Lecture Notes in Computer Science 4970

Commenced Publication in 1973
Founding and Former Series Editors:
Gerhard Goos, Juris Hartmanis, and Jan van Leeuwen

Editorial Board

David Hutchison
  Lancaster University, UK
Takeo Kanade
  Carnegie Mellon University, Pittsburgh, PA, USA
Josef Kittler
  University of Surrey, Guildford, UK
Jon M. Kleinberg
  Cornell University, Ithaca, NY, USA
Alfred Kobsa
  University of California, Irvine, CA, USA
Friedemann Mattern
  ETH Zurich, Switzerland
John C. Mitchell
  Stanford University, CA, USA
Moni Naor
  Weizmann Institute of Science, Rehovot, Israel
Oscar Nierstrasz
  University of Bern, Switzerland
C. Pandu Rangan
  Indian Institute of Technology, Madras, India
Bernhard Steffen
  University of Dortmund, Germany
Madhu Sudan
  Massachusetts Institute of Technology, MA, USA
Demetri Terzopoulos
  University of California, Los Angeles, CA, USA
Doug Tygar
  University of California, Berkeley, CA, USA
Gerhard Weikum
  Max-Planck Institute of Computer Science, Saarbruecken, Germany

Manfred Nagl   Wolfgang Marquardt (Eds.)

Collaborative and Distributed Chemical Engineering

From Understanding to Substantial Design Process Support

Results of the IMPROVE Project

 Springer

Volume Editors

Manfred Nagl
RWTH Aachen
Institut für Informatik 3
Ahornstrasse 55, 52074 Aachen, Germany
E-mail: nagl@informatik.rwth-aachen.de

Wolfgang Marquardt
RWTH Aachen
Lehrstuhl für Prozesstechnik
Turmstraße 46, 52056 Aachen, Germany
E-mail: Wolfgang.Marquardt@avt.rwth-aachen.de

Library of Congress Control Number: 2008931577

CR Subject Classification (1998): J.2, I.6, J.6, D.2

LNCS Sublibrary: SL 2 – Programming and Software Engineering

ISSN	0302-9743
ISBN-10	3-540-70551-1 Springer Berlin Heidelberg New York
ISBN-13	978-3-540-70551-2 Springer Berlin Heidelberg New York

This work is subject to copyright. All rights are reserved, whether the whole or part of the material is concerned, specifically the rights of translation, reprinting, re-use of illustrations, recitation, broadcasting, reproduction on microfilms or in any other way, and storage in data banks. Duplication of this publication or parts thereof is permitted only under the provisions of the German Copyright Law of September 9, 1965, in its current version, and permission for use must always be obtained from Springer. Violations are liable to prosecution under the German Copyright Law.

Springer is a part of Springer Science+Business Media

springer.com

© Springer-Verlag Berlin Heidelberg 2008
Printed in Germany

Typesetting: Camera-ready by author, data conversion by Scientific Publishing Services, Chennai, India
Printed on acid-free paper   SPIN: 12438235   06/3180   5 4 3 2 1 0

Preface

Collaborative Research Center (Sonderforschungsbereich) 476, "Information Technology Support for Collaborative and Distributed Design Processes in Chemical Engineering" (IMPROVE), is a large joint project of research institutions at RWTH Aachen University, including different groups in engineering, in particular process systems engineering, plastics processing, labor research, and different groups in informatics, namely, communication, information systems, and software engineering. It is funded by the German Research Foundation (Deutsche Forschungsgemeinschaft, DFG), with more than 20 scientists collaborating continuously in a long-term research effort since 1997.

In this volume we summarize the results of IMPROVE after 9 years of cooperative research work. Together with master's theses and the contribution of technical personnel we report on the total effort of more than 200 person years of scientific work. This includes research work done in dissertation projects.

The 9-year period of the CRC has been and will be continued by technology transfer activities from mid 2006 to mid 2009 (Transfer Center 61 "New Concepts and Tools for Chemical Engineering Practice"), also mostly funded by the DFG. The main activities of this transfer center are also described in this volume.

The focus of IMPROVE is on understanding, formalizing, evaluating, and, consequently, improving *design processes* in chemical engineering. In particular, IMPROVE focuses on conceptual design and basic engineering, where the fundamental decisions concerning the design or redesign of a chemical plant are undertaken. Design processes are analyzed and evaluated in collaboration with industrial partners.

The design of a plant for producing Polyamide-6 is used as the *reference scenario* within IMPROVE. Based on this scenario, novel concepts, models, and tools for supporting integrated forms of design processes in chemical engineering have been developed.

The kernel of the approach is a *formal process/product model* for design, which is regarded on different levels, covering domain models on the top to internal models used for the realization of tools at the bottom. Infrastructure projects introduce research results on distributed platforms for computer processes and data integration as part of the architecture of the overall environment of supporting tools. The formal product/process model has already gained a certain degree of maturity. However, further research work will be needed to present it in a formalized way. We invite other research groups to join our activities and to contribute to this challenging task.

The IMPROVE approach in particular deals with *collaborative* and *distributed* design *processes* across different organizations. Cross-company collaboration takes place every day. It is still a big problem for all engineering disciplines with respect to in-depth understanding and a suitable tool support. In order to improve the state of the art, not only organizational boundaries have to be considered. Rather, knowledge across different specific domains and design disciplines is needed and, therefore, has to be uniformly presented.

The goal of IMPROVE can also be seen from the *tool integration perspective*. There, we follow an approach which is based on existing tools and aims at enriching their functionality (bottom-up or a-posteriori strategy). Hence, integration does not only refer to system integration in the sense that all existing tools work coherently on top of a distributed platform. In contrast, new tool functionality is added to support and, therefore, improve distributed design across different persons, roles, disciplines, or companies. For this purpose, new informatics concepts are introduced to facilitate the design process, e.g., by reusing developers' experience, controlling consistency on a fine-grained level, direct multimedial communication of results, and by reactive project management. Furthermore, the combination of these new functionalities is studied providing further synergistic functionalities. This comprehensive tool integration approach is unique and not yet covered by any available book on tool integration.

Another unique feature is the derivation of new *tool functionality* from explicit *models* of the *chemical engineering domain* including the organization of design processes, thereby embedding existing design support tools. The analysis of industrial design processes is the basis of the formalization. The new tool integration functionality is derived from the resulting formal models. Thus, we do not build a new integrated environment and check afterwards whether it matches the needs of the design processes of interest. Rather, the functionality is derived from validated practical models. This vertical and formalized integration aspect demonstrates the close cooperation between chemical engineering and informatics within IMPROVE.

Although chemical engineering is the application domain for which we produced conceptual and practical results in the form of models and software, a substantial part of the research results contained in this volume hold true for other engineering domains as well. Some results are directly valid in other domains, others relate to approaches which can be carried over to other domains without revision. Hence, this book can also be seen as dealing with *engineering design processes* and their *integrated support in general*.

The results of IMPROVE have been reviewed four times by peers at the beginning and after every 3 years. Furthermore, evaluation by the industry has been implemented by affiliating a number of industrial partners with the CRC. Transfer activities will strengthen the relations between IMPROVE and the industry in the next few years. Both peer reviews and cooperation with the industry imply a *spiral research approach*: Goals have been checked and revised after a period of 3 years. Finally, the progress of one period is based on the results of the preceding period.

This book intends to summarize the results of 10 years of interdisciplinary research and future plans for an additional 2 years to present them to the international community. The results are applicable to different design processes including chemical engineering, mechanical engineering, electrical engineering, or computer science. Thus, the book addresses the whole community of people involved in the improvement of design processes in different engineering disciplines, either in academia or in industry.

Hence, a specific goal of this book is to broadcast our results across different disciplines. As informatics is the main focus, publication in the *Lecture Notes in Computer Science* series is an obvious objective. However, engineers from many disciplines are addressed as well.

There is no interdisciplinary joint project on design process modeling and support of a comparable breadth and depth known to the editors. Therefore, the approach and the results of the CRC received broad international recognition in both the chemical engineering and informatics communities.

Results of IMPROVE have been published in many scientific journals or conference and workshop proceedings, as compiled in the bibliography. The aim of this book is to give a complete and uniform description of the corresponding results with a quality of coherence which comes close to that of a monograph.

Our thanks go to different institutions: The German Research Foundation has given and will give us remarkable support during the last 10 years and the 2 years to come. RWTH Aachen University and the Ministry of Science and Research of the State North-Rhine Westphalia have also provided us with additional funding. The funding sums to about 11 million euro. Without these generous donations IMPROVE could not have been started. Section 1.3 gives more detailed figures. Additional grants have been given to the research groups participating in IMPROVE by the DFG, the Federal Ministry of Research in Germany, or by the European Community for other and related projects. Their results have been the basis for IMPROVE or they will continue the research described in this book.

We are also indebted to several persons: Peers have visited us four times to evaluate our proposals and to make remarks for improving the project. Their engagement is especially acknowledged. Members of IMPROVE (see Appendix A.2) have worked hard to achieve the results described in this book. Furthermore, many master's degree students contributed to the project during their thesis work.

Finally, Mrs. Fleck and Mr. Haase spent a lot of effort and time getting the layout of this book in shape.

April 2008

Manfred Nagl
Wolfgang Marquardt

Contents

Preface V

Part I Overview

1 Goals, Approach, Functionality of Resulting Tools, and Project Structure 1

    1.1 A Model-Driven Approach for A-posteriori Tool Integration .. 3
              W. Marquardt and M. Nagl
    1.2 A Scenario Demonstrating Design Support in Chemical Engineering 39
              R. Schneider and B. Westfechtel
    1.3 The Interdisciplinary IMPROVE Project 61
              M. Nagl

Part II Technical Results

2 Application Domain Modeling 81

    2.1 An Introduction to Application Domain Modeling 83
              J. Morbach, M. Theißen, and W. Marquardt
    2.2 Product Data Models 93
              J. Morbach, B. Bayer, A. Yang, and W. Marquardt
    2.3 Document Models 111
              J. Morbach, R. Hai, B. Bayer, and W. Marquardt
    2.4 Work Process Models 126
              M. Eggersmann, B. Kausch, H. Luczak, W. Marquardt,
              C. Schlick,
              N. Schneider, R. Schneider, and M. Theißen

2.5 Decision Models 153
M. Theißen and W. Marquardt
2.6 Integrated Application Domain Models for Chemical
Engineering 169
J. Morbach, M. Theißen, and W. Marquardt

3 New Tool Functionality and Underlying Concepts 183

3.1 Using Developers' Experience in Cooperative Design
Processes 185
M. Miatidis, M. Jarke, and K. Weidenhaupt
3.2 Incremental and Interactive Integrator Tools for Design
Product Consistency 224
S. Becker, M. Nagl, and B. Westfechtel
3.3 Multimedia and VR Support for Direct Communication of
Designers 268
A. Schüppen, O. Spaniol, D. Thißen, I. Assenmacher,
E. Haberstroh,
and T. Kuhlen
3.4 An Adaptive and Reactive Management System for Project
Coordination 300
M. Heller, D. Jäger, C.-A. Krapp, M. Nagl, A. Schleicher,
B. Westfechtel,
and R. Wörzberger

4 Platform Functionality 367

4.1 Goal-Oriented Information Flow Management in Development
Processes 369
S.C. Brandt, O. Fritzen, M. Jarke, and T. List
4.2 Service Management for Development Tools 401
Y. Babich, O. Spaniol, and D. Thißen

5 Integration Aspects 431

5.1 Scenario-Based Analysis of Industrial Work Processes 433
M. Theißen, R. Hai, J. Morbach, R. Schneider, and
W. Marquardt
5.2 Integrative Simulation of Work Processes 451
B. Kausch, N. Schneider, S. Tackenberg, C. Schlick, and
H. Luczak
5.3 An Integrated Environment for Heterogeneous Process
Modeling and Simulation 477
L. von Wedel, V. Kulikov, and W. Marquardt

5.4 Design Support of Reaction and Compounding Extruders 493
M. Schlüter, J. Stewering, E. Haberstroh, I. Assenmacher,
and T. Kuhlen

5.5 Synergy by Integrating New Functionality 519
S. Becker, M. Heller, M. Jarke, W. Marquardt, M. Nagl,
O. Spaniol,
and D. Thißen

5.6 Usability Engineering 527
C. Foltz, N. Schneider, B. Kausch, M. Wolf, C. Schlick, and
H. Luczak

5.7 Software Integration and Framework Development 555
Th. Haase, P. Klein, and M. Nagl

6 Steps Towards a Formal Process/Product Model 591

6.1 From Application Domain Models to Tools: The Sketch of a
Layered Process/Product Model 593
M. Nagl

6.2 Work Processes and Process-Centered Models and Tools 605
M. Miatidis, M. Theißen, M. Jarke, and W. Marquardt

6.3 Model Dependencies, Fine-Grained Relations, and Integrator
Tools 612
S. Becker, W. Marquardt, J. Morbach, and M. Nagl

6.4 Administration Models and Management Tools 621
R. Hai, T. Heer, M. Heller, M. Nagl, R. Schneider,
B. Westfechtel,
and R. Wörzberger

6.5 Process/Product Model: Status and Open Problems 629
M. Nagl

Part III Transfer and Evaluation

7 Transfer to Practice 641

7.1 Industrial Cooperation Resulting in Transfer 643
R. Schneider, L. von Wedel, and W. Marquardt

7.2 Ontology-Based Integration and Management of Distributed
Design Data 647
J. Morbach and W. Marquardt

7.3 Computer-Assisted Work Process Modeling in Chemical
Engineering 656
M. Theißen, R. Hai, and W. Marquardt

XII   Contents

- 7.4 Simulation-Supported Workflow Optimization in Process Engineering 666
  B. Kausch, N. Schneider, C. Schlick, and H. Luczak
- 7.5 Management and Reuse of Experience Knowledge in Extrusion Processes 675
  S.C. Brandt, M. Jarke, M. Miatidis, M. Raddatz, and M. Schlüter
- 7.6 Tools for Consistency Management between Design Products . 696
  S. Becker, A. Körtgen, and M. Nagl
- 7.7 Dynamic Process Management Based upon Existing Systems . 711
  M. Heller, M. Nagl, R. Wörzberger, and T. Heer
- 7.8 Service-Oriented Architectures and Application Integration .. 727
  Th. Haase and M. Nagl

8 Evaluation 741

- 8.1 Review from a Design Process Perspective 743
  W. Marquardt
- 8.2 Review from a Tools' Perspective 753
  M. Nagl
- 8.3 Review from an Industrial Perspective 764
  W. Marquardt and M. Nagl
- 8.4 Review from Academic Success Perspective 774
  M. Nagl

Part IV Appendices, References

Appendices 781

- A.1 Addresses of Involved Research Institutions 781
- A.2 A.2 Members of the CRC 476 and TC 61 783

References 785

- R.1 Publications of the IMPROVE Groups 785
- R.2 External Literature 817

Author Index 851

1

Goals, Approach, Functionality of Resulting Tools, and Project Structure

This chapter consists of three sections.

Section 1.1 gives a detailed problem *analysis* of design processes in Chemical Engineering and the deficits of tools available in practice to support these processes. Then, we describe the overall goals and *approach* of IMPROVE, namely to introduce novel process aspects from the Engineering and the Informatics side to get a better support for designers. The resulting key problem is to *formalize* design *processes* and their products. From the *tool* perspective the task is to build up an integrated environment for the cooperation of different designers.

In Sect. 1.2 we present one practical result of IMPROVE, namely one *version* of an integrated *environment* of existing, extended, and new tools, built to support different facets of a cooperative and distributed design process in Chemical Engineering and Plastics Processing. The results are shown by giving a demo in form of a *guided tour*. We concentrate on those development steps where the novel IMPROVE concepts induce a remarkably better collaboration of developers.

In Sect. 1.3, finally, we explain the IMPROVE *project structure*. The project addresses the problem of Sect. 1.1 and produces results like that of Sect. 1.2. The section also characterizes the predecessor projects on which IMPROVE is based and gives an overview of the structure of this book. Finally, this section contains figures about the funding of IMPROVE.

1.1 A Model-Driven Approach for A-posteriori Tool Integration

W. Marquardt and M. Nagl

Abstract. The following section discusses the *long-term approach* of IMPROVE[1]. We start with a careful analysis of current design[2] processes in chemical engineering and their insufficient support by tools. Then, we sketch how new ideas from the engineering and informatics side can considerably improve the state-of-the-art. The resulting tools according to these new ideas are sketched in the next section.

It should be remarked that both, the long-term goals and the principal approach of IMPROVE [343], were not changed within the whole project period, i.e. – if also counting preliminary phases of the project – within a duration of 11 years. This *stability* of *goals* and *approach* shows that we have addressed a fundamental and hard problem which cannot be solved in a short period. It, furthermore, shows that the undertaken approach went in the right direction. Therefore, it was not necessary to revise the approach within the whole project period, neither in total nor in essential parts.

1.1.1 Chemical Engineering Design Processes and Their Computer-Assisted Support

We first describe the application domain the IMPROVE project has been focussing on. The next sections provide first a coarse characterization of the design process in chemical engineering. A case study is introduced next for illustration and benchmarking purposes. The major properties of industrial design processes are reviewed and supporting software tools are described. The practice of chemical design processes is critically assessed to identify weaknesses and shortcomings of current computer-assisted support.

The Chemical Process Design Process

A *chemical process system* comprises of a set of interconnected physical, chemical, biological and even information technological unit processes. The functional task of a chemical process system is the transformation of the type, properties and composition of given raw materials to result in a desired *material product* with given specifications. This transformation process is implemented in a *process plant* which consists of various pieces of equipment dedicated to one or more related unit processes in order to achieve the sequence of

---

[1] Collaborative Research Center 476 IMPROVE (in German Sonderforschungsbereich, abbr. SFB) is funded by German Research Foundation (Deutsche Forschungsgemeinschaft, abbr. DFG) since 1997.

[2] In this volume, we mostly speak of design processes. In other disciplines outside Chemical Engineering they are called development processes. Analogously, we use the term designer or developer.

the desired transformations. The equipment is interconnected by pipelines and signal lines. Pipelines transfer material and associated energy streams between apparatuses, while information flows are transferred between process control devices, respectively. Hence, the chemical process system relates to a conceptual and functional representation of the manufacturing process whereas the chemical process plant refers to its concrete technical realization [559].

The objective of a *design process* in chemical engineering is the development of a manufacturing process to produce a newly developed or an already known material product. We distinguish between *grassroot* and *retrofit design*, if a chemical process plant is designed from scratch or if an existing plant is modified or partly replaced to match new requirements of the changing economical environment. The design process constitutes of all the activities and their relations to solve the design problem in a team effort [1047]. It starts from a rough description of the design problem and it ends with a complete specification of the chemical process plant which comprises all the detailed information necessary to build and commission the plant.

Processes and *products* (and in subsequent chapters the models for their representation) have to be distinguished on *two* different *levels*. The term process can either refer to the chemical process or plant or to the design process. Likewise, the product can either refer to the result of the design process or to the material product manufactured in the chemical plant. In most cases, the meaning of the terms will be obvious from the context they are used in. In order to eliminate any degree of ambiguity, we will often refer to a chemical process and a design (or more generally a) work process and likewise to a material product and to the product of the design process.

Like in any other engineering domain, the chemical process design process can be subdivided in a number of distinct phases. In the first phase, the requirements for the material products to be produced by the chemical plant are fixed on the basis of an analysis of the market demand and the current and expected supply by competitors. These requirements emphasize the business objective of the design project and form the basis for a first description of the design problem. The design problem formulation is further refined in *conceptual design*, where the major conceptual decisions on the raw materials, on the chemical synthesis route, on the process structure, and even on the strategy for plant operation are taken. These conceptual considerations are refined in *front-end engineering* and detailed and completed during *basic engineering* [559, 638, 906].

At the end of *basic engineering* all major *design data* of the plant are fixed. They typically include

- all raw materials and their origin,
- the chemical reaction pathways for chemical synthesis,
- the main and side products including their achievable product quality,
- the waste streams into the environment,
- the process units and their couplings in a process flowsheet,

- the nominal operating point which is given by flow rates and compositions of all process streams as well as by the temperatures and pressures in all the process units,
- the decision on the operational strategy for the plant and the plant-wide control system structure,
- the selection of suitable equipment to implement the functional tasks of process units,
- the major engineering design data for any piece of equipment,
- a rough plant layout, and
- an estimate of the total capital and operational cost per unit mass of produced material.

Basic engineering is followed by *detail engineering*, where the design data fixed during basic engineering are used to specify all pieces of equipment of the plant including all the instrumentation and control systems in full detail. The design process is completed at the end of detail engineering. The result (or product) of the design process is a set of complete specifications which are on a level of detail to allow *procurement* of all parts and *construction of the plant* on site in the subsequent plant lifecycle phases.

The focus of our investigations on computer-assisted support of process design processes in the IMPROVE project has been *focussing* on the *early phases* of the design process, in particular on conceptual design and front-end engineering. The late phases of basic engineering or detail engineering have only been considered to a lesser extent during the project. In the first place, a restriction on certain phases of the design process is mandatory to pragmatically limit the complexity of the application scenario to benchmark the information technology concepts and novel tools to a manageable level.

The *decision* to concentrate on the early phases of the design process has been based on the following more specific *considerations* and *arguments*:

- Though only a small fraction of the total investment costs is spent during the early phases of the design process, the *impact* of the *decisions* taken on the economical performance of the plant during its total lifetime (i.e. the cost of ownership of the plant) is most significant. The results of the early design process phases not only form the basis for a subsequent refinement during basic and detail engineering, but rather fix about 80 % of the total production costs. Furthermore, an investment decision is typically taken at some time during front-end engineering, which emphasizes the importance of this phase of the design process.
- The engineering activities performed during the early phases of the design process reveal a *high* level of *complexity* because different aspects and objectives have to be considered simultaneously to facilitate a holisitc design of the process system with a balanced trade-off between potentially conflicting goals.
- The level of complexity can only be mastered if mathematical models of the process system are used on various scales of resolution to support the

synthesis of the process system and to analyze the performance of a certain design alternative. Since only certain tasks during the design process can be mastered by means of *model-based* approaches they have to be smoothly integrated with *manual design activities* in the overall design process.
- While largely routine and therefore well-understood design activities are more typical during detailed engineering, the design activities during the early phases of the *design* process are of a *creative nature*. Since creative work processes are usually poorly understood, they can neither be properly captured and formally represented nor properly planned in advance. Therefore, this kind of design processes constitute a major challenge for any information technological support.

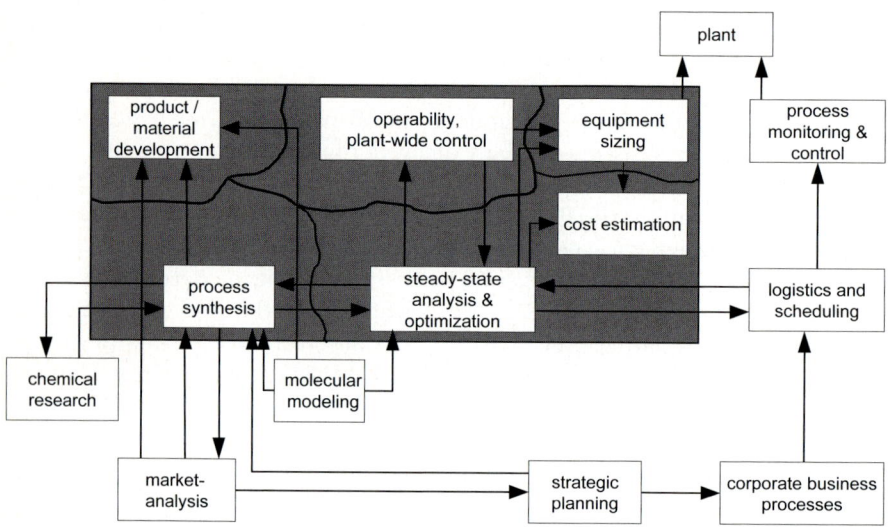

Fig. 1.1. Development tasks and their interdependencies

The *design activities* which have been *considered* in more detail during the IMPROVE project are depicted in Fig. 1.1. They include in particular the development of a favorable material, the synthesis of the process flowsheet, the determination of an economically optimal nominal operating point, the sizing of the major equipment, the estimation of operating and investment costs, the analysis of process dynamics and operability, as well as the synthesis of a plantwide control system structure. Most of these major design activities are closely related to each other. In particular, design decisions taken in one of them will affect the decision making in another. These *interdependencies* are often not transparent or even not well understood and are therefore only handled incompletely during the design process [1016].

The Polyamide Process as Case Study

A *scenario-based* research *approach* has been used in IMPROVE in order to identify the requirements, based on a concrete chemical process design case study. A case study from the polymerization domain has been chosen because there are much less mature design support tools as compared to petrochemical processes. Therefore, tool integration and work process support are of considerable interest in both, the end user as well as the software vendor industries.

The selected scenario comprises the conceptual design of a polymerization process for the *production* of *Polyamide-6* (PA6) from caprolactam [99, 104]. PA6 is a thermoplastic polymer with a world production capacity of more than 4 million tons per year (as of 2006). The most frequent use of PA6 is the production of fibers, which are used in home textiles, bath clothing and for carpet production. In addition, PA6 is used as an engineering construction material if high abrasion resistance, firmness, and solvent stability are required. Glass-fiber reinforced and mineral material-filled PA6 is a preferred construction material if a combination of rigidity, elasticity and refractory quality characteristics are required.

A complete *process chain* for moulded parts made of PA6 is shown in Fig. 1.2. PA6 is synthesized in a polymerisation step from the monomers caprolactam and water. The production of caprolactam itself can be traced back to benzene which is isolated from crude oil or formed from crude oil in a cracking process. The polymer melt is degassed from unreacted monomers and then further processed to condition it for the manufacturing of semi-finished products and of moulded parts. The polymer material in the moulded parts may be eventually recycled and reused in the polymerisation after a polymer degradation reaction has been carried out.

The figure 1.2 also emphasizes three distinct *engineering domains*, namely chemical engineering, polymer processing, and material recycling. The design problems in these domains are highly interrelated but typically treated in almost complete isolation in different organizations. Any computer-assisted design support system has to bridge these gaps to fully exploit the economical potential of the design.

The PA6 molecule has two different end groups, namely an amide end group and a carboxyl end group, which can react with each other to form longer polymer chains. PA6 can be formed via an anionic or a hydrolytic reaction route. The anionic mechanism is mainly used for special polymers whereas the hydrolytic one is more often applied industrially. The *case study* considers the design of a process to produce 40.000 t/a of PA6 via the hydrolytic route with tight quality specifications. For example, impurity constraints are formulated for the residues of caprolactam, cyclic dimer and water in the polymer. A viscosity specification is added to guarantee favorable properties during the following polymer processing step. The molecular weight is specified to account for desirable mechanical properties of the moulded parts.

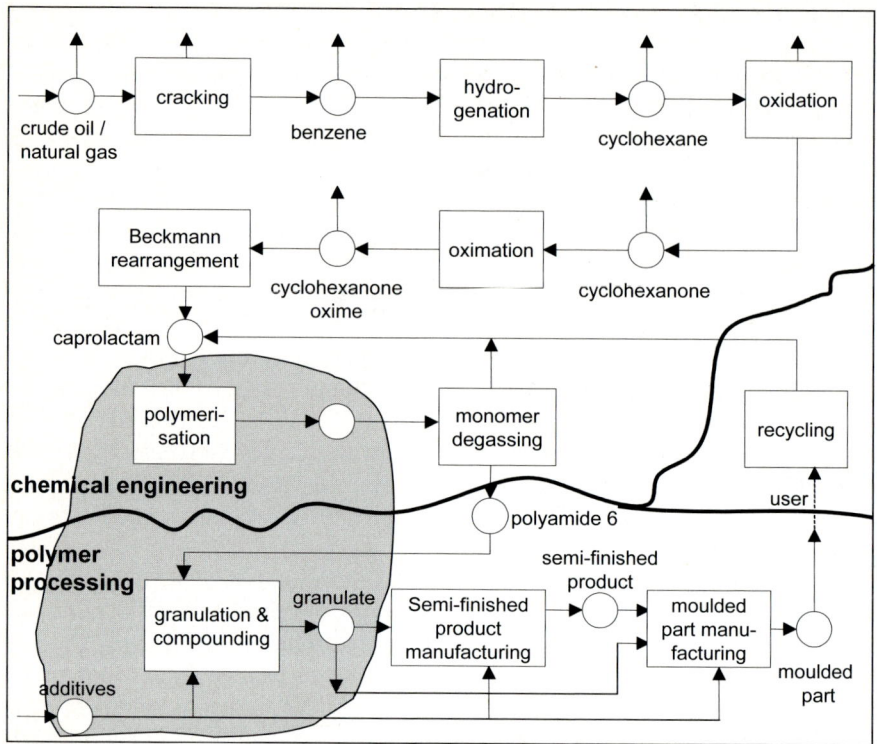

Fig. 1.2. Process chain for Polyamide-6 (the section considered in CRC 476 is shaded grey)

Figure 1.3 shows a *possible flowsheet* of a PA6 process. The process can be decomposed into three sections, the reaction, separation, and compounding sections. The monomers water and caprolactam are converted in a series of polymerization reactors to the PA6 polymer. The monomer in the polymer melt leaving the last reactor has to be separated from the mixture to meet the quality requirements of the final product. There are various alternatives to accomplish this separation step. The flowsheet in Fig. 1.3 shows a separation alternative where caprolactam is leached from the polymer pellets formed from the melt in the upstream granulation unit by means of hot water. The residual water is removed from the polymer pellets in a subsequent dryer. The separation is followed by a compounding step, in which fibers, color pigments, or other additives are added to the polymer to adjust the polymer properties to the specific customer product specifications. The extruder is not only used for compounding but also for degassing of the monomer traces remaining in the melt. The monomer is recycled from the compounding extruder to the polymerization reactors.

The design process starts with *requirements* the final product, e.g. the moulded part, has to fulfill. The mechanical properties of the moulded part determine the properties of the polymer material, which have to be achieved in the PA6 process by means of the integrated reaction, separation, and compounding steps briefly introduced above. In particular, a certain molecular weight, a specific melt viscosity, and a limited monomer residual have to be guaranteed in the polymer product.

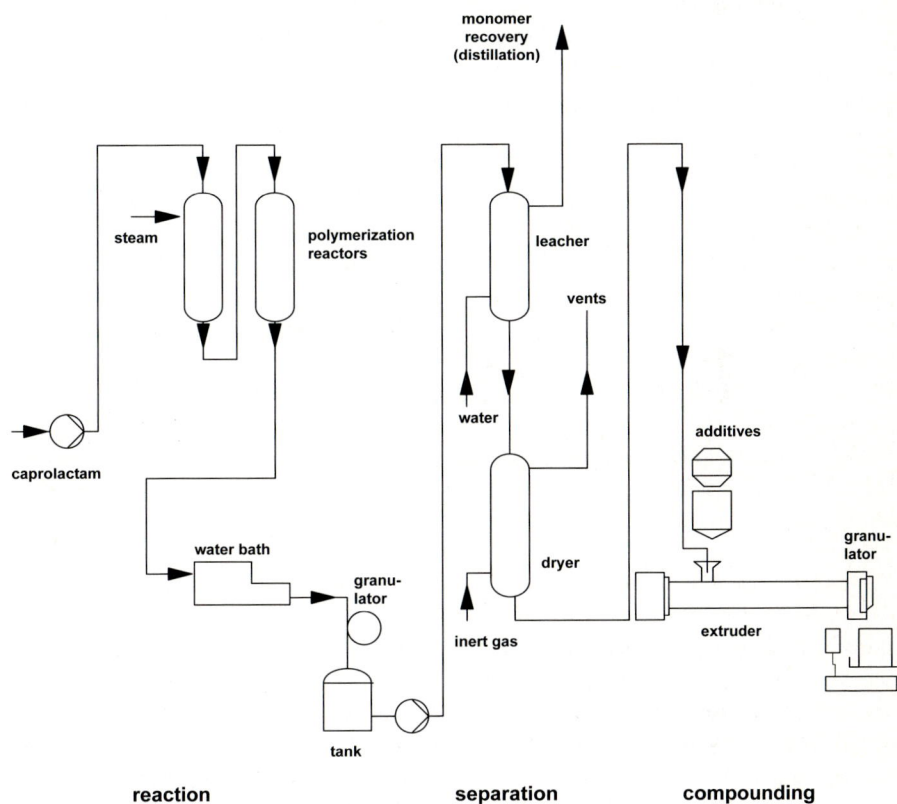

Fig. 1.3. Flowsheet of the Polyamide-6 process

The *conceptual design* is accomplished in series of related design activities which are described in detail elsewhere [99] following the methodology suggested by Douglas [638]. First, the decision for a hydrolytic polymerization in a continuous process is taken, based on qualitative design rules reflecting domain experience. According to Douglas [638], the input-output structure and the recycle structure are fixed in the next design activities. The results form the basis for the design of the reactor network. The polymer reaction can be carried out in stirred tank or plug flow reactors or in a combination thereof, with or without an intermediate separation of the condensate and of the unconverted monomer. The design of the reaction section largely fixes the quality of the reaction products which itself determines the design of the separation section. The alternatives for the separation include devolatilisation in a wiped-film evaporator [840, 935], the removal of low molecular-weight components by leaching of polymer pellets followed by a drying step [930], the degassing in the compounding extruder, or a combination thereof.

The *close relation* between the *design problems* related to the three process sections (polymer reaction, separation, and compounding) becomes obvious if the flowsheet alternatives are reflected in more depth. The design of the separation section is not independent of the design of the polymer processing section because a decision has to be taken regarding the split of work for monomer separation between the extruder and the upstream separation unit. Further, polymer reaction and compounding are not independent, because the polymer chains can be further grown if a reactive extrusion process [313, 734] is considered as an alternative to the process flowsheet in Fig. 1.3, where part of the polymer reaction is carried out simultaneously with polymer compounding in the extruder. The activities during the design process have to be coordinated across disciplinary boundaries in order to account for these interdependencies. In particular, the separation of part of the monomer from the melt and the increase of the mean molecular weight in the extruder allow a reduction of the capacity of the upstream separation units and polymerization reactors which consequently results in less investment cost. However, the decisions on the most favorable process alternative are also determined by product quality which is achieved by the synergistic interaction between all the process units.

All these design steps require *different types* of mathematical *models* which are preferably implemented in different types of modeling environments. The models are used to analyze the performance of a certain design alternative and to optimize the steady-state operating point [99]. A preliminary engineering design calculation has to be carried out by dedicated software to get a rough idea of the equipment sizes as a basis for the subsequent estimation of the investment cost. For example, MOREX [147] can be employed to identify a favorable structure and geometry of the extruder screw and the type and dimensions of the processing segments in the extruder. A rough description of degassing and even polymer reaction processes is also covered. Finally, dynamic simulations are carried out to analyze the process dynamics and to assess the operability of the process in order to build up the process understanding required to design a plantwide control system structure.

The design support *software tools* employed in the case study are of a completely *different nature*. They include commercial as well as legacy tools. Examples are Microsoft Excel, various process simulators such as Polymers Plus from Aspen Technology, gPROMS from PSE, MOREX, BEMFlow and BEMView from Institut für Kunststoffverarbeitung at RWTH Aachen University, the project database Comos PT from innotec, the document management system Documentum from EMC^2 as well as the platform CHEOPS for run-time integration of different types of simulators, the repository ROME for archiving mathematical models, and the modeling tool ModKit, all from Lehrstuhl für Prozesstechnik at RWTH Aachen University.

Typically, despite the significant interdependencies, polymerization, separation, and compounding are designed in *different organizational units* of the same or even different corporations using different methodological approaches and supporting software tools. An integrated solution of the design problem of the PA6 case study has to *overcome* the traditional *gap* between polymer reaction engineering, separation process engineering and polymer processing with their different cultures as well as the problem of incompatible data and software tools. Hence, the scenario poses a challenge for any integrated methodology for conceptual design and its supporting software environment.

Characteristics of Industrial Design Processes

The case study presented in the last subsection can also be taken to summarize the *characteristics* of *industrial design processes*. This subsection abstracts from the concrete example. The reader, however, may also relate the following general statements to the polymerisation case study.

Industrial design processes are always carried out by a *team* of multidisciplinary experts from different organizational units within the same or different companies. The team is formed to carry out a dedicated *project*, it is directed by a project manager. Usually, a number of consultants, who have expert knowledge on a very specific technological topic, are contributing to the design activities in addition to the team members. All team members are typically part of more than one team at the same time. Often, the team operates at different, geographically distributed *sites*. The *duration* of a single project may range from weeks to years with varying levels of activity at a certain point in time. Hence, the team and the status and assignments of its members may change over time, in particular in case of long project duration. Inevitably, there is no common *understanding* about the design problem at the beginning of the project. Such a common understanding, called shared memory by Konda et al. [791], has to *evolve* during collaborative work.

The design process constitutes of all the related activities carried out by the team members while they work on the design problem [1047]. This multidisciplinary process shows an immense complexity. It has to deal with the culture and paradigms from different domains. Complicated multi-objective *decision making processes under uncertainty* are incorporated in the design. They rely on the typically incomplete information produced in the current and previous design activities. In particular, conceptual design processes show a *high degree of creativity*, they are of an inventive nature and do not just apply existing solutions. Creative conceptual design processes are hardly predictable and, therefore, can only be pre-planned on a coarse-grained level. An at least coarse-grained work process definition is mandatory to establish simultaneous and concurrent engineering to reduce the total time spent on a design.

The lack of precise planning on a medium-grained level inevitably results in *highly dynamic* work processes. They show *branches* to deal with the assessment of alternatives and to allow for a simultaneous work on only loosely related subtasks. *Iterations* occur to deal with the necessary *revision of previous decisions and solutions*. In the first place, they are due to inevitable uncertainties during decision making because of lacking or incomplete information. While the design process is carried out, this uncertainty can be continuously reduced because of additional information becoming available. It is either collected from various available but not yet exploited resources or it is generated while the design process progresses. Additional information always gives rise to new insight to either address a problem which has not yet been recognized, to exploit an identified potential for improving an existing solution, or to even evolve the design requirements. A strict and prescriptive definition of the work process (as accomplished in many administrative business processes, e.g. [671]) is not only impossible but also highly undesirable in the context of a design process. It would largely constrain the creativity of the designer with obviously undesirable consequences for the quality of the resulting design.

The team of experts typically uses a *multitude of resources* in the various phases of the design process. For example, web-based *text retrieval and browsing systems* are used to search the scientific and patent literature or internal archives for information on materials or processing technologies. *Lab-scale or pilot-scale experiments* allow the investigation of specific questions related to physical properties, kinetics, scale-up of equipment, or the accumulation of impurities in recycles and their impact on the process behavior. All kinds of *software tools* with diverse and often overlapping functionality have been increasingly used in the last two decades to support different design activities.

In the course of the design process, a *complex configuration of different types of information* is created. This information appears in multiple ways. There are, for example, standardized documents including equipment specification sheets or design reports, informal texts like e-mail or telephone notes, or input or output files of certain software tools containing problem specifications or result summaries in a formal syntax. More recently, audio and video clips may be included in addition. This information is typically held in a decentralized manner in the local data stores of the individual software tools, in document management systems, or in project databases. Typically, the relationship between the various information units is not explicitly held in the data stores. Information is exchanged in the design team by means of *documents*, which aggregate selected data relevant to a certain work process context.

Though a large amount of information is created and archived in some data store during the design process, there is typically no complete *documentation* of all the alternatives considered during the design. However, a full documentation of the final conceptual design has to be compiled from the information created during the design process. Typically, this documentation is handed over to an engineering contractor to continue the design process during basic and detail engineering.

Available Supporting Software Tools

Industrial design processes are supported by different kinds of software tools. We do not want to elaborate on the functionality of an individual tool, but try to *classify* various kinds of *tools* and to *relate* them to the process *design process* described above. Largely, there are two different groups of software tools in use in chemical process design.

The first group comprises *generic software tools* which are not specific to chemical engineering but which are used in all industries to support collaborative work processes and administrative business processes. They include word processing and spreadsheet systems, enterprise resource planning (ERP) [781, 961], project management tools [777], workflow management systems [762], decision support tools [915], document management systems [1064], and groupware systems [654] which are completely independent of a specific application domain and hence are established in all industrial segments. The functionality of these tools is not driven by the needs of the (chemical process) design process. Rather, these tools address the *domain-independent* needs of a (much) larger customer base. Hence, such tools do not reflect the peculiarities and the particular needs of the ill-defined, complex and dynamically changing, creative work processes, which typically occur in any (engineering) design process. Furthermore, these tools do not account for appropriate interfaces to relate activities, project schedules, or documents produced by individuals in a design process in a transparent manner [355].

In contrast to these generic tools, the second group of *domain-specific software tools* addresses certain tasks during the design process in the chemical engineering domain. They can roughly be classified as data retrieval, synthesis, and analysis tools. A typical example for data retrieval tools are physical property systems [1048, 1053]. Synthesis tools include flowsheet synthesis [951, 1039], plant layout and pipe routing [955], model development [54, 558] or even solvent selection [737]. The most prominent examples of analysis tools are process simulators for steady-state [518, 556, 1046] and dynamic simulation [288, 518, 916].

Though most simulators provide a library of standard models for process units, there is only *limited support* for very *specific units* such as those typically occurring in polymerization processes. Equation-oriented [895, 916] and object-oriented [54, 1002] process modeling environments are suitable for the development of customized models for non-standard units such as for the leacher in the PA6 process. Complex transport problems involving fluid dynamics as well as other kinetic phenomena (i. e. chemical reaction, nucleation and growth of particles, interfacial heat and mass transfer etc.) can be treated with computational fluid dynamics codes and their extensions. Such codes are also available for polymer processing (e.g. [860, 889]).

Though these tools are still largely used in a stand-alone manner, there are some *attempts* toward their *integration*. For example, Aspen Plus and Aspen Custom Modeler, a steady-state and a dynamic simulator, have been integrated by AspenTech to transfer models from one into the other simulation environment. Run-time integration of different simulation tools has been recently addressed by means of a generic framework [252, 409] which facilitates the integration of block-oriented and equation-oriented simulators for steady-state and dynamic simulation. All major vendors integrate their tools into some design environment with interfaces to selected tools from cooperating vendors to address particular aspects in the design process (e.g. AspenTech's Engineering Suite or Intergraph's Smart Plant). In particular, these tools are at least partly linked to engineering or product data management systems [907] such as Aspen Zyqad, Comos PT or SmartPlant Explorer which persistently store the information used and generated during the design process.

Most of the vendors offer more general interfaces to import and export data between a product data management system and some software application. XML is the preferred exchange format. The data schema is typically based on the STEP modeling methodology [503] and on the STEP models standardized as ISO 103030 and ISO 15026. None of these schemas covers the complete design process. The collection of data exchange schemas does not comprise a complete data model, since the relations and dependencies between the data are not captured. A *unifying data model* across platforms offered by different vendors is currently not available [506].

Some vendors also offer solutions for controlling and archiving the data exchanged between applications, which are based on the idea of a *central data warehouse*. Examples include TEF [746] of Intergraph, which builds on the data warehouse SmartPlant Foundation [747] or the data warehouse VNET [520] of AVEVA. However, these solutions only support the proprietary XML models of the vendor.

There is *no* (or very little) *integration*, neither conceptually nor technically, between the *two groups of tools* supporting the administrative domain-independent and the technical domain-specific business processes in the design process in chemical engineering. In particular, *workflow management* or *groupware systems* do not provide any interfaces to the dedicated chemical engineering tools. In more general terms, there are no tools available, which integrate the design process and the design product perspectives in a sound manner.

An Assessment of Current Design Practice

The analysis of current *design practice* reveals a number of *weaknesses* which have to be overcome in order to successfully establish design process excellence [791, 1047]. The most import issues are the following:

- There is *no* common *understanding* and terminology related to the design process and its results. Communication in the project team is largely informal. A misinterpretation of the communicated data and information often occurs due to the lack of common understanding.
- Creative design *processes* are *not* properly *understood*. Neither routine tasks nor the experience of senior developers is properly captured. The complex coordination tasks on the management level in a multi-project and multi-team environment are not made explicit. The dynamics inherent to creative design processes is not captured. Therefore, the design process is not transparent for the project team. Intuition and experience have to compensate the lack of a properly defined work process and of a systematic reengineering of the work process for continuous improvement.
- Design *processes* and their *results* are *not* sufficiently well *documented*. This lack of documentation prevents tracing (i) of ideas which have not been pursued further for one or the other reason, (ii) of all the alternatives studied, (iii) of the decision making processes, and (iv) of the design rationale.
- A systematic *analysis* of all the relevant *alternative solutions* is not carried out.
- The creation of knowledge through *learning* from previous experience is *not* systematically *supported* by information technologies.
- There is neither a systematic evolution of requirements and nor an assessment of the design objectives with respect to the requirements during the design process. There is no systematic management of conflicts between design information or change propagation mechanism between design documents. The *quality* of the *reconciliation* of different *parts* of a *design* with each other or with possibly changing requirements and specifications depends on the care of the designer.
- A *coherent configuration* of all the design data in the context of the work process is *not available*. Time spent for searching and interpreting information on a certain design in the course of the plant lifecycle is enormous. Often, it is less effort to repeat a task rather than to retrieve the already produced results from some source. Reuse of previous solutions and experiences at a later time in the same or a similar design project is not supported.
- There is no systematic *integration* of *design methodologies* based on mathematical models of the chemical processes with the overall design work process.

In addition to these work-process oriented deficiencies, there are also serious *shortcomings* with respect to the *software tools* supporting the design process [783, 834, 1043, 1047, 1062]. Some important considerations are as follows:

- Tools are determining the design practice significantly, because there has been largely a technology push and not a market pull situation in the past. Tool functionality has been constrained by technology, often preventing a

proper tailoring to the requirements of the design process. Usually, the tools are providing *support functionality* for only a *part* of a design task or a set of design tasks. Tool support is limited to determine and present the result of some task in the design process. Relations to the work process are not captured.
- Design *data* are represented differently in the various tools. There are not only *technical*, but also *syntactic* and *semantic mismatches* which prevent integration.
- There is a lack of managing *relations* between *data* and *documents* produced by different tools in different design activities.
- The implementation of the tools relies on proprietary information models and software platforms. Therefore, *integration* between *tools* is quite *limited* and largely focuses on those of a single vendor or its collaborating partners. The integration of legacy tools of an end-user into such a design environment or the integration of the design environment into the software infrastructure of the end-user is often difficult to achieve and induces high development and maintenance cost.
- Tool *integration* is largely accomplished by data transfer or data integration via a central data store, *neglecting* the *requirements* of the *work processes*.
- Project *management* and administration software is *not* at all *integrated with* engineering *design* support software. Hence, proper planning and controlling of creative design processes is difficult.
- *Communication* in the design team is only supported by generic tools like e-mail, video conferences, etc., which are *not integrated* with engineering design tools.
- The *management* of creative design processes is *not* supported by means of *domain specific* tools.
- The heterogeneity of the software environment *impedes cooperation* between *organizations*.

These two lists clearly reveal the *high correlation* of the *work processes* and the *supporting software tools*. Both have to be synergistically improved and tailored to reflect the needs of the design process in a holistic manner. We believe that a work process-oriented view on design and the required information technology support is a major prerequisite to achieve design process excellence. In particular, efficient computer-assisted support of complex engineering design processes has to address the integration problem in different dimensions in order to overcome isolated methodologies and informatics solutions currently available [537].

These *integration dimensions* include

- the different application *domains* and *disciplines* such as polymer reaction engineering, separations, polymer processing and materials engineering in the PA6 case study,

- the different design *tasks* such as raw materials processing, material product development, process synthesis, control system synthesis, or waste reduction and recycling,
- the *participants* with different disciplinary and cultural *backgrounds* in a project team consisting of developers, manager, customers, and consulting experts,
- the work processes of all the *team members*, and
- the various software *tools* employed during the design process.

1.1.2 Long-Term Goals and Solution Approach of IMPROVE

This subsection summarizes the key ideas of IMPROVE on the conceptual level. After a short motivation, we introduce design process innovations from the Engineering and the Informatics side, on which we later base our ideas for novel design support. We then concentrate on the question, how subprocesses of the work-process interact, or how their results are integrated. The formalization of such results is called *process/product model*, in short PPM. We show that this model has to be layered. Finally, we summarize all the requirements, we can find for such a PPM.

Motivating a New Approach

The global aim of any support of a design process in chemical engineering or in any other discipline is to *improve efficiency* and *quality* of the subprocesses and their interaction as well as the integration of their results within the overall work-process. The same holds true for the product.

This improvement is enforced by *market pressure* due to globalization [675, 721]. The pressure extorts either a more efficient production of chemical bulk products or of specialty products [650, 651, 728]. Both are based on improvements in the corresponding design processes. Especially, for polymer products, there are many possibilities to adjust product properties [678].

Therefore, chemical process design is facing big *challenges* [704]. A prerequisite for addressing these challenges is to better *understand* the nature of the design processes, to *improve* these processes by integratively regarding all inherent aspects [293], and to *accelerate* these processes. This is required for all integration dimensions referred to above. Especially, tool support has to be shaped to bridge the gaps between different persons, their roles, design disciplines, and companies.

The IMPROVE approach addresses the problems corresponding to the three steps understand, improve, and accelerate, which have to be applied to development processes: (a) We *avoid* – wherever possible – *manual* developer *activities* by transferring the corresponding activities to process-supporting tools. (b) We aim to understand, document, and reuse *design results*, design *knowledge*, and the underlying design *experience*. (c) Design processes are accelerated by including all relevant aspects and by relying on both, *concurrent* [494, 521, 539, 729] as well as *simultaneous engineering* [708].

Pragmatically, we want to use *existing* industrial *tools*, wherever possible. However, these tools have to be *extended* to better support the design process or cooperation within this process. In some cases, *new tools* have to be built, if the required functionality is not available.

All tools have to form an *integrated design environment*, supporting all essential activities of the design process. Therefore, developers are enabled to mostly interact within this environment. The realization of this environment is much *more than* a *systems-engineering integration* of existing tools. As a consequence, a nontrivial software development process has been incorporated into the IMPROVE project, dealing with extending existing tools, building new tools, and integrating all these tools to a coherent design environment.

Design Process Innovations

The computer-assisted support of design processes, as presented in this book, is based on a number of *novel aspects* which have been identified prior to the start of the IMPROVE project in related research areas. The following innovative aspects have been *identified* in various research *projects* by the *collaborators* of IMPROVE.

The Engineering Perspective

1. *Model-Based Process Design*: Model-based procedures are increasingly used in the chemical industry. They give an evaluation of a design decision on one side and they reduce the costs by avoiding experiments in chemical process development on the other. The development of mathematical models, necessary for model-based design on different degrees of detail, usually causes high costs. These costs can be reduced by shortcut calculations [13, 467] or by introducing computer-based modeling tools to derive detailed models [53]. Systematic and formalized mathematical models as well as corresponding model development processes [277, 303, 401] are necessary.

2. *Comprehensive View on Conceptual Design*: Chemical process design can be divided into different phases, each of them consisting of synthesis, analysis, and evaluation of process alternatives. Every result influences succeeding phases. Often, backtracking is necessary to achieve optimal designs, which is sometimes skipped due to cost reasons. Furthermore, an integrated treatment of different aspects in design [54, 98, 104] results in better solutions [106]. Examples include the integration of process and control strategy design [1], or an early consideration of safety, oparability, and economic aspects.

3. *Processes Analysis in Plastics Engineering*: The compounding of plastics materials requires a basic understanding of fluid dynamics and heat transfer inside the extruders to realize the requested polymer modifications

[313, 314]. The implementation of these effects is done by selecting and placing screw and barrel elements in order to configure the compounding extruders' flow channels. To optimize the screw configuration in critical functional sections, the polymer flow is analyzed by numeric simulation to gain process understanding in order to apply design modification on the detailed level.

4. *Ergonomic Analysis and Evaluation*: When developing new groupware systems and adjusting them to a specific application field, neither the precise characteristics of the application field nor the mechanisms, rendering the application of the groupware system profitable, are known at the beginning of the development process. Therefore, it is necessary to collect, understand, and document work and communication processes in a repetitive manner. This has to be done such that technical and organizational design measures can be identified gradually and precise requirements of an informational support [119, 395] can be deduced. For this purpose, an adequate modeling methodology [221] is as essential as a corresponding analysis procedure [283, 284, 482], a close interlocking between software development [281, 486] and iterative evaluation [285], as well as an introduction model adjusted to the business situation [178].

The Informatics Perspective

5. *Direct Process Support of Human Actors* [21, 93, 94, 189, 191, 194–196]: Such support is achieved by investigating process traces of developers [366, 469] and by offering best-practice traces by means of process fragments [21, 156, 371]. Specifically, process actors are supported during decision making and during negotiating to gain an agreement [192, 193, 458]. For that, heuristics are offered how to proceed. Basically, direct process support is directed to facilitate or to standardize acting, and to temper cooperation problems, both through best practices by using the experience of developers. This experience is being incorporated in tools. Direct process support is usable for all roles in the development process (developers, manager etc.).

6. *Product Support for Achieving Consistency* [15, 26, 27, 33, 36, 37, 134, 229, 254, 334]: This kind of support is used to assure structural and value constraints of development products on the language or methodology level. This holds true for single documents but, even more, for consistency handling between different documents, usually produced by different developers. Corresponding constructive tools for such consistency maintenance between different documents, called integrators, are offered for change processes, where different developers are involved. These integrators help to assure or to reinstall consistency across different documents. These tools provide an essential support for modifications in a development process. Actors are not directly regarded and their activity is not restricted.

7. *Informal Multimedia Communication of Developers* [317, 420, 453, 456]: This type of spontaneous collaboration of developers is used for clarifying

a problem, for discussing intermediate results, for preparing decisions or agreements. Formal cooperation cannot substitute spontaneous communication "in between" which makes use of the ability of human actors to grasp high-level information, ideas, suggestions from descriptions, especially in the form of graphical representations. Spontaneous communication usually also influences later formal communication and can be a step of formal cooperation, e.g. within a decision meeting. Multimedia communication is valuable, especially if offered in an application-dependent form.

8. *Reactive Administration* [30, 160, 162, 174, 211, 355, 475, 476]: Most process approaches for administration neglect the fact and the resulting problem that design processes are reactive, i.e. have to be permanently adapted during the process itself. Adaptation is due to decisions, details in results, upcoming problems, necessity of changing the proceeding, introduction of method patterns [390], reuse rules, and the like. All these adaptations are consequences of creative steps in the design process. Work processes, products, and resources [391, 477] are not regarded in all details, when managing design processes. A specific problem, addressed in IMPROVE, is the management of distributed[3] design processes [208].

Key Topic: Process/Product Model

Quality and efficiency improvement cannot be achieved by merely integrating existing tools on a system-technical level [295]. Instead, additional *functionality* has to be offered which is adjusted to cooperative development and which is able to bridge the above mentioned gaps [296, 298, 343, 344, 346, 352]. This requires that the interaction of subprocesses and the integration of subproducts is well-understood and also formalized. So, there is an urgent need for a comprehensive and *formal process/product model.*

A process/product model has *various advantages*. It can be used to better understand the cooperation between developers, to document current or already finished design processes, to know how to support such processes by tools, and to promote reuse of partial designs. Especially, such a model is necessary for building an integrated design environment essentially supporting the whole design process.

Design Processes and Products

In a design process different subprocesses interact. Any such subprocess results in a subproduct which has to be integrated in the overall design product. The interaction of subprocesses can vary. Consequently, we find different forms of product integration. In the following, we discuss such aspects of interaction and integration.

Basically, there are *different granularities* of *subprocesses*: A subprocess may be an essential part of the work process of a developer. A subprocess may

---

[3] also called cross-company processes or interorganizational processes

enclose the collaboration of two developers. Finally, a subprocess may require a team of different developers. An essential task of that team collaboration – again being a subprocess – is the coordination of team members.

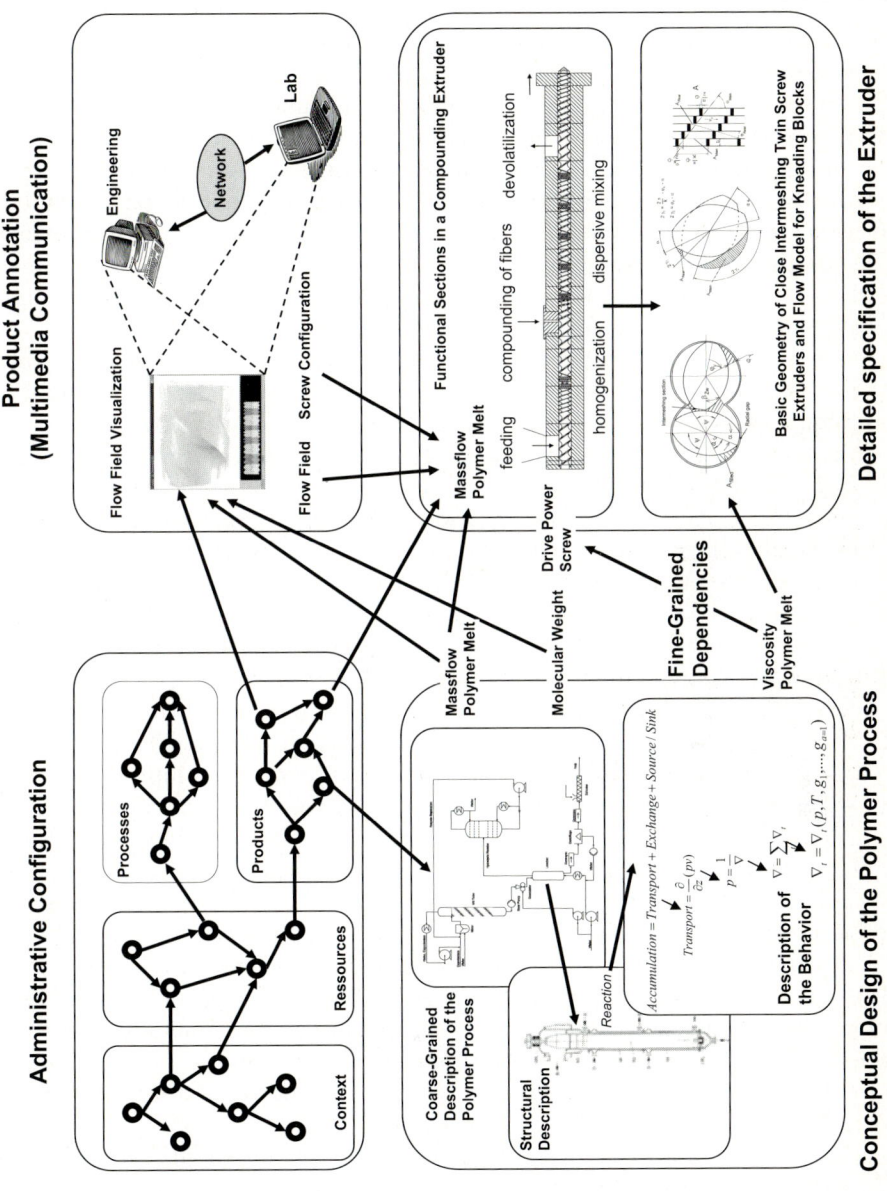

Fig. 1.4. Overall configuration of the development product (cutout)

Subprocesses yield results in the form of subproducts. More often than constructing new results, modifications have to be made due to mistakes, changed decisions, or requirements. Furthermore, *subprocess interaction* is not only that one subprocess constructs or changes a result which is needed, elaborated, or detailed by the next. We also find that one subprocess determines for another one, how to proceed, how the resulting products looks like, which specific aspects have to be obeyed, and the like.

The result of a complex design process is a complex product, called *overall configuration*(cf. Fig. 1.4). This configuration contains the various results of all designers together with their mutual relations (technical configuration). Logically separable parts of a moderate size, each usually constructed by one designer, are called *documents* (e.g. a flowsheet for a certain part of a plant). The overall configuration contains *subconfigurations*, e.g. for synthesis, analysis, and evaluation of an essential part of a plant, together with corresponding design alternatives and a design rationale.

Documents have a rich *internal structure* with mutual relations between their parts, called increments. For example, the behavioral description of an apparatus consists of terms with mutual syntactic and semantical relations.

Especially, there are many *fine-grained relationships between* increments of *different documents* (see again Fig. 1.4). E.g., parts of the behavioral description of a plant are in close relation to the structural description. These fine-grained relations are needed to estimate the effort of changes, to carry out changes, to check consistency of the changes, etc.

An essential part of the overall configuration is the so-called *administrative configuration*, the information of which is used for coordinating the collaboration of developers. Here, we find information corresponding to the design process, its product, the corresponding resources, but also of the context (department, company etc.) in which the process is carried out. The information is coarse-grained: It is only interesting that a subprocess or subproduct exists, in which state it is, but not how it is carried out or structured internally.

Figure 1.4 shows a cutout of an *example* for an *overall configuration*. In the administrative configuration we find all the information necessary to coordinate the corresponding group of designers. Especially, we find information for product management. The biggest part of the configuration consists of technical documents, produced by different designers. Between those documents there are many fine-grained relations which refine the coarse-grained relations of product management data. As a result of direct communication between designers, there are various visualizations and there are further annotations of design products to denote decisions, identified problems and mistakes, hints or references to indicate additional investigations etc.

In case of distributed design processes, the overall configuration is produced by and is spread between different companies. The different parts of different companies are not completely disjoint, as any developer, subteam, or company has to see the context, in which the corresponding part of the product has to be developed. This *distributed* and partially redundant *overall configuration* has to be kept *consistent*, as if the design process were carried out within one single enterprise.

Above, we have argued only from the product perspective of a design project. We could also and *analogously* have *argued* from the *process perspective*. Then, we would have found subprocesses which have a rich internal structure, are built hierarchically, have different forms of complex interactions, and so on.

Processes and products are dual to each other as any process has a result, and a result needs a corresponding process. However, they contain different information. Therefore, in the process/product model we are investigating, we do not express one aspect implicitly by the other. Instead, we regard both *processes* and *products explicitly*.

Interaction, Integration, and Reactivity

As already stated, Fig. 1.4 is a cutout of the results of a design process in chemical engineering. It solely shows the *product* of the process, the *process* itself cannot be found. So, we do not see how a manager creates the administrative configuration, how a designer synthesizes the flowsheet, how different developers annotate products in a spontaneous discussion etc. For any document or any bundle of relations between two documents there must be a corresponding subprocess for that document or bundle.

Every *designer* does only see a *specific part* of the overall configuration which is important for his subprocess. This part consists of the results the designer is responsible for, but also the necessary *context* – produced by others – namely necessary input, rules to obey, proceedings to follow, explanations for the task to fulfil etc. These specific parts of the configuration are produced, changed, and stored in a distributed manner.

The elimination of errors, the modification of a chemical process design, the extension of the design product, the adaptation according to new requirements, or the use of reuse techniques induce significant, often structural changes within and of the overall configuration. To support those changes, the *subproducts* of different designers have to be *integrated*.

Modifications of the overall configuration are changes of and within documents. Even more, there are induced changes due to modifications in other documents enforced by fine-grained consistency constraints between documents. Both, changes and induced modifications, have to be regarded transitively. Hence, we get *change chains* (in the product) or change *traces* (of the process). These changes are coordinated using the information contained in the administrative configuration. Hence, any change of the overall configuration is a *cooperative* and *distributed* process.

Changes can seldomly be carried out automatically. *Subprocesses* for producing documents are *creative*, as there are many ways to solve a subtask resulting in different solutions. In the same way, a designer can give different answers how to modify his document, as a consequence of changes done by other developers. Therefore, in cooperative and distributed development processes we have *interaction* of *subprocesses* of different developers. This interaction can have different forms, as already discussed.

The structure of a design process and its product is in most cases only determined in the process itself, as design decisions predetermine following processes and products. In the same way, made modifications or specific error repairs imply what and how to do. Hence, the process can only be partially fixed before it starts. We call this the *dynamics* within the process.

The *results* of a completed design process can be *used* in the following one. They are just used and need not be developed. This drastically *changes* the *overall structure* of this next process. *Knowledge* about processes and products can be introduced in form of templates. Their use within the process, again, evidently changes this following design process. Also, predeterminations of the process influence future subproducts.

A subprocess produces a result, e.g. a flowsheet of the essential part of the plant. The flowsheet determines the following subprocess, e.g. the analysis of the design by simulation. We call this mutual relationship between subproducts of one subprocess and later subprocesses *reactivity*. This can also happen backwards: One may detect an error in a resulting document of a subprocess. The error, however, has been made in a preceding subprocess, which has to be taken up again.

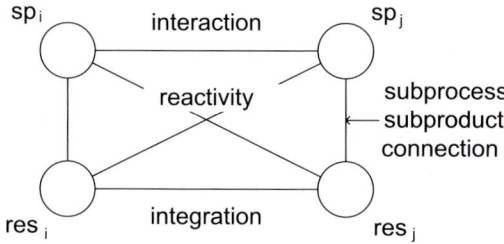

Fig. 1.5. Dynamic interplay of processes (interaction, integration, reactivity)

Figure 1.5 shows the *interplay* of *subprocesses* (interaction) sp_i and sp_j, the *subproducts* res_i and res_j which have to fit together (integration), as well as how a subproduct influences the following subprocesses or vice versa (reactivity). Furthermore, subprocesses and subproducts are dual to each other and have to be seen as two parts of a whole.

As already explained, the interaction of subprocesses can serve different purposes. In the same way, the subproducts to be integrated may express different things (hierarchy, parallel views, fixing the structure of other and following subproducts etc.). Therefore, in a design process we have dynamic *interaction* of different subprocesses yielding different results to be *integrated* to a *consistent* whole.

Improvement of *quality* and *efficiency* can now be achieved by a better interaction of subprocesses, by a better integration of subproducts, by understanding and making use of the duality of subprocesses and subproducts, or of the reactivity induced by their mutual relation. This is one of the key messages of IMPROVE.

Layers of the Process/Product Model

Processes as well as products have to be regarded and *modeled* on *different* levels, which we call modeling *layers*. The simplicity and uniformity of layers as well as of the transformation between layers is a *prerequisite* for a simple and uniform process/product model. From this model we later "derive" the functionality of the integrated environment as well as its implementation (cf. Fig. 1.6).

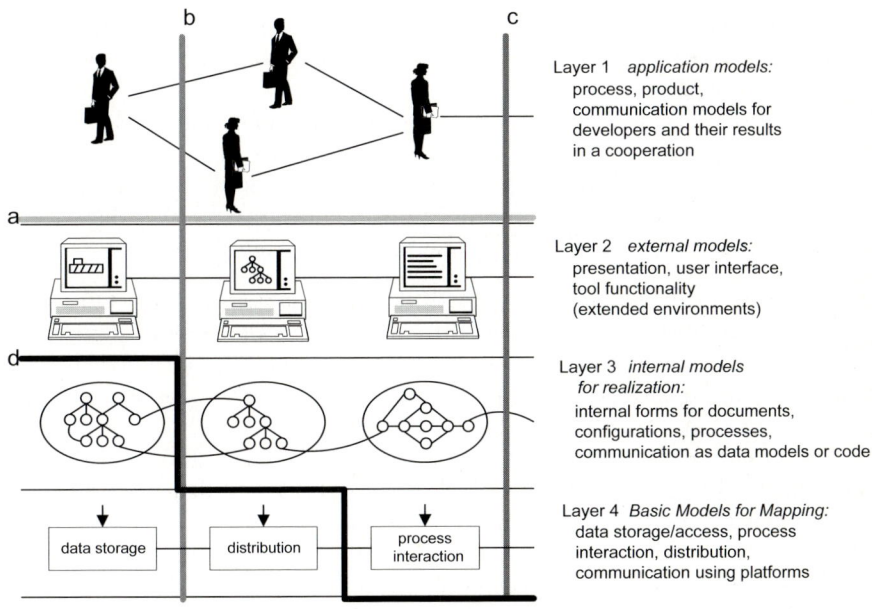

Fig. 1.6. Layered process/product model

Layers

- On layer 1 of this figure we find the *application domain models* for the process, its product, but also the required forms of collaboration and communication. Even more, to be seen later, we find models to prestructure the application domain and its underlying knowledge. We find organizational

knowledge of the process, carried out in a specific subdomain, company, cooperation of departments, etc.
- The support of the design process by tools should be as close as possible to these application models. This facilitates that the process is carried out within the design environment and not outside of it. On layer 2, we, therefore, find *external models* of *tools* for different kinds of users, where the presentation of these tools corresponds to the languages and methods used within the design process. Analogously, the user interfaces fit the application domain, the organization of a process, its context, and the like.
- On layer 3 we find *internal models* of *tools* which represent subconfigurations, process structures, collaboration and communication structures etc. on an abstract level, but containing all necessary details for tool construction. These internal models are either represented by suitable data models within the tools of the environment, or these models act as a description from which equivalent tool software is later derived. In any case, these models have to guarantee that tools offer the functionality and user interface of layer 2. Typically, these tools offer a suitable cutout view for a certain user.
- Internal models have to be mapped onto services of distributed platforms. In order to keep the realization of tool extensions or of new tools independent of the specifics of the underlying platforms, we find *basic models* on layer 4. They are used for tool implementation and they have to be transformed to a chosen and specific proprietary platform.

We give some *examples of what* should be found *on which layer*: On layer 1 we find a formalization of incremental and iterative development processes, incremental changes within results, or the integration of models for chemical engineering and plastics processing. On layer 2 we have to offer tool behavior and user interfaces to allow, support, and facilitate these new application concepts, thereby regarding that users have different roles needing different tool support. Layer 3 models tool internal structures – efficient data access due to efficient structure-, attribute-, relation-changing commands – which contain a lot of further details compared to layer 2. Layer 4 refers to the internal structure of tools consisting of various computer processes and data components to be mapped onto a given distributed platform, like CORBA.

A Layered PPM

Models for processes, products, collaboration/communication, subprocess interaction, subproduct integration etc. *occur on every* of these four *layers* and have to be mapped onto the next layer. So, we find the *same basic notions* on every layer. Let us explain this by taking the notion of a process as example: On level 1 we speak of work processes of designers, containing subprocesses, which again contain process steps. On level 2 we speak of support of an essential step by a complex command of a tool, if the process is interactive, or of an automatic tool if the subprocess can be carried out automatically. On layer 3 we structure complex commands, we organize the complex intermediate states of tools, we organize how complex models are to be presented to the user and how this is done (e.g. following the model-view-controller paradigm). On level 4 we organize how complex tool components can be mapped on operating system processes.

In the same way, we find *hierarchies* on *every* of the four *layers*. Application processes are in hierarchical order, from life cycle structures to work processes and their steps. Tools on layer 2 have to support all hierarchical levels. On layer 3 these hierarchies have to be organized, stored, and retrieved. On layer 4 automatic parts of that process hierarchy or command execution are mapped on computer processes.

Hence, we see that the *process/product model* appears on *any* of the four *layers*. What we have argued for processes, we could have done for the other basic notions of above (products, collaboration, etc.) as well. In addition, the specific forms of the process/product model on a layer have to be *mapped* downwards to the next.

A great challenge for IMPROVE is to handle *distributed development processes* across different departments of a company or even between companies. In this case, Fig. 1.6 has to be seen in multiple occurrences. Different application, external, internal, or basic models have to be integrated. Specific problems arise, as *different model universes* can exist on both sides of cross-company processes (chemistry, chemical engineering, plastics processing, etc.), or different organizational structures, design process habits, underlying languages/methods (e.g. a strict sequential process on one side, simultaneous engineering on the other). Nevertheless, the different layered process/product models on each side of the cooperation have *to be integrated*.

Requirements for a Comprehensive Process/Product Model

The *process/product model* for distributed design is *no static* and *fixed determination*, as we have learned above when discussing dynamics of the process and the various interaction forms for its subprocesses. We now discuss the *requirements* this model should fulfill.

Requirements

The PP Model has to be open, extensible, adaptable, and dynamical. Here are *some* of the *reasons*:

- A design process in chemical engineering looks different, depending on the specific chemical process, the applied standards, the available knowledge and experience, the used apparatus etc.
- The design subprocesses and subproducts are different. Hence, they cannot be described and formalized equally well.

- Departments, companies, and combines of companies structure and coordinate their design processes differently. Furthermore, the creativity of designers also enforces freedom and variety.
- The structure of design processes in most cases is only determined at process runtime. The development process can rarely be structured and prescribed before it starts.

Therefore, the *comprehensive process/product* model for development has to *describe* in a *formal way*

- different forms of subprocesses (granularity, structure),
- different interactions of these subprocesses,
- different subproducts (granularity, structure),
- different integration mechanisms for these products,
- the duality and connection between subprocesses and products,
- the necessary reactivity mechanisms,
- the mapping from one layer to the next,
- parameterization mechanisms to handle the huge variety of underlying process structures, and
- the cooperation of different processes to one distributed process.

Hence, if we talk about a *uniform* process/product model we think that there has to be a comprehensive assembly of *modeling constructs* to express variety of processes, interaction forms, ..., difference and similarity of cooperating processes within one cross-company process. So, uniformity applies to modeling constructs. The corresponding *models* for different processes, evidently, are *different* and not uniform.

We have to recognize that design *processes* are *structured* and *carried out by human actors* which are supported by more ore less intelligent tools. It is obvious that subprocesses and overall processes cannot be completely formalized, as creativity is beyond formalization. Furthermore, even if a process were completely formalizable, this formalization can only be done piecewise, as the process is developing and changing at process runtime. The above modeling constructs have to be flexible enough to express the variety of processes, having human actors as creative elements, and evolving at runtime.

A Nontrivial Task

The development of a comprehensive and uniform process/product model is a *nontrivial modeling task*, since

- there were no formal notations for subprocesses, subproducts, interaction, integration, and reactivity mechanisms at hand with precise syntax, semantics, and pragmatics fulfilling the above requirements,
- consequently, there is no methodology for these notations,
- language definition, method definition, and modeling has to be started in parallel,

- there is no uniform modeling methodology available (to use modeling constructs in a clean and approved way) within the layers, nor to describe the transformation between layers,
- up to now nobody has regarded the breadth between application models and operating system models to be a part of one big model,
- commonalities by structuring a model, use of common and general substructures, introduction of generic models and use of generic instantiations is not state-of-the-art, not even on one of the above layers,
- in case of distributed and cross-company development we have the additional problem, that equal or similar circumstances may be structured differently: Modeling should be able to deal with similarity, difference, and correspondences of similar models.

There was little to be found in literature regarding this ambitious problem of a comprehensive, uniform, and layered process/product model . There were some ontology approaches [540, 617] or some coarse partial models [14] on the application level. Also, there were some internal models available for tool construction, e.g. to formalize tool behavior [334, 350]. So, *setting* the ambitious *goal* of a *process/product model* is another key message of IMPROVE. This general model must have a tight *connection* from *application* models down to the *realization* of tools.

What we have achieved within IMPROVE are only *partial solutions* to this big problem. They are described in Chap. 6. There is plenty of room for continuing research.

1.1.3 Design Support by an Integrated Software Environment

Available tools are far from providing a complete and comprehensive support for all designers of a cooperative project. We find *gaps* where no support or no valuable support is available, and we find *islands* of good support. Especially, we find areas, where the support of tools should be *closer related* to the nature of design activities or to the patterns of the cooperation between designers. Only experience or intuition of developers and organizational as well as management rules give a chance that development processes nowadays are successful.

In this subsection we argue that new integration functionality transforms existing tools to personalized environments which facilitate cooperation. Such cooperative personalized environments are then integrated to build up an integrated environment for the cooperative design team. The integrated environment is based on an open framework and makes use of advanced reuse techniques.

Gaps of Support and Cooperative Environments

Gaps of Tool Support

There is usually a big *gap* between the *application* layer and the layer of external *models* of *tools* (see again Fig. 1.6; separation line a): Tools only support minor or primitive parts of the work processes of developers, support does not fit to the task to be solved, the presentation of results is not evident, or the communication between developers is done outside of tool environments (phone, meetings, etc.). Work processes are mostly manual and tedious.

Especially, there is little support for the *cooperation* of *different developers* (separation line b): Results of one developer are interpreted by another developer to infer what to do. In the same way, project coordination and design are not integrated.

In the case of *distributed* and cross-company *design processes*, the situation is even worse (line c): There is at most data exchange in the form of files, eventually in some intermediate or standard format. Cooperation is remarkably hindered, because there may be different cultures in the partner organizations.

Finally, the *realization* of *environments* may be completely different (see separation line d): There is no standard structure for design environments nor for their integration. One environment may directly use files, another a data base, a third may use a standard for product data on top of a data base system etc.

Bridging the Gaps by New Functionality

Developers and Tool Support: Processes of human actors are supported by new semantical tool functionality, concentrating on major design steps and/or eliminating bookkeeping effort.

As IMPROVE starts by investigating and evaluating industrial processes from engineering and labor research perspectives before tool development is started, there is a high probability that the new functionality is accepted by the design team.

Approved action patterns (experience) are used for process support. If there are structural conditions for results, tools may support to obey these structures or even generate these structures.

Different Developers or Developers and Management: Integrator tools help to assure fine-grained consistency relations between increments of different documents produced by different developers. Action patterns can also be used for actions of different persons. Both can be used for simultaneous as well as concurrent engineering.

As argued above, there is much spontaneous communication among developers "in between". The results of these discussions can be stored as annotations to element objects of products. Project coordination can support to find the right partner for spontaneous communication, and it can help to organize this communication.

Most dynamic effects result from details or problems, elaborated or detected during design. Tools can extract the corresponding information from master documents, describing the product of the design process, in order to support dynamic coordination.

Different Subprojects of Distributed Development: It should be possible to parameterize tools for a specific context in an application domain, or for well-defined or approved structures of processes or documents. The corresponding adaptation should be possible from one project to the next, or even better during a running project.

Subprojects are independent on one side and have to be integrated to one project on the other, especially in cross-company design. This enforces interactions between both, the designers' and the management level.

Extended Personal Environments

New Concepts and A-posteriori Integration

The new informatics concepts of IMPROVE (see 1.1.2) do *fit* with *a-posteriori integration*, i.e. to use existing tools (cf. Fig. 1.6):

Direct process support analyzes traces of developers, determines well-behaving traces and chunks, and offers them in the form of complex commands. Hence, we have additional functionality on top of existing tools.

Integrator tools for assuring fine-grained *interdocument relations* support to keep results of different designers in a consistent state. They can also be applied between designers and manager. These tools use views of documents, elaborated by conventional tools. These conventional tools are still necessary to produce the designers' results.

Multimedia communication replaces communication outside the design environment (phone, meetings). Available tools are extended to support this advanced communication. The results of communication are added as annotations to usual documents. Again, we use existing tools.

Reactive administration is for coordinating dynamic design processes. There is no functionality at hand on top of which we could offer these reactive mechanisms. Hence a new "generalized workflow system" was built. A similar system can also be built on top of classical management tools (cf. Chap. 7).

Not only the new informatics concepts can *extend* available *functionality* of existing tools. The same holds true for the *engineering innovations* we discussed above (see 1.1.2): Introducing model-based design and supporting it by further tools can also be seen as an a-posteriori extension, as conventional tools will be necessary as well. The same holds true for the comprehensive view on conceptual design by regarding additional aspects as costs, security, etc.

Cooperative Personal Environments

An essential part of an integrated environment to support the whole chemical design process are the *personal environments* for specific subprocesses of the overall process. The informatics concepts can now be used to prepare available tools/environments for this cooperative overall process. This is done by introducing interfaces for new *cooperative functionality* and by adding this new cooperation functionality. Figure 1.7 shows the different parts of that extension and explains their use.

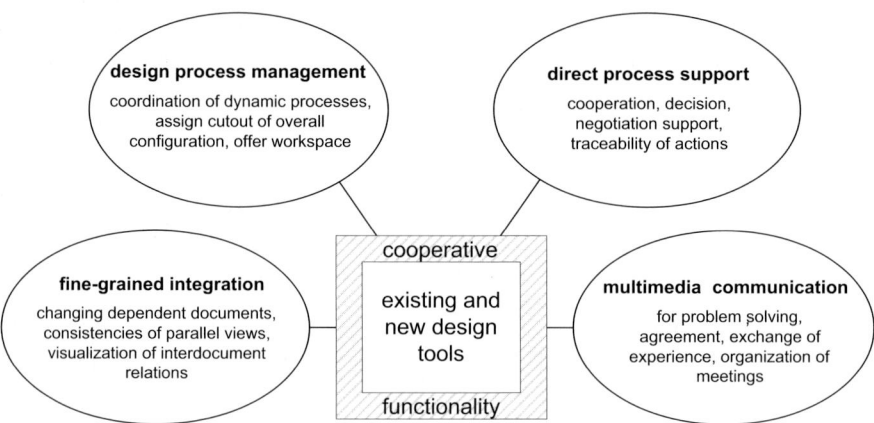

Fig. 1.7. Cooperative environments around given or new tools

The aim of these cooperative environments is to support cooperation of human process actors such that they need not use auxiliary means outside their personal environment. Therefore, the cooperative environment should be *complete* w.r.t. all *means* necessary *for cooperation* within the design process.

An even more important goal of the extension of a given environment to form a cooperative personal environment is to *specialize* this environment to *fit* the corresponding *personal* and *cooperative subprocess* of the actor. All the above new informatics concepts offer valuable support of a new kind (see Fig. 1.7): Direct process support introduces experience and best-practice for cooperation, decisions, and negotiations. Product consistency support facilitates to keep dependent documents or parallel views consistent to each other and visualizes fine-grained dependencies. Multimedia communication supports discussions to find out if and where there is a problem and to sketch the "right" way to solve that problem. Reactive administration allows to handle project coordination even if there are severe changes in the design process. So, by these new functionalities, any given software environment (for synthesis, simulation etc.) gets a new quality for being usable in the cooperative process.

New *engineering concepts* according to 1.1.2 yield *new environments* providing necessary, specific design functionality, usually not available or broadly used today, as e.g. specific environments for mathematical modeling and the like. Of course, these new environments can also be *extended* to cover the *cooperation* aspects outlined above. Fig. 1.7 holds true for standard tools or for new engineering design functionality. For both, additional cooperation functionality is useful.

It should be mentioned that in a complete design environment for the whole design process we do not only find technical environments for specific technical tasks (synthesis, simulation etc.) or coordination management, possibly extended by collaborative functionality. There are also specific *environments for other purposes*: (1) an environment to define process fragments extracted from the designers' experience, (2) an environment to introduce specific process/product types or process knowledge on the coordination level, (3) an environment to handle changes of the underlying platform, and many others.

Furthermore, and of another quality, there are *specification environments* which we use to build tools. The latter environments are not part of the integrated environment to support the design processes in chemical engineering. Instead, these environments belong to the tool development process. Therefore, they are not discussed here but in later chapters of this book.

Realization of the Integrated Environment by an Open Framework

The approach of IMPROVE is not to add specific environments to given ones in order to get a more or less complete coverage for all activities in design. Instead, we aim at getting a *common* and *uniform approach* for a-posteriori integration. Thereby, we expect that the corresponding results are (at least in parts) also applicable to other engineering application domains [353].

Uniform A-posteriori Integration Approach

Existing environments for technical tasks in the design process are extended, as they have not been developed for integration using an integrated process/product model [296, 298, 346]. Possibly, they have to be modified before [88]. *Extensions* of environments thus consist of *two parts*: (a) to introduce interfaces to couple new functionality, and (b) to offer this new and cooperative functionality. (This new functionality can later be integrated again and, therefore, offers further synergy [348].)

Finally, the extended environments are to be integrated again to an integrated but distributed overall environment for the whole design process. As we aim at *tight* integration, the goal of a *common* a-posteriori *solution* is a scientific *challenge*.

Our *approach* essentially *differs* from other tool *integration approaches*, e.g. [795, 950], but also from standardization approaches [751, 859], other tool projects in chemical engineering [505, 524, 928], as well as approaches to support specific parts of a design process [541, 622, 931, 952].

All these, but also other approaches, do not regard the *specific requirements* for cooperative, distributed, cross-company software environments for engineering design applications, namely loosely interacting subprocesses, integration of subproducts, reactivity etc., thereby also regarding technical aspects as security, different views, various specific tools etc.

Integrated and Distributed Environment of Design Environments

Figure 1.8 shows the *architecture* of the *integrated overall environment* in a simplified form (for a pragmatic understanding of architectures see [331, 336]). Usual architectural sketches are even coarser [950]. The existing specific environments for developers to be integrated are of different quality and completeness. They, also, have quite different purposes in the development process but are regarded as having the same rank on an architecture level.

Some *remarks* are to be made before we start the discussion of the architecture: The architecture reflects layer 3 and 4 of Fig. 1.6 and, in addition, has a layer for existing platforms to be used. User interface handling does not appear in the architecture, as our architecture is too sketchy. Therefore, layer 2 of Fig. 1.6 is missing. Layer 1 does not appear, as layer 2 should reflect the application models with corresponding and suitable user interface functionality. A further simplification of the architecture is that the distribution of the overall environment is not shown. Finally, the specific environments of the last subsection (for extracting developers' experience, defining consistency rules, or adapting a process, etc.) are also not shown.

The overall environment consists of different *parts* ordered in different *architectural layers*, which we are going to discuss next:

- Given specific *technical environments* (for synthesis, simulation, etc.) are to be integrated (white parts (1) of Fig. 1.8). The code of these environments usually handles proprietary data structures. Code and data usually use a specific platform.
- Specific environments behave as personal and cooperating environments by an interface and by additional cooperation functionality (see above). For that purpose, given environments are connected via *technical wrappers* (2.a). Furthermore, data and functionality of existing tools have to be offered in a homogeneous form which is done by so-called *homogenization wrappers* (2.b).
- New functionality is offered by an extension of existing tools. For that purpose, new *internal data models* (3.a) are introduced to allow this new functionality (document structures, process descriptions, fine-grained relations, annotations, administration information). To be more precise we do not find the data descriptions and their states in the architecture but the *software components* by which those representations are defined, retrieved, or updated using some underlying and structured data. Furthermore, as already sketched, new cooperation functionality (3.b) can again be integrated, yielding further synergistic functionality (3.c).
- Components and data for new functionality as well as program and data components of existing environments are mapped onto distributed plat-

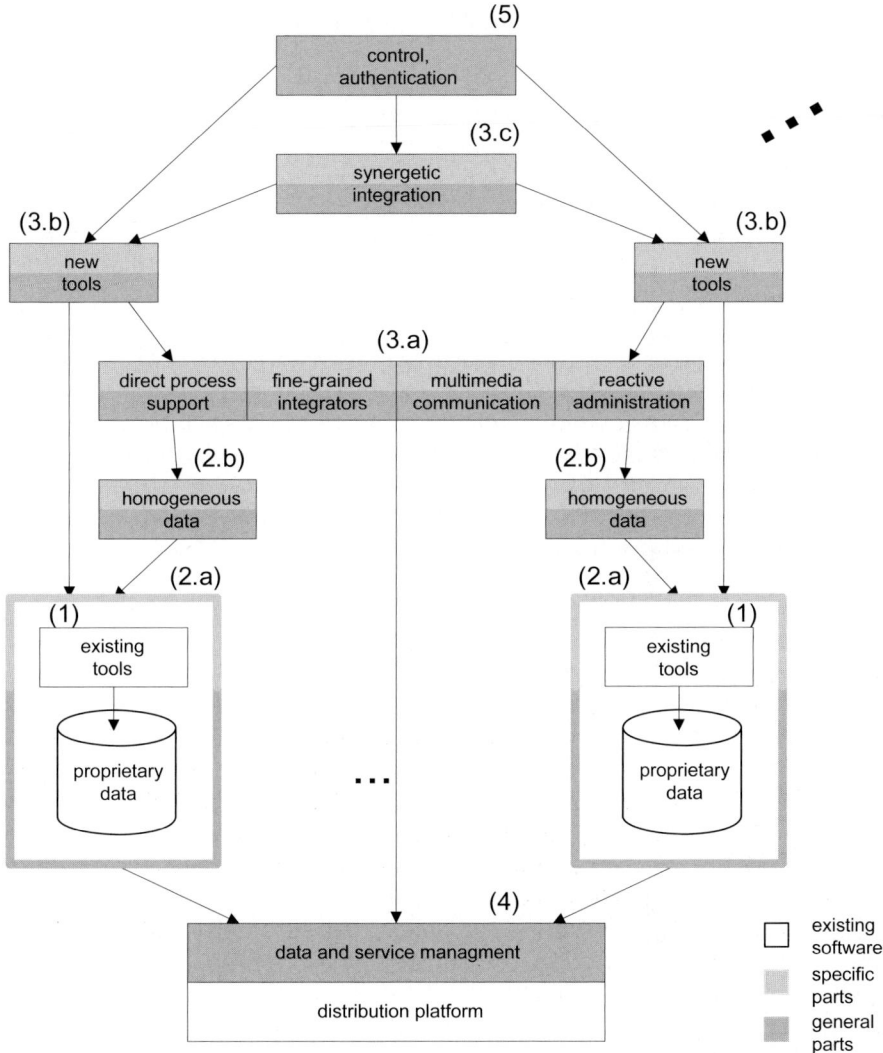

Fig. 1.8. Integrated environment of design tools: architectural sketch

form services. Thereby, we use existing platforms like CORBA [877] or COM [846]. At the moment, the use of such platforms comes along with a lot of details: On which computer are processes or data, how to reorganize in case of bottlenecks or in case of breakdowns etc.? Furthermore, some platform services are still missing (e.g. protocols for multimedia communication, security services [76]). Therefore, we have an additional architectural layer (4) in our architecture. The interface of that layer offers all necessary services, the realization maps these interfaces onto services

of existing platforms. Basic data models are introduced [262, 273, 421] to administrate platform details as well as mechanisms to extract data [11, 190].
- Figure 1.8 shows the extension of two given environments and their integration. Of course, the integration prototype, to be discussed in the next section, integrates much more specific environments. Any extended and integrated environment has some component to start/end a user dialog (5). In case of distributed and cross-company cooperation different integrated environments, each for one location, have to be integrated to a distributed and overall environment.

It should be said that integration of existing, extended, and newly built environments for one location and their further integration to a distributed overall environment results in a big software development process and project for tool construction/extension/integration within IMPROVE. The underlying question is practical as well as big and, therefore, needs software engineering and architectural knowledge. The solution we have achieved leaves the size and lucidity of textbook examples.

Framework and Reuse Aspects

Open Framework for A-posteriori Integration

Universality of IMPROVE's tool integration approach means that we have to *show*, (a) that the solutions can be adapted to a specific scenario in chemical engineering, (b) that the prototypical solution of IMPROVE is complete w.r.t. covering all occurring scientific and practical problems, and (c) that the fundamental concepts of the solution can also be applied outside of chemical engineering to other engineering design processes.

Due to the realization effort of the challenging approach of IMPROVE this can only mean that we do not aim at case-dependent solutions. Instead, we are looking for a *framework* of reusable components (dark grey parts of Fig. 1.8) and a carefully *structured software development process* to get a distributed overall environment by making use of the framework.

In the following we discuss the *reusable components* of that framework: Parts of the technical wrappers to enclose existing tools belong to the framework (2.a). Technical wrappers also contain specific parts, not to be reused. Mapping (layer 4) is completely reusable and, therefore, is a part of the framework. Homogenization wrappers (2.b) as well as new data structures (3.a) contain reusable parts. New tools (3.b) also are partly reusable (e.g. command cycle elaboration) as well as the components for synergy (3.c). Finally (5) is completely independent.

The framework is called *open* as arbitrary existing environments can be integrated. Thereby, the framework components remain unchanged. The framework *avoids inhomogeneity* of environments, at least corresponding to new code for new environments, integration functionality, synergy, etc. (see separation line d of Fig. 1.6).

Integration approaches [950] often cite *integration dimensions* [1038], namely control, presentation, data, and platform integration. This classification is too narrow, as constituents for these dimensions can be found on *every layer* of Fig. 1.8. To give the arguments for the data integration dimension, which is mostly understood to put data into the "same" repository, or to use the same data modeling (e.g. relational) approach: On the external level (not contained in Fig. 1.8), we find corresponding functionality to build up and consistently change data of the overall configuration of a design process. These overall configurations have to fit the application models (again not shown in Fig. 1.8). On the internal level, uniformly modeled corresponding data structures (syntax, semantics) are necessary to achieve tight integration. On the mapping level the data have to be stored in different data containers or retrieved from them. On the platform level they are transported in a common standard format.

Distributed systems [572, 573, 648, 665, 738, 980] in the form of bound and coupled programs are *different from cooperating design processes* supported by distributed overall environments. For design processes we need loose mechanisms for interaction, integration, reactivity, etc. which, nevertheless, have to be as tight as possible. We will probably never be able to describe a design process by a program, a statement [892] which raised some controversy. On the other hand there are some similarities between distributed systems and cooperating processes and their environments, which have to be detected.

Reuse Applied to Framework and Overall Environment Realization

Reusable components of the framework were not used from the very beginning in IMPROVE. We needed some experiments and corresponding experience to find commonalities, generic templates, general mechanisms, and the like. Reuse within development of new cooperative tool functionality was mostly detected and applied by the corresponding partner, offering and realizing this new functionality. Nevertheless, some nice results corresponding to *architecture transformations* in order to structure and in order to adapt the overall architecture have been found (see Sect. 5.7).

Technical or homogenization wrappers appear in many occurrences in an overall architecture. Therefore, some effort was spent in order to *mechanize* the *construction* of those *wrappers*. Again, the corresponding results are described in Sect. 5.7.

The underlying *general platform* and its mapping to services of existing platforms also constitutes a major part of the reusable framework. The corresponding research work is described in Chap. 4.

All remaining components (not being components of existing tools, not being completely reusable) are indicated in light grey in Fig. 1.8. They are not part of the framework. These *specific components* are *realized differently* in IMPROVE: Some of them have been manually coded with or without a clear methodology. In other cases, these specific components are specified and the specification is directly interpreted. Finally, in some cases generators are used to generate code from a specification. Thereby, generally usable components were detected (interpreters, generators) which, however, are not a part of the overall environment, but belong to the tool construction/extension/integration process.

1.2 A Scenario Demonstrating Design Support in Chemical Engineering

R. Schneider and B. Westfechtel

Abstract. The IMPROVE demonstrator is an integrated research prototype of a novel design environment for chemical engineering. Throughout the IMPROVE project, it has been considered essential to evaluate concepts and models by building innovative tools. Moreover, an integrated prototype was seen as a driving force to glue research activities together. Two versions of the demonstrator were implemented. The first version was demonstrated after the first phase of the project (from 1997 to 2000). The second demonstrator built upon the first one and was prepared at the end of the second phase (lasting from 2000 to 2003).

This section describes the second demonstrator, which shows interdisciplinary cooperation (between chemical and plastics engineering), interorganizational cooperation (between a chemical engineering company and an extruder manufacturer), and synergistic tool integration.

1.2.1 Introduction

From the very beginning, it has been considered crucial to address both concepts and tools in the IMPROVE project. Elaborating concepts and models alone bears the risk of doing fundamental research which cannot be put into practice. Conversely, building tools without studying and developing the underlying concepts and models ends up in software development activities which do not advance the state of research.

For this reason, it has been decided to build demonstrators on a regular basis. The demonstrators would provide feedback to evaluate the research on concepts and tools. In the first place, demonstrators have to be built in the individual subprojects to evaluate their respective contributions. However, individual demonstrators were not considered sufficient, since integration has been a key goal of the IMPROVE project. Integration has been addressed both at the level of modeling (integrated process and product model for chemical engineering) and at the level of implementation. The role of the *IMPROVE demonstrator* is to bundle research activities in the IMPROVE project and to show the added value of an integrated environment for chemical process design which is based on innovative concepts and models.

Throughout the course of the IMPROVE project, building tools has been a constant activity. Milestones in tool development were demonstrated after the completion of the first and the second phase, respectively. This section describes the second demonstrator, which was successfully shown at the project review that took place in April 2003. This demonstrator builds upon its predecessor, which was demonstrated at the end of the second phase in May 2000. Since the second demonstrator extends the functionality of the first one in several areas, while it retains its most essential capabilities, only the second demonstrator is discussed here.

The presentation focuses on the functionality and the user interface of the IMPROVE demonstrator. It follows the lines of the demonstration given for the project reviewers. Thus, the core part of this section consists of a tour through the demonstrator, whose user interface is illustrated by several screenshots. Before starting the tour, we summarize the most essential contributions of the demonstrator. We will briefly sketch the underlying concepts and models, which are described in depth in other sections of this book. Likewise, the architecture of the demonstrator is discussed elsewhere.

The rest of this section is structured as follows: Subsection 1.2.2 provides an overview of the demonstrator. A guided tour through the demonstrator follows in Subsect. 1.2.3 and 1.2.4. Finally, a short conclusion is given in Subsect. 1.2.5.

1.2.2 Overview

Case Study

The demonstration refers to a *case study* [17] which has been used in the IMPROVE project as a reference scenario. The case refers to the conceptual design of a plant for the production of *Polyamide-6* (PA6), as introduced in Subsect. 1.1.1. By means of this case study, the workflow of industrial design processes is examined in order to identify weak points and to define requirements for the development of new tool functionalities or even new tools. The case study therefore serves as a guideline for the tool design process and hence constitutes a common basis for research in the IMPROVE project. All tools developed in IMPROVE are evaluated in the context of the case study. Furthermore, these tools are integrated to a common prototype demonstrating the interactions between the different tools and their support functionalities. Following this working procedure, it is possible to evaluate whether the tools really fulfill the defined requirements and contribute significantly to an improvement of design processes in chemical engineering.

Figure 1.9 shows different kinds of *flowsheets* which mark the start and the end of the part of the overall design process which is covered by the case study. At the beginning, the chemical process is described by an *abstract flowsheet* which decomposes the process into basic steps without considering the equipment to be used (upper part of Fig. 1.9). The process consists of three steps: reaction of caprolactam and water, separation of input substances which are fed back into the reaction, and compounding, which manipulates the polymer produced in the reaction step such that the end product meets the requirements. The lower part of Fig. 1.9 shows a *process flowsheet* which consists of chemical devices and therefore describes the chemical plant to be built – still at a fairly high level of abstraction. The process flowsheet serves as input for detail engineering, which is beyond the scope of the case study.

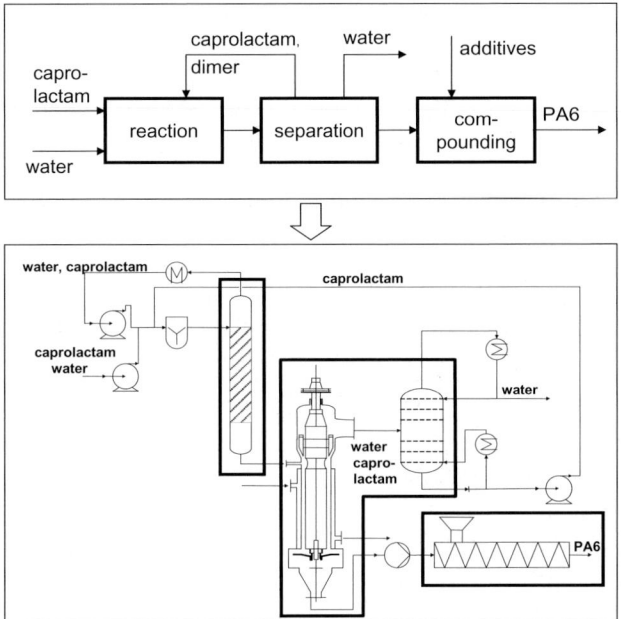

Fig. 1.9. Flowsheets for the case study

A description of the *design process* [169] is given in Fig. 1.10 using the notation and modeling concepts of C3, which will be introduced in Sect. 2.4. This process description is based on three sources: study of the literature, self observation, and interviews. Self observation refers to design activities that were performed by the engineering partners participating in the IMPROVE project. Furthermore, interviews were conducted with several industrial partners. Based on these three sources, the design process was structured. On the one hand, the design process was constructed such that it reflects industrial practice and the scientific state of the art. On the other hand, we incorporated innovative concepts which are intended to advance the current state of the art, e.g. with respect to the collaboration between chemical and plastics engineering.

Figure 1.10 merely shows a simplified cutout of the overall design process serving as a reference scenario within the IMPROVE project. Furthermore, the diagram should be viewed as a *trace* rather than as a normative description. That is, it shows a partially ordered set of steps parts of which will be covered by the guided tour (shaded rectangles). Therefore, the figure serves as a map for the guided tour. Of course, a process model for design processes in chemical engineering has to be defined on a much more general level since it has to cover a class of design processes rather than a specific instance.

The design process was modeled in WOMS [400], a modeling tool which has been developed in one subproject of IMPROVE (see Sect. 5.1). The nota-

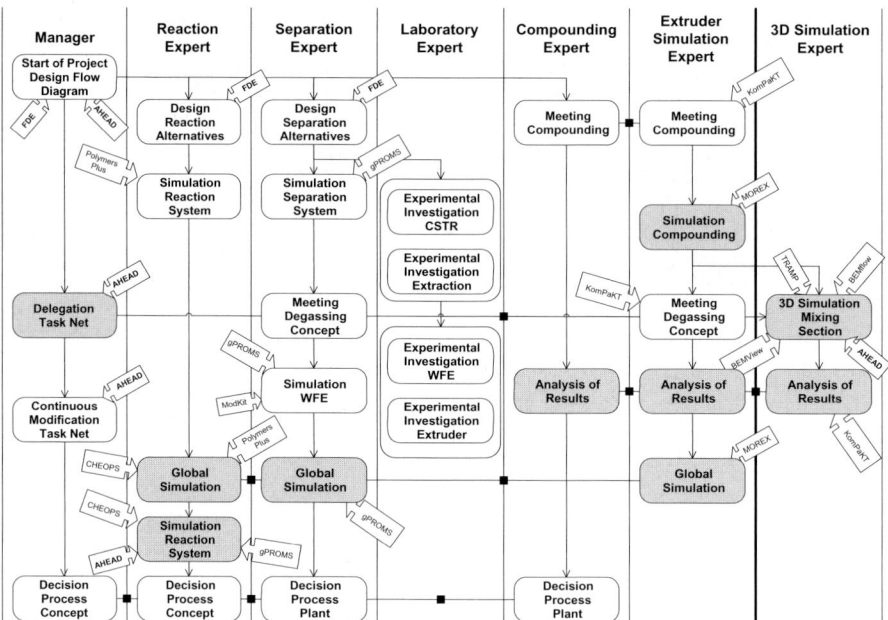

Fig. 1.10. Overview of the design process

tion used to describe the design process is the *C3 formalism* [221], a modeling language for the notation of work processes. The abbreviation C3 stands for the three aspects of workflow modeling which are represented in this formalism: cooperation, coordination, and communication. The elements of C3 are roles (e.g. reaction expert), activities (e.g. design reaction alternatives), input/output information (not shown in this figure), control flows (solid lines), information flows (also not shown), and synchronous communication (represented by black squares and horizontal lines). For each role, there is a swimlane which shows the activities executed by that role. A rounded rectangle which contains nested rectangles corresponds to an activity whose subactivities can be performed in any order. Finally, the notes attached to activities represent the supporting tools. For example, a flow diagram editor (FDE) is used to create flowsheets for reaction alternatives.

Different organizations contribute to the overall design process. The chemical engineering company is responsible for the overall process design, including reaction, separation, and compounding. Only the compounding cannot be handled in the chemical engineering company alone. Rather, an extruder manufacturer has to assist in designing the extruder. Altogether, this results in an *interorganizational design process*. Up to a certain level of complexity, the compounding expert and the extruder simulation expert (of the chemical engineering company) may take care of the extruder design. The simulation

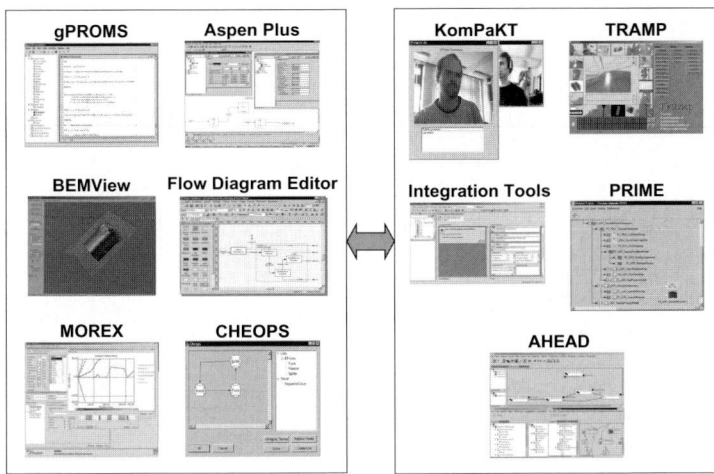

Fig. 1.11. Tools of the demonstrator

expert of the extruder manufacturer is consulted only to solve those problems which require specific expertise.

Besides the interorganizational cooperation, the tight *interdisciplinary cooperation* between chemical and plastics engineering constitutes an innovative key contribution of the case study. In this respect, the case study goes far beyond the current state of industrial practice (and also advances the state of research). First, the design of the extruder is considered in an early phase of the design process by elaborating the interactions between the separation step and the compounding step of the chemical process. Second, the compounding expert and the extruder simulation expert of the chemical engineering company closely cooperate with the 3D simulation expert of the extruder manufacturer in designing the extruder. Third, the extruder is included in a simulation of the overall chemical process.

Tool Support

The demonstrator is composed of a set of tools which are classified into two categories (Fig. 1.11):

- The left-hand part of the figure shows *technical tools* which support engineers in design and simulation activities. With the help of these tools, engineers create flowsheets, prepare simulation models, run simulations, analyze simulation results, etc. The flow diagram editor [21] is used to create abstract and process flow diagrams. Steady-state and dynamic simulations of the chemical process are performed in the commercial tools Aspen Plus and gPROMS, respectively. MS EXCEL (not shown) is employed for storing simulation results (instead of using a plain text file). 1D

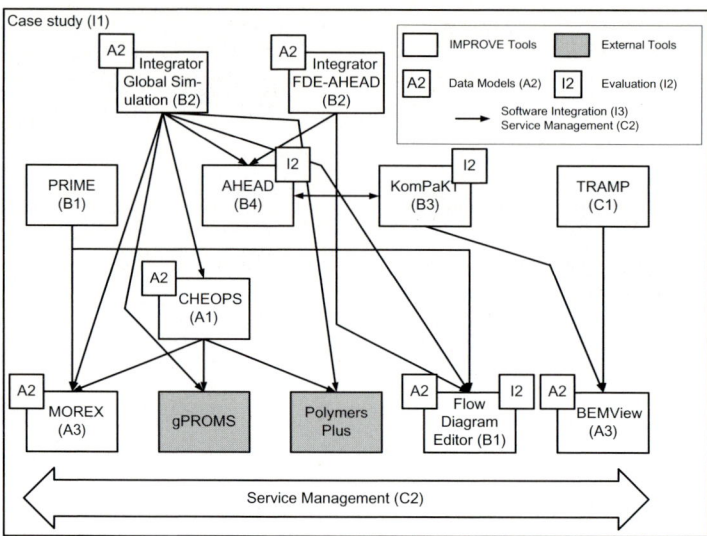

Fig. 1.12. Synergistic tool integration

simulations of the extruder are carried out with the help of MOREX [147]. The commercial tool BEMFlow is used for 3D simulations. The results of these simulations are visualized in BEMView [145]. Finally, simulations of the complete plant are run in CHEOPS [462], which couples heterogeneous simulation tools for individual steps of the chemical process. The subproject names (e.g. A1, B1, and I2) are explained in the next subsection.

- The right-hand part of the figure displays *novel tools* which provide added value by new functionality, relying on and integrating technical tools (see 1.1.2). All of these tools have been developed in the IMPROVE project. AHEAD [169, 355] is a management system which supports the coordination of design activities and provides for coarse-grained tool integration (tools are launched via the work environment of the AHEAD system). KomPaKt [456] supports synchronous cooperation in distributed multimedia work sessions. PRIME [371] is a process engine which adds fine-grained process support to other tools such as e.g. the flow diagram[4] editor. TRAMP [188] is used to record and organize product and process data in a multimedia data warehouse. Finally, the demonstrator also includes various integrator tools to be explained below.

From the perspective of computer science, the key contribution of the demonstrator not only comprises *novel tool functionality* [348], but also its *synergistic integration*, to form an overall design environment. This environment, sketched in Fig. 1.12, provides added value which goes beyond the use of the individual

---

[4] The terms flowsheet and flow diagram are used as synonyms in this book. Also, we use flowsheet editor and flow diagram editor in parallel.

tools. Technical tools are located at the bottom, synergistic tools are placed on top of the technical tools. The tools are connected by *use relationships*, some of which are omitted (e.g., the use relationships between AHEAD and technical tools are not shown). (This picture is more detailed than Fig. 1.8, as it contains the existing tools of the scenario. It is more abstract, as it ignores wrappers.) Let us briefly explain the use relationships, proceeding from the top to the bottom and from left to right:

- CHEOPS couples different kinds of simulators to perform plant-wide simulations and therefore uses MOREX, gPROMS, and Polymers Plus.
- The process engine PRIME is used to extend the functionality of the flow diagram editor by providing executable process fragments. Furthermore, PRIME calls MOREX to launch 1D simulations of the extruder.
- AHEAD calls KomPaKt in order to initiate multimedia conferences. Conversely, KomPaKt relies on managerial data provided by AHEAD, e.g., data about participants of a conference.
- TRAMP calls BEMView, a tool built on top of BEMFlow, to visualize simulation data.
- The integrator for plant-wide simulations composes a simulation model for CHEOPS by querying the flowsheet for components to be simulated, retrieving the simulation documents for components via AHEAD, and including them into the set of files for CHEOPS.
- Finally, the integrator between the flow diagram editor and AHEAD examines the flowsheet to update managerial data in the AHEAD system (such as updating of a task net after structural changes to the flowsheet have been performed).

In addition to the tools and their synergistic integration, Fig. 1.12 also shows other contributions which were indispensable for building the demonstrator:

- An IMPROVE subproject was responsible for the case study, which was modeled in WOMS. Thus, this subproject served as a global coordinator which integrated the contributions of individual subprojects into a coherent design process.
- Another subproject contributed through the development of data models for chemical engineering, which were collected and integrated in the overall CLiP data model [491].
- Another subproject developed the service management layer [438], which provides a communication platform for tool integration.
- Finally, the software architecture of the overall design environment was also developed in a subproject, which provided the organizational framework for software integration [26].

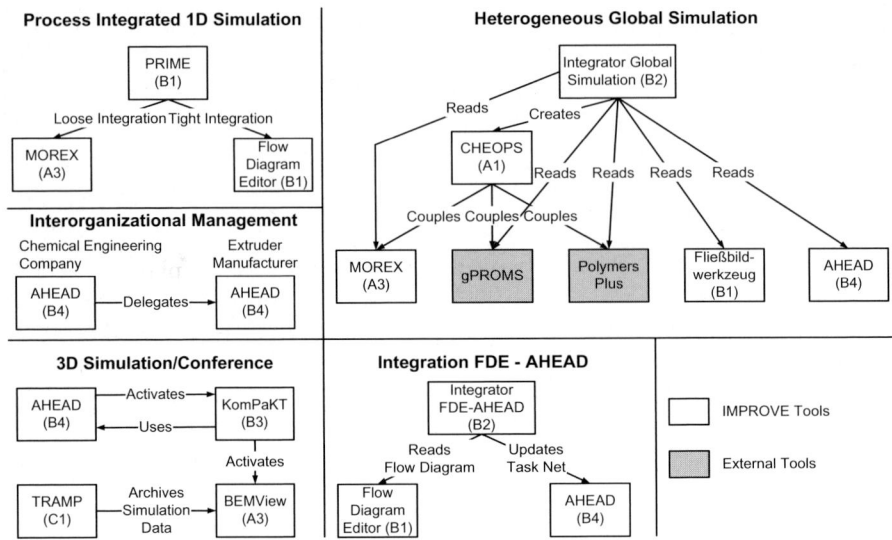

Fig. 1.13. Key parts of the demo

Key Parts of the Demo

Figure 1.13 summarizes the key parts of the demo (see novel process aspects of 1.1.2). These highlights are explained briefly below, following the order in which they appear in the demo.

Process Integrated 1D Simulation

The design of the extruder is considered early in the design process. In order to integrate the extruder design into the overall design, the extruder simulation expert uses the flow diagram editor to model the extruder in terms of its functional zones. The process engine PRIME is used to provide process fragments to support the simulation expert above the level of basic commands. To achieve this, PRIME is tightly integrated with the flow diagram editor (e.g., PRIME extends command menus, queries and controls user selections, etc.). From such a process fragment, the simulation tool MOREX is called, which is integrated with PRIME in a much looser way by means of wrappers.

Interorganizational Management

The cooperation between the chemical engineering company and the extruder manufacturer is supported at the managerial level by the AHEAD system, which is used to delegate the 3D simulation of the extruder. Both companies run their own instance of the AHEAD system each of which accesses a local database. In this way, the use of a central database is avoided. The management data are kept consistent by a runtime coupling which is based on the exchange of events. Thus, the chemical engineering company is kept informed about the state of execution of the delegated subprocess.

3D Simulation and Conferencing

The simulation expert performs the 3D simulation in BEMFlow (not shown in the demo). Before running the simulation, he uses TRAMP to retrieve data from previous simulations which may give hints on setting up the simulation parameters. The simulation results are visualized in BEMView and are stored and annotated in TRAMP in order to augment the experience database. The results are discussed in a conference, which is performed with the help of KomPaKt. The conference tool is launched via the AHEAD system. Conversely, KomPaKt queries the AHEAD system for various kinds of data such as the participants to be invited and the documents to be presented. BEMView is used in the conference to visualize and discuss the simulation results.

Heterogeneous Plant-Wide Simulation

A plant-wide simulation is performed with the help of CHEOPS, which couples different kinds of simulators (MOREX, gPROMS, and Polymers Plus) at run time. The input data for CHEOPS are generated by an integration tool which queries the flowsheet for the components and their connections, retrieves the respective simulation models via the AHEAD system, generates an input file for CHEOPS, and also passes the simulation models to CHEOPS. The integrator cannot run completely automatically, since the user still has to select the simulation models to be used from sets of candidate models.

Integration between the Flow Diagram Editor and AHEAD

At the end of the demo, the requirements are changed with respect to the properties of the chemical product (PA6). An integrator tool is used to assist in propagating these changes. However, user interaction is required to determine the consequences of the changes and the activities to be performed. This interactive process results in an updated task net determining which new activities have to be executed and which old activities have to be restarted. Major decisions still rest (intentionally!) with the manager. For example, the affected parts of the flowsheet cannot be determined automatically. Rather, they have to be marked manually.

1.2.3 Demonstration

This subsection offers a guided tour through the IMPROVE demonstrator 2003. The demo was also documented by a set of video clips which may be viewed online (http://se.rwth-aachen.de/sfbdemo).

Process Integrated 1D Simulation

The demo starts when the reaction and the separation have already been designed in parallel. In the sequel, the compounding step of the chemical process is addressed. At this stage, it has already been decided that compounding is performed with the help of an extruder. The polymers fed into the extruder are melted. Furthermore, glass fibers are added, and monomers are degassed.

The overall design process is managed with the help of the *AHEAD system*, which will be shown later. AHEAD provides a *management environment* which represents the design process as a task net. In addition, AHEAD supports the management of products (of the design process) and resources (the design team). The products are represented as versioned documents such as flow diagrams, simulation models, simulation results, etc. These documents are created with the help of the technical tools which were introduced in Subsect. 1.2.2.

In addition to the management environment, AHEAD provides a *work environment* for designers which displays a personalized agenda of tasks. When a task is selected in the agenda, a work context is displayed containing the relevant documents (input, output, and auxiliary documents). The work context is used to start tools operating on these documents. All tools for performing technical tasks (i.e., design and simulation tasks in the demo process) can be activated via the work environment of the AHEAD system.

The following steps are concerned with the design of the compounding step. They are supported by the *flow diagram editor* and the simulation tool *MOREX*. The extruder simulation expert models the compounding process as a part of the overall chemical process with the help of the flow diagram editor. The respective part of the flow diagram is used to derive a model for 1D simulation in MOREX.

Both the flow diagram editor and MOREX are integrated with the *PRIME process engine*. The user of the respective tool is not aware of the process engine, which operates transparently. Rather, process fragments appear as high-level commands which bundle basic commands of the tools to be invoked manually without process support. Furthermore, the process engine may trace user interactions for different purposes (event-based activation of process fragments or recording of user interactions, see below).

Process integration was performed *a posteriori* for both tools: The flow diagram editor is based on MS Visio, a commercial drawing tool. From the perspective of the process engine, MOREX can be considered as a legacy system. In both cases, the source code of the tool was not modified to perform process integration. The achievable level of process integration was constrained by the interfaces provided by the tools:

- For the flow diagram editor, *tight process integration* was implemented. E.g., by modifying the command menus, it was possible to launch process fragments directly from the editor.

A Scenario Demonstrating Design Support in Chemical Engineering

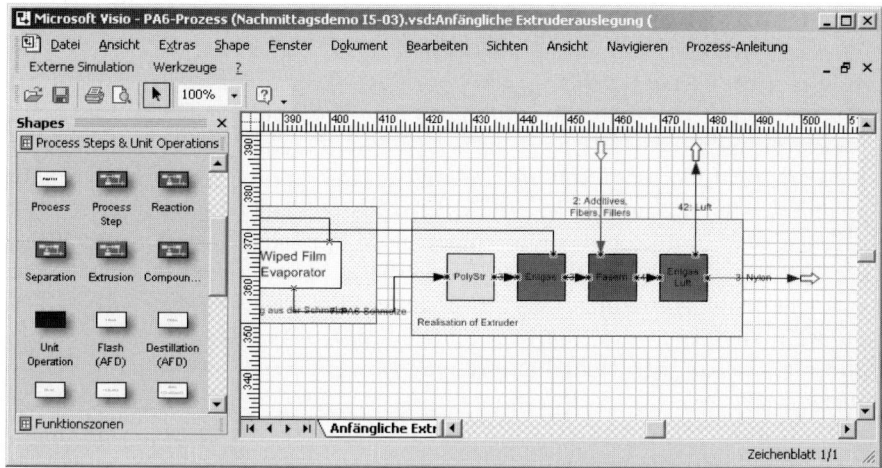

Fig. 1.14. Flow diagram editor

- Only *loose process integration* could be realized with MOREX. Basically, PRIME may only trace the actions of MOREX and record these traces. In addition, process fragments may be launched in an event-based fashion, but it is not possible to activate process fragments directly via a command offered in MOREX.

The flow diagram editor is invoked via the work environment of the AHEAD system on the current flow diagram, which decomposes the chemical process into the basic steps reaction, separation, and compounding. To model the internals of the compounding step, the compounding expert creates a refining subdiagram. The compounding process in the extruder is decomposed into functional zones: simple polymer flow, degassing of monomers, addition of glass fibers, and degassing of air (Fig. 1.14). Furthermore, parameters relevant for the simulation are specified (e.g., the screw speed is defined as 300/min). Finally, the compounding expert passes the flow diagram to the extruder simulation expert for performing a 1D simulation in MOREX. Please note that the addition of glass fibers cannot be simulated in MOREX; this will be handled later in BEMFlow (3D simulation).

The 1D simulation expert opens the flow diagram using the flow diagram editor. He activates a command for generating a simulation model for the extruder. The simulation model contains the functional zones defined in the flow diagram. The simulation model is opened in MOREX, a 1D simulator which is used to simulate mass flows and heat flows in an extruder. Before the simulation is run, the simulation expert refines the simulation model by adding functional zones and enriching the functional zones with screw elements. The flow diagram contains only sketchy information; for the simulation, the actual geometry of the extruder is required.

MOREX is coupled loosely with the process engine PRIME. In this coupling mode, PRIME can merely trace the actions performed in MOREX, but cannot control user interactions actively. The traces observed by PRIME are stored persistently and can be analyzed later. But traces can also be used to support the user of MOREX in an active way. For example, the simulation expert adds a functional zone for mixing and compounding processes. This user interaction is observed by PRIME, which recognizes a context in which previously acquired process knowledge can be applied. PRIME opens FZExplorer, a tool for exploring functional zones (Fig. 1.15). The simulation expert may select one of the realizations for the functional zone, which is added to the simulation model via the COM interface of MOREX. Thus, even in the case of loose process integration the user may be supported actively to a certain extent[5].

The simulation expert still has to add material and process parameters to the simulation model. After the simulation model has been completed, the simulation expert runs the simulation. The results, which are displayed graphically (Fig. 1.16), include e.g. temperature and pressure profiles.

After the simulation has been completed, the modified extruder configuration is propagated back into the flow diagram. Thus, the functional zone which was added to the extruder configuration for the mixing and compounding processes is also added to the subdiagram refining the extruder. Furthermore, the functional zones are extracted from the extruder configuration and are written to a database which is accessed by the FZExplorer. In this way, experienced-based process support is provided.

With the help of the 1D simulation performed in MOREX, some process parameters cannot be determined. In particular, this refers to the mixing quality which is achieved when glass fibers are added. Therefore, it is decided to examine the flow conditions by means of a 3D simulation in BEMFlow.

Interorganizational Management

The 3D simulation cannot be performed locally: This task requires specific expertise which goes beyond the capabilities of the staff of the chemical engineering company. Therefore, 3D simulation is *delegated* to the extruder manufacturer. Delegation is performed with the help of the *AHEAD system*. In general, any connected subprocess may be delegated to an external organization. Here, the subprocess consists of a single task which is embedded into its context. The contractor refines the task locally to decompose the contract into manageable activities.

Delegation is performed in multiple steps. First, the client *exports* the subprocess by activating a respective command offered by the management environment (Fig. 1.17). The delegated process is written into an XML document which is transferred to the contractor. Next, the contractor *imports* this

---

[5] Please note that PRIME cannot modify the command menu of MOREX.

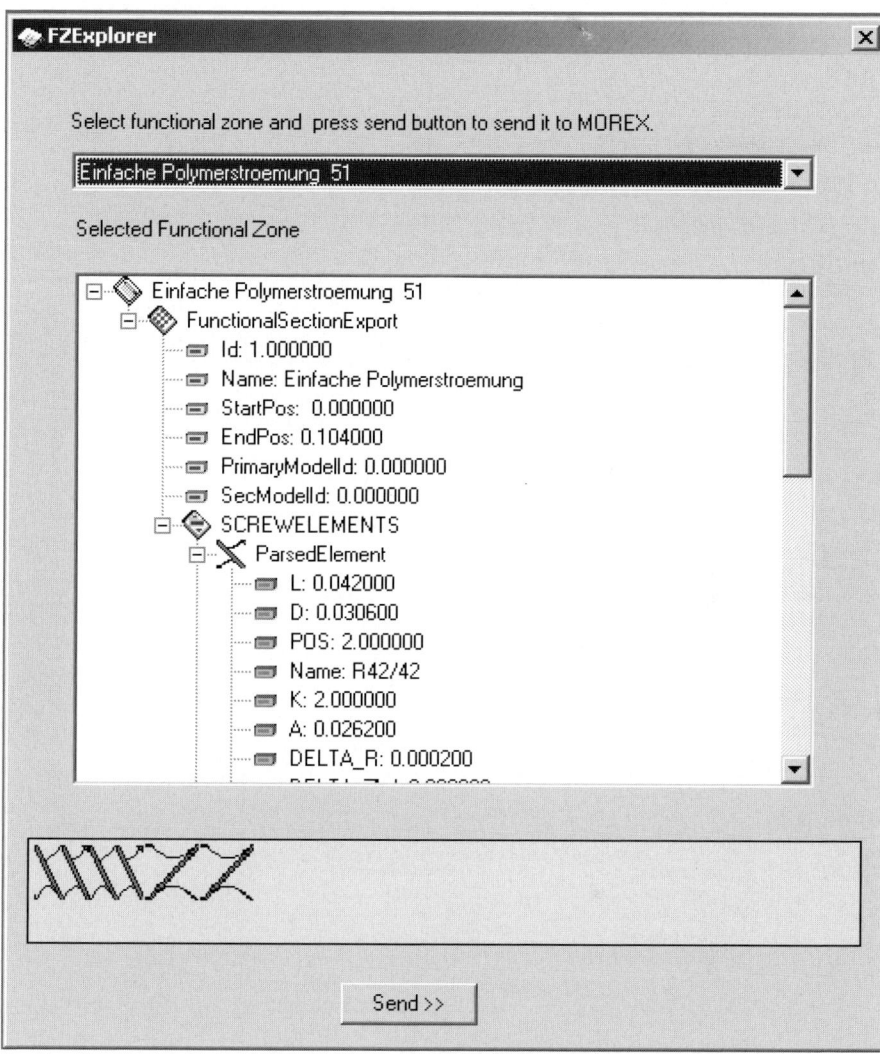

Fig. 1.15. Function zone explorer

document into his local instance of the AHEAD system. Now, the database of the contractor contains a local copy of the delegated subprocess, including the context into which the subprocess is embedded. Finally, both the client and the contractor connect to a *communication server* which is used to synchronize the local databases. Messages which are relevant to the communication partner are transmitted via the communication server. In this way, the local copies are kept consistent. When an instance disconnects from the communication server, a message queue is maintained which is flushed on re-connect.

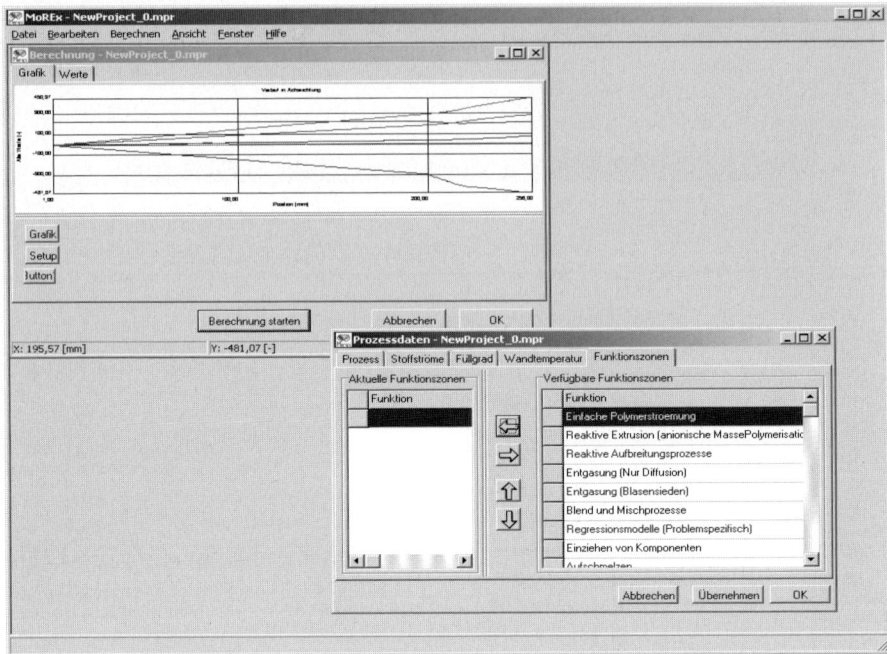

Fig. 1.16. 1D simulation in MOREX

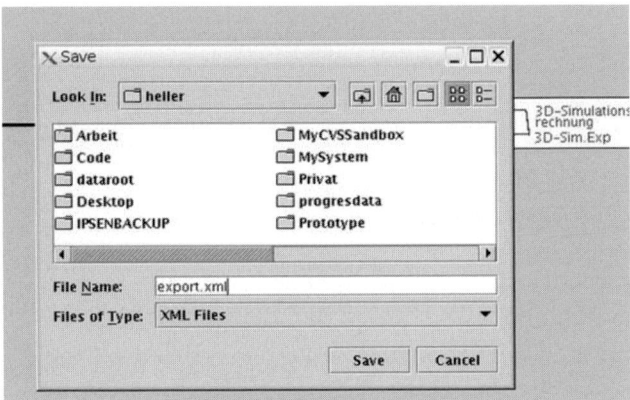

Fig. 1.17. Export of a task to a subcontractor

The 3D simulation expert, who also plays the manager role for this small subprocess, starts the task for 3D simulation. The corresponding state change is propagated immediately via the communication server. Thus, the client keeps informed about the operations performed by the contractor (and vice versa). Subsequently, the 3D simulation expert refines the simulation task into a sequence of subtasks for analyzing the results of 1D simulation, generating

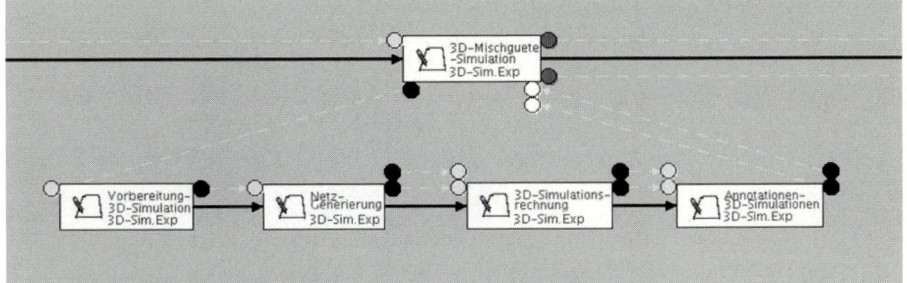

Fig. 1.18. Refinement of the imported task

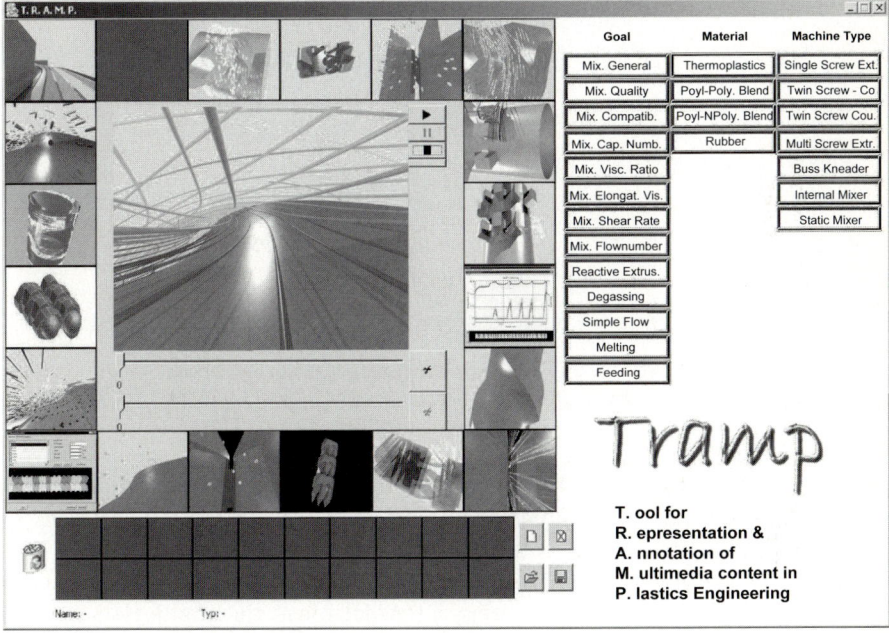

Fig. 1.19. Analyzing simulation histories in TRAMP

the mesh required for the 3D simulation, performing the actual simulation by solving a set of equations, and analyzing the results of 3D simulation. The refining task net, which is shown in Fig. 1.18, is not visible for the client, who may monitor only the public parts of the delegated subprocess.

1.2.4 3D Simulation and Conference

After having planned the subprocess for 3D simulation, the simulation expert starts by analyzing the results of 1D simulation, which have been transmitted via a file generated by MOREX. In the next step, he creates the sim-

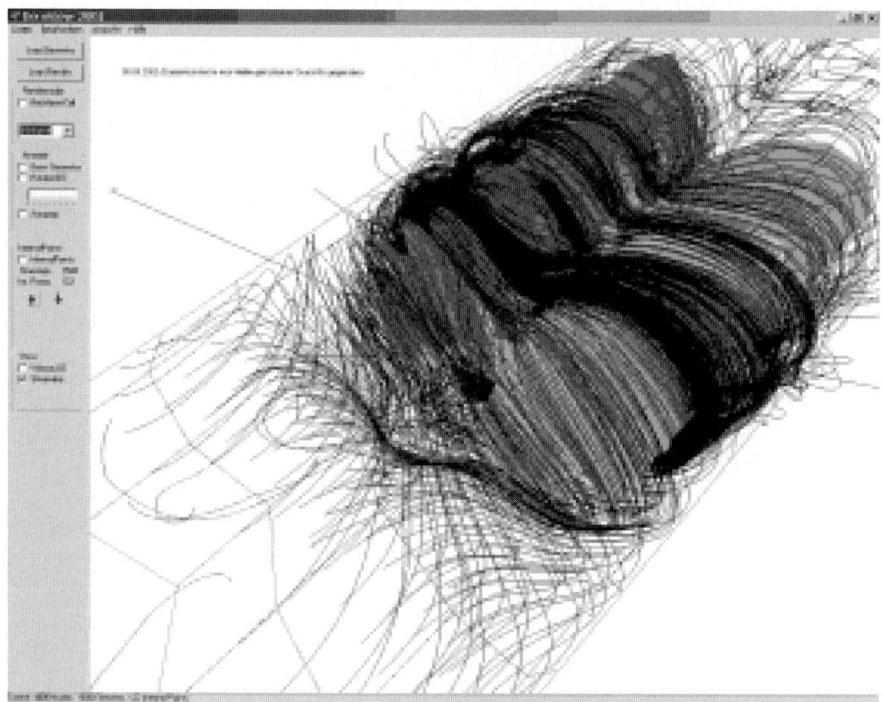

Fig. 1.20. Visualizing simulation results in BEMView

ulation model. To this end, he makes use of *historical data* gathered from previous simulation runs. This is supported by *TRAMP*, which maintains a multi-media database of animated simulation results. TRAMP organizes its database by domain-specific categories. When the simulation experts selects the category "mixing quality", the simulations matching this category are displayed by TRAMP. In the next step, TRAMP plays the simulation selected by the expert (Fig. 1.19). After having investigated several simulations stored in the database, the simulation expert creates the simulation model for BEMFlow.

The simulation is performed in *BEMFlow* off-line. Afterwards, the simulation results are visualized with the help of *BEMView*, which animates the flow conditions in 3D (Fig. 1.20). The simulations have been run with different screw speeds in order to compare the respective flow conditions. The simulation expert, who took advantage of historical simulation data stored in TRAMP, provides his own contributions by recording animations as video clips, categorizing and annotating them, and storing the annotated videos in the TRAMP database.

Subsequently, the simulation results for different screw speeds are discussed in a *conference*. The conference serves as an example of *synchronous*

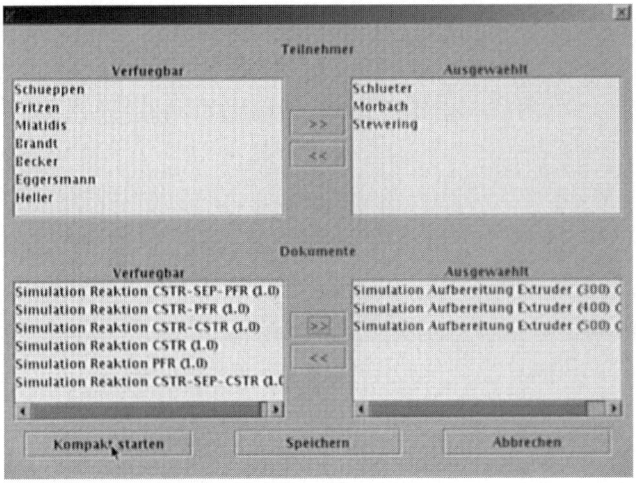

Fig. 1.21. Launching a conference from the AHEAD system

inter-organizational cooperation because both the client and the contractor participate.

Since the conference was planned in advance, it is represented as a task in the *AHEAD system*. The compounding expert, who is assigned to this task, initiates the conference via the work environment of the AHEAD system. To this end, he starts the task and opens its work context, from where external tools may be activated. When the compounding expert activates the command for initiating a conference, a dialog is displayed in which the participants of the conference are selected from a menu of available participants (Fig. 1.21). Likewise, all document versions contained in the work context are offered as subjects of the conference. Here, the compounding expert selects the animated simulation results for different screw speeds which have been prepared by the 3D simulation expert.

The conferencing tool *KomPaKt* is started via a CORBA wrapper. KomPaKt accesses managerial data provided by the AHEAD system via the CORBA wrapper. First, KomPaKt determines the participants to be invited. When all participants are available, the compounding expert invites them to join the conference. Each participant is informed by an invitation window and accepts the invitation. Subsequently, the simulation results are discussed in a joint working session. To this end, the respective documents are retrieved from the management database of the AHEAD system via a CORBA wrapper. Please note that *wrappers* play a crucial role in a posteriori integration.

For supporting joint working sessions, KomPaKt offers a sharing mode called *event sharing*. Traditional application sharing requires to transmit high volumes of data over the network since the actual contents of the screen has to be sent to each participant. In contrast, KomPaKt passes more high-level events to each application instance. This significantly reduces the amount

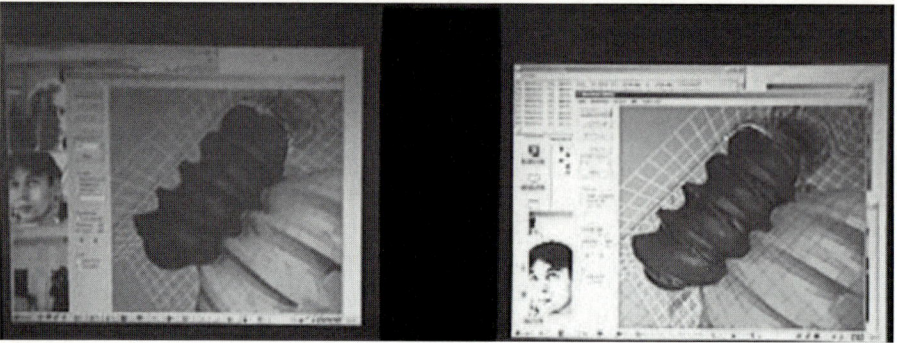

Fig. 1.22. Sharing BEMView in KomPaKt

of data traveling over the network. In the demo, event sharing is applied to BEMView, which is used to visualize the simulation results. Figure 1.22 demonstrates sharing of BEMView among two participants of the conference. However, event sharing works also for more than two participants. In the demo, three experts take part in the conference from the very beginning. Then, another expert from the extruder company, who is connected via a low bandwidth telephone line, is invited ad hoc during the conference. Event sharing also works for such a low bandwidth of 28.8 kbit/sec.

The participants of the conference agree on a screw speed of 400/min, which differs from the screw speed proposed originally (300/min). The increased screw speed guarantees an improved mixing quality. The compounding expert updates the flow diagram accordingly. Subsequently, the 1D simulations are performed once again for the updated screw speed.

Heterogeneous Plant-Wide Simulation

After having simulated the components of the chemical process individually, a simulation of the overall process is performed. To this end, *different kinds of simulators* have to be coupled to form a heterogeneous simulator. We have shown that MOREX has been used for 1D simulation of the extruder. Furthermore, reaction and separation simulations have been performed with the help of the commercial simulators Polymers Plus and gPROMS, respectively.

The simulator framework *CHEOPS* couples different simulators at runtime. CHEOPS requires an input file which describes the simulation models and simulators for the components of the overall process, as well as their dependencies. Based on this information, CHEOPS starts the simulators in the correct order. In the case of feedback loops. CHEOPS iterates the simulation runs until a steady state is reached. The input file for CHEOPS is created by an *integrator tool* which operates on multiple data sources (product management database, integration documents, and the flow diagram).

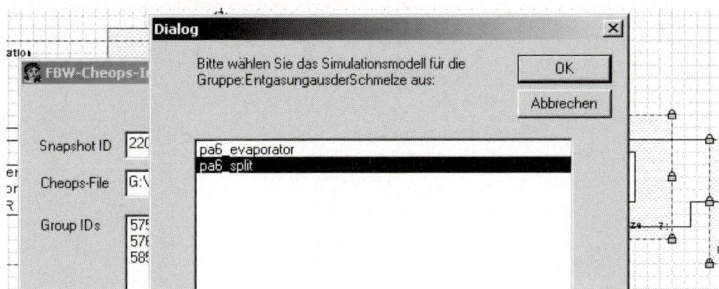

Fig. 1.23. Integration tool for creating an input file for CHEOPS

As the first step, the flow diagram is opened in the flow diagram editor. Subsequently, the simulation expert selects the regions to be included into the simulation. Then, he starts the integrator tool for creating the CHEOPS input file from the flow diagram editor. For the selected regions, the integrator tool determines the simulation models which have been created for these regions. This is done by using both coarse-grained dependencies stored in the product management database and fine-grained dependencies stored in integration documents. For each region, the integration tool presents a list of candidate models to the simulation expert, who selects an appropriate model from the list (Fig. 1.23). Subsequently, the selected models are combined according to the mutual connections defined in the flowsheet. For performing the global simulation, the parameters of external streams have to be defined, as well. Finally, the input file for CHEOPS is created as an XML document.

To initiate the global simulation, the CHEOPS server as well as the wrappers of all participating simulators are started. Next, the simulation expert specifies the input file for CHEOPS and runs the simulation. CHEOPS solves the simulation in a sequential-modular mode: Each simulator is run in turn, where the simulation results are passed from one simulator to the next. In the case of feedback loops, this process is iterated until a steady state is reached (i.e., the global simulation converges).

During the simulation run, the active region is highlighted in the flow diagram editor. Furthermore, the respective simulation model is displayed, as well. In Fig. 1.24, the reactor is highlighted in the flow diagram (background window), and the corresponding simulation model is shown in Polymers Plus (foreground window). After several iterations, the simulation converges, and the plant-wide simulation run is terminated. The simulation results are written into an MS Excel spreadsheet.

Integration between the Flow Diagram Editor and AHEAD

For the final part of the demo, we assume that the requirements to the product of the chemical process are changed: Both the molecular weight and the purity of the polymer product have to be increased. As a consequence, certain tasks

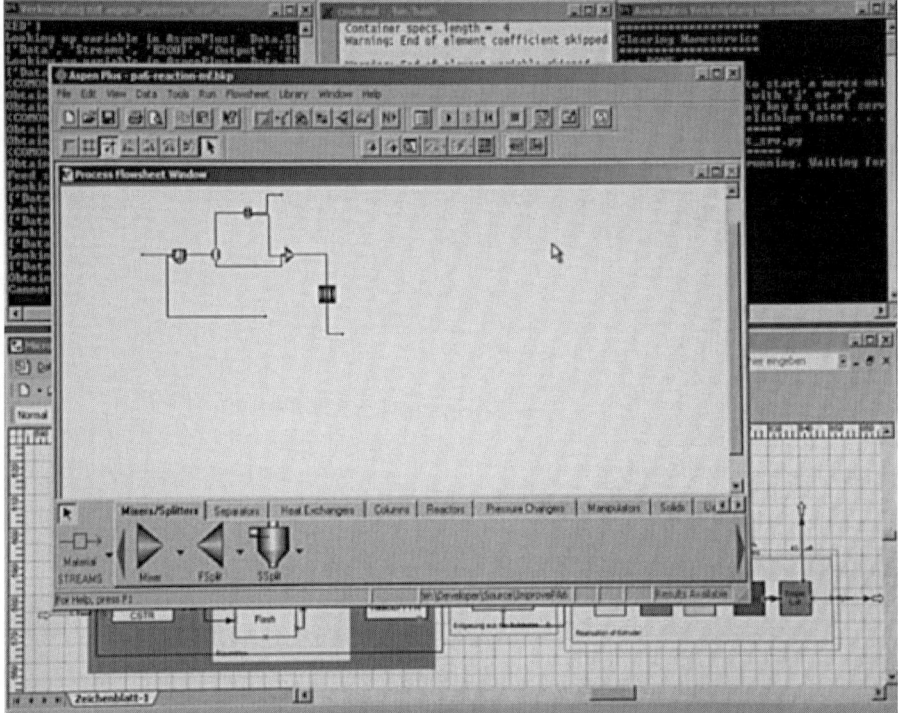

Fig. 1.24. CHEOPS at runtime: coupling heterogeneous simulators

have to be iterated, and certain documents have to be revised. Unfortunately, it is difficult to assess the implications of the changed requirements. In particular, it is impossible to determine these implications completely automatically. In contrast, the chief designer is supported by an interactive tool which employs both managerial data maintained in the AHEAD system and technical data represented in the flowsheet. This tool is called *FSE-AHEAD integrator*[6] because it assists in propagating (estimated) changes of the flowsheet into the task net maintained by AHEAD. The updates of the task net determine which actions have to be taken in response to the changed requirements.

To cope with the changed requirements, the reaction expert, who also plays the role of the chief designer, inspects the flowsheet and identifies devices which have to be investigated further. He concludes that both the reactor and the extruder may be affected, and marks these devices. Next, the chief designer starts the FSE-AHEAD integrator from the flow diagram editor. The FSE-AHEAD integrator receives the selected devices as inputs and determines the regions in which they are contained. Subsequently, all documents are searched which refer to these regions. This is performed in a similar way as it

---

[6] FSE stands for flowsheet editor

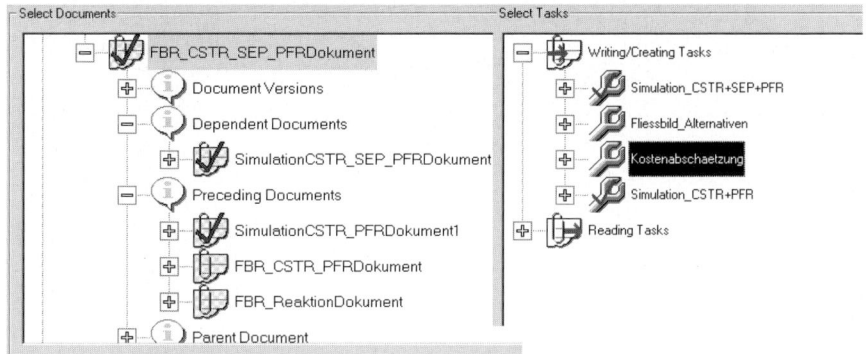

Fig. 1.25. Determining affected documents and tasks

is done in the integrator tool for plant-wide simulation described in the previous subsection (i.e., both coarse- and fine-grained dependencies are queried). All potentially affected documents are displayed to the chief designer, who selects those documents he assumes to be actually affected (Fig. 1.25, left).

Subsequently, the FSE-AHEAD integrator determines all tasks which created the documents selected in the previous step. These tasks are presented to the chief designer, who selects those tasks he considers actually affected (Fig. 1.25, right). The FSE-AHEAD integrator composes an agenda of tasks which have to be re-activated. The chief designer annotates the agenda, which is written by the FSE-AHEAD integrator into an XML document.

The agenda created by the chief designer does not become immediately effective. In contrast, it is inspected by the project manager, who makes the final decision which tasks are going to be performed in response to the requested changes (Fig. 1.26). When the project manager has released the agenda, the FSE-AHEAD integrator creates a script which is sent as a batch command file to the AHEAD system. After the script has been executed, AHEAD displays the changed task net. Subsequently, the project manager would assign responsibilities to the tasks to be executed. However, the demo tour ends at this point. The reader may imagine on his own how to carry on.

1.2.5 Conclusion

In this section, we have taken a short tour through the demonstrator prepared at the end of the second phase of the IMPROVE project (in 2003). More detailed information about the tools contributing to the demonstrator is given elsewhere in this book. The demo tour presented here primarily serves to demonstrate the tight integration of heterogeneous tools (given, extended, or new) making up a novel design environment for chemical engineering. The key message here is *synergy*: The interplay of the components adds value which goes beyond the value of the individual components alone.

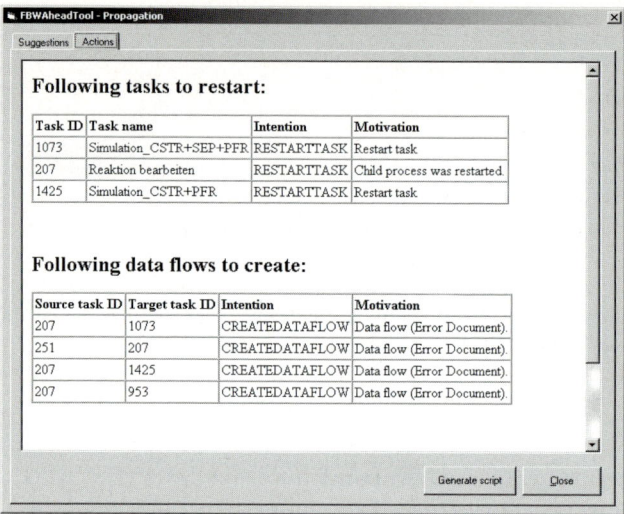

Fig. 1.26. Deciding on the actions to be performed

The second contribution refers to the underlying design process, which differs from traditional design processes in two respects. First, the design process is characterized by tight *interdisciplinary cooperation*, namely between chemical and plastics engineering. Second, the overall design process involves *interorganizational cooperation*, as well (between a chemical engineering company and an extruder manufacturer). This cooperation may be used for iterations and studies of alternatives within design processes, but also for error situations and changing requirements.

Acknowledgments

The authors are deeply indebted to all participants of the IMPROVE project who contributed to the demonstrator, directly or indirectly.

1.3 The Interdisciplinary IMPROVE Project

M. Nagl

Abstract. As already introduced in Sect. 1.1, IMPROVE is a *joint project* between *different disciplines*, namely Chemical Engineering and Informatics. Even more, Plastics Engineering and Labor Research are also involved. Informatics itself is represented by three groups of different research fields.

Whereas Sect. 1.1 dealt with the problem to be solved and Sect. 1.2 discussed a resulting integrated prototypical environment, this section discusses the project and the *research and development process* to solve these problems.

The *section* is structured as *follows*: First, the IMPROVE project structure is discussed, next its iterative and practical approach is introduced. Two further characterizations are given, namely the funding of IMPROVE and the predecessor projects. Finally, we give a survey of this book's structure.

1.3.1 Project Structure and Policy

There are 6 *research groups* involved in the IMPROVE project the contact data of which are given in Appendix A.1. Every *subproject* of this project is carried out by exactly one research group.

Project Areas

In Fig. 1.27, the *coarse structure* of the IMPROVE project is presented as three technical project areas and one area for integration. Any of these areas is divided into several subprojects. Furthermore, there is a common research topic dealing with the key problem of the CRC, namely to understand and to evaluate design processes and, especially, to define a formal process/product model. This topic is elaborated by all research groups and, therefore, by all subprojects. Last but not least, there is a subproject Z for organization, administration, and documentation of the results of the CRC. The latter is not further discussed in this book. In the following, we discuss the research areas and their corresponding subprojects.

Technical Project Areas

In area A *Development Processes in Chemical Engineering* on top of Fig. 1.27 we find the definition of problems to be solved, the specification of requirements for new tool support, and the formalization of underlying models for processes and products on the application level. Three subprojects, A1–A3, are included.

Area B *New Methods and Tools* deals with innovative concepts for supporting development processes, the synergistic and integrative use of these concepts, a methodology for implementing these new support concepts in the

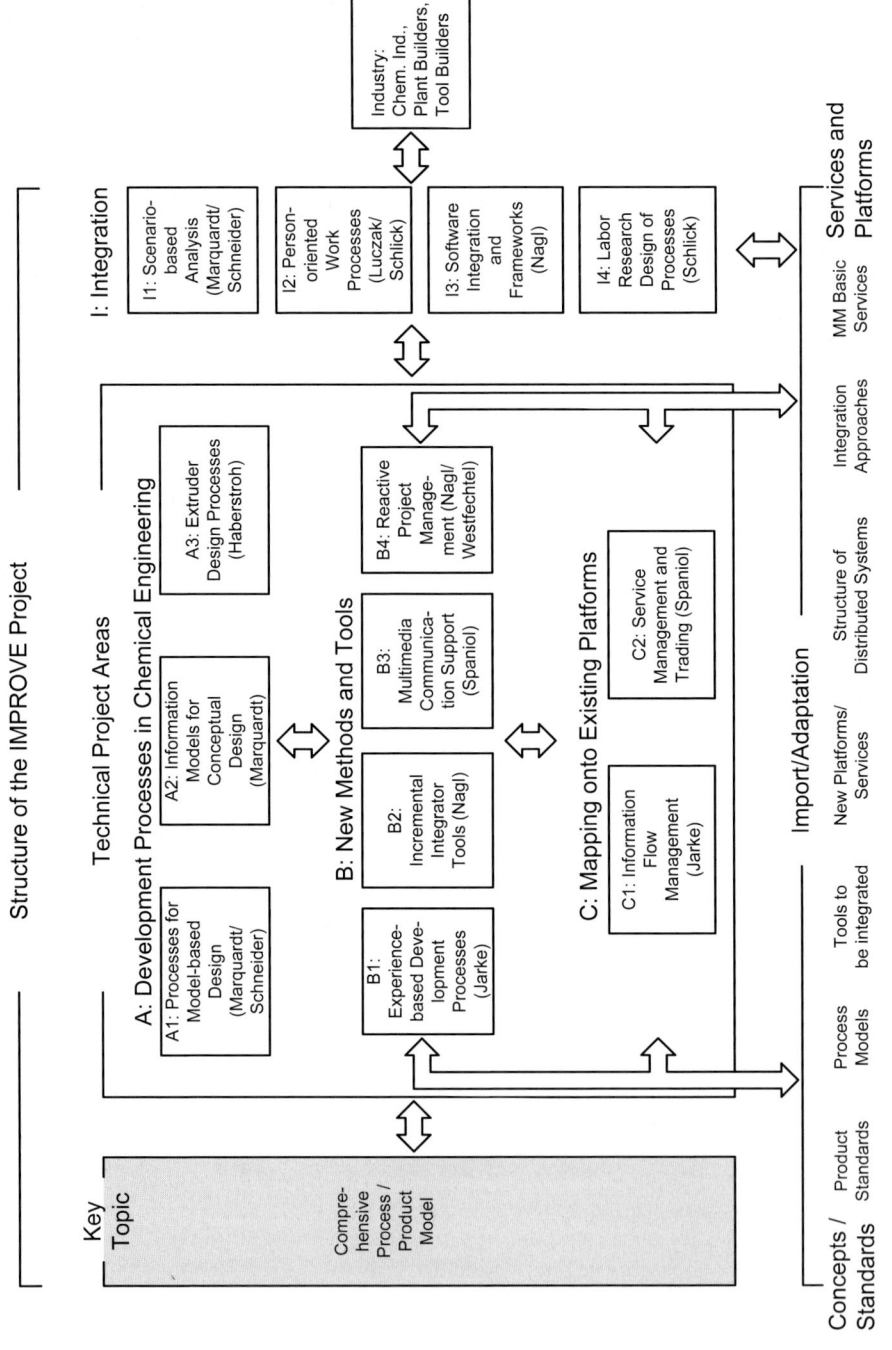

Fig. 1.27. Project structure of IMPROVE

form of tools or tool extensions, thereby applying reuse techniques for tool implementation. This project area consists of four subprojects, B1 to B4.

Finally, project area C *Mapping onto New and Existing Platforms* summarizes unification, generalization, and use of different tool platform approaches. The result is a generalized layer for data access and computing processes being developed by the two subprojects C1 and C2.

Corresponding to the *relations* between the above *project areas* the following remarks hold true:

1. The relation between areas A and B is *more than specifying* tool *functionality* in area A and realizing tools in area B. Instead, tools are realized in a close cooperation between the engineering and computer science projects in areas A and B, respectively. Furthermore, the engineering projects also developed some tools to be described later.
2. Relation B and C: In most cases, the realization of engineering tools relies on a specific underlying platform. To avoid this situation, project area C defines an interface which *unifies* existing platform services. Furthermore, there are functionalities not offered by existing platforms which, however, are necessary for the realization of advanced tools. Hence, the platform interface also gives an *extension* of existing platform functionalities.

Integration Project Area

Integration here has *three* different *semantics*:

- Subproject I1: Existing industrial *design processes* are analyzed and evaluated; the corresponding results serve as an input for IMPROVE. Therefore, real world problems are brought into IMPROVE. Conversely, new concepts and new tool support of IMPROVE are made applicable to industry. This mutual connection to industry is done by using different and graduated scenarios as already described in Sect. 1.2. Therefore, I1 is, among others, responsible for installing and maintaining relations between industry and IMPROVE.
- Subprojects I2 and I4 deal with *labor research aspects*. In I2, work processes are studied, requirements for tools are postulated, and new tools of IMPROVE are evaluated, altogether corresponding to labor research aspects. In I4, design processes are planned and simulated under the labor research perspective, in order to make a prognosis about their time and effort behavior, before they are carried out eventually.
- Subproject I3 deals with the *software development process* connected to tool realization. On the one hand, the subproject gives architectural advice for tool constructors such that newly developed tools or extended existing tools later fit together. This was especially important during the development of the integrated prototypes. On the other hand, the subproject collects and administrates all reusable components of the integrated prototypes we have realized so far. These reusable components are collected in a framework. One specific topic, thereby, is to incorporate existing tools.

Relations to the Outer World

These *relations* (see right side of Fig. 1.27), are mostly to *industry*, from which problems, addressed in the IMPROVE project, are received and discussed. On the other hand, results are transferred to industry.

Thereby, *industry* has *three different appearances*: (a) The users of chemical process technology and of corresponding tools in the chemical companies, (b) the constructors and builders of chemical plants and, finally, (c) tool builders supporting the process of developing chemical plants. In Chap. 7 we explain our efforts to transfer results of the IMPROVE project to the industry of these three forms.

The second relation to the outside world is *import* of results to be used in the IMPROVE project. Here, also, import has different meanings: concepts to be used/adapted, standards to be followed, platforms to build upon, and tools to be used as a part of our realization. At the bottom of Fig. 1.27 the different forms of imports are enumerated.

Comprehensive Process/Product Model

The most ambitious scientific problem to be solved by IMPROVE – in other words the key topic of this project – is to develop a comprehensive and formal *process/product model* for design processes. The importance of this model, its appearance across different layers, its hierarchical structure in every layer, and the mapping between different layers were already introduced in Sect. 1.1.

This key topic had to be addressed by all research groups of IMPROVE. There were no special projects solely addressing this problem. Instead, *every project* gave its *specific contribution* according to the layer structure introduced in Fig. 1.6.

Projects A1, A2, A3, I1, I2, and I4 mainly contributed the *application layer* of this model. B1, B2, B3, and B4 addressed the *internal* and conceptual *models* for tool development, and projects C1 and C2 contributed to the *platform* layer. Of course, there was also a cooperation of projects on any of these layers in order to discuss the models on each layer. Even more, there was a cooperation in order to discuss the transitions of the models from one layer to the next, from top to bottom.

For *example*, A1, I1, B1, and B4 cooperated for elaborating *process models* on the fine-grained technical as well as on the coarse-grained management level. This was done on the application side, the tool construction side, and the transition between both. Cooperations like this are typical for model elaboration.

The contributions for the process/product model came from *four different directions*, namely processes, products, cooperation, and communication. Some projects are rather specific in their contribution, others deliver results corresponding to more than one of these four aspects.

The current state of the layered process/product *model* is described in Chap. 6. A brief *summary how far we got* is as follows: For specific topics (fine-grained process models, fine-grained product models, coarse-grained product and process models) we have some nice results. They deal with, how the models look like on different layers and how the transition between these layers has to be described and, especially, how they can be derived from application domain models. However, there are still a lot of open problems.

1.3.2 Practical Relevance

This subsection characterizes the IMPROVE project by regarding different perspectives: focus, procedure, role of demonstrators, and cooperation with industry.

Focus and Procedure

Focus and Importance

As already argued, design processes have to be improved with respect to quality and efficiency. This is especially important in a developed industrial country, like Germany, with high salaries. The *early phases* of the development process, furthermore, have great *economical impact*. In scientific terms, this part of the process is a particularly *challenging*.

The current state of design processes can essentially not be improved by making only small steps. Instead, a new approach is necessary. Thereby, we face *principal questions* and *nontrivial problems*. We find new questions and corresponding problems by coherently and uniformly modeling the application domain and by defining new and substantial tool functionality. The layered process/product model is a scientific question which – even in a long-term project like IMPROVE – can only be answered partially.

Finally, the *realization* of an integrated design *environment* with new, integrated, and synergistic functionality on top of existing tools was specifically difficult, as we aimed at tight integration and as we also tried to apply *reuse* techniques for tool implementation.

By taking existing tools and by asking fundamentally new questions, the IMPROVE project tried to *balance* between *relevance for industry* on the one and *scientific focus* for research on the other hand. Usability is specifically addressed by carefully regarding application-specific aspects and by the ergonomic evaluation of intended concepts and resulting tool functionality. Model and tool integration is still a hot topic of practice and research. As shown in Chap. 7, results of the IMPROVE project can be transferred to industrial practice.

66   M. Nagl

Evolutionary and Iterative Procedure

The procedure for getting solutions for the ambitious problems described above *cannot* just be *top-down*. Neither the comprehensive process/product model nor the know-how to structure this model were at hand when the IMPROVE project started. In the same way, deep knowledge how to realize an integrated environment with new functionality on top of existing tools was not available.

Instead, it was necessary to start *yo-yo*: First steps for the process/product model were made top-down and first steps towards the realization of an integrated environment were made bottom-up. Furthermore, a *step-by-step* approach was necessary, as difficult problems can only be clearly stated when trying to start the realization of new tools. Fortunately, as to be explained in Subsect. 1.3.4, the involved research groups had specific experience and deep knowledge with respect to certain aspects and problems addressed by IMPROVE.

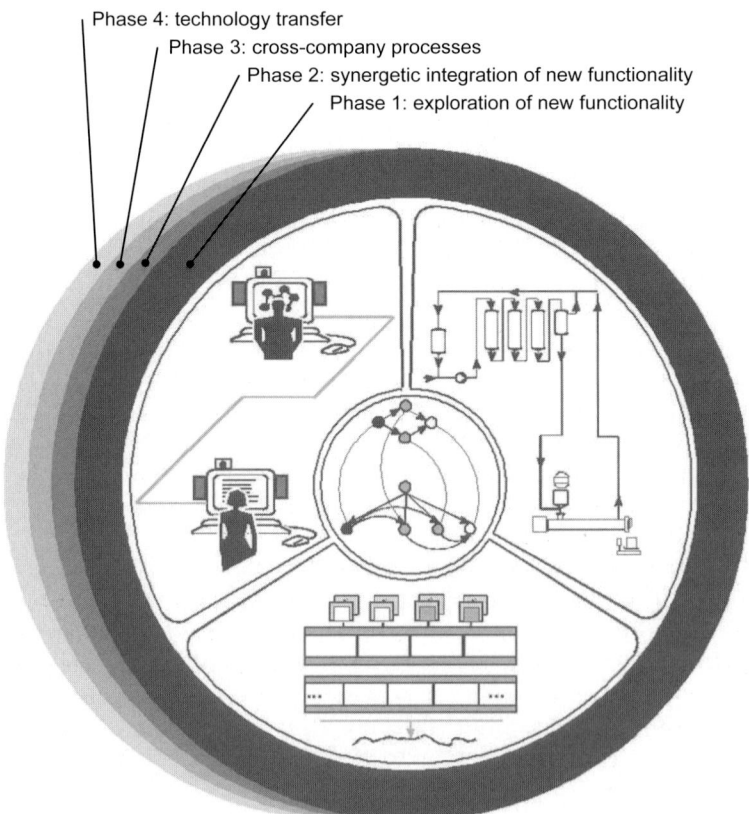

Fig. 1.28. Logo symbolizing the spiral approach of IMPROVE

Therefore, we took an *evolutionary* and *iterative* proceeding for the long-lasting IMPROVE project. This is to be seen by the IMPROVE logo (cf. Fig. 1.28). Four cycles were planned, three of them are already finished. Below, we describe the foci of cycles 1 to 4. By the way, the cycle periods correspond to the funding periods, as money was only given by DFG for three years, and a continuation is only granted after a successful peer review, evaluating the results of the last funding period as well as the detailed research plan for the next.

In every *cycle* the *following steps* were taken (see again the logo): We regard the current situation of design processes in chemical industry (right sector of the wheel), we derive new tool functionality (left sector), and we realize this new functionality (bottom sector) making use of reuse results, if available. Then, we start the next cycle. In the middle of the logo we find the hard problem to be solved, namely the process/product model.

So, in every cycle we got an *extension* of the tool results and of the integrated development environment of the last cycle together with underlying *conceptual results*. *Extending* the *environment* means (a) regarding an extended scenario by taking further tools into account, (b) having an extended version of the process/product model and a deeper understanding of development processes, (c) getting an improved knowledge of tool integration, (d) also improving the reuse machinery for building new tools on top of existing ones within an integrated environment, and (e) generalizing and unifying results of the last cycle.

Key-Notes of Iterations

The key-note of the first cycle, from July 97 to June 2000, was the *exploration of new concepts* on the engineering and informatics side (see Sect. 1.1). Their use was demonstrated by a prototypical first integrated environment shown to peers in the spring of 2000, who evaluated our results of the first phase and our plans for the second. This *first prototype* showed the use of *new functionality* for supporting new steps of design processes within suitable steps of the design process. These steps were carefully selected from both, an application-specific as well as an informatics perspective.

The key-note of the second cycle, from July 2000 to June 2003, was to demonstrate *synergistic integration* of this new functionality and to show its use for design processes, having a different form than state-of-the-art processes of industry. Again, a *prototype* of an integrated environment was demonstrated to peers in the spring of 2003. The corresponding review was to evaluate the second cycle and the proposal of the third cycle. A *round trip* for this prototype has been described in Sect. 1.2.

The third cycle, from July 2003 to June 2006, had the motto of considering cross-department and *cross-company development processes*. We showed integrated and synergistically cooperating new tools to bridge the gaps between different departments or companies using different development tools, different cultures, organizational forms etc. The peer review took place in March

2006 evaluating the results of that third cycle and our plans for the next fourth one. No new integrated prototype was demonstrated, as we spent our effort for preparing the next transfer cycle. However, different but *separate demonstrators* of the involved groups have shown the use of cross-company support.

The last cycle, from July 2006 to June 2009, follows the key-note *technology transfer*. The corresponding goals and work packages are to be described in more detail in Chap. 7. Not every subproject of IMPROVE has a transfer phase. There are 6 transfer subprojects which got through the review process of DFG. Another CRC subproject transfers its research results on the basis of other funding. The Collaborative Research Center 476 has been finished in June 2006. The new Transfer Center (TC 61) was established for this fourth cycle.

This book gives a *survey* on the *results* of IMPROVE in the *past* three *cycles*. It also describes ongoing work on transfer in Chap. 7.

The Role of Scenarios and Demonstrators

In Sect. 1.1 we explained that the IMPROVE project studies the problems of better understanding, formalizing, and supporting design processes in Chemical Engineering, by taking a Polyamide-6 plant as a *case study*. This case study defines the frame of IMPROVE's exemplary view: early phases of design process, hydrolytic polymerization, the necessity of cross-company design, concrete requirements for an improved and innovative design support etc.

More specific than the case study is the *overall scenario* of the IMPROVE project. It defines which tools have to be incorporated into the design environment, it describes which runs of an exemplary design process are regarded, and how cooperation is carried out within one company or between different companies. Of course, the overall scenario is based on the case study. The scenario was the basis for an advanced environment for the first phases of a design process in connection with the case study example. Clearly, the overall scenario changed from project cycle to project cycle.

The overall *scenario* is *characterized* by the following *attributes*: Experiments with cross-company cooperation by including a plastics expert into the design process of a chemical company, extending the simulation of single devices by an overall simulation of the whole plant, showing the necessity and use of incremental and iterative changes within the design process.

As already mentioned, the IMPROVE project gave a *demonstration* of an *integrated prototype* three times when being evaluated by peers: In the first evaluation 1997 this was the prototype of the predecessor project SUKITS. After the first cycle the prototype showed the first version of an integrated environment demonstrating the ideas of a new kind of support. After the second cycle, the demonstrator showed the integrated environment discussed in Sect. 1.2, dealing with synergy of new support concepts and already dealing with cross-company development. Demonstrations were worked out according to this overall scenario.

After phases 1 and 2 a *demonstration scenario* showed a tour through the corresponding available, integrated environment. The walk through consisted of a reasonable and exemplary part of the design process and its corresponding support. For this demonstration scenario a detailed script was elaborated. Subsection 1.2.3 describes this tour.

Specific demos of the IMPROVE research groups gave and give a detailed look on the corresponding research results. They usually deepen corresponding parts of the demonstration scenario or they deal with possible extensions of this scenario which could not be shown in an overall demonstration scenario due to time reasons.

Cooperation with Industry

There were a lot of *industry workshops* having been organized by IMPROVE. Their corresponding auditorium consisted mostly of chemical companies. However, also plant building companies as well as tool builders and vendors have been cooperating with us.

Chemical companies went through a turbulent period in the last twenty years: mergers, buy-outs, restructuring of the product portfolio etc. changed the cooperation relationships. These events rendered cooperations with the IMPROVE project more difficult. Nevertheless, there was a stable core of partners for discussion which consisted of Bayer, BASF, and Degussa.

Cooperation with *engineering and construction companies* was not simple as well. Market pressure made it difficult for these companies to think in mid- or long-term goals. Nevertheless, we found cooperation partners for the IMPROVE project. Stable partners were Uhde and Linde.

Even harder were the last 20 years for *tool builders*. Tool builders of our partner Bayer are now with Bayer Technology Services. Aspen swallowed Hyprotech and Icarus, among others resulting in internal re-organization, which did not favor our collaboration. The German company innotec, however, has been cooperating with us for many years.

Section 7.1 gives a *report* about *industry cooperation* within the last 10 years of the IMPROVE project. There were a lot of activities and corresponding results.

For the planned transfer projects new industrial partners have been found, or the cooperation with previous partners has been intensified. Chapter 7 gives more details.

1.3.3 Funding and Total Effort

IMPROVE was installed in 1997 as a so-called *Collaborative Research Center* (in German Sonderforschungsbereich, in short SFB, in the US similar to a center of excellence), namely *SFB 476* by German Research Foundation (in

German Deutsche Forschungsgemeinschaft, in short DFG). As already told, IMPROVE ran through three periods of financing. The corresponding proposals have been peer-reviewed before any of these periods. The fourth period is carried out in form of a technology *Transfer Center* (in German Transferbereich, in short TB), namely as *TB 61*.

Funding

Most of the *funding* of the research activities of IMPROVE, as described in this book, has been given by DFG. However, other sources have also been involved. This subsection gives the *figures* of funding given by different sources

Funding of the CRC IMPROVE

The following table gives the *funding* in kEuros for the Collaborative Research Center IMPROVE in the three 3year-periods *from* mid *1997* to mid *2006*.

Table 1.1. Funding of CRC 476: Different Sources in kEuros (1€ ≅ 1,3US$)

periods of CRC 476	funding by DFG	extra funding by RWTH Aachen University	funding by Ministry of Research of North-Rhine Westphalia	total
1st period mid 97 – mid 00	2.498	98	185	2.781
2nd period mid 00 – mid 03	2.910	64	19	2.993
3rd period mid 03 – mid 06	2.585	85	115	2.785
	7.993	247	319	8.559

It should be noted that DFG funding is only given if the applying research groups also invest a remarkable amount of money. We have no detailed figures about the sum which was given by the IMPROVE research groups themselves. It should be in the order of but less than one half of the money given by DFG. This additional sum of approximately further 3 Mio Euros should be added to the above total amount.

Funding of the Transfer Center

The fourth phase of IMPROVE was organized in form of a transfer center. It is financed by similar sources as the CRC 476. In addition, also money from industry was necessary to be given. The remark on an own contribution of the research groups, given for CRC 476 above, also applies here.

It should also be remarked, that there is a further transfer project, which is financed by other sources.

Table 1.2. Funding of TC 61 in kEuros

TC 61	funding by DFG	extra funding by RWTH Aachen University	funding by industry	total
mid 06 – mid 09	1.500	76	appr. 800	2.300

Total Sum and Acknowledgements

Altogether, IMPROVE CRC 476 from 1997 to 2006 and TC 61 from 2006 to 2009 has spent/will spend the amount of 10.1 Mio Euros funding. If we also count the contributions of industry and that of the involved groups, we end up by more than *15.0 Mio* Euros.

It is this long-term research funding instrument of Collaborative Research Centers which makes such a fundamental and long-lasting project possible. *Thanks* go to the sponsors, special thanks go to the *DFG*. We are deeply indebted to the above mentioned organizations for their help

Total Effort

We did not really precisely accumulate the *total human* effort. However, 3 different *rough calculations* come to the same result of person years of fully-paid scientists: (a) looking at the above total sum and regarding the average salary, (b) counting the number of involved scientists of Appendix A.2 thereby regarding which of them had looser connections and, therefore, should only be counted part-time, and (c) looking on the number of finished Ph.d. projects of Sect. 8.4.

We get an approximate number of 200 person years of *fully-paid scientists*. Let us now regard that any of these persons is the advisor of master projects, let's say two a year. We now calculate the workload of a student doing a Master's Thesis by half of a year and multiply this by 0.5, as he/she is not too experienced. Then, we get about additional 100 person years for the contribution of *Masters' students*.

So, finally, we end up with the estimation that IMPROVE and the following transfer center comprise a *total personpower* effort of about *300 p.y.*, including work of Masters's students. This is a remarkable size for a university project.

1.3.4 Predecessor Projects

The specific know-how needed for the topics, problems, and solutions addressed by some subprojects was gained by predecessor projects. In the following, we sketch these *predecessor projects* by explaining their goal, their solutions, and in which way they influenced the IMPROVE project. All these predecessor projects are of a reasonable size.

SUKITS

The topics of the *SUKITS project* [110, 351, 352] (**S**oftware **u**nd **K**ommunikation **i**n **t**echnischen **S**ystemen) was on tool integration. SUKITS was a joint project between two groups of mechanical engineering and two groups of computer science at RWTH (Informatics 3 and 4, both being also involved in IMPROVE). The project was funded as a DFG researchers' group from 90–97. Counting also the Masters' Theses, SUKITS was a project of about 80–100 person years.

SUKITS was also on *a-posteriori tool integration*. Thereby, tool integration was done by a project management system as a coarse-grained integration instance and using a common generalized communication platform. To speak in terms of IMPROVE, the topics of subprojects B4 and C2 were involved. The two engineering projects dealt with modeling requirements aspects of coarse-grained processes and products and on integrating specific mechanical engineering design tools into the environment.

The *focus* of SUKITS was *more narrow* than that of IMPROVE. The project management system was to coordinate a development project in which classical tools are used for the specific roles of mechanical engineers. So, the SUKITS integration approach only took one integration concept of IMPROVE into account. Furthermore, there were no plans of developing an integrated process/product model or to study cross-company development. There were nice results published in a book [352].

IPSEN

The *IPSEN project* [226, 329, 330, 332–334] (**I**ntegrated Software **P**roject **S**upport **En**vironment) was a long-lasting research project financed by different sources; the biggest contribution has been given by DFG. IPSEN was on novel support of software development, so solely within informatics. The project was carried out from 1981–1996 with a total personal effort of about 110 person years at Chair Informatics 3 (Software Engineering), one of the groups involved in IMPROVE.

The *topic* of IPSEN was *a-priori integration*. This means that IPSEN built new tools for being integrated. Integration, therefore, is very tight. The new tools were for requirements engineering, architectural design, programming, documentation, and project management, and for making the transitions in between.

Three *integration dimensions* were addressed. (a) Tight integration on *one document*: This was shown by a tight integration of editor, analysis, execution, and monitoring tools for the programming task [108, 385]. (b) Also, tight integration *between different documents* was discussed by offering *fine-grained* integrator tools between the above mentioned working areas. For example, integrator tools were built between different views of requirements engineering [229], from requirements engineering to architectural descriptions [184, 255], from architectural descriptions to program modules [260] etc. Finally, (c) integration between *fine-grained* and *coarse-grained descriptions* were explored: within management, objects and relations were studied on a fine-grained level. From management to outside, only coarse-grained structures were regarded.

In current terminology, IPSEN was on *model-driven development.* Thereby, models were not regarded on an outside level but only on a *tool construction level.* All tools were specified by graph transformations [328, 350, 363, 412, 492]. From these specifications, tools were either manually but methodologically derived, or they were generated. In IPSEN, also the specification environment PROGRES and corresponding code generators were developed, both being an integral part of the tool production machinery.

From IPSEN the IMPROVE project *borrowed* several aspects: (a) On one side the idea of integrator tools [33] was taken and also the way how the prototypical versions of such tools are realized [350, 363]. However, the discussion within IMPROVE on integrator tools is more elaborate, the application field is now chemical engineering, and the machinery to generate these tools has been improved. (b) The I3 subproject took over the architectural discussions of IPSEN on tools as well as on tool integration. However, the discussion within IMPROVE is not on a-priori but on a-posteriori integration [136]. Finally, (c) IMPROVE also took over the idea of tightly integrating management into a development process [475]. The ideas about how this management system is structured and which functionality it offers, is more from SUKITS than from IPSEN.

ConceptBase

In cooperation with the University of Toronto, the Chair of Informatics 5 (Information Systems, Prof. Jarke) had been developing an adaptable, logic-based conceptual modeling language called Telos since 1986. Telos was designed to provide a uniform description of the different design artefacts within an information systems engineering process, such as requirements models, designs, and implementations, as well as their static interdependencies and process relationships. Telos was initially developed in the context of ESPRIT project DAIDA [199, 202] and elaborated for the special case of requirements management in ESPRIT Basic Research Project NATURE [200] (see below). Outside the Entity-Relationship and UML language families, Telos has become the most-cited conceptual modeling language of the 1990's. A complete description of Telos together with its formal semantics was first given in [327].

Within the ESPRIT Basic Research Project on Computational Logic (COMPULOG), Manfred Jeusfeld was able to show in his doctoral thesis [203] that a slightly simplified version of the Telos object model could be elegantly mapped to Datalog with stratified negation which provided the basis for an efficient implementation. From this result, a rather efficient and very expressive metadata management system called ConceptBase was developed [189]. ConceptBase has since been deployed in more than 350 installations worldwide for teaching, research, and industrial applications such as reverse engineering, view management, or heterogeneous information systems integration [358, 425]. Several of these applications as well as some of the implementation techniques have also influenced the design of ontology languages and XML databases.

Within CRC IMPROVE, ConceptBase has been used as the basis to describe the interrelationships between heterogeneous design artefacts and some aspects of the semantics of design process guidance, mostly at the fine-grained level [193]. In the course of the CRC, a large number of improvements were made, including novel compilation and optimization techniques and new interfaces, e.g. for XML. A recent system overview can be found in [205].

DWQ

The goal of the ESPRIT Long Term Research Project DWQ "Foundations of Data Warehouse Quality" was to improve the design, the operation, and most importantly the long-term evolution of data warehouse systems [192]. In the years from 1996 to 1999, researchers from the DWQ partners – the National Technical University of Athens (Greece), RWTH Aachen University (Chair of Informatics 5; Germany), DFKI German Research Center for Artificial Intelligence (Germany), the INRIA National Research Center (France), IRST Research Center in Bolzano (Italy), and the University of Rome – La Sapienza (Italy) had cooperated on these goals.

The primary result of the DWQ project consists of a neutral architectural reference model covering the design, the operation, the maintenance, and the evolution of data warehouses. The architecture model is based on the idea, that any data warehouse component can be seen from three different perspectives: The conceptual perspective, the logical perspective, and the physical perspective. In the design, operation, and especially evolution of data warehouses, it is crucial that these three perspectives are maintained consistent with each other.

Placed in a completely different and far more fixedly structured domain, the results of the DWQ project were not directly transferable to the CRC 476 IMPROVE. For the support of creative design processes, different approaches were – and are – necessary. Yet the approach of applying methods of meta modeling in domain and application models was expected to succeed there as well. Especially the use of ConceptBase, as described before, was used to achieve a sound conceptual and ontological basis for the modeling aspects, the process extensions to Data Warehousing as researched in the subproject C1, and the meta process and product repository of the subproject B1.

NATURE

The ESPRIT Basic Research Project NATURE (Novel Approaches to Theories Underlying Requirements Engineering, 1992–1995) has investigated central problems of Requirements Engineering (RE), especially the selection, visualization, and reuse of requirements, the management of non-functional requirements, and the transformation of requirements from natural language into formal semantics [187, 201]. Researchers from the City University London (England), ICS-Forth Heraklion on Crete (Greece), SISU Kista in Stockholm (Sweden), the Université Paris 1 Pantheon-Sorbonne (France), and from cooperation partners in Canada and the US have worked under the leadership of the Chair of Informatics 5 at the RWTH Aachen University. The aforementioned problems have been treated from the viewpoints of domain analysis, process guidance, and formal representation.

The role of domain theory is mainly placed in facilitating the identification, elicitation, and formalization of domain knowledge. Furthermore, concepts for supporting experience reuse were developed, based on similarity mappings and classifications.

As part of the process guidance, systems were designed to support software design by context-based and decision-guided tools. The influence of domain knowledge onto actions like validation and view integration was also researched.

A framework for domain analysis and process guidance was created. The views of the system, the usage, and the subject world had to be integrated into a common formal representation. This included the formal representation of characteristics of informal language like ambiguity, incompleteness, inconsistency, simplification, or redundancy.

NATURE does not define a methodology, but rather a "meta-level" framework that defines languages for thinking about, and organizing work in, RE. Concrete specializations can then be designed to improve existing methodologies. To integrate and evaluate the theories developed based on the different view points, they were represented in the formal modeling language Telos, implemented in a prototype based on ConceptBase, and evaluated by exemplary application scenarios.

The conceptual framework for Requirements Engineering in software development represented by the NATURE project heavily influenced the CRC 476 IMPROVE, as many of the approaches could be adapted from software RE to the elicitation, formalization, and analysis of requirements in chemical process engineering. The knowledge representation framework based on Telos and ConceptBase was used for the domain modeling in the subprojects of area A. The design of a data and meta data repository for all aspects of the design activities formed the basis of the Process Data Warehouse as developed in the C1 subproject. The experiences from requirements traceability were employed for the experience reuse in B1 and C1. With respect to (fine-grained) process guidance, the modeling and guidance aspects of NATURE were applied in the B1 subproject, including the adaption of the process-centered Requirements Engineering development environment PRO-ART into the PRIME environment.

1.3.5 Survey of This Book's Structure

The *structure* of the *book* can easily be explained by taking the project structure of Fig. 1.27 and having a look on the iterations of IMPROVE's evolutionary approach of Fig. 1.28. The structure of this book as well as the interdependencies of its sections are discussed here.

Survey

We are now at the end of the introductory Part I of this book. The following *main part* of this book about technical results can be *derived* from IMPROVE's *project structure*:

- Chapter 2 describes the results of the engineering partners (upper layer A of Fig. 1.27),
- Chapter 3 gives the main contributions of the middle layer B about new support concepts and their implementation,
- Chapter 4 explains our bottom layer C on platforms for tools,
- Chapter 5 on integration discusses various integration aspects (cf. right column of Fig. 1.27) and, finally,
- Chapter 6 gives the results we have gained so far for the major topic process/product model (left column of Fig. 1.27).

Chapters 2 –6 describe the status of the IMPROVE project after the third project cycle. This *description*, for simplicity reasons and for giving a clear view on the scientific outcome, is *organized along* the corresponding *subprojects* of IMPROVE, essentially one section per subproject. This section collects the results of the first three cycles of an IMPROVE subproject, if the results are not overridden by later results. So, the book is not structured along the timeline. Instead, it is structured along results for specific topics elaborated by subprojects.

The fourth cycle, as already stated, deals with *transfer* of results to industry. Transfer is not only one-way directed. Also, new questions, problems, and experiences are brought in by industrial partners. Not all subprojects of IMPROVE are represented by a *transfer project*. Most of the transfer projects are financed by the Transfer Center 61, funded by DFG. The transfer projects, as described in Chap. 7, are given in the common *order* of areas (from A to B and C) and in the order within a project area, as shown in Fig. 1.27.

Part III of this book gives a description of current/future activities on the one and an evaluation of achieved results on the other hand. Chapter 7 describes the transfer project plans/activities (to be) carried out in the fourth cycle. The plans are rather concrete, as we have already gone through 21 months of this transfer period. Chapter 8 of this book gives an *evaluation* of IMPROVE's results from different perspectives: How we contributed to a better understanding of design processes, how we contributed to an improved and elaborated support for design processes, why and how we contributed to an improvement of the state-of-the-art in industry, and how we influenced the scientific community.

The book contains an extended *bibliography*. This bibliography is split into two parts, namely publications of IMPROVE and literature outside IMPROVE. This is done to give a quick survey on the scientific literature output for the reviewing peers and our science companions.

Structure and Dependencies of This Book

We now discuss the structure of this book, which is given as a *structure* and *dependency graph* in Fig. 1.29. The graph does not contain all nodes.

Tree Structure

The tree of Fig. 1.29 is the *structure part* of this graph.

The chapters and sections are given as an *ordered tree* (the root node 'book' with five successors is not shown). The tree in *preorder* visit of its nodes delivers the table of contents, i.e. the order of reading, if a reader goes completely through the text, from its beginning to its end.

Part II and Part III are the *main parts* of the book describing the achieved results, evaluations, and transfer still going on. So, the graph for this part contains all the nodes down to sections.

Dependencies

There are many *dependencies* in the book, i.e. that one chapter or section contains information necessary for the reading of another part.

There are *trivial* dependencies: Part I is a prerequisite for all chapters and sections of the main part of this book. Analogously, every section of this book contributes to the references.

Also, there are *simple* and being mostly 1:1 *dependencies*: A section of the main Part II on technical results is a prerequisite for a corresponding section in Part III of this book on future transfer activities, if this project is involved in transfer. For example, there is a dependency between Sect. 3.2 on integration and the corresponding transfer of integrator results in Sect. 7.6.

All these relations are either only given on a coarse level (as from Part II to Part III) or not shown at all. When discussing the dependencies of this book, we *concentrate* on the *main part* of the book, consisting of Parts II and III. Furthermore, we only explain some *examples*, one per chapter.

The schema within *Chap. 2* is rather simple: The introduction prepares for all the following sections. Furthermore, the Sects. 2.2 to 2.5 are prerequisites for understanding the Sect. 2.6, which summarizes all the discussed models.

For *edges* to be discussed, we now use an *abbreviation*: The target node of an edge is given, the corresponding source nodes are all the nodes written down on the edge with a corresponding section or chapter number.

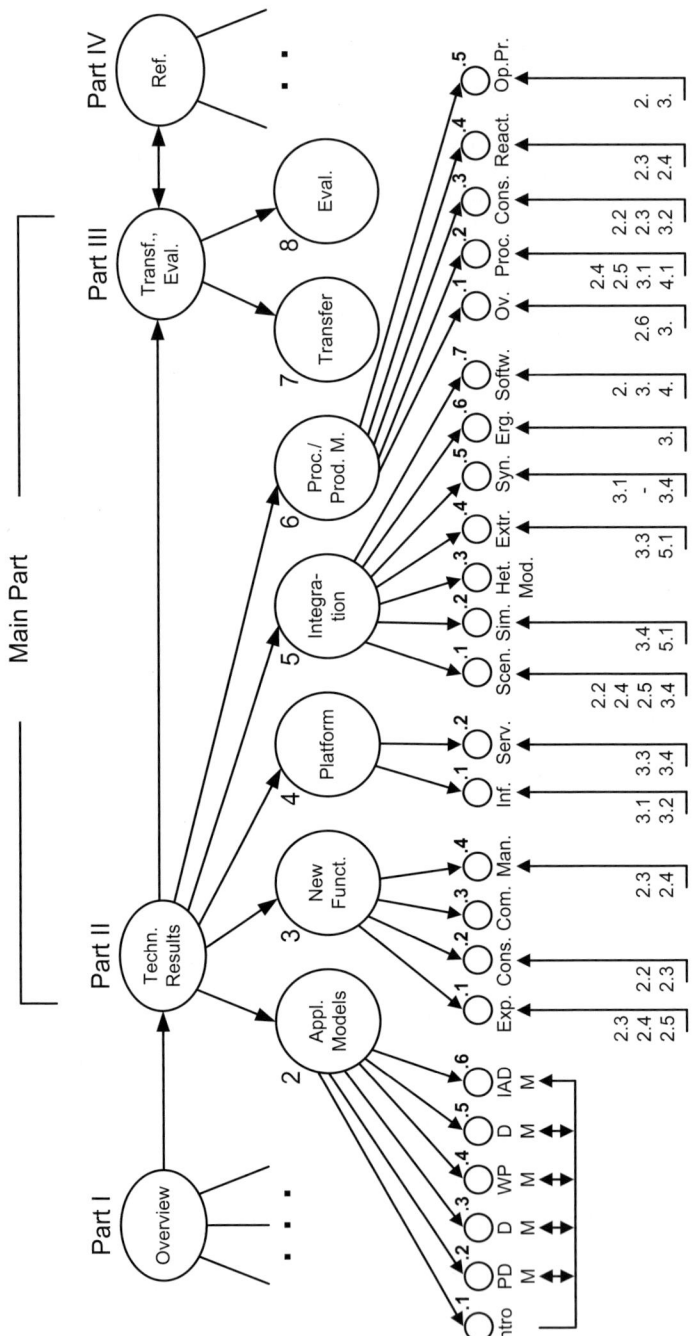

Fig. 1.29. A dependency graph describing the structure of this book

Let us take the incoming edges to node 3.1 of *Chap. 3* as an example. We see that Sects. 2.3 on document models, Sect. 2.4 on work process models, and Sect. 2.5 on decision models are prerequisites for Sect. 3.1 on experienced-based design process support.

For Sect. 4.2 on service management, as example of *Chap. 4*, Sect. 3.3 on communication and Sect. 3.4 on management are the prerequisites.

The scenario-based analysis of work processes of *Sect. 5.1* needs Sect. 2.2 on product data models, Sect. 2.4 on work process models, and Sect. 2.5 on decision models. To see the relations to management, we also need input from Sect. 3.4 on management on development processes.

For *Sect. 6.2* on process/product models for experienced-based tools we need again Sect. 2.4 and Sect. 2.5 on work processes and decisions, but also Sect. 3.1 on experienced-based support, and Sect. 4.1 on the process data warehouse.

2

Application Domain Modeling

This chapter deals with *application domain modeling*. It presents the various models for the application domain of chemical process engineering, which have been developed in the CRC 476. As already indicated and to be demonstrated by the following chapters, application domain modeling is the input of the tool construction processes.

The *chapter* is organized as *follows*: Section 2.1 gives an introduction to application domain modeling. The subsequent two sections deal with the products of the application domain, whereas Sects. 2.4 and 2.5 discuss the process and decision perspective. Section 2.6 discusses the integration of product models and process models into a coherent framework and gives a brief review of the related work in this area.

2.1 An Introduction to Application Domain Modeling

J. Morbach, M. Theißen, and W. Marquardt

Abstract. This section serves as an introduction to application domain modeling. Firstly, we will motivate the objectives of modeling. Next, we will suggest definitions for different types of application domain models. Finally, a brief survey of the modeling languages applied throughout this chapter will be given.

2.1.1 Objectives of Modeling

Different types of application domain models can be distinguished according to the objective of the model. *Information models* are created to support the development of software systems. Information models can be differentiated into *conceptual models*, *design models*, and *implementation models*, which form a series of consecutively refined models. Moreover, *ontologies* are to be mentioned here, whose aim is to provide an explicit specification of a conceptualization of the application domain. In the following, more precise definitions of these different types of models will be given.

Information Models

Information modeling is a commonly used method for the analysis and formalization of information structures and work processes as the basis for software design [862]. Three types of information models can be distinguished, which are consecutively developed through refinement:

- The *conceptual model* describes the major entities of the domain of interest and their interrelations on a conceptual level, irrespectively of a particular application or implementation. Its function is to familiarize with the vocabulary of the domain of interest and to establish a common understanding of its key concepts. In the areas of software engineering and database design, conceptual models are used as a preliminary draft or template, based on which more specific models can be created. In these communities, the terms 'domain model' or 'analysis (object) model' are often used instead of 'conceptual model' (e.g., [510, 673, 761]).
- Next, the conceptual model is transformed into a *design model*. It serves the needs of the software engineer during software development. A design models defines the architecture and the internal data structures and determines the user interfaces of the software system.
- Finally, the design model is implemented by means of some programming language, resulting in the *implementation model*. Thus, an implementation model is the realization of a design model in form of software.

Ontologies

'Ontology' is originally a term in philosophy, indicating a study field that strives to provide *a systematic account of existence*. In the last two decades, this term has been adopted by computer science, firstly used in the field of artificial intelligence and recently in other areas, as well. While the meaning of this term in computer science still appears controversial, the definition given by Gruber [705] has been the one most widely cited: "*An ontology is an explicit specification of a conceptualization*". As further explained by Studer et al. [988], "*conceptualization* refers to an abstract model of some phenomenon in the world by having identified the relevant concepts of that phenomenon. *Explicit* means that the type of concepts used and the constraints on their use are explicitly defined". To this aim, an ontology will necessarily "include a vocabulary of terms and some specification of their meaning. This includes definitions and an indication of how concepts are inter-related, which collectively impose a structure on the domain and constrain the possible interpretation of terms" [1007].

As explained by Uschold and Grüninger [1006], an ontology can be

- *informal* if expressed in natural language;
- *semi-informal* if expressed in a restricted and structured form of natural language;
- *semi-formal* if expressed in an artificial and formally defined language; and
- *rigorously formal* if it provides meticulously defined terms with formal semantics, theorems, and proofs of properties such as soundness and completeness.

In computer science, the notion of ontology is often restricted to the latter two subtypes.

Relationship between Ontologies and Information Models

Ontologies and information models should not be regarded as mutually exclusive. They have various characteristics in common. In particular, both conceptual models and ontologies provide conceptualizations of the problem domain. Moreover, formal ontologies can be used directly by computer agents – in this respect, they resemble implementation models.

The major difference between information models and ontologies results form their different objectives. Ontologies emphasize the sharing of domain knowledge, which requires the explicit definition of the meaning of terms. Information models, aiming at the design and implementation of software systems, do not need such definitions, since the semantics of the modeled entities is implicitly known to the modeler.

In accordance with the above definitions, a conceptual model as well as an implementation model can be considered as an ontology if it explicitly specifies the meaning of its classes and relations. On the other hand, an implementation model stated in a formal ontology language, such as the schema of a deductive database, is not necessarily an ontology. Again, the necessary criterion to qualify as an ontology is the availability of an explicit specification of the meaning of terms.

2.1.2 Modeled Entities

In this subsection, definitions for different types of application domain models, distinguished by the type of the modeled entities, will be given. We will differentiate between *process models* describing work processes, *product models* describing the products (i.e., the outcomes or results) of the work processes, and *decision models* describing the decisions made during a work process. Product models can be further subdivided into *product data models*, dealing with the elementary data items, and *document models*, describing the aggregation of product data into documents in a certain work context.

Product Data Models

In engineering disciplines, the term *product data* is used to denote data about a *technical product and its characterizing properties* [19]. We will extend this definition to the data representing *the products (i.e., the outcomes or results) of a design process*. This agrees well with the conventional definition of product data, since engineering design processes aim at the specification of a technical product, such that their results characterize – either directly or indirectly – a technical product. Thus, product data is defined as *data about a technical product and its characterizing properties as well as the data created or used during some design process*.

Consequently, a *product data model* defines the form and the content of product data [839]. It classifies, structures, and organizes the product data and specifies their mutual relationships.

In the context of chemical engineering, both the chemical plant and the chemical substances produced by the plant can be regarded as products. Thus, the following examples qualify as product data: the dimensions of a plant equipment (e.g., volume, diameter), the operating conditions of a process step (e.g., pressure, temperature), or the physical properties of a chemical compound (e.g., density, boiling temperature).

Document Models

Product data can be aggregated to documents like reports, data sheets, or process flowsheets, each of which represents a specific *view on data*. During project execution, documents are used as carriers for the data they hold [940]. Thus, documents can be defined as *carriers of data*.

Physically, documents may come either in form of a hard copy (i.e., a stack of paper) or in form of a soft copy (i.e., a data file within a computer system).

A comprehensive document model should describe the different types of documents used in a given application domain, their interrelations and contents, as well as the dynamic behavior of documents over time [946].

Work Process Models

In contrast to product data and documents, the term *process* and its precise meaning have not been agreed on in the literature. This is largely due to its origin in different, at least partially unrelated disciplines. We therefore look at different kinds of processes before we come up with the definition of a work process model.

Business Process and Work Process

The term *business process* became a buzzword in the last decade of the past century, but even today there is no widely accepted definition. The term is used in particular by the *business process reengineering* and *workflow management* communities.

According to Hammer and Champy [714], business process reengineering (BPR) is the '*fundamental rethinking and radical redesign of business processes to achieve dramatic improvements in critical, contemporary measures of performance, such as cost, quality, service and speed.*' In the BPR literature, rather general definitions of a *business process* are proposed. Davenport [626], for instance, defines it as a '*structured, measured set of activities designed to produce a specified output for a particular customer or market.*'

Workflow management (WM) refers to the '*automation of a business process, in whole or part, during which documents, information or tasks are passed from one participant to another for action, according to a set of procedural rules*' [1058]. The Workflow Management Coalition, a group of vendors, users, and researchers of workflow management systems, defines a business process as a '*set of one or more linked procedures or activities which collectively realize a business objective or policy goal, normally within the context of an organizational structure defining functional roles and relationships*' [1058].

Despite the universality of the definitions proposed in the BPR and WM literature, BPR and WM typically focus on recurrent processes on the operative level which are completely determined and therefore can be planned in a detailed manner. Examples include the processing of a credit transfer in a bank or of a claim in an insurance company. As a result, the term *business process* is often used in this restricted sense. Other processes, such as the design of an engineering artifact or the production of a physical good, are commonly not considered to be business processes although they actually meet the definitions cited above.

To overcome the ambiguities associated with the term *business process*, we introduce the term *work process* and define it as follows: *A work process is a collection of interrelated activities in response to an event that achieves a specific product for the customer of the work process.* As this definition was originally proposed by Sharp and McDermott [962] for business processes, it becomes clear that there is no fundamental difference between business processes and work processes. Though, a business process in the colloquial meaning of the term is a special type of a work process.

Design Process

Mostow [858] characterizes *design* as a process whose *'purpose is to construct a structure (artifact) description that satisfies a given (perhaps informally) functional specification.'* Hence, a design process is a special type of work process aiming at the creation of a special product, that is, the specification of an engineering artifact by means of documents which contain product data describing the artifact.

Design processes in chemical engineering comprise all the activities related to the design of a new product and the associated production plant including the process and control equipment as well as all operation and management support systems [299]. One of these design processes is the *process design process*. According to Westerberg et al. [1047], its target is the *'creation or modification of a flowsheet capable of manufacturing a desired chemical.'*

The focus of CRC IMPROVE is on the early stage of the chemical process design process, called *conceptual process design*. The main product of conceptual process design is a process flow diagram – or a set of several diagrams – specifying the basic layout of a plant or its parts, including important process data such as mass or energy streams. Further documents are created during conceptual process design, for instance, mathematic models of the chemical process for the evaluation of different design alternatives. However, in general not all of these documents are provided to the customer of the design process, in particular if the customer is located at another organizational unit or enterprise.

Work Process Models

A *work process model*, in the following referred to as *process model* for short, contains the steps performed in a work process as well as their interdependencies, for example a sequence of several steps or their parallel execution. In this contribution, the steps are referred to as *activities*; in contrast to other terms which can be found in literature (e.g., *action* [880], *function* [949]), this term emphasizes the temporal dimension of an activity which can itself be decomposed in several sub-activities which form a work process in their own right. Depending on their purpose, process models can contain further information beyond the activities, for instance about the products of activities, the people performing the activities, or the tools required.

Process models can be *descriptive* or *prescriptive* [372]. The former describe a work process as it is or has been performed by the involved actors (*as-is models*). Descriptive models can serve as part of the documentation of a particular design project. In CRC IMPROVE, descriptive models of design processes are also used as an intermediate step for the creation of prescriptive models which incorporate best practices for future design processes (*to-be models*) and can serve as guidelines for designers or as a starting point for implementation models required for automated process support.

Decision Models

In this contribution, we use the term *decision model* for both a representation of *design rationale* [856], i.e., the decisions taken by a designer during a design process that led to a particular artifact, and for a representation of some rules or methods which can guide a designer confronted with a decision problem or which can even solve a decision problem algorithmically.

Models of design rationale comprise at least the design problem under consideration (e.g., *Which reactor type?*), the alternatives considered (e.g., *VK-tube* or *CSTR*), and arguments for or against the alternatives. Similar to *as-is* process models, descriptive design rationale models can serve the documentation of past design projects – although this is rarely done in practice – and they can be used for the identification of approved decision strategies to be incorporated in a prescriptive model for future decisions.

2.1.3 Modeling Languages

For the representation of the different application models presented in this chapter, various modeling languages have been used, which provide graphical as well as lexical notations. In the following, a short introduction to those languages and some references for further reading will be given.

The modeling languages can be divided in two groups. The first group comprises generic modeling languages used to represent domain or application models; these models, in turn, can be considered as application-specific modeling languages, which constitute the second group[7]. For instance, the elements of the Unified Modeling Language UML can be used to define the vocabulary of the Decision Representation Language DRL, which can then be employed to describe a particular case, such as a design decision.

Generic Modeling Languages

UML

The Unified Modeling Language UML (e.g., [673]) is a graphical modeling language, which consists of a number of different modeling diagrams that

---

[7] Note that this agrees well with the four-layer metamodel hierarchy proposed by the Object Management Group (OMG) [882]: The generic modeling languages correspond to the M2 layer, whereas the application-specific modeling languages correspond to the M1 layer.

allow the graphical notation of all aspects of a software development process. Of these different modeling diagrams, only the class diagram is used within this chapter.

Within UML, domain entities are represented through *classes* and their instances (*objects*). Classes can be hierarchically ordered and further characterized by means of attributes. Additionally, binary relations, called *associations* in UML, can be introduced between classes. Associations can be either uni-directional or bi-directional. The number of objects participating in an association can be indicated by cardinality constraints. Additionally, UML introduces two special types of associations: aggregation and composition.

O-Telos

O-Telos [327] is a conceptual modeling language that supports basic deductive reasoning capabilities to assist schema development and maintenance. O-Telos is implemented in the ConceptBase system [189], a deductive object base intended for conceptual modeling.

An outstanding property of O-Telos is its capability for meta modeling: The language supports not only the definition of *classes* but also the definition of *meta classes* (and even *meta meta classes*); this leads to a layered model structure where the entities on each level are instances of the entities on the level above. Rules and logical constraints can be defined at any abstraction level. Constraints specify conditions on the specialization and instantiation of classes. Rules make implicit information explicit and thus derive new information from the asserted facts. Within each level, binary relations between the model entities can be indicated through so-called *links*, and the entities can be further characterized by attributes.

DAML+OIL and OWL

The *OWL Web Ontology Language* [546] and its predecessor *DAML+OIL* [609] are ontology markup languages that have been developed for publishing and sharing ontologies in the Web. Their syntax is based on existing Web markup languages, the most prominent of which is XML [1060]. By now, DAML+OIL has been largely superseded by its successor OWL, which has been endorsed as a *W3C recommendation*[8]

As OWL is derived from DAML+OIL, it shares most of its features (a listing of the differences between the two languages can be found in Appendix D of [546]). Therefore, only OWL will be discussed in the following.

Model entities can be represented through *classes* in OWL; their instances are called *individuals*. Classes can be hierarchically ordered in OWL, thereby allowing multiple inheritance. Complex class definitions can be constructed

---

[8] A W3C Recommendation is the final stage of a ratification process of the World Wide Web Consortium (W3C) concerning a standard for the Web. It is the equivalent of a published standard in many other industries.

by means of the following language primitives: set operators (union, intersection, complement of classes), equivalence of classes, disjointness of classes, and exhaustive enumeration of the instances of a class.

Classes can be further characterized by means of attributes. Additionally, binary relations can be introduced between classes. In OWL, attributes and binary relations are represented through the same language primitives, the so-called *properties*; their ranges are different, though (data types and individuals, respectively). Properties can be hierarchically ordered, and their usage can be restricted through cardinality constraints and type constraints. Moreover, two properties can be declared to be equivalent or the inverse of each other. Finally, additional logical information can be indicated about binary relations, namely the transitivity and the symmetry of a relation.

The OWL language provides three increasingly expressive sublanguages, called *OWL Lite*, *OWL DL*, and *OWL Full*. Each of these sublanguages is an extension of its simpler predecessor, both in what can be legally expressed and in what can be validly concluded [971]. Throughout this chapter, only the OWL DL subset is used for modeling. This sublanguage is compatible with a particular type of description logic (DL) called SHOIN(D) [736]. As a consequence, the models represented in OWL DL can thus be processed with standard DL reasoners.

VDDL

The *VeDa Definition Language VDDL* [12] is a self-defined frame-based language, which was specially developed for the representation of the VeDa Data Model (cf. Subsect. 2.2.2). Combining features from object-oriented modeling and description logics, VDDL is a highly expressive general modeling language providing many different, specialized features. For sake of simplicity, only the most important language elements will be discussed here.

Entities are represented through *classes*, which can be ordered by means of *superclasses* and *metaclasses*. Class definitions can include *attributes*, possibly restricted through *facets*, as well as *methods* and *laws*. Different types of attributes are distinguished. Among those, so-called *relational attributes* can be used to represent binary relations between classes. Attributes values can be restricted through both facets and laws. Methods represent numerical or symbolical functions that allow to derive complex properties of an object from known (simple) ones by executing some procedural piece of code.

Application-Specific Modeling Languages

C3

The *C3* language has been developed within the CRC subproject I4 for the representation of weakly structured work processes. In particular, C3 allows the adequate capture of issues related to *C*oordination, *C*ooperation, and *C*ommunication which are characteristic for design processes.

C3 is based on the state diagram and the activity diagram of UML. As these diagrams do not cover the requirements for modeling cooperative work processes [221], additional elements and concepts have been added to C3, such as the blob element of the Higraphs [719], a modified subset of Role-Function-Activity Nets [872], and some concepts of Task Object Charts [1069].

C3 allows for participative modeling, that is, the language is intentionally kept simple to allow for an easy application and to enable the participation of all stakeholders in a work process modeling session, independent from their distinctive backgrounds and experiences in process modeling.

IBIS and DRL

Issue Based Information Systems (*IBIS*, [798]) and the Decision Representation Language (*DRL*, [808]) are both semi-formal modeling languages for the representation of design rationales. The basic elements provided by IBIS are the **Issue** denoting a problem to be decided on, the **Position** representing an alternative, and the **Argument**, which supports or denies a **Position**. Thanks to its ease of use, IBIS can be seen as a participative modeling language; its proponents even advocate its use during meetings of design teams in order to structure the evolving discussions [607].

In contrast to IBIS, DRL is characterized by a considerably richer set of modeling elements, the most important extension being the **Goal** that allows for a more adequate representation of the constraints to be met by the final specification of an artifact.

Computer systems for cooperative modeling have been created for both IBIS and DRL [608, 807], which demonstrate the usability of the languages in academic and industrial settings and thus their empirical validation. Whereas several applications of IBIS have been reported in literature (e.g., [54, 525]), no results seem to be available for DRL. Nevertheless, we estimate the expressiveness of DRL as indispensable for the appropriate representation of design decisions during conceptual process design.

Notation and Naming Conventions

There are considerable similarities between the components of the different generic modeling languages applied in this chapter. In particular, all of them provide language primitives for classes, relations, attributes, and instances (though named differently within the respective formalisms). For sake of clarity, we will use the term *class* to refer to classes as well as to concepts, frames, or similar language constructs (however named) in the remainder of this chapter. Similarly, the term *relation* is used to denote (inter)relations, properties, slots, and associations; the term *instance* subsumes instances, individuals, and concrete objects alike.

Throughout this chapter, **UpperCamelCase** notation in sans-serif font is used to denote identifiers of classes, and likewise **lowerCamelCase** notation to represent relations. Instances are denoted by sans-serif font, yet without using a special notation. Finally, *italicized sans-serif font* refers to structural elements of a model, such as modules, packages, or layers.

In figures, the notation of UML class diagrams is used. Boxes represent classes, lines represent relations, and dashed arrows represent instantiation. Aggregation relations are represented through a white diamond-shaped arrowhead pointing towards the aggregate class. Similarly, composition relations are indicated by a black diamond-shaped arrowhead.

2.2 Product Data Models

J. Morbach, B. Bayer, A. Yang, and W. Marquardt

Abstract. This contribution summarizes the results of more than a decade of product data modeling at the authors' institute. In this effort, a series of consecutive product data models has been developed for the domain of chemical engineering, starting with the chemical engineering data model VeDa, followed then by the conceptual information model CLiP and finally by the domain ontology OntoCAPE. The evolution of these different models is described, and their similarities and differences are discussed with regard to scope, structural design, conceptualization, representation language, and intended usage.

2.2.1 Introduction

Due to the large number of domain concepts, ever-changing application areas, newly arising insights, and the continuous update of technological capabilities for model representation, the development of product data models for a complex domain is deemed to be an evolutionary process. Since the early nineties, several product data models have been developed by the authors and their former colleagues, which model various aspects of the chemical engineering domain. The results are *VeDa* [10, 55], *CLiP* [14, 19], and *OntoCAPE* [325, 326], the second version of which has been released lately.

While OntoCAPE is a formal ontology as defined by Uschold and Grüniger [1006], CLiP and VeDa were historically called "conceptual models" (cf. Subsect. 2.1.1 for a definition of these terms). However, since they agree with Uscholds and Grünigers definition of semi-formal ontologies, they can be regarded as ontologies, as well.

In the following, the different models will be characterized, and their similarities and differences with regard to scope, structure, conceptualization, representation language, and intended usage will be discussed.

2.2.2 VeDa

The development of VeDa began more than a decade ago, when the subject of conceptual information modeling received first attentions in chemical engineering. It was initiated by Marquardt [289–291] to support the mathematical modeling of chemical processes. VeDa provides concepts for the description of mathematical models, the modeled objects, and – for documentation purposes – the activities and the decisions taken during model building. Thus, VeDa describes not only product data but to some extent work processes, as well.

Model Structure

VeDa relies on *general systems theory* (e.g., [578, 784, 901, 1015]) as a general organization principle. General systems theory has been successfully applied to represent complex structured systems in various application areas (e.g., systems theory was used by Alberts [499] and Borst [562] as an organizing principle for the construction of large engineering ontologies). In VeDa, mathematical models as well as the modeled objects (i.e., the objects that are to be represented by the mathematical models) are described as systems.

A technique for complexity reduction that is widely used in systems engineering is the adoption of a *viewpoint*. A viewpoint is an abstraction that yields a specification of the whole system restricted to a particular set of concerns [744]. Adopting a viewpoint makes certain aspects of the system 'visible' and focuses attention on them, while making other aspects 'invisible', so that issues in those aspects can be addressed separately [529]. In VeDa, viewpoints are used to organize the description of mathematical models: A mathematical model is considered from two distinct viewpoints, originally called *complexity coordinates* [291]. The *structural coordinate* describes the structural elements of the model and their interrelations, and the *behavioral coordinate* describes the behavior of the model.

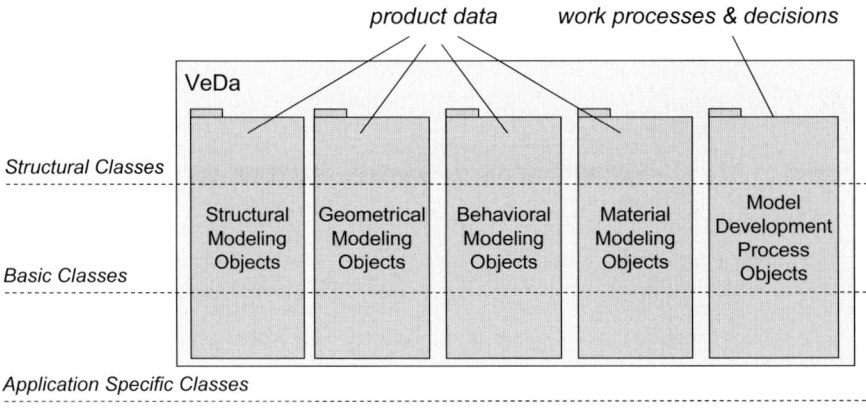

Fig. 2.1. Model structure of VeDa

VeDa is structured into three horizontal *layers*, separating so-called **Structural Classes**, **Basic Classes**, and **Application Specific Classes** (cf. Fig. 2.1). The design of each layer follows the principle of "minimal ontological commitment" [705], meaning that each layer holds only those concepts that are essential for its function; concepts that are more specific than the layer's function are placed on a lower layer.

Vertically, VeDa is partitioned into five *partial models*, representing thematically distinct areas. The partial models **Structural Modeling Objects** and **Behavioral Modeling Objects** describe mathematical models from a structural and behavioral viewpoint, respectively. The description of mathematical models is complemented by the remaining partial models, which represent geometrical concepts, materials, and the process of model building.

Representation Language

Since no prevalent general modeling language was available in the early nineties, a frame-based language was developed for the representation of VeDa. This language, called *VDDL* (Veda Data Definition Language), has been formally specified in [12]. Combining features from object-oriented modeling and description logics, VDDL is a highly expressive general modeling language, which supports the definition of classes, metaclasses, and instances, as well as different types of attributes and relations. Class definitions can include methods and laws. Methods represent numerical or symbolical functions that act upon the objects. Laws restrict the possible instances of classes and attributes by logic expressions.

Since VDDL does not provide a graphical representation, parts of VeDa have been additionally represented through UML class diagrams for reasons of clarity.

Scope and Content

The partial model **Structural Modeling Objects** [417] describes mathematical models from a structural viewpoint. According to an organization principle introduced by Marquardt [289, 292] and later refined by Gilles as 'Network Theory' [693], a mathematical model can be decomposed into submodels for *devices* and for *connections*. Device models have the capability for the accumulation and/or change of extensive physical quantities, such as energy, mass, and momentum; they represent for example process units, or thermodynamic phases. Connection models describe the interactions and fluxes (e.g., mass, momentum, or energy fluxes) between the devices; they represent for instance pipes, signal lines, or phase boundaries. **Structural Modeling Objects** also describes the mereotopological relations between these different types of models. Furthermore, the partial model provides concepts for the representation of some modeled objects (e.g., various reactor types). A modeled object is described as a system, which is considered from both a functional and a structural viewpoint.

The partial model **Geometrical Modeling Objects** [418] is concerned with the geometrical description of the structural modeling objects. It provides concepts to represent surfaces, shapes, and spatial coordinate systems.

Behavioral Modeling Objects [55, 56] characterizes mathematical models from a behavioral viewpoint. It introduces concepts to represent mathematical equations, physical dimensions, and units.

Material Modeling Objects [461] represents the intrinsic characteristics and the thermodynamic behavior of materials.

Finally, *Model Development Process Objects* [248] describes the (work) process of model development, introducing concepts for the description of activities, goals, and decisions.

As indicated in Fig 2.1, the first four partial models are concerned with the representation of product data; they can be considered as the origin of the product data models of the CRC, which are presented in the following subsections. The fifth partial model, in contrast, addresses work process modeling; it forms the basis of the later-developed work process and decision models presented in Sects. 2.4 and 2.5. The partial models of VeDa's product data part are closely related to each other, and their mutual relations are explicitly represented in the model. The process part, on the other hand, is somewhat detached from the rest of the model as the interdependencies between the product part and the process part have not been explicitly covered by VeDa.

Usage

VeDa is intended as a modeling language for the domain of process engineering, which can be used for the representation of specific mathematical models of chemical process systems in a model library as well as for the representation of knowledge about models and the modeling process in a knowledge-based modeling environment.

Based on VeDa, implementation models for the model repository ROME [463] and the modeling environment ModKit [52, 54], both described in Sect. 5.3, have been developed. In this context, VeDa was refined and extended. While most of these changes were implementation-specific and therefore did not become part of the conceptual model, some were incorporated in VeDa, like for example detailed taxonomies of physico-chemical phenomena and physical properties.

2.2.3 CLiP

In succession to VeDa, the conceptual information model CLiP has been developed to provide a conceptualization for the domain of chemical process design. CLiP covers the product data produced during the design process, the mathematical models used in the various model-based design activities, the documents for archiving and exchanging data between designers and software tools, as well as the activities performed during process design.

Except for the partial model *Geometrical Modeling Objects*, CLiP comprises all areas covered by VeDa and extends beyond its scope. Key notions from

VeDa were reused in CLiP; however, there are essential differences with regard to conceptualization due to the different objectives of the models. While VeDa's focus is on mathematical modeling, CLiP puts its main emphasis on the *products* of the design processes, which VeDa covered only marginally, reducing them to the role of modeled objects. In terms of representing chemical process systems, VeDa focuses on its behavioral aspect to meet the need of mathematical modeling, whereas CLiP highlights the function and realization of process systems, as well, since they are important in the context of design. As for work process modeling, the area of interest has been extended from work processes targeting the development of mathematical models to work processes targeting chemical process design.

Model Structure

Meta modeling has been used as a structuring mechanism to allow for an efficient representation of recurrent model structures (Fig. 2.2). This way, the coarse structure of the information model can be fixed, and a simple categorization of the most important modeling concepts becomes feasible. We distinguish the *Meta Meta Class Layer*, which introduces the concept of a general system and its aspects, the *Meta Class Layer*, which introduces different kinds of systems and their specific properties, and the *Simple Class Layer*, which defines concepts related to different tasks in the design process and therefore corresponds roughly to VeDa's *Basic Classes Layer*. Unlike VeDa, CLiP has no layer for application-specific classes.

The open model structure of CLiP is achieved by assembling thematically related concepts on the *Simple Class Layer* into partial models (cf. Fig. 2.3). In comparison to VeDa, their number has been significantly increased, resulting in a more fine-grained partition of the domain. This improves the extensibility and maintainability of the model, as the partial models can be introduced and maintained largely independently from each other. The individual partial models can usually be associated with design tasks, which are typically addressed independently during the design process. However, since the same real object is often referred to in different design tasks from different viewpoints with differing degree of detail, overlap, partial redundancy, conflicts, and even inconsistency can hardly be avoided. Existing relationships between classes are explicitly indicated. To reduce the specification effort and the complexity of the resulting information model, only those relations are modeled that are of relevance for the design process. This principle of systematic, task-oriented decomposition and subsequent selective reintegration is considered an essential prerequisite to successfully deal with the inherent complexity of a large product data model.

Representation Language

CLiP is implemented by means of different modeling formalisms. Both *Meta Layers* and some parts of the *Simple Class Layer* have been implemented in

ConceptBase [189]. This system supports meta modeling and offers a logic-based language and basic deductive reasoning capabilities to assist schema development and maintenance. The concepts on the *Simple Class Layer* are represented by means of UML class diagrams (e.g., [673]). This has been done, since the a graphical notation like UML is better suited for the representation and management of large and complex product data models than the frame-based O-Telos language of ConceptBase. On the other hand, the UML does not provide the flexibility of introducing several meta levels as they were needed for the conceptual framework of CLiP. Therefore, the framework has been developed with ConceptBase and the detailed class models have been modeled in UML. There is no formal integration between the two models given; rather, consistency is ensured by an overlap of major concepts on the *Simple Class Layer*.

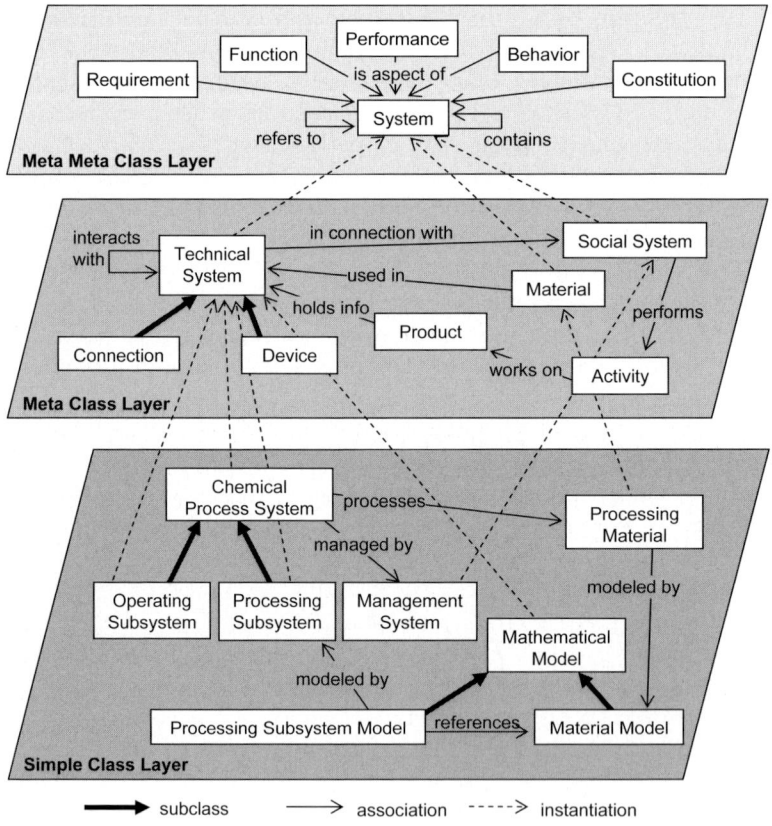

Fig. 2.2. Model structure of CLiP

Scope and Content

CLiP is structured according to the ideas of general systems theory. In comparison to VeDa, CLiP takes a more systematic approach by representing the principles of systems theory explicitly on the *Meta Meta Class Layer* (Fig. 2.2): The root class, from which all other classes can be derived, is the System, consisting of interacting Subsystems and characterized by distinct Properties and their different Values (not shown in Fig. 2.2). The class Aspect (and its subclasses) represent a System that is considered from a particular viewpoint. An Aspect is again a System, which represents those components of a System that are relevant to a particular viewpoint. Five major Aspects are of importance in the context of system design: In addition to functional and structural aspects already established in VeDa (represented through the classes Function and Constitution in CLiP), the aspects Behavior, Requirement, and Performance are introduced.

Different kinds of Systems can be distinguished on the *Meta Class Layer* (cf. Fig. 2.2). TechnicalSystems represent all kinds of technical artifacts. They can be decomposed and do interact with other technical systems. TechnicalSystems are either Devices or Connections. Similar to the idea of device and connection models in VeDa, Devices hold the major functionality and are linked by Connections. Furthermore, Material and SocialSystem are introduced

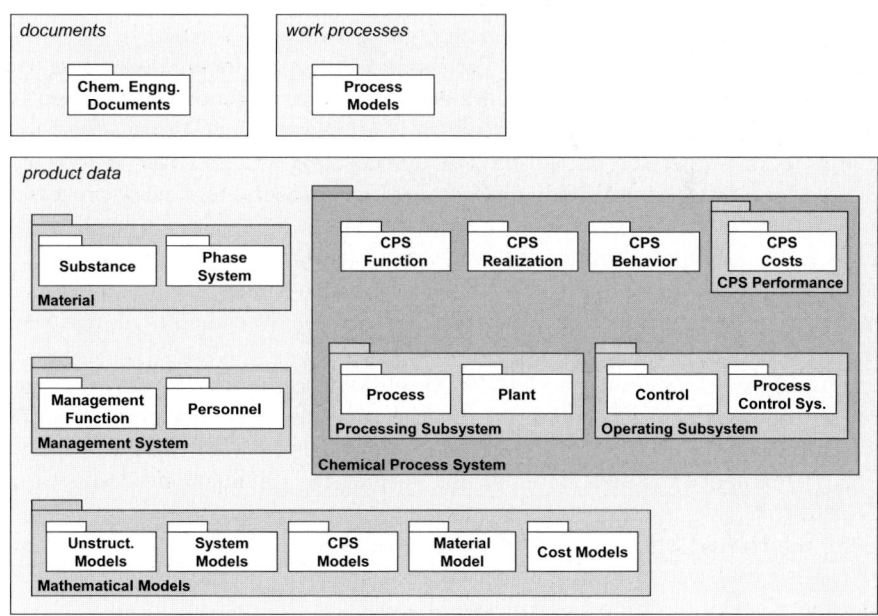

Fig. 2.3. Partial models on the *Simple Class Layer* of CLiP

as instances of System. Material abstracts matter and substances, whereas SocialSystem is the superordinate class for both a single person and a group of people. SocialSystems perform Activities, which work on Products. A Product represents a document version, or a part thereof, which may hold different sorts of information, for example about a Technical System as indicated in Fig. 2.2. Thus, a first attempt has been made to model the dependencies between product data, documents, and work processes (see also [18]). This matter is further elaborated in Sects. 2.3, 2.4, and 2.6.

On the *Simple Class Layer*, the TechnicalSystem class is instantiated and thus further refined to ChemicalProcessSystem, which can be decomposed into two distinguished parts: the ProcessingSubsystem and the OperatingSubsystem. The ProcessingSubsystem holds the functionalities of material processing, including chemical, physical, or biological procedures and their realization in a plant in order to produce some specified product; the OperatingSubsystem comprises the technology to operate the plant (i.e., observation and control). The ChemicalProcessSystem is managed by the ManagementSystem, which represents the staff working on the chemical plant as well as the different management functions performed by the staff. ProcessingSubsystem and OperatingSubsystem are instances of a TechnicalSystem, whereas the ManagementSystem is an instance of a SocialSystem.

There are two different instantiations of Material on the *Simple Class Layer*: the ProcessingMaterial, which is processed in order to get a chemical product, and the ConstructionMaterial (not shown in Fig. 2.2) used to build the chemical plant system. The behavior of these materials can be described by MaterialModels. These are referenced by ProcessingSubsystemModels describing the ProcessingSubsystem. Both MaterialModels and ProcessingSubsystemModels are refinements of MathematicalModel, which is an instance of TechnicalSystem.

Figure 2.3 shows the partial models on the *Simple Class Layer*. Three major areas can be distinguished: *product data*, *documents*, and *work processes*. Packages of the UML are used to represent the partial models in all three areas. Each specific System as well as each of its Aspects are described in an individual partial model, with the aspect packages being part of the system package. Only the product data part of CLiP will be described here; the other parts will be discussed in the following sections.

The *Chemical Process System* (CPS in short) describes the chemical process and its different aspects, ranging from the function (*CPS Function*), the realization (*CPS Realization*), the behavior (*CPS Behavior*) to the performance (*CPS Performance*), which includes the cost of the chemical process system (*CPS Costs*).

As mentioned before, a ChemicalProcessSystem can be decomposed into a ProcessingSubsystem and an OperatingSubsystem. The partial models *Process* and *Control* describe the functional aspect of the respective subsystems, whereas the partial models *Plant* and *Operating System* represent the constitution (i.e., technical realization) of the subsystem. The associated *Management System* is also considered under the aspects function and constitution

(*Management Function* and *Personnel*, respectively). The latter represents the staff working at and with the chemical process system, while the former describes the management functions or roles exerted by the personnel.

The partial model *Material* is subdivided into the partial models *Substance* and *Phase System*. *Substance* describes the intrinsic characteristics of materials that do not alter in mechanical, thermo-physical, or chemical processes; *Phase System*, on the other hand, holds those properties that change according to the given physical context [491].

Reusing parts of VeDa and of the VeDa-based implementation models of the afore mentioned research prototypes ROME and ModKit (cf. Sect. 5.3), the partial model *Mathematical Models* comprises both generic concepts for mathematical modeling and specific types of mathematical models. Within the partial model *Unstructured Models*, models are described from a mathematical point of view. Concepts of *Unstructured Models* can be used to specify the equations of *System Models*, which model *Systems* in a structured manner. Models for the description of ChemicalProcessSystems (as defined in *CPS Models*) are examples for system models. Such models employ concepts from *Material Model* to describe the behavior of the materials processed by the chemical process system. Finally, the partial model *Cost Models* introduces methods to estimate the cost for construction, procurement, and operation of chemical process systems.

The CLiP partial models concerned with mathematical models are not as detailed as their counterparts in VeDa. They introduce only the high-level concepts required for mathematical modeling. Moreover, the representation of units and physical dimensions has not been adopted by CLiP. An improvement in comparison to VeDa is the strict distinction between partial models describing the language of mathematics in terms of equations and partial models describing the physical interpretation of these equations [152], which enhances the reusability of the model.

Usage

CLiP provides a conceptualization of the most important entities and their relations in the domain of chemical process design. However, CLiP is not intended as a comprehensive, detailed product data model. Rather, it is understood as a conceptual model, based on which specialized implementation models can be developed, such as data schemata for domain-specific databases and tools, or neutral representation formats for the exchange of product data between heterogeneous software systems. Within the CRC, several implementation models for domain-specific software tools have been elaborated on the basis of CLiP.

- Section 4.1 presents an extended product reuse and method guidance system, called Process Data Warehouse (PDW). An earlier version of the PDW, which is described in Subsect. 4.1.3, has been implemented in ConceptBase directly based on the CLiP partial models [96]. Note that the

current implementation of the PDW makes use of OntoCAPE instead of CLiP, as will be explained in the next subsection.
- In Sect. 3.2, an integrator tool for the coupling of the process simulator Aspen Plus [518] and the CAE system Comos PT [745] is described. The integration is driven by integration rules, which have been partially derived from the CLiP partial models *Processing Subsystem* and *Mathematical Model*, as explained in [15]. See also Sect. 6.3, where a more recent approach to derive such integration rules is described.
- The database schema of the CAE system Comos PT has been extended according to the partial models describing the chemical process and the plant, resulting in a design database for the lifecycle phase of conceptual process design [22]. Some prototypical tool integrations have been realized around that design database in order to demonstrate CLiP's capability for the integration of existing data and data models [24].
- The implementation model of ModKit+ [151] is based on the partial models *Unstructured Models* and *System Models*.

In addition to the above applications, CLiP can be utilized as an integration framework to support the integration of existing data models [14, 23]. To this aim, the classes of the data models that are to be integrated must be related to high-level classes of CLiP, either through instantiation or subclassing relationships. By identifying those classes that belong to the same superordinate class or meta class, a loose connection is established between them, which is a first step towards a more tight integration.

Finally, CLiP can be used in the sense of an ontology: it provides a common vocabulary for the domain of chemical engineering, which promotes a shared understanding and facilitates the communication between people and across organizations.

2.2.4 OntoCAPE

Starting in 2002, the formal ontology OntoCAPE [325, 326] has been developed, combining results of VeDa and CLiP. Structure and terminology of CLiP have been adopted to the extent possible, especially for the part of representing chemical process systems. Besides, the VeDa partial models *Geometrical Modeling Objects* and *Behavioral Modeling Objects*, which are not fully incorporated in CLiP, have been revisited and included. Several additional areas not covered by the previous models, such as numerical solvers, have been newly conceptualized in OntoCAPE.

Unlike CLiP and VeDa, OntoCAPE is solely concerned with product data; it does not attempt to describe documents and work processes. However, document models describing the types, structures, and contents of documents, as well as the interdependencies between documents, can be integrated with OntoCAPE (cf. Sect. 2.3). Moreover, OntoCAPE can be combined with formal models of work processes and decision-making procedures, as presented in Sects. 2.4 and 2.5. The integration of these different types of application models is discussed in Sect. 2.6.

OntoCAPE is represented in a formal ontology language, which allows for a precise definitions of both the meaning of terms and the interrelations between the individual classes, as well as for the indication of axioms and constraints that restrict (and therefore clarify) the usage of the vocabulary. By exploiting the potential of the language, OntoCAPE is able to incorporate a significantly larger amount of domain knowledge than CLiP and VeDa.

The development of OntoCAPE started in the EU-funded COGents project [70], which addressed the automatic selection of mathematical models and numerical solvers for chemical process systems through cooperating software agents. Within this project, version 1.0 of OntoCAPE has been established as a shared communication language for human and computer agents. Due to the focus of the COGents project, the 1.0 version of OntoCAPE particularly emphasizes the areas of process modeling and simulation. Yet its overall model structure, which was developed in close cooperation with the CRC 476, has explicitly been designed for a later extension towards other areas of chemical engineering.

After completion of the COGents project, the further development of OntoCAPE was taken over by CRC 476. In 2007, version 2.0 of OntoCAPE has been released [324], which, in addition to mathematical modeling and simulation, also covers the design of chemical process systems. In the following, OntoCAPE 2.0 is presented.

Model Structure

As shown in Fig. 2.4, OntoCAPE is organized by means of three types of structural elements: layers, partial models, and modules. Subsequently, the particular functions of these elements will be discussed.

Similar to VeDa and CLiP, OntoCAPE is subdivided into different levels of abstraction, referred to as *layers*, in order to separate general knowledge from knowledge about particular domains and applications:

- The topmost Meta Layer is the most abstract one[9]. It introduces domain-independent root terms, such as Object or N-aryRelation, and establishes an approach to represent mereotopological relations between objects. Furthermore, generic *Design Patterns* (e.g., [682]) are introduced, which define best-practice solutions for general design problems.
- The Upper Layer introduces the principles of general systems theory as the overall design paradigm according to which the domain ontology is organized; thus, it is comparable to the Meta Meta Class Layer of CLiP.

---

[9] Conceptually, the Meta Layer is not a genuine part of OntoCAPE, but represents a meta level on top of the actual ontology.

104   J. Morbach et al.

- On the *Conceptual Layer*, a conceptual model of the CAPE domain is established. It corresponds to the *Basic Classes* layer of VeDa and to the *Simple Class Layer* of CLiP.
- The subsequent layers refine the conceptual model by adding classes and relations that are of practical relevance for certain tasks and applications. The idea is similar to VeDa's layer of *Application Specific Classes*, yet split

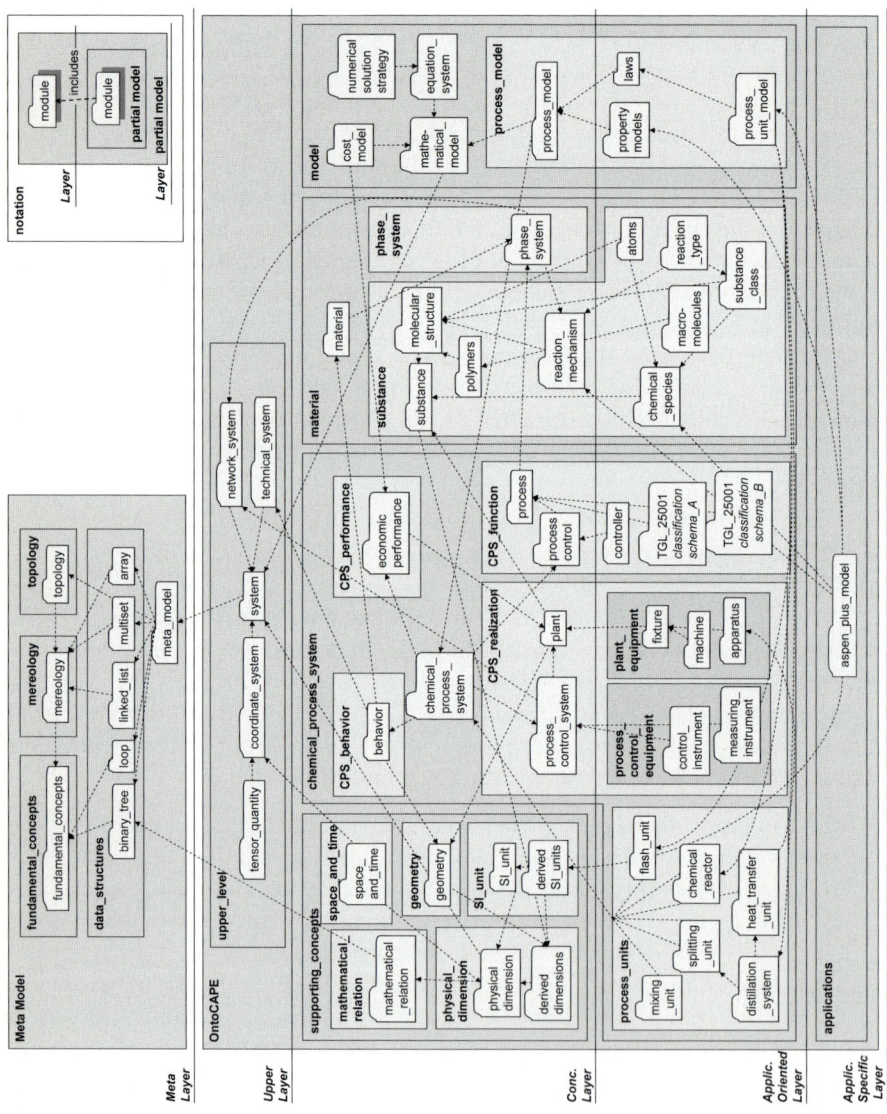

Fig. 2.4. Model structure of OntoCAPE

up into an *Application-Oriented Layer* and an *Application-Specific Layer*. The former describes the diverse application areas in general, whereas the latter provides specialized classes and relations for concrete tasks and applications.

The notion of partial models as used in VeDa and CLiP has been refined in OntoCAPE: here, *modules* and *partial models* are distinguished. Modules assemble a number of classes that cover a common topic, relations describing the interactions between the classes, and some constraints defined on them. A module can be designed, adapted, and reused to some extent independently from other parts of an ontology [987]. Modules that address closely related topics are grouped into partial models. The partial models constitute a coarse categorization of the domain. Unlike modules, partial models may stretch across several layers. While the boundaries of the modules are chosen for practical considerations (e.g., to minimize interaction with other modules or to facilitate their handling with respect to the application at hand), the boundaries of the partial models reflect the "natural" boundaries of the domain. In the course of ontology evolution, the partial model structure is therefore supposed to remain relatively stable, whereas the number of modules as well as their content and interrelations are likely to change over time in order to adapt to new applications.

Representation Language

Two different specifications of OntoCAPE have been constructed: A *formal specification* represents the ontology in a formal ontology language, such as DAML+OIL [609] or OWL [546]. A supplementary *informal specification* presents the ontology in a human-readable form and provides additional background information that cannot be coded within the ontology language.

An important function of the informal specification is to provide the development team of OntoCAPE with a shared understanding of the domain before the formal encoding of the ontology begins. Therefore, the informal specification can be considered a conceptual model, based on which implementation models, such as the formal specification of OntoCAPE, can be developed. The formal specification can be regarded as an implementation model as it is encoded in a formal ontology language and can therefore be directly processed and employed by software tools. Further, the informal specification serves as a user manual, which supports new users in familiarizing with the ontology and its design principles and understanding the intended usage of the individual concepts. To this aim, all ontology modules have been represented by UML-like class diagrams, which provide a graphical view of the main interrelations between classes and their instances. In addition to graphical illustrations, the informal specification comprises natural language descriptions of the meaning of classes, attributes, relations, and instances.

In the course of creating the formal specification, modeling errors were detected that had not been noticed before. On the one hand, the usage of a formal ontology language compels the modeler to define the meaning of classes more exactly and thus to revise the somewhat sloppy verbalizations given in the informal specification and in its predecessors VeDa and CLiP. On the other hand, as the ontology language is based on formal logic, it is possible to check the consistency of the model statements through a reasoner. That way, remaining inconsistencies within the model can be detected automatically.

The formal specification of OntoCAPE was initially represented by means of the ontology language DAML+OIL, the state-of-the-art ontology modeling language at the time when version 1.0 of OntoCAPE was developed. The formal specification was created using the ontology editor OilEd [545] and verified by the reasoner FaCT [735]. For the representation of version 2.0, the formal specification was converted into the Ontology Web Language OWL (OWL DL in particular, [546]), which has replaced DAML+OIL as a standard ontology modeling language. This laborious task of conversion was supported by a converter [2], which handled most of the translation effort automatically. However, as OWL does not offer equivalents for all language elements of DAML+OIL, some parts of the ontology had to be remodeled manually within the ontology editor Protégé [979]. For the verification of the OWL version of the ontology, the reasoner RacerPro [918] has been used.

In the formal specification of version 2.0, ontology modules are manifested through different XML namespaces [1060], each of which is stored in a single OWL file; connections between the modules are realized by using the import mechanism provided by OWL. The partial models correspond to identically named directories in the formal specification. That way, they establish a directory structure for managing the OWL files.

Scope and Content

The *Meta Layer* on top of OntoCAPE holds two partial models: the *Meta Model* and the *Core Ontology* (the latter is not shown in Fig. 2.4). The *Meta Model* introduces fundamental modeling concepts and design guidelines for the construction of the OntoCAPE ontology. Its function is (1) to explicitly represent the underlying design principles of OntoCAPE and (2) to establish some common standards for the design and organization of the ontology. Thus, the *Meta Model* supports ontology engineering and ensures a consistent modeling style across the ontology. Details about the *Meta Model* can be found in Subsect. 2.6.2.

The *Core Ontology* integrates the different ontologies that are required for some application. Different applications require different core ontologies – for instance, the ontology-based Process Data Warehouse described in Subsect. 4.1.5 relies on a core ontology that links OntoCAPE with ontologies for the description of documents, work processes, and storage systems (cf. Fig. 4.6). Generally, a core ontology has the function (1) to retrieve the concepts that are relevant for the respective application from the different ontologies, (2) to define how these concepts are to be used (i.e., interpreted) by the application, and (3) to introduce additional top-level concepts, which cannot be retrieved from the existing ontologies.

The module *system* is the major ontology module on the Upper Layer. It introduces important systems-theoretical and physicochemical primitives such as System, Property, Value, PhysicalQuantity, Unit, etc., and specifies their mutual relations: A System is characterized by its Properties, each of which can take numerous Values. Furthermore, the *system* module introduces concepts to represent mereological and topological relations between Systems. It also establishes the notion of an AspectSystem, which represents a System considered from a particular viewpoint.

To distinguish the different Values of a certain Property, the concept of a *backdrop*, as suggested by Klir [784], is introduced. Adapting Klir's definition to the terminology of OntoCAPE, a backdrop is some sort of background against which the different Values of a Property can be observed. In OntoCAPE, the Values of any Property can act as a backdrop to distinguish the Values of another Property. Time and Space are typical choices of such distinguishing Properties: For example, the different Values of a Temperature[10] arising in the course of an observation can be distinguished by mapping each Temperature Value to a Value of Time. Thus, the backdrop concept enables the representation of dynamic system behavior as well as of locally distributed systems, and it may be used to model any other type of distribution (e.g., particle size).

The *system* module is supplemented by four ontology modules: The module *tensor_quantity* allows for the representation of vectors and higher-order tensors. In *coordinate_system*, the concept of a coordinate system is introduced, which serves as a frame of reference for the observation of system properties. The ontology module *network_system* establishes the principles of network theory; network theory has already been applied in VeDa, but is now defined in a more general and systematic manner. Finally, the *technical_system* module introduces the class TechnicalSystem as a special type of a System which is developed in an engineering design process. In the design lifecycle of a TechnicalSystem, five viewpoints are of particular interest, which are represented as special types of AspectSystems: the system Requirements, the Function of the system, its Realization, Behavior, and Performance. While these classes have already been introduced by CLiP, OntoCAPE defines their meaning more precisely and clarifies their interrelations.

On the Conceptual Layer, the CAPE domain is represented by four major partial models: The central *chemical_process_system* represents all those concepts that are directly related to materials processing and plant operating – just like the identically named partial model in CLiP. The partial model *material* provides an abstract description of materials involved in a chemical process, whereas *model* defines notions required for a description of models and model building. Finally, the partial model *supporting_concepts* supplies

---

[10] a subclass of Property

fundamental notions such as space, time, physical dimensions, SI-units, mathematical relations etc., which do not directly belong to the CAPE domain but are required for the definition of or as supplements for domain concepts. This partial model is only rudimentarily elaborated, as it is not the objective of OntoCAPE to conceptualize areas beyond the scope of the CAPE domain.

The organizing principle of general systems theory, which has been introduced on the *Upper Layer*, is taken up on the *Conceptual Layer*: With the exception of *supporting_concepts*, each of the aforementioned partial models holds a key module describing the main System as well as supplemental modules holding AspectSystems which reflect a particular viewpoint of the main System: For instance, the module *chemical_process_system* holds the main system ProcessUnit, while the module *process* contains the class ProcessStep reflecting the functional aspect of the ProcessUnit.

The *Application-Oriented Layer* extends the *Conceptual Layer* by adding classes and relations needed for the practical usage of the ontology. Some exemplary modules are shown in Fig. 2.4: *substance*, which analogously to CLiP describes the intrinsic characteristics of materials, is supplemented by chemical species data for atoms, molecules, and polymers. The *process* module is extended by two alternative classification schemata for unit operations [20] based on the national standard TGL 25000 [995]. The partial model *process_units* contains descriptions of typical process units (i.e., modular parts of a chemical process system); exemplarily shown are the modules *mixing_unit*, *splitting_unit*, *flash_unit*, *chemical_reactor*, *heat_transfer_unit*, and *distillation_system*. The module *process_model* is extended by three ontology modules: *laws*, which specifies a number of physical laws that are applicable in the context of chemical engineering (e.g., the law of energy conservation); *property_models*, which provides models that reflect the behavior of designated physicochemical properties (e.g., vapor pressure correlations or activity coefficient models); and *process_unit_model*, which establishes customary mathematical models for process units, such as ideal reactor models or tray-by-tray models for distillation columns.

The Application-Specific Layer holds classes and relations that are required for particular tasks or applications. Figure 2.4 shows exemplarily a module for the description of model files that are available in the format of the process simulator Aspen Plus [518].

Usage

So far, OntoCAPE has been used in a number of software applications for process modeling and design [479]:

- In the COGents project [70], OntoCAPE forms part of a multi-agent framework, which supports the selection and retrieval of suitable process modeling components from distributed model libraries. Within this framework,

OntoCAPE serves as a communication language between interacting software agents, and between the software agents and the human users. Concepts from OntoCAPE are used to formulate a modeling task specification, which is then matched against available process modeling components also described through OntoCAPE.
- OntoCAPE has been applied for the computer-aided construction of process models following an ontology-based approach [489, 490]. This approach suggests constructing a mathematical model in two successive steps, namely conceptual modeling and model generation. In the first step, a human modeler constructs a conceptual model of the chemical process; this is accomplished by selecting, instantiating, and connecting appropriate concepts from OntoCAPE that reflect the structural and phenomenological properties of the chemical process. Based on these specifications, the mathematical model can automatically be created by a model generation engine: The engine selects and retrieves appropriate model components from a library of model building blocks; these model components are subsequently customized and aggregated to a mathematical model, according to the specifications of the conceptual model. Note that the conceptual modeling tool is an adaptation of the aforementioned ModKit+ [151], which itself is a reimplementation of ModKit [54] (see also Sect. 5.3).
- The Process Data Warehouse (PDW) described in Sect. 4.1 acts as a single point of access to all kinds of design information within an organization, regardless of the format and storage location of the original data sources. The implementation of the PDW is based on loosely connected ontology modules (cf. 4.1.5), some of which are provided by OntoCAPE [62–64]. Within the PDW, concepts from OntoCAPE are used to annotate electronic documents and data stores (cf. Subsect. 2.3.4). That way, one obtains a consistent, integrated representation of the contents of these heterogeneous data sources. These content descriptions can be processed and evaluated by the semantic searching and browsing functions of the PDW, which support the navigation between and retrieval of resources.
- Like CLiP, OntoCAPE can be used to derive integration rules for the integrator tools developed by subproject B2 (cf. Sect. 3.2). Section 6.3 describes the approach in detail.

2.2.5 Conclusion

A series of product data models has been developed for the domain of chemical engineering. Model evolution occurred as a consequence of adopting new modeling insights, encountering new applications, and taking advantage of novel modeling languages. For the time being, OntoCAPE 2.0 represents the final synthesis of this effort, incorporating the experience and consolidating the results of more than a decade of product data modeling.

The scope of OntoCAPE 2.0 comprises both the area of process design previously covered by CLiP and the area of process modeling previously covered by VeDa and OntoCAPE 1.0. Like CLiP and VeDa, OntoCAPE 2.0 is based on general systems theory, yet defining the underlying theoretical concepts more explicitly than the preceding models.

As for the modeling language, the expressiveness of OWL, which is employed for the representation of OntoCAPE 2.0, is comparable to VDDL, the self-defined language used to represent VeDa, and to ConceptBase, the deductive database language used to represent the upper levels of CLiP. At the same time, OWL has the advantage of being an international standard language, as it is the case with the less expressive UML used to represent the class level of CLiP. Thanks to OWL's roots in description logics, reasoning support, which had already been experimentally used when developing certain parts of VeDa [51, 61, 318, 384] and CLiP [20], could now be regularly applied for the development and refinement of OntoCAPE.

Regarding structural design, the strategy of modularization, successfully applied in the preceding models, is continued and refined in OntoCAPE: The ontology has been partitioned into nested partial models, which are further subdivided into ontology modules. Complementary, the concepts of the ontology have been allocated across five layers, thus combining and improving the structural layers of CLiP and VeDa: The meta modeling approach first applied in CLiP is further elaborated, now clearly distinguishing between the generic **Meta Layer** and the systems-theoretical **Upper Layer**. The layer of application classes originally introduced by VeDa is adopted and split into two, with the aim of further differentiating between application-oriented and application-specific concepts.

The presented product data models have been developed as part of a continuous information modeling effort. In this effort, we have witnessed the necessity of making various changes to existing data models or ontologies, as the consequence of adopting new modeling insights or encountering new applications. While the preceding models already incorporate some organizational mechanisms to support such evolutionary changes, OntoCAPE 2.0 further enhances these mechanism, thus allowing for an easy extension and reuse of the ontology. This is of special importance, since OntoCAPE — unlike the conceptual models VeDa and CLiP — is intended for direct use by intelligent software systems; consequently, the ontology must be reconfigured regularly in order to meet the requirements of new applications.

For future work, two major challenges remain to be addressed: (1) validating the ontology's ability to describe and manage large amounts of real-world data, and (2) further improving its capabilities for extension and reuse. To this aim, we are going to apply OntoCAPE in an industrial project dealing with the integration and management of technical data throughout the lifecycle of a chemical plant (cf. Sect. 7.2). Within this project, which will be run in close cooperation with two industrial partners, we are able to test the ontology in industrial practice and, if necessary, adapt it to real-world requirements and needs. The ultimate goal is to obtain a validated, comprehensive ontology that is easily applicable to various applications in the process engineering domain.

2.3 Document Models

J. Morbach, R. Hai, B. Bayer, and W. Marquardt

Abstract. In this contribution, a comprehensive document model is presented, which describes the types and dependencies of documents as well as their internal structures and dynamic behavior. Additionally, the contents of documents can be indicated by integrating the document model with a product data model. Due to these properties, the document model provides an adequate basis for a novel class of software tools that support efficient document handling in development processes.

2.3.1 Introduction

Documents play a major role in chemical engineering work processes. Data are often handled not as single instances but in the form of documents, which act as carriers for the data during project execution. These documents differ widely in form and content, ranging from informal documents containing natural language text, over formal ones like tables or simulation files with well-defined syntax and semantics, to documents containing graphical data or even multimedia contents. Documents such as flowsheets and equipment specification sheets constitute the human perspective on the plant and the plant data [914]. Executable documents like simulation files are used a means for information exchange between users and/or software tools.

Various interrelations and dependencies exist between the different types of documents – for instance, a Process & Instrumentation Diagram is a master document that determines the content of peripheral documents such as equipment lists. Further, it needs to be considered that documents and their contents evolve over time: Each document has a lifecycle of its own, comprising creation, editing, review, approval, and release processes. During a document lifecycle, different versions of a document are created that need to be managed.

To obtain a complete description of documents, Salminen et al. [946] have identified three important aspects about documents that should be reflected by the document model:

- A static description of the documents given by a classification of the *document types* together with a description of their *interrelations*.
- A description of the *dynamic behavior* of documents over time.
- A description of the *document content* by describing the internal structure (i.e., the syntax) of the document, for example through a document template.

While we agree with this approach in general, the method for content description suggested by Salminen and coworkers is, in our opinion, not sufficient. The expressiveness of a structural description is rather restricted. Only the syntax of a document and of its contents can be given that way, but no information about the semantics of the contents can be provided. A semantic description of the contents can be achieved by linking the single elements of which the document is composed to corresponding classes of a product data model, as we have suggested before [19].

In the following, we present an approach to document modeling that allows to describe the three aspects postulated by Salminen et al. on a coarse-grained level within a single representation language. Firstly, we will discuss how individual document types and their dependencies can be described. Next, we will introduce a method to describe the dynamic behavior of documents. Finally, the modeling of syntactic as well as semantic content will be addressed.

2.3.2 Document Types and Dependencies

For the identification of the individual document types existing in a given application area, document-oriented work process models can be exploited – either generic work process models (e.g., [851]) or models of particular work processes (e.g., [17] or Sect. 5.1 of this book). Other possible information sources are articles and textbooks that cover the handling of documents in engineering projects (e.g., [853, 957, 1004, 1027]), technical rules and standards (e.g., [634]), or project execution manuals as they are available in various chemical engineering companies.

Document types can be represented as UML classes. Figure 2.5 shows a taxonomy of important document types involved in a chemical engineering design project, which has been derived from some of the above mentioned sources. The class **ChemEngDesignDoc** subsumes all documents created or used in such design projects. Four different types of **ChemEngDesignDocs** can be distinguished: reference documents (**ReferenceDoc**), which are static resources of information; technical documents (**TechnicalDoc**), which are created by the chemical engineering department to specify or calculate technical information about the chemical process and/or plant to be developed; supplementary documents (**SupplementaryDoc**), which are created by other functions than the chemical engineering department but are processed by chemical engineers; and comprehensive documents (**ComprehensiveDoc**), which summarize the content of other documents. These classes can be further refined as shown in Fig. 2.5. The semantics of the individual classes is explicated in Table 2.1, which can be found at the end of Sect. 2.3.

It should be noted, though, that the collection of the given document types is not exhaustive, and the above classification schema is not the only feasible one. In an analogous manner, alternative taxonomies can be developed to represent the document types used in a certain domain or by a particular organization.

Between the different types of documents, various dependencies as well as other types of relations exist. In this context, dependency between documents means that, if document B depends on document A, the content of B

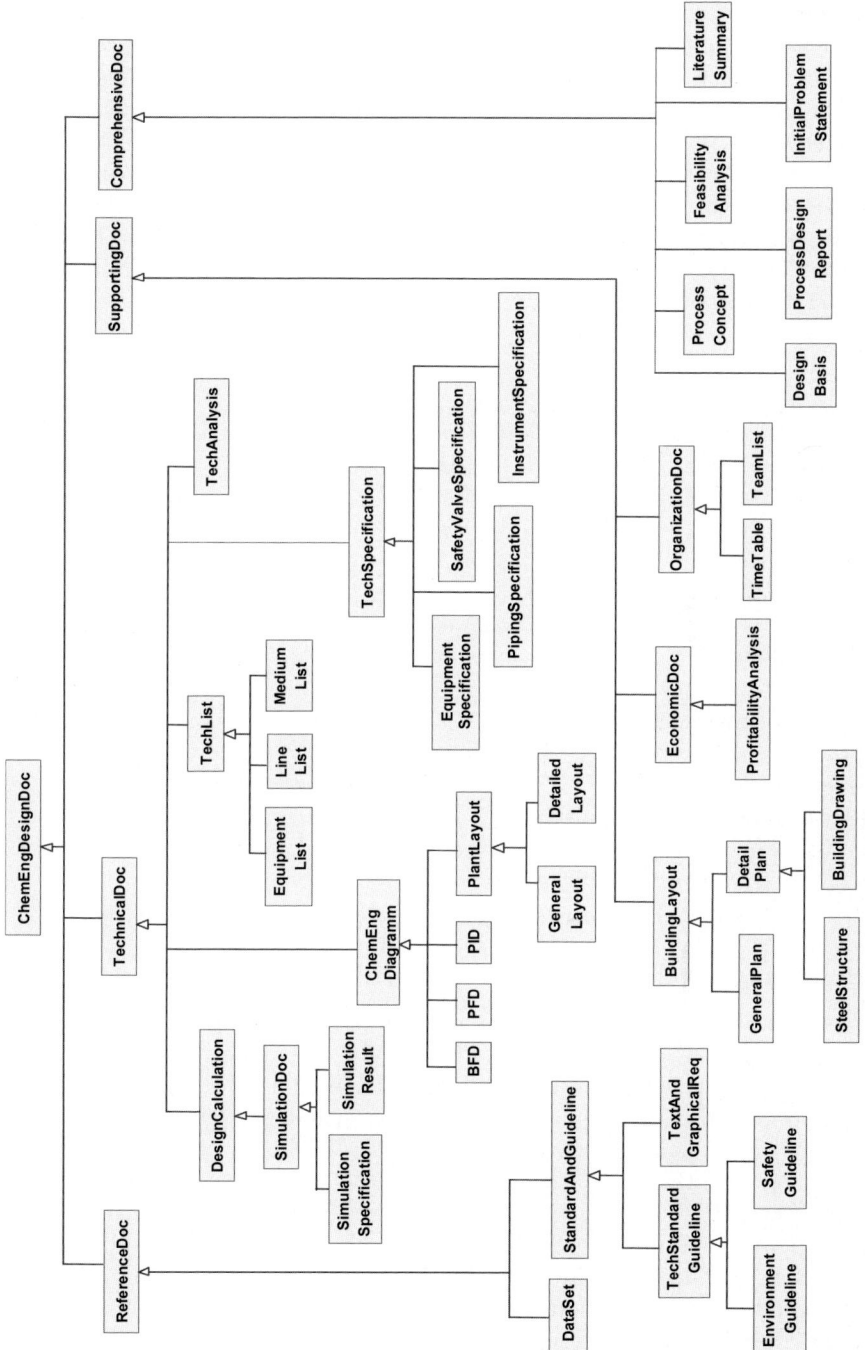

Fig. 2.5. A hierarchy of document types in chemical engineering design projects

is contingent on the content of A. Hence, changes in document A may require consequent changes in document B. Figure 2.6 shows exemplarily some dependencies between the document classes introduced in Fig. 2.5. Most of these dependencies are explicated in [851]. Dashed arrows represent unidirectional dependencies, solid lines denote mutual dependencies.

As can be deduced from the large number of depending documents, the process & instrumentation diagram (PID) is the central document of the design process. It collects information from various sources, such as the DesignBasis or the PFD, and acts as master document for others, such as the EquipmentList or the PlantLayout.

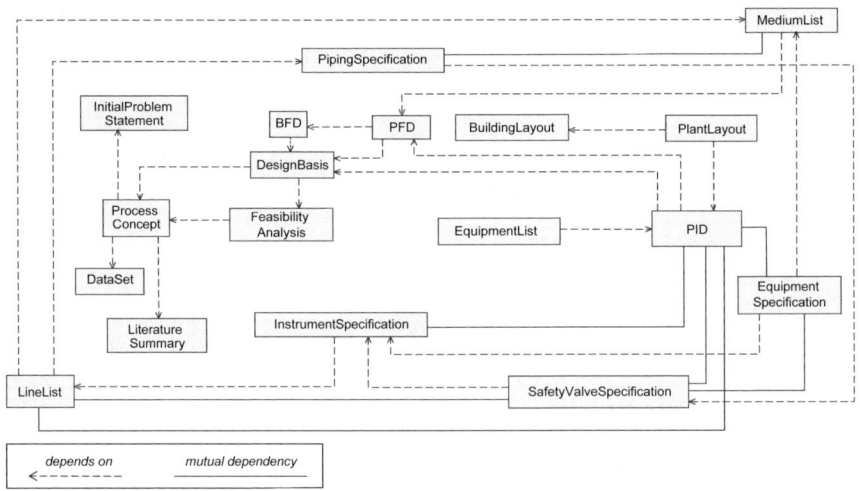

Fig. 2.6. Dependencies between the individual document types

Moreover, it can be observed that numerous mutual dependencies (e.g., between EquipmentSpecification and PID) as well as circular dependencies (e.g., between classes LineList, InstrumentSpecification, and SafetyValveSpecification) exist between the document classes. Usually, such dependencies arise if several versions or revisions of the documents are created during project execution. For instance, in the case of the mentioned circular dependencies, a first version of the LineList is created, based on which the InstrumentSpecification can be developed. The SafetyValveSpecification in turn requires some input from the InstrumentSpecification (namely the sizing information of the control valves). The SafetyValveSpecification, however, has an impact on the LineList, as all piping connected to a safety valve has to be sized according to the specified relief rates, such that a second version of the LineList needs to be established.

In order to represent such dependencies between document versions explicitly in the model, the dynamic behavior of documents needs to be taken into account. This issue is discussed in the next subsection.

2.3.3 Dynamic Behavior

To describe the dynamic behavior of documents, their different versions and configurations evolving over time need to be captured. In the following, we will introduce a model that combines such a description with the static description of document types and their dependencies introduced in the preceding Subsect. 2.3.2. In addition, the model reflects the structural composition of documents and document versions. Figure 2.7 shows the top-level concepts of this *Document Model*, which is represented both in UML and in OWL. It is consistent with the *Product Area* of the PDW Core Ontology, which is presented in Subsect. 4.1.5.

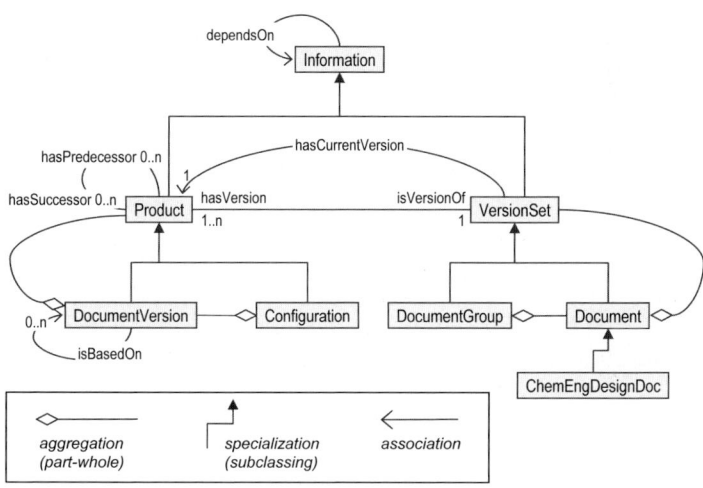

Fig. 2.7. Top-level concepts of the Document Model

The **Information** is the root class of the Document Model. Through the relation **dependsOn**, generic dependencies between **Information** can be expressed. Refinements of this relation can be introduced to denote special types of dependencies, such as **isMasterDocumentOf**.

The class **Product** denotes all kinds of information objects existing in form of a hard copy or a soft copy; for instance, a **Product** might be a part of or even an entire text document, a (part of a) binary file, or even a particular view on a database.

Different **Products** may represent different versions of the same information object. In this case, the different versions can be linked to (and thus managed by) a single **VersionSet**. The class **VersionSet** represents the information object from a static perspective, while **Product** represents the different versions of the information object that evolve over time. **Products** and the corresponding **VersionSet** are connected through the relation isVersionOf and its inverse hasVersion, respectively. Moreover, one of the **Products** is identified as the current version of the **VersionSet**.

A **DocumentVersion** is a **Product** that represents a particular version of an entire **Document**. A **DocumentVersion** may be decomposed into several **Products**, as it is indicated by the part-whole relation between these two classes. Similarly, a **Document** may be aggregated of several **VersionSets**. **Document** can be refined through detailed taxonomies like the one shown in Fig. 2.5 to represent the types of documents used in a certain domain or by a particular organization. Analogous to the CoMa model [473] developed by subproject B4, **DocumentGroup** and **Configuration** are introduced to denote clusters of interdependent **Documents** and **DocumentVersions**, respectively.

As mentioned before, all former versions as well as the current version of a document can be indicated by instantiating the class **DocumentVersion**. An example is provided in Fig. 2.8, within which the following notation is used: The white boxes represent instances of the classes introduced above. Their names are given on top of the respective boxes, while the names of the instantiated classes are given within squared brackets on the bottom of each box. A diagonal slash denotes a class/subclass-relation (e.g., Document/LineList).

Two **DocumentVersions** of a particular line list are shown in Fig. 2.8, namely **Line List PA6 Plant v1.0** and **Line List PA6 Plant v2.0**, respectively. Both are connected with **Line List PA6 Plant**, an instance of the document type **LineList**, via the hasVersion-isVersionOf relation. The actual version is additionally marked through the hasCurrentVersion relation. The two versions are connected by

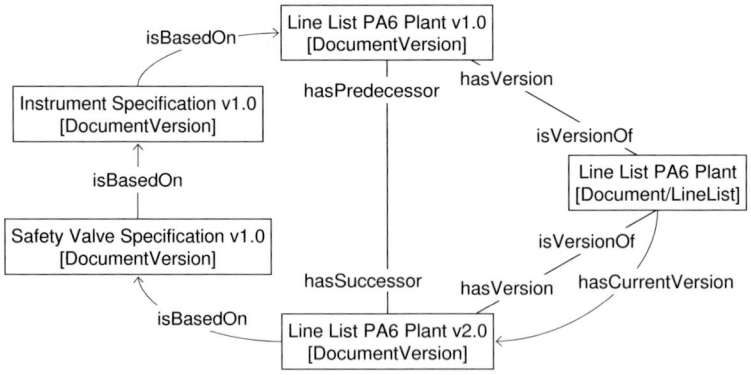

Fig. 2.8. Relations between instances of **Document** and **DocumentVersion**

the relation hasSuccessor and its inverse hasPredecessor, respectively. That way, the version history of a document can be retraced.

The model allows to express dependencies not only between document types but also between versions of documents. As it is shown in Fig. 2.8, the aforementioned dependencies between documents of types LineList, InstrumentSpecification, and SafetyValveSpecification (cf. Subsect. 2.3.2) can now be represented explicitly through the isBasedOn relation[11]: First, the Instrument Specification v1.0 is created, based on which the Line List v1.0 can be established. Next, the Safety Valve Specification v1.0 can be prepared, which in turn is the basis for the generation of the Instrument Specification v2.0.

2.3.4 Document Content

As defined in Subsect. 2.1.2, a document is an aggregation of data and acts as a carrier of product data in a certain work context. One possibility to represent models of document contents is the use of the eXtensible Markup Language (XML) [1060] and its Document Type Definitions (DTDs), as suggested by Bayer and Marquardt [19]. Within a DTD, the structure of a specific document type can be described. Figure 2.9 shows the (incomplete) DTD of a specification sheet for vessels. Here, the different data items like temperature or construction material are indicated to specify the piece of equipment. However, the expressiveness of such document type definitions is rather restricted. A DTD specifies only the syntax of the document content but not its semantics. One possibility to enrich the DTD with meaning is to relate the single DTD elements to the classes and attributes of a product data model. This is exemplarily shown Fig. 2.9, where relations between some DTD elements and the corresponding classes of the product data model CLiP (cf. Subsect. 2.2.3) are indicated.

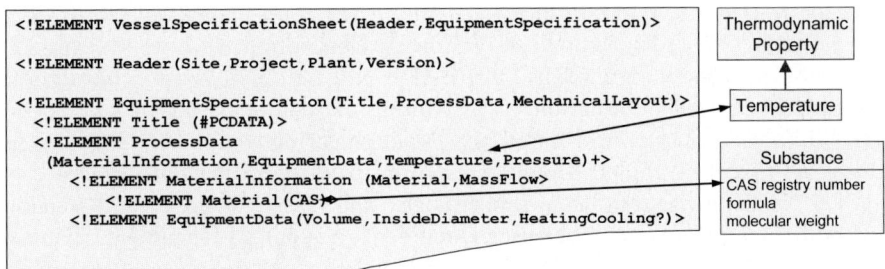

Fig. 2.9. DTD of an equipment specification sheet and corresponding classes in CLiP

---

[11] The isBasedOn relation is a refinement of the dependsOn relation that is solely applicable between DocumentVersions.

A drawback of the approach is that it cannot be represented within a single modeling language. Therefore, the CASE tool Rational Rose had to be used to manage the relations between the product data model and the DTD elements. In order to obtain a model represented within a single formalism, the Document Model introduced in Fig. 2.7 has been extended, as shown in Fig. 2.10. Newly introduced classes and relations are emphasized in bold print.

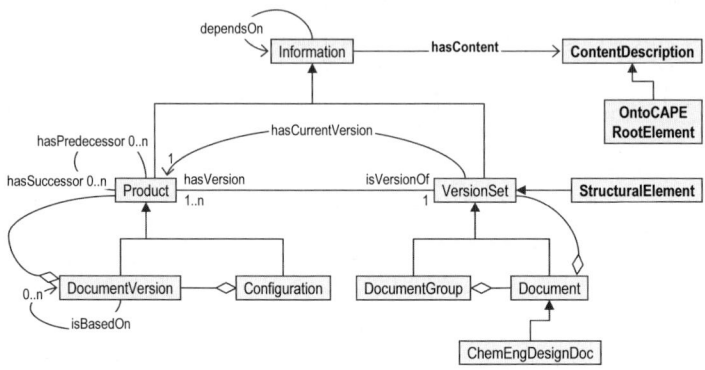

Fig. 2.10. Content description of documents on the class level

The class ContentDescription is introduced, which is connected with the Information via the relation hasContent. ContentDescription acts as an extension point, where a qualified product data model suitable for describing the content of a Information can be integrated into the Document Model. Integration is realized by declaring the classes of the product data model to be subclasses of ContentDescription. For example, by defining the root element of OntoCAPE (cf. Subsect. 2.2.4) as a subclass of ContentDescription, as it is indicated in Fig. 2.10, all classes defined in OntoCAPE can be employed to characterize the content of some Information. If the content of documents lies in another domain than that of computer-aided process engineering covered by OntoCAPE, OntoCAPE can be replaced by a more appropriate product data model.

The relation hasContent is inherited by the classes Product and VersionSet and their respective subclasses DocumentVersion and Document. Thus, it can be employed to describe the content of both documents and document versions. If the content of a document remains the same for all its versions, a ContentDescription can be associated with a Document directly; if, on the other hand, the content may change from one version to another, the different ContentDescriptions should be linked with their corresponding DocumentVersions.

Moreover, the **hasContent** relation can be used for both coarse-grained and fine-grained content descriptions: For a coarse-grained description, a **ContentDescription** may be directly associated with a **Document** or **DocumentVersion**. If a more fine-grained description is needed, a **Document** or **DocumentVersion** should firstly be decomposed into its constituents (i.e., into **Products** and **VersionSets**, respectively); next, individual content descriptions can be specified for each constituent via the **hasContent** relation.

So far, only the semantic content of documents has been addressed. In order to provide some means for specifying the syntax of a document, as well, the class **StructuralElement** is introduced to indicate the format of a particular document element. **StructuralElement** can be refined to introduce special format types: For instance, a text document could be composed of structural elements like **Heading**, **Paragraph**, and **Table** whereas a block flow diagram could be decomposed into **Blocks** and connecting **Streams**. However, this part of the model has not been elaborated in detail, as the model does not aim at providing fine-grained format definitions. Other technologies exist for this purpose, including DTD and XML Schema.

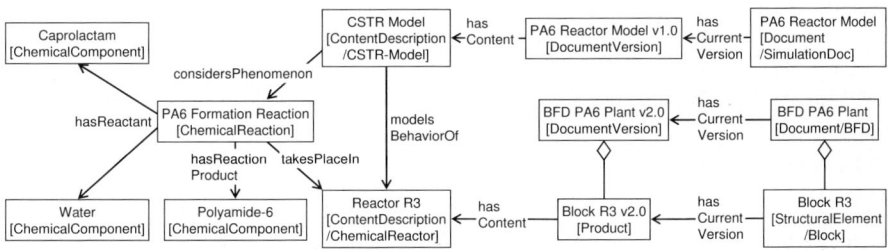

Fig. 2.11. Exemplary description of document content on the instance level

An example of a content description is given in Fig. 2.11. Two instances of **Document** are shown on the right hand side of the figure, namely **PA6 reactor model** of type **SimulationDoc** and **BFD PA6 Plant** of type **BFD**. Below, one particular **StructuralElement** of **BFD PA6 plant** is shown, namely **Block R3** of type **Block**. The current versions of **PA6 Reactor Model** and **Block R3** (i.e., **PA6 Reactor Model v1.0** and **Block R3 v2.0**) are associated with **ContentDescriptions**: **CSTR Model** is an instance of the OntoCAPE class **CSTR-Model**, while **Reactor R3** is an instance of the OntoCAPE class **ChemicalReactor**.

The content descriptions are further specified through additional classes from OntoCAPE, as shown on the left hand side of Fig. 2.11: The **CSTR model** considers the **PA6 Formation Reaction** that takes place in **Reactor R3**. Moreover, the **ChemicalComponents** involved in the **ChemicalReaction** are indicated: **Caprolactam** and **Water** are the reactants and **Polyamide-6** is the reaction product.

In addition, the relation **modelsBehaviorOf** establishes a direct connection between the content descriptions **CSTR-Model** and **Reactor R3**. That way, concise *semantic relations* between documents can be specified via their content descriptions. The semantic relations clarify the character of document dependencies, which could only be indicated through the rather general **dependsOn** relation before.

2.3.5 Usage

Document models, as the one presented above, provide the basis for efficient computer support in various fields of application: Document models are a first step towards the *improvement of business processes* and the solution of problems related to *document handling* and *information services* [656]. For the development of *document management* facilities, it is necessary to consider the different document types and their dependencies (Subsect. 2.3.2) together with their dynamic behavior (Subsect. 2.3.3) [14]. Models of document contents and their integration with product data models (cf. Subsect. 2.3.4) can support the *automatic generation of documents* [553], the *retrieval of information from a document base* ranging from single data items over parts of documents to entire documents [1064], as well as the *exchange of information between dependent documents* [26].

The applicability of the above Document Model to the areas of information integration and document management has been practically demonstrated by utilizing the Document Model in two software development projects, as shortly discussed next:

- Integrator tools manage fine-grained interdependencies between documents that arise in the course of a design project. Their goal is to keep the vast amount of inter-document relations consistent and automatically propagate changes through the network of dependent documents. Without such integrators, these fine-grained dependencies need to be managed manually by developers without appropriate tool support - a tedious and error-prone task, especially in case of concurrent and simultaneous engineering.

  Model-based integrators, as those developed by subproject B2 (Sect. 3.2), rely on formal *tool models* that describe the entities of which the documents are composed as well as the relations between these entities. Such tool models can be derived from the Document Model presented here. In Sect. 6.3, the transition from document to tool models is described in detail.

- The Process Data Warehouse (PDW) described in Sect. 4.1 collects, enriches, and manages the various information created during engineering design projects. The implementation of the PDW is based on interrelated ontology modules, which are built around a central Core Ontology (cf. Subsect. 4.1.5). The Document Model presented here forms part of this Core

Ontology, where it provides the concepts required for the management of documents[12]. Within the PDW, a document is represented through metadata describing its structure, version history, and contents, as well as its dependencies and semantic relations with other documents. These metadata are created by instantiating the Document Model, as explained in the previous subsections. To enhance user acceptance and usability of the PDW, converters are needed that automatically derive the required metadata from legacy documents. Some prototypical converters have been built to demonstrate the feasibility of this approach (see [3] for the description of a converter that maps XML data on a given OWL ontology).

2.3.6 Conclusion

In development processes, information are typically handled by means of documents, which act as carriers for the data during project execution. To better support the needs of the developers, efficient tools for document handling are required, which rely on formal models of documents.

In this section, a comprehensive Document Model has been presented, which describes the types and dependencies of documents as well as their internal structures and dynamic behavior. Moreover, by integrating the Document Model with a product data model, the contents of documents and their semantic relations can be indicated.

In order to validate the model and to prove its applicability to real-world problems, it has been tested (and thus practically validated) in two software development projects targeting the areas of information integration and document management. Further testing will be performed in the near future by applying the model in a large-scale industrial project dealing with the management and integration of design information across the project lifecycle of a chemical plant (cf. Sect. 7.2).

Table 2.1. Semantics of the document classes shown in Fig. 2.5

Class Name	Definition
BFD	The Block Flow Diagram is a schematic representation of the overall process. Block or rectangles represent a unit operation or groups of unit operations. The blocks are connected by straight arcs, which indicate material and utility transfers between the various units.
BuildingDrawing	Detailed drawings of buildings.
BuildingLayout	Documents provided by civil engineers and used by chemical engineers as a basis for arrangement of process units or equipments.

---

[12] Note that the class **Information** is named **ProductObject** within in the Core Ontology.

Table 2.1. Semantics of the document classes shown in Fig. 2.5 *(continued)*

Class Name	Definition
ChemEng-DesignDoc	All documents used, created, or processed by chemical engineers during conceptual design, basic engineering, or detail engineering.
ChemEng-Diagram	A graphical representation of a process, a chemical plant, or of parts of them, viewed from a chemical engineering perspective.
Comprehensive-Doc	This class denotes detailed reports based on documents of type ReferenceDoc, TechnicalDoc, SupplementaryDoc and/or other ComprehensiveDoc. They are usually written at the end of a developing stage to outline all the important procedures and the results, and to establish the foundation for the next developing stage.
DataSet	A document in which some raw data (i.e., data without context) is embodied, for example a cost database.
Design-Calculation	Documents of this type contain design calculations of processes, units and/or equipments. The calculations can be done manually or by using software such as Aspen and Excel.
DesignBasis	The design basis describes the feedstocks, the product specifications, and the utilities available. In principle, all the project-specific requirements have to be listed here. Also a project-specific numbering and codification system will be listed in the design basis. The document is very central in the further execution of the project, and the information stated in the design basis is essential for preparing the flowsheet [851].
DetailedLayout	The detailed layout specifies the locations of main equipment and equipment elevations.
DetailPlan	A detailed drawing of buildings, structures, foundation, etc.
EconomicDoc	A document containing a cost estimation or an economic analysis of the chemical process at different stages of the development process.
Environment-Guideline	A guideline that describes the environmental requirements on a chemical process.
EquipmentList	A list of all the equipments shown on the PID.
Equipment-Specification	The equipment specification contains the process requirements to the equipment and the basic mechanical requirements. Details like orientation of nozzles are left to the detail designer. The main objective of an equipment specification is to find vendors, who can bid for the equipment [851].
Feasibility-Analysis	The feasibility analysis is an important result of the conceptual study, which gives technical and economic assessments of the process, evaluations of different concepts, and a statement of license application [1027].

Table 2.1. Semantics of the document classes shown in Fig. 2.5 *(continued)*

Class Name	Definition
GeneralLayout	The general layout specifies the location of process units, utilities units, roads, building and fencing.
GeneralPlan	A drawing that gives an overview on the functional zones of the entire facility.
InitialProblemStatement	The initial problem statement, also called primitive design problem statement [957], describes the point of origin of the design project. It expresses the current situation and provides an opportunity to satisfy a societal need [957]. A typical initial problem statement is given by [957]: "An opportunity has arisen to satisfy a new demand for vinyl chloride monomer, on the order of 800 million pounds per year, in a petrochemical complex on the Gulf Coast, given that an existing plant owned by the company produces 1 billion pounds per year of this commodity chemical. Because vinyl chloride monomer is an extremely toxic substance, it is recommended that all new facilities be designed carefully to satisfy governmental health and safety regulations".
LineList	A list of all the pipes shown on the PID. Each pipe will be identified with a unique number, which is printed on the PID as well. Different projects have different requirements to the actual content of the information in a line list. As a minimum, the operating and design pressure and temperature, the nominal diameter and the pipe class have to be listed. In some projects a pressure-drop budget figure will also appear [851].
LiteratureSummary	A survey of project-related internal documents and publications in the open literature.
MediumList	A listing of the different media involved in the process. Along with the medium, which is usually designated with a two letters code, normal operating temperature and pressure and design temperature and pressure are listed. The last field is material selection [851]
OrganizationDoc	A document concerned with the work organization but not the work content of chemical engineers.
PFD	The Process Flow Diagram (PFD) shows the basic layout of the unit. A heat and mass balance is included in the flowsheet. Some flowsheets will show the basic control philosophy [851].
PID	The Piping and Instrumentation Diagram (P&ID) shows in detail all the equipment, how it is connected with piping, where and what type of instrumentation and valves are used, the location of temperature, pressure, level and flow controllers (and indicators). Further it will show all the safety devices and alarms. The PI&Ds give a comprehensive overview over the plant or the unit and it is a common document for mechanical, process, and instrumentation engineers in the downstream activities [851].

Table 2.1. Semantics of the document classes shown in Fig. 2.5 *(continued)*

Class Name	Definition
Piping-Specification	This document specifies the piping material and defines the different piping classes and to be used in the project. A piping class is a project specific standard of a piping material that meets the mechanical requirements in a certain range temperature wise, pressure wise and corrosion wise. To complete piping specification all possible fittings and valves to be used within the piping class are specified [851].
PlantLayout	The plant layout specifies the arrangement of the units, buildings, equipments etc.
ProcessConcept	A conceptual report of possible solutions of the primitive problem. It indicates the resource of raw materials, the scale of the process, an approximate location for the plant, and other restrictions [957].
ProcessDesign-Report	This report gives the design details of a process and its profitability. A recommendation on whether or not to invest further in the process should be also included [957].
Profitability-Analysis	A document supplying information on the profit potential of the final plant. Several standard methods such as return on investment, venture profit, payback period, and annualized cost are used to calculate the expected profitability [957].
ReferenceDoc	Static resources of information, which are independent of a particular project and are not changed during project execution. Reference documents within chemical engineering are for example material data sheets, patents, and technical articles.
SafetyGuideline	A guideline that specifies the safety requirements on a chemical process.
SafetyValve-Specification	A series of documents specifying the sizes of the safety valves in the plant. The reason for not just merging this group of documents into the instrument specifications is that the sizing calculations of the safety valves actually require the sizing of the control valves [851].
SimulationDoc	A document created or used by simulation software.
SimulationResult	A document that contains the numerical results of the simulation.
Simulation-Specification	The input file for a simulator, specifying the simulation model and/or the parameters.
StandardAnd-Guideline	Standards and guidelines for developing a chemical process.
SteelStructure	A detailed drawing of the steel structure of the plant.
SupportingDoc	Supplementary documents created by other functions or departments than chemical engineering that directly support the work of chemical engineers.

Table 2.1. Semantics of the document classes shown in Fig. 2.5 *(continued)*

Class Name	Definition
TeamList	A list of staff members, their responsibilities, and their contact information.
TechAnalysis	An analysis of a chemical process system from a technical perspective.
TechList	A listing that summarizes all technical items of specific type involved in a particular chemical process.
TechnicalDoc	Documents created and processed by chemical engineers at different stages of the chemical process design containing important technical information.
Tech-Specification	A document that specifies some technical details about a the chemical process or plant to be developed.
TechStandard-Guidline	Standards and guidelines for technical calculation and construction [1004].
TextAnd-GraphicalReq	Standards for text and graphics [1004].
TimeTable	The time table includes the particular project definition, all the different projected activities, the periodical review points, the change points in project management, the requirement for introduction of additional support teams, and the emphasis on specific efforts [853].

2.4 Work Process Models

M. Eggersmann, B. Kausch, H. Luczak, W. Marquardt, C. Schlick,
N. Schneider, R. Schneider, and M. Theißen

Abstract. Empirical studies are a prerequisite for creating meaningful models of work processes, which can be used to analyze, improve, and automate design processes. In this contribution, a modeling procedure is presented, which comprises the creation of semi-formal models of design processes, their analysis and improvement, and finally the formalization of the models as a prerequisite for the implementation of supportive software tools. Several modeling languages have been created for representing design processes, including the C3 language for participative modeling of design processes on a semi-formal level and a Process Ontology for the formal representation of design processes.

2.4.1 Introduction

According to current studies in German research and development departments, a substantial part of research and development expenditures is not used efficiently [724]. Also, in [575] a *significant potential to reduce development times* is identified. In order to exploit this potential, a better coordination of the developers involved in design processes is required, parts of the design process are to be automated, and proven knowledge from previous design processes is to be reused in further projects. To this end, work process models, providing the relevant information about design processes in a concise way, are essential.

Design processes in chemical engineering are highly creative in nature, and thus detailed models prescribing all the activities that might be performed during a design process are infeasible. Thus, current modeling approaches for design processes rather aim at a coarse-grained description of the process, for example by defining the milestones of a design project and enumerating some activities to reach these milestones (e.g., [638, 1019]). In few exceptional cases, such as best practice guidelines, the focus is on a more detailed description of single tasks, comprising a small number of activities within a design process. However, knowledge about design processes on a fine level of granularity is scarce. Therefore, meaningful and sound models of design processes require empirical studies addressing exemplary design processes [125]. Several research groups emphasize the value of knowledge about past design processes for future design projects [574, 1047].

A research project prior to IMPROVE has addressed a special type of design processes: A methodology for the *creation of mathematical models* has been elaborated and implemented in the modeling environment ModKit [54, 277] (cf. Subsect. 5.3.4). Part of the support offered by ModKit is based on the tool's workflow functionality. Thus, profound knowledge about mathematical modeling processes was essential for the success of the ModKit project. This knowledge was initially derived from case studies; subsequent steps included an iterative enrichment and formalization of the process knowledge.

The *modeling approach* pursued in the ModKit project has been *elaborated and generalized* in the IMPROVE projects A1 and I1 to cover the modeling, improvement, and implementation of work processes during the conceptual design of chemical processes. In Subsect. 2.4.2, this modeling procedure is presented in detail. Practical application of the procedure relies on the availability of adequate modeling languages to describe complex design processes. In Subsect. 2.4.3, we give a review of typical languages for work processes and show that they do not comply with some fundamental requirements. In consequence, *new modeling languages* had to be developed during IMPROVE.

In Subsect. 2.4.4, the first of these languages, called *C3*, is introduced. *C3* is an easy-to-use semi-formal language suited for the early phases of the modeling procedure, when a fundamental understanding of a design process must be reached. In Subsect. 2.4.5, we get back to some issues related to the modeling procedure and make use of the C3 language.

The implementation of computer support for design processes requires *formal representations* of the work processes. To this end, the conceptual data model CLiP (cf. Subsect. 2.2.3) has been extended with a partial model covering design processes. The experiences gained with this model have led to its revision and re-implementation in form of a Process Ontology. Structure, content, and usage of both approaches are described in Subsect. 2.4.6.

2.4.2 A Procedure for Modeling and Improving Design Processes in Chemical Engineering

Fig. 2.12 shows the generic procedure for modeling, improving, and implementing design processes in chemical engineering, which was elaborated during the IMPROVE projects A1 and I1 [104, 437]. Three roles are involved in the execution of the procedure:

- the *designer*, who has the best (tacit rather than explicit) knowledge about a particular design process,
- a *work process engineer*, a person who has some background on interview techniques and the domain of interest, chemical engineering in our case,
- and a *work process modeler*, who is familiar with different modeling approaches for work processes.

The first step is to define the *scope of the subsequent modeling activities* (1). Depending on the intended usage of the model and the character of the design process in consideration, it must be decided which aspects of the process are to be incorporated in the model. For instance, if an analysis aiming at a reduction of the execution time is envisaged, the time need for all activities should be included, whereas it can be neglected in case of a model intended solely for computer support of the design process.

128 M. Eggersmann et al.

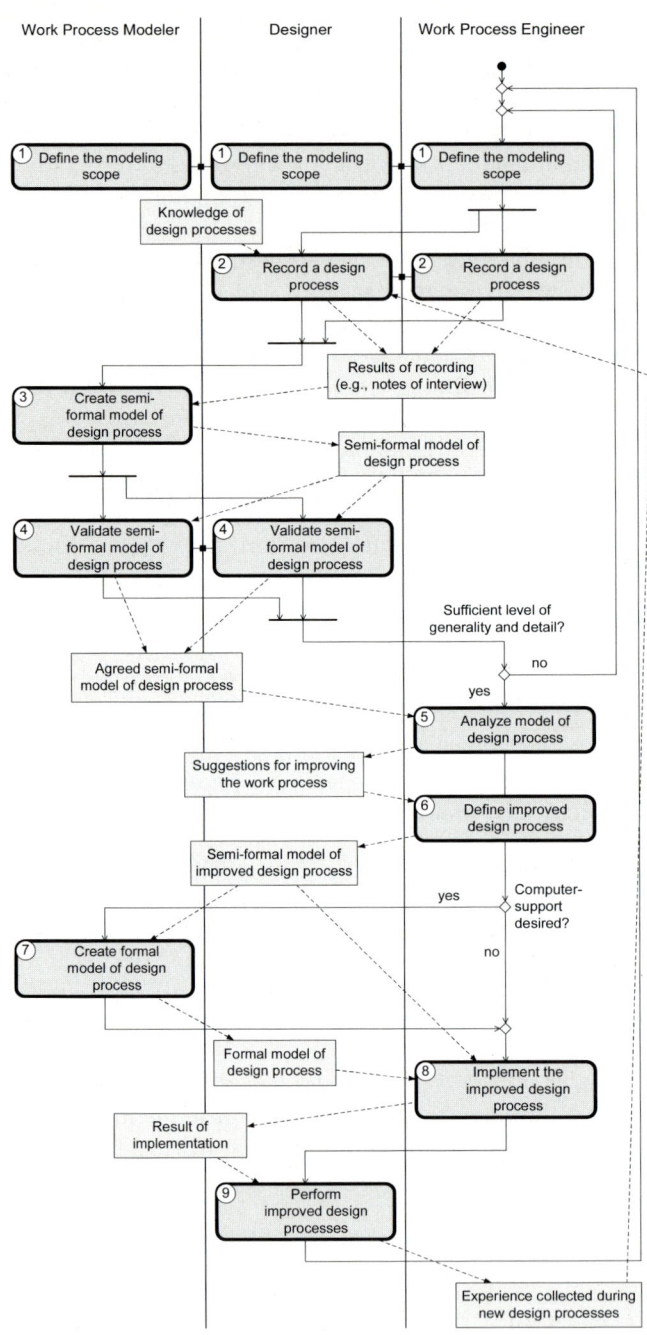

Fig. 2.12. Procedure for modeling and improving design processes (in C3 notation)

Subsequently, an *exemplary concrete design process* as performed by a designer is recorded (2). The recording is supported by the work process engineer, who interviews the designer after the completion of (part of) the design process. The roles of the designer and the work process engineer may coincide; in this case the designer personally protocols what he/she is doing during the design (*self observation*).

The work process modeler transforms the usually informal results of the recording phase into a *semi-formal work process model* (3). As this model must be *validated* by the designer in a subsequent step in order to eliminate misunderstandings (4), it must be easily understandable for the designer, who cannot be expected to be an expert of work process modeling. Simultaneously, the model must allow to represent the main characteristics of design processes in chemical engineering. To this aim, the C3 language for the semi-formal representation of work processes is used (cf. Subsect. 2.4.4). Note that also the depiction of the modeling procedure in Fig. 2.12 makes use of this intuitively comprehensible notation.

If the work process model is intended to support further design processes, it must be both sufficiently general, i.e., it must cover possible variants of the originally recorded process, and it must comprise details relevant for its analysis and reuse. Given the complexity of industrial design processes, it is infeasible to create a model with adequate generality and detail in a single step. Instead, an *iterative procedure* is advisable until a model with the required properties is built. Possibly, the scope of the model has to be reconsidered and adapted during the iterations. For example, a coarse-grained model describing the milestones of a project should not be overloaded with all software tools ever used during its execution. In contrast, a model focusing on the activities of single designers should cover such tools.

During an analysis of the semi-formal work process model (5), the work process engineer may *identify problems and shortcomings of the design process* and make suggestions for their improvement. Some examples for a manual analysis are given in [99, 105]. The analysis can also be supported by tools like discrete-event simulation systems (cf. Sect. 5.2).

From the analysis of a certain number of design processes, it should be possible to *define improved standard processes* (6), which have proven to be effective and can serve as some kind of template for future projects. Such standard processes could be directly implemented (8) without the use of information technology: A project manager, for example, can use standard work processes to organize his project, or a designer can employ a certain design procedure to design a process unit.

Obviously, work processes can be even better supported by the application of computer tools. A software tool which captures information about various best practice work processes could guide a designer by suggesting appropriate activities and advising him how to proceed. This requires the creation of a *detailed and sufficiently formal conceptual model of the design process* to be supported (7), which would then be used for the creation of a design model required for the *implementation* (8) of the software tool (cf. Subsect. 2.1.1). Due to the complexity of this conceptual model, it must be created by the work process modeler.

After the application of the *new design processes* (9), their success has to be evaluated, again starting with the recording of the work process. In Fig. 2.12, this iteration is indicated by the feedback flow of the experience collected during the new design process to the recording step (1) at the beginning of the procedure.

The remainder of this section addresses different modeling languages for work processes, which have been developed by the IMPROVE subprojects A1, I1, and I4, taking into account the requirements imposed by the modeling procedure and the characteristics of design processes in chemical engineering. In Subsect. 5.1.2, we discuss some issues related to the practical implementation of the procedure, focusing on its implementation in industrial settings.

2.4.3 Modeling Approaches for Work Processes

The availability of adequate *semi-formal and formal modeling languages for design processes* is an indispensable precondition for applying the modeling procedure presented above. Research in fields like software engineering, workflow management [1058], and business process reengineering [714] has led to a wide range of modeling approaches with different objectives. In the following paragraphs, we will give a review of exemplary approaches. The subsection concludes with an evaluation of the existing approaches with respect to their suitability for design processes.

Business Process Modeling

The probably best-known modeling language for business processes is the activity diagram of the *Unified Modeling Language* (*UML*, [560]). The roots of UML are in software engineering, but nevertheless it is widely used in many domains. UML 2.0 [880] comes with a rich set of concepts for a detailed description of the behavior of software. The user is free to extend UML with the concepts needed for a similar expressiveness for other process types. According to [568] such extensions, denoted as *profiles* in UML, are in particular required for business processes modeling. An exemplary extension for business process is described in [884]

Both graphical representation and semantics of the *Business Process Modeling Notation* (*BPMN*, [873]) are similar to that of UML activity diagrams. BPMN addresses the creation of process models for workflow execution and automation; for this purpose mappings from BPMN to executable languages like the *XML Process Definition Language* (*XPDL*, [1059]) and the *Web Services Business Process Execution Language* (*WS-BPEL*, [887]) have been defined.

Event Driven Process Chains (*EPC*, [842, 949]) also address the analysis and execution of business processes. As the expressivity of an EPC in its basic form is rather limited, several variants have been developed. *Extended Event-Driven Process Chains* (*eEPC*, [958]), for example, cover organizational units.

Task Object Charts (*TOC*, [1069, 1070]) are meant for the design of interactive software systems and user interfaces. They provide a modeling element called *blob*, which allows to include several activities in a model without restricting their execution order. The *blob* has been adopted in C3 and is discussed in Sect. 2.4.4. *CoCharts* [221], an enhancement of TOC, also cover coordination mechanisms for groups of activities.

Diagrams like *Gantt* and *PERT* charts (e.g., see [777]) are tools for project management and resource allocation. Consequently, their focus is on the representation of resources and roles.

Mathematical Formalisms

In addition to the approaches described so far, mathematical formalisms like *Petri nets* [770], *state charts* [718], and *process algebra* [551] have been developed which allow to describe the behavior of concurrent systems in an unambiguous, mathematically sound way, which makes up their suitability for computer-based applications. Petri nets, in particular, are an appropriate basis for work process automation by means of workflow systems [1014] or analysis by means of process simulation (cf. Sect. 5.2).

However, in their basic form Petri nets are unsuitable for modeling, planning, and documenting complex work processes, since even relatively simple facts result in highly complex Petri nets (see also [939]). Thus, in order to better exploit the advantages of Petri nets, mappings from some of the languages described above have been defined (e.g., [986])[13]. A recent development is *Yet Another Workflow Language* (*YAWL*, [1063]), whose semantics is formally defined based on Petri nets. YAWL claims to incorporate all workflow patterns provided by any workflow system and thus, the language is characterized by a large expressivity.

Requirements for Modeling Languages for Design Processes

The characteristics of creative design processes induce several requirements for adequate modeling languages. These requirements provide a basis for the evaluation of the existing approaches in the next paragraph.

First of all, an adequate modeling language must be sufficiently expressive to enable the modeling of all aspects relevant to documenting, analyzing, and implementing design processes. These aspects comprise the *activities* in a

---

[13] In case of executable languages, these mappings aim primarily at process analysis. Implementations of the languages are typically not based on Petri nets.

design process and their interdependencies, the *information* produced during the activities or needed for their execution, the *roles* or organizational units of the actors, and finally the *tools* required for performing the activities [102, 103]:

- *Information and information flow.* Typically, the execution of activities produces information (e.g., in form of documents, data, or oral messages), which is then used in subsequent activities. The explicit representation of information and information flows is crucial as the availability of input information imposes restrictions on the order of activities.
- *Activities and their interdependencies.* Due to their creative character, design processes are *weakly structured*: In general, it is not possible to give a detailed descriptions of all activities *a priori*. For instance, the necessity to perform certain activities can depend on criteria which cannot be formulated before some intermediate results of the design process are available.[14] In addition to conventional modeling concepts for predetermined procedures, which comprise activities in a well-defined order or with explicit conditions for their execution, a modeling language for design processes must provide powerful concepts for temporal and structural abstraction. *Temporal abstraction* refers to the freedom to omit information about the temporal relations between activities (e.g., by permitting a partial overlap of two subsequent activities). An even stronger *structural abstraction* allows to create process models which do not determine whether an activity is performed at all.
  A frequent pattern encountered in design processes is that of *distributed activities*: Several simultaneous activities require a considerable exchange of information between the actors involved (e.g., between several managers, engineers, and technicians at a project meeting). Even if the exact nature of this information is neglected in a model – for instance, if it is unknown *a priori* – the strong interdependencies between the activities should be recorded.
- *Roles and organizational units.* Current research results emphasize the importance of roles and organizational structure in design processes. The coordination of activities in different units depends substantially on their spatial distribution, their fields of activity, and their competencies [981]. A modeling language for design processes must be able to represent work processes across the borders of the organizational units involved.
- *Resources and tools.* The efficient execution of activities relies on the technical resources and tools used by the actors. In case of chemical engineering design, these tools are typically software tools, but they can also comprise physical resources like lab equipment.

---

[14] In contrast to this terminology, the workflow community uses the term *structured workflow* to denote workflows that are built according to certain rules (e.g., see [778]).

Table 2.2. Expressivity of business process modeling languages
+: satisfied, o: satisfied with restrictions, -: not satisfied

	temporal/ structural abstraction	distributed activities	information	information flow	resources (tools)	roles/ organizations
UML activity diagram	-	-	+	+	-	+
BPMN	-	-	+	+	-	+
EPC/eEPC	o	-	+	+	-	+
TOC	o	-	+	+	-	o
Gantt Chart	-	-	-	-	+	+

A further requirement are *flexibility* and *extensibility* of the modeling language. Depending on the intended purpose of a process model (e.g., analysis and improvement, automation, compilation of best practices), the importance of different aspects of a design process can vary. Hence, a modeling language must allow to bring the relevant aspects into focus. This concerns the possibility to enrich the basic elements of a modeling language with relevant attributes if need arises, but also the freedom to omit information which is considered as irrelevant in a certain modeling context.

Evaluation of Existing Modeling Languages

A wide range of modeling languages and notations for work processes exists, but none of them fulfils all requirements imposed by the characteristics of design processes in chemical engineering [221]. Table 2.2 gives an overview of the *expressivity of the most prominent of the modeling languages* discussed above with respect to the requirements. The most critical shortcoming common to all approaches is their inability to represent work processes on different levels of structural and temporal abstraction (except for TOC) and their limited support for distributed activities; the existing languages are tailored to predetermined and well-structured work processes. Some requirements are met to different degrees by several languages. In consequence, new modeling languages had to be developed within IMPROVE, building as far as possible on proven techniques and extending them with missing concepts.

2.4.4 C3 – A Semi-Formal Language for Participative Modeling of Design Processes

In order to receive detailed and reliable information about the different activities of designers and other experts in the entirety of the organizational units involved in a design process, representatives from all roles must participate in both the creation of the work process model and its validation (cf. steps 2 to 4 of the modeling procedure). In general, these people cannot be expected to be experts of work process modeling. Thus, a prerequisite for *participative modeling* is a rather simple semi-formal modeling language whose concepts and elements are easily understood across the borders of the involved disciplines [694]. In addition, a small learning effort to get acquainted with the modeling language reduces both the time and the costs for its introduction in an organization and increases its acceptance among planners, decision makers, and the designers themselves.

Building on a language developed earlier at IAW [424], the C3 language for participative modeling of work processes has been developed with a special focus on the requirements imposed by the creative character of chemical engineering design processes as discussed above. The term *C3* is an acronym for *Cooperation, Coordination,* and *Communication*; it indicates the characteristics of design processes which are in particular addressed by the language:

- *Cooperation* is a collective effort in working towards a common goal. Decisions are made collectively, individual goals are subordinated to the overall goal of the group, collective plans are developed, and the achievement of the group as a whole is evaluated.
- *Coordination* refers to the management of such cooperation.
- *Communication* is a person-based, goal-oriented as well as non-anonymous exchange of information.

C3 is intended for users below expert-level. The language is not meant for fine-grained models, but rather to describe unstructured and partially structured activities and work processes. C3 inherits from other languages, in particular UML activity diagrams, Task Object Charts, and state charts. Also, a modified subset of *Role-Function-Activity nets (RFA nets,* [872]) for the representation of information flows and cooperation partners has been adopted.

C3 diagrams contain two types of symbols, *elements* and *connectors*. In the interest of usability, C3 provides a rather restricted number of core elements and connectors, which are essential for the expressiveness demanded above. Elements and connectors can be further specified by *attributes*. Whereas some attributes should always be provided, such as a meaningful name and an ID allowing for unambiguous reference, the adequacy of other attributes depends on the purpose of the work process model and the users' needs. For the sake of illustration, some possible attributes are given in the following discussion of the core elements.

Core Elements

As shown in Fig. 2.13, C3 provides four core elements: *activities, information* elements, *tools,* and *roles*.

Activities are the central element of any work process model. They describe what has been done or has be done. Some exemplary attributes for activities are

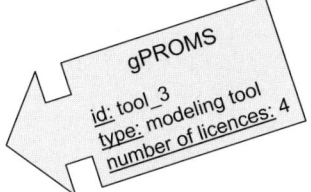

Fig. 2.13. Core elements *activity*, *information*, *tool*, and *role* with some exemplary attributes

- the qualifications required for executing the activity or
- the execution time.

Information items can represent documents such as flowsheets or mathematical models, but also ideas, messages, or any information which is produced during activities and further processed in other activities. Typical attributes of information comprise

- a textual description of the information or
- the file format in case of an electronic document.

Tools are used during activities. In a C3 diagram, tool elements are attached to the activities they are used for. The representation of tools may contain attributes like

- the type of the tool (e.g., modeling tool) or
- the number of available tools (e.g., the number of licenses in case of a software tool).

Roles can represent organizational units (e.g., *R&D department*), but also people or groups with certain functions or qualifications (e.g., *project leader* or *chemical engineer*). Like in UML activity diagrams, roles in C3 are depicted as vertical swimlanes, in which the activities are placed. Thus, C3 diagrams provide a clear view on the assignment of activities to roles. Typical attributes comprise

- the number of persons performing in the role or
- the qualifications of these persons.

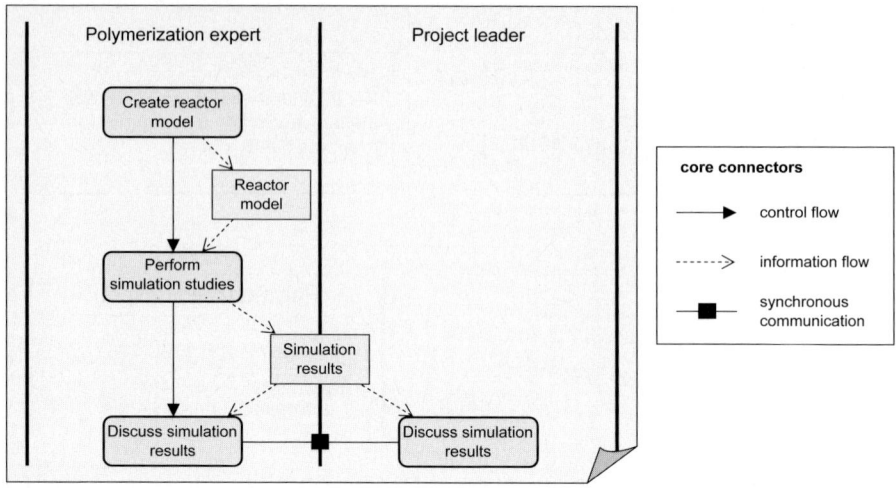

Fig. 2.14. Core connectors *control flow, information flow,* and *synchronous communication*: symbols and usage

Core Connectors

Three core connectors are available for representing interrelations between the core elements (see Fig. 2.14).

The *information flow* shows the interaction between information elements and activities. Information required for an activity is marked by an information flow from the information item to the activity. Similarly, the information produced by an activity is linked by an information flow from the activity to the information.

In its basic form, the *control flow* represents a temporal sequence of two activities. Further uses of the control flow are discussed below.

The third core connector is the *synchronous communication*; it connects activities whose execution requires communication between the actors involved (e.g., during project meetings).

Synchronization Bar

More complex relations between activities can be modeled by means of the *synchronization bar*. One possible usage is depicted in Fig. 2.15 a). Here, two synchronization bars indicate the beginning and ending of two independent control flows to be followed in parallel. Both activities A1 and A2 are executed.

Conditional branchings are modeled by attaching conditions to the outgoing control flows of a synchronization bar. In Fig. 2.15 b), these conditions are mutually exclusive (either *reaction rates available* or *no reaction rates available*). Exactly one of the two activities A1 and A2 is executed. In general,

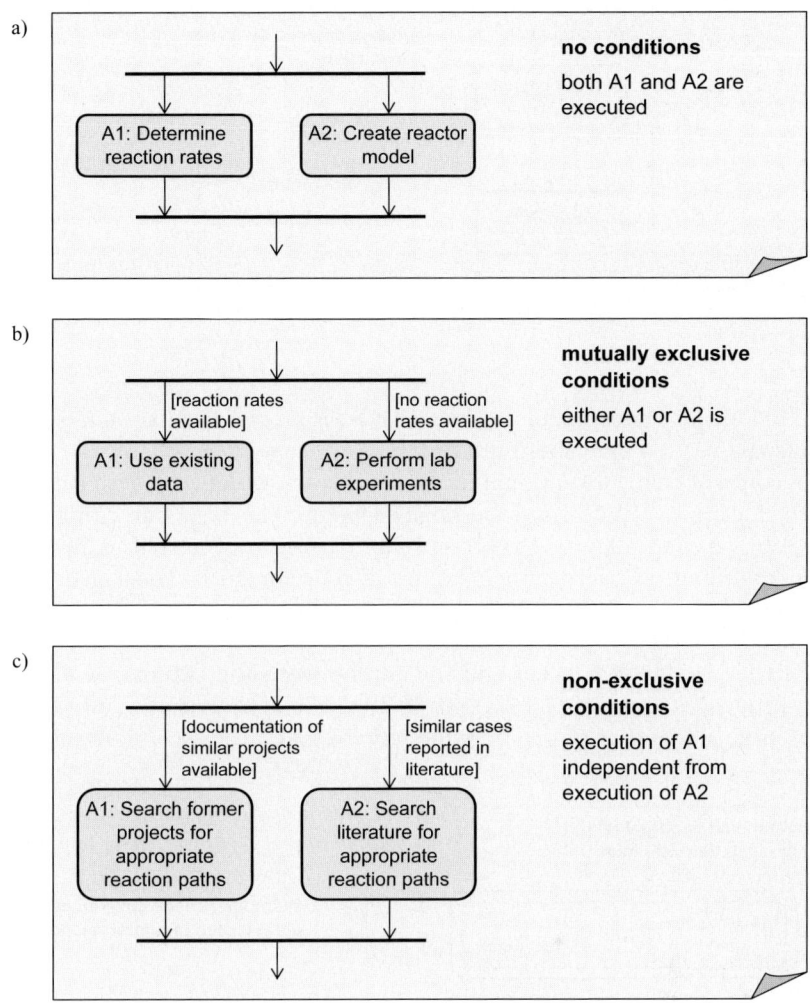

Fig. 2.15. Usage of the synchronization bar

the conditions for outgoing control flows can be independent of each other (non-exclusive conditions). An example is given in Fig. 2.15 c): For each of the activities A1 and A2, a separate condition for its execution is given.

Modeling Concepts for Temporal and Structural Abstraction

Due to the creative and dynamic character of complex design processes, temporal relations between activities are often undefined before a certain stage in a design process has been reached. However, if some knowledge of a design process is available *a priori*, it should be represented in a model. C3 offers

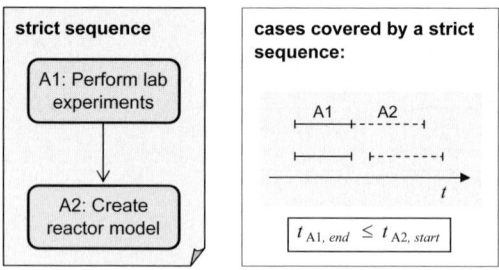

Fig. 2.16. Strict sequence of activities

some modelings concepts for representing *relations between activities* without forcing the user to give a definite order.

In contrast to a conventional (strict) control flow between activities, which states that one activity must be finished before the next one can start (see Fig. 2.16), a *blob* allows to represent the existence of activities in a design process without restricting their temporal relation. In its most general form as depicted in Fig. 2.17, a blob is equivalent to a representation using synchronization bars.

Further modeling concepts fill the gap between the two extremal cases of strict sequences and completely unrestricted sets of activities in a blob. A frequent situation is that of *overlapping activities*, i.e., the second activity starts

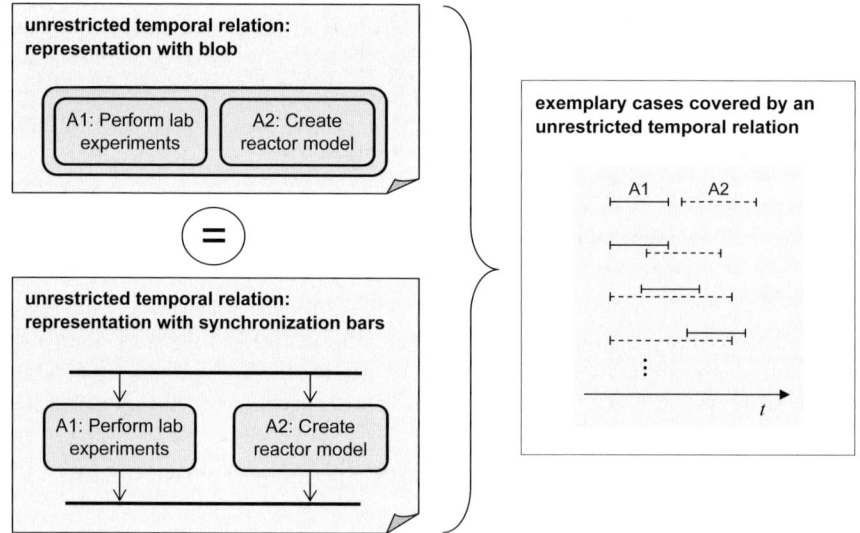

Fig. 2.17. Unrestricted temporal relation between activities

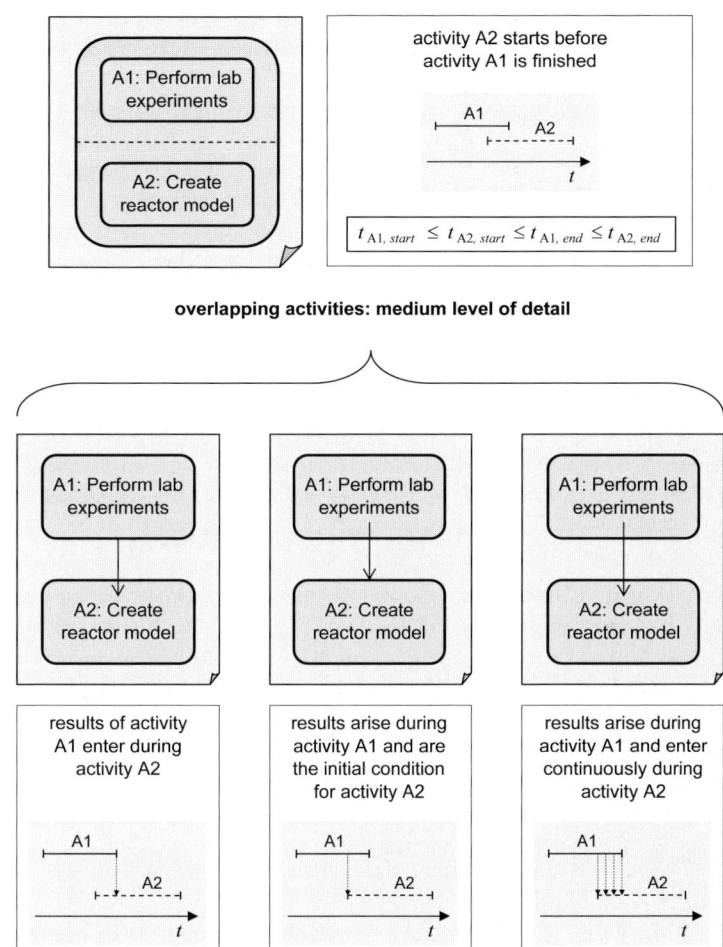

Fig. 2.18. Overlapping activities on different levels of detail

while the first activity is still performed. This can be modeled by splitting a blob in horizontal segments as depicted in the upper part of Fig. 2.18. In case a more detailed representation is wanted and if the required process knowledge is available, several modifications of the control flow allow to distinguish between different cases of overlapping activities, which discriminate by the information transfer between the activities (see lower part of Fig. 2.18).

Fig. 2.19. Optional elements and shortcoming element

Further Modeling Concepts

An exhaustive description of the C3 modeling language can be found in [221]. Here, we confine ourselves to briefly introducing two further concepts: Activities and information items can be declared as *optional elements* (indicated by a shadow behind the symbol, see Fig. 2.19) if their existence in a work process is uncertain at the time of modeling or depends on factors which go beyond the scope of the model. A *shortcoming* element can be placed anywhere in a work process to indicate possible problems.

2.4.5 Two Dimensions of Design Process Modeling

For the creation of C3 models, the prototypical Workflow Modeling System WOMS has been developed at LPT (cf. Sect. 5.1). By means of WOMS, it has been possible to pursue several academic and industrial case studies addressing the modeling and improvement of design processes. These case studies have revealed the necessity to adapt the modeling procedure under certain circumstances in order to cope with the inherent complexity of design processes. Before discussing these issues, we introduce a two-dimensional space in which different models of a design process can be located. These two dimensions are

- the *level of generality* referring to the variety of potential design processes covered by a model and
- the *level of detail* indicating the amount of information captured in a model.

These levels should not be understood as some quantitative measures but rather as an orientation guide helping a work process modeler – and other persons involved – to focus on the relevant issues during the modeling process.

Level of Generality

The level of generality of a model refers to the number of systems which are represented by the model. We adopt a definition given in [888], which states that a model is more general than another if it applies to more real-world systems. In case of work processes, we deal with non-material systems, both in the past and in the future, which arises the question of what should be considered a real-world system. Here, we regard a work process model B as more general than a model A if B applies to all work processes covered by A and if B applies to additional work processes which are not covered by A.

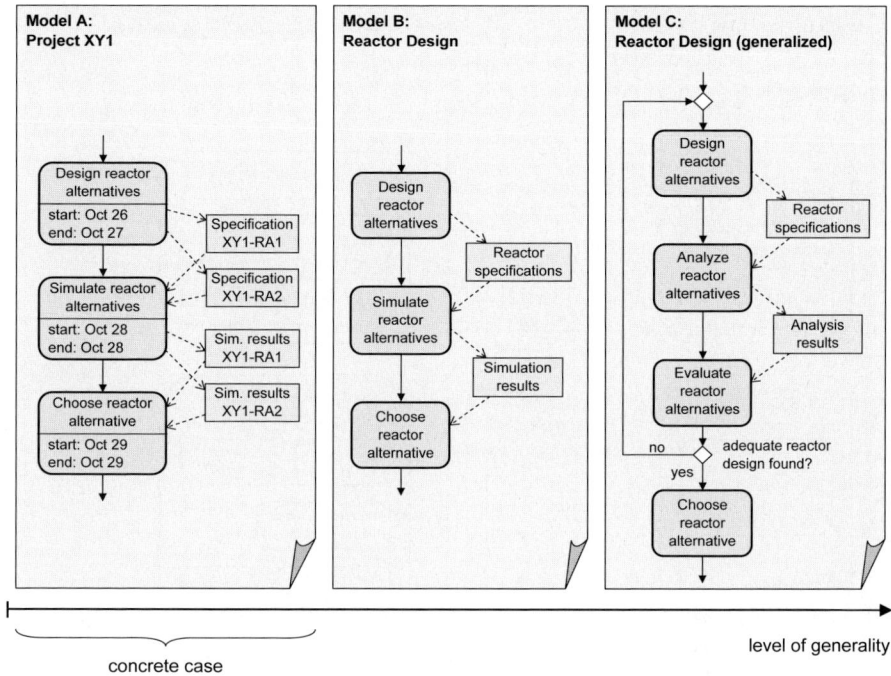

Fig. 2.20. Reactor design on different levels of generality

Figure 2.20 shows three work process models, which are arranged according to increasing level of generality. *Model A* is a simple model of a concrete *Project XY1* addressing the design of a chemical reactor. The model is essentially a sequence of several activities representing the design of some reactor alternatives, their subsequent simulation, and finally the choice of the most adequate alternative. Obviously, *Model A* is restricted to the concrete *Project XY1* as it contains information specific to this project, i.e., information which cannot be valid for any other design project (such as the project number, the dates when the activities were performed, and the concrete information items produced during the work process).

In contrast, the start and end dates of the activities in *Model B* are not specified. Compared with *Model A*, the references to concrete reactor specifications and simulation results have been replaced by the general information items *Reactor specifications* and *Simulation results*. Whereas *Model A* is restricted to a single design process, the more general *Model B* applies to any reactor design process in which an adequate reactor is chosen after first designing and then simulating some alternatives.

Model C is even more general than *Model B*. The simulation of different alternatives has been replaced by a more general analysis, which can comprise a numerical simulation, but also other analysis activities such as checking the

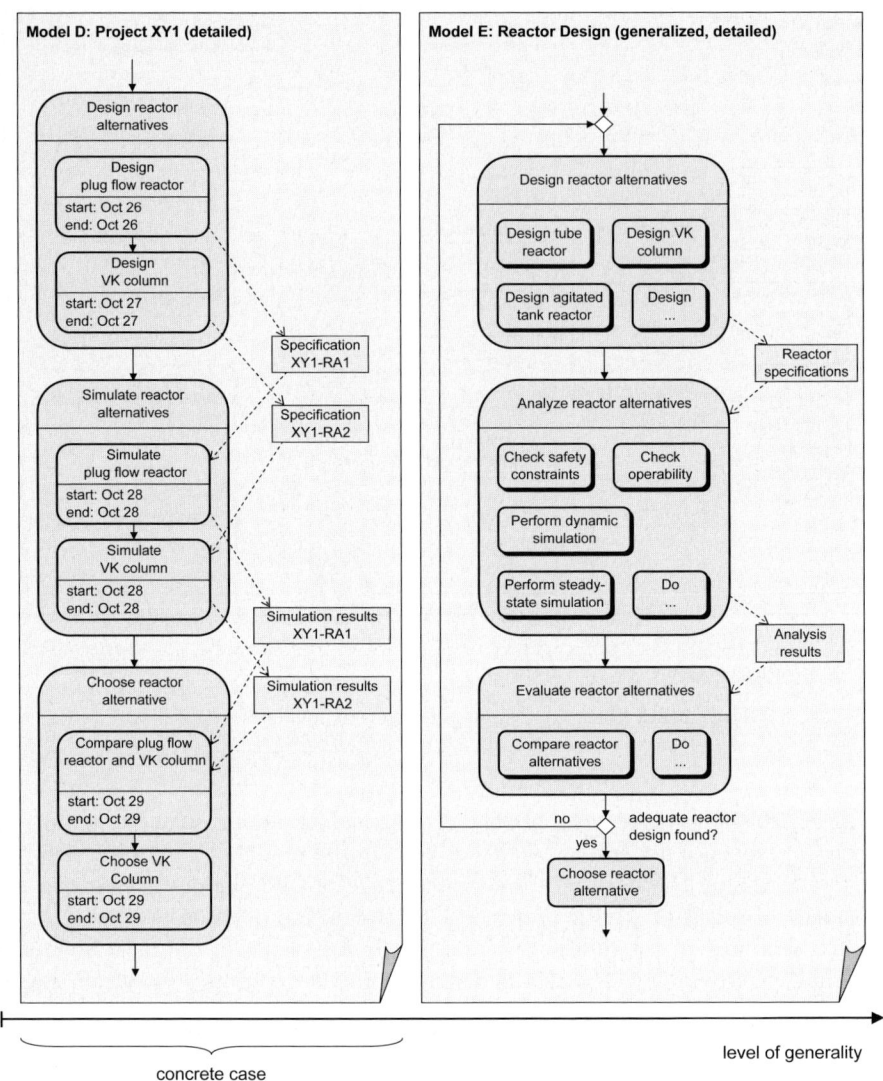

Fig. 2.21. More detailed models of the reactor design process

compliance of reactor alternatives with safety regulations. Instead of the choice of a reactor alternative after the analysis – which requires that an adequate alternative has already been designed at this stage in the process – *Model C* contains an evaluation activity and a subsequent branching of the control flow depending on whether an adequate alternative exists; thus, *Model C* also covers work processes with several design, analysis, and evaluation phases before an alternative is chosen.

Level of Detail

To a certain extent, the three models in Fig. 2.20 differ with respect to their level of detail. These differences result from the necessity to remove or to add information in order to generalize them, i.e., make them applicable to more work processes. A more critical issue is the incorporation of more information while keeping the level of generality.

Fig. 2.21 shows two further models of the reactor design process. *Model D* is a more detailed representation of the concrete design project in *Model A*; for each of the three activities in *Model A* – represented by blobs in the new model – some sub-activities are given.

In *Model E*, a more detailed view of the generalized *Model C* is given. In a similar way, the activities of the less detailed models have been replaced by blobs containing sub-activities; the sub-activities themselves are marked as *optional*, because in a typical reactor design process not all of the sub-activities would be performed.

In these examples, the level of detail is increased by decomposing coarse-grained activities. In an analogous manner, details could also be added by decomposing other model elements such as information items or by including new aspects such as the tools used for the activities. The following discussion is restricted to the *decomposition of activities*, as this is the most critical issue when sufficiently detailed and general models aiming at the support of future design processes are to be created.

Implications for Design Process Modeling

Except for a concrete case, decomposing an activity practically always reduces the generality of a model because there may be an almost infinite number of different sub-activities that could be applied. However, in *Model E* the generality of *Model C* is more or less kept by incorporating some placeholder activities (e.g., *Do ...*) symbolizing possible sub-activities, which are not modeled explicitly. In case of design processes, this approach is acceptable in order not to restrict the designers' creativity. Nevertheless, there may be situations when such indeterminateness is inappropriate, for instance when describing the process of checking the compliance of a design product with safety regulations.

Whereas a more detailed version of a model with low generality – such as a model of a concrete case – is rather uncritical, this is not true for more general models. We briefly discuss two issues, which typically arise when activities in a general work process are decomposed.

Choice of Adequate Discriminators

Typically, there are several possibilities to discriminate sub-activities of a given activity. For instance, in *Model E* the design activity is decomposed in sub-activities addressing the design of different reactor types, whereas the analysis activity is decomposed in several analysis methods which can each be applied to any reactor type (see Fig. 2.21). The choice for a useful discrimination is of minor importance when a single model like *Model E* is considered, but it becomes crucial when the sub-activities themselves are refined, i.e., when they are modeled as sub-processes. Decomposing an activity should be made in a way such that the sub-processes can be easily modeled and applied. In case of complex processes, it is recommendable to model some representative concrete cases in a first step, and then to analyze them in order to find out their similarities. If significant differences are identified, it is preferable to create several generalized models for different groups of the representative cases rather than trying to include all cases in a single overloaded model.

Modeling Design Decisions

The generalized and detailed *Model E* in Fig. 2.21 comprises several sub-activities for the design of reactor alternatives, but it does not provide any criteria for the selection of an adequate sub-activity in the context of a particular project. In fact, the problem of selecting an appropriate sub-activity for the design activity is actually the design problem of selecting an appropriate reactor type. In some well-defined cases, it is possible to give best-practice procedures for such decisions; for instance, some guidelines for the design of separation systems are proposed in [530, 531]. However, these best-practices are essentially lists of questions to be answered in order to exclude inappropriate separation alternatives. A similar example is given in [95] for selecting the mode of operation of a chemical plant.

In general, design decisions are too complex to be represented by a set of conditional branchings in a work process. Instead, a modeling approach tailored to decision modeling should by pursued (cf. Sect. 2.5).

2.4.6 Formal Models of Design Processes

Step 7 of the modeling procedure for design processes addresses the formalization of work process models (see Fig. 2.12). Two models for the formal representation of design processes in chemical engineering have been created. The first one, *Process Models*, is part of the conceptual information model CLiP (cf. Subsect. 2.2.3). Taking into account the lessons learned from this model, a Process Ontology has been developed. In the following, both models are discussed.

Partial Model *Process Models* of CLiP

The focus of *Process Models* [95] is on concepts which are relevant for the computer-based support of design processes. *Process Models* is in part based on a modified version of NATURE, originally developed for requirements and software engineering [201] (see also Subsect. 1.3.4), and *IBIS* (*Issue-Based*

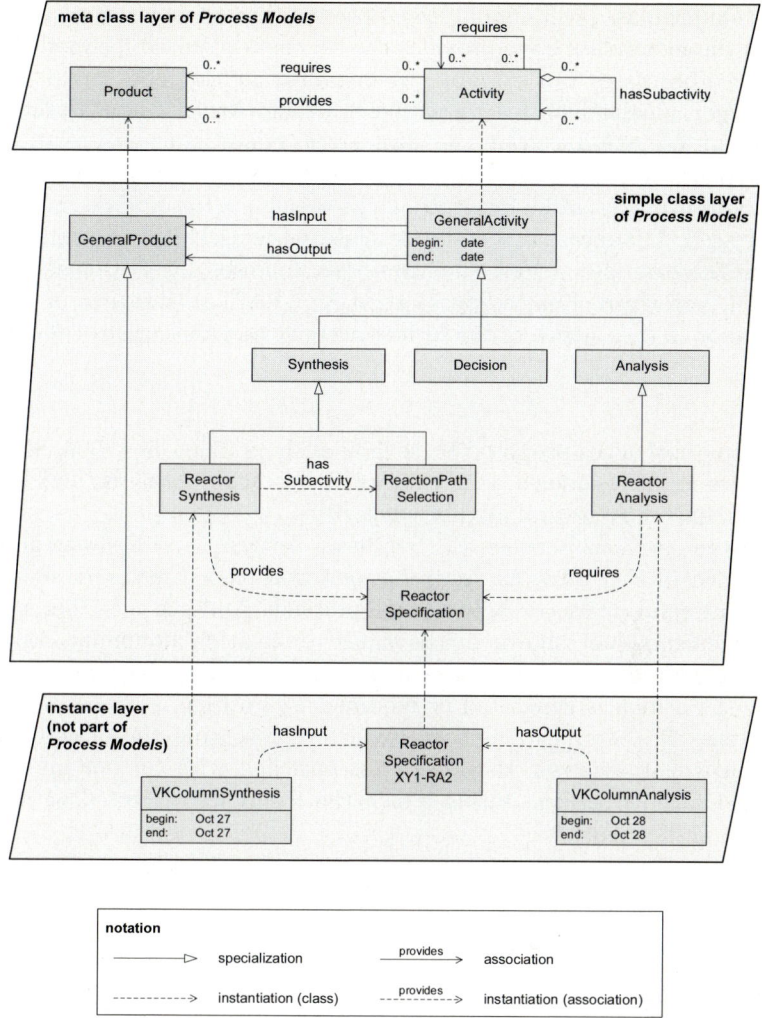

Fig. 2.22. Structure of *Process Models*

Information Systems, [798]), a modeling approach addressing discussions and argumentations related to 'wicked problems' such as design problems (see also Sect. 2.5).

Structure

As shown in Fig. 2.22, *Process Models* comprises a *meta class layer* and a *simple class layer*. The *instance layer*, on which concrete work processes are represented, does not form part of *Process Models*.

The simple class layer enables the representation of generalized processes, whereas the meta class layer provides the concepts required for modeling the generalized processes. For example, by instantiating the meta classes Activity and Product, classes like ReactorSynthesis, ReactorAnalysis, and ReactorSpecification can be introduced on the class layer. Furthermore, as relations like requires and provides are introduced on the meta layer, the fact that a ReactorSynthesis provides and a ReactorAnalysis requires a ReactorSpecification can be expressed. Instances of the simple classes themselves represent concrete cases; in the example, a VKColumnSynthesis, followed by VKColumnAnalysis, has been performed. ReactorSpecification XY1-RA2, an instance of ReactorSpecification is the output of the former activity and the input of the latter.

Content

Similar to approaches proposed by several authors (e.g., [564, 569, 852]), design processes are modeled as iterations of Synthesis, Analysis, and Decision activities, linked by several auxiliary activities.

The interrelations between the Synthesis, Analysis, and Decision activities are depicted in Fig. 2.23. An Issue is a question to be answered, such as an appropriate solution for a reactor design problem. An Issue is further specified by Requirements (not shown in the figure) indicating boundary conditions such as a desired annual production or product purity. During a Synthesis activity, a Position is generated or modified. A Position is a possible answer to the Issue. It describes a GeneralProduct representing the design solution itself. During an Analysis, the Position is enriched with further information required for the Decision, during which the Position is selected or rejected.

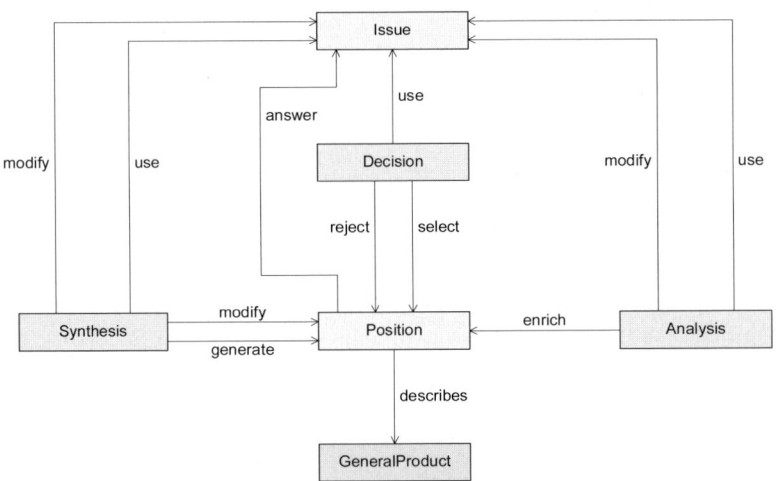

Fig. 2.23. Basic meta-model of the activity representation

During both **Synthesis** and **Analysis**, the **Issue** may be modified, for instance when **Requirements** turn out to be infeasible.

In [97, 98], typical sub-activities of **Synthesis** (e.g., *generating* an artifact such as a reactor), **Analysis** (e.g., *calculating* the product purity reached by a reactor), and **Decision** (e.g., *comparing* the product purities of different reactors) are described. These sub-activities do not form part of *Process Models* as they should not be understood as a formal classification, but rather as in informal description which facilitates the assignment of an activity to one of the types. They are meant as an aid to understand and structure design processes and should help their planning within the framework of a support environment. The feasibility of this approach has been demonstrated by modeling the design process of a separation system for the recovery of ethylene and byproducts from the steam pyrolysis of light hydrocarbons [98].

A *role* in C3 corresponds to an **Actor** in *Process Model*. The actor model of *Process Model* [100] has been adopted in the Process Ontology; it is therefore discussed in the next subsection.

Usage

The partial model *Process Models* extends CLiP's range of use (see Subsect. 2.2.3) to cover process-oriented aspects. Parts of *Process Models* are implemented in the prototypical support environment *COPS* (*Context Oriented Process Support*, [247]), which is integrated with the modeling environment ModKit (cf. Subsect. 5.3.4). COPS supports the creation of mathematical models by suggesting appropriate, context-dependent activities and guiding the modeler through a predefined workflow.

In subproject B1, the *PRIME* system (*PRocess-Integrated Modeling Environments*) has been developed for the experience-based process support at technical workplaces. The usage of PRIME for a certain application domain requires the definition of executable process models relevant to the domain. Generalized work processes from chemical engineering design, represented in *Process Models*, were used as a basis for the creation of such executable models. This transition is described in Sect. 6.2.

However, as parts of *Process Models* have been developed before or during the elaboration of the modeling procedure described in Subsect. 2.4.2, some difficulties can occur when *Process Models* is used for the formalization of C3 models. The modeling concepts of *Process Models* differ substantially from those of C3. For instance, *Process Models* does not provide modeling concepts equivalent to the synchronization bar in C3; in consequence, it must be paraphrased using the available concepts. Thus, describing the formal content of a generalized C3 model can be awkward or even impossible in some cases, even before a formalization of the informal content of a C3 model – such as textual annotations – is addressed.

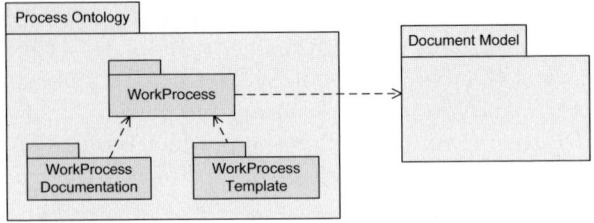

Fig. 2.24. Structure of the Process Ontology.

Process Ontology

The limitations of *Process Models* motivated its revision and led to the development of an ontology for the representation of work processes. As CLiP's successor OntoCAPE (cf. Subsect. 2.2.4) is specified in the web standard OWL (*Web Ontology Language*, [546]), it was decided to use OWL also for the Process Ontology. Thus, OntoCAPE and the Process Ontology can easily be combined in order to get an ontology covering both the work processes and the products in chemical engineering design (see Subsect. 2.6.2 for details on the integration). A further advantage of OWL is the existence of reasoners which can check the internal consistency of the Process Ontology as well as the consistency of process models represented by means of the classes of the ontology.

The scope of the Process Ontology is beyond that of *Process Models*. Its focus is not only on the support and (semi-)automation of work processes in chemical engineering design, but also on the documentation of concrete work processes and on their analysis and generalization.

Structure

Other than in *Process Models*, both generalized models and models of concrete cases are represented on the instance layer. This is in part due to the fact that OWL does not allow to represent graph-like structures – such as a network of interrelated activities in a work process model – on the class layer. Besides, this approach has some advantages with respect to the modeling procedure. First, the transition from concrete to generalized cases is simplified as no change of the modeling approach is required. Secondly, compared to the approach in *Process Models*, which supports only two levels of generality (the concrete case and the general case), it is possible to create a hierarchy of generalized models.

The topmost module *WorkProcess* (Fig. 2.24) provides classes and relations similar to the modeling elements of C3, extended by some additional concepts. The semantics of the classes in **Work Process** is refined in two further modules, *WorkProcessDocumentation* for the documentation of concrete cases and *WorkProcessTemplate* for the definition of generalized work processes. This

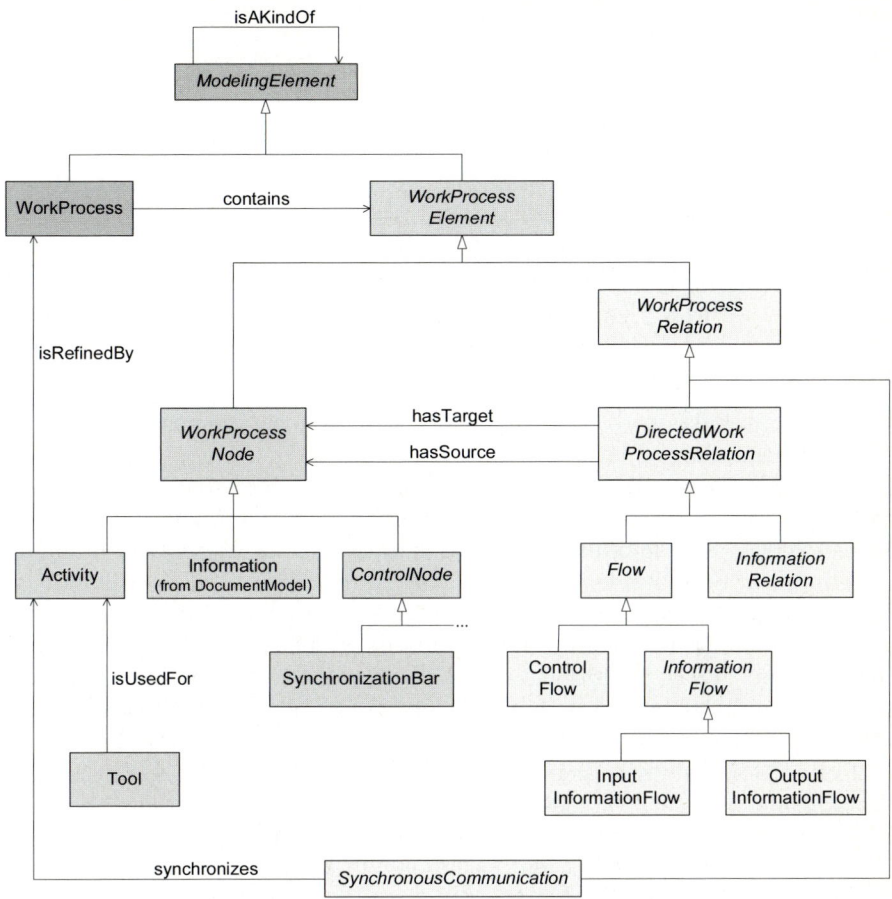

Fig. 2.25. Main classes of *WorkProcess*

distinction is necessary as the semantics of most modeling elements can vary depending on whether they refer to a concrete or generalized work process. For example, the assignment of an Actor to an Activity in a concrete case states that the Activity has been performed by the Actor, whereas a similar assignment in a generalized model recommends or prescribes the involvement of the Actor in the Activity.

Content

As shown in Fig. 2.25, all classes of in the *WorkProcess* module are derived from ModelingElement. That way, the isAKindOf relation is defined for all classes. As demonstrated in a modeling example below, isAKindOf fulfills the function of the instantiation in the partial model *Process Models* of CLiP (see

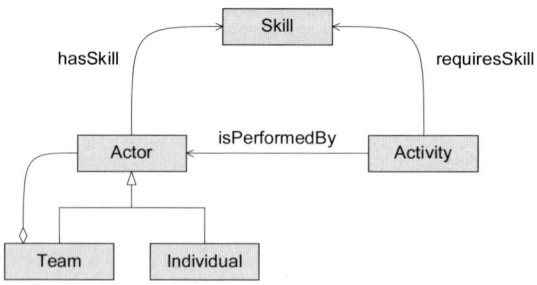

Fig. 2.26. Actor and Skill model of *WorkProcess*

Fig. 2.22): it links work process models and their components on different levels of generality.

A WorkProcess contains WorkProcessNodes like Information items and Activities which are connected by WorkProcessRelations like ControlFlow and InformationFlow. The Information class is imported from the document model introduced in Sect. 2.3.4. An Activity can be further refined by linking it to another WorkProcess. All modeling concepts of C3 find their match in the Process Ontology. In general, C3 elements map to classes and C3 connectors to relations.[15] Thus, the gap between semi-formal and formal representations is closed, and the transition from C3 models to models represented in the Process Ontology becomes simple and straightforward. Once the formal content of a C3 model (such as interrelations between activities) is transformed, the informal content, for instance textual annotations describing the skills required for activities, can be integrated in the formal model.

Fig. 2.26 shows the Actor and Skill model in *WorkProcess*, which has been adopted from *Process Models* [100]. An Actor, corresponding to a *role* in C3, is an abstract entity which can be refined to be either an Individual or a Team (i.e., an aggregation of Actors). Individuals are characterized by the Skills they provide. The set of Skills possessed by a Team is the union of the Skills of its members. In addition, Activities can be linked to the Skills required for their execution. The Actor and Skills model allows the assignment of an Activity to an Actor who has the required skills.

For the reasons discussed in Subsect. 2.4.5, the Process Ontology does not provide classes corresponding to Issue and Requirement within *Process Models*. Design decisions are now covered by a separate decision model (cf. Sect. 2.5).

The InformationRelation is introduced as an abstract base class for any relation between Information items. Its usage is exemplified in the section on decision modeling.

---

[15] Technically, C3 connectors map to classes derived WorkProcessRelation, a generic relation class. Modeling relations as relation classes permits to declare attributes, like it is done for the connectors in C3.

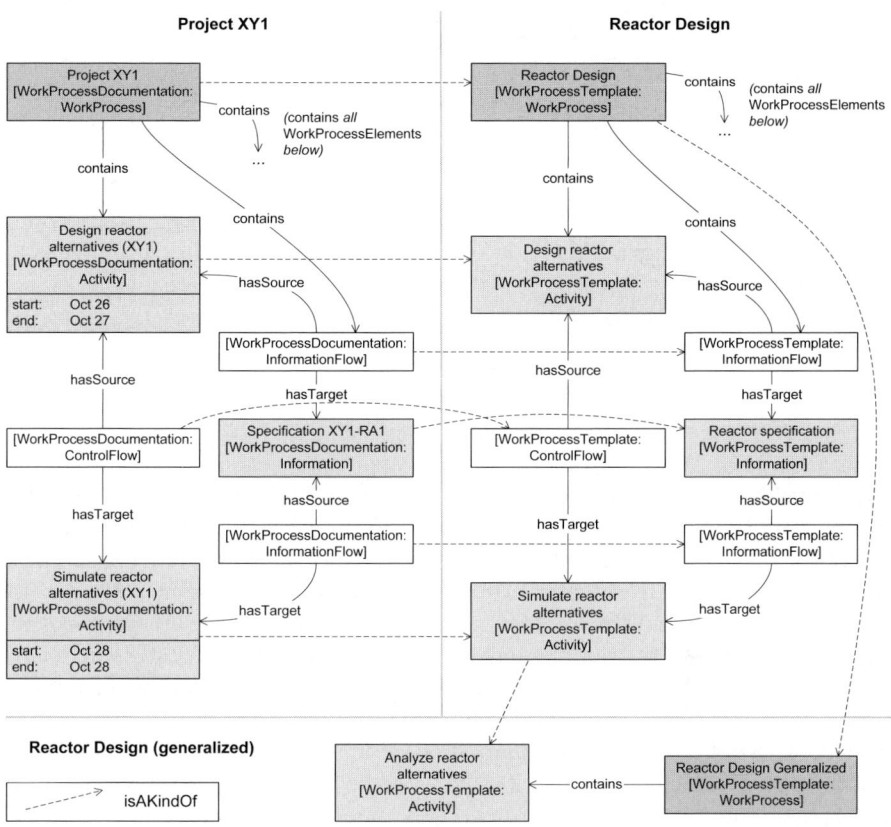

Fig. 2.27. Modeling a concrete case and generalized processes on the instance layer.

As pointed out above, the *WorkProcessDocumentation* and *WorkProcessTemplate* modules adapt the classes in *WorkProcess* for the modeling of concrete cases and generalized work processes, respectively. Further attributes and restrictions are introduced, as required for each of the model types. For instance, an Activity in *WorkProcessDocumentation* can be characterized by a start and end time, and it must not be the target of an isAKindOf relation.

Fig. 2.27 shows some fragments of the models discussed in Subsect. 2.4.5 (see Fig. 2.20), now represented by means of the process ontology. Project XY1, a WorkProcess of the WorkProcessDocumentation module, represents a particular case. It is linked to its generalization Reactor Design, an instance of WorkProcessTemplate:WorkProcess, via isAKindOf. The latter process is a specialization of Reactor Design Generalized. The work processes are characterized by the elements they contain as well as their interrelations. Specialization relations between the elements of different processes are represented by further instances of isAKindOf.

As argued in [95], the definition of specialization relationships between Activities without considering their context can lead to problems, like very flat hierarchies without practical use. In the approach proposed here, such specializations are established while keeping their context within a work process. The isAKindOf relation between Simulate reactor alternatives (within the Reactor Design process) and Analyze reactor alternatives (as part of Reactor Design Generalized) does not mean that a simulation is always an appropriate analysis method. Simulation is rather the analysis method recommended within the exemplary Reactor Design process. In other specializations of the Reactor Design Generalized process, alternative techniques for analysis may be scheduled.

Usage

Currently, no appropriate modeling tool exists that supports the process ontology to its full extent, which has inhibited its intensive application. WOMS, originally developed for C3 modeling, can be used for creating work process models covering activities, information items, and their interrelations. These models can be converted to OWL files, which are then finalized manually by means of the OWL editor Protégé. Also, a simple work around has been realized which allows to include some further aspects in a WOMS model by means of textual annotations in a prescribed format. This additional information is kept during the conversion. So far, the available converters allow to use the process ontology as a domain module for the Process Data Warehouse described in Subsect. 4.1.5.

2.4.7 Conclusion

A procedure for modeling, improving, and implementing design processes in chemical engineering has been elaborated. Two modeling languages – C3 and the Process Ontology – addressing the requirements imposed by the characteristics of creative design processes have been created. Case studies in cooperation with industrial partners (cf. Sect. 5.1) have been performed to ensure practical relevance and usability of the scientific results.

The modeling procedure and the Process Ontology are planned to be further elaborated in a future project described in Sect. 7.3. The project aims at a generalization of the approach to cover different types of work processes in the chemical industries. It is performed in cooperation with four major players of the process industries, thus indicating the increasing importance that is attached to work process modeling by industrial users.

2.5 Decision Models

M. Theißen and W. Marquardt

Abstract. In this contribution, the last of the four application domain models is presented. The model, implemented as an ontology, covers the rationale underlying the decisions in design projects. The ontology supports the representation of concrete decisions for documentation purposes as well as generalized decision templates, which can serve as guidelines for designers and help to reduce the effort for documenting concrete decisions.

2.5.1 Introduction

During design processes, engineers do not only create technical specifications and auxiliary documents such as flowsheets and mathematical models; they also produce *design rationale*, i.e., *'reasoning that goes into the design of the artifact'* [643], including evolving cognition of the requirements the artifact must fulfill, the creation of design alternatives for the artifacts, the disclosure of arguments for and against the alternatives, and finally the decision to choose one alternative.

The benefits of documenting design rationale are manifold. *Explicit representations of design rationale* support a consistent view among the stakeholders involved in a design project, they help to keep track of possible effects when requirements for an artifact change, and they can improve later design projects when similar problems are to be solved.

However, design rationale is rarely captured in industrial projects. Producing any kind of *documentation* is often seen as an unacceptable overhead by designers (e.g., [810]), while the beneficiaries of the documentation are typically found later in the design process or even later in the life cycle of the artifact [706], for example on the occasion of a retrofit of an existing plant. A further obstacle is the reluctance to explicitly document discarded alternatives as they might be conceived as an indication of not working efficiently. Furthermore, unsuitable design decisions can be easily tracked at a later stage with possibly undesirable consequences for the designer.

Some aspects of design rationale, such as the work processes during the creation of a design artifact or the dependencies between different artifact versions, are covered by the domain models presented in the previous sections. This section most notably focuses on *decision rationale*, a term coined by Lee and Lai referring to the evaluations of alternatives, the arguments underlying the evaluations, and the criteria used for the evaluations [808].

In chemical engineering, several methodologies and procedures are applied for decision making during design processes. Representative examples include design heuristics, shortcut methods, decision analysis methods, and mathematical optimization.

In the early stages of conceptual design, simple *design heuristics* permit to exclude inadequate alternatives without detailed knowledge about the process. For instance, [530, 531] propose some rules for the design of separation systems for liquid and gas/vapour mixtures.

Approximations or boundaries of process variables can often be calculated by means of *shortcut methods*. Like heuristics, such methods allow to identify unsuitable alternatives based on a minimum of information. The *rectification body method* (*RBM*), for example, allows to compute the minimum energy demand of distillation processes [13].

Formal *decision analysis methods*, in particular *Multi-Criteria Decision Analysis* (*MCDA*, see [960] for an overview) such as *Utility Analysis* and *Analytic Hierarchy Process* (*AHP*), seek to formally assess the importance of several criteria and the grade to which the criteria are respected by different alternatives, to detect inconsistencies in the assessments, and finally to recommend the best fitting alternative. Several applications in the domain of chemical engineering are reported in the literature, including the design of urban wastewater treatment plants [672] and separation systems for hydrocarbons [621].

Mathematical optimization seeks to minimize or maximize a real function by choosing its arguments (real or integer variables or functions) from within an allowed set [649, 692]. Optimization can be applied at different stages of a design project for finding out optimal solutions for continuous (e.g., reactor size) or discrete (e.g., number of stages in a distillation column) parameters or for structural design decisions (e.g., the configuration of several distillation columns).

Such methods are typically not applied independently of each other, but their outcomes may interact; an example addressing the combination of shortcut calculations with the rectification body method and rigorous optimization is given in [231].

The section is organized as follows. In Subsect. 2.5.2, we give an overview of notations and modeling approaches for design and decision rationale. The requirements to be met by a decision model for chemical engineering design projects are discussed in Subsect. 2.5.3. Building on existing approaches, a *decision ontology* has been created; its structure, content, and usage are described in Subsect. 2.5.4.

2.5.2 Approaches to Design and Decision Rationale

The first and probably best-known notation for design rationale is *IBIS* (*Issue-Based Information Systems*), developed in the early seventies of the past century [798]. IBIS provides a rather restricted set of modeling elements: *Issues* are questions to be answered. Possible answers are represented by *Positions*, which are backed or disapproved by *Arguments*. IBIS was originally developed for structuring and supporting discussions dealing with *wicked* problems in the field of social policy. Among other things, IBIS' author characterizes wicked problems as ill-defined and open to an infinite set of potential solutions [936]. As this also applies to many engineering design problems, IBIS has been used for recording design rationale in several engineering disciplines. A prominent example is the *Knowledge Based Design System (KBDS)* for chemical engineering design [524]. Based on an IBIS representation of design decisions, KBDS offers supportive functions like automatic report generation and backtracking the effects of changes in specifications [525]. In a similar way, the modeling environment *ModKit* makes use of IBIS for the documentation of decisions during the creation of mathematical models [54] (see also Subsect. 5.3.4).

Given its intended usage for capturing ongoing discussions, ease of use and a small set of modeling elements were key design criteria for IBIS. A plethora of further notations and models has been developed which focus on different aspects of decisions and decision making procedures beyond those in IBIS. *PHI (Procedural Hierarchy of Issues*, [835]), for instance, introduces sub-issues as a means for structuring argumentations. The *Potts and Bruns* model [913] emphasizes the *derivation* of detailed artifacts from more general predecessors. Exhaustive presentations of design rationale approaches and systems are given in [644, 857, 927]. In the following, we shortly describe two notations which are of importance for our work on decision modeling.

The *QOC* notation was proposed as an auxiliary for *Design Space Analysis*, an analysis style that places an artifact in a space of alternatives and explains why the particular artifact was chosen [826]. QOC stands for *Questions, Options, and Criteria*, thus indicating the main elements of the notation. In contrast to the more general *Issues* in IBIS, *Questions* are a means to structure the design *Options*. The most notable difference to IBIS is the explicit modeling of the *Criteria* which are the basis for *Assessments* of *Options*. *Reusable Rationale Blocks* (*RRB*, [733]) are a generalization of QOC: Whereas a QOC model embodies the rationale for a concrete decision, an RBB describes a generic design problem in conjunction with its available solution options and their effects. When applying an RBB to a concrete problem, the designer must evaluate the options in the context of the problem on hand.

The *Decision Representation Language* (*DRL*, [808, 809]) is a notation for decision rationale. Its top-level element is the *Decision Problem*, equivalent to the *Question* in QOC. *Alternatives* and *Goals* in DRL correspond to *Options* and *Criteria*, respectively. The outstanding characteristic of DRL is that *Claims*, i.e., statements which may be judged or evaluated, can be linked in a way that enables to represent complex argumentations in a straightforward manner.

2.5.3 Requirements for a Decision Ontology for Chemical Engineering Design

In this subsection, we discuss the requirements to be fulfilled by a decision rationale model adequate for chemical engineering design. This model is referred to as *Decision Ontology* in the remainder of the section. The requirements are as follows:

- Obviously, concepts corresponding to the elements of IBIS (*Issue, Alternative, Evaluation*) are indispensable. In addition, we regard requirements and constraints (denoted as *Criteria* in QOC and *Goals* in DRL) as fundamental elements of a Decision Ontology. To a large extent, design decisions depend on the relevant goals. Vice-versa, decisions are only comprehensible if the goals are represented explicitly. Furthermore, goals might evolve during the course of a project. For instance, the desired purity of a chemical product may change during a design process due to changes in the market. Also, the requirements for mathematical models (e.g., precision, process variables considered, implementation tool) typically evolve during a design project.
- In order to keep the Decision Ontology as simple and general as possible, the detailed representation of these basic elements (in particular *Alternatives* and *Goals*) should rely on the product data and document models presented in Sects. 2.2 and 2.3.
- Decisions are the result of decision making processes. The temporal dimension of decisions should be reflected by a tight integration of the Decision Ontology with the Process Ontology (Subsect. 2.4.6).
- Like for work process models, any successful application of the Decision Ontology depends on the ability of its intended users (i.e., chemists, chemical engineers, technicians) to use it without considerable learning effort. Thus, at least the top-level concepts of the ontology should be kept as simple as possible. For these top-level concepts, an intuitive graphical representation should be provided (similar to the main concepts of the Process Ontology, which can be represented in a C3 model, cf. Subsect. 2.4.4). More complex features of the ontology, which are nevertheless required for advanced applications, should be optional extensions.
- Similar to the ontology for work processes, the Decision Ontology should allow both the *documentation of concrete design decisions* in a project as well as the representation of *generalized decision templates*, which can serve as guidelines for a design team confronted with a certain type of decision problem. A further advantage is that parts of decision templates can be incorporated in models of concrete cases. Thus, the overhead of modeling and documenting decisions during design projects can be considerably reduced [434].
- Decisions concerning the different types of artifacts created during a design project should be represented in a consistent – and thus easy to apply –

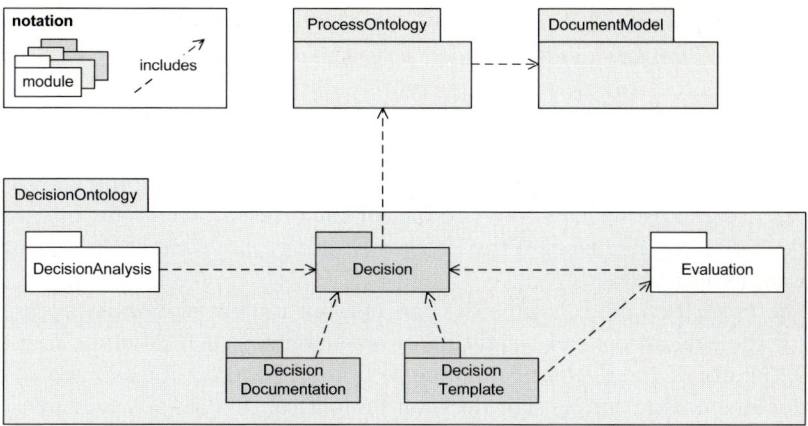

Fig. 2.28. Structure of the Decision Ontology

manner. These artifacts comprise the main products of a design project, i.e., specifications of chemical plants or their parts, but also auxiliary products like mathematical models.
- The Decision Ontology should *represent the different methodologies and procedures* for evaluating and discriminating alternatives (e.g., heuristics, decision analysis methods) in a consistent way.

2.5.4 Decision Ontology

The necessity to incorporate goals makes notations like QOC or DRL a good starting point for a Decision Ontology. We have opted for the more expressive, but also more complex DRL. However, QOC can be regarded as a subset of DRL; there is no necessity for a decision modeler to apply the entirety of DRL concepts when a simpler QOC-like representation is sufficient.

DRL is adequate for modeling and documenting concrete decisions, whereas its applicability for decision templates is limited. By an approach similar to Reusable Rationale Blocks for QOC, we have extended DRL to cover also decision templates.

For similar reasons as for the Process Ontology (Subsect. 2.4.6), OWL (*Web Ontology Language*, [546]) has been chosen as implementation language. That way, the integration of the Decision Ontology with the other application domain models is simplified, and description logics reasoners such as RacerPro [918] can be used for consistency checks.

Structure

The structure of the Decision Ontology shown in Fig. 2.28 is analogous to that of the Process Ontology. Common concepts are defined in a top-level module (*Decisions*); this module is imported by the *DecisionDocumentation* and *DecisionTemplate* modules which serve the representation of the decisions made in concrete projects and the representation of generalized guidelines, respectively.

Similar to artifact specifications in form of documents, also the decision-related information created during a design process (e.g., arguments for and against design alternatives) is a product of the process. This tight integration is reflected by the inclusion of the Process Ontology in *Decisions*; it is discussed in more detail below.

The *Evaluation* module provides concepts for a more elaborate representation of the evaluation and weighting of arguments. When documenting decisions, the use of the *Evaluation* module is optional and can be avoided to keep the documentation simple. For decision templates, it is usually indispensable.

The *DecisionAnalysis* module demonstrates the extension of the Decision Ontology with concepts required for representing a group of decision making methodologies.

Content

Main Modules

The classes of the *Decision* module are shown in Fig. 2.29. All classes are derived from the abstract DecisionObject. We first discuss the six classes located in the left part of the figure; instances of these classes form the nodes of a decision model. Gray shapes inside the class boxes depict the graphical representation of their instances in the modeling examples below.[16]

A DecisionProblem is a design problem requiring a Decision. This can include design decisions related to the final artifact (such as the specification of a chemical plant) like the choice of the mode of operation of a chemical plant or the selection of an appropriate reactor type, but also decisions referring to auxiliary artifacts like mathematical models. By linking a DecisionProblem to another DecisionProblem by means of the IsASubdecisionOf relation, it can be stated that solving the latter problem requires solving the former. The semantics of IsASubdecisionOf does not require the sub-decisions to be independent of each other nor does it require the entirety of sub-decisions to describe the complete DecisionProblem.

Alternatives are options meant to solve a decision problem. For instance, batch and continuous mode as well as their combinations would be Alternatives for the DecisionProblem to choose a mode of operation. Alternatives can be linked to DecisionProblems by two different relations. The first one, isAnAlternativeFor, does not express any evaluation or even preference of the Alternatives. It is an auxiliary to capture any Alternatives which possibly solve the DecisionProblem, before a suitable Alternative for the problem is eventually evaluated by means of the second relation, IsAGoodAlternativeFor.

---

[16] The representation of instances complies largely with the notation proposed by the authors of DRL [808].

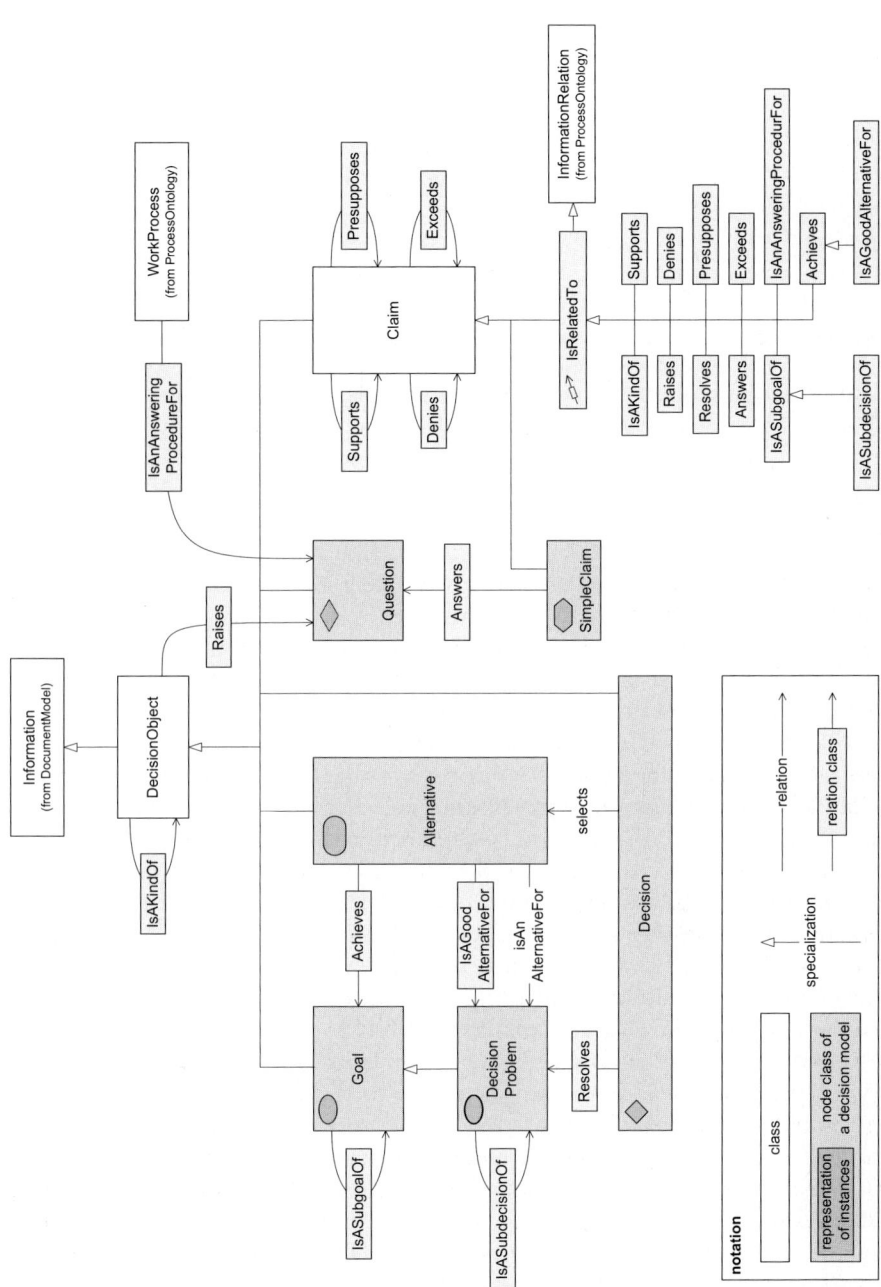

Fig. 2.29. Module *Decision* of the Decision Ontology

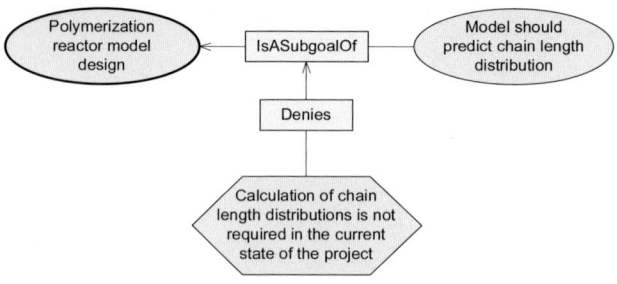

Fig. 2.30. A SimpleClaim which Denies another Claim

Goals describe desired properties, requirements, and constraints to be fulfilled by a suitable Alternative for a DecisionProblem. Annual capacity and product purity are examples for Goals of a plant design DecisionProblem. The ability to predict chain length distributions can be a Goal for the DecisionProblem to select a mathematical model of a polymerization reactor. Goals can be assigned to another Goal by IsASubgoalOf; this relation states that reaching the sub-goal helps reaching the superordinate goal. Like for IsASubdecisionOf, sub-goals are not necessarily independent, and a set of sub-goals does not need to describe the entire super-goal. Alternatives are evaluated with respect to Goals by means of the Achieves relation.

It should be noted that DecisionProblem is a subclass of Goal: A DecisionProblem is a special type of Goal for which an Alternative is sought. The evaluations of an Alternative with respect to subgoals – represented by Achieves – influence the evaluation of the Alternative with respect to the superordinate DecisionProblem, represented by IsAGoodAlternativeFor.

Questions are issues to be considered in the context of a decision problem. For instance, the DecisionProblem to choose a mode of operation Raises the Questions whether solids or strongly exothermic chemical reactions occur; the answers of Questions can influence the evaluation of Alternatives. Similarly, the selection of a suitable reactor model may depend on the Question whether trustable parameter data are available. WorkProcesses describing procedures for answering Questions can be assigned by IsAnAnsweringProcedureFor.

Finally, a Decision represents the selection of one Alternative that is meant to resolve a DecisionProblem.

Any statement in a decision model which may be subject to uncertainty or to disaccord, or, in general, may be evaluated, is a Claim. Claims are either SimpleClaims or relation classes derived from IsRelatedTo. Most of the relations introduced above are subclasses of IsRelatedTo, and thus they are actually Claims. Statements which cannot be represented as a relation between two DecisionObjects are modeled as SimpleClaims, typically qualified by a textual annotation.

Modeling relations as **Claims** offers the possibility to argue for or against them. For example (see Fig. 2.30), the adequacy of an **IsASubgoalOf** relation between the **Goal** *Model should predict chain length distribution* and the **DecisionProblem** *Polymerization reactor model design* can be doubted, for instance because the *Calculation of chain length distributions is not required in the current state of the project*. Such argumentations are modeled by means of four additional subclasses of **IsRelatedTo**:

- A **Claim** can **Support** the validity of another **Claim** (*argument*).
- A **Claim** can **Deny** the validity of another **Claim** (*counter argument*).
- A **Claim Presupposes** another **Claim** if the validity of the other **Claim** is a precondition for validity of the **Claim** itself.
- A **Claim Exceeds** another **Claim** if the first one is more valid than the second one.

All classes described so far – including the relation classes derived from **Claim** – are subclasses of **DecisionObject**, which is derived from the **Information** class of the *DocumentModel*. Thus, the statements represented in a decision model – such as considering a **Goal** as a constraint to be respected or finding out that an **Alternative IsAGoodAlternative** for a **DecisionProblem** – are seen as potential products of the **Activities** in a work process.

Integrating the work process and the decision model allows to remedy the difficulties in defining subclasses of **Activity** as they were encountered for the partial model *Process Models* of CLiP (called sub-activites there; see Sect. 2.4.6). The relation class **OutputInformationFlow** defined in the Process Ontology does not impose any restrictions on the type of **Information** created in an **Activity** (see Fig. 2.31), whereas a **DecisionActivity** is required to produce at least one **DecisionObject**. Subclasses of **DecisionActivity** are characterized by

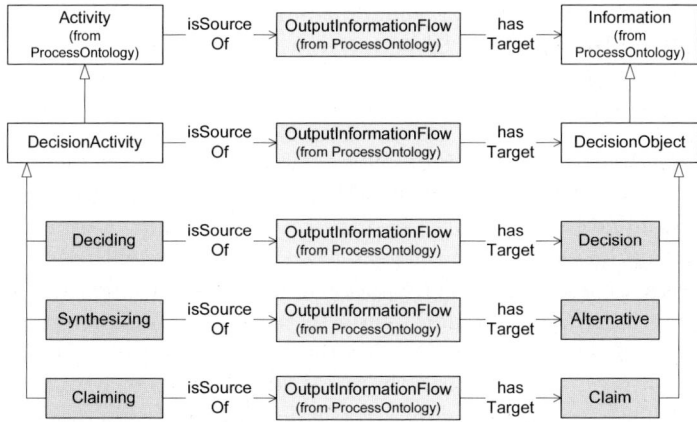

Fig. 2.31. Some subclasses of **Activity** defined in the Decision Ontology

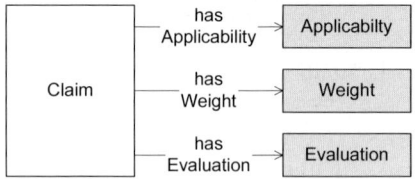

Fig. 2.32. Module *Evaluation* of the Decision Ontology

the subclasses of DecisionObject created during their execution. For instance, Claiming is a DecisionActivity whose output information comprises at least one Claim, and Deciding must produce at least one Decision.

Evaluation Module

The optional *Evaluation* module defines additional concepts for more subtle evaluations of Claims (see Fig. 2.32).

The Evaluation is an overall measure for the applicability and importance of a Claim. The Evaluation class is not further specified in the *Evaluation* module. Instead, tailored measurement units can be defined depending on the nature of the decision problem in consideration. For instance, in the early stages of a design process, a coarse-grained evaluation of Claims is sufficient because not much knowledge about relevant aspects of a design problem is available and heuristics as well as screening and short-cut methods are applied.

Often, in particular in case of decision templates, the evaluation of a Claim cannot be given because the applicability of the Claim is unknown. Thus, Applicability and Weight are introduced. The Weight reflects the Evaluation of the Claim provided that the Applicability is high. The usage of Evaluation, Applicability, and Weight is demonstrated in the modeling examples below.

DecisionAnalysis Module

The *DecisionAnalysis* module exemplifies the extension of the Decision Ontology with concepts relevant for a certain group of decision making methods.

A decision model represented by means of the Decision Ontology contains much information that is also relevant when an MCDA method is used (e.g., the hierarchy of the Goals). The application of such methods would be simplified if this information could directly be used in an MCDA tool. However, MCDA methods impose restrictions on the relations between the DecisionObjects. For instance, AHP requires the sub-goals of a goal to be independent of each other. These restrictions are formally defined within the *DecisionAnalysis* module.

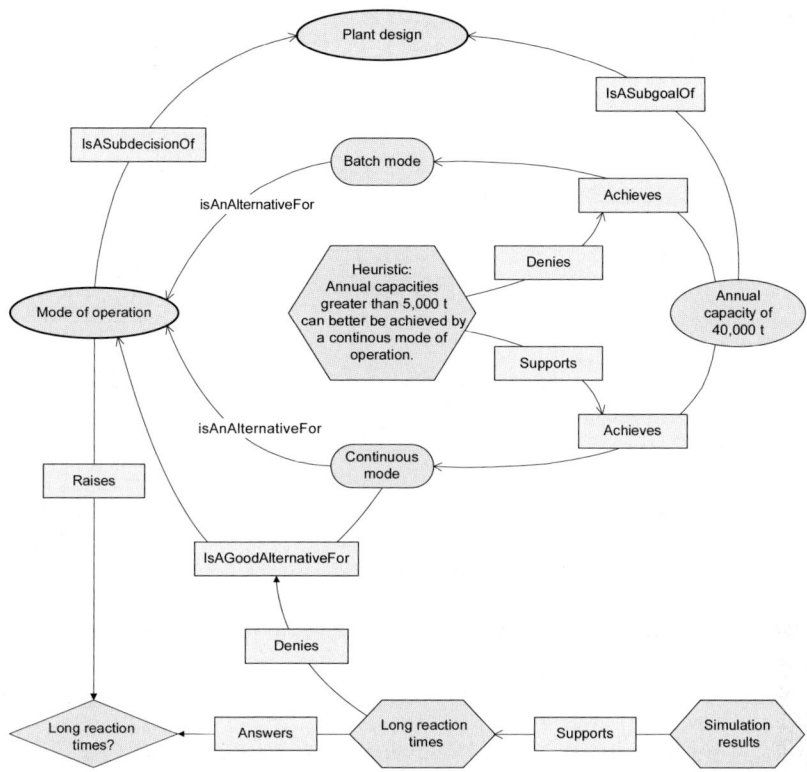

Fig. 2.33. Documenting parts of the argumentation for selecting a *Mode of operation*

Modeling Examples

Decision Documentation

Figure 2.33 gives a simple example for using the *DecisionDocumentation* module. The model describes part of the argumentation for selecting the *Mode of operation* of a chemical plant. This problem IsASubdecisionOf the overall *Plant design* problem. To reach an *annual capacity of 40,000 t* IsASubgoalOf the overall problem. Two Alternatives (*Batch mode* and *Continuous mode*) are mentioned for the *Mode of operation*. A heuristic, modeled as a SimpleClaim, states that for annual capacities greater than 5,000 t, a continuous mode is preferable in general. Thus, the heuristic Supports that the *Continuous mode* Achieves the capacity *Goal*, and simultaneously it Denies that the *Batch mode* Achieves the capacity. However, the choice of the *Mode of operation* requires more issues to be considered, for instance the Question whether *Long reaction times?* occur. *Simulation results* Support that there are in fact *Long reac-*

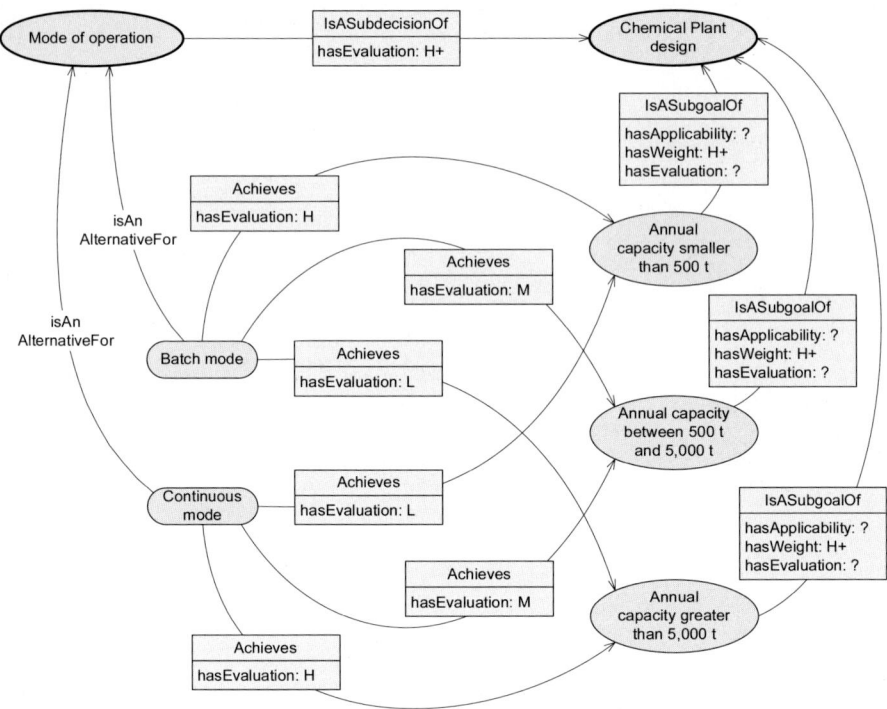

Fig. 2.34. Part 1 of a decision template for selecting a *Mode of operation*

tion times, which **Denies** that the *Continuous mode* **IsAGoodAlternative** for the *Mode of operation*.

Decision Template

An example for a decision template, generalizing the above documentation model, is shown in Figs. 2.34 and 2.35. The template has been split in two parts to keep it more readable: The first part exemplifies the generalization of the evaluation of **Alternatives** with respect to **Goals**, whereas the second part shows the generalization of **Questions** and their answers, which can affect decisions. In the example, we use a simple measure for **Evaluations** ranging from very high ($H+$) via high (H), medium (M), and low (L) to very low ($L-$).

Three sub-goals are given for the overall *Chemical plant design* Decision-Problem (Fig. 2.34), each representing a possible range for the annual capacity of the plant. The applicability of the three instances of **IsASubgoalOf** is unknown. When the template is applied in a concrete plant design problem, a high applicability would have to be chosen for one **IsASubgoalOf** instance – depending on the capacity actually required for the concrete plant. The applicability of the remaining **IsASubgoalOf** instances would be low. The weight of each IsASubgoalOf is very high. Thus, the evaluation of an IsASubgoalOf should be very high if it is considered as applicable, reflecting the importance of the annual capacity when designing a plant. The *Mode of operation* is assigned to the plant design problem as a sub-goal. The IsASubgoalOf relation has a very high evaluation; it is considered for granted that choosing a *Mode of operation* is inevitable when designing a plant. The template proposes two Alternatives for the mode of operation[17]. Each of the Alternatives is linked to each of the Goals representing different annual capacities; these Achieves relations are evaluated according to the same heuristic rule already used in Fig. 2.33.

The second part of the template (Fig. 2.35) lists some of the Questions arising when deciding on the *Mode of operation* of a plant. Furthermore, some SimpleClaims are shown, representing different answers to the Questions. For instance, the SimpleClaim *Strongly exothermic reactions* is a possible answer of *Strongly exothermic reactions?* The applicability of the Answers relation is unknown and can only be decided in the context of a particular decision problem. *Strongly exothermic reactions* Deny that *Batch mode* IsAGoodAlternativeFor the *Mode of operation*. Also for the Denies relation, the applicability and thus the evaluation are unknown. However, the weight is considered as high, which means that *Strongly exothermic reactions* would be an important argument against *Batch mode*.

Usage

As argued in Subsect. 2.4.5, the Process Ontology – like any modeling language for work processes – is not adequate for representing complex decision rationale, in particular decision templates. The concepts introduced in the Decision Ontology, complementary to those of the Process Ontology, permit to apply the modeling procedure described in Subsect. 2.4.2 for the improvement of those parts of design processes, in which complex decisions are made [434]. When performing design processes, engineers gain tacit knowledge about both their work processes and the rationale underlying their design decisions. These two aspects are recorded in an explicit *decision process model (DPM)*, an integrated model of both the design process and the decision rationale, which is meant to support the original design process, subsequent phases in the lifecycle of the same artifact, and other similar design processes in various ways.

Supporting the Original Design Process

The DPM helps to improve the communication between different stakeholders on the status of the design project. Team members add criteria, alternatives, and evaluations of the alternatives with respect to the criteria to the DPM. For

---

[17] It should be noted that this does not prevent a user of the template from adding other alternatives for a concrete case, such as combinations of batch and continuous mode.

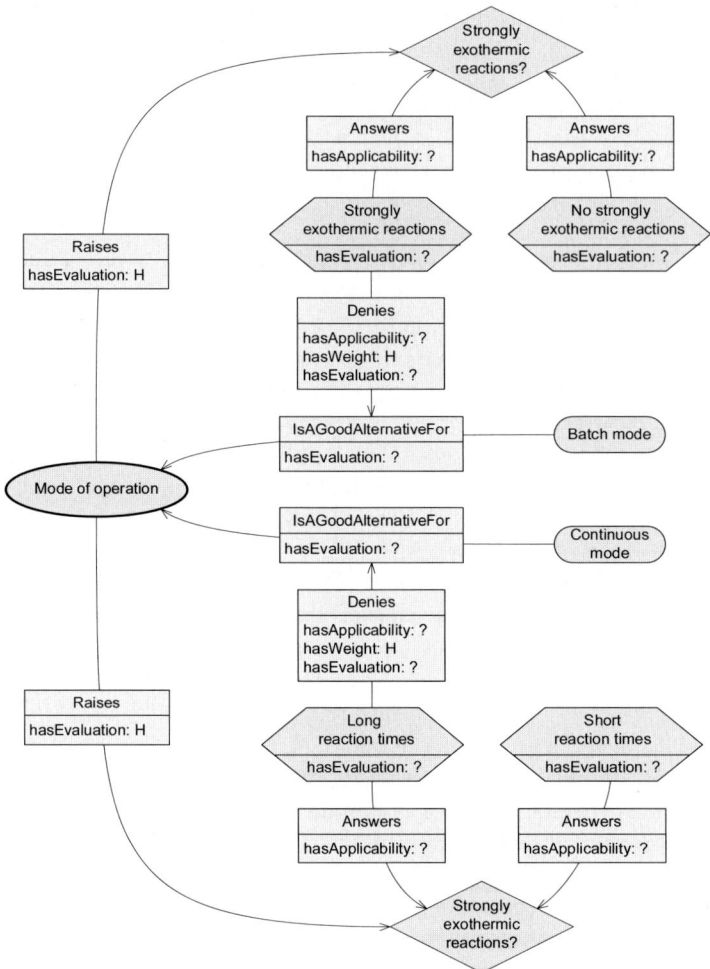

Fig. 2.35. Part 2 of the decision template for selecting a *Mode of operation*

this purpose, easy access to the DPM must be provided to all team members. In order to simplify the use of methods like decision analysis and mathematical optimization, the information contained in the DPM must be easily reusable in suitable software tools.

Supporting Subsequent Phases in the Lifecycle of the Same Artifact

The DPM constitutes a valuable asset for later phases in the life cycle of the artifact. For instance, assume a damaged reactor in a plant which needs to be replaced. Due to technical progress and changes in the market since the design and construction of the plant, there may be better alternatives than the installation of a reactor of the same type. However, normally only a technical specification of the old reactor is available at best, but there is no documentation of the requirements to be met by the reactor, which would be necessary for the selection of a better reactor.

Supporting Further Design Processes

Knowledge from previous design projects can be reused in similar design tasks. Concerning the rationale aspect, a DPM contains information about constraints to be respected and questions to be posed which otherwise might be neglected or forgotten. A crucial issue for this kind of support is the provision of adequate retrieval mechanisms for relevant DPMs.

However, each DPM is specific to a particular project, and the information relevant for a new project may be scattered among several models. Therefore, knowledge from completed projects which is considered to be important for other design projects can be collected, generalized, and finally represented by work process and decision templates which provide for simpler retrieval than a set of concrete DPMs. As the relevant parts of a decision template can directly be incorporated into the new DPM, the effort for documenting decision rationale in a new project is considerably reduced.

Even better support for both decision making and decision documentation can be provided if parts of the design process are automated. As the implementation of software tools for such support requires the involvement of experts from other domains than chemical engineering, empirically proven expert knowledge about chemical engineering design must be made available to the developers of the tools. In addition to the process models mentioned in the original modeling procedure, this knowledge transfer is simplified by decision templates.

2.5.5 Conclusion

We have presented a Decision Ontology which enables the representation of both concrete design decisions and generalized decision templates. As for the *documentation* of design processes and decisions, our approach enables the integrated modeling of the process and rationale perspective (as well as product data and documents, see Subsect. 2.6.2). As for *templates*, our work has addressed the separate modeling of process templates (cf. Subsect. 2.4.6) and decision templates. Future work will have to cover the *integration of process and decision templates*.

The *representation of decision models on the instance level* is still an open issue. The simple modeling examples in Figs. 2.33 – 2.35 suggest that increasing complexity hinders the readability of larger and more realistic models, as indicated by a set of decision models addressing the reactor choice in the IMPROVE reference scenario (cf. Subsect. 1.2.2) [228]. Decomposition of decision models into clearer parts is not straightforward due to the complex network of relations that can exist between its nodes. Also, alternative representations may be adequate. Parts of DRL have been implemented by its authors in SYBIL [807], a system using two-dimensional matrices in addition to the graphical representations. However, no experiences concerning the usability of this approach are reported.

Like for the Process Ontology, no modeling tool is available that fully supports the Decision Ontology (apart from general-purpose tools like the OWL editor Protégé [979]). However, as all **DecisionElements** are derived from **Information**, the Workflow Modeling System WOMS (cf. Subsect. 5.1.3) can be used as a modeling tool for DPMs to a certain extent. The incorporation of decision templates in models of concrete decisions is still restricted to simple copy and paste. The converter already mentioned for the Process Ontology also supports the Decision Ontology; hence, the Decision Ontology can be used as a domain module for the Process Data Warehouse (Subsect. 4.1.5), which provides the necessary access and retrieval functions.

The *pivotal obstacle* to decision rationale *documentation* in industrial practice is the overhead for its production. We claim that this effort can be reduced significantly by means of decision templates. The transfer project described in Sect. 7.3 aims at the further elaboration of our modeling approach for work processes and decision rationale, the implementation of a suitable modeling tool (based on the experiences gained with WOMS), and the validation of the approach in several case studies in cooperation with industrial partners.

2.6 Integrated Application Domain Models for Chemical Engineering

J. Morbach, M. Theißen, and W. Marquardt

Abstract. A comprehensive summary of the application domain models presented in this chapter is given, and their integration into a common framework is discussed. Other existing application domain models of comparable scope are reviewed and compared to the models presented herein.

2.6.1 Introduction

A major research objective of the IMPROVE project is the development of an integrated process/product model, which enables a comprehensive and formal description of the products and processes of the application domain. As previously argued in Subsect. 1.1.2 and later elaborated in Sect. 6.1, the application domain models presented in this chapter jointly constitute the upper layer of such a process/product model (cf. Figs. 1.6 and 6.1). This, of course, requires the formal integration of the different application domain models into a common framework.

While some aspects of model integration have already been discussed earlier in this chapter, a coherent summary of the dependencies and interrelations between these different models is still missing and shall be given in this section. First, the integration of the individual application domain models is discussed in Subsect. 2.6.2. Next, we briefly review the related work on application domain modeling and contrast it with our modeling efforts (Subsect. 2.6.3). The section concludes with a detailed comparison of our integrated model against the benchmark of the ISO 15926 information model (Subsect. 2.6.4).

2.6.2 A Comprehensive Model for the Chemical Engineering Domain

Four different types of models have been presented in the preceding sections, each describing a particular aspect of the application domain: product data models (Sect. 2.2), document models (Sect. 2.3), work process models (Sect. 2.4), and decision models (Sect. 2.5). These models were not developed independently of each other, but were designed in such way that they may be easily combined into a comprehensive model of the application domain. Hereafter, we denote this comprehensive model as **C^2EDM** (abbr. for Comprehensive Chemical Engineering Domain Model); the individual application domain models that constitute the C^2EDM are referred to as *submodels* of the C^2EDM.

As a prerequisite for a formal integration, all four submodels must be represented in a common modeling language. The Web Ontology Language OWL

[546] has been chosen for this purpose. Consequently, only the most recent, OWL-based versions of the individual submodels are considered here: that is, OntoCAPE 2.0 (cf. Subsect. 2.2.4) as a submodel for product data, the latest version of the Document Model (cf. Subsect. 2.3.4), the Process Ontology (cf. Subsect. 2.4.6) for the representation of work processes, and the Decision Ontology (cf. Subsect. 2.5.4) for the modeling of design decisions. The integration of some of the earlier versions of these submodels has been presented elsewhere: [19] describes the integration of product data and documents within the conceptual information model CLiP (cf. subsection 2.2.3); in [96] and [299], the interrelations between product data, documents, and design activities in CLiP are discussed.

The remainder of Subsect. 2.6.2 is organized as follows: firstly, the general dependencies between the individual submodels are described; next, the interrelations between the submodels' top-level concepts are clarified; and finally, the function of the superordinate Meta Model is discussed.

Submodel Dependencies

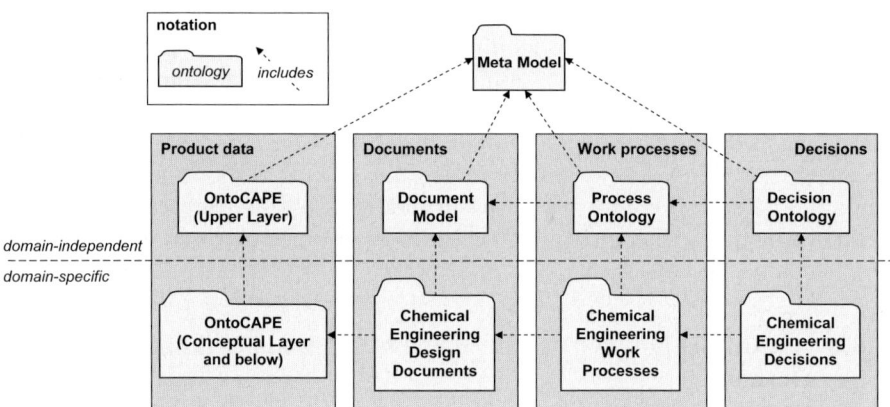

Fig. 2.36. General dependencies between the individual submodels

Figure 2.36 displays the structure of the four submodels of the C^2EDM as well as their interdependencies on a coarse-grained level. As indicated in the figure, each submodel can be split into a domain-independent and a domain-specific part. Since the individual parts are represented as OWL ontologies, inclusion relations[18] can be established between them by means of the OWL

---

[18] Inclusion means that if ontology A includes ontology B, the ontological definitions provided by B are included in A. Inclusion is transitive – that is, if module B includes another module C, the definitions specified in C are valid in A, as well.

import mechanism. As can be seen in Fig. 2.36, the Meta Model is situated on top of the inclusion lattice; it constitutes a common design framework for the submodels, as will be explained in Subsect. 2.6.2.

Furthermore, the following conclusion can be drawn from the figure:

- OntoCAPE is independent of the other submodels.
- Similarly, the core of the Document Model is self-contained. However, it relies on a product data model to describe the contents of documents (cf. Subsect. 2.3.4). Thus, the domain-specific extension of the Document Model imports OntoCAPE as a means to characterize the documents created or used in an engineering design project.
- The Process Ontology requires concepts from the Document Model to describe the handling of (domain-independent) documents as input and output information of activities. Consequently, its extension specific to chemical engineering (*Chemical Engineering Work Processes*) includes the domain-specific extension of the Document Model. That way, the document types specific to engineering design processes can be referenced in a work process model.
- As argued in Sect. 2.5, the elements of a decision model (e.g., alternatives, goals ...) are conceived as information which is created in a work process; thus, the Decision Ontology imports the Process Ontology and (implicitly) the Document Model. By importing *Chemical Engineering Work Processes*, the *Chemical Engineering Decisions* model extends the Decision Ontology with concepts required for the enrichment of the generic model elements. For instance, an alternative can be linked to the chemical engineering document which contains a detailed specification of the alternative.

Interrelations between Top-Level Concepts

To illustrate how the four submodels are integrated on the class level, Fig. 2.37 shows the interdependencies between the essential top-level concepts of the C^2EDM[19]; the affiliation of a concept to a particular submodel is indicated by the grey-shaded boxes in the background of the figure.

From the Process Ontology (module *WorkProcessDocumentation*), the following concepts are shown: The class **Activity** represents a single step of a work process. **Activities** are performed by **Actors**, often with the help of some (software) **Tool**. A connection to the Document Model is established via the **Information** class. **Information** can be both input or output of an **Activity**.

Within the Document Model, **Information** subsumes the classes **Product** and **VersionSet**; the former denotes all kind of information objects, the different versions of which can be logically bundled (and thus managed) by the latter. Since **Products** are typically not handled in an isolated manner, they can be assembled into documents, which act as carriers of **Products** during the

---

[19] Note that some of the concepts described here also form part of the Core Ontology of the Process Data Warehouse, which is presented in Subsect. 4.1.5.

work process. The class DocumentVersion denotes the individual versions of a document that arise in the course of a project, which can again be bundled through a (logical) Document.

An Information can be further characterized through a ContentDescription. For the domain of chemical engineering, classes (or composite expressions) from the OntoCAPE ontology can be used as such.

In the Decision Ontology, special types of Information are defined, such as the DecisionProblem, the Alternative, the evaluation of an Alternative with respect to a DecisionProblem (modeled by means of the IsAGoodAlternativeFor relation class), and the Decision to select a certain Alternative. Also, special Activities are introduced, which are qualified by the type of Information produced during their execution. For instance, during an Evaluating activity, it may be found out that an Alternative is a good alternative for a DecisionProblem. The information created during Deciding is the Decision to select an Alternative.

Meta Model

The *Meta Model* is defined on top of the C^2EDM. Like the submodels of the C^2EDM, it is implemented in OWL to enable a formal model integration.

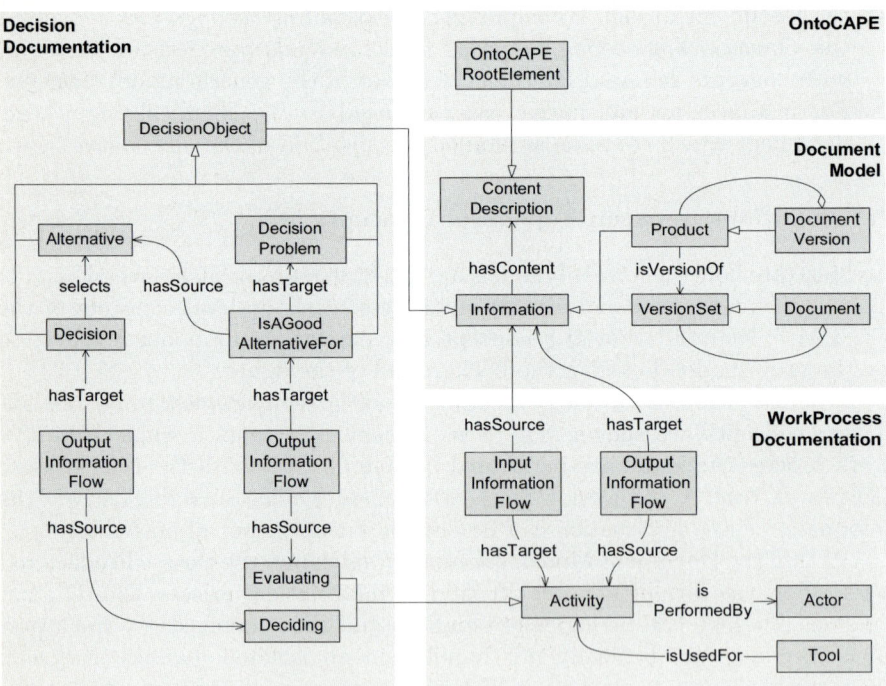

Fig. 2.37. Interdependencies between the essential top-level concepts of C^2EDM

The Meta Model is imported by all four submodels, as indicated in Fig. 2.36. That way, the ontological assertions of the Meta Model are included in the submodels.

Complementary to the mechanisms discussed above, the Meta Model is not concerned with the integration of the submodels. Rather, it serves as a general design framework, establishing common standards for the construction and organization of the four application domain models, thus ensuring a consistent modeling style. This is achieved (1) by introducing *fundamental concepts* from which all the classes and relations of the individual submodels are derived (either directly or indirectly), and (2) by providing *design patterns* that define best-practice solutions to common design problems.

- The *fundamental concepts* are the basic classes and relations of the Meta Model. They form the topmost layer of the concept hierarchy; all other classes and relations – those in the Meta Model as well as those in the submodels – are specializations (i.e., subclasses or subrelations) of these fundamental concepts[20]. RelationClass is a good example of such a fundamental concept: The class is a means for representing n-ary relations in OWL (by default, the OWL language provides language primitives for binary relations only). Specializations of RelationClass are utilized in all four submodels of the C^2EDM.

  By linking a submodel concept to a fundamental concept, its role within the model is characterized. That way, a user or a software program is advised how to properly treat that particular concept. To give an example: submodel classes that are derived from RelationClass are obviously auxiliary constructs for the representation of n-ary relations. Consequently, instances of such classes do not need to be given meaningful names (cf. [869]). Thus, if a subclass of RelationClass is instantiated, a user or an intelligent software program can conclude that the instance can be labeled automatically, according to some standard naming convention.

- A *design pattern* is a best-practice solution to a commonly occurring modeling problem. It is formed by a set of interconnected classes and relations, which jointly define a sort of template that can be applied within a submodel of the C^2EDM. That way, patterns encourage a consistent, uniform modeling style across the individual submodels. An example for the application of a pattern is the representation of mereological relations (part-whole relations), which appear numerous times in the submodels: In the Document Model, mereological relations are used to assemble Documents and DocumentVersions from VersionSets and Products, respectively; in the Process Ontology they are utilized to compose WorkProcesses from WorkProcessElements; in OntoCAPE, mereological relations are, amongst

---

[20] Conceptually, the application domain concepts should be derived from the meta model concepts via instantiation. However, as current OWL reasoners do not support this kind of meta modeling, the application domain concepts are currently linked to the meta model concepts by specialization.

others, applied to decompose a **System** into **Subsystems**. As a common basis for these different cases, a design pattern defines a standard way of modeling this relation type, which is then adopted by all submodels.

While the Meta Model is highly useful during model design, it is less relevant for practical applications. Under certain conditions, the Meta Model may even prove harmful, as its highly abstract concepts may confuse an inexperienced user rather than support him/her. For that reason, the interconnectivity between the Meta Model and the C^2EDM needs to be minimized such that they can be separated easily, once the design has been completed. To this end, the classes and relations defined in Meta Model are not directly used within the individual submodels of the C^2EDM. Rather, they are redefined and subsequently linked to the original concepts in the Meta Model. Thus, only the links to the Meta Model need to be disconnected if a stand-alone usage of the C^2EDM is desired.

2.6.3 A Review of Related Work

In the area of chemical engineering, various application domain models have been developed over the last decades. Some of these are conceptual models, intended to clarify the interrelations between domain entities and to obtain a better understanding of the domain; others are implementation models that are targeted at specific applications, such as information storage or data exchange. In the following, the most outstanding of these models will be reviewed. Only application domain models of a scope comparable to our C^2EDM are considered here; that is, models which enable a comprehensive description of product data, documents, work process, and decision-making procedures. Models focusing on single aspects like work processes (Sect. 2.4) or decision rationale (Sect. 2.5) are discussed in the respective sections. Also, since an extensive review of application domain models in chemical engineering has already been published elsewhere [18], the present review focuses solely on the more recent developments: It discusses the later changes of the models covered in [18] and reports on newly evolved models.

ISO 10303

The international standard ISO 10303, widely known as *STEP* (STandard for the Exchange of Product data) proposes a series of data models for the exchange of engineering data between computer systems. The STEP model framework has a modular architecture: Basic modules introduce generic classes, methods, and building blocks; from these, domain-specific models, the so-called Application Protocols (AP), can be constructed. For the domain of process engineering, the following APs are of major interest[21]:

---

[21] There are further APs that cover more peripheral aspects of chemical engineering, such as the AP 212 *Electrotechnical Design and Installation* or the AP 230 *Building Structural Frame: Steelwork*.

- AP 221 *Functional Data and their Schematic Representation for Process Plant* [754] is being developed for the file-based exchange of process design information between large data warehouses. The model describes process-plant functional data and its 2D schematic representations, such as piping and instrumentation diagrams and datasheets. Its main focus is on the identification and description of the plant equipment. Additionally, two types of work process activities can be described, namely operating activities that transform or transport processing materials and design activities that create or modify some plant equipment. Currently (as of 2007), the AP 221 is still under development. Its completion is delayed due to the need for harmonization with the ISO 15926 data model (see below).
- AP 227 *Plant Spatial Configuration* [755] focuses on the exchange of the spatial configuration information for plant equipment with an emphasis on piping systems. A specification of the material streams and the operation conditions, as it is needed for the design of a piping system, is also included.
- The objective of AP 231 *Process Engineering Data: Process Design and Process Specifications of Major Equipment* [752] is to establish a common neutral data exchange format for conceptual process design and basic engineering. AP 231 describes unit operations, process simulations, stream characteristics, and design requirements for major process equipment. AP 231 also covers (experimental) material data, thermodynamic data, and chemical reaction data. Further, process flow diagrams, detailed process and plant descriptions, and basic control strategies can be specified. Currently (as of 2007), work on the AP 231 has stopped, and the project has been withdrawn by the responsible subcommittee TC184/SC4 of the ISO organization.

Based on the activity models of these application protocols, the *PIEBASE* consortium (Process Industries Executive for Achieving Business Advantage Using Standards for Data Exchange, [909]) has developed a generic model describing the design and production processes in the process industries, called *PIEBASE activity model* [908]. The model is represented in the NIST standard IDEF0 for functional modeling [788]; additional constraints are imposed on the usage of the IDEF0 standard by introducing a general process template with which any specific work process needs to comply. The template distinguishes three different types of activities (*Manage*, *Do*, and *Provide Resources*), which exchange certain types of information in a predetermined way.

While STEP is an established standard in some industrial sectors like the automotive industry, it is less accepted in the chemical process industries (with the exceptions of the AP 221 and its companion standard ISO 15926 in the oil and gas industries). There are several reasons for the poor acceptance of STEP: A major problem is the sheer complexity of the standard, which complicates its application [14, 726, 815]. Yet in spite of its magnitude, the data scope of the model is not sufficient; important areas are not covered by the available APs [18, 401, 726, 898]. Furthermore, the individual APs have too much overlap with each other and are not sufficiently harmonized [401, 726, 909].

ISO 15926

ISO 15926 is an evolving international standard that defines information models for the integration and the exchange of lifecycle data. Even though its title, *Industrial automation systems and integration – Integration of life-cycle data for process plants including oil and gas production facilities*, suggests a focus on the oil and gas industry, the data model is rather generic and allows to represent the lifecycle data of all kinds of process plants. The standard consists of seven parts; the first two parts have already been released [756, 757], while the others are still under development [758–760]. Eventually, the standard will comprise information models for the representation of product data, documents, and activities. These models describe the physical objects that exist in a process plant (materials, equipment and machinery, control systems, etc.) as well as the design requirements for and the functional descriptions of these objects; they cover the lifecycle stages of development, construction, operation, and maintenance. In the original ISO publication, the models are represented in the EXPRESS modeling language [753]; an alternative representation in the Web Ontology Language (OWL, [546]) is under development [994], which will be discussed later in this subsection.

The development of ISO 15926 originated from the STEP AP 221 (see above). It was motivated by a new approach to information sharing between software tools, which differs from the one followed by STEP: The STEP approach advocates the exchange of data files at designated points in the plant lifecycle. By contrast, the ISO 15926 propagates the continuous information sharing via a common database, which contains all the information emerging over the lifetime of the plant. This requires an information model that is able to represent the evolutionary changes to a plant over its lifecycle – an objective that is outside the scope of STEP [806]. Thus, a new information model has been developed for ISO 15926, the core idea of which is the *4D approach* for change representation: According to this paradigm, which is formally founded in [983], objects are extended in space as well as in time, and they may be decomposed both spatially and temporally[22]. While the spatial parts represent the physical constituents of the object, the temporal parts represent segments of the object's life time. An example is the conversion of a steel bar into a pipe, where steel bar and pipe are modeled as different temporal parts (or states) of the same object [542]. The 4D approach also supports the definition of so-called "replaceable parts" [1045], which have functional, rather than material continuity as their basis for identity [542].

The ISO 15926 standard is organized in a layered architecture:

---

[22] The 4D paradigm is based on a philosophical theory known as *perdurantism*. The perdurance theory is opposed to *endurantism*, which assumes that a (physical) object is wholly present at every moment of its existence.

- On top, a high-level *Data Model* (cf. part 2 of ISO 15926) introduces generic classes like **physical object**, **activity**, and **event**, and defines generic relations, such as composition, connection, containment, and causality. Also, the aforementioned 4D approach is established here. The Data Model is domain-independent and contains roughly 200 classes.
- The Data Model is extended by the *Reference Data Library*, RDL (cf. part 4 of ISO 15926). The RDL establishes the different terminologies (i.e., the Reference Data) required for the individual application domains. Particularly for the chemical process industries, taxonomies for the description of materials, plant equipment, physical properties, and units are introduced by refining the classes of the Data Model. The RDL is to be harmonized with the STEPlib library of the AP 221 (cf. Annex M of ISO 10303-221). Currently (as of 2007), merging of these originally independent libraries is still in progress. So far, about 15,000 classes have been defined in the RDL. It is expected that the RDL will contain up to 100,000 standard classes in the end.
- On the next lower layer, approximately 200 *Templates* are introduced (cf. part 7 of ISO 15926). A Template retrieves classes from the Data Model and correlates them via n-ary relations; that way, it defines a configuration of interconnected classes, which jointly represent some generic modeling concept. Thus, Templates are comparable to the Design Patterns described in Subsect. 2.6.2.
- The subsequent layer introduces specialized Templates, the so-called *Object Information Models*, OIM (cf. parts 4 and 7 of ISO 15926). An OIM refines a Template by replacing the generic classes from the Data Model involved in the Template by more specific classes from the RDL.

In addition to the above layers, which are specified by the ISO 15926, the standard allows for user-specific extensions, thus allowing user organizations to configure the model to their individual requirements. Besides specialized Reference Data and specialized OIMs, an organization may define its corporate *Document Types*. A Document Type is a model for a particular document; it specifies the document's structure and places constraints on its contents. The structure is indicated by decomposing the document into its structural elements, called Document Cells. The contents are specified by referring from a Document Cell to some Reference Data or OIM, which characterizes the information content of the respective Document Cell. Thus, the approach to document modeling taken by the ISO 15926 is very similar to the Document Model presented in Sect. 2.3: Both models decompose a document into its structural constituents and define the semantic contents by linking the constituents to corresponding classes defined in an (external) product data model. As a further similarity, both models have a similar understanding of the concept of a document: Any collection of data is considered as a document; valid examples are a simulator input file or (parts of) an entire database

Upper Ontology for ISO 15926

The objective of the *OMPEK* (Ontologies for Modeling Process Engineering Knowledge) project [799] is to provide foundation ontologies that can be extended for use in knowledge-based applications in the process engineering domain. These ontologies shall eventually cover the areas of substances, physicochemical processes, production plans and operations, processing equipment, human systems, and value-chains. Ontology development is based on an *Upper Ontology*, which defines general-purpose concepts and theories (such as mereotopology and causality) and acts as a foundation for more specific domain ontologies. An earlier version of the Upper Ontology was based on the SUMO ontology [865] and was represented in the DAML+OIL language [609]. A more recent version, which is described in [542, 543], is an OWL [546] implementation of the ISO 15926 Data Model ([757], see above). To this end, the EXPRESS code [753] of the ISO 15926 has been translated into OWL, and some axiomatic definitions and constraints have been added, which could not be represented in the original model because the EXPRESS modeling language lacks the necessary expressiveness. Also, some parts of the Upper Ontology (e.g., the part concerned with the modeling of physical quantities) differs from the original Data Model, as the OWL language enables a more advantageous conceptualization of the model contents. Presently (as of 2007), the Upper Ontology comprises about 200 concepts.

Two extensions of the Upper Ontology have been published so far, which go beyond the scope of the ISO 15926. Suzuki, Batres, and co-workers [990] have added a theory of causality to represent and query knowledge about plant accidents and hazards and operability studies (HAZOP). Fuchino, Takamura, and Batres [679] introduce concepts for representing IDEF∅ activity models and integrate the PIEBASE Activity Model [908] with the Upper Ontology.

POPE Ontology

Zhao and co-workers are developing an information management system to support the design of pharmaceutical products and processes, which is based on an ontology named POPE (Purdue Ontology for Pharmaceutical Engineering) [1021, 1065–1067]. Within the suggested framework, part of the information created by the different application tools is to be stored in shared ontologies (represented in OWL) to enable the information exchange between tools and the systematic storage and retrieval of knowledge.

Some domain ontologies have been developed, which so far cover only a small portion of the chemical engineering domain: They enable the representation of material properties, experiments and process recipes, as well as the structural description of mathematical models. In parallel, an ontology for the modeling of work processes and decisions has been built. Yet the representation of work processes is confined to deterministic guidelines, the input/output information of activities cannot be modeled, and the acting persons are not considered. Moreover, the approach to decision representation is somewhat limited, as it presumes that the decision criterion can be explicitly modeled by means of a logical expression – an assumption that, based on our experience, does usually not hold true. Summarizing, the POPE ontologies specialize in particular aspects of pharmaceutical products and processes, and are consequently not as generic and broadly applicable as the application domain models presented herein.

To date (as of 2007), a detailed documentation of the POPE ontologies has not been published, but from the available information it can be concluded that there is no common model framework for the different ontologies. Instead, the ontologies are being developed independently and are only partially integrated. Thus, consistency between the different ontologies is not guaranteed.

2.6.4 A Comparative Evaluation

From all the reviewed models, the ISO 15926 information model bears closest resemblance to the C^2EDM; in the following, we use it as a benchmark against which our model is compared. To this end, the two models will be contrasted with respect to model quality, scope, level of detail, structure and organization, representation of temporal changes, and modeling language. As for the Upper Ontology for ISO 15926, it basically constitutes a reimplementation of the ISO standard itself; thus, the following considerations apply to it, as well.

Model Quality

Generally, the ISO 15926 can be criticized for being of mediocre quality, at best. In particular, Smith [970] lists a number of systematic defects of the ISO Data Model, which cause it to be both unintelligible and inconsistent. Opposing the ISO 15926, Smith states some general principles for model quality, which should be satisfied by any good information model. The first four of these principles are rather generic: The author advises to supply intelligible documentation, provide open access, keep things simple, and reuse available resources. These are well-known targets for information models (e.g., [538, 589, 705, 944]), which we were aiming to achieve right from the start of the model development process (e.g., [18, 303, 326]). The remaining principles are more specific, suggesting best practices to avoid the deficiencies of the ISO model, particularly its lack of terminological coherence and its unintelligible and faulty concept definitions. Since most[23] of these best practices have been applied in the C^2EDM, we may claim that the C^2EDM constitutes a model of superior quality.

---

[23] Excluded are principles no. 8 and 12, which state rules that govern the construction of compositional concept names; they do not apply to the C^2EDM, as we do not utilize such compositional terms. Principle no. 13 is debatable: It advices to avoid ambiguous words in concept definitions, such as 'which may', 'indicates', 'characterizes', etc. While we certainly tried to define concepts as precisely as possible, words like 'characterize' and 'indicates' have been used regularly, in order

Scope

While both models have a similar coverage of documents, their scopes are not identical: The lifecycle phases of plant operation and maintenance, which are addressed by the ISO 15926, are only partially considered in the C^{2}EDM; however, coverage of these areas will be enhanced in the near future (cf. Sects. 7.2 and 7.3). On the other hand, the subject of mathematical modeling, a major focus of OntoCAPE, is not well covered by the ISO 15926.

As for process modeling, while the ISO 15926 allows to model activities, the other components of a work process model (i.e., actors/roles, information ...) are not provided. Decision rationale, which we consider an important type of information created during design processes (complementary to product data), is also not covered by ISO 15926.

Level of Detail

The ISO 15926 standard defines an implementation model intended for shared databases and data warehouses as well as for electronic product catalogues. Consequently, it gives a fine-grained and highly detailed description of the domain, resulting in a very complex data model that incorporates a large number of specialized classes. A drawback of this approach is that the ISO 15926 model is only accessible to modeling experts, who are willing to spend a considerable amount of time to get acquainted with it.

In contrast, we propagate less complex process and product models, which can be easily understood and applied by less experienced practitioners. To this end, we have tried to keep the models as simple and intuitive as possible. In consequence of this principle, it is sometimes necessary to trade accurateness and precision against usability, which leads to a coarse- to mid-grained description of the application domain. However, this level of detail has proven to be sufficient for many applications like those presented in the later chapters of this book.

Structure and Organization

In spite of its complexity, the ISO 15926 model is not sufficiently organized. While the layers constitute at least a coarse model structure, the concepts within each layer are organized by subclassing only, which is certainly not sufficient to manage ten thousands of classes. A complementary organizing principle is provided through the Templates, which are comparable to the design patterns defined in our Meta Model; but again their large number

---

to define relations that actually have the *function* of indicating/characterizing a particular class. Also, the phrase 'which may' is often utilized to denote a cardinality of $0..n$. If used this way, we do not consider these words to be ambiguous, thusly not violating the above principle.

makes it difficult for the user to find and apply a desired Template. A possible improvement would be the grouping of thematically related concepts into modules, partial models, and finally submodels, as it is done in our approach. Moreover, introducing the idea of different perspectives, as realized in OntoCAPE through the notion of aspect, might further improve the organization of the ISO 15926 Data Model.

Representation of Temporal Changes

An important difference in conceptualization between the ISO model and the C^2EDM is the representation of temporal changes: The ISO 15926 advocates the perdurantistic (or 4D) view, while the C^2EDM takes the endurantisitc (or 3D) perspective. Generally, the representation of temporal persistence is subject of controversial debate in philosophy – a summary of the different arguments is, for example, given in the Stanford Encyclopedia of Philosophy [725, 867]. Also in ontology engineering, the issue is disputed: For example, the SUMO project of the IEEE has finally (and only after some discussion) adopted the 3D paradigm [865].

In the end, the decisive factor is always the practicability of the chosen approach: Due to its intended use as a schema for shared databases, the ISO 15926 sets a high value on the representation of temporal changes, for which the 4D perspective seems advantageous. By contrast, temporal changes are less relevant for the applications targeted by the C^2EDM, such that we consider the 3D paradigm more practicable for our purposes: It has the advantage of being more intuitive, which supports our goal of an easily usable ontology. Note that the submodels of the C^2EDM provide alternative mechanisms for the representation of temporal changes: OntoCAPE introduces the backdrop concept for the description of dynamic system behavior; the Document Model allows for the versioning of documents; procedures and work processes are modeled by means of the Process Ontology.

Modeling Language

The original ISO 15926 standard utilizes the EXPRESS modeling language for the formal representation of the data model. EXPRESS provides language features that go beyond the expressiveness of OWL; particularly, class definitions may include functions and procedures, which allow formulating complex statements with local variables, parameters, and constants just as in a programming language. On the other hand, EXPRESS is a rather uncommon language, and consequently there is a lack of available tools supporting it. Moreover, having been developed in the late 80s, the language has limited compatibility with modern Web technologies.

OWL is less expressive than EXPRESS, but it is still sufficient for most applications. Particularly OWL DL, the DL-based language subset used for the representation of our C^2EDM, constitutes a favorable compromise between expressiveness and computational scalability (cf. Subsect. 2.1.3). This allows supporting the modeling process through reasoning services like consistency checking and automatic classification. Thanks to its status as a Web-enabled standard endorsed by the W3C, OWL has received wide attention both in academia and industry. Although it has only been released lately (in 2004), there are already a number of compatible software tools available, commercial as well as open-source.

Due to these advantages, a reimplementation of the ISO 15926 in OWL is currently in progress. However, judging from the preliminary samples presented in [542] and [668], the ISO models will not be realized in OWL DL, but in the more expressive sublanguage OWL Full (cf. Subsect. 2.1.3). OWL Full supports such advanced language features as metamodeling (i.e., instantiation across multiple levels) or augmenting the meaning of the pre-defined language primitives, yet at the cost of loosing scalability and compatibility with DL reasoners.

Concluding Remarks

The comparison against the benchmark of the ISO 15926 information model demonstrates that the C^2EDM makes a valuable contribution to the field of application domain modeling. Nevertheless, the model cannot be considered to be complete. Information modeling is an incessant effort, demanding the continuous improvement and adaptation of the model according to prevailing conditions and requirements. Our next goal is the application and evaluation of the C^2EDM submodels in industrial practice, as explicated in Sects. 7.2 and 7.3. That way, the models will be validated against industrial requirements and, if necessary, adapted to practical needs.

3 New Tool Functionality and Underlying Concepts

Four *new informatics concepts* have been introduced in IMPROVE to support chemical engineering design processes in a novel way (cf. Sect. 1.1 for their introduction and 1.2 for their usage in a prototype). This chapter gives their *detailed description*.

In the corresponding sections, not only the new *concepts* are discussed but, moreover, their *realization* resulting in *new* tool *functionality*. This new functionality is mostly implemented on top of given tools, following the bottom-up approach of IMPROVE. We are going to learn that in most cases sophisticated *reuse* techniques have been used for the implementation of this new functionality. Reuse forms include tool specification.

Tool specification has to do with modeling. It contributes to the layered process/product model which is the overall goal of IMPROVE. The corresponding contributions to this formal process/product model are discussed in Chap. 6.

Section 3.1 introduces process chunks to support and process traces to keep track of the experience of developers in a design process. Section 3.2 discusses how links between the contents of different documents can support change processes. These links are automatically introduced by interactive tools, called integrators. Direct communication of designers is an essential part of any development process and complements organized communication and cooperation. Section 3.3 introduces corresponding audio and video tools on one hand and virtual reality tools on the other hand. Finally, Sect. 3.4 introduces novel management tools to organize the collaboration in a design process. These tools regard changes within the design process, parametrization in order to meet a specific context, and interorganizational coordination.

3.1 Using Developers' Experience in Cooperative Design Processes

M. Miatidis, M. Jarke, and K. Weidenhaupt

Abstract. The process industries are characterized by continuous or batch processes of material transformation with the aim of converting raw materials or chemicals into more useful and valuable forms. The design of such processes is a complex process itself that determines the competitiveness of these industries, as well as their environmental impact. Especially the early phases of such design processes, the so-called conceptual design and basic engineering, reveal an inherent creative character that is less visible in other engineering domains, such as in mechanical engineering. This special character constitutes a key problem largely impacting final product quality and cost.

As a remedy to this problem, in cooperation with researchers and industrial partners from chemical and plastics engineering, we have developed an approach to capture and reuse experiences captured during the design process. Then, fine-grained method guidance based on these experiences can be offered to the developer through his process-integrated tools. In this section, we describe the application of our approach on the case study of the IMPROVE project. We first report on experiments made with a prototypical implementation of an integrated design support environment in the early project phases, and successively describe how it has been reengineered and extended based on additional requirements and lessons learned.

3.1.1 Introduction

Undoubtedly, engineering design is a process of major importance for the production lifecycle of a chemical product. Its main concern is the investigation and application of novel, state of the art methodologies on the product design process, in order to increase its expected quality in a profitable way. Especially the early stages of the development and reengineering of chemical processes, the so-called conceptual design and the basic engineering are of particular interest, since they already predetermine to a large extent the competitiveness of the final product (see Sect. 1.1).

As a consequence, the computer-based process support of early phases of chemical engineering design has drawn considerable attention from the research society. Various paradigms, frameworks and environments have been developed in order to address the need for design process improvement and excellence. Their support can be provided at two granularity levels. At a coarse-grained level, it cares for the efficient project planning and coordination of the whole design project in order to establish Concurrent/Simultaneous Engineering and reduce the total design time [299]. At a fine-grained level, on the other hand, the interest shifts to the *direct process support* of the developers. Especially in the chemical engineering domain, fine-grained process support has to adequately address the inherent dynamics of design.

The second variation of process support constitutes the central topic that the subproject B1 *"Experience-Based Support of Cooperative Development Processes"* deals with. In the subproject B1, in cooperation with researchers and industrial partners from chemical and plastics engineering, we have developed the idea to provide direct process support based on the reuse of experiences gained during design processes. Process support is offered to the developer in the form of methodical guidance while interacting with process-integrated software tools. Because of design creativity and uncertainty, such guidance is only possible for certain well-understood working and decision steps that tend to repeatedly occur across several design processes. On the other hand, our approach is able to capture interesting experiences from the process execution, and provide them to the developer for situation-based reuse in analogous design processes in the future.

This section pinpoints the quest for our research and details the solutions we have provided. In Subsect. 3.1.2, we briefly identify the creative nature of engineering design and elaborate the key ideas behind our approach for its direct experience-based support. In Subsect. 3.1.3, we outline the prototypical flowsheet-centered design support environment that we pursued during the first phase of the project. The feedback given by domain experts on our first prototype provided us with additional requirements for improvements and extensions (Subsect. 3.1.4). The consideration of these requirements led to a significant reengineering of our approach undertaken during the last two phases of the project. The reengineered environment and its underlying key ideas are detailed in Subsect. 3.1.5. In Subsect. 3.1.6, we observe the current state of practice concerning the various aspects of our approach and make a brief comparison. Finally, in Subsect. 3.1.7, we draw some conclusions and provide an outlook to future work.

3.1.2 Motivation

In this section, we provide the motivation for our research. First, we pinpoint the high degree of creativity of chemical engineering design and next, we outline the solution ideas behind our approach for its direct experience-based support.

The Creative Character of Engineering Design

Engineering design, as not only in chemical engineering, is a highly *cooperative* process (see Subsect. 1.1.1). It involves extensive communication and collaboration among several developers working synergistically for the definition, analysis and evaluation of the possible alternatives for the design of a product. Each developer is assigned a set of responsibilities (tasks) in the context of the process, according to the roles he carries. Developers synergistically working for a shared goal are usually grouped together in interdisciplinary teams, possibly geographically distributed across different company sites.

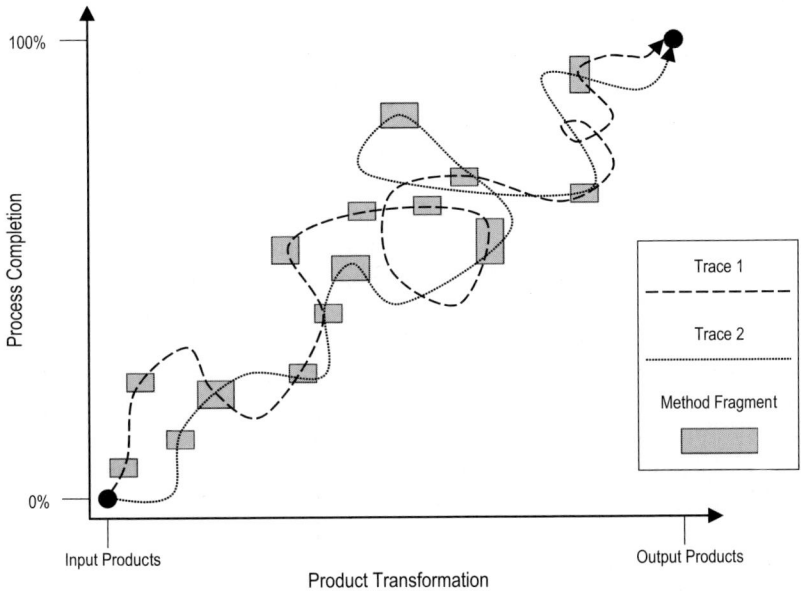

Fig. 3.1. Two arbitrary traces of the same design process

A design process constitutes all the tasks assigned to team members while working on a design problem. Throughout these tasks, developers are challenged to tackle a broad range of unexpected design complications in order to meet last-minute changes in plans and requirements. Thus, they often have to invent a solution "on the fly" appropriate to the current problem, or heavily tailor an Existing Solution According To The Problem'S Individualities Rather than simply reuse it. In either cases, the inherent *human problem solving* nature of the design process is revealed. Then, the developer is demanded to exploit his knowledge background in order to come to the optimal solution. The knowledge background concentrates mental stimulus from previous experience, education, know-how and guidelines stemming from the company, or even pure instinct.

In such highly *knowledge-intensive* settings, developers may interpret the design in different ways and thus, employ diverse "ways of working" in order to resolve a design problem. Some of these ways of working might work efficiently and deliver good results in reasonable time, while others can perform less effectively than expected. Thus, design shows an inevitable *high degree of creativity*. This creativity disturbs the planning at a fine-grained level and makes the prediction of the followed ways of working hard.

Figure 3.1 shows the traces of two arbitrary instances of the same design process projected on an orthogonal Cartesian system. Its two axes represent the design progress and design product transformation dimensions respectively. Along the horizontal axis a number of input products are being processed in order to deliver a set of final products. Along the vertical dimension, the completion of the process increases until the design is 100% complete. Each instance starts at a specific point (initial state) where a number of initial design products are available. By the end of the design (final state), the final design products are delivered.

Both processes continue in a mostly unpredictable and poorly-structured way, as a consequence of their creative nature. Nevertheless, there exist certain process chunks which tend to appear with a relatively high frequency across several design processes and are therefore well-understood and can be adequately defined. These might be parts of the process where a single way of working tends to be followed by the overwhelming majority of the developers, either because it is the only known way to produce acceptable results at the specific situations, or because it constitutes a prescribed company expertise asset that is enforced to the employees. We call these well-understood ways of working *method fragments* [94, 919].

Approach to Direct Process Support

Many contemporary process support systems are driven by explicit process models that capture in a precise way the whole lifecycle of the enacted process. The prominent focus of such systems is the complete automation of the work through the provision of prescriptive guidance to the worker. This kind of support is mainly suitable for domains with complete, rigid and well-structured process patterns that can be thoroughly predetermined in advance, such as business processes, and has been widely adapted by workflow management systems [1014]. Such complete guidance of engineering design would severely restrict the choices of the developers and thus, obscure the creativity of the overall process.

Based on this consideration, we have developed the approach to provide direct experience-based support based on the reuse of captured experiences during design processes. On one hand, captured experiences (*traces*) can be directly reused in similar design processes. On the other hand, certain method fragments that appear with high frequency among several instances of a design process can be formalized and guide the performer, when it is applicable. In the following, we outline the three basic ideas behind our approach, sketched in Fig. 3.2.

Situated Method Guidance

Method guidance is intended to influence the way that a design process is performed based on the reuse of best practices. The support provided to the developer can take two forms with regards to its way of application [642]. At one extreme, guidance strives for a strict conformity of the developer's actions to the process prescriptions through *enforcement*. Such enforcement does not

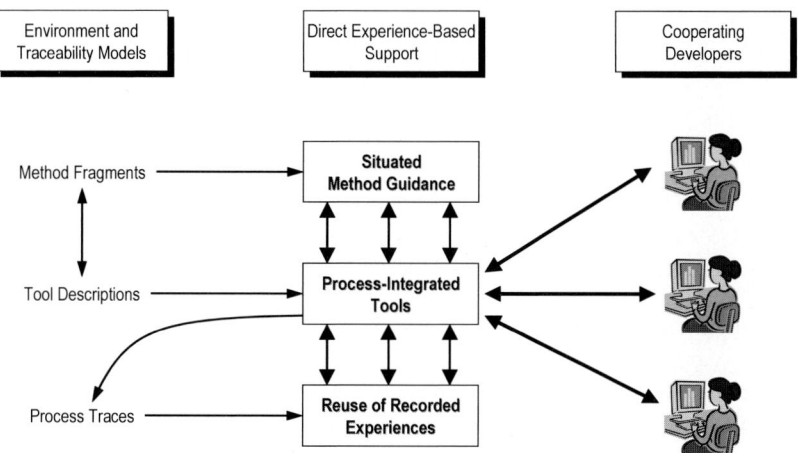

Fig. 3.2. Basic structure of direct, experience-based process support

mix well with the nature of engineering design, as it stands in the way of human creativity. On the other hand, guidance can be provided in the form of *advice*, and it depends on the judgement of the developer whether he follows it or not. According to our approach, only limited guidance is provided to the developer in the form of advice, inside well-understood process parts where a method fragment can methodically guide him. Thus, a prerequisite for such guidance is the identification of the process *situation*. When a valid situation with eligible method fragments for enactment is addressed, the developer gets notified, and can selectively request their enactment.

Process-Integrated Tools

A prerequisite for method guidance is the effective dissemination of the method guidance information to the developer. Traditionally, this kind of support has been given in the form of *handbooks*. Since handbooks are totally decoupled from information systems, *separate guidance tools* have been built providing task manager interfaces notifying the developer about the next task to be performed (e.g. [870]). Nevertheless, their guidance is typically coarse-grained and there is no way, during a method fragment enactment, to check the conformance of the developer's actions with the method definitions. We argue that the limitations of separate guidance tools can be avoided by providing integrated method guidance inside the software tools used during the design process (i.e. CAD, simulation and modeling tools), through the *process integration* mechanism. A process-integrated tool can trace a situation where method guidance can be provided, and directly inform the developer about the applicable method definitions through dynamic adaptation its user interface and exposed services (*process sensitivity*). Further, a process-integrated tool can automatically provide feedback information concerning the developer's actions. This information is necessary for the adjustment of the enactment state (i.e. situation detection and loading of applicable method fragments) according to the current process performance inside the tool.

Reuse of Process and Product Traces

The feedback information provided by a process-integrated tool is organized according to a concrete *traceability metamodel* adjusted to our project-specific needs [368]. According to this model, the recorded information provides evidence of the dependencies between the *design products* (flowsheets, simulation models etc.), the so-called *supplementary products* (goals, decisions etc.) and the *process observation data* concerning the changes that the products have undergone (i.e., process steps). Consequently, captured traces closely monitor the developer's design history (i.e., design process trace depicted in Fig. 3.2). Whereas prescriptive process definitions are biased by perceptions, captured traces provide an objective abstraction of the reality and faithful evidence of design evolution and improvement. Thus, observation and comparative analysis of the captured traces can set the foundation for experience-based explication and dissemination of design knowledge: in the short term, for empowering the developer to reuse best practices abstracted from design experiences; in the long run, for providing accountability of good and bad practices, as well as lessons learned.

3.1.3 The Prototypical Flowsheet-Centered Design Support Environment

The work of the first project phase (1997–2000) concentrated on the support of developers in small groups inside the same company. The support has been provided through embedding systematic method guidance in the design process on the basis of detailed method fragments, and through the process integration of design support tools. The mechanical interpretation of method fragments resulted in the partial automation of certain activities inside the process-integrated tools and the dynamic adaption of their user interface according to the actual process state.

The PRIME[24] approach developed at our chair[25] has been used as the integration framework for our process-integrated design support environment [367, 370]. PRIME fulfills the requirements for our design environment through four solution ideas:

1. The explicit definition of method fragments using the NATURE situation-based process metamodel [366, 937]. A *process engine* can mechanically interpret method definitions and, based on their enactment, provide situated support.

---
[24] PRIME: PRocess-Integrated Modelling Environments
[25] The initial version of PRIME has been developed in the frame of the DFG project 445/5-1 "Process Integration of Modelling Workplaces".

2. The integration of the method definitions with tool models into the so-called *environment metamodel* that lays the foundation for the process integration of tools.
3. The recording of design history according to a concrete traceability structure capturing traces along the three orthogonal dimensions of *specification, representation* and *agreement* [365, 366].
4. The definition of a generic object-oriented implementation framework for the interpretation of environment model definitions by the process-integrated tools and the dynamic adaptation of their interactive behavior.

Originally, PRIME has focused on the support of software engineering design processes, and facilitated only the a-priori process integration of newly-implemented tools. Nevertheless, in the context of the IMPROVE project, we have been interested in the a-posteriori integration of existing tools. As implication, we were confronted with the challenge of defining the criteria that an existing tool should fulfill in order to become process-integrated, and the need to design a generic mechanism for their criteria-wise integration.

In this section, we outline the prototypical PRIME based design support environment we developed during the first phase of the project. First, we describe the generic mechanism for the a-posteriori process integration of existing tools. Next, we demonstrate how it has been applied for the process integration of a flowsheet editor. In the end, we present the coupling of the flowsheet editor with other domain-specific tools, complemented by generic tools for the documentation of design rationale and visualization of traces, and illustrate their interplay on a demonstration example.

A-posteriori Process Integration of Existing Tools

The PRIME process metamodel for the definition of method fragments is organized around the situation-based NATURE process metamodel, originally proposed for the requirements engineering domain.

The NATURE process metamodel explicitly represents situations and intentions. A *situation* describes the subjectively perceived state of the process and is based on the individual states of the *products* undergoing development. An *intention* reflects the goal that the human actor has in his mind. The process knowledge of how to reach a specific intention in a given situation is represented by a context that can be refined into three kinds of chunks:

- *Executable contexts* describe pieces of the process that can be automated and are usually applied by tool actions.
- *Choice contexts* capture the most creative process parts where a decision among several alternatives is demanded. For each alternative, supporting or rejecting arguments can be provided.
- *Plan contexts* define workflow strategies and systematic plans and can recursively contain contexts of all three types.

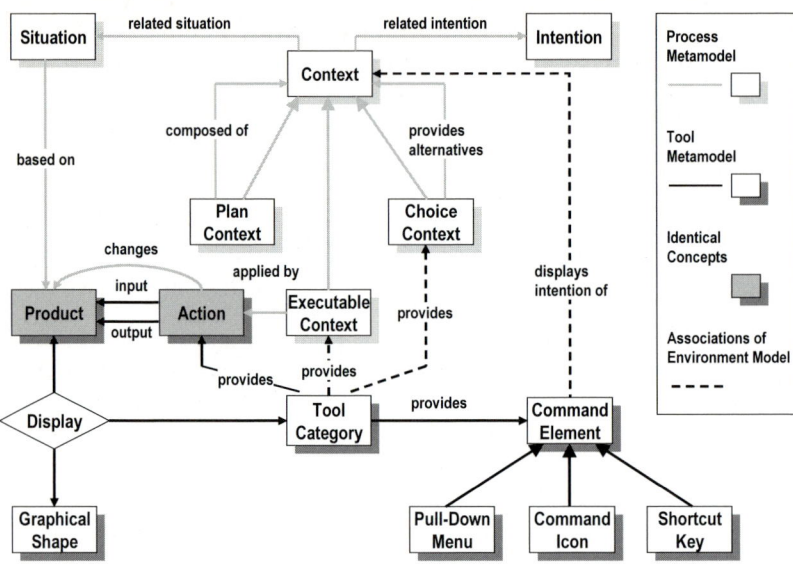

Fig. 3.3. Integrated environment metamodel

Process integration in PRIME is based on the integration of the contextual description of the design process steps with descriptions of the tools responsible to perform these steps. A tool model is constructed in a tool modeling formalism describing its capabilities (i.e. services provided) and GUI elements (menu items, tool bars, pop-up menus etc.). Process and tool metamodels are integrated within the so-called *environment metamodel* (Fig. 3.3). The interpretation of environment models enables the tools to adapt their behavior to the applicable process definitions for the current process state. Thus, the user is able to better understand and control the process execution.

The explicit definition of a tool metamodel and its integration with the contextual process metamodel, allows the formalization of the following six requirements on the APIs exposed by a tool in order to be fully process-integrable (cf. Fig. 3.4, [369, 469]):

A1 A service invocation API required for triggering the services provided by the tool.
A2 A feedback information API required for keeping track of the results obtained from executing a tool service.
A3 A command element API for introducing additional command elements in the tool's user interface.
A4 A product display API for highlighting the actual situation product parts.
A5 A selectability API for adapting the tool's user interface to the feasible alternatives when a choice context is active.
A6 A selection API for obtaining notification about user selections of products or command elements.

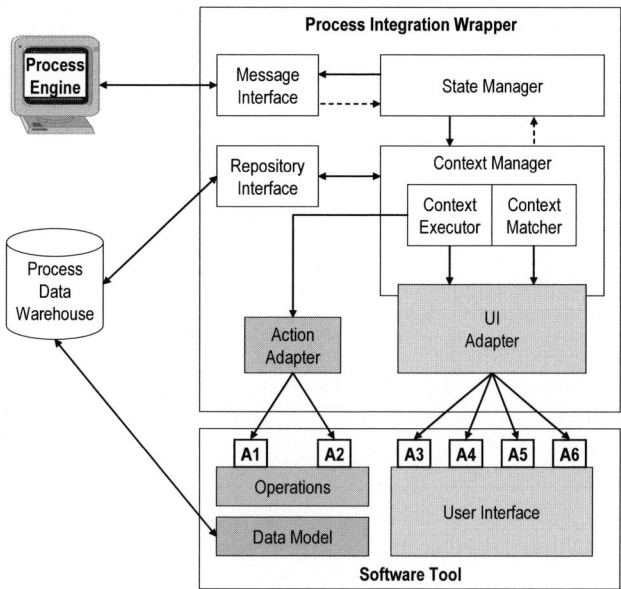

Fig. 3.4. Generic tool wrapper architecture

A *generic tool wrapper architecture* (Fig. 3.4) defines the basic guidelines for the construction of wrappers that mediate the interactions between the process engine and process-integration tools. A tool wrapper has three responsibilities:

1. It hides the heterogeneity of the above sketched APIs.
2. By interpreting the relevant parts of the environment model, it is able to restrict the interaction capabilities of the tool, and notify the process engine of human-triggered requests for guidance through the tool's GUI elements.
3. It is responsible for the communication with the process engine according to a strict *interaction protocol* [469].

Some parts of the generic tool wrapper architecture have been reused from the original PRIME process integration specification. The *message interface* and the *state manager* are responsible for the communication with the process engine and the exchange of messages. The *context manager* cares for the process definition conformed behavior of the integrated tool. It uses adapter modules specialized for each of the tool-specific API interfaces. Method fragments, product object schemas and tool models are maintained in a *Process Data Warehouse (PDW)* (see Sect. 4.1).

The Process-Integrated Flowsheet Editor

Within the IMPROVE project various commercial tools like CAE tools, simulators and model builders are employed for the support of the design and construction of a chemical plant. As a prototypical example of a-posteriori process integration, we have developed a fully process-integrated flowsheet editor extending a commercial CAE tool [21]. In the following, we present the motivation behind its integration, as well as some details of its technical realization.

The Flowsheet as Cornerstone of Chemical Engineering Design

During the development of a chemical process, many information pieces are created which have to be maintained and kept easily accessible. Among these documents, *flowsheets* (a graphical representations of the structure of a plant) play a prominent role. The importance of flowsheets is not only stressed in various text books about chemical engineering (e.g. [559, 638]), but can also be observed in chemical engineering practice.

Indeed, one of the findings of a workshop we conducted with developers and managers from a large chemical engineering department was that the flowsheet reflects in a natural manner the assumptions made by the various stakeholders (chemical engineers, material engineers, costing people, safety engineers, managers etc.) about the current state of plant design [767]. Thus, the flowsheet acts as the main communication medium across several organizational units, and throughout the often decade-long lifecycle of a chemical plant or chemical production process. Moreover, the flowsheet is used as an anchoring point and structuring device for information pieces such as simulation specifications and results, cost calculations, design rationales, safety considerations etc.

The manifold relationships of the flowsheet to other information units are illustrated on Fig. 3.5, which additionally depicts the tools used in the IMPROVE demonstration scenario.

Realization of the Process-Integrated Flowsheet Editor

In current practice, flowsheets are frequently created using drawing tools or CAD systems. These tools provide no specific support for chemical engineering, and often confront the user with superfluous functionality. In competition to these pure drawing tools, dedicated tools for chemical engineering, such as block-oriented simulators (Aspen Plus, PRO/II) have been augmented by flowsheet user interfaces. This pragmatic trend reflects the close relationship between flowsheet design and mathematical models, and provides the user with considerable support as long as he does not have to leave the boundaries of the tool. The enrichment of simulation programs with flowsheet functionality has indeed led to monolithic, hardly maintainable software systems, which rarely provide open interfaces for extensions. As a consequence of such "islands

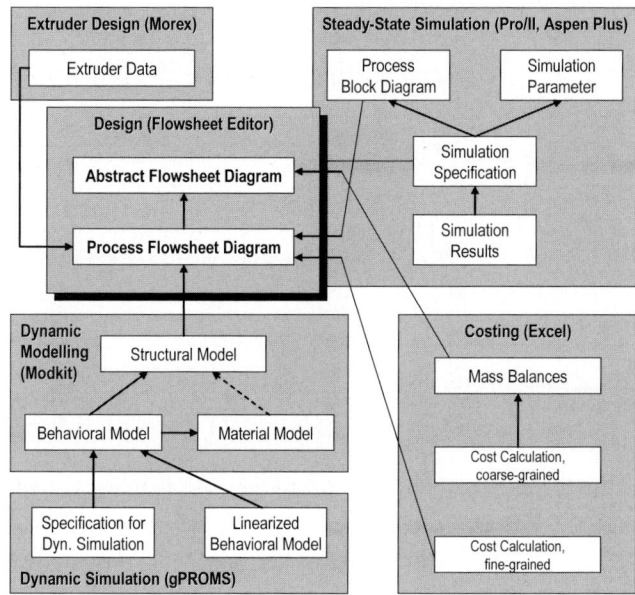

Fig. 3.5. Role of the flowsheet in the overall IMPROVE scenario

of automation", flowsheet information is repeatedly entered manually, e.g. if different simulators are used within one project. Moreover, as flowsheet editors of common simulators do not allow to annotate individual flowsheet elements with, e.g. cost calculations or safety remarks, isolated flowsheet documents emerge in these working environments, too.

The development of a completely new flowsheet editor is a formidable task, and the result is unlikely to be able to compete with the rapid advances in chemical engineering design tools. Our process-integrated flowsheet editor has therefore been realized according to the a-posteriori philosophy of PRIME, on top of an existing tool. None of the commercially available flowsheet tools fulfilled the requirements concerning flowsheet refinement and extensible type systems for flowsheet components. Thus, the existence of open interfaces, which allow the addition of the above-mentioned functionality (besides the process integration), became the most important criterion during the choice of the tool. We finally decided for Microsoft VISIO [845], a widely used tool for creating technical drawings. VISIO's strengths lie in its comprehensive and extensible symbol libraries and its add-on mechanisms based on COM interfaces. These interfaces provide to external extensions a fine-grained access to VISIO's internal object model.

Figure 3.6 depicts the coarse-grained architecture of the flowsheet editor and a snapshot of its user interface. The VISIO based flowsheet tool fulfils the

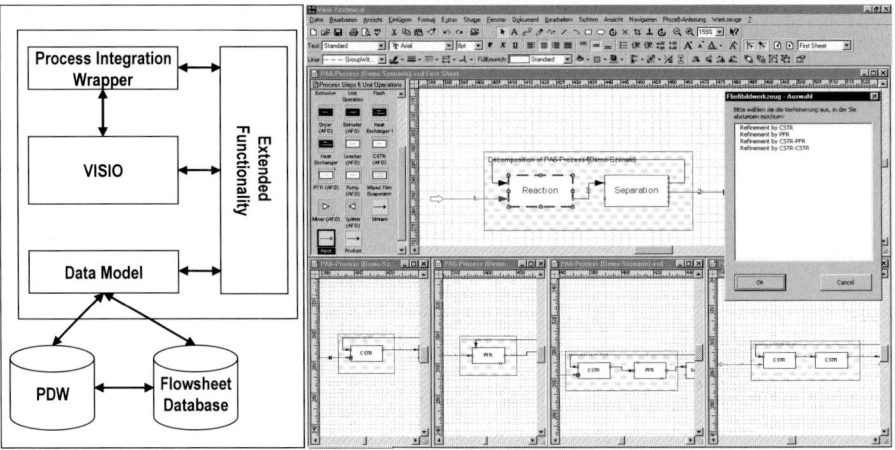

Fig. 3.6. Architecture and user interface of the flowsheet editor

following essential functionalities from a chemical engineering perspective for constructing and maintaining flowsheets:

Complex Refinement Structures. Flowsheets are refined across various hierarchical abstraction levels and depicted in specific representation formats. In the context of the IMPROVE project, we are mainly interested in two of the three commonly used flowsheet variants: the abstract flow diagram (AFD) and the process flow diagram (PFD). Thus, the flowsheet editor supports the full hierarchy of complex stepwise refinement operations from AFD to PFD and assures a set of consistency constrains in order to preserve their semantic correctness.

Rich and Extensible Type System. The organization of flowsheet components in an expressive type system is an indispensable prerequisite for guiding consistent flowsheet refinement. However, the richness and the rapid growth of knowledge goes beyond the abilities of typical object-oriented type systems: there are many conceivable process steps and possible stream types. Thus, the type system of the flowsheet provides a semantically rich characterization of individual flowsheet components that is extensible. The chemical engineer is empowered to define his own flowsheet components and characterize them semantically in order to promote their reuse in future projects.

The data model of the flowsheet editor is closely related to the partial models Chemical Process and Plant of the conceptual IMPROVE product data model (see Sect. 2.2). Its design was also influenced by emerging data exchange standards in chemical engineering (e.g. the process data exchange interface standard PDXI in the STEP context [623]).

VISIO's user interface and the data model layer are used by the enhanced functions modeled as method fragments, for the realization of operations such as creating a new refinement within the same flowsheet, or navigating inside existing hierarchical refinement structures.

The Integrated Flowsheet-Centered Architecture

The fact that the flowsheet plays a prominent role in manifold development activities (Fig. 3.5) leads to two important requirements for a design support environment. From a developer's perspective, the flowsheet editor should be seamlessly integrated into the design lifecycle (e.g. simulation, costing and safety engineering), and the corresponding tools used in each design step. Of course, tight coupling between tool functionality of different working domains should be avoided in order to remain sufficiently flexible and not to fall in the trap of monolithic solutions. Further, in order to support the central role of the flowsheet during engineering design, the flowsheet editor should serve the role of a unified interface for the exchange of data across other design support tools.

Process integration offers the potential to couple different tools more flexibly and to provide high quality support for the developer at the same time. Explicit, easily modifiable method fragments guide the developer during activities across multiple tools, while the tools themselves only cover a limited scope. In this way, process integration complements data integration mechanisms, which maintain structural consistency between documents of different tools, and component-based approaches such as the CAPE-OPEN approach [72].

In order to support the central role of flowsheets during engineering design, we have implemented a prototypical flowsheet-centered architecture (Fig. 3.7). The central element of this architecture is the VISIO based flowsheet editor that, based on the PRIME process integration mechanism, is operationally linked to other domain-specific tools and acts as the prominent communication medium among developers.

Within the database of the flowsheet editor, only prominent information about the flowsheet elements is stored, including characteristic parameters of process steps and equipment items as well as stream and piping information. This information needs to be shared for the development of further mathematical models inside modeling tools, the cost analysis of different process parts, as well as the specification of steady-state dynamic simulations in simulators.

To this end, we have partially process-integrated the generic Microsoft Excel application [844] for the user-triggered calculation of the mass flow in streams of flowsheet process groups. In a similar manner, a partial integration between the flowsheet editor and the MOREX dedicated simulation tool [147] has been realized for the transfer of information concerning the simulation of compounding extruders.

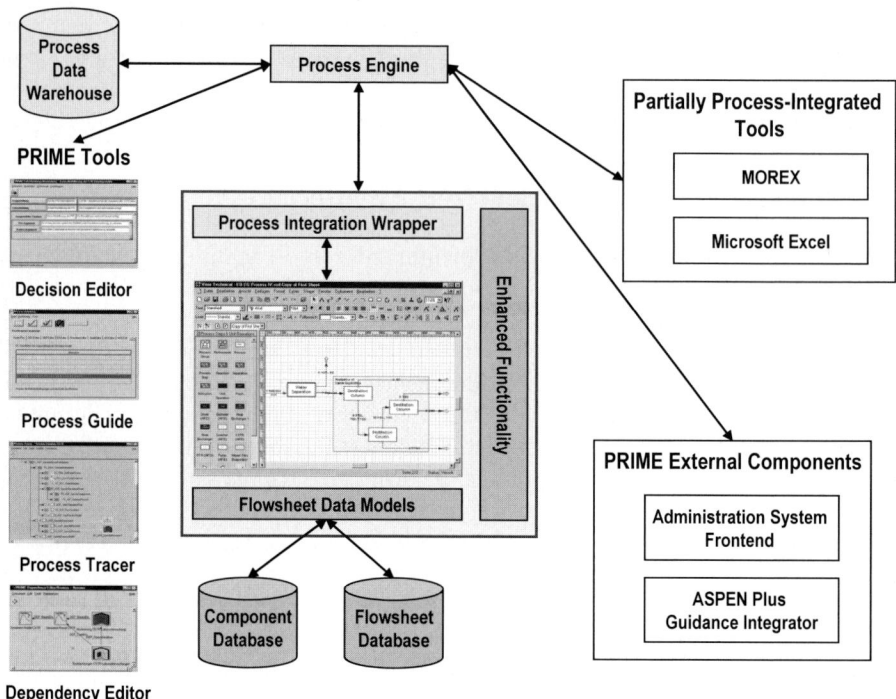

Fig. 3.7. Integrated flowsheet-centered architecture

Some domain-specific tools, do not provide at all open interfaces for the exposure of their functionality for a potential process integration (e.g. the Aspen Plus block-oriented simulator [516]). For these cases, we have used integrators for exchanging information and ensuring consistency between data storages and documents [27]. Specialized integrators have also been used for the definition of mapping rules between the constructs of the process-integrated flowsheet editor and the AHEAD administration system [212]. Both of these cases of integration involve extensive synergy with other subprojects, and are described in Sect. 5.5.

The overall environment is complemented by a number of generic PRIME interactive tools for several other supplementary aspects (see [469] for a detailed description):

1. A decision editor allows the capture of the developer's design rationale though documenting the decisions taken during choice contexts together with the arguments that led to them (cf. Sect. 2.5). More specifically, the developer can set an initial issue, one or more positions about it, and support or reject each of them through pro or contra arguments.

2. A *process guide* gives advice to engineers how to use poorly integrated tools, and allows them to document the use of such tools with respect to conceptually defined process guidance.
3. A process tracer allows the developer to explore the complete history of a process, organized chronologically and hierarchically according to the decomposition of contexts and with linkage to the objects created.
4. A dependency editor allows local analysis and editing of dependencies of product objects with other process objects.

Contributions to the IMPROVE Scenario

The benefits of the direct process support have been validated on the basis of the Polyamide-6 conceptual design scenario used in the IMPROVE project (cf. Sect. 1.2). In the following, we briefly deal with three parts of the overall scenario.

Guidance through Flowsheet Refinements

During the creation of a flowsheet diagram, many variants have to be considered, e.g. the reaction part can be realized through a stirred tank reactor, or a tubular reactor, or the interconnection of these reactor types with an intermediate separator. Method fragments can guide the process by automatically generating the chosen alternatives.

The upper part of Fig. 3.8 shows the first refinement level of the Polyamide-6 design process in the flowsheet editor. The reaction expert expresses his decision to refine the Reaction process group by selecting it and activating the "Refine" menu item from the menu bar. Then, a suitable choice context becomes active that retrieves from the PDW the four possible refinement alternatives, and the "Guidance" menu gets automatically adapted to display their intention. Additionally, two items with the respective intentions "Generate all Alternatives" and "Quit Refinement" are displayed. Lets suppose that the developer selects the menu item to generate all alternatives. In accordance with the definitions of the method fragment, automatically four groups of refinements for the Reaction process group are inserted into the flowsheet editor window (lower left part of Fig. 3.8). As a final step, the method fragment requests the developer to document which of these alternatives should be first examined (lower right part of Fig. 3.8).

Tool-Spanning Method Fragment

Process integration offers the potential to couple other tools participating in the design with the central flowsheet editor. We show, how a method fragment can guide the export of flowsheet process group information to the Excel tool for the calculation of the mass balance of the process.

Figure 3.9 shows the enactment of a method fragment involving multiple tools. In the flowsheet editor (left), the developer has selected the refinement

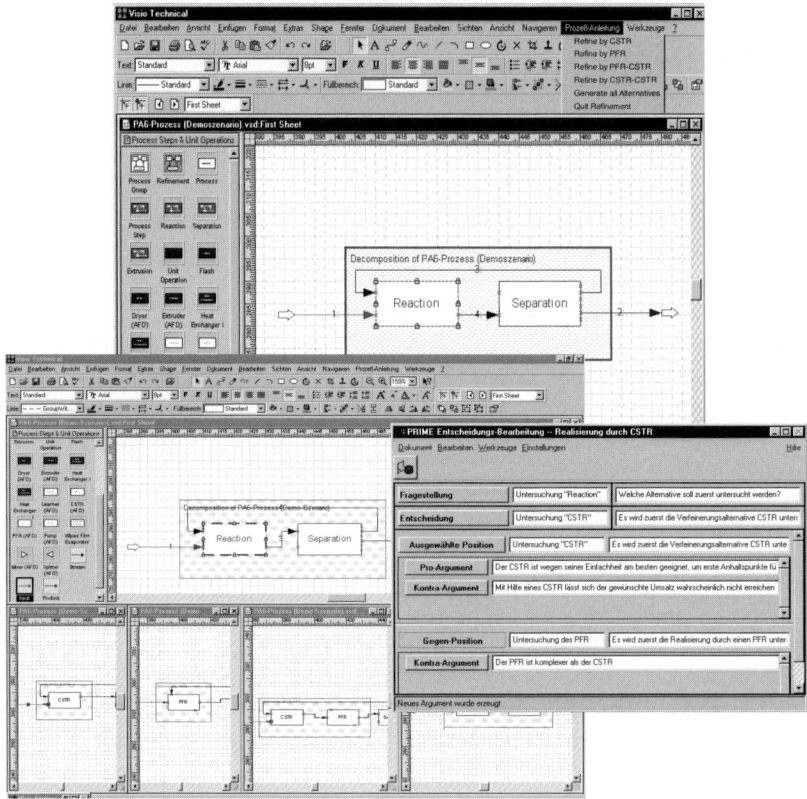

Fig. 3.8. Method fragment guided flowsheet refinement

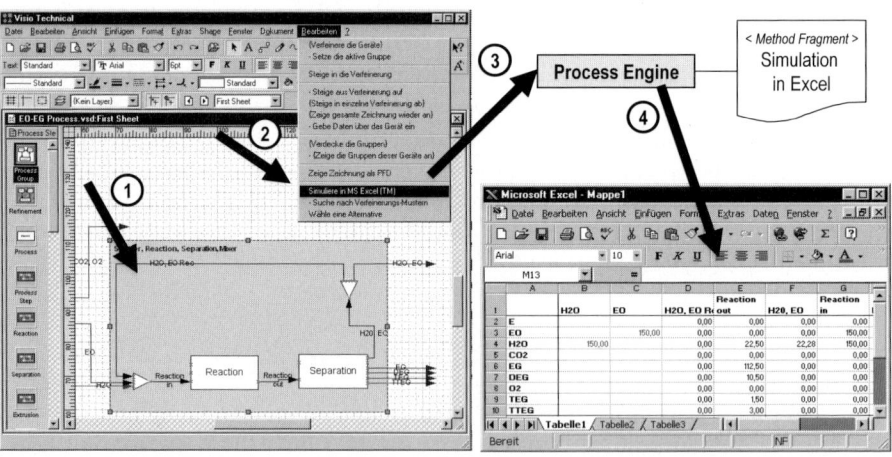

Fig. 3.9. Method fragment guided tool interoperability

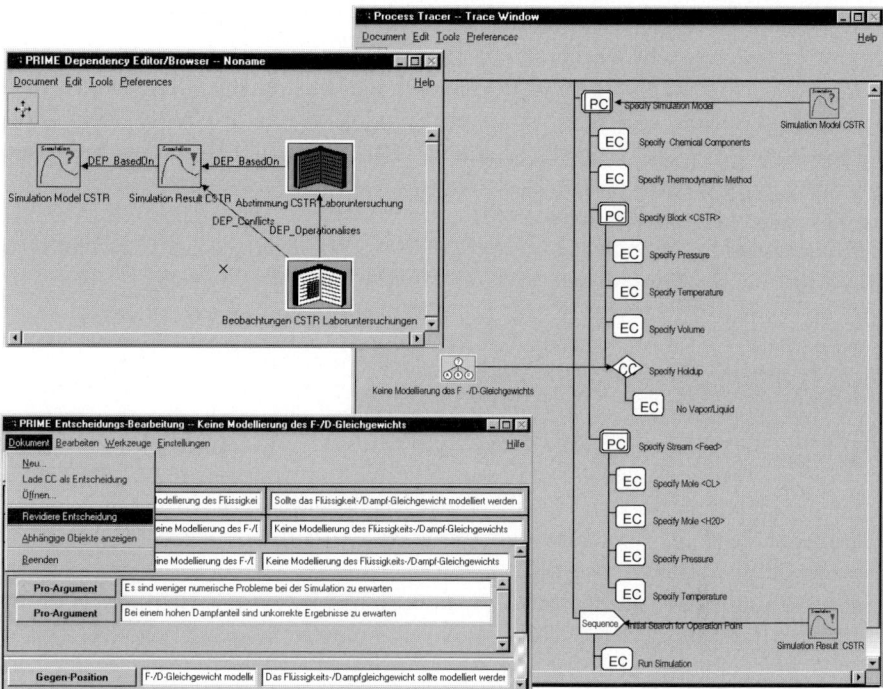

Fig. 3.10. Traceability of process steps and decisions

of a process group (step 1). Instantly, the situation based on this product is matched by the process engine, and menu items that show the tool actions as well as method fragments that can be applied to the selected situation are enabled. The developer chooses the menu item "Simulation in Excel" (step 2). As this menu item is not related to an action provided by the flowsheet editor, a corresponding method fragment is enacted by the process engine (step 3). According to the definition of the method fragment, the process engine invokes Excel (right) and creates a spreadsheet which mathematically calculates the mass balance of the selected process group.

Use of Process Traces

The documentation of major decisions taken during the design contributes to the better understanding of the overall design. We show, how captured process traces can help to improve the unsuccessful modeling decision for the simulation of a stirred tank reactor.

After the simulation of the reaction process through a continuous stirred tank reactor (CSTR), the laboratory expert is requested to make experiments on the simulation results. At the end of the experiments, the expert realizes that discrepancies exist between the results of the laboratory and the simulation. In order to find the possible false decisions that led to this deviation, he traces back to the steps and decisions taken during the simulation. In the dependency editor, he selects the icon "Simulation Result CSTR" (left upper part of Fig. 3.10) and selects the menu item "Show Trace". Then, a process tracer window opens that displays the hierarchical decomposition of the respective trace chunk (right part of Fig. 3.10). By browsing through the trace chunk elements, the expert can recognize the wrong decision that has been taken during the simulation, and he can use the decision editor (lower left part of Fig. 3.10) to revise it and provide pro and contra arguments.

3.1.4 A Critique

The first experiences gained by the application of our approach on the IMPROVE Polyamide-6 scenario suggested that the direct process support offered significant assistance to the work of the developer. However, some shortcomings showed up that provided the starting point for extensions and improvements in the next two phases.

Our initial approach for process support has been provided across multiple design tools intended to be used by a single human actor (developer or groups of developers) at a single technical workplace. This hypothesis does not match well with the highly cooperative nature of engineering design. Thus, two prominent requirements were posed for the integration of multi-user aspects to the existing infrastructure. From an implementation perspective, multi-user support means the exploitation of new standards and technologies allowing distribution of the process support across multiple technical workplaces. From a modeling perspective, process models should get extended in order to capture the user dimension and the context of the information that is allowed to be transferred from one user to the other, based on enterprise rules and policies.

The main research goal of the first period had been the provision of direct process support by means of an a-posteriori process integration of interactive design tools. The developed approach was illustrated through the full process integration of the VISIO based flowsheet editor. However, our experience showed that the realization of the VISIO wrapper has been a very labor-intensive and lengthy task of high expenditure. Moreover, our generic tool wrapper architecture could not sufficiently accommodate tools only partially supporting the six APIs required for full process integration. The integration of such tools, is especially vital for the thorough support of the demonstrative Polyamide-6 scenario, where the complexity lies not only in the flowsheet design during the synthesis phase, but also in the exchange of information with the analysis steps taking place in various simulation tools.

Last but not least, the trace visualizer developed during the first phase displayed in a chronologically and hierarchically organized way the sequence of process actions traced during the lifecycle of a design process. It was mainly intended to be used by certain specialists for the observation of operational sequences of actions of developers and trace of design failures and inconsistencies. However, domain experts pinpointed us the great benefit of the use of such kind of tools also by the designer himself. Due to the creativity of design, some parts of it are not clearly-defined and sometimes, especially for the inexperienced developer, the path to be followed is ambiguous and full of assumptions. To this end, specialized tools can be developed for the selective retrieval and exploration of best practices in product design from the past, complementing the process support based on well-known method fragments.

3.1.5 The Reengineered Design Support Environment

During the rest of the project (2000–2006), we have improved and extended the original concepts of our approach based on the above considerations. In this section, we detail the key ideas behind the reengineered design support environment, and we show how they have been validated on extended case studies in the IMPROVE context.

Cooperative Extensions to Direct Process Support

Engineering design involves an intricate interplay of conceptual synthesis of alternative requirements and design configurations, preliminary impact analysis of these alternatives using complex simulations, and human decision making. Such activities exhibit a highly multidisciplinary character and thus, cannot be carried out by a single person or organization. Instead, they employ numerous human performers carrying diverse knowledge backgrounds and heterogeneous skills.

Initially, each developer has a set of goals in his mind that reflect the strategic contributions of his assigned role to the complex interplay of design activities. These goals can be shared with other developers (e.g. belonging to the same engineering team), or even depend on the achievement of other goals by others (e.g. during cross-disciplinary activities). In the course of design, cooperation arises inside or across functional teams and company boundaries in order to synergistically attain common goals that otherwise would have been too complex and time-consuming to be attained.

Different empirical studies have shown that an engineer typically spends as much as 35-50% of his whole daily time while cooperating with his colleagues [221]. Nowadays, cooperation has become even more intense due to increased global competition that forces manufacturing companies to develop even increasingly complex products in even decreasing times. As a response to these demands, today's companies require even more cooperative work across cross-functional teams along their supply chains resulting to the establishment of extended enterprises. Thus, the importance of cooperation in modern design processes cannot be underestimated. Indeed, cooperation and teamwork has been extensively covered by several systematic approaches for design excellence and improvement ranging from the Total Quality Management paradigm to the more modern Participatory Engineering and Concurrent/Simultaneous Engineering approaches that try to bring together design and manufacturing activities [576].

As a consequence, cooperation inside the multidisciplinary settings of engineering design is inevitable, and support is demanded for the efficient and effective dissemination of information among developers. In order to adequately capture the cooperative character of design, we have to take into consideration the various facets of process information and goals from each individual developer's viewpoint. These facets provide answers to questions of the type *what* work has to be done, *who* is going to do it, *when* should it be done and *how* should it be done. Based on these considerations, a modeling formalism capturing cooperative work should underlie the following aspects [619]:

- *functional* aspects, describing what process steps are to be followed, and what flow of information will take place during them;
- *behavioral* aspects, describing how something has to be done (i.e. routing information of process steps and their flow of information), as well as when it can happen (i.e. preconditions and criteria);
- *informational* aspects, detailing what kind of data will be produced, consumed or transformed by a process step;
- *organizational* aspects, outlining where and by whom a process step is going to be realized.

A modeling language (i.e. metamodel) that combines the above four aspects, as well as their domain specific interdependencies and constraints, is able to consistently describe the design process. Our existing NATURE based environment metamodel captures the ways of working that support the enactment of developers while cooperating. In order to touch cooperative work, it will have to be extended with further elements of a *cooperation metamodel* that describes the above aspects. The cooperation metamodel is compatible with the basic elements of the C3 formalism for the description of cooperative work that has been developed in the IMPROVE project (see Subsect. 2.4.4). The extended environment metamodel in shown in Fig. 3.11.

The logical pieces of work that require the support of human and machine resources for their execution are captured in the *task*. A task represents the elementary activity level inside which the real work happens. High level tasks corresponding to management activities refined until elementary tasks are only interesting for systems of the administration level (see Sect. 3.4) and are, thus, not considered.

Each task requires a number of *resources* in order to be carried out. We distinguish two kinds of them: *human actors* that perform a task and the engaged software *tool categories*. Human actors are indirectly associated to their assigned tasks via their *roles*. Roles are distributed according to the knowledge background, skills and assigned responsibilities of the human actors. During

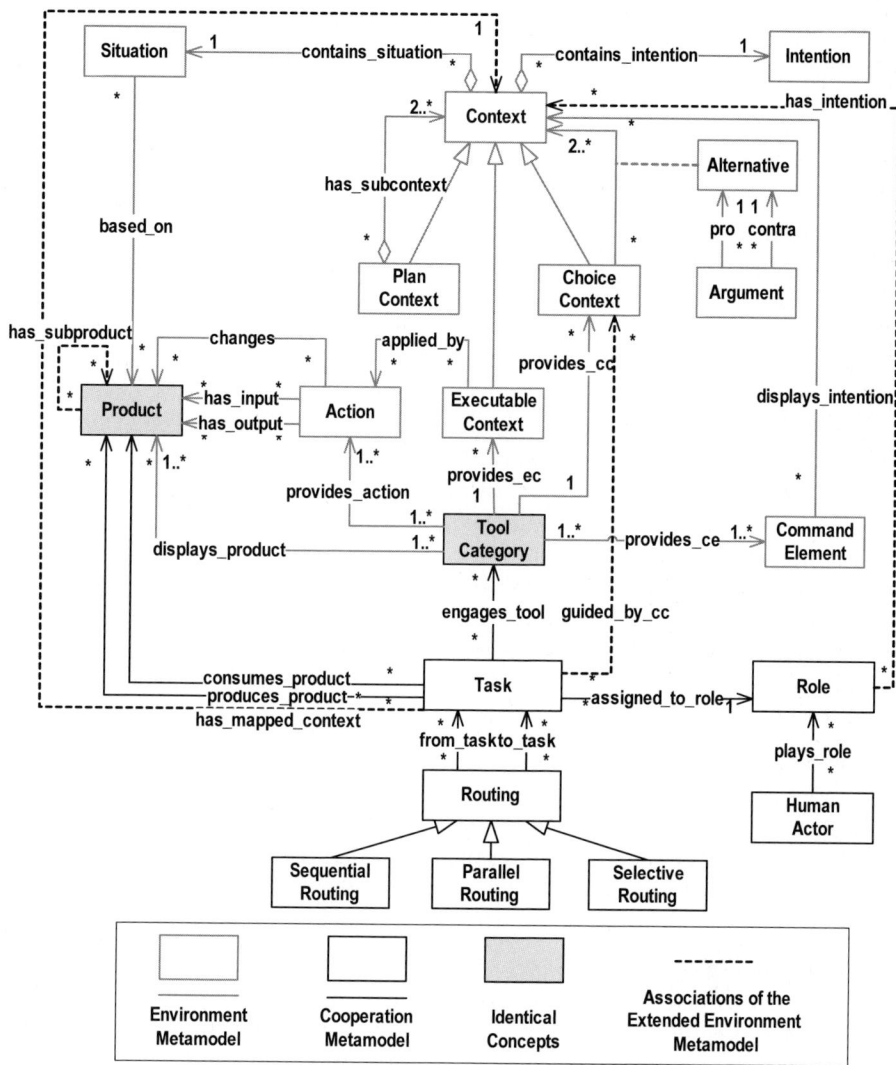

Fig. 3.11. Extended environment metamodel

the task enactment, human actors and computer agents work on objects of the real world (e.g. project specifications, flowsheets, simulation results and mathematical models). All these objects are captured as *products* produced or consumed by a task.

The *routing* element models the different ways of control flow organization among tasks. Three basic routing types (specializations) are distinguished: *sequential routing* when the tasks are carried out one after the other, *parallel routing* when more than one tasks can be active at the same time (i.e. *AND-*

split and *AND-join*) and *selective routing* for the situation when a specific task among several others has to be selected based on the evaluation of preconditions (i.e. *OR-split* and *OR-join*).

The integration of the NATURE based environment metamodel with the cooperation metamodel has been built upon the identical concepts of product and tool category, and three additional relationships:

- Each developer is provided with method guidance while enacting his assigned tasks. This method guidance occurs through the enactment of specific methodologies described as contexts by the NATURE process metamodel. Thus, a *has_mapped_context* relationship represents the possible contexts that guide the enactment of a task.
- At the cooperation level, products are usually considered as pieces of information at the level of document. Method guidance, on the other hand, requires a further refinement of the product to its comprising parts that are changed by tool actions. Thus, there was a need to bring the product decomposition relationship *has_subproduct* to the metamodel level in order to distinguish the two cases.
- The intention of the NATURE process metamodel represents the goal that a developer wants to attain. In order to bring these goals to the cooperation level and capture the decisions of developers while cooperating, we added the *has_intention* relationship between a human actor and the context of his intention.

To sum up, the extended environment model presented above is based on the integration of a number of concepts that model selected aspects of the cooperative work among several technical workplaces, and the existing PRIME environment metamodel for fine-grained and flexible method guidance inside each individual technical workplace. The integration of the two metamodels has been realized through three prominent associations that preserve:

- *Data integration*: the products worked upon by a task are related to the products transformed by the tools engaged during the task.
- *Control integration*: an external coordination system can directly influence the work at a technical workplace by directly requesting the enactment of a method fragment mapped to the actual task.
- *Consistency management*: monitoring information can flow from process-integrated tools to the coordination level.
- *Confidentiality*: the plan contexts that the method guidance provided to a developer is based on, are restricted to the allowed ones (i.e. plan contexts whose intention is associated with him).

A Comprehensive Framework for Process Integration

As mentioned earlier, tools can offer different degrees of accessibility to their internal structures and functions though the APIs they expose. At one extreme, some of them might not allow at all their external control and thus,

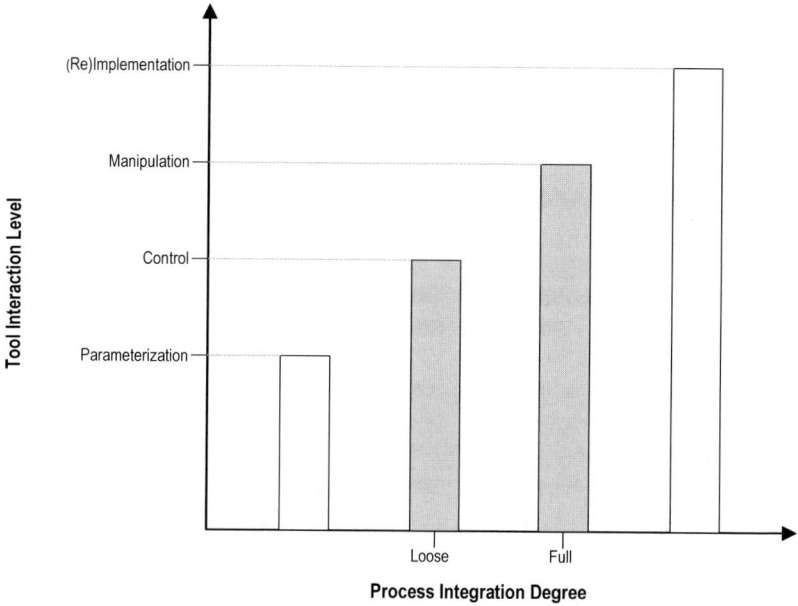

Fig. 3.12. Possible degrees of process integration

their process integration potential is restricted to a simple starting of the tool with command-line parameters (like the one supported by the WFMS reference metamodel). At the other extreme, though not so often, a tool might provide facilities for external invocation of its services and changes of its user interface, or send back notifications concerning its internal state. Tools of this last class are required by the original PRIME process integration mechanism. The most interesting and usual case though, is that of a tool that lies somewhere in the middle of the spectrum. Such a tool, does not offer all the necessary APIs for process integration but only a subset of them.

In order to address the problem of process-integrating the latter case of tools that do not fully comply to all PRIME requirements, we extended the original process integration framework with ideas towards a more flexible and easily adaptable mechanism for the integration of a broader range of tool categories. To this end, we have identified a number of discrete interaction levels that can be established with external tools and assigned to them corresponding degrees of process integration (cf. Fig. 3.12). The horizontal axis represents the process integration spectrum from non existing and loose, up to the extreme of full. The vertical axis represents the four possible levels of interaction with a tool, inspired by [999].

At the very first level of tool interaction (*parameterization*), a tool is embedded in the process as a black box. The only possible ways of interacting with it is by providing it with input information when it starts up. Such input

Table 3.1. Correspondence between tool interaction levels for process integration and PRIME APIs.

Interaction Level/API	A1	A2	A3	A4	A5	A6
Parameterization						
Control	x	x				
Manipulation	x	x	x	x	x	x
Reimplementation	x	x	x	x	x	x

is usually given in the form of command line parameters that either initialize the internal tool state, or activate specific execution modes, or even load a specific product. After the tool has been started, it provides no means for its external control (e.g. triggering of a service inside it), or retrieval of notification concerning the interaction of the user with its user interface. Thus, the potential of such degree of process integration is limited to the on-time opening of the tool. When the tool uses an open product format, then specific product information can be loaded inside it too.

In order to offer method guidance to such a poorly-integrated tool, albeit externally, a specialized guidance tool that is part of the generic PRIME architecture called *ProcessGuide* can be employed. The idea behind ProcessGuide is similar to that of agenda managers widely used by workflow management systems. The interface of ProcessGuide provides the user with information concerning the available contexts that he can manually execute inside the tool. It is, then, the responsibility of the user to correctly execute the context inside the tool, and document his action back in the ProcessGuide. Since the ProcessGuide has been developed with process-integration in mind (a-priori), the process engine can virtually control the consistency of the guidance inside the tool and automatically trace the user's actions. Nevertheless, experience has shown that this kind of process support is highly complex and error-prone because it depends on the user to firmly and on-time report his execution progress [371].

With the second level of tool interaction (*control*), we introduce the notion of *loose* process integration. Loose process integration corresponds to the most often addressed case of commercial tools that restrict their external interactions to those of service invocation and notifications for tool events. Such tools provide service invocation (A1) and feedback information (A2) APIs (Table 3.1), but they provide no means for extending their user interface with new command elements (A3), or increasing their process sensitivity by highlight product shapes or command elements and returning feedback concerning their selection (A4, A5 and A6).

Thus, a loosely process-integrated tool is not able to provide by itself integrated method guidance based on method definitions. The guidance of the user inside the tool is restricted by the interaction patterns prescribed by its tool vendor. Nevertheless, the loose process integration of a tool can bring benefits in the following two cases:

- *Tool-spanning* method fragments (plan contexts) offered as guidance alternatives inside a fully process-integrated tool, can couple it with a loosely-process integrated one, and allow control and/or information flow from one to the other.
- Traces of specific tool actions captured from a loosely process-integrated tool can be used as signals of situation matchings that, then, initiate the execution of plan contexts.

The third degree (*manipulation*) captures fully process-integrated tools like the VISIO based flowsheet editor, that can be fully manipulated both in terms of interactions with tool services and command elements. A fully process-integrated tool exploits all six APIs (Table 3.1) and is able to faithfully follow the process definitions and accordingly guide the user when he demands it. Obviously, full process integration brings the highest potential level of integrated method guidance, albeit requires ideal tools that are not so often found in the market.

For completeness reasons, the last degree (*(Re)implementation*) captures the tools that have been implemented from scratch, or have been largely reimplemented in order to obey in a hardwired way their intended process-aware behavior (a-priori process integration). (Re)Implementation guidelines for the a-priori process integration of tools, as well as external guidance techniques for tools allowing limited interaction at the level of parameterization have been studied in the PRO-ART framework that preceded PRIME [366]. Nevertheless, in the context of the IMPROVE project we are not interested in this case.

The described process integration framework is complemented by a new generic tool wrapper architecture. Our goal was to reengineer the existing one in order to reduce the effort expenditure of the wrapper creation process and make it flexible enough to accommodate both fully and loosely process-integrated tools. The UML 2 component diagram [880] shown in Fig. 3.13 gives an overview of the major pluggable components comprising the new wrapper architecture.

In the new architecture, the burden of maintaining context-specific information has moved to the process engine. Wrappers are able to comprehend strictly tool-specific information. Thus, each time that context information has to flow to a tool, a module inside the process engine is responsible to convert it to tool-specific calls and forward them to the appropriate tool. The state manager is responsible for the maintenance of the wrapper's internal state based on the exchange of messages with the process engine using precise interfaces. The action adapter maintains mappings between actions in the tool

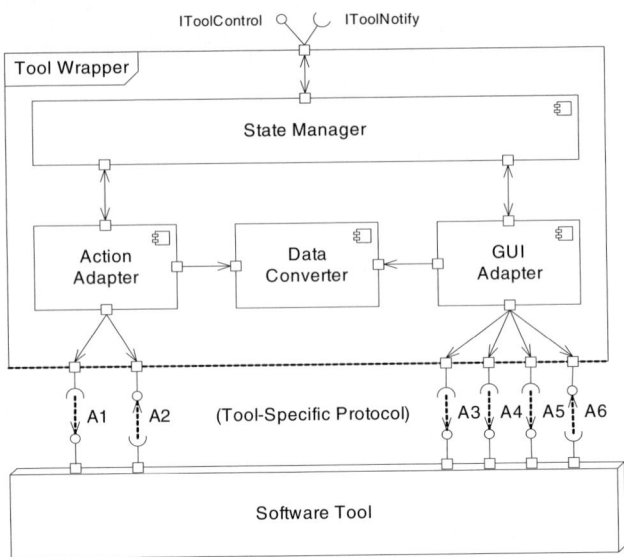

Fig. 3.13. New component-based tool wrapper architecture

model and services of the wrapped tool for the dispatching of service invocations and receival of feedback information (A1 and A2 APIs). Similarly, the GUI adapter uses the appropriate APIs provided by the wrapped tool in order to send or receive messages concerning changes in its user interface (A3-A6 APIs). The data converter is responsible for the conversion of the internal tool product format to the neutral one used in the PDW, and vice-versa. Obviously, the architecture for the case of a loosely process-integrated tool can be reduced to the first two or three components.

Improvement-Oriented Reuse of Captured Experiences

So far, we have contended that because of the high degree of creativity in engineering design, it cannot be fully prescribed in advance. Our PRIME based design support environment tackles this problem by providing effective process support based on the *explicit knowledge* of well-understood method fragments where an agreed way of working dominates engineering practice. The PRIME process support chain can be summarized as follows (Fig. 3.14):

1. The process support chain starts with the knowledge acquisition phase. Here, domain experts (i.e. process engineers) define the specifications, requirements, and design methodologies of the addressed design scenarios. This domain knowledge is, then, formalized as NATURE based method fragments that are stored in a process repository.

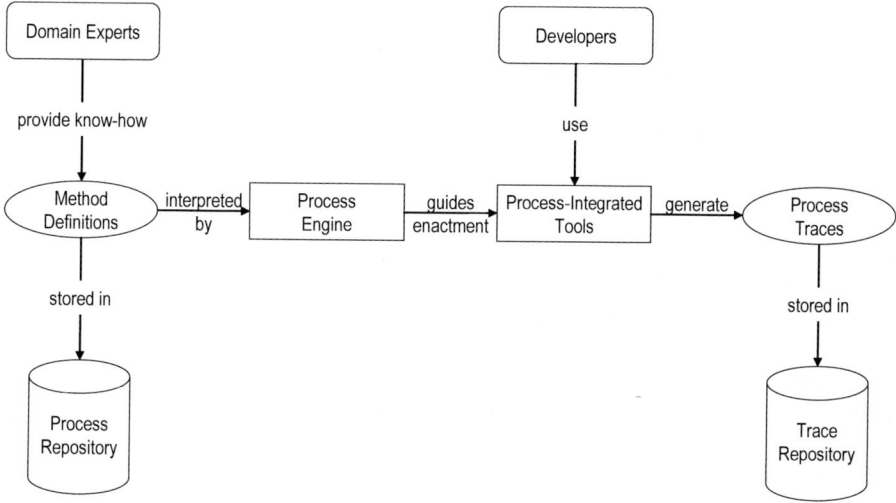

Fig. 3.14. PRIME process support chain

2. Method definitions are integrated with tool models in environment models. These environment models are enriched with runtime semantics (e.g. Petri-Net extensions) and can get mechanically interpreted by the process engine, when method guidance is demanded.
3. The interpretation of environment models can control the enactment at the technical workplace through the process-integrated tools that the human actor interacts with.
4. Design history inside process-integrated tools is automatically recorded and organized in trace chunks using a concrete traceability model. Finally, the captured trace chunks are stored in a trace repository.

The method definitions stored in the process repository of PRIME represent the capitalized knowledge of the company that is provided to its workers on demand. This knowledge constitutes a valuable strategic asset, that is disseminated to the designers in appropriate situations through their process-integrated tools. Further, the stored trace chunks at the end of the chain describe the real design lifecycle in a systematic way. Except for the prescribed process parts, they further capture the "unrestricted" parts where no well-known way of working exists and the human actor's decisions are solely based on his experience background.

As a consequence, PRIME can be seen as a knowledge-based engineering framework that facilitates the computer-based *internalization* and *externalization* knowledge conversion processes, according to the SECI model of Nonaka and Takeuchi [866]. The internalization of explicit knowledge (method definitions) is supported via the method guidance provided through process-integrated tools. On the other hand, externalization of human choices, men-

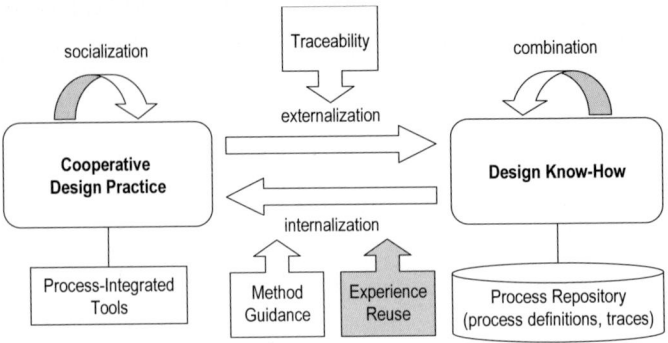

Fig. 3.15. Cooperative knowledge management in engineering design practice and PRIME contributions

tal models and technical skills (the so-called *tacit* knowledge) is empowered through traceability. Concerning the other two knowledge creation processes, the *socialization* process is outside the scope of PRIME, since it requires some kind of social network and collaboration environments. The explicit knowledge maintained by PRIME is provided through external processes based on interviews, observations and discussions between domain experts. Thus, the process of *combination* is neither supported.

Figure 3.15 puts the SECI model in engineering design context, using concepts from the cooperative information systems framework described in [186]. We distinguish the cooperative design practice domain where the actual design process is being enacted, and the design know-how domain that captures the formalized company knowledge. The stored traces at the end of the internalization process, reflect the experiences of developers while working with their software tools (operational knowledge).

The developer can greatly benefit from reusing knowledge extracted directly from traces of the past that apply to his current context. Observation and comparative analysis of the captured traces can set the foundation for experience-based dissemination and reuse of design knowledge: in the short term, for empowering the human actor to reuse best practices abstracted from design experiences and improve his mental models and technical skills; in the long run, for providing accountability of good and bad practices, as well as lessons learned.

As a result, we were challenged to provide support to the developer facilitating the context-based internalization of design knowledge clustered in experiences from the past (captured traces). To this end, we extended the original PRIME support chain according to a reuse infrastructure illustrated on Fig. 3.16. Specialized reuse interfaces constitute the cornerstone of this infrastructure, which can provide solution or advice for the actual situation on demand , by utilizing knowledge of recorded experiences from the past. The

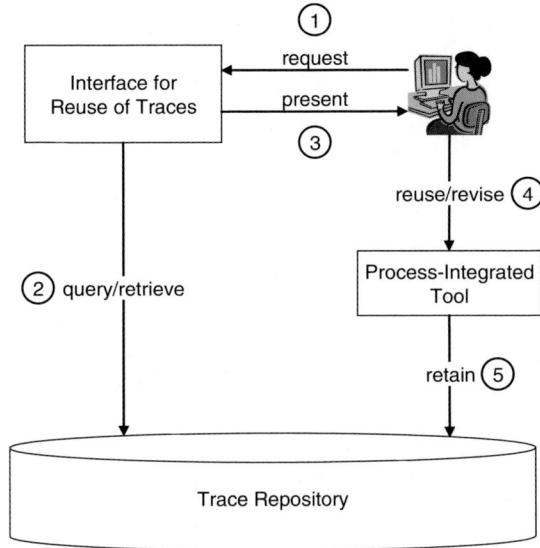

Fig. 3.16. Infrastructure for the reuse of captured traces

very essence of a recorded experience is represented using the basic NATURE formalism:

- A *situation* describes the process state in which the experience was captured.
- An *intention* reflects the sought goal.
- A *context* encapsulates the intended solution.
- A *product* represents the assets that the solution worked upon.

Whenever the developer feels like he would benefit from experience-based support, he can trigger the appropriate reuse interface of the infrastructure on demand. Then, support is provided following the workflow of five steps shown in Fig. 3.16:

1. *Request*: Initially, the developer requests the reuse interface for support. The interface is embedded inside PRIME and, is thus able to retrieve the actual situation of the developer's work that is an important criterion for the identification of the relevant stored traces.
2. *Query/Retrieve*: The reuse interface queries the trace repository and retrieves the stored traces that fit the situation at hand. The query is processed and executed by an analysis module having direct access to the trace repository (not shown in Fig. 3.16)
3. *Present*: The retrieved results (if any) are presented to the developer through the reuse interface using an appropriate representation formalism. If the developer is interested in process experiences, the respective

chunks are organized chronologically and hierarchically according to their contextual decomposition. Product experiences are displayed in a user-friendly hierarchical way showing the product decomposition along with a graphical visualization (when applicable).
4. *Reuse/Revise*: The developer can, then, decide to reuse the traces or revise them inside his process-integrated tool. In the case that he decides to revise a proposed solution, he can optionally document his decision using the PRIME decision editor.
5. *Retain*: Any previous experiences reused (without or after revision) are automatically recorded by the process-integrated tool and their trace is stored in the trace repository. From now on, they also constitute pieces of recorded experience that can be provided for support to analogous situations in the future.

Validation

In the following, we illustrate the application of the new PRIME ideas on some demonstration examples from the IMPROVE project.

Traceability across Workplace Boundaries

By integrating the cooperative work perspective to our NATURE based meta-model, we have been able to extend the PRIME process support in order to support the cross-functional interplay among developers. More precisely, PRIME is able to remotely trace the process enactment at several workplaces and transfer information from one to the other, when needed, using a shared ontology that obeys company policies and rules. In order to show the effectiveness of the distributed process support, we have implemented a tool called cooperation console that is able to visualize the workflow of elementary activities of the design process, provide information concerning its execution status, and to unfold the design history for each developer.

Figure 3.17 shows a snapshot of the PRIME cooperation console. The simplified workflow shown for demonstrative reasons, is described using the C3 formalism (cf. Subsect. 2.4.4). The tasks with black border have already been delegated to the developers, whereas the ones with gray border are still inactive. The highlighted task "1D-Simulation of Compounding Alternatives" is currently active. By double clicking on any task, a pop-up window opens that displays the design history of the developer during that specific task (lower left part of the figure). In order to protect confidential data, only process traces of specific tasks are shown to specific classes of developers who use this tool, according to a classification schema of information. For example, a developer assigned with a central managerial role, can observe the design history of the whole process, whereas a group leader can only access the design history of his own group members.

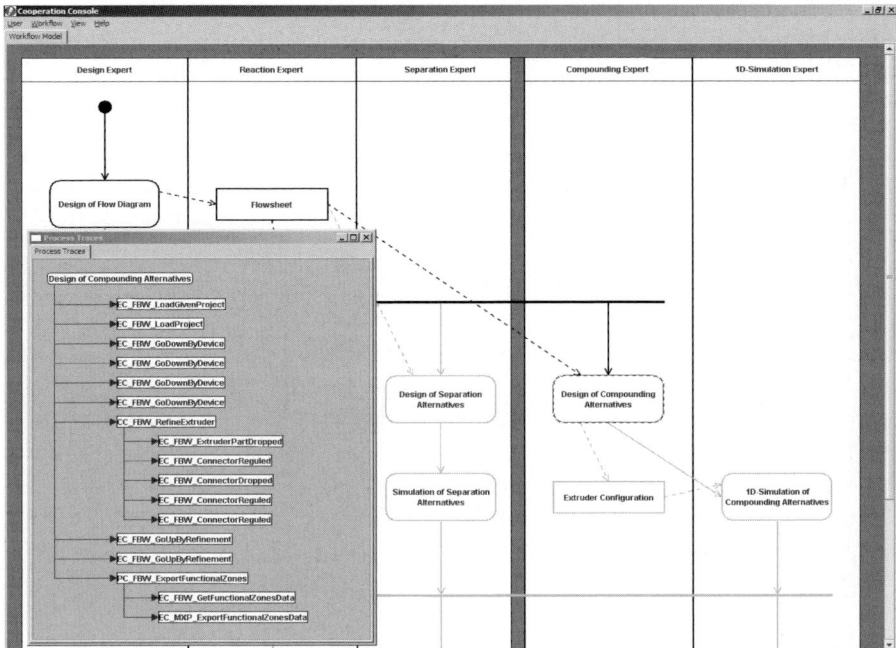

Fig. 3.17. User interface of the cooperation console

Loose Process Integration of MOREX

The concept of loose process integration has been validated with the prototypical integration of the one-dimensional simulation tool MOREX. Concerning the requirements for process integration, MOREX, as typically the case for a broad range of commercial tools, exposes only the A1 (service invocation) and A2 (feedback information) APIs using COM interfaces. Based on these APIs, we have built a MOREX wrapper employing on our new generic wrapper architecture with little effort. Since MOREX maintains its product information in XML files, the data exchange of the wrapper with the process engine has been realized based on a standardized XML interface format. Simultaneously with the process integration of MOREX, we have extended the original hierarchical model of the compounding extruders inside the flowsheet editor with in intermediate level of *functional zones* that are refined by screw element types, which can then be simulated inside MOREX.

The contribution to the overall scenario is summarized through the three process steps shown in Fig. 3.18. Suppose that the developer has finished processing the initial version of the extruder functional zones in the flowsheet editor. This information is sent, under the control of the process engine, to MOREX (step 1). MOREX receives the extruder configuration, along with material and process parameters via XML. The engineer, then, starts the

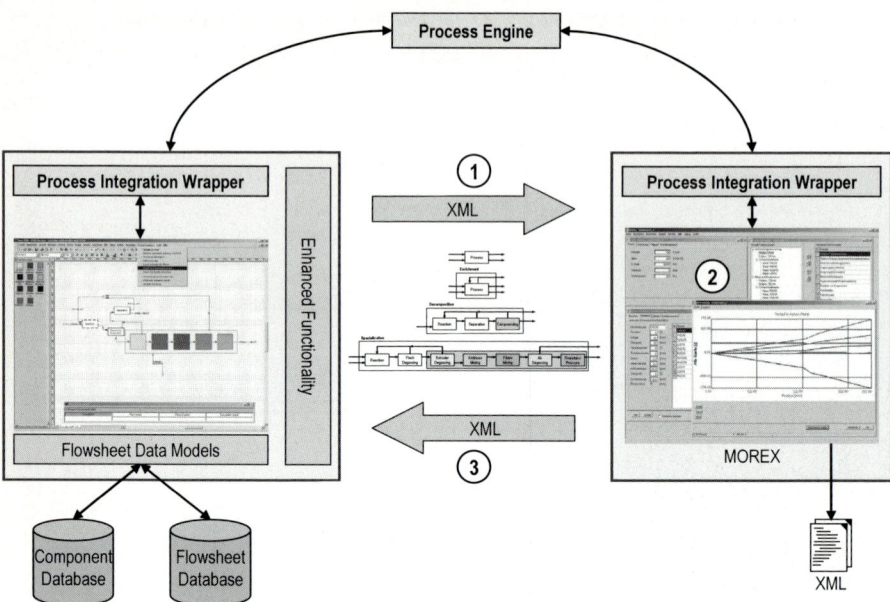

Fig. 3.18. Process-integrated interactions between the flowsheet editor and MOREX

analysis by running simulations and creating visualizations and diagrams of the results (step 2). Based on the results, he can change and optimize accordingly the original extruder configuration in MOREX. When the optimized configuration appears satisfactory, the developer can send it back to the flowsheet editor (step 3). In both cases of information exchange, guidance based on the corresponding method fragments is requested through menu items of the fully process-integrated flowsheet editor.

Domain-Specific Reuse of Product Traces

The introduction of the intermediate level abstraction of functional zones can empower the reuse of product traces with two goals: on the one hand for the effective composition of an extruder configuration from functional zones and, on the other hand, for the reuse of experience-based knowledge of successful realizations of functional zones in real devices. Being unable to integrate such reuse support inside the existing MOREX simulation tool, a special loosely process-integrated tool has been realized to help the engineers selectively reuse mappings between the (shared) functional zones flowsheet and the (discipline-specific) screw configurations associated with one or more adjacent functional zones. We call this domain-specific reuse tool FZExplorer (Functional Zones Explorer).

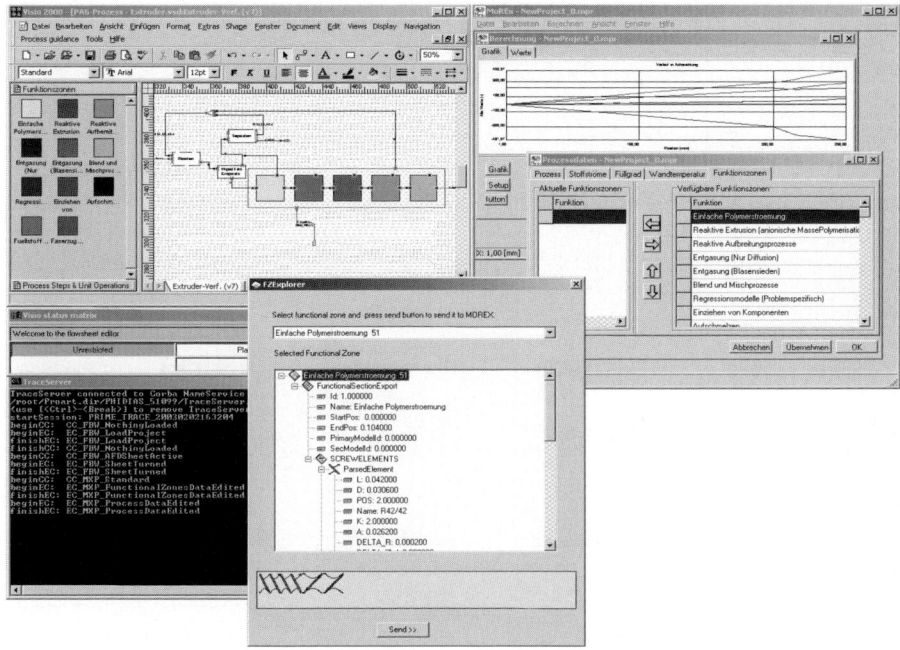

Fig. 3.19. Experience-based reuse of functional zones in MOREX

Figure 3.19 sketches a small scenario demonstrating the FZExplorer usage. At the upper left part, the window of the flowsheet editor is shown and next to it, is the MOREX tool that has just received the functional zones from the flowsheet editor. In MOREX, the functional zones will be added to a new project process data configuration, and the received process and geometry parameters will be loaded. Possibly, existing functional zones should be enriched with screw elements, or new ones zones should be added. To help this reuse task, the FZExplorer is invoked by a PRIME process context (center of the figure). After querying the PRIME traces repository for the requested information, the FZExplorer retrieves it and displays it in a user-friendly hierarchical way, along with a graphical visualization of the screw elements to the user. The user can then select the appropriate configuration and import it back to MOREX. At the lower left part is shown the text window of the automated tracing of the execution of the whole sequence of described actions (EC stands for executable context and CC for choice context).

Method Advice Based on Recorded Process Traces

Whereas method definitions are biased by perceptions, process traces provide an objective abstraction of reality: they further describe the experience that was acquired by a developer in a specific problem context. Due to creativity

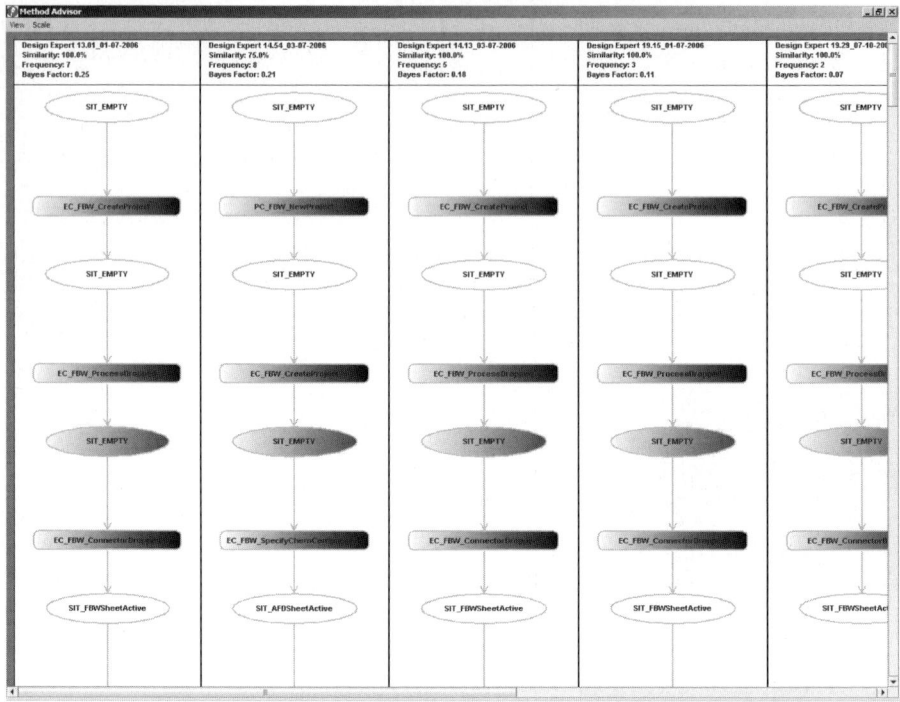

Fig. 3.20. Petri-Net based visualization of retrieved trace chunks in the method advisor's user interface

of design, the process traces might considerably deviate from the method definitions and, thus, provide faithful evidence of design change and evolution.

Thus, the developer can benefit from method advice that exploits captured traces of the PRIME environment, generalizes knowledge from the experiences indirectly captured inside them that apply to the current process situation, and mediates them for reuse. Then, it remains in the hand of the developer to evaluate the appropriateness of each one of the retrieved proposals to his actual context, and decide which one, if any, fits better. A chosen proposal, either in its original or altered form, can be manually enacted by the developer in his process-integrated tools.

The above strategy of method advice is implemented by a specialized module of the PRIME architecture that executes the FIST algorithm for finding relevant process traces, developed in [80]. The returned trace chunks are presented to the developer through the method advisor reuse interface. The method advisor continually traces the current situation and previous actions of the developer. At any time during his task (e.g. while interacting with MOREX, where no method guidance is offered), the developer can request the method advisor for method advice giving the preferred values for the FIST algorithm parameters. The process traces returned (if any) are visualized in the Method Advisor's user interface using either a Petri-Net based (Fig. 3.20) or a tree-like structure (not shown here).

The Petri-Net based visualization depicts a sequential, flat view of the trace chunks consisting of the addressed situations and the contexts executed at each situation. The displayed trace chunks are sorted according to their Bayes factor calculated at the final step of the FIST algorithm, and annotated with values of their timestamp, similarity, frequency and Bayes factor. Moreover, the displayed situations that are similar to the current one are highlighted, and their exact similarity to the current situation can be shown by clicking on them.

3.1.6 Comparison with the State-of-the-Art

Contributions that are relevant to our work can be identified in various research areas. The direct, fine-grained process support relates to approaches from the areas of computer-based process support and tool integration. For experience-based reuse, research results about traceability and reuse support infrastructures from the disciplines of software and chemical engineering are applicable.

Process Support Systems

Numerous process support systems have been developed for the computer-based coordination of processes. Traditionally, they mainly fall into two broad categories. Process-centered (software) engineering systems are often used to support (software) engineering design processes [501, 628], whereas workflow management systems care mainly for the support of business processes [954, 1014]. Both of these categories of systems strive for the effective and efficient coordination of work among several participants based on the interpretation of formal models, usually employing Petri-Net based languages. Nevertheless, most of the identified approaches are suited for deterministic processes that are well-known in advance. They do not fulfill the requirements for the support of open, creative and knowledge-intensive processes like engineering design. A partial solution has been given with the introduction of runtime exception handling and workflow adaptability [586, 984] that, nevertheless make the models more complex. In addition, the exceptions have to be modeled in advance, resulting to an anaemic treatment of eventuality.

In contrast, our solution follows a situation-based approach and models the context only in situations where a well-defined method fragment exists. When such a situation occurs, the user is offered the opportunity to, on demand, follow the methodical guidance of the process engine, without being restricted to do so. Further, most of the existing approaches use process modeling formalisms with a coarse- or medium-grained representation of the process. Thus, they are more appropriate for project administration purposes and they provide no real support in terms of fine-grained method guidance at the technical workplace, like PRIME does.

Tool Integration

In the literature, there have been several discussions on the benefits of tool integration as part of a system development environment [768, 950, 998]. Most of the proposed solutions center around the five kinds of integration initially identified by Wasserman [1038]: platform integration, concerned with environment services; presentation integration, concerned with the offering of uniform interface; data integration, concerned with the exchange of data; control integration, concerned with the sharing of functions; and process integration concerned with the process-guided behavior. According to our approach, process integration plays a first citizen role among all other kinds of integration, as it provides the means to provide integrated method guidance to the user through the increased process sensitivity of his tool. In the context of the IMPROVE project, our research on process integration is complemented by that of other subprojects that investigate data integration (cf. Sect. 3.2), control integration (cf. Sect. 4.2), and platform integration (cf. Sects. 4.1 and 4.2).

Several traditional approaches emphasize only control and data integration aspects of tools. For example, some researchers propose component-oriented approaches for control integration in open tool environment [948, 1052], or combine them with emerging standards like XML Metadata Interchange (XMI) for the integration of tools in the context of conflicting data models [624]. Other paradigms are exemplified by agent-based approaches for the orchestration of tools in heterogeneous distributed environments, encapsulating them into agents [614]. In the workflow management arena, the Workflow Management Coalition (WfMC) has defined, as part of its workflow reference model, a standardized interface for the communication of workflow enactment services with external applications ("invoked applications") [1057]. This interface mainly supports the starting and stopping of external tools according to the Workflow Application Programming Interface (WAPI). In practice, tools are integrated with workflow aspects of the process using middleware technologies like CORBA and DCOM (e.g. SAP Workflow).

The modern horizontal business landscape has forced companies to find new ways to integrate their business processes across organizational boundaries [837]. This new reality had a great impact on the traditional tool integration trends that have been reworked and shifted towards ubiquitous integration. The resulting movement of *application integration* strives to capture the heterogeneity of systems, applications and legacy tools, in order to integrate them inside business processes and increase the competitiveness, agility and market responsiveness of a company [819]. Service-Oriented Architectures (SOA) play a key role in application integration by providing guidelines for the loose-coupling of heterogeneous applications in a coherent whole that exposes selected functionality to the outside world as services [659, 740]. Although there already exist several mature technologies for the implementation of SOAs, the most recent trends converge on the use of internet technologies for communication employing platform-independent Web Services [500, 814]. The reengineered PRIME framework employs SOA principles for the process integration of tools using Enterprise Java Beans and CORBA. Specifically, services of PRIME process-integrated tools are uniformly described (using the environment metamodel semantics), and orchestrated by the process engine.

Experience-Based Reuse and Improvement

Reuse of lifecycle experiences and products has drawn a lot of attention in the (software) engineering domain, as a means of increasing final product quality and productivity, while at the same time reducing overall cost of operations [818]. Several improvement paradigms have been proposed that employ reuse through feedback loops in order to improve quality for various processes at different levels [533, 741]. As a special form of experience-based reuse, traceability in (software) product families engineering has recently drawn a lot of attention [544, 912]. Traceability, facilitates the development and maintenance of a product family, while at the same time offering the potential for discovering commonalities and variabilities of the product family artifacts, and learning from them.

In the domain of engineering design, experience reuse has drawn a lot of attention among researchers. The KBDS system attempts to capture design rational of chemical engineering and evaluate it by tracking the interrelations between design alternatives, objectives and models inside a flowsheet tool (similarly to our VISIO based flowsheet editor) [524]. The n-dim approach promotes the collaborative character of design and proposes a classification schema for the organization of diverse types of information from multiple developers along multiple dimensions that allow the capture of evolution and reuse [989]. More recent research efforts concentrate on the experience reuse inside the frame of specific design support tools and employ the CBR (Case-Based Reasoning) paradigm for the retrieval, reuse and adaption of design cases from the past [588, 805].

Our approach differs from the above identified contributions with respects to some of its key ideas. Most of the other approaches implement external experience repositories decoupled from the running system, and the user has to manually provide a description of the problem and query them. In contrast, the PRIME reuse interfaces exploit the process integration mechanism in order to continually monitor the actual situation at the technical workplace and, on demand, directly present to the user the relevant experiences from the past. Moreover, while the experience gathered by most other CBR systems captures a pure description of the problem and its related solution, PRIME exploits whole trace chunks that further capture the temporal dimension of the solution in the form of precise steps that other users followed (the interested reader is referred to [848] for a similar work). Last but not least, in our approach, the user does not have to explicitly document the revision of a solution that he might reuse. A reused solution, in an original or revised form, is automatically recorded by process-integrated tools without requiring user intervention.

3.1.7 Conclusions and Outlook

The chemical engineering domain brings out some problems in large-scale design which are perhaps less clear in the more widely discrete domains such as mechanical or software engineering. In this section, we focused on the question how to bring knowledge to the engineering workplace as directly as possible (i.e. in a process-integrated manner) even if the knowledge is complex, and undergoes continuous change.

Our solution to this challenge has been exemplified through the flowsheet-centered design support environment we have developed, based on the PRIME approach. PRIME was originally able to provide integrated method guidance based on the a-priori process integration of tools. Empowered by its solution ideas and driven by project requirements, we extended the original PRIME approach towards the a-posteriori process integration of commercial tools, cross-functional support among developers, and situated visualization and reuse of process and product traces.

In further informal experiments with our environment by chemical and plastics engineering researchers and practitioners, the efficiency gain of integrated method guidance in supporting their tasks has been quite substantial. Together with the automated capture of process traces, according to their evaluation, a major step towards the reuse of design experiences has been accomplished. Reported key success factors that distinguish our approach include:

- the promotion of the contextualized nature of design through our NATURE based metamodel that can be exploited both for the purposes of guidance and traceability;
- the provision of methodical guidance directly from inside the domain-specific tools, increasing their process sensitivity;
- the integration of several experience reuse tools empowering learning based on the comparison of accumulated experiences from the past.

The experience with the IMPROVE case study has further shown that full and loose process integration, while conceptually building on the same framework, should not necessarily use the same implementation techniques. The PRIME object-oriented implementation framework for tight process-integration was originally implemented in C++, using COM [846] and CORBA [877] for integrating tools. The succeeding reengineered framework that further incorporated loose process integration was implemented as a three-tier service-oriented architecture employing EJB and JMS [504] for service orchestration, and COM/CORBA for the communication with loosely process-integrated tools. In order to take advantage of subsequent standardization efforts, but also to enable better support for the security aspects of internet-based cross-organizational integration, we are planning to employ Web Services [500] for loose process integration. This implementation will also be strongly based on ontology specification languages like OWL [546].

Until now, we have focused on the experience reuse on a project basis. Equally important is the experience reuse on a corporate basis in order to feed and improve the existing company know-how. To this end, we are planning in the near future to develop a methodology for the bottom-up capitalization of knowledge by abstracting from individual experiences across several project instantiations (high level cases). Inadequate methodologies can then be detected through the identification of certain ways of working that perform poorly according to qualitative metrics, whereas new ones can elaborate with a statistical analysis of frequent process violations and discrepancies.

The current implementation of our environment provides a client/server model for process enactment. One central enactment server is used to support many developers who are connected to it. In real world scenarios, a design process can span among several geographically distributed engineering teams. In such a case, our central enactment mechanism might become the bottleneck. Thus, more than one enactment mechanisms should be provided to ensure adequate performance, each one residing at different company sites. Only well-defined interfaces for the exchange of specific process and product information will be available for their intercommunication, resulting to a better preservation of confidentiality constraints.

3.2 Incremental and Interactive Integrator Tools for Design Product Consistency

S. Becker, M. Nagl, and B. Westfechtel

Abstract. Design processes in chemical engineering are inherently complex. Various aspects of the plant to be designed are modeled in different logical documents using heterogeneous tools. There are a lot of fine-grained dependencies between the contents of these documents. Thus, if one document is changed, these changes have to be propagated to all dependent documents in order to restore mutual consistency.

In current development processes, these dependencies and the resulting consistency relationships have to be handled manually by the engineers without appropriate tool support in most cases. Consequently, there is a need for incremental integrator tools which assist developers in consistency maintenance. We realized a framework for building such tools. The tools are based on models of the related documents and their mutual relationships. Realization of integrators and their integration with existing tools is carried out using graph techniques.

3.2.1 Integrator Tools for Chemical Engineering

Introduction

Development processes in different engineering disciplines such as mechanical, chemical, or software engineering are highly complex. The product to be developed is described from multiple perspectives. The results of development activities are stored in *documents* such as e.g. requirements definitions, software architectures, or module implementations in software engineering or various kinds of flow diagrams and simulation models in chemical engineering (cf. Sect. 1.1). These documents are connected by mutual dependencies and have to be kept consistent with each other. Thus, if one document is changed, these changes have to be propagated to dependent documents in order to restore mutual consistency.

Tool support for maintaining inter-document consistency is urgently needed. However, conventional approaches suffer from severe limitations. For example, batch converters are frequently used to transform one design representation into another. Unfortunately, such a transformation cannot proceed automatically, if human design decisions are required. Moreover, batch converters cannot be applied to propagate changes incrementally. Current tool support in chemical engineering is mainly characterized by numerous software tools for *specific purposes* or *isolated parts* of the design process. However, a sustainable improvement of the design process can only be achieved by the integration of single application tools into a comprehensive design environment [548]. During the last years, commercial environments like Aspen Zyqad [517] or Intergraph's SmartPlant [605] have been developed. They are mainly restricted to the tools of the corresponding vendor. The adaptation of the tools to specific work processes of developers within a company or the integration of arbitrary tools, especially from other vendors, are unsolved issues.

In this paper, we present *incremental integrator tools* (integrators) which are designed to support concurrent/simultaneous engineering and can be tailored to integrate any specific interdependent documents from any application domain. The only restriction is that documents have to be structured such that a graph view on their contents can be provided.

The key concept of our tools is to store fine-grained relationships between interdependent documents in an additional *integration document* which is placed in between the related documents. This integration document is composed of *links* for navigating between fine-grained objects stored in the respective documents. Furthermore, these links are used to determine the impact of changes, and they are updated in the course of change propagation. Changes are propagated and links are established using an extension of the triple graph grammar formalism originally introduced by Schürr [413].

Integrator tools are driven by *rules* defining which objects may be related to each other. Each rule relates a pattern of source objects to a pattern of target objects via a link. Rules may be applied automatically or manually. They are collected in a rule base which represents domain-specific knowledge. Since this knowledge evolves, the rule base may be extended on the fly. The definition of rules is based on domain knowledge. Rules for our integrators for chemical engineering design processes are defined using the model framework CLiP [20] (cf. Subsect. 2.2.3).

Integrator tools for specific documents are built based on a universal *integrator framework* and by specifying a corresponding rule base. Additionally, some tool-specific extensions, like wrappers for the tools to be integrated, have to be implemented.

There has been a tight *cooperation* of this subproject with the CLiP project at the department of process systems engineering (LPT) [15]. All chemical engineering examples used throughout this paper have been elaborated in cooperation with the LPT and our industrial partner innotec [745].

Motivating Example

We will use the sample scenario in Fig. 3.21 to illustrate how integrator tools assist the design team members. The scenario deals with the integration of *process flow diagrams* (PFD) and *simulation models*. A PFD describes the chemical process to be designed, while a simulation model serves as input to a tool for performing steady-state or dynamic simulations. Different tools may be used for creating flowsheets and simulation models, respectively. In the following, we assume that the flowsheet is maintained by Comos PT [745] and simulations are performed in Aspen Plus [516], both of which are commercial tools used in chemical engineering design.

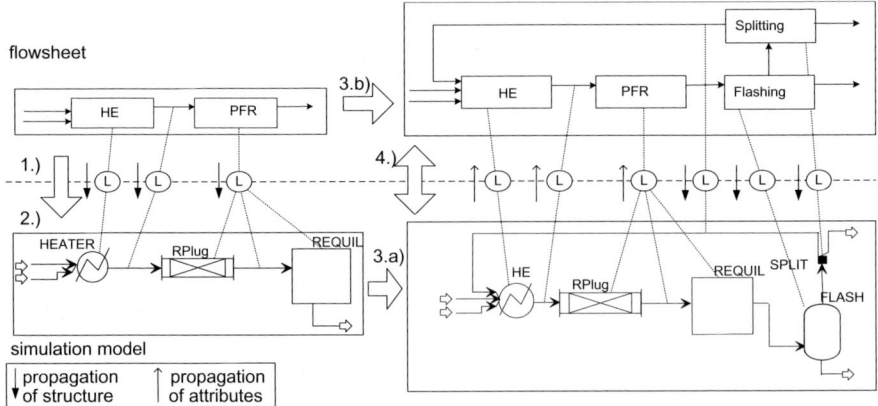

Fig. 3.21. Sample integration scenario: integration of PFD and simulation model

PFDs act as central documents for describing chemical processes. They are refined iteratively so that they eventually describe the chemical plant to be built. Simulations are performed in order to evaluate design alternatives. Simulation results are fed back to the PFD designer, who annotates the flowsheet with flow rates, temperatures, pressures, etc. Thus, *information* is *propagated back* and *forth* between flowsheets and simulation models.

Unfortunately, the *relationships* between both results are *not* always *straightforward*. To use a simulator such as Aspen Plus, the simulation model has to be composed from pre-defined blocks. Therefore, the composition of the simulation model is specific to the respective simulator and may deviate structurally from the PFD.

The chemical process taken as example produces ethanol from ethen and water. The PFD and simulation models are shown above and below the dashed line, respectively. Two *subsequent versions* of *both models* are depicted side by side. The integration document for connecting both models contains links which are drawn on the dashed line[26]. The figure illustrates a design process consisting of four steps:

1. An initial *PFD is created* in Comos PT. This PFD is still incomplete, i.e., it describes only a part of the chemical process (heating of substances and reaction in a plug flow reactor, PFR).
2. The integrator tool is used to *derive a simulation model* for Aspen Plus from the initial PFD. Here, the user has to perform two decisions. While the heating step can be mapped structurally 1:1 into the simulation model, the user has to select the most appropriate block for the simulation to be performed. Second, there are multiple alternatives to map the PFR. Since the most straightforward 1:1 mapping is not considered to be sufficient, the

---

[26] This is a simplified notation. Some details of the document and integration model introduced later are omitted.

user decides to map the PFR into a cascade of two blocks. These decisions are made by selecting among different possibilities of rule applications which the tool presents to the user.
3. a) The simulation is performed in Aspen Plus, resulting in a *simulation model* which is augmented with simulation *results*.
   b) In parallel, the *PFD* is *extended* with the chemical process steps that have not been specified so far (flashing and splitting).
4. Finally, the integrator tool is used to *synchronize* the parallel *work* performed in the previous step. This involves information flow in both directions. First, the simulation results are propagated from the simulation model back to the PFD. Second, the extensions are propagated from the PFD to the simulation model. After these propagations have been performed, mutual consistency is re-established.

An integrator tool prototype has been realized to carry the design process out in this example. This was part of an industrial cooperation with innotec [745], a German company which developed Comos PT.

Requirements

From the motivating example presented so far, we derive the following requirements:

Functionality An integrator tool must manage *links between objects* of interdependent documents. In general, links may be m:n relationships, i.e., a link connects m source objects with n target objects. They may be used for multiple purposes: *browsing, correspondence analysis,* and *transformation.*

Mode of operation An integrator tool must operate incrementally rather than batch-wise. It is used to *propagate changes* between interdependent documents. This is done in such a way that only *actually affected parts* are *modified*. As a consequence, manual work does not get lost (in the above example the elaboration of the simulation model), as it happens in the case of batch converters.

Direction In general, an integrator tool may have to work in *both directions*. That is, if a source document is changed, the changes are propagated into some target document and vice versa.

Integration rules An integrator tool is driven by *rules* defining which *object patterns* may be related to each other. There must be support for defining and applying these rules. Rules may be interpreted or hardwired into software.

Mode of interaction While an integrator tool may operate automatically in simple scenarios, it is very likely that user *interaction* is *required*. On the one hand, user interaction can be needed to resolve non-deterministic situations when integration rules are conflicting. On the other hand, there can be situations where no appropriate rule exists and parts of the integration have to be corrected or performed manually.

Time of activation In single user applications, it may be desirable to propagate changes eagerly. This way, the user is informed promptly about the consequences of the changes performed in the respective documents. In multi user scenarios, however, *deferred propagation* is usually required. In this way, each user keeps control of the export and import of changes from and to his local workspace.

Traceability An integrator tool must *record a trace* of the rules which have been applied. This way, the user may reconstruct later on, which decisions have been performed during the integration process.

Adaptability An integrator tool must be adaptable to a *specific application domain*. Adaptability is achieved by defining suitable integration rules and controlling their application (e.g., through priorities). In some cases, it must be possible to modify the rule base on the fly.

A-posteriori integration An integrator tool should work with *heterogeneous tools* supplied by different vendors. To this end, it has to access these tools and their data. This is done by corresponding wrappers which provide abstract and unified interfaces.

Not every integrator tool has to fulfill all of these requirements. E.g., there are some situations where incrementality is not needed. In other situations, the rule base is unambiguous such that there will never be user interaction. In such cases, it has to be decided whether a full-scale integrator tool based on the integrator framework is used anyway, or some other approach is more suitable. In IMPROVE, we implemented both types of integrator tools for different parts of our overall scenario. For instance, in addition to the integrator tool described in the motivating example, we created a tool that generates the input file for heterogeneous process simulation with CHEOPS ([409], see Subsect. 3.2.6). This tool only uses small parts of the integrator framework and most of its behavior is hand-coded instead of being directly controlled by rules. Other tools have been realized using completely different approaches like XML and XSLT [567, 602]. Even for the realization of these simpler tools, the experiences gained with the full-scale integrator tools were helpful.

Organization of This Paper

The rest of this paper is structured as follows: In the next Subsect. 3.2.2, we give a short overview of our integrator framework. Our modeling formalism for integration is explained in Subsect. 3.2.3. Subsection 3.2.4 addresses the integration algorithm and its relation to the triple graph grammar approach. We support two approaches for the implementation of integrators, which are introduced and compared in Subsect. 3.2.5. In our project, furthermore, some integrators have been realized following a modified or an entirely different approach. They are sketched in Subsect. 3.2.6. Subsection 3.2.7 compares our approach to other integration R&D work. In Subsect. 3.2.8, we give a summary and an outlook on open problems.

3.2.2 Integrator Framework

Architecture Overview

In each application domain, e.g. in chemical engineering, there are a lot of applications for integrator tools. As a consequence, the *realization* of a specific integrator tool has to require as *little effort* as possible.

We are addressing this by two means: First, our approach allows to define *rules* for integrator tools based on already-existing domain models (cf. Sect. 6.3) and to *derive* an *implementation* for such tools (process reuse within the integrator development process). Second, we provide a *framework* for integrator tools that offers most parts of the integrator functionality in predefined and *general components* (product reuse) [27, 251]. To create a specific integrator tool, only some additional components have to be implemented and integration rules have to be defined.

Figure 3.22 provides an overview of the *system architecture* for integrator tools. It shows an integrator tool between Comos PT (source, lower left corner) and Aspen Plus (target, lower right corner) as example. Note that the terms "source" and "target" denote distinct ends of the integration relationship between the documents, but do not necessarily imply a unique direction of transformation. For each pair of related Aspen and Comos documents, their fine-grained relationships are stored as links in an integration document. The structure of the integration document is the same for all integrator tools realized using the framework and will be discussed below.

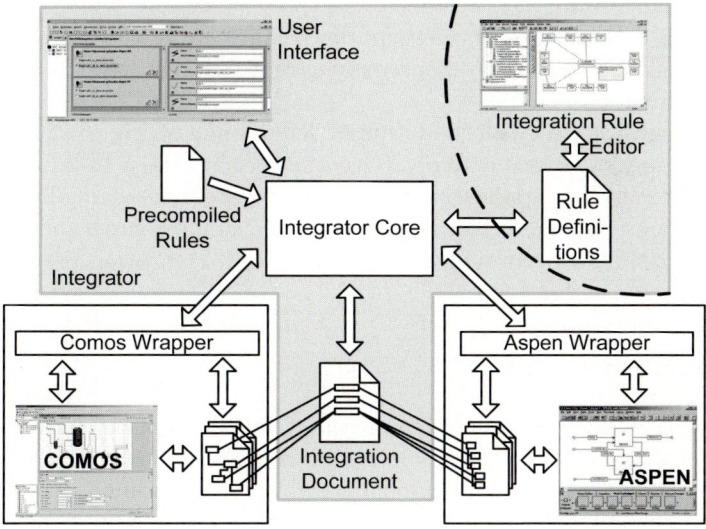

Fig. 3.22. System architecture of an integrator

The integration is performed by the *integrator core*. It propagates changes between source and target documents and vice-versa and modifies the links in the integration document. The integrator core includes an implementation of the integration algorithm which will be explained in detail in Subsect. 3.2.4. It is a universal component that is used by all framework-based integrator tools.

The core does not access source and target tools directly but uses *tool wrappers*. These wrappers provide a standardized graph view on the tools' documents to keep the integrator code independent of the tools' specific interfaces. Additionally, they provide functions for launching tools and locating specific documents. For each tool, a new wrapper has to be implemented. To minimize the required programming efforts, the model-based wrapper specification approach described in [136] and in Sect. 5.7 of this book can be used.

During integration, the integrator core is controlled by *integration rules*, which can be provided following *different approaches*:

First, they can be defined using the integration rule editor (upper right corner of Fig. 3.22), be exported as rule definition files, and then be executed by a *rule interpreter* being part of the core. The formalism for defining integration rules will be explained in Subsect. 3.2.3. Depending on whether the integrator supports the definition of integration rules on the fly, the rule editor is considered either a part of the framework or a part of the development environment for integrators.

Second, they can be *implemented manually* in a programming language, compiled, and linked to the integrator tool in order to be executed. This can lead to a better performance during execution and allows to define rules whose functionality goes beyond the rule definition formalism and the predefined execution algorithm.

Third, as a *combination*, rules can be specified using the rule editor, then translated automatically into source code, and then compiled and linked to the integrator tool.

The integrator *user interface* (upper left corner of Fig. 3.22) is used to control the integrator interactively. Here, the user has the possibility to choose between different rule applications or to manipulate links manually. Although there is a generic user interface implementation, in most cases an application-specific GUI should be provided to facilitate the integration process for the user.

Integration Document

An integration document contains a set of *links* which represent the relationships mentioned above. Each link relates a set of syntactic elements (*increments*) belonging to one document with a corresponding set belonging to another document. A link can be further structured by adding *sublinks* to a link. A sublink relates subsets of the increments referenced by its parent link and is created during the same rule execution as its parent.

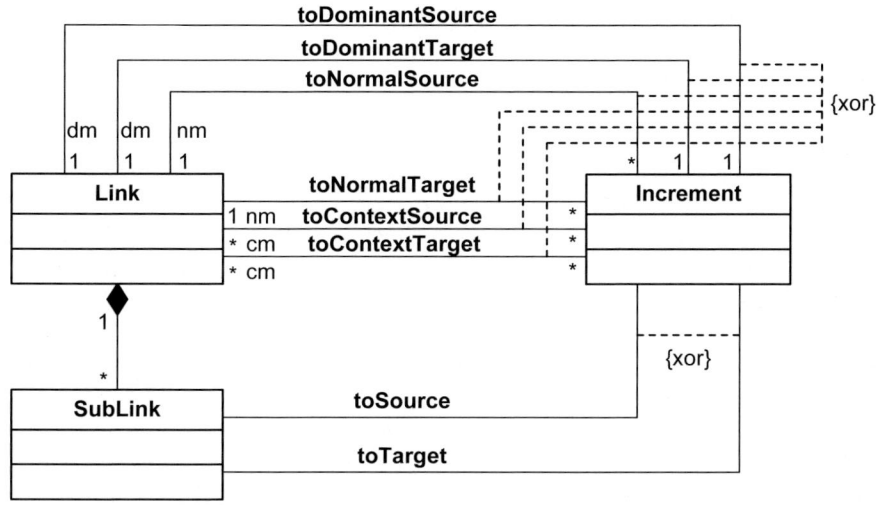

Fig. 3.23. Link model

Figure 3.23 shows the *structure of links* in a UML class diagram [560]. Most constraints needed for a detailed definition are omitted, only examples are shown. An increment can have different roles w.r.t. a referencing link: Increments can be *owned* by a link or be referenced as *context* increments. While an increment can belong to at most one link as owned increment, it can be referenced by an arbitrary number of links as context increments. Owned increments can be created during rule execution, whereas only existing increments can be referenced by new links as context increments.

Context increments are needed when the execution of a rule depends on increments belonging to an already existing link that was created by the application of another rule. Context is used for instance to embed newly created edges between already transformed patterns. Owned increments can be further divided into *dominant* and *normal* increments. Dominant increments play a special role in the execution of integration rules (see Subsect. 3.2.4). Each link can have at most one dominant increment in each document. A link can relate an arbitrary number of normal increments.

There is additional *information* stored *with a link*, e.g. its state and information about possible rule applications. This information is needed by the integration algorithm but not for the definition of integration rules.

3.2.3 Definition of Integration Rules

Overview: Different Levels for Modeling Rules

To create a specific integrator, a set of integration rules specifying its behavior is needed. Therefore, we provide a modeling formalism for such rule sets. The *explicit modeling* of rules has several *advantages* over implicitly coding them within an integrator: First of all, it is much easier to understand rules and their interdependencies if they are available as a human readable visual model. Additionally, our modeling approach is multi-layered and allows consistency checking between the layers. This cannot guarantee the complete correctness of rule sets but at least prevents some major mistakes and thereby ensures the executability of rules.

Another advantage is that the *source code* of integrators is *independent* of specific rules or – if rules are linked to the integrator (see above) – dependencies are limited to certain sections of the code. This *facilitates* the *adaptation* of integrators to new applications or changed rules. If integration rule sets are interpreted using the rule interpreter of the integrator framework, even learning new *rules on the fly* is possible.

In most application domains, *domain models* already exist or are at least under development. Consequently, the information has to be used when defining integration rules. For instance, in another project of IMPROVE the product data model framework CLiP [20] was defined (see Sect. 2.2). Such domain models normally are not detailed enough to allow for the derivation of integration rules or even integrator tools. Nevertheless, they can serve as a starting point for integration rule definition [15]. Together with some company-specific refinements, they can be used to identify documents that have to be integrated and to get information about the internal structure of the documents as well as about some of the inter-document relationships. Of course, the models have to be further refined to be able to get executable integration rules. The process of refining domain models to detailed definitions of the behavior of integrator tools is described in detail in Sect. 6.3. In this section, the focus is on defining integration rule sets without a strong relation to domain models.

For the definition of integration rules, we follow a *multi-layered approach* as postulated by OMG's meta object facility (MOF) [874], based on the Unified Modeling Language (UML) [560]. Figure 3.24 provides an overview of the different modeling levels and their contents for the running example. Using MOF as meta-meta model, on the meta level the existing UML meta model is extended. Additional elements are added that form a language to define models of the documents to be integrated and to express all aspects concerning the documents' integration.

The *extension* on the *meta level* comprises *two parts*: First, *graph-* and *integration-related* definitions are added. These are used by all integration rule sets. Second, *domain-specific* extensions can be made. They can facilitate the definition of integration models when being combined with domain-specific

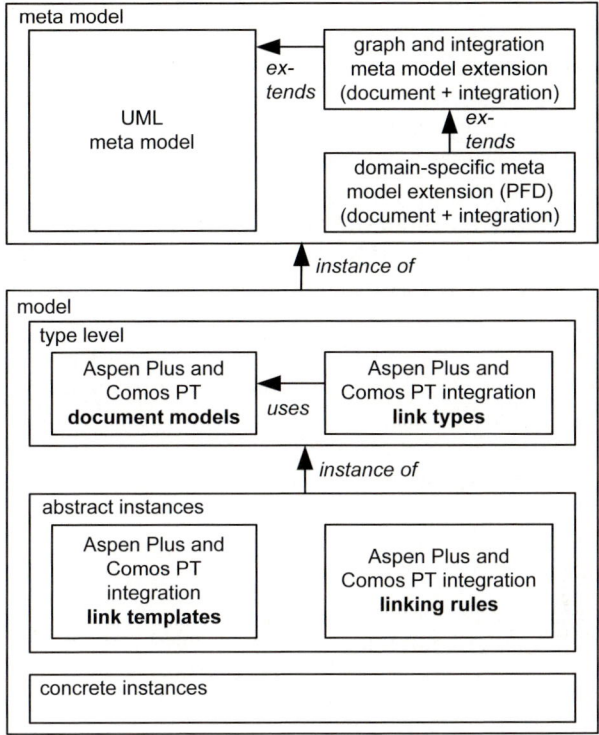

Fig. 3.24. Levels of modeling

visualization. To express the relation between meta model and model, we use the approach sketched in [391], where UML stereotypes are used on the *model level* to express the instance-of relationship between meta model elements and model elements. A more detailed description of the meta level can be found in [26].

On the *model level*, we distinguish between a type (or class) level and an instance level, like standard UML does. On the *type level*, *document models* for specific types of documents are defined. They are expressed as class hierarchies describing the documents' underlying type systems. In our example, documents containing simulation models for Aspen Plus and flowsheets for Comos PT are defined. To be able to perform an integration of these documents, *link types* that relate classes contained in the documents' class hierarchies are defined. All occurrences of links in further definitions on lower levels are instances of these link types and are thereby constrained by these types.

The *instance level* is divided into an abstract and a concrete level. On the *abstract instance level*, *link templates* and *linking rules* are specified using collaboration diagrams. Link templates are instances of link types relating a pattern (which is a set of increments and their interrelations) that may exist in one document to a corresponding pattern in another document. A link template only defines a possible relation between documents. It is not imposed that the relation always exists for a given set of concrete documents.

Link *templates* can be *annotated* to define executable linking rules. The annotations provide information about which objects in the patterns are to be matched against existing objects in concrete documents and which objects have to be created, comparable to graph transformations. Linking *rules* are *executed* by integrators and are thus also called integration rules. Rule execution is described in detail in Subsect. 3.2.4.

While on the abstract instance level only patterns are defined that may appear in source, target, and integration document, on the *concrete instance level*, concrete existing documents and integration documents can be modeled. The resulting models are snapshots of these real documents, which can be used as examples, e.g., to document rule sets. As this is not vital for defining integration rules, models on the concrete instance level are not further described here.

In the following, selected aspects of the integration rule model are described in more detail. For further information, we refer to [26, 39, 40]. The modeling approach presented here is based on work dealing with a purely graph-oriented way of specifying integration rules and corresponding tools [131–134].

Type Level Modeling

Before integration rules for a given pair of document types can be defined, *document models* describing the documents' contents have to be created. In our current modeling approach, this is done by providing class hierarchies defining *types* of *entities* relevant for the integration process (increment types) and the possible *interrelations* of increments being contained in a document. To facilitate the definition of integration rules, it is planned to use more advanced document models that address further structural details (see Sect. 6.3). Here, simple UML-like class diagrams are used to express the type hierarchies.

To illustrate our modeling approach, we use excerpts of the underlying rule base of the motivating scenario presented in Subsect. 3.2.1. Figure 3.25 shows a part of the *Aspen Plus* type hierarchy. The figure is simplified, as it does not show stereotypes, cardinalities, and association names. It only shows an *excerpt* of the *simulation document model*. The type hierarchy does not reflect the whole Aspen Plus data model as offered by Aspen's COM interface. Instead, it is the model offered by our Aspen tool wrapper which supplies only information relevant for our integration between simulation models and flowsheets.

On the lowest layer on the left side of the type hierarchy we find increment *types* for some of the simulation *blocks* and *streams* that are predefined

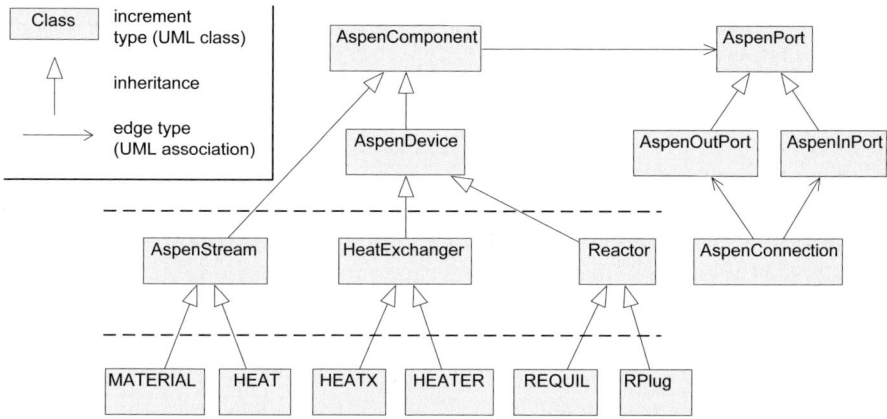

Fig. 3.25. Aspen Plus document model (UML class diagram)

in Aspen Plus (e.g. **RPlug** modeling the behavior of plug flow reactors and **MATERIAL**, the type of stream transporting substances between blocks).

One layer higher, the blocks are further classified by grouping them into *categories* that correspond to the tabs holding the blocks in the Aspen user interface. All blocks and streams inherit from **AspenComponent** (top layer). Each *component* can have an arbitrary number of ports (**AspenPort**) which can be further refined regarding their orientation (**AspenInPort** and **AspenOutPort**). **AspenConnection** is used to express that two ports of opposite orientation are connected by referencing them via associations.

In Fig. 3.21, blocks are represented as *rectangles*, streams are shown as *arrows* inside source and target document. Connections and ports are not represented explicitly (rather, they may be derived from the layout), but they are part of the internal data model.

The *document model* for *Comos PT* is not presented here, since it is structured quite similarly. This similarity is due to two reasons: First, both types of documents, simulation models and PFDs, describe the structure of chemical plants by modeling their main components and their interconnections. Second, technical differences between the internal models of both applications are eliminated by tool wrappers. Nevertheless, the remaining mapping between simulation models and PFDs is not straightforward, as we have already discussed above.

As a first step for defining integration rules, *link types* are modeled. They are used for two different purposes: First, each link that is contained in an integration document or occurs in an integration rule on the instance level has to be an *instance* of a link *type*. As a result, it is possible to *check* the *consistency* of integration rules against a set of link types. This cannot ensure the semantical correctness of integration rules but facilitates the definition process by eliminating some mistakes.

Second, it is very likely that for a new integrator a basic set of integration *rules* is defined *top-down using domain knowledge* [15]. For instance, in our running example it is common domain knowledge that devices in a PFD are simulated by predefined blocks in Aspen Plus. More concrete statements can be made as well: A reactor in the PFD could be simulated by some reactor block in Aspen Plus. Each of these statements can be translated easily in a link type definition. Thereby, our modeling formalism provides a means of formalization and communication of domain knowledge concerning the relationships between documents. The resulting link types can be further refined and, finally, on the abstract instance level, integration rules can be defined accordingly.

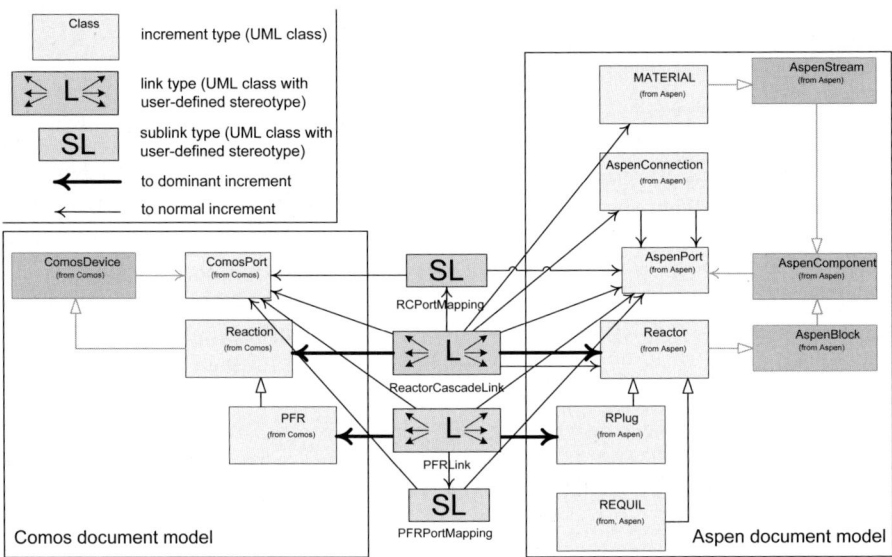

Fig. 3.26. Reactor link types (UML class diagram)

Figure 3.26 shows an example class diagram defining two *link types* concerning the relation between *reactors* in *PFDs* and simulator *blocks* in *Aspen Plus*. The left side of the figure contains an excerpt of the Comos document model, the right side one of the Aspen document model. In between, link types and sublink types are depicted. While the documents' increment types are shown as plain UML classes, link and sublink types are presented using graphical stereotypes. This can be applied to increment types as well if domain-specific elements are defined in the meta model together with corresponding graphical stereotypes. For instance, all reactor increment types could be displayed with a reactor symbol. The usage of graphical stereotypes facilitates the readability of models for domain experts having little UML knowledge.

In the figure, a link type called PFRLink is defined. This *example link type* expresses that exactly one instance[27] of the specific reactor type PFR (plug flow reactor) can be simulated by exactly one instance of the specific reactor block RPlug in the simulation model. The reactors in both documents can have an arbitrary number of ports which are mapped by the link as well. To assign corresponding ports, the link may have sublinks each mapping pairs of ports[28]. Both reactors are referenced as dominant increments. This link type is rather specific, as it forms relatively tight constraints for all of its instances.

In general, it is not always possible to simulate a reactor in the PFD by one single simulator block, but rather a cascade of reactor blocks can be necessary. This is the case in our running example. Therefore, *another link type*, namely for mapping reactor devices to reactor block cascades is defined (ReactorCascadeLink in Fig. 3.26). It assigns *one* Reaction (or one of its subtypes') *instance* to *multiple instances* of Reactor subtypes[29]. These instances are connected via their ports and connections with MATERIAL streams transporting the reacting substances. Again, sublinks help identifying related ports.

The ReactorCascadeLink *type* is *rather generic* compared to the PFRLink. For instance, it does not specify the number of reactor blocks used in the simulation, nor does it specify their concrete types. Even how they are connected by streams, is not further specified. To get an executable rule, a lot of information is still missing, which is supplied on the abstract instance level (see below).

The type level definitions of the rule set for our running example comprise much more link types. Some of them are more concrete, others are more generic than those discussed above. Even from the link types described so far, it can be seen that the definition of the relations between given document types is quite ambiguous. This reflects the complexity of the domain and the need for engineers' knowledge and creativity during integration. Thus, our integration *models* only *provide heuristics to support* the engineers at their work.

Abstract Instance Level Modeling

In general, the information modeled on the type level is not sufficient for gaining executable integration rules. Link types constrain links but do not fully specify the structures to be related by links. Therefore, on the abstract instance level, a detailed definition of the corresponding *patterns related by a link* is made. This is done by defining so-called *link templates* in UML collaboration diagrams. These instances are abstract because they do not describe concrete documents but situations that *may* occur in one concrete document at runtime and how the corresponding situation in the other document could

---

[27] Cardinalities are not shown in the figure.
[28] To keep the figure simple, it is not distinguished between ports of different orientation, as it is done in the real model.
[29] The Reactor class is abstract, thus no instances of it are allowed.

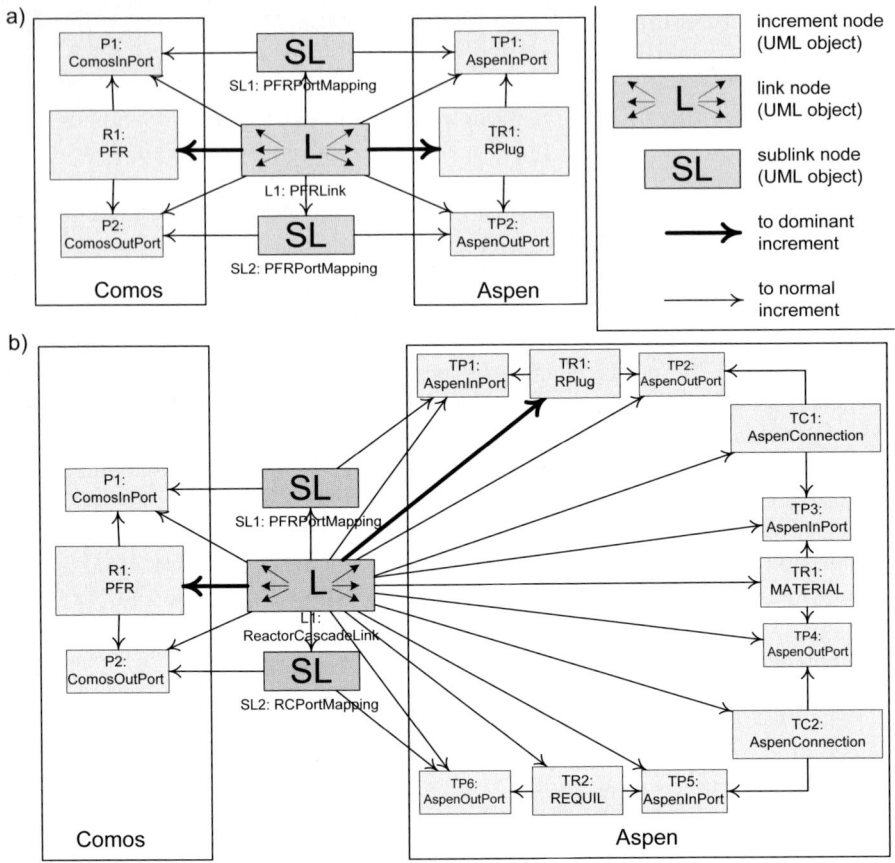

Fig. 3.27. Reactor link templates (UML collaboration diagram)

look like. From link templates operational integration rules that describe actions like "if the described pattern was found in the source document, create a corresponding pattern in the target document" can be derived easily.

To illustrate the definition of link templates, we again use the rule set of our running *example*. Figure 3.27 shows two UML *collaboration diagrams* defining link templates which are instances of the link types introduced in the previous subsection.

The *link template* in Fig. 3.27 a) *relates* a plug flow *reactor* (PFR) with two ports *to a* RPlug *block* with two ports in the simulation model. All increments are referenced by the link L1, which is refined by the sublinks SL.1 and SL.2 to map the corresponding ports. The port mapping is needed later to propagate connections between ports (see below). The link L1 is an instance of the PFRLink type. It could be an instance of the more generic link type ReactorCascadeLink as well, with the cascade just consisting of one reactor block.

However, a link should always be typed as instance of the most concrete link type available that fits its definition. Link types reflect domain knowledge and the more concrete a link type is the more likely it is that its instances are supported by domain knowledge.

In the running example, the simple link template of Fig. 3.27 a) was not applied, because a single simulator block does not suffice to study the reaction of the plug flow reactor. Instead, the **PFR** was mapped to a *reactor cascade*. Figure 3.27 b) contains the *link template* that is the basis of the corresponding integration rule. The PFD-side pattern is the same as in Fig. 3.27 a). On the simulation side, a cascade of a **RPlug** and a **REQUIL** block are defined. Substances are transported by a **MATERIAL** stream, whose ports are connected to the reactors' ports. Again, all increments in both documents are referenced by the link, which is an instance of **ReactorCascadeLink**, and the external ports of the cascade are associated with the **PFR** ports by sublinks.

The link types and templates discussed so far and some other rules in our running example address the mapping of devices and blocks together with their ports. If rules are derived from these templates and are then used, e.g. to transform the devices of a PFD into a simulation model, it is necessary to *transform* the *connections between* the *ports* as well. This is done with the help of the definitions in Fig. 3.28. Part a) of the figure contains the link template for mapping a **ComosConnection** to an **AspenConnection**. While the mapping

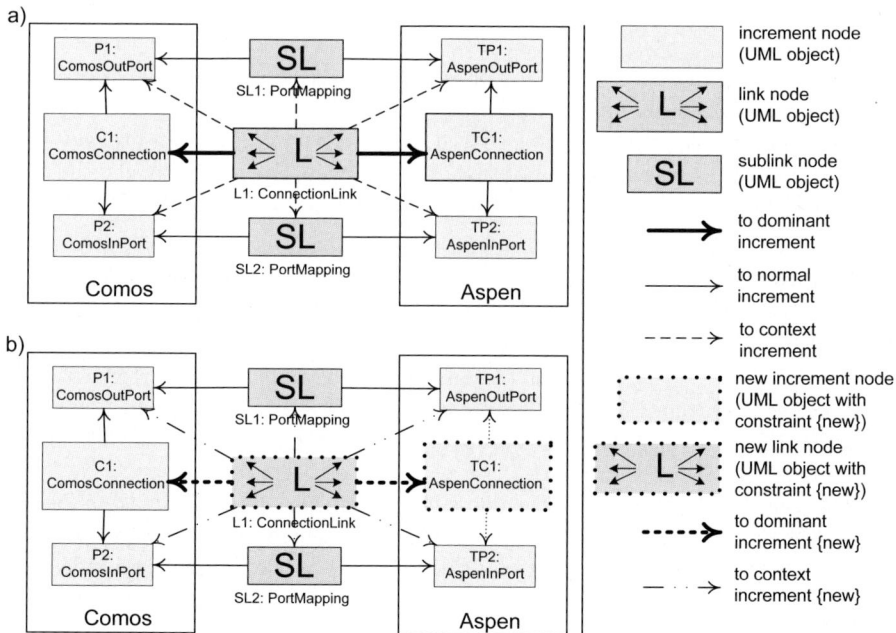

Fig. 3.28. Connection link template (a) and forward rule (b)

is quite simple with regard to its 1:1 structure, there is a particularity about this link template: To be able to embed the edges leading from the newly created connection in the target document to the correct ports, the source document ports have to be already mapped to their corresponding ports in the target document. As a result, the ports and their mapping are referenced as context information by the link L1. This ensures that the rule transforming connections is applied only after the ports have been mapped by applications of other rules and thus it can be determined which ports to connect. Then, the edges from the newly created connection to the ports can be created easily.

Modeling Operational Rules

The link templates described so far are *purely declarative* and just describe which patterns could be related by fine-grained inter-document links. They do *not* contain *operational* directives for transforming a document into another one or for establishing links between documents. Following the triple graph grammar approach [413], operational integration rules can be easily derived from link templates.

For each *link template, three integration rules* can be derived[30]:

- *Forward* transformation *rules* look for the context in source, target, integration document, and the non-context increments in the source document, as well as for all related edges. For each match, it creates the corresponding target document pattern and the link structure in the integration document.
- *Backward* integration *rules* do the same but in the opposite direction from target to source document.
- *Correspondence* analysis *rules* search the pattern in source and target document including the whole context information. For each match, the link structure in the integration document is created.

The derivation of a forward transformation rule from a link template is illustrated in Fig. 3.28, as an example, using the rule to transform a connection. Part b) shows the *forward rule* corresponding *to* the *link template* in part a). All dotted nodes (L1 and TC1) and their edges are created when the rule is executed. To determine whether the rule can be applied, the pattern without these nodes is searched in the documents. Here, the already related ports and the connection in the PFD are searched and the corresponding connection in the simulation model is created.

The notation of integration rules can be compared to *graph transformations* [328, 652] with left-hand and right-hand sides compressed into one diagram. The dotted nodes and edges are contained on the right-hand side only and thus are created during execution. The other nodes and edges are contained on both sides and are thus searched and kept.

---

[30] Additional rules can be derived if consistency checking and repairing existing links are taken into account.

So far, only the structural aspects of link templates and rules were addressed. In practice, each increment is further defined by some attributes and their values. Using a subset of the OCL language (Object Constraint Language [879], part of the UML specification), *conditions based on* these *attributes* that further constrain the applicability of the resulting integration rules can be added to link templates.

To deal with the consistency of attributes, link templates can be enriched with different *attribute assignment statements* using a subset of the OCL language as well. An attribute assignment can access all attributes of the increments referenced by a link. There are different situations in development processes in which an integration is performed. Depending on the situation, an *appropriate* attribute assignment is *chosen*. For instance, for each correspondence (i.e., for the set of resulting rules) there is one attribute assignment for the initial generation of the simulation model, one to propagate the simulation results back into the flowsheet, etc.

Rule Definition Round-Trip

Figure 3.29 shows the interrelations between the different *parts* of the *modeling formalism* from a practical point of view. The meta model serves as basis both for the implementation of integrator tools and the rule modeling process. It is defined according to domain-specific knowledge like, in our case, the information model CLiP [20] for chemical engineering and the requirements concerning integration functionality.

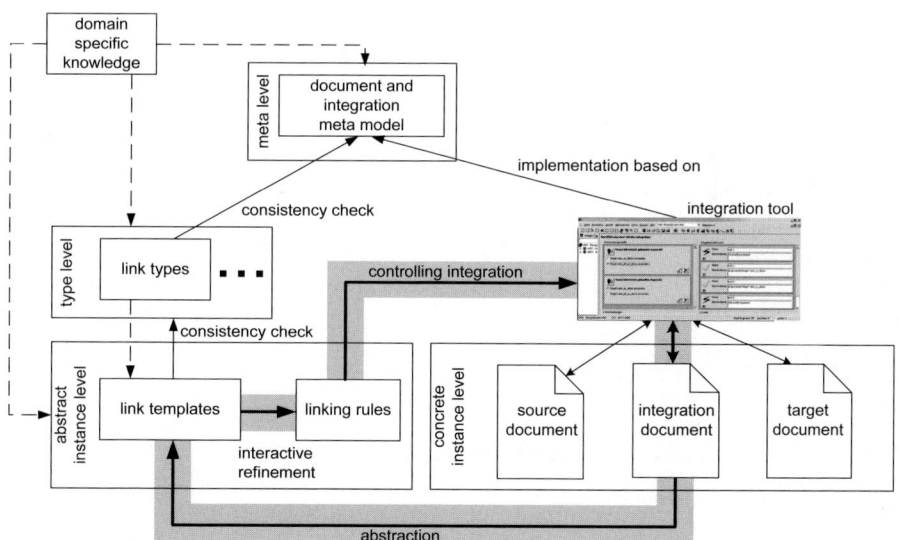

Fig. 3.29. Layers of modeling and modeling round-trip

Basically, there are two ways to define integration rules: *top-down*, before the integrator tool is applied, or *bottom-up*, based on situations occurring during the usage of the tool. It is most likely that in practice first a basic set of rules is defined top-down by a *modeling expert* and then the rule base is extended bottom-up by the *engineer using* the integrator tool.

Before any further definitions can be made, the *documents* to be integrated have to be *modeled* on type level as described above, which is not shown in this figure. Next, *link types* have to be *defined* on type level that declare types for possible correspondences on the abstract instance level. Again, for both tasks domain-specific knowledge has to be used.

Following a top-down approach, *link templates* on the abstract instance level are modeled based on the link types of the type level. These are then refined to *linking rules*. The resulting set of rules is *used by* the *integrator* to find corresponding parts of source and target document and to propagate changes between these two documents. The corresponding document parts are related by *links* stored in the *integration document*. If no appropriate rule can be found in a given situation, the chemical engineer performing the integration can *manually modify* source and target document and add links to the integration document.

To extend the rule base *bottom-up*, the links entered manually in the integration document can be automatically *abstracted to* link *templates*. Next, a consistency check against the link types on the type level is performed. If the link templates are valid, the engineer is now guided through the interactive *refinement* of the link templates to *linking rules* by a simplified modeling tool. The rules are *added* to the rule base and can be *used* for the following integrations.

This can be *illustrated* by an *extension* of the *scenario* presented above: Initially, there is no rule for mapping a plug flow reactor to a cascade of two reactors. Instead, the first time the situation occurs, the mapping is performed manually: The user creates the reactor cascade in the simulation model and adds a link to the integration document. From this link, the link template in Fig. 3.27 b) is abstracted. The link template is consistency-checked against the available link types. It is detected that the link template fits the ReactorCascadeLink type from Fig. 3.26 and, therefore, it is permanently added to the rule base and applied in further runs of the integrator.

3.2.4 Rule Execution

In this subsection, the *execution algorithm* for integration *rules* is presented. First, the triple graph grammar approach which serves as the basis for our approach is briefly sketched. Furthermore, it is explained how our work relates to this approach and why it had to be extended. Second, an overview of our integration algorithm is given, using a simple and abstract scenario. Third, the individual steps of the algorithm are explained in detail, using the integration rule for a connection as an example. In this subsection, the execution of the algorithm with PROGRES [414] is considered. Please note that the execution with PROGRES is only one approach for building integrator tools (cf. Subsect. 3.2.5).

Triple Graph Grammars and Execution of Integration Rules

For modeling an integration, the source and target documents as well as the integration document may be modeled as graphs, which are called *source graph*, *target graph*, and *correspondence graph*, respectively. If the tools operating on source and target documents are not graph-based, the graph views can be established by tool wrappers (cf. Subsect. 5.7.4). Moreover, the operations performed by the respective tools may be modeled by graph transformations.

Triple graph grammars [413] were developed for the high-level specification of graph-based integrator tools. The core idea behind triple graph grammars is to specify the relationships between source, target, and correspondence graphs by *triple rules*. A triple rule defines a coupling of three rules operating on source, target, and correspondence graph, respectively. By applying triple rules, we may modify coupled graphs synchronously, taking their mutual relationships into account. In the following, we give a short motivation for our integration algorithm only. For a detailed discussion of the relation between our approach and the original work by Schürr, the reader is referred to [37].

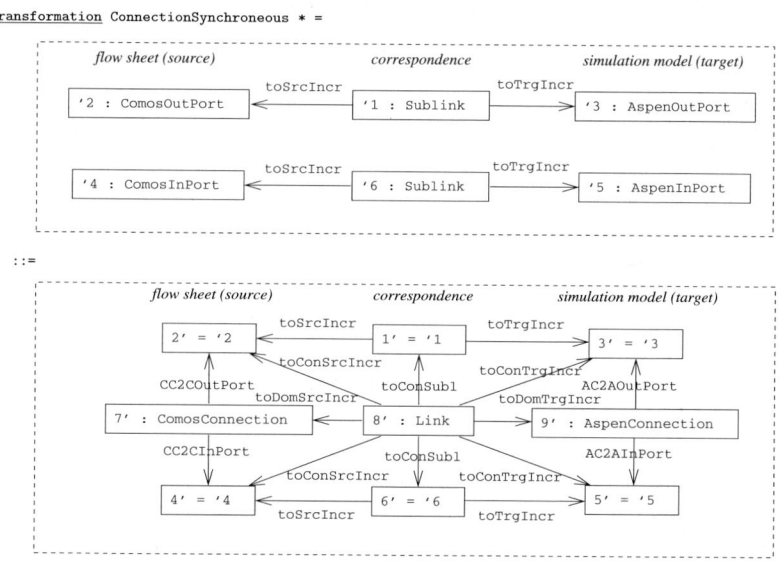

Fig. 3.30. Triple rule for a connection

An *example* of a *triple rule* is given in Fig. 3.30. The rule refers to the running example to be used for the integration algorithm, namely the *creation* of *connections* as introduced in Fig. 3.28 a). Here, the triple rule is presented as a graph transformation in PROGRES [414] syntax.

A *graph transformation* consists of a left-hand and a right-hand side, which are displayed on top or bottom, respectively. Each side contains a graph pattern consisting of nodes and interconnecting edges. When a graph transformation is applied to a graph, the left-hand side pattern is searched in the graph (pattern matching) and replaced by the right-hand side pattern. All nodes on the left-hand side and new nodes on the right-hand side are further specified by giving their type. The nodes on the right-hand side that appear on the left-hand side as well are related to the corresponding left-hand side nodes by their node numbers.

Both sides of the triple rule `ConnectionSynchronous` span all participating subgraphs: the source graph (representing the PFD) on the left, the correspondence graph in the middle, and the target graph (for the simulation model) on the right. The triple rule can be seen as a different representation of the link template in Fig. 3.28 a). Its *left-hand side* is composed of all context nodes of the link template: It contains the port nodes in the source and target graphs, distinguishing between output ports and input ports. Furthermore, it is required that the port nodes in both graphs correspond to each other. This requirement is expressed by the nodes of type `subLink` in the correspondence graph and their outgoing edges which point to nodes of the source and target graph, respectively.

The *right-hand side* contains all nodes of the link template: All elements of the left-hand side reappear on the right-hand side. New nodes are created for the connections in the source and target graph, respectively, as well as for the link between them in the correspondence graph. The connection nodes are embedded locally by edges to the respective port nodes. For the link node, three types of adjacent edges are distinguished. `toDom`-edges are used to connect the link to exactly one dominant increment in the source and target graph, respectively. In general, there are additional edges to normal increments (not needed for the connection rule). Finally, `toContext`-edges point to context increments.

Figure 3.30 describes a *synchronous graph transformation*. As already explained earlier, we cannot assume in general that all participating documents may be modified synchronously. In the case of asynchronous modifications, the triple rule shown above is not ready for use. However, we may derive asynchronous forward, backward, or correspondence analysis rules as explained in Subsect. 3.2.3. Figure 3.31 shows the *forward* rule for a connection from Fig. 3.28 b) in PROGRES syntax. In contrast to the synchronous rule, the connection in the PFD is now searched on the left-hand side, too, and only the connection in the simulation model and the link with its edges are created on the right-hand side.

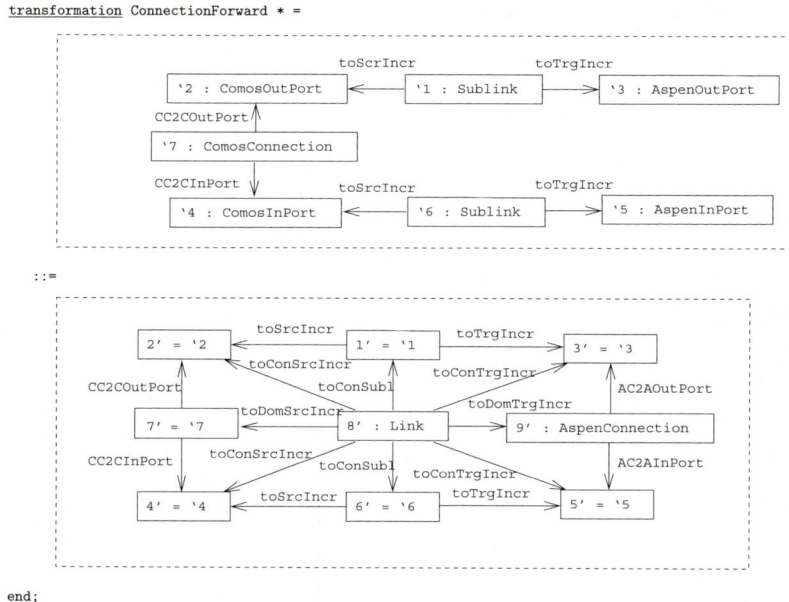

Fig. 3.31. Forward rule for a connection

Unfortunately, even these rules are not ready for use in an integrator tool as described in the previous section. In case of non-deterministic transformations between interdependent documents, it is crucial that the user is made aware of conflicts between applicable rules. A conflict occurs, if multiple rules match the same increment as owned increment. Thus, we have to consider all applicable rules and their mutual conflicts before selecting a rule for execution. To achieve this, we have to *give up atomic rule execution*, i.e., we have to decouple pattern matching from graph transformation [33, 255].

Integration Algorithm

An integration rule cannot be executed by means of a single graph transformation. To ensure the correct sequence of rule executions, to detect all conflicts between rule applications, and to allow the user to resolve conflicts, all integration rules contained in the rule set for an integrator are automatically translated to a set of graph transformations. These rule-specific transformations are executed together with some generic ones altogether forming the *integration algorithm*.

While the algorithm supports the concurrent execution of forward, backward, and correspondence analysis rules, we *focus* on the *execution* of *forward rules* here. Also, we present the *basic* version of the algorithm only, without optimizations. A full description of all aspects can be found in [29] which is an extended version of [33].

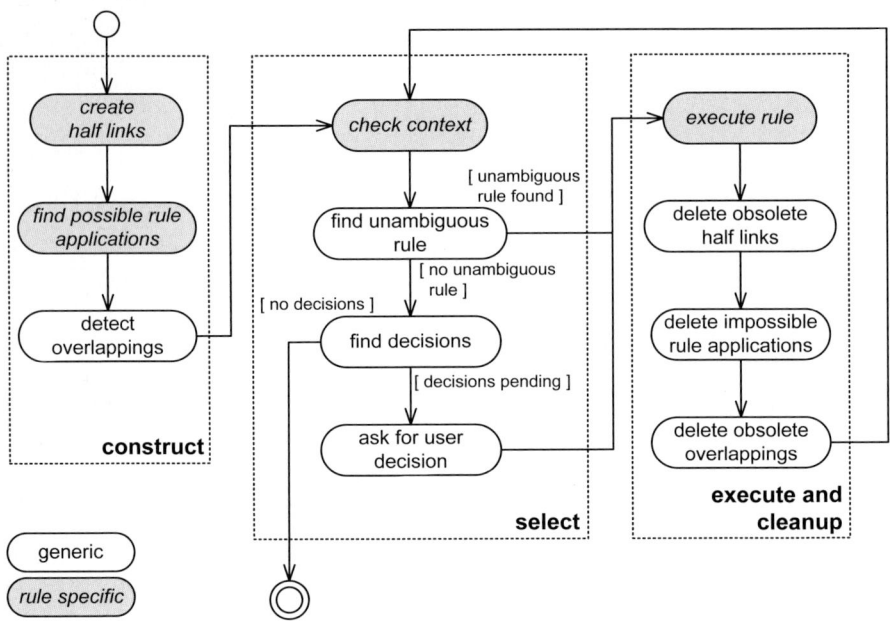

Fig. 3.32. Integration algorithm

The execution of integration rules is embedded in an *overall algorithm* which can be briefly sketched as follows: When performing an integration, first all links already contained in the integration document are checked for consistency. Links can become inconsistent due to modifications applied to the source and target documents by the user after the last run of the integrator tool. In this case, they can be interactively repaired by applying repair actions proposed by the integrator tool or fixed manually by adding or removing increment references to the link or by simply deleting the link. For an initial integration the integration document is empty, so this applies only to subsequent integrations. After existing links have been dealt with, rules are executed for all increments that are not yet referenced by links. In case of a subsequent integration, these increments have been added by the user to source and target documents since the last run of the integrator tool.

Figure 3.32 shows a UML *activity diagram* depicting the integration algorithm. To perform each activity, one or more graph transformations are executed. Some of these graph transformations are generic (white), others are specific for the integration rule being executed (grey and italics). Thus, the algorithm is composed of all *generic* and *rule-specific graph transformations*, the latter for all integration rules contained in the rule set. The overall algorithm is divided into *three phases*, which are described informally in the following using the example of Fig. 3.33. The example is rather abstract and is not related to specific rules of our scenario.

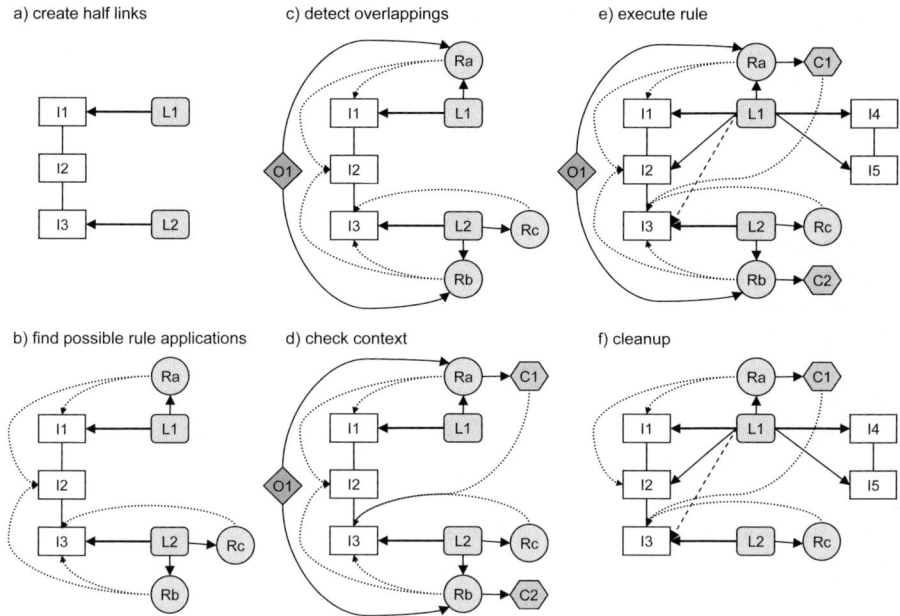

Fig. 3.33. Sample integration

During the *first phase* (construct), all possible rule applications and conflicts between them are determined and stored in the graph. First, for each increment in the documents that has a type compatible with the dominant increment's type of any rule, a *half link* is *created* that references this increment. In the example, half links are created for the increments I1 and I3, and named L1 and L2, respectively (cf. Fig. 3.33 a).

Then, for each half link the possible *rule applications* are *determined*. This is done by trying to match the left-hand side of forward transformation rules, starting at the dominant increments to avoid global pattern matching. In the example (Fig. 3.33 b), three possible rule applications were found: Ra at the link L1 would transform the increments I1 and I2; Rb would transform the increments I2 and I3; and Rc would transform increment I3.

Here, two types of conflicts can be found. First, the rules Rb and Rc share the same dominant increment. Second, the rules Ra and Rb share a normal increment. Both situations lead to conflicts because each increment may only be transformed by one rule as normal or dominant increment. To prepare conflict-resolving user interaction, *conflicts* of the second type are *explicitly marked* in the graph by adding an edge-node-edge construct (e.g. O1 in Fig. 3.33 b).

In the *second phase* (select), the *context* is *checked* for all possible rule applications and all matches are stored in the graph. Only rules whose context has been found are ready to be applied. In the example in Fig. 3.33 d), the context for Ra consisting of increment I3 in the source document was found (C1). The context for Rb is empty (C2), the context for Rc is still missing.

If there is a possible rule application, whose context has been found and which is not involved in any conflict, it is *automatically selected* for execution. Otherwise, the user is asked to *select* one rule among the rules with existing context. If there are no executable rules the algorithm ends. In the example in Fig. 3.33 d), no rule can be automatically selected for execution. The context of Rc is not yet available and Ra and Rb as well as Rb and Rc are conflicting. Here, it is assumed that the user selects Ra for execution.

In the *third phase* (execute and cleanup), the selected *rule* is *executed*. In the example (Fig. 3.33 e), this is the rule corresponding to the rule node Ra. As a result, increments I4 and I5 are created in the target document, and references to all increments are added to the link L1. Afterwards, *rules* that cannot be applied and *links* that cannot be made consistent anymore are *deleted*. In Fig. 3.33 f), Rb is deleted because it depends on the availability of I2, which is now referenced by L1 as a non-context increment. If there were alternative rule applications belonging to L1 they would be removed as well. Finally, *obsolete overlappings* have to be deleted. In the example, O1 is removed because Rb was deleted. The cleanup procedure may change depending on how detailed the integration process has to be documented.

Now, the *execution goes back* to the select phase, where the context check is repeated. Finally, in our example the rule Rc can be automatically selected for execution because it is no longer involved in any conflicts, if we assume that its context has been found.

In the following, some of the rule-specific and generic graph transformations needed for the execution of the connection rule will be explained in more detail.

Construction Phase

In the construction phase, it is determined which *rules* can be possibly applied to which *subgraphs* in the source document. Conflicts between these rules are marked. This information is collected once in this phase and is updated later *incrementally* during the repeated executions of the other phases.

In the first step of the construction phase (create half links), for each increment, which type is the type of a *dominant increment* of at least one rule, a link is created that references only this increment (*half link*). Dominant increments are used as anchors for links and to group decisions for user interaction. Half links store information about possible rule applications; they are transformed to consistent links after one of the rules has been applied.

To *create half links*, a *rule-specific* PROGRES *production* (not shown) is executed for each rule. Its left-hand side contains a node having the type of the rule's dominant increment, with the negative application condition that there is no half link attached to it yet. On its right-hand side, a half link node is created and connected to the increment node with a `toDomSrcIncr`-edge. All these productions are executed repeatedly, until no more left-hand sides are matched, i.e., half links have been created for all possibly dominant increments.

The second step of the construction phase (find possible rule applications) determines the integration rules that are possibly applicable for each half link. A rule is *possibly applicable* for a given half link if the source document part of the left-hand side of the synchronous rule without the context increments is matched in the source graph. The dominant increment of the rule has to be matched to the one belonging to the half link. For potential applicability, context increments are not taken into account, because missing context increments could be created later by the execution of other integration rules. For this reason, the context increments are matched in the selection phase before selecting a rule for execution.

Figure 3.34 shows the PROGRES transformation for the example forward rule for a connection of Fig. 3.28 b). The left-hand side consists of the half link ('2) and the respective dominant increment ('1), as all other increments of this rule are context increments. In general, all non-context increments and their connecting edges are part of the left-hand side. The link node is further constrained by a *condition* that requires the attribute `status` of the link to have the value `unchecked`. This ensures that the transformation is only applied to half links that have not already been checked for possible rule applications.

On the right-hand side, a rule *node* is created to *identify* the possible *rule application* (4'). A transfer is used to store the id of the rule in its attribute `ruleId`. A `possibleRule`-edge connects it to the half link. A role node is inserted to explicitly store the result of the pattern matching (3'). If there are more increments matched, role nodes can be distinguished by the `roleName`-attribute. The asterisk (*) behind the production name tells PROGRES to

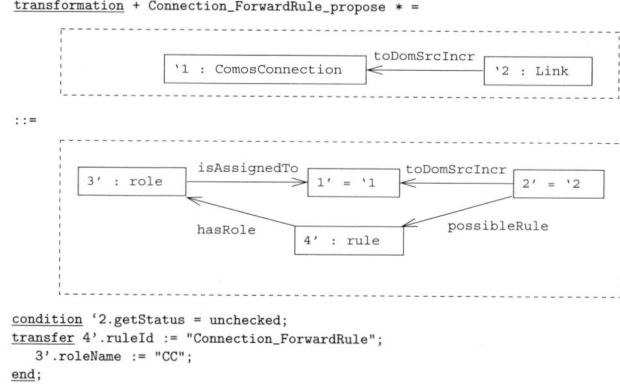

Fig. 3.34. Find possible rule applications

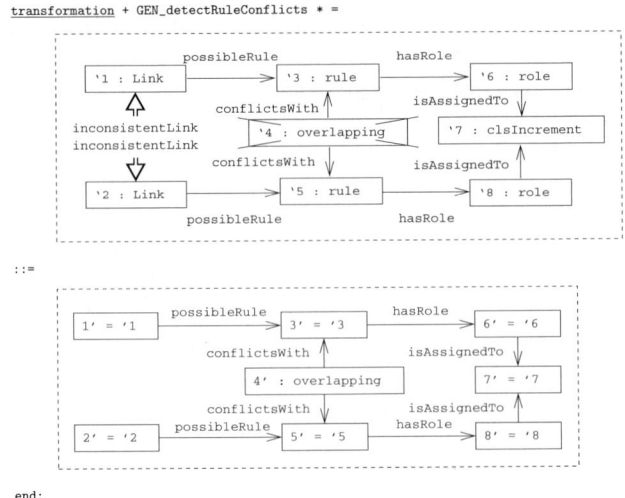

Fig. 3.35. Detect overlappings

apply this production for each possible matching of its left-hand side. When executed together with the corresponding productions for the other rules, as a result all possibly applicable rules are stored at each half link. If a rule is applicable for a half link with different matchings of its source increments, multiple rule nodes with the corresponding role nodes are added to the half link.

In the selection phase, for each link that is involved in a conflict all possible rule applications are presented to the user who has to resolve the conflict by selecting one. Thus, these conflicts are directly visible. *Conflicts* where possible rule applications share the *same normal increment* are *marked* with cross references (hyperlinks) between the conflicting rule applications belonging to different links. This is done with the help of the generic PROGRES production in Fig. 3.35. The pattern on the left-hand side describes an increment ('7) that is referenced by two roles belonging to different rule nodes which belong to different links. The negative node '4 prevents the left-hand side from matching if an overlap is already existing and therefore avoids multiple markings of the same conflict. The arrows pointing at the link nodes, each labeled inconsistentLink, call a PROGRES *restriction* with one of the link nodes as parameter. A restriction can be compared to a function that has to evaluate to true for the restricted node to be matched by the left-hand side. The definition of the restriction is not shown here. It evaluates to true, if the link's attributes mark it as being inconsistent.

On the *right-hand* side, the *conflict* is *marked* by adding an overlap node (4´) is inserted between the two rule nodes. Again, this production is marked with an asterisk, so it is executed until all conflicts are detected. Besides detecting conflicts between different forward transformation rules, the depicted production also detects *conflicts between forward, backward, and correspondence analysis rules* generated from the same synchronous rule. Thus, to prevent unwanted creation of redundant increments, it is not necessary to check whether the non-context increments of the right-hand side of the synchronous rule are already present in the target document when determining possible rule applications in the second step of this phase.

Selection Phase

The goal of the selection phase is to *select* one possible *rule* application for execution in the next phase. If there is a rule that can be executed without conflicts, the selection is performed *automatically*, otherwise the *user* is *asked* for his decision. Before a rule is selected, the contexts of all rules are checked because only a rule can be executed whose context has been found.

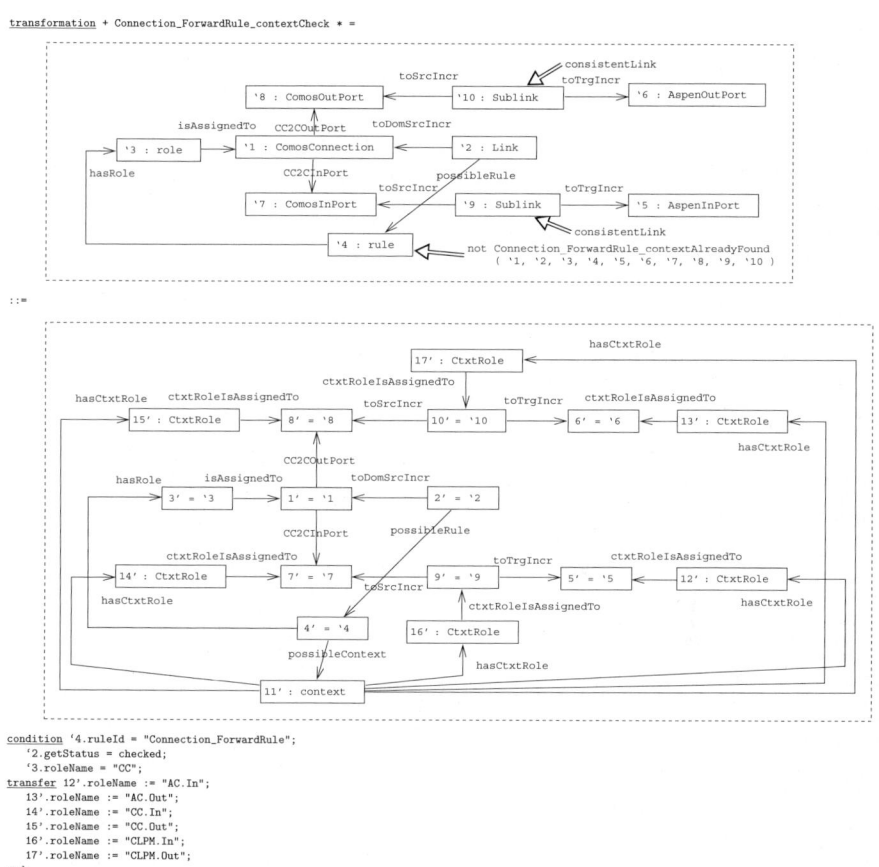

Fig. 3.36. Check context

The *context check* is performed in the first step of this phase. The context is formed by all context elements from the synchronous rule. It may consist of increments of source and target documents and of links contained in the integration document.

Figure 3.36 shows the PROGRES *production checking the context* of the example integration rule. The left-hand side contains the half link ('2), the non-context increments (here, only '1), the rule node ('4), and the role nodes ('3). The non-context increments and their roles are needed to embed the context and to prevent unwanted folding between context and non-context increments. For the example rule, the context consists of the two ports connected in the source document ('7, '8), the related ports in the Aspen document ('5, '6), and the relating sublinks ('9, '10). The restrictions make sure that the sublinks belong to a consistent link.

On the right-hand side, to mark the matched context, a new *context node* is created ('11). It is *connected* to all nodes belonging to the context by *role nodes* (12', 13', 14', 15', 16', 17') and appropriate *edges*. If the matching of the context is ambiguous, *multiple* context nodes with their roles are created as the production is executed for all matches.

As the selection phase is executed repeatedly, it has to be made sure that each *context match* (context node and role nodes) is *added* to the graph only *once*. The context match cannot be included directly as negative nodes on the left-hand side because edges between negative nodes are prohibited in PROGRES. Therefore, this is checked using an additional graph test which is called in the restriction on the rule node. The graph test is not presented here as it is rather similar to the right-hand side of this production[31].

The *context* is *checked* for all possible *rule applications*. To make sure that the context belonging to the right rule is checked, the rule id is constrained in the condition part of the productions. After the context of a possible rule application has been found, the rule can be applied.

After the context has been checked for all possible rule applications, some rules can be applied, others still have to wait for their context. The next step of the algorithm (find unambiguous rule) tries to find a *rule application* that is *not* involved in any *conflict*. The conflicts have already been determined in the construction phase. As any increment may be referenced by an arbitrary number of links as context, no new conflicts are induced by the context part of the integration rules. The generic PROGRES production in Fig. 3.37 finds rule applications that are not part of a conflict. On the left-hand side a rule node is searched ('1) that has only one context node and is not related to any overlap node. It has to be related to exactly one half link ('2) that does not have another rule node.

For forward transformation rules, a rule node belongs to one link only, whereas nodes of correspondence analysis rules are referenced by two half

---

[31] In the optimized version of the integration algorithm, the context check is performed only once for each rule, thus this test is avoided.

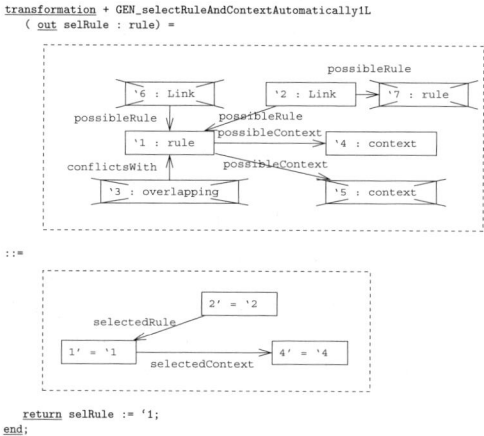

Fig. 3.37. Select unambiguous rule

links. Therefore for *correspondence* analysis *rules*, *another production* is used which is not shown here.

A rule node is not selected for execution if there are *conflicting rules*, even if their *context* is still *missing*. As the context may be created later, the user has to decide whether to execute this rule and thereby making the execution of the other rules impossible.

If a *match* is *found* in the host graph, the *rule* node and the context node are *selected* for execution by substituting their referencing *edges* by `selectedRule` and `selectedContext` edges, respectively (cf. right-hand side of production in Fig. 3.37). The rule node is returned in the output parameter `selRule`. Now, the corresponding rule can be applied in the execution phase.

If *no rule* could be selected *automatically*, the *user* has to decide which rule is to be executed. Therefore, in the next step (find decisions), all conflicts are collected and presented to the user. For each half link, all possible rule applications are shown. If a rule application conflicts with another rule of a different half link, this is marked as annotation at both half links. Rules that are not executable due to a missing context are included in this presentation but cannot be selected for execution. This information allows the user to *select* a rule *manually*, knowing which other rule applications will be made impossible by his decision. The result of the user interaction (ask for user decision) is stored in the graph and the selected rule is executed in the execution phase.

If *no rule* could be selected automatically and there are no decisions left, the *algorithm terminates*. If there are still half links left at the end of the algorithm, the user has to perform the *rest* of the integration *manually*.

Execution Phase

The *rule* that was selected in the selection phase is *executed* in the execution phase. Afterwards, the *information* collected during the construction phase has to be *updated*.

Rule *execution* is performed by a *rule-specific* PROGRES *production*, see Fig. 3.38. The *left-hand side* of the production is nearly identical to the right-hand side of the context check production in Fig. 3.36. The main difference is that, to identify the previously selected rule, the edge from the link ('2) to the rule node ('4) is now a `selectedRule` edge and the edge from the rule node to the context node ('11) is a `selectedContext` edge. The `possibleRule` and `possibleContext` edges are replaced when a rule together with a context is selected for execution either by the user or automatically in the previous phase of the algorithm (see above).

On the *right-hand side*, the new *increments* in the *target* document are *created* and embedded by *edges*. In this case, the connection (18') is inserted

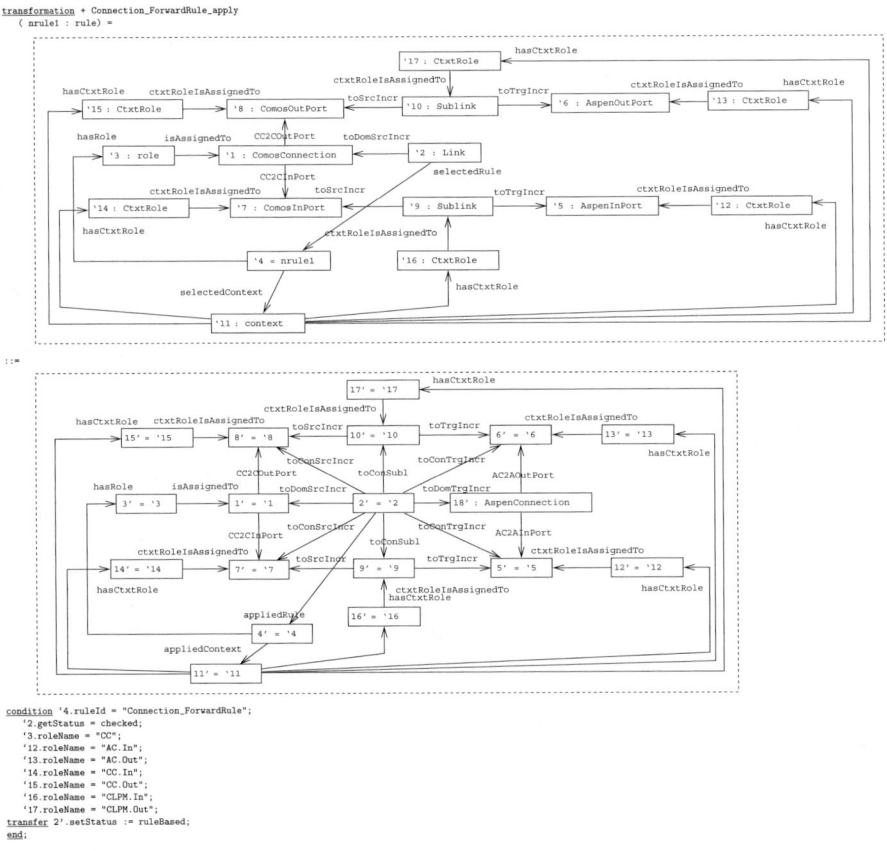

Fig. 3.38. Execute rule

and connected to the two Aspen ports (5', 6'). The half *link* (2') is *extended* to a full link, referencing all context and non-context increments in the source and the target document. The information about the applied rule and roles etc. is kept to be able to detect inconsistencies occurring later due to modifications in the source and target documents.

The last steps of the algorithm are performed by generic productions not shown here that update the information about possible rule applications and conflicts. First, obsolete *half links* are *deleted*. A half link is obsolete if its dominant increment is referenced by another link as non-context increment. Then, potential *rule applications* that are no longer possible because their potentially owned increments are used by another rule execution are *removed*.

3.2.5 Implementation

Besides realizing integrators completely on an ad-hoc hardwired basis, there are *four* different *alternatives* for implementing a given set of integration rules (see Fig. 3.39).

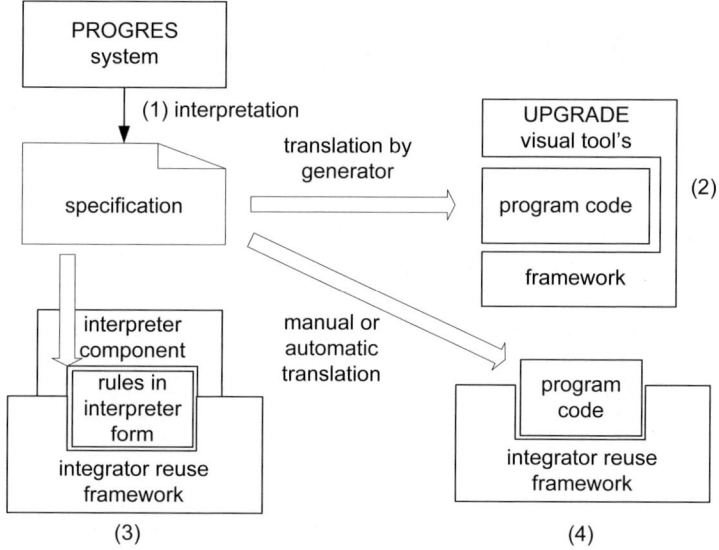

Fig. 3.39. Four different possibilities for realizing integrators based on a specification, (2) and (3) being discussed in detail below

Different Realization Strategies

Two of them are based on the academic PROGRES system which allows for defining and executing graph transformation systems [414]. Both have in common that the set of integration rules has to be available as a PROGRES graph transformation specification, which can be generated as described in the previous subsection. Following alternative (1), the *specification* is *interpreted* within the PROGRES system. As the PROGRES interpreter does not offer domain- and application-specific visualization and graph layout, this alternative is not considered in this paper.

The second alternative (2) is based on first *generating* program *code* from the PROGRES specification and then compiling it together with the *UPGRADE* visual tool's framework [49]. This results in a PROGRES-independent prototype with graphical user interface.

Alternatives (3) and (4) are based on a specific *framework* designed for the realization of *integrator tools* by making use of reuse. These alternatives follow a more industrial approach. So, they do not need the academic platform used for alternatives (1) and (2). Following alternative (3), integration rules are *interpreted* at runtime by a specific interpreter component. Alternative (4) is to include program *code* for all integration rules – either generated automatically or written manually – into the framework.

In the following, we are focusing on alternatives (2) and (3) which are compared in Fig. 3.40. We give a short overview and explain the common ground as well as the differences of both approaches. In the following subsections, we will present each of them in more detail.

Both implementation approaches rely on a set of integration rules as having been described in Subsect. 3.2.3. For *modeling* these *rules*, we provide a special *editor* which is shared by both approaches. To reduce the implementation efforts, the rule editor uses the commercial case tool Rational Rose [743] as a basis, since it already provides basic modeling support for all types of UML diagrams needed by our modeling approach. To supply modeling features *specific* for *integration* rules, a plug-in implemented in C# (about 8600 lines of code) was created.

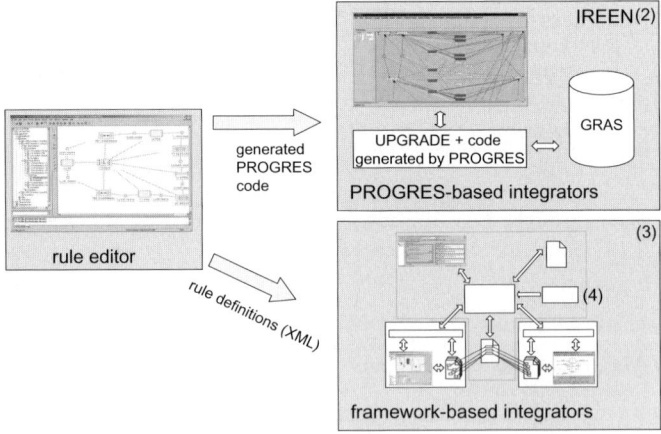

Fig. 3.40. Different implementation approaches

The integration rule editor comprises *all layers* of our *modeling approach* which are organized in different model packages: The meta layer package contains class diagrams for the generic integration meta model, which are fixed, as well as for domain-specific meta model extensions, which can be added and modified. In the type layer package, class diagrams for document type definitions and class diagrams for link type definitions can be created. In the abstract instance package, link templates and rules can be defined using UML collaboration diagrams.

The plug-in provides consistency *check* support *between* the *layers*: the type layer is checked against the meta layer and the instance layer is checked against the type and meta layers. For both checks, all detected inconsistencies are listed in the user interface and the affected model elements are highlighted. This helps the user to remove the inconsistencies. Additionally, the plug-in is able to derive forward, backward and consistence analysis rules from a given link-template. After the model is checked for its consistency and integration rules are derived, integration rules are exported in the XML dialect for graphs GXL [567, 732].

Both implementation approaches *apply* the *integration algorithm* presented in Subsect. 3.2.4. They differ in how the algorithm is executed.

Following the approach in the upper half of Fig. 3.40 (2), *PROGRES code* is derived for the rule-specific steps of the algorithm from the GXL rule definitions and combined with a predefined specification of the generic ones. Then, using PROGRES' *code generation* capabilities and the UPGRADE framework [49], an integrator prototype with a GUI is derived. Up to now, integrators realized by this approach are not connected to existing real-world applications. Instead, they are used for the quick evaluation of integration rules and of the integration algorithm itself. This realization method is called IREEN (Integration Rule Evaluation ENvironment). Current work at our department aims at providing a distributed specification approach for PROGRES [50] and interfaces to arbitrary data sources for UPGRADE [46]. Forthcoming results of this work could be used in the future to connect integrators realized by this approach to existing tools.

Up to now, integrator tools to be used in an industrial context and integrating existing applications are realized differently. *Integration rules* contained in GXL files are *interpreted* at runtime by a *C++-based integrator framework* (lower half of Fig. 3.40, (3). This approach was already sketched in Subsect. 3.2.2 (cf. Fig. 3.22). Besides interpreting rules, which is done for most rules, pre-compiled rules can be linked to the integrator as well (4). Up to now, these rules have to be hand-coded, but a C++ code generation comparable to the PROGRES code generation could be realized. The integrator is connected to the existing applications by tool wrappers which provide graph views on the tools' data.

Realization with PROGRES and Prototyping

Figure 3.41 gives an overview of how a PROGRES-based *integrator prototype* including a graphical user interface can be derived from a given integration rule set, following the *IREEN method*. First, from each synchronous triple graph rule being contained in the integration rule set to be executed a forward, a backward, and a correspondence analysis rule is derived as explained in Subsects. 3.2.3 and 3.2.4.

As mentioned before, the integration algorithm for rule execution consists of rule-specific and generic graph transformations. The *rule-specific graph transformations* are automatically derived from the UML collaboration diagrams containing all forward, backward, and correspondence analysis rules using a code generator. The generator output is an incomplete PROGRES graph transformation system [412, 414, 481] containing all rule-specific transformations for all rules.

To obtain a complete and executable specification, the partial specification has to be *combined* with three generic *specification parts*: One specification contains the static integration-specific parts, which are the integration graph scheme, the overall integration algorithm control, and the generic transformations for the algorithm. Additionally, for both source and target document there is a specification containing the document's graph scheme and some operations allowing the user to modify the document. Currently, the specifications for source and target documents are constructed manually. In general, it is possible to – at least partially – derive these specifications from UML models as well.

The *complete specification* is *compiled* by the PROGRES system resulting in C code which is then *embedded* in the UPGRADE framework [49, 206]. This

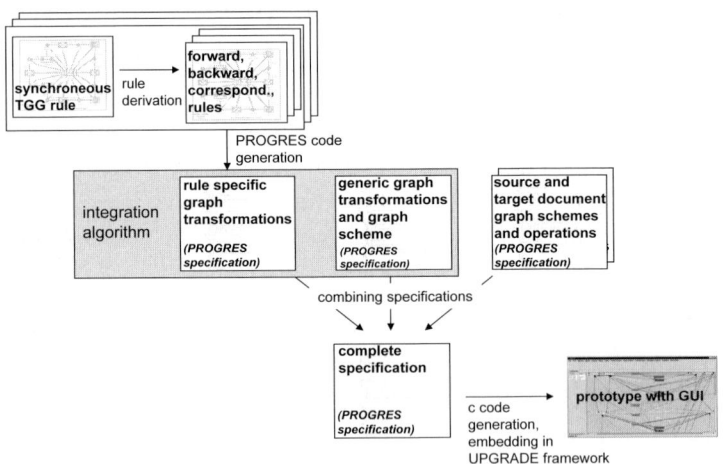

Fig. 3.41. Composition of PROGRES specification

leads to a *prototype* with a graphical user interface which allows construction and modification of source and target documents as well as performing runs of the integrator tool. All documents reside in the underlying graph database GRAS [8]. Some additional coding is required for application-specific layout algorithms and user interfaces. However, these efforts can be kept small because the resulting prototypes are not targeted at the end user. Instead, they are intended to serve as proof of concept, i.e., for the evaluation of integration rules without having to deal with real applications.

Industrial Realization

For the practical realization of integrators, i.e. for demonstrators in industry, an *integrator framework* is used. The framework is implemented in C++ and comprises about 14.000 lines of code. We focus here on a sketchy survey, because an overview of the system components of this framework has already been given in Subsect. 3.2.2 (cf. Fig. 3.22),

The *architecture* of the integrator framework is displayed in Fig. 3.42. The package IntegratorCore contains the execution mechanism and the overall control of the integrator. The package IntegrationDoc provides functionality for storing, retrieving, and modifying links in the integration document. DocumentWrapper consists of interface definitions that all tool wrappers have to implement. As there are no generic wrappers, there is no implementation in this package. The packages mentioned so far correspond to the main components in the system architecture of Fig. 3.22.

There are two additional packages: First, IntegrationGraphView provides an *integrated graph view* on source, target, and integration document. Therefore, it uses the corresponding document packages. Second, GraphPatternHandling supplies graph *pattern matching* and *rewriting functionality*. This functionality is rather generic and could be used for arbitrary graph rewriting tasks.

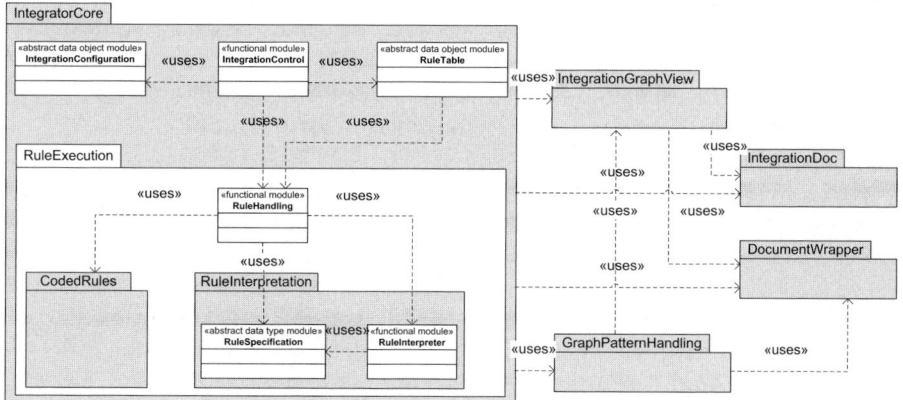

Fig. 3.42. Integrator framework

Indeed, some of the pattern matching algorithm optimizations incorporated in PROGRES have been re-implemented in the package. But as the graph rewriting formalism for integration rules is limited, so far the package only supplies the features needed for integrators.

The module IntegrationControl in the integrator core provides the *overall control* for enacting integrations. For instance, its interface offers methods that are used by the GUI to start or restart the integrator. To customize a run of the integrator, administrative settings are stored in IntegrationConfiguration and read by IntegrationControl. At the start of the integrator, all integration rules are stored in the module RuleTable. The sub-package RuleExecution implements the integration algorithm introduced in Subsect. 3.2.4. Rule-independent steps of the algorithm are implemented in RuleHandling. Rule-specific steps are either implemented directly in the sub-package CodedRules or executed by the rule interpreter (sub-package RuleInterpretation). The module RuleHandling serves as a "router", either calling a rule-specific piece of code in CodedRules for coded rules, or handing the execution over to the rule interpreter. For either type of rule, the realization of the algorithm steps is mostly based on graph pattern handling. But unlike the PROGRES-based implementation, some algorithm steps can be implemented by calling methods of the integration document package directly to provide a more specific and, thereby, more efficient implementation.

Prototype Demonstrator

Our integration approach described so far has been *applied* in a *cooperation* with our industrial partner innotec GmbH. Innotec is a German software company and the developer and vendor of the integrated engineering solution Comos PT. In our cooperation, the integrator for Comos PT process flow diagrams and Aspen Plus simulation models as described in the motivating example (cf. Subsect. 3.2.1) has been implemented.

The *integrator realization* is based on an early version of the C++ integrator framework which is interpreting integration rules at runtime. Integration rules are modeled using the formalism described in Subsect. 3.2.3 with Rational Rose, the rule modeling plug-in is used to export the rules to XML files. Figure 3.43 shows the graphical user interface of the integrator.

The *user interface* is integrated into the Comos PT environment as a plug-in. It is divided into *two parts*: On the left-hand side, all pending decisions between alternative rule applications are listed. The user has to choose a rule before the integration can proceed. On the right-hand side, all links in the integration document are shown. Symbols illustrate the links' states and for each link a detailed description showing the related increments can be opened. The integrator realized so far only addresses the integration of PFDs and simulation models. Future work aims at providing an integration platform for Comos PT for arbitrary integration problems (cf. Sect. 7.6).

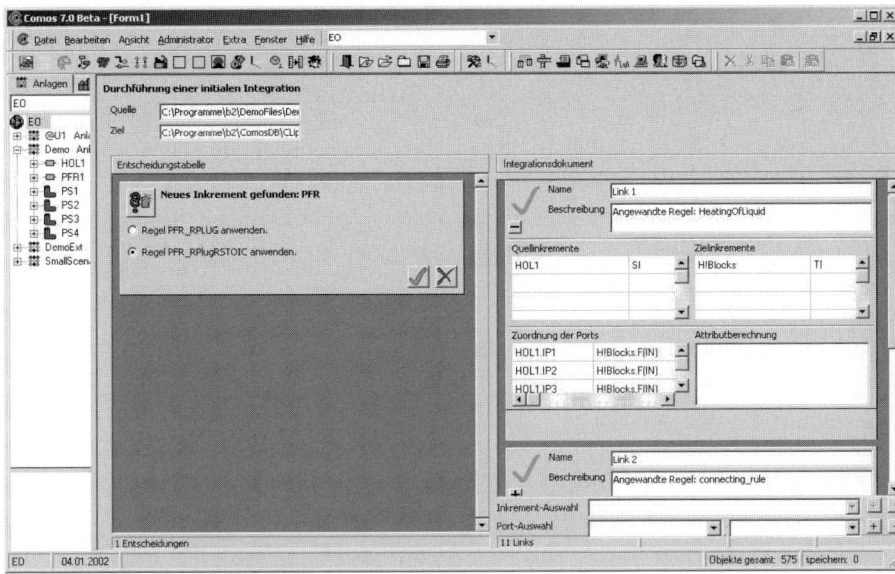

Fig. 3.43. User interface of the integrator for Comos PT PFDs and Aspen Plus simulation models (in German)

3.2.6 Additional Integrators

Apart from integrators described so far, some *further integrators* have been realized in IMPROVE and for predecessor projects that only partially make use of the described concepts and infrastructures. In this section, we will provide a short overview of these tools. Some tools will be described in more detail in Sect. 5.5 and Sect. 7.6.

CHEOPS Integrator

In the IMPROVE scenario (cf. Sect. 1.2), the tool CHEOPS [409] (cf. Subsect. 5.3.5) is used for the *simulation* of the *overall* chemical *process*. This simulation consists of multiple simulation models for different heterogeneous simulation tools. The task of CHEOPS is to perform partial simulations with the appropriate tools and to exchange simulation results between them.

An *integrator* was implemented to generate the *XML file* that is used to control CHEOPS for a specific process. This integrator differs from the ones described so far. Unlike other integrators, it deals with *more* than two *documents*. It unidirectionally generates the XML document out of multiple input documents: It reads the top-level PFD, the AHEAD product model (cf. Sect. 3.4) and the single simulation models. Additionally, links contained in integration documents between PFDs and the simulations are exploited. Another difference is the kind of user interaction taking place. There are *no conflicts* as explained in Subsect. 3.2.4. Instead, the user has to *select* which simulation *model* to *use* if more than one is available for a part of the process and to supply initial values for external input streams.

Therefore, the integrator was *implemented manually* using only parts of the integrator framework. The rules that describe how the XML file has to be integrated are hard-coded into the prototype. This does not lead to problems here as the rules are simple and static. Additionally, *rule execution* is much *simpler* as in other integrators because of the lack of real conflicts. This integrator is described in more detail in Sect. 5.5.

Integrating Technical and Design Level

Most integrators integrate the results of technical activities in the development process, like PFDs and simulation models. Unlike that, the PFD-AHEAD integrator performs the *integration* of a *technical master document*, the PFD, with the administrative configuration of the development process in the AHEAD system (cf. Sect. 3.4) being some *organizational master document*.

As the PFD is a master document that serves as overview of the whole chemical plant to be designed, it can be used to provide an *interface* to the administration of the development process. The PFD-AHEAD integrator does this to help the *chief engineer* in a development process to *determine* the *consequences* of *changes* that are made to a part of the plant design. To do so, the chief engineer marks components in the PFD that have to be redesigned due to a change. After that, the integrator interactively *determines* how project *coordination* in AHEAD has to be adapted to contain tasks that deal with these changes. The *process manager* reviews the changes to the AHEAD process and either modifies them or directly applies them.

For this integrator as well, a *realization* approach *different* from the one for normal integrators has been applied. The main reason for this are the peculiarities of the user interaction needed: Instead of selecting between conflicting rule applications, the chief engineer *annotates* the *source document* (PFD) to make his decisions. Later, he interactively *refines* his *annotations* with the help of the integrator.

Additionally, user *interaction* is performed by two *different roles* in the development process. The first is the chief engineer, who uses a PFD-related user interface. The second is the project manager, whose user interface is closely related to the AHEAD system. As a result, the integrator was implemented manually. Nevertheless, the experience with other integrators was quite helpful as some concepts of the original approach could be applied resulting in a clean architecture and a straight-forward integration algorithm. This integrator is also described in more detail in Sect. 5.5.

Other Integrators within and Outside of IMPROVE

Some additional integrators have been built which are listed below. Due to space restrictions, they are explained very briefly only.

- An integrator collecting *data from production control* has been realized in an industrial cooperation with the German company Schwermetall [956]. It integrates a large number of heterogeneous data sources into a centralized database with the mapping being quite simple. This integrator is not based on the framework sketched above.
- Two *XML-based integration tools* have been built in the area of *process management*. The first translates AHEAD (cf. Sect. 3.4) process definitions into the petri net dialect used as workflow language by the commercial workflow management tool COSA [615]. The second handles the import of workflow definitions made with WOMS (cf. Sect. 2.4) into AHEAD. Both made use of the XML transformation language XSLT [602].
- During the first phase of IMPROVE, an integrator between *Aspen Plus* and the *flowsheet editor* (FBW) of IMPROVE (cf. Sect. 3.1.3) and one between the modeling tools *gPROMS* and *ModKit* have been implemented manually [84]. The experience gained with their implementation was important for the design of the integrator framework.
- After the first version of the framework was implemented, the integrator between FBW and Aspen Plus has been reimplemented to evaluate the framework.
- At our department, *integrator tools for other domains* have been built: In the ConDes project, an integration between a *conceptual* model of a building with the *concrete building architecture* is performed. Additionally, different ontologies modeling common knowledge about building architecture are integrated [234, 241].

  In the CHASID project, written *text* is integrated with a *graph structure* describing its *contents* [128].

  In the domain of *reverse-* and *reengineering*, triple graph grammars have been applied to integrate different *aspects* [81–83, 88, 89].

  The integration of different logical documents was first studied for development processes in *software engineering* [109, 260]. For instance, the relationship between requirements engineering and software architecture has been studied [74, 184, 185, 254]. Theses studies have been broadened during the IPSEN project [334] dealing with a tightly integrated development environment [229, 256–259].
- In our cooperation with innotec, we currently develop an integrator tool between the *data* structure *definition* of Comos PT and corresponding *UML models*.

3.2.7 Related Work

Our approach to the specification of incremental and interactive integrator tools is based on triple graph grammars. Therefore, we will discuss the relationships to other research on *triple graph grammars* in the next subsection. Subsequently, we will address competing approaches to the *specification* of integrator tools which do not rely on the triple graph grammar approach.

Related Work Based on Triple Graph Grammars

The triple graph grammar approach was invented in our group by Schürr [413], who gave the theoretical foundations for building TGG-based integrator tools. The work was motivated by *integration problems in software engineering* [349]. For example, [259] describes how triple graph grammars were applied in the IPSEN project [334], which dealt with integrated structure-oriented software development environments.

Lefering [255] built upon these theoretical foundations. He developed an early framework for building integrators which was based on triple graph grammars. The framework was implemented in C++, rules had to be transformed manually into C++ code to make them operational. The framework was applied to the integration of requirements engineering and software architecture documents, but also to integrate different views of requirements engineering [229].

Other applications of triple graph grammars have been built using the PROGRES environment. In our *reengineering* project REFORDI [88], synchronous triple rules were transformed manually into forward rules (for transforming the old system into a renovated one being based on object-oriented concepts). The PROGRES system was used to execute forward rules – which were presented as PROGRES productions – in an atomic way.

Our work on rule execution differs from systems such as REFORDI (or, e.g., VARLET [766] from another department) inasmuch as a single *triple rule* is executed in *multiple steps* to detect and resolve conflicts, as originally introduced by Lefering.

Our work contributes the following improvements:

- We added detection, persistent storage, and resolution of *conflicts* between integration rules.
- We provide a precise formal *specification* of the *integration algorithm*. In [255], the algorithm was described informally and implemented in a conventional programming language.
- Likewise, rules had to be hand-coded in Lefering's framework. In contrast, synchronous triple rules are *converted automatically* into specific rules for execution in our approach.
- We *used* the *specification* in *two ways*: First, IREEN was constructed by generating code from the formal specification (Fig. 3.41). Second, an implementation designed for industrial use was derived from the formal specification (Fig. 3.42).

To conclude this subsection, we will briefly discuss related work on triple graph grammars:

The PLCTools prototype [528] allows the *translation* between *different specification formalisms* for programmable controllers. The translation is inspired by the triple graph grammar approach [413] but is restricted to 1:n mappings. The rule base is conflict-free, so there is no need for conflict detection and user interaction. It can be extended by user-defined rules which are restricted to be unambiguous 1:n mappings. Incremental transformations are not supported.

In [786], *triple* graph *grammars* are *generalized* to handle *integration* of *multiple documents* rather than pairs of documents. From a single synchronous rule, multiple rules are derived [787] in a way analogous to the original TGG approach as presented in [413]. The decomposition into multiple steps such as link creation, context check, and rule application is not considered.

In [579, 1033], a plug-in for *flexible* and *incremental consistency management* in Fujaba is presented. The plug-in is specified using story diagrams [670], which may be seen as the UML variant of graph rewrite rules. From a single triple rule, six rules for directed transformations and correspondence analysis are generated in a first step. In a second step, each rule is decomposed into three operations (responsibility check, inconsistency detection, and inconsistency repair). The underlying ideas are similar to our approach, but they are tailored towards a different kind of application. In particular, consistency management is performed in a reactive way after each user command. Thus, there is no global search for possible rule applications. Rather, modifications to the object structure raise events which immediately trigger consistency management actions.

Other Data Integration Approaches

Related areas of interest in computer science are (*in-*)*consistency checking* [975] and *model transformation*. Consistency checkers apply rules to detect inconsistencies between models which then can be resolved manually or by inconsistency repair rules. Model transformation deals with consistent translations between heterogeneous models. Our approach contains aspects of both areas but is more closely related to model transformation.

In [658], a *consistency* management approach for *different view points* [669] of development processes is presented. The formalism of distributed graph transformations [992] is used to model view points and their interrelations, especially consistency checks and repair actions. To the best of our knowledge, this approach works incrementally but does not support detection of conflicting rules and user interaction.

Model transformation recently gained increasing importance because of the model-driven approaches for software development like the *model-driven architecture* (MDA) [876]. In [689] and [776] some approaches are compared and requirements are proposed.

In [977], an approach for non-incremental and non-interactive *transformation between domain models* based on graph transformations is described. The main idea is to define multiple transformation steps using a specific meta model. Execution is controlled with the help of a visual language for specifying control and parameter flow between these steps.

In the *AToM project* [627], modeling tools are generated from descriptions of their meta models. Transformations between different formalisms can be defined using graph grammars. The transformations do not work incrementally but support user interaction. Unlike our approach, control of the transformation is contained in the user-defined graph grammars.

The *QVT Partner's proposal* [509] to the QVT RFP of the OMG [875] is a relational approach based on the UML and very similar to the work of Kent [498]. While Kent is using OCL constraints to define detailed rules, the QVT Partners propose a graphical definition of patterns and operational transformation rules. These rules operate in one direction only. Furthermore, incremental transformations and user interaction are not supported.

BOTL [565] is a transformation language based on UML object diagrams. Comparable to graph transformations, BOTL rules consist of an object diagram on the left-hand side and another one on the right-hand side, both describing patterns. Unlike graph transformations, the former one is matched in the source document and the latter one is created in the target document. The transformation process is neither incremental nor interactive. There are no conflicts due to very restrictive constraints for the rules.

Transformations between documents are urgently needed, not only in chemical engineering. They have to be incremental, interactive, and bidirectional. Additionally, transformation rules are most likely ambiguous. There are a lot of transformation approaches and consistency checkers with *repair actions* that can be used for transformation as well, but none of them fulfills all of these requirements. Especially, the detection of conflicts between ambiguous rules is not supported. We address these requirements with the integration algorithm described in this contribution.

3.2.8 Summary and Open Problems

In this section, we presented the results of the IMPROVE subproject B2. The *main contributions* of this section are the integration algorithm defined in the PROGRES specification of IREEN, the specification method for integration rules, and the integrator framework. Our framework-based integrator prototypes realized so far could be implemented with considerably lower effort than those that were built from scratch. The explicit specification of integration rules helped improving the quality of the resulting tools.

First practical experiences have been gained in a *cooperation* with our industrial partner innotec. The cooperation will be continued in a DFG transfer project, see Sect. 7.6.

Another important aspect of *methodolodical* integrator *construction* is the step-wise refinement of coarse-grained domain models or ontologies to fine-grained specifications defining the behavior of operational tools. In this section, this topic has only been sketched, focusing mostly on the fine-grained definition of integration rules. The relationship to domain models will be discussed in more detail in Sect. 6.3.

Besides evaluation in industry, current and future work will address some major *extensions* to the integration approach. For instance, more language constructs of graph transformations, e.g. paths and restrictions, are to be incorporated into the integration rule language. Additionally, the framework will be extended to offer repair actions for links that have become inconsistent due to modifications of documents. Further research will be conducted to support the integration of multiple documents considering complex multi-document dependencies.

3.3 Multimedia and VR Support for Direct Communication of Designers

A. Schüppen, O. Spaniol, D. Thißen, I. Assenmacher, E. Haberstroh, and T. Kuhlen

Abstract. The development of design processes in chemical engineering and plastics processing requires close cooperation between designers of different companies or working groups. A large number of *communication relationships* is established, e.g. for the clarification of problems within single tasks, or within project meetings covering the discussion about interim results. With the ongoing development of a process, different types of communication relationships will occur, as the required communication form as well as the extent of communication are depending on the work task. For an efficient support of the communication between designers, support tools are needed to enable cooperative computer-aided work tailored to an actual task and to obtain a speed-up of the development of a process.

This section discusses several activities to improve the communication between engineers by *multimedia and Virtual Reality tools and protocol mechanisms* for supporting *new forms of cooperative work* in the design of a process. To ease the usage of those tools and protocols within the work processes, they are integrated into the normal working environments of the designers. For communication processes involving geographically distributed designers, the *communication platform KomPaKt* was developed, which integrates the new communication and cooperation tools for different purposes in a single, configuration-free, and intuitive user interface. As a specially interesting case of cooperation between designers, the technology of *immersive Virtual Reality for simulation sciences* was examined in more detail using the example of compound extruders, as Virtual Reality technology plays a key role in interdisciplinary communication processes.

3.3.1 Direct Communication

A design process in the considered application domain comprises several *different forms of communication* and cooperation. To give only a few examples, video conferences can be planned for the discussion about interim results of a design step between geographically distributed designers, shared document editing helps in document manipulation and presentation in distributed work groups, and informal communication could be necessary if a question arises which can only be answered by a remotely located project partner.

For an *efficient support* of all communication needs, certain *requirements* have to be met:

- Currently available tools that support communication only integrate some simple functionalities like e-mail or address books, or they are realized as independent and possibly incompatible products for special tasks. This can cause a huge amount of time that a designer has to spend on learning how to efficiently use these communication tools. Ultimately, this may lead

to a complete refusal of such supporting tools. To avoid such problems, a designer should only need to use a *single interface* comprising *all* different *functionalities* for *communication* in a design process. This interface should provide intuitive usability, and should free the designer from configuration tasks.
- To facilitate a better exchange of ideas in a communication process between geographically distributed engineers, a *combination of documents and video/audio data* is needed. Only this combination enables a discussion about results.
- Tasks on which several engineers work together have to be supported by functionalities of common, *simultaneous manipulation of design documents*. This should be realized on one hand by *enhancing* existing tools for document manipulation, but on the other hand also by new interactive environments, like *Virtual Reality*.
- *Changes in design documents* produced during a communication session should be *restored in the document management system*. To do so, an integration of the tools with the product model is needed.
- Results of one design step, or of a whole design process have to be discussed in a group of geographically distributed designers. For this purpose, *group communication* is necessary, not only comprising multicast communication but also conferencing control.

To meet these requirements, a *communication platform* was developed which enables a designer to use multimedial communication mechanisms in his work process (see Subsect. 3.3.2). This platform integrates synchronous and asynchronous communication tools into one graphical interface. A major design goal was the intuitive usability of the communication platform. Through integration with the administration system (described in Sect. 3.4), contact information of cooperation partners as well as the actual documents can be aggregated without any additional configuration work.

The platform makes it possible to *access several* communication *tools*; some stand-alone tools were integrated, and some new tools and communication mechanisms were developed. Most important in this context are the so-called *event sharing* for cooperative document manipulation in distributed work groups together with the required mechanisms for the adaptive transfer of multimedia data (see Subsect. 3.3.3), as well as *Virtual Reality* (see Subsect. 3.3.4 and 3.3.5) as a new way to represent results.

For the control of a cooperative work session, the *Scalable Conference Control Service (SCCS)* was developed as a signalling protocol (see Subsect. 3.3.6).

A description of work related to the communication platform and the tools and services (Subsect. 3.3.7) closes the section.

3.3.2 The Communication Platform KomPaKt

When introducing new forms of cooperation, it is important to get acceptance of the mechanisms by the target user group. Thus, the communication platform Kommunikationsplattform für die Prozessentwicklung und -analyse in der Kunststoff- und Verfahrenstechnik (KomPaKt) was developed [456]. This platform integrates several communication tools and offers the designers a single interface for accessing them.

The *tasks of developing the platform* not only comprised the set-up of an interface and the integration of communication tools, but also the development of new tools for supporting the steps in a design process on one hand, and of an infrastructure for conference management on the other. The new tools and the conference management infrastructure are covered in the following subsections. This subsection only focuses on the communication platform's functionality and its integration.

Conception of the Communication Platform

To integrate synchronous and asynchronous communication tools into distributed work sessions, a framework was created which modularly incorporates different tools into one interface. In this context, *synchronous communication* comprises all those services which support *multimedia conferencing*. This includes not only the real-time transmission of audio/video data, but also the usage of locally installed design tools by the whole team.

In synchronous communication, all involved team members have to be available at the same time. So, all services belonging to this category need some mechanisms for ensuring the *communication availability* of all participants. The typical conferencing scenario in the given application domain is a project meeting of a group of designers. In the usual working environment it cannot be expected that special conferencing rooms with audio/video equipment are available for such a group at any time it would be needed. Instead, it should be possible for the designers to join a *distributed conference during their work* from their workstations, i.e. standard PCs, equipped with common audio/video devices.

In contrast, *asynchronous communication* does not require the availability of all participants at the same time. The most popular functionality here is e-mail. Additionally, an analogous service can be thought of for voice: By allowing the generation of audio messages which can be sent to the communication partner, a functionality like a phone mailbox is provided, allowing the communication partner to replay it when he has time to do so. Such functionalities are of use for delegating tasks, exchange of results of process steps, or questions which do not need to be answered immediately. The advantage using asynchronous communication is its simplicity, as no coordination of the involved project partners is necessary.

The *means for synchronous and asynchronous communication* are described in Fig. 3.44, showing all communication functionality considered in KomPaKt. These functionalities were realized by developing and implementing new tools (for audio/video transmission [183] and event sharing [317]),

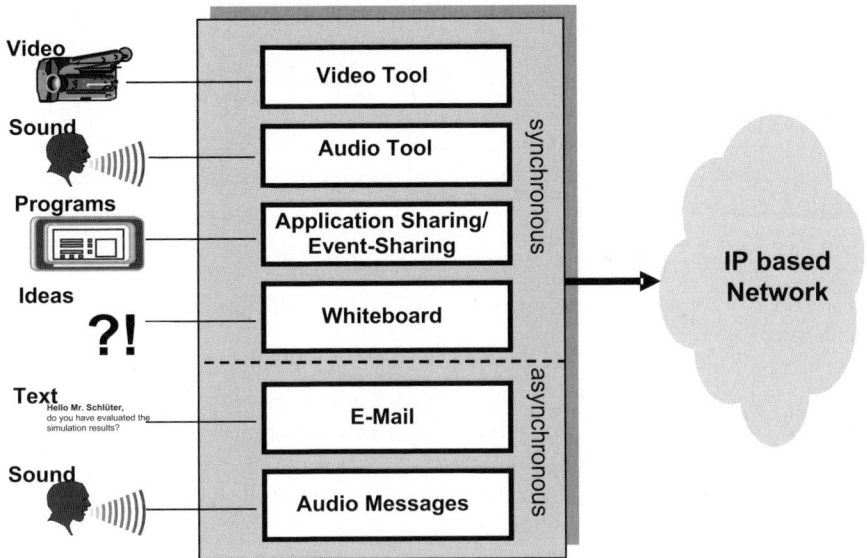

Fig. 3.44. Functionalities of the communication platform

or by integrating commercial products (for application sharing, shared whiteboard, e-mail, and audio messages). A scenario for the usage of those functionalities was defined in cooperation between project partners from communication systems and plastics processing.

The minimum functionality for a distributed work session is the exchange of audio and video information. Data streams can be sent independently; if a coordinated interaction is needed in a conferencing, a so-called *floor control* as part of a conference management realizes a coordinated access to the communication channel. Additionally, one participant (usually the initiator of the distributed meeting) manages the conferencing session. Communication between the participants considers certain social interaction rules [1036], e.g. signalling of the wish to ask a question before sending it (as audio/video data). Furthermore, *secure communication* is to be provided; authentication and conference access assure that only authorized users can participate in a conference. Thus, in addition to floor control, conference access control is provided for the communication tools.

Figure 3.45 shows the *protocol stack* upon which the communication tools are based. The Internet protocols TCP and UDP in combination with RTP [902] are used for data transmission. The audio/video tools were developed for transmission over the Internet considering possible congestions; thus, they realize an adaptive transmission where the transmission rate is adapted to currently available capacities [304], [465]. Both, the video and audio tool, use RTP for transmission and the capabilities of the corresponding RTCP

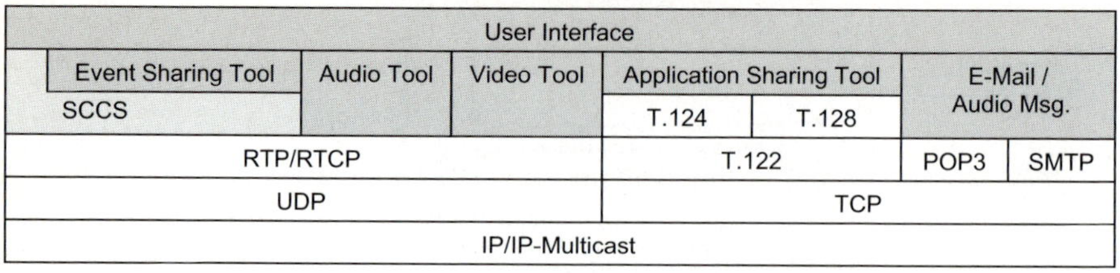

User Interface						
Event Sharing Tool	Audio Tool	Video Tool	Application Sharing Tool		E-Mail / Audio Msg.	
SCCS			T.124	T.128		
RTP/RTCP			T.122		POP3	SMTP
UDP			TCP			
IP/IP-Multicast						

Fig. 3.45. Protocol stack of KomPaKt

to exchange state information about Internet traffic, thereby discovering free capacities of the networks. Based on the RTCP messages, the compression scheme of the audio/video data can be changed to adapt the required capacity, keeping the quality of the transferred data as high as possible. In case of only two participants, IP is used as a network protocol, for conferences comprising more participants, IP multicast is used. The same holds for the so-called *event sharing* used for distributed work on a single document. Additionally, a transmission channel for sending the event sharing data is provided here by the *Scalable Conference Control Service (SCCS)*. SCCS implements a signalling protocol which handles the conference and floor control mechanisms for synchronous communication.

Similar to the event sharing, but also suited for other application scenarios, is the so-called *application sharing* [850]. We selected application sharing provided by Microsoft's NetMeeting [841]. It is based on the ITU-T T.120 protocol family. While T.128 defines the data exchange, T.124 provides conference management functionality. Both use T.122 for transferring application and management data. T.122 is independent of the used network by avoiding to use IP multicast. It maps point-to-multipoint transmission onto a set of point-to-point connections using TCP. Clearly, this network independence causes additional transmission costs.

Before discussing more details of the communication tools and services, an *introduction* to the *capabilities* provided for the engineers is given, focussing on the user interface and the relation to other supporting functionality.

User Interface and Functionality

A simple and *intuitive interface* for the *users* [410] is an important yet often neglected point. Multimedia communication tools are perceived as just an additional support in process design, but not as absolutely necessary for solving a task. Thus, the usage has to be simple in order to motivate the designer to use the additional functionality. Consequently, the design of the user interface has been carried out in cooperation with the project partners of the process engineering and ergonomics departments.

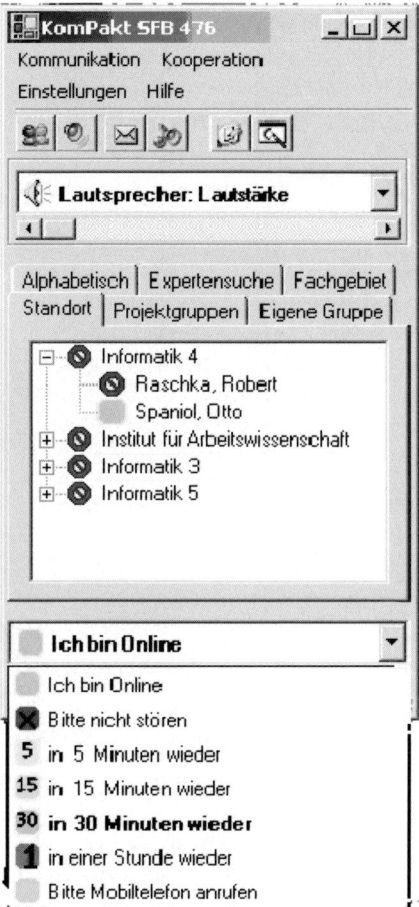

Fig. 3.46. User interface of KomPaKt

As a result, the user interface of KomPaKt was designed, implemented, and evaluated. The *evaluations* have shown that KomPaKt is very well suited for sporadic usage in the context of a design process. All test persons, experienced as well as unexperienced computer users, could solve all tasks for initiating and performing a communication without problems. More details on these results are described in Sect. 5.2.

The user interface manages information about all persons involved in a design process. This enables the presentation of a clear list of possible communication partners relevant in a work context. The willingness of all planned participants is a pre-requisite for initiating a synchronous multimedia conference. Thus, an *awareness functionality* has been integrated into KomPaKt.

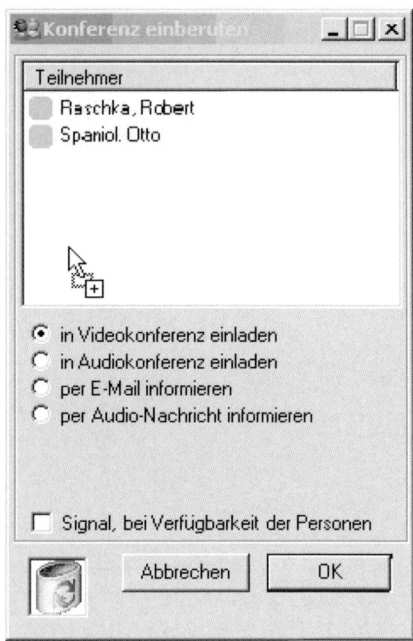

Fig. 3.47. Invitation window of KomPaKt

In principle, it would be possible to automatically detect the availability of a user by simply detecting keyboard or mouse input. Yet, this would not only violate security guidelines, but would also only indicate an information about availability, though availability does not necessarily signal willingness to communicate. The awareness functionality enables a user to signal his willingness for communication: It allows to indicate whether he is available for communication, whether he does not want to be disturbed, whether he is away, and possibly when he will be back, see Fig. 3.46. This information is visible for all other users and can help planning a conference. Additionally, each user can leave a file with contact data for urgent situations.

The first step in the *initiation of a communication relation* is to choose the communication type (synchronous or asynchronous). An invitation window opens, to which participants from the address book can be added by drag and drop (see Fig. 3.47).

It is not necessary to deal with any configuration details (e.g. providing the IP address of the communication partner) before starting a conference. The corresponding address information about project partners from the address book, as well as documents for a work session are transparently provided through a connection to the administration system AHEAD, which is described in Sect. 3.4. This connection is one of the reasons why it was not possible to just use commercial products for multimedia support. Some tools and services had to be newly developed to allow for the necessary connections to other supporting tools developed in IMPROVE.

Connection to the Administration System AHEAD

The *administration system* AHEAD manages the resources of a design process, which also include the representation of the involved project members. By implementing a *connection* to AHEAD, information about designers and their contact data is transparently gathered by KomPaKt. Additionally, the documents used in a work session can be identified and used in a conferencing session.

At this point it is important to make a distinction between two types of multimedia conferences. *Planned conferences*, e.g. meetings for discussing about intermediate results obtained in the design process, can be modeled as tasks in the process management of AHEAD. Thus, KomPaKt can be initiated, like any other tool, through the work context for a task, directly gathering all documents and contact information of the assigned project members for this task.

However, KomPaKt can also be used for *spontaneous conferencing*, for instance, if an engineer wants to discuss problems with partners. In this case, KomPaKt uses the project team management of AHEAD to determine the possible participants of a conference.

Last but not least, *results of a* simulation *conference* (e.g. protocols, video recordings, annotated documents, etc) can be stored by using the document management of AHEAD. On the technical layer, the connection to the administration system is based on CORBA.

For the integration with AHEAD, the *communication process* is independent from the process models described in Chapter 2. The communication process adds communication and cooperation possibilities to the C3 model (see Sect. 2.4). Detailed information about synchronous communication for certain tasks is provided. On the one hand, this is done by specifying requirements in terms of Quality of Service for the transmission. Figure 3.48 shows the parameters necessary for different communication types.

On the other hand, *four abstraction layers* have been considered to model communication processes. First, the description layer gives information about given input or output functionality, and if the data are persistent. On the next abstraction layer, data sources and sinks are combined to transmission channels, characterized by the media coding used. In general, an application will use several channels, e.g. one for audio and one for video transmissions. To integrate these channels into a communication process, contexts between channels are created on the third layer. So far, the whole specification is independent from communicating applications located on the users' hosts.

	Audio	Video		Data
		uncompressed	compressed	
Data rate	16-64 Kbit/s	~ 100 Mbit/s	~1.5 Mbit/s	~ 0.2 -10 Mbit/s
End-to-end delay	~ 250 ms	~ 250 ms	~ 250 ms	~ 1.000 ms
Maximum jitter	~ 10 ms	~ 10 ms	~ 1 ms	~ 1 ms
Maximum loss rate	10^{-2}	10^{-2}	10^{-11}	10^{-11}

Fig. 3.48. Communication requirements for different media

Thus, the fourth layer locates the involved components. The layered model is linked to the model of AHEAD, to allow for modelling multimedia conferences as own tasks, but also to enable KomPaKt to obtain configuration information from AHEAD.

3.3.3 Communication Tools

The user interface of KomPaKt only gives a unified way of accessing synchronous and asynchronous communications. This interface integrates some useful existing communication tools. For the considered application domain, also the development of some *new tools* was necessary. These tools are described in this subsection.

Audio and Video Transmission

Video conferencing needs the transmission of both audio and video data. Since a designer should be able to use the functionality from his own PC, no assumptions about the underlying network can be made. It can be an Ethernet with 100 MBit/s, or it can be based on ISDN with up to 2 MBit/s. For video and audio transmission, therefore, there is a need to adapt to the network capabilities. Products like NetMeeting were designed to work over low-capacity networks, with reduced video quality. Thus, dedicated *video and audio transmission tools* were developed, better *adapting to the network*.

The *video transmission tool* is able to adapt to available capacities of the network by changing the video codec and/or the compression ratio before sending the data [304]. As shown in Fig. 3.45, this approach uses the Real-Time Transmission Protocol RTP for a connectionless data transmission without error handling. RTP also provides a control protocol named RTCP, by which the receiving party can give feedback information about the received data and the error ratio. The sending party uses this feedback information to estimate the available network capacity and correspondingly changes coding and compression of the video data stream. Integrated error detection and correction, also using information from RTCP, prevents the error propagation in the received video stream, which is typical for commercial products like NetMeeting [306], [307]. A corresponding *tool* for transferring *audio* data was also developed [465], [464] and *integrated* into the KomPaKt environment.

While these video and audio transmission methods were realized as separate tools, together offering a mechanism for video conferencing, in a later project phase the functionality was generalized to offer an *independent multimedia data transmission service* which can be used in conjunction with a broader spectrum of tools. The communication service which arose from this effort is presented in Subsect. 3.3.6.

Event Sharing

In a typical cooperation process, not only video and audio information are to be exchanged. Also, designers discuss about documents, and maybe these documents are modified. Thus, in addition to the conferencing some way of shared document editing is needed. A well-known technology is *application sharing* [850].

Application sharing simply means that one application is started locally on one host and shared with a number of other hosts. The graphical *output* of the application is *distributed* to all participants. Also, the *input* can be *switched* between all parties, allowing one participant at a time to take over the control of the application. If a document is opened in the application, everyone can take over control and modify the document. The modifications are immediately visible to all others.

The disadvantage of this technique is the *large volume of screen data* which is to be transferred to all partners to display the current state simultaneously for all users. The scenario in question was the cooperative work of a distributed team regarding 3D-animations using BEMView in plastics processing (see Subsect. 5.4.6). The amount of data to be transferred for shared working with an animation sequence is intolerably high.

Thus, as a first solution, 3D-animation of simulations of streams in an extruder were *recorded as a video sequence* and streamed to all participants. This method reduces the influence of congestions in the network compared to application sharing, but does *not allow* to *modify* the document. The chance for spontaneous interaction in the evaluation of animation sequences is lost. Even simple things, like having a look at the extruder from another angle is impossible - one would have to wait for the sequence to end and to produce a new video sequence. This makes a cooperative evaluation very uncomfortable and will hardly be accepted by any user.

An *alternative* approach was developed, which enables the transmission of a visualized extruder simulation as a 3D-world. The *compressed data* files that are storing the simulation results for usage with BEMView contain all

Application Sharing:

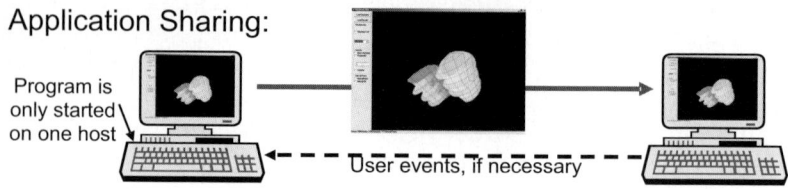

Event-Sharing:

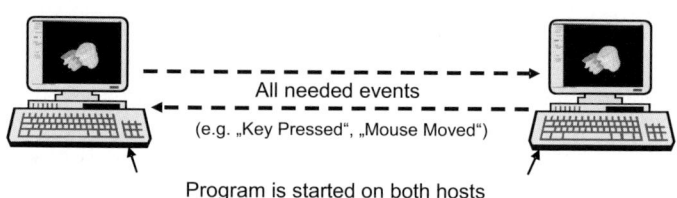

Fig. 3.49. Comparison of application and event sharing

information about the 3D-animation, and at the same time are only slightly larger than the recorded video sequences. Only user information which is to be provided during the loading of the file are not available (angle of view, distance, start time of simulation).

To address these problems the so-called *event sharing* was developed [317]. While in application sharing the screen information is transferred to all participants, event sharing needs to *run the application on all hosts* and only distributes events which are causing a change in the application's state. This way, an *input* of one participating host is sent to all partners and processed by the locally running application instances. *All instances* are kept in the *same state*, such that all users have the same view, like in application sharing. Such events can be user inputs (mouse, keyboard) or system inputs (system time, data). Figure 3.49 compares application and event sharing.

Figure 3.50 shows the *architecture* of the *event sharing tool*: The component **Conference** is responsible for conference management and for the management of the data transfer channels. As these tasks are not specific to event sharing but are needed for conferencing in general, they are managed by the control service SCCS, which is described in Subsect. 3.3.6.

The **Controller** is the interface for initializing the event sharing and controls the interaction of the other components. For example, first it checks whether the conference service is available. If so, the **Synchronizer/Sender** is invoked which provides a messaging interface for mapping events to a data channel provided by the conference service. The **Synchronizer/Sender** enables synchro-

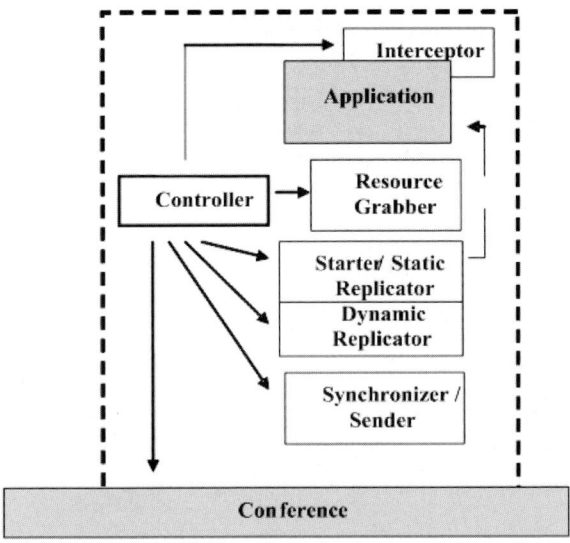

Fig. 3.50. Components of the event sharing architecture

nization by forwarding new events only when all participating hosts are in the same state. Subsequently, the other components are started.

The Interceptor grabs the events that change the state of the shared program. How to implement the Interceptor depends on the operating system; most operating systems offer mechanisms for stopping or adding events. Only one user has control over the program; his host copies and distributes the events to the other parties. On all other sites events are blocked and replaced by the received ones. The Resource Grabber recognizes which resources are used by the program and makes reservations on all other sites.

The latter two components are responsible for distributing documents needed in the shared work process. The Starter/Static Replicator takes over that task upon startup; it distributes the necessary documents to all parties and starts the shared program. During the work session, the Dynamic Replicator tracks the program run and distributes newly loaded documents to all sites. As a brief comparison, Fig. 3.51 summarizes the most important characteristics of application and event sharing.

The plastics processing's application *BEMView* was used for *realizing* and *evaluating* the event sharing approach. A clear reduction of necessary transmission capacity could be observed. Additionally, as a side effect, a *new* type of *interaction* with a shared application was found: By loosening the processing or blocking of events, it is possible to realize a *loose coupling*, thus allowing each user his own view at the displayed simulation results within the same document.

	Application Sharing	Event Sharing
Method	Screen output is transferred	Events which change the state are transferred
Number of running application instances	One (only on initializing host)	Many (one on each involved host)
Target applications	Any program	Programs with extensive graphical output
Platform independence	Yes (theoretically)	No
Needed transmission capacity	High	Low
Loose Coupling	Impossible	Possible

Fig. 3.51. Application sharing vs. event sharing

Unfortunately, the approach has also disadvantages *compared to application sharing*: Only users who have installed the shared application, can participate in shared work. Additionally, the event sharing mechanism has to be adapted to an application in order to be able to grab all events necessary for synchronization. Thus, the installation process causes much more effort than for application sharing. It depends on the application area whether application or event sharing is more useful. Event sharing is not to be seen as a replacement of application sharing, but as an *alternative*.

Nevertheless both, application and event sharing, have shown a disadvantage in the application domain of plastics processing: extruder simulations consist of large, complex sets of data, which are hard to be handled by the engineers even with tools like BEMView. For this application scenario, *Virtual Reality* was promising to be a much more effective communication medium, although it needs special hardware and thus cannot be integrated with KomPaKt. Because of the extensive examinations regarding this technology, an *own subsection* is spent to present the results (see Subsect. 3.3.4).

Other Functionalities

Not all communication tools integrated in KomPaKt have been newly developed, also some *existing* and well-functioning *tools* were *integrated*. For synchronous communication, a shared whiteboard delivered together with NetMeeting was integrated for providing a mechanism for drawing sketches or writing down some ideas in spontaneous meetings. For asynchronous communication, an e-mail service was integrated, also supporting the sending of audio messages.

Furthermore, it was planned to develop a general *shared editor* which would be able to store *annotations* to documents to provide a mechanism for informal comments on design documents. Due to the complexity of such a tool it was decided not to pursue this as one function. Instead, several mechanisms for specific document types were developed.

Most important here was the *annotation of video sequences*. As the recording of video sequences had already been implemented as predecessor of the event sharing tool, video sequences of simulation runs or screenshots of those sequences can be stored. The tool TRAMP (see Sect. 4.1) provides the functionality of storing and also annotating such sequences.

3.3.4 Virtual Reality in Multimedia Communication

In the following, *Virtual Reality* is introduced as an effective instrument to interpret application data and to communicate on these interpretations.

Virtual Reality for Interdisciplinary Communication

Virtual Reality, as a technology, aims at presenting synthetic worlds with which a user can interact in realtime with his natural senses. Despite this rather abstract definition, the additional value in concrete application examples comes from the naturalness of perception and presentation of abstract data. As such, Virtual Reality technology plays a *key role* in *interdisciplinary communication* processes (cf. Fig. 3.52). An important observation in interdisciplinary projects is, that people from different fields of science or engineering often use a very different language when discussing about the same topic. In Virtual Environments, the participants often start talking about the very same object of visualization they see, in conjunction with gestures and full body maneuvering. This obviously allows an easier access to the ideas of the conversational partner, which indicates the usage of virtual reality for communication.

Although this is already possible in room mounted Virtual Environments, such as CAVE-like or workbench environments, these setups are typically optimized for single users. As such, they have a little drawback for the cooperative work of many users. More possibilities result from the idea of *Collaborative Virtual Environments* (CVE), where more than one Virtual Environment is coupled on an application level and users interact remotely through teleconferencing mechanisms, avatar representations, and specific interaction metaphors.

Computational Fluid Dynamics Post-Processing in Virtual Environments

In the last few years, simulation of technical and physical processes has become an important pillar in engineering. In particular, the simulation of flow phenomena – also known as *Computational Fluid Dynamics* (CFD) – is nowadays an indispensable and essential tool for the development of, e.g., airplanes,

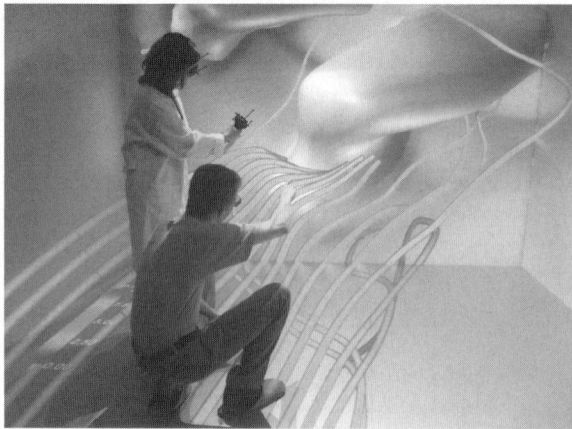

Fig. 3.52. Discussing engineers in a CAVE-like environment

cars, combustion engines, turbines etc. [883]. Even in medicine, CFD is going to play an important role in the analysis of flow within blood vessels and the development of artificial blood pumps, or in order to understand the air flow within the human respiratory organs. Today, CFD is the preferred technique to assess flow fields while the classical experimental approach is mainly used to validate the simulation results. Due to the considerable costs of experiments, flow simulations continuously push forward.

CFD simulations produce *numerical data* which cannot be interpreted by the engineer without further treatment. Efficient post-processing approaches are needed to extract structures and features from these raw data. Scientific *visualization* methods provide a comprehensive overview of underlying datasets by converting the data into geometrical objects that can be rendered on computer displays.

Although mathematical and physical models for the description of flows have been developed, for a long time only rather simple flows could be simulated numerically. This situation changed dramatically with the availability of high performance computing (HPC) systems. The enormous growth of computing power was also for the benefit of CFD. Today, flow *phenomena* are increasingly *simulated* in *three* instead of only two *dimensions*, based on very fine grids containing up to several million cells. In addition, researchers are investigating unsteady flow phenomena, where the flow field changes over time, resulting in huge datasets, especially when used in combination with highly refined grids.

Due to the high complexity of the simulated phenomena, the *analysis* procedure of the resulting datasets becomes more *explorative*. In an explorative analysis, the hypotheses about the characteristics of a flow phenomenon still have to be made during the analysis procedure, resulting in a trial-and-error process. This is a contrast to the *confirmative* approach of data analysis, where regions of interest or specific parts of the simulation are selected for a visualization beforehand. During the explorative analysis session, the engineer continuously defines parameter values to extract flow features which are thereafter often rejected because of unsatisfying results. Then, the parameters are modified for a renewed feature extraction. This *iterative* scheme is applied until a comprehension of the flow characteristics will be attained. Thus, an explorative analysis relies heavily on the *interactivity* of the underlying system.

All in all, researchers are going to examine physical phenomena of such a high complexity that traditional methods of post-processing, like producing static images or at best animations, are no longer neither an effective nor an efficient approach to understand the simulated flow fields. Instead, engineers demand interactive exploration of their data in 3D space, eventually leading to the *use* of *Virtual Reality* technology.

This requirement comes along with the statement made by [1010] that the size of computed simulation results increases faster than the possibilities of data processing and data analysis. In the long term, they expect that only artificial intelligence techniques will solve this problem by offering a completely automatic pre-processing of raw data. Then, the user would only be confronted with pre-structured, handy quantities of prepared data. For the near future, they propose the employment of *Immersive Virtual Reality* (IVR) that combines interactive visualization with immersive sensation. Appropriate systems have to ensure the complete integration of users into virtual, computer-generated worlds in order to enable an interactive investigation of phenomena located in simulation datasets.

After all, we identify the following *potentials* of *IVR* in comparison to traditional post-processing and visualization:

- *3-D viewing.* It is obvious that a 3D simulation can be understood much more intuitively when visualized in 3D space, using stereoscopic projection and head tracking to produce a quasi-holographic representation of the three-dimensional flow phenomena. In comparison to 2D or at best $2\frac{1}{2}$D visualizations, the one-to-one spatial relationship between the real world and VR reduces the mental workload considerably. This is especially true for CFD visualizations, because unlike in, e.g., architectural applications, there hardly exist any psychological clues for 3D viewing of abstract data. Thus, the user has to rely even more on the physiological clues, like stereo and motion parallax provided by viewer centered projection, e.g., when following the course of a path line in 3D space.
- *Navigation.* In an animation-based visualization, a full animation sequence has to be generated, in case it becomes necessary to assess a dataset from a different viewpoint. This makes this approach completely unsuitable for an explorative analysis. In VR, the user can navigate through the dataset in real-time, actually allowing an explorative analysis within a reasonable time. Besides interactivity, a further benefit of IVR is that – depending

on the number of screens installed – a multiple of information can be presented at once as compared to a monitor-based presentation with its limited field of view. As a consequence, the engineers cannot only physically walk around the flow phenomena or just position themselves in the middle of the dataset, but can also examine single features without loosing track of the whole dataset. Unlike in a monitor-based solution, intuitive orientation within the data is guaranteed even when zooming into interesting details.

- *Interactivity.* In VR, interactivity is not only limited to navigation, but it also includes the manipulation of a virtual scene or virtual objects. For that reason, a VR-based post-processing should by definition provide functionalities for interactive feature extraction and for variation of visualization parameters, thus inherently supporting the trial-and-error procedure in an explorative analysis.
- *3D, multimodal interaction.* A lot of publications exist which approve that 3D interaction techniques significantly increase a user's performance and acceptance when positioning tasks or manipulative tasks have to be accomplished in 3D space. These general findings can be directly transferred to VR-based post-processing. For instance, the setting of path- and streamlines is an important and often executed task during an explorative analysis. Making use of 3D input devices, positioning and orienting of seed points can be intuitively and precisely achieved even in complex 3D datasets. Besides three-dimensionality, multimodality is another attribute of VR. Possibly, a multimodal interface including acoustics, haptics, or speech input may have the potential to improve or accelerate a CFD analysis further.
- *Communication.* A 3D presentation of the data representing simulated flow phenomena, is considered to be much easier to understand, especially for non-experts. As a consequence, the design of a newly developed turbine, motor or in our case an extruder can be communicated to customers, project partners etc. Also, interdisciplinary discussion of CFD results within a company, which today becomes more and more important in the product development process, can be improved by means of Virtual Reality.

Virtual Reality for the Analysis of Flow Phenomena in Compound Extruders

In plastics processing, polymers are usually processed within twin *extruders*. Due to the complex and unsteady flow phenomena within such extruders, their *design* and *optimization* is a complicated task in which people from different disciplines are involved.

In particular, engineers are going to combine 1D *simulations* with *FEM* and *BEM* methods in the *3D* domain throughout the optimization process, which by ViSTA and its framework Viracocha all had to be integrated into one exploration tool.

As the numerical simulation of flows inside a compound extruder is a complex task, engineers usually do not simulate complete units but concentrate on *functional zones*. Due to the modularity of a compound extruder, a wide variety of *configurations* is possible for any functional zone. 1D simulations calculate averaged functions or determine an overview of a specific configuration on an empirical level. They are used to select a specific extruder setup to be simulated with finite element (FEM) or boundary element (BEM) methods.

Even though in principal, the visualization of 1D simulations in a three-dimensional space is not too meaningful, the visualization of a number of 1D-functions with exact geometrical placement can be an interesting alternative. In addition to that, the simple and natural interaction metaphors that are used in VEs can *ease* the task of *configuration construction*. Figure 3.53 depicts the prototype we developed for the interactive configuration of twin screw extruders that uses fast 1D simulations to prepare functional zones for expensive FEM simulations.

Fig. 3.53. PME Screenshot

As in other application areas that examine flow phenomena, in plastics processing numerical simulations replace the common model-based experiment. With increasing complexity, the *requirements* on the methods for the *visualization* rise. Traditionally, visualization software allows the simple animation of transient data sets. This is not enough for the interactive exploration of complex flow phenomena, which is, in contrast to a confirmative analysis, comparable to an undirected search in the visualization parameters for a maximum insight into the simulation. In a worst case scenario, important features of a flow are not detected. Due to this fact, the interactive explorative analysis in a *real-time virtual environment* is demanded by scientists.

It turns out that the interactive setting of *seed points* and *visualization of particle traces* is the most adequate technique to understand the flow inside an extruder. Thus, the application particularly profits from the parallel calculation of pathlines in Viracocha as well as from their efficient rendering by Virtual Tubelets (see Fig. 3.54). A special challenge in the application is to find intuitive *representation* metaphors for *physical* and *chemical processes* that directly *depend* on the *flow characteristics* inside an extruder.

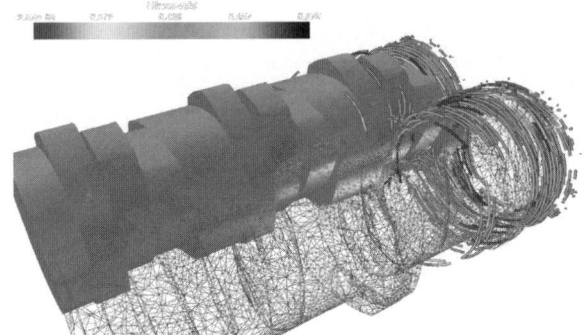

Fig. 3.54. Pathlines inside a simulated twin extruder, rendered as Virtual Tubelets

3.3.5 A Framework for Flow Analysis in Virtual Environments

As a comprehensive application of VR in the application scenario, in the following we will introduce a *comprehensive tool* under development, where among many other features, minimum system response time and a maximum frame-rate are respected, in order to allow for an explorative analysis of complex, unsteady flow phenomena in *IVRs*. The main *components* of this tool are ViSTA, ViSTA FlowLib, and Viracocha. ViSTA covers basic VR functionality like navigation, interaction, and scenegraph handling, whereas ViSTA FlowLib [386] provides special features for flow visualization. While ViSTA and ViSTA FlowLib both run on the graphics computer, a framework called Viracocha [130] is running on a different computer – preferably the simulation host or another HPC system – where parallel feature extraction and data management are held. In Fig. 3.55, the overall architecture is depicted.

Immersive Display Technology

Nowadays, immersive multi-screen displays like CAVEs^{TM} are driven by off-the-shelf PC clusters with consumer graphics cards instead of multipipe, shared memory graphics computers. This reduces the costs for IVR infrastructure dramatically. VR toolkits supporting PC clusters must inherently have a distributed software architecture, and data synchronization is an issue in such frameworks. Besides a *client-server* approach, where the scenegraph is distributed over the cluster nodes, a *master-slave* approach is most often

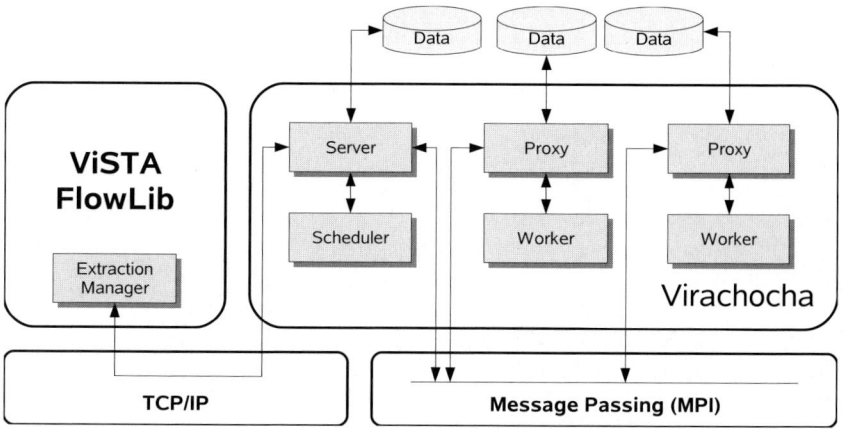

Fig. 3.55. Architecture and communication concept of ViSTA FlowLib and Virachocha

used. Here, copies of the same application run on every node, and events are propagated via network from one dedicated master node to the slave nodes. Details about these concepts will be explained in Subsect. 3.3.5.

Our framework represents a combination of both approaches. In principle, ViSTA follows the master-slave concept. In the context of CFD postprocessing however, the master requests and gets data from the simulation host, which then has to be propagated to the slaves in order to be rendered. To combine both design patterns, a so-called *data tunnel* has been implemented into ViSTA, working in clustered and non-clustered environments.

In the *clustered case*, the master data tunnel consists of a dispatcher forwarding received data packets to all connected render nodes. Additionally, slave data tunnels exist which ignore all computation request to the HPC backend. Consequently, only requests by the master are processed. The hybrid architecture is depicted in Fig. 3.56.

The following section will explain the very basics of *Virtual Reality applications* in a *distributed setting*. Parts of the technology are applied for the realization of PC cluster setups in order to drive large VR displays. On a technological level, much of the applied methods also count for collaborative applications. As a consequence we will illustrate this technology and then extend them to the collaborative setting.

Virtual Reality in Distributed Environments

Distributed environments that are used for collaboration usually suffer from a number of *problems*, and a huge number of remedies exist. The most basic problems arise from distributed working memory of the participating nodes and the need for sophisticated data locking strategies over network connections.

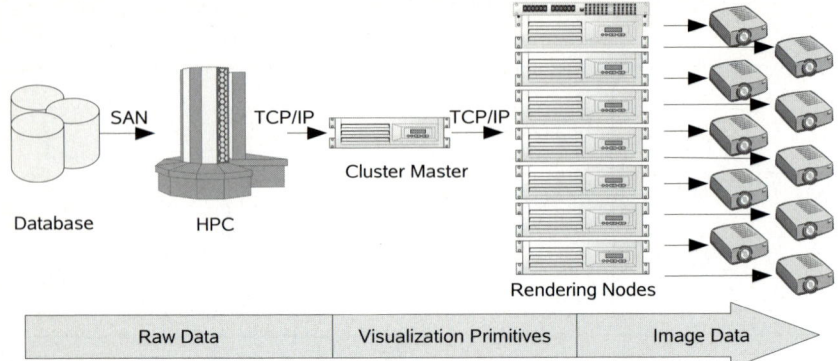

Fig. 3.56. Hardware setup for the distributed visualization system

In the field of Virtual Environments, this is a severe constraint, as these applications require a real-time environment to run in. This can usually not be fulfilled and is stated in the *Consistency-Throughput Tradeoff* [966]: "It is impossible to allow dynamic shared state to change frequently and guarantee that all hosts simultaneously access identical versions of that state". As such it states that a dynamic VE can not support dynamic behavior and consistency across all connected sites simultaneously.

One approach to this situation is to research *low latency update mechanisms* across a group of participants, either by sophisticated communication schemes or the sheer reduction of information that has to be synchronized. A Virtual Reality application usually processes a lot of data for the visualization. In comparison to that, *interaction events* that actually steer the application are both, small in size and less frequent in comparison to video interrupts from the graphics hardware. A simple, but working approach is to share interaction events between collaborative applications in order to ensure data locking. This approach can be used for the synchronization of PC cluster based large displays and is depicted in the following section.

Event Sharing in PC Clusters

A complex large display driven by a PC cluster architecture needs tight synchronization for an immersive sensation in a VR application. The following passages describe a *software-based solution* to the *data-* and *swap locking problem* that arise in this task. It will explain the difficulties and necessary terms in detail and close with a presentation of an abstracted framework that can deal with data flow and synchronization issues in a more general way. All software described is completely implemented in the ViSTA VR toolkit.

As stated above, room mounted multi-screen projection displays are nowadays driven by off-the-shelf PC clusters instead of multi-pipe, shared memory machines. The topology and system layout of a PC cluster raises fundamental differences in the software design, as the application suddenly has to respect *distributed computation* and *independent graphics* drawing. The first issue introduces the need for data sharing among the different nodes of the cluster, while the latter one raises the need for a synchronization of frame drawing across the different graphics boards.

Data locking deals with the question of sharing the relevant data between nodes. Usually, nodes in a PC cluster architecture do not share memory. Requirements on the type of data differ, depending whether a system distributes the scene graph or synchronizes copies of the same application. Data locked applications are calculating on the same data, and if the algorithms are deterministic, are computing the same results in the same granularity.

An important issue especially for Virtual Reality applications is *rendering*. The frame drawing on the individual projection screen has to be precisely timed. This is especially true for active stereo displays where a tight timing between the activation of the shutter glasses and the presentation on the display is vital to see a stereoscopic image. *Tight frame locking* usually is achieved with specialized hardware, e.g. genlocking or frame locking features, that are available on the graphics boards. However, this hardware is not common in off-the-shelf graphics boards and usually rather expensive. In addition to this, while exact synchronization is mandatory to the active stereo approach, it is not that important when using *passive stereo*, where frames for the left and the right eye are given simultaneously on the projection surface. Ideally, a software-based solution to the swap synchronization issue would strengthen the idea of using non-specialized hardware for Virtual Reality rendering thus making the technique more common.

Additionally, the *scalability requirement* is important as well. This means, that a programmer and the application should not need to be aware of the fact that it is running in a distributed environment or as a stand-alone application. Ideally, the same application can run in a cluster environment as well as on a laptop setup with minor modifications.

Data locking deals with the issue of distributing knowledge between applications. In Virtual Reality applications, it is usually distinguished between two types of knowledge, the graphical setup of the application (the scenegraph) and the values of the domain models that define the state of the application.

Distributing the scenegraph results in a simple application setup for cluster environments. A setup like this is called a *client-server* setup, where the server cluster nodes provide the service of *drawing* the scene, while the client dictates what is drawn by providing the initial scenegraph and subsequent modifications to it. The client node performs all *user interaction* and is usually not a part of the rendering environment. As a consequence, the client node should provide enough computational resources to deal with the additional complexity of user input dispatching and calculations. The server nodes need only enough graphics performance, as they do not do additional calculations.

The client-server technique is usually embedded as a low level infrastructure in the scenegraph API that is used for rendering. Alternatives to the

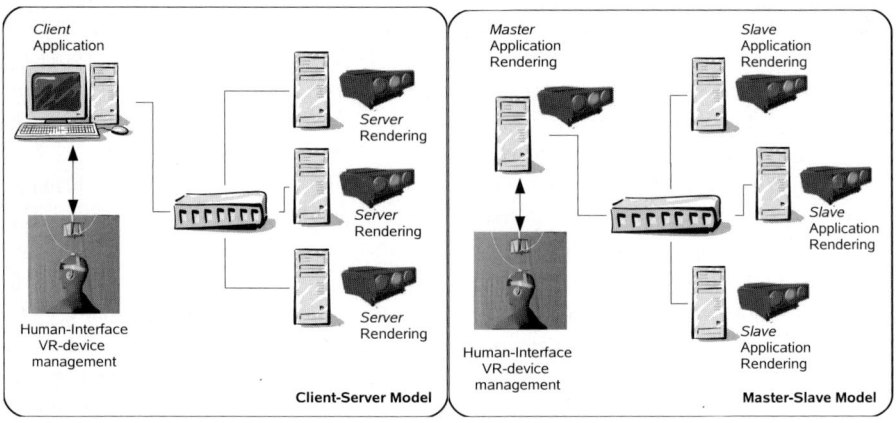

Fig. 3.57. Client-Server and the Master-Slave model for PC cluster applications

distribution of the scenegraph can be seen in the distribution of pixel-based information over high bandwidth networks, where all images are rendered on in a high performance graphics environment [677], or the distribution of graphics primitives, e.g., parameters of OpenGL primitive calls.

A totally different approach respects the idea that a Virtual Reality application that has basically the same state of its domain objects will render the same scene, respectively. It is therefore sufficient to distribute the state of the domain objects to render the same scene. In a multi-screen environment, the camera on the virtual scene has to be adapted to the layout of your projection system. This is a very common approach and is followed more or less, e.g., by approaches such as ViSTA or NetJuggler [978]. It is called the *master-slave*, or *mirrored application* paradigm, as all slave nodes run the same application and all user input is distributed from the master node to the slave nodes. All input events are replayed in the slave nodes and as a consequence, for deterministic environments, the state of the domain objects is synchronized on all slave nodes which results in the same state for the visualization. The master machine, just like the client machine in the client-server approach, does all the user input dispatching, but as a contrast to the client-server model, a master machine can be part of the rendering environment. This is a consequence from the fact that all nodes in this setup merely must provide the same graphical and computational resources, as all calculate the application state in parallel.

Figure 3.57 depicts both architectures in their principal layout. This is a *higher-level approach* than the distribution of the scenegraph, as the state of domain objects is usually depending on user interaction and to some extent on non-determinism, e.g. time and random number calculations. One can select between the distribution of the domain objects or the distribution of the influences that can alter the state of any domain object, e.g. user input. Domain objects and their interactions are usually defined on the application level by the author of the VR application, so it seems more reasonable to distribute the entities of influence to these domain objects and apply these influences to the domain objects on the slave nodes of the PC cluster.

A common model for interactive applications is the *frame-loop*. It is depicted on the left side of Fig. 3.58. In this model, the application calculates its current state in between the rendering of two consequent frames. After the calculation is done, the current scene is rendered onto the screen. This is repeated in an endless loop, until the user breaks the loop and the application exits. A single iteration of the loop is called a *frame*. It consists of a calculation step for the current application state and the rendering of the resulting scene. Our solution assumes that user interaction is dispatched after the rendering of the current scene and any state change of an interaction entity is propagated to the system and the application using *events*. An event indicates that a certain state is present in the system. E.g. pressing a button on the keyboard represents such a state. All events are propagated over an *event-bus*.

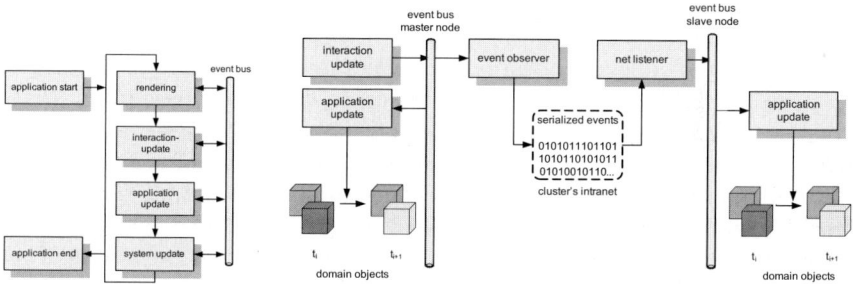

Fig. 3.58. Left: A typical endless loop of a VR application. Right: Data locking by event sharing.

We can see that the distribution of user interaction in the form of events is a key component to the master-slave approach. As a consequence, in a well-designed system it is sufficient to mirror application events to a number of cluster nodes to transparently realize a master-slave approach of a clustered virtual environment. The task for the framework is to *listen* to the *events* that run over the event bus during the computational step in between the rendering steps of an application frame, and *distribute* this *information* across a network. The following paragraphs will focus on the *recording, serializing* and *reproducing* of application events.

In an interactive Virtual Reality application, a frame rate of at least 30 scene renderings per second is a fundamental requirement. This enforces any calculation and time for the actual rendering to happen in less than 33 ms. For a cluster setup this includes graphics rendering, application calculation, and network handling. In VR programs, it is a typical idiom to delegate i/o handling to separate threads to give more time to the rendering and calculational part of the application. This will relax the model of a PC cluster setup to a thread that handles network communication, buffering, deserializing and intermediate object creation, and the VR core part that is basically the same as in the stand alone version. Between two render requests, an arbitrary number of events can occur at a high frequency. Furthermore, in a networked environment, it is better to send few large informational chunks instead of a high number of small ones. The latter approach increases communication overhead and reduces the chances of a successful buffering scheme.

As a consequence, we introduce the term of an *application frame*. It consists of the ordered sequence of real world events that happen in between two rendering steps. A master node thus collects an application frame, chooses a proper encoding, and sends the frame as a complete collection to all the slave nodes, where the sequence will be reconstructed and injected to the local event bus. As shown above in Fig. 3.58, this will result in the same state of the domain objects right before rendering.

Event Sharing for Collaborative Virtual Environments

A central part of all collaborative systems is the *communication* between *participating users*. Any VR application reacts mainly on user input that is derived from a number of sensors. The types of input are more versatile than input from desktop PC devices like mouse or keyboard. This is easily seen in the higher number of degrees of freedom for typical VR devices and the essential requirement on Virtual Reality applications to demultiplex user input from different modalities at the same time. However, the occurrence of user input can be represented by events that have to be handled whenever the user triggers a sensoric device for a specific modality. As such, a number of events, or in a very simple case, the complete stream of events can be used as input for distributed VR applications, once they are transmitted over a network interface.

A small *example* will *illuminate* this *setting*. Users that participate in a collaborative session in an immersive VE are usually tracked and use mouse-like input devices that provide six degrees of freedom and additional buttons for specific commands that can be used to interact with the system. As stated above, the collaborative part of this setting is that more than one user interact in the same shared virtual world. That means that all users have access or at least the possibility to interact with the presented objects at the same time, usually with a mouse-like device, gestures or speech recognition. E.g., the system detects the push of a button on the 3D-mouse over a virtual object and utters an event to the application that has then the possibility to interpret the push of the button over the object as a try to select that object for further manipulation, e.g., dragging around. In a different implementation, the user application does not see the push of the button as such, but is presented an event that indicates the selection of an object within the world and can react on that.

No matter what granularity is chosen for the concrete implementation, the important point is that at some stage a *selection* of an object is *detected* and *rendered* as an *event*. By promoting this event to all participants in the collaborative session over the network interconnect, a (delayed) synchronicity can be achieved by simply replaying the event on the remote sites.

A complete solution is not as simple, as any participant in a collaborative session is basically a loosely coupled system over a shared space, and *conflicts* might arise when two or more *users* try to *manipulate* the *same object*. In order so solve this, additional *locking conflict* solving *strategies* are needed, but are beyond the scope of this article.

Data Streaming

The shift of the post-processing to an HPC system, parallelization strategies, and the innovative data management mainly aimed at the reduction of total runtime. On the other hand, these approaches are not sufficient to fulfill the demand of short system response times needed for interactive exploration tasks. Because of the size of today's datasets, it is not possible to meet this criterion fully since the speed-up of an algorithm cannot be increased significantly. However, a *fast representation* of first *temporary* or *approximate* results leads to a considerable reduction of noticeable delays by simply decreasing the latency time of an algorithm.

This is the motivation and the main goal for the integration of *streaming functionalities* into Viracocha. Normally, the term streaming is used to describe a special mode of data transmission where the incoming data is processed right after reception. For example, the streamed parts of a digitized movie can be played back long before the transmission finally terminates. During the transmission process, the data are completely known to the sender but not to the receiver. In our notion, streaming describes the process of transferring intermediate or approximate results to the visualization system during an ongoing computation.

Using streaming schemes, *meaningless* extraction *processes* can be *identified* early during the execution. Then, running jobs can be *discarded* immediately in order to continue the investigation at another point. Thus, streaming supports the trial-and-error process of an explorative analysis and contributes to a higher user acceptance.

3.3.6 Communication and Management Services

After the presentation of tools for supporting communication and cooperation of designers, some *technical topics* are still open: A framework for the transmission of multimedia data streams, a conference control service, and security services.

Framework for Transmission of Multimedia Data Streams

Based on the adaptive audio and video transmission described above, a *streaming protocol* was developed as a generalization. It adapts video data to be transferred taking into account the current network situation [59, 60].

Figure 3.59 shows the *framework* for this *service*. The **Video Data Adapter** regulates the data stream depending on the network situation. **Congestion Control (CC)** tries to detect overload situations by monitoring its own sender buffer and the loss rate in the network. From this information a target value for the transmission capacity is computed and transferred to the **Video Data Adapter**. **Jitter Control (JC)** estimates the delay on the transmission path as well as the buffer load at the receiver side. Based on this estimation, the sending of further packets is slowed down or sped up. **Loss Control (LC)** recognizes packet losses and schedules retransmissions if the ration of lost data is important for avoiding error propagation. **Transport** is not specified in the framework. It can be any protocol able to provide fast transmission as well as statistics about the network state. Again, RTP is the best choice here.

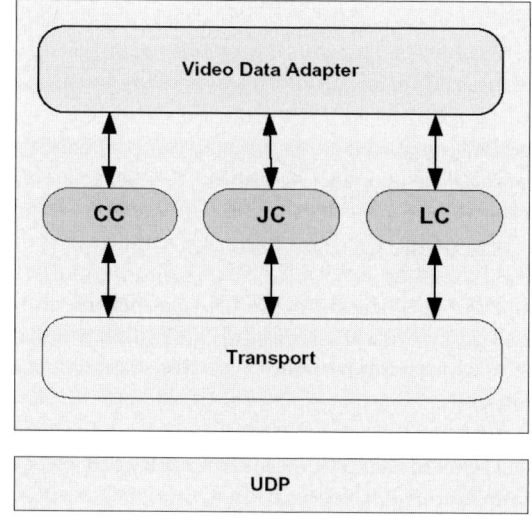

Fig. 3.59. Video transmission framework

Conference Management

A *conference management system* is an important mechanism for formalizing interactions between communicating parties. Only such a system enables a controlled flow of a synchronous communication.

The basic functionality that is to be provided by a conference management is the *announcement* and *initialization* of a conference. To this end, a conference mediation functionality as well as a conference announcement channel are required. In addition, it is necessary to have a function for *inviting* other parties to the conference after conference setup. Last but not least, moderated conferences are often desirable, where single users get special rights, e.g. for excluding certain users from the conference, and performing access control.

To serve these needs, the *Scalable Conference Control Service (SCCS)* [450], [451] was developed and integrated into KomPaKt. SCCS differs from common conference control systems by offering a *flexible control structure*, independent from the real network topology, creating a tree-like structure of control connections, see Fig. 3.60. Thus, a directed and efficient signalling between the involved parties is possible [454], [452]. If the current control structure becomes inefficient, e.g., through introduction of new users during runtime, a re-configuration of the signalling channels is done to speed-up the signalling. This re-configuration is done without changes to the transmission quality [455], [457].

Furthermore, for conferences to be secure, *access rights* and *authentication* are necessary. In SCCS this is done using a password for each user and an additional acknowledgement by the conference leader. Thus, unknown users do not have access to the conferences. Also, an encryption of data is optional by using the Multicast Transport Protocol (MTP) [959] for point-to-multipoint transmission [453].

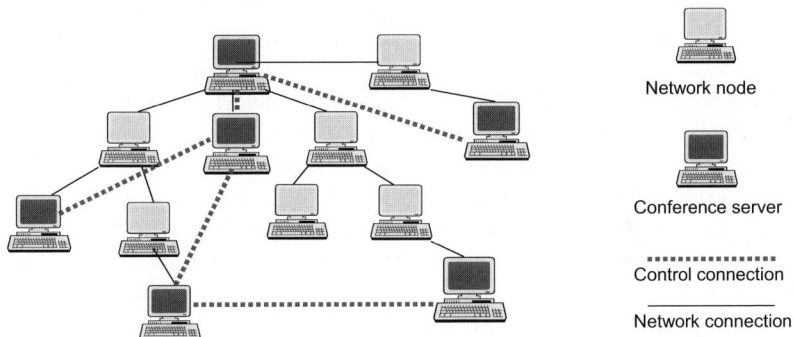

Fig. 3.60. Control structure of the conference management system SCCS

Security Aspects

Security aspects were briefly mentioned before: Only certain users should get access to a conference, which is controlled by SCCS together with a conference leader. Also, it is important to encrypt data which are exchanged if team members from different companies are having a conference with confident information. This functionality is included in MTP, which is used by SCCS for multicast communication.

One more security aspect also concerns the cooperation between companies. If no third party should know about the cooperation partners, methods to guarantee *privacy and anonymity* in the Internet are necessary.

This can be done by using so-called *mix servers* [58], [57]. A mix server makes tracing of data packets nearly impossible. It receives a packet (which has to be encrypted), re-codes it, and forwards it with some delay. This delay causes several packets received by the mix server to be permuted in their order in the sending process. The re-coding is done with an asymmetric encryption method. The sender encrypts the packet with the public key of the mix server, the mix server decrypts it with the corresponding private key and forwards it. It is also possible to create a chain of mix servers for increased security. Due to the delay caused for decrypting and also for forwarding the packets in a different order, this techniques is unsuited for synchronous communication.

3.3.7 Comparison with Related Work

In general, no tools exist which provide *all functionalities* presented in this section. Typically, comparable systems offer only a part of the functions necessary for a comprehensive support of cooperative work. Furthermore, a common and simple usage of these systems is not possible. Furthermore, such environments do not *integrate tools* needed in a *design* process. Only with such an integration, communication tools can take special requirements of the design process into account and are able to hide technical details from the users.

For instance, most systems have a focus on improving the synchronous communication. Reference [549], for example, provides a framework for the development of conferencing systems with high quality by proposing several optimizations on the technical layer. Yet, only video and audio transmission as well as application sharing are considered; the *handling* of *shared documents* and *tools* with support of more complex information structures (annotation mechanisms, change management) as well as an *integration* into a certain *work process* are neglected.

Approaches for the realization of such additional functionality nevertheless exist. Reference [1037], for example, presents an approach for an improved support of distributed work sessions by offering a simple mechanism for initiating a communication session and by providing a *session-oriented management* of meetings with an *integrated document management*. Furthermore, a shared editor is realized, which enables the *presentation* of semantic constructs, necessary for a simple capturing of meeting results. The latter aspect is of particular importance, as it can happen with handwritten notes that important details are missing. Not everything can be written in time, or short comments are misinterpreted later on [1050]. However, the mechanisms for conference initiation and document management are isolated and proprietary. In KomPaKt, an integration with AHEAD is done, allowing for the reuse of existing information from the design process. Additionally, KomPaKt offers a broader range of cooperation functionalities by providing event sharing.

Also, several other approaches exist for the *recording* of video sequences. Reference [982] records presentations and presenter. Commercial products like NetMeeting also offer such mechanisms. Reference [690] sketches an architecture for the management of *shared work processes* and the control of documents. Synchronous as well as asynchronous activities are supported. The core aspect of the presented system is to *capture information* and to integrate it with already existing information. For this purpose, meetings are recorded as video sequences to provide an information base about past meetings [932]. Furthermore, the system comprises an annotation mechanism for PowerPoint presentations. The central problem with these approaches is the handling of recorded, possibly annotated sequences. Again, work process integration is missing; it would allow for a feedback of results or changes into the work process itself.

A central aspect of KomPaKt is the *awareness functionality* of the user interface. Contacting a project partner is easier if that partner will probably not be disturbed by the contact attempt. Thus, awareness of the *willingness* to communicate simplifies initiating a spontaneous communication [561]. Early work regarding that awareness is described in [639]. Video cameras are used to record persons at their work place, and the *recordings* are available on a web page with access restricted to a certain work group. Observing the current activity of a person, it can be decided if that person could be disturbed by a communication attempt. Clearly, the acceptance of this system is low because it has the possible potential for complete work monitoring [640], [698] – on this reason such techniques are highly illegal. Another approach would be the *monitoring* of user *activity* at his computer, e.g. by monitoring keyboard activity [698]. Using e-mail, an activity notification can be sent to all project members. However, this approach implies a willingness for communication, independent of the current task. Better mechanisms with more expressive descriptions about a user's communication willingness are offered by *messengers* like ICQ. Those mechanisms only offer the sending of short messages and of files, other communication tools are not integrated with the awareness functionality.

The *event sharing tool* was newly developed in this project. Commercial tools only implement application sharing. The implementation of event sharing was questioned in the literature [850]. Implementation approaches had not led to products which could be used in the design process. Sometimes only prototypes were developed, which are not available, or implementations are only for a certain type of programs, e.g. Java Applets. Hewlett Packard had developed two prototypes to share 3D CAD/CAM applications in an X window environment. While the Reduced Event Set Application Sharing only transmits a restricted set of event, the input of keyboard and mouse [717], the Event Distribution Application Sharing needs a central server for delivering all events to the connected clients [716]. Both approaches were not further considered by Hewlett Packard in ongoing developments. A working approach, the Java Applets Made Multiuser, was developed at the Virginia Polytechnic Institute [547]. Yet, this solution is not usable for general Windows or Unix/Linux applications.

Regarding the *transmission* of audio and video data *over networks with limited capacity*, a framework was developed to adapt the amount of data transmitted to the currently free capacity. One possible way of reaction to a bottleneck causing packet loss is to use Forward Error Correction [903]. To avoid jitter unacceptable for a continuous playout, [666] and [802] cover the optimization of transmission and playout of multimedia data streams. Furthermore, [1017] gives an overview of several other adaptive techniques in multimedia data transmission. The approach realized in IMPROVE partly uses these technologies but additionally covers the applicability in wireless networks for supporting mobile project partners.

3.3.8 Conclusions

This section describes the multimedia and VR support for the communication of designers from chemical engineering and plastics processing. First, a *communication platform* for supporting the cooperative work between chemical and plastics processing engineers was developed. This communication platform integrates several forms of synchronous and asynchronous communication in *one graphical interface*. An automated interaction with the administration system AHEAD *frees* the user from *configuring* the communication tools before starting a cooperation, and at the same time it enables the *provisioning* of all *documents* related to the cooperative task. Partly, existing communication tools were integrated, but also some new tools had to be developed because the user should be able to perform the communication at his normal working place. Important in this context were an audio/video *transmission* tool and an *event sharing tool* for cooperative work on documents over limited bandwidth.

As a special form of cooperation, *collaborative Virtual Reality environments* were introduced. VR communication can be considered as an additional component in the tool chain for the support of communication of designers. However, due to the *complex nature* of the analysis process, *special technology* as discussed in this section has to be applied. Thus it is impossible to integrate this technology so tightly within KomPaKt as it was possible for the other forms of direct communication.

Collaborative Virtual Reality environments are needed for such complex tasks as *designing and optimizing an extruder* by studying the flow phenomena in detail. Although visualization is much more complex in this case, it was shown that event sharing can be used also in distributed VR applications. Studying the impact on and the *possibilities of VR in specific development processes* is a rather new field of research in engineering and computer science.

Last but not least a communication framework was set up together with a *conference management* service to handle cooperative work sessions.

Still, several *improvements* have to be made. First, it is only possible to obtain and modify all task-related documents for *planned* communication sessions. In case of *spontaneous cooperations* the administration system AHEAD is not involved and cannot provide related documents. Here, a mechanism for document retrieval and integration needs to be developed that also supports a spontaneous communication session. However, not only this interaction with AHEAD needs to be improved; a closer *integration* with the *process integration* framework PRIME (Sect. 3.1) or the *integrator tools* (Sect. 3.2) would also be of further use.

Also, tools and services already developed can be further refined. The event sharing mechanism is a good addition to the well-known application sharing approach. As event sharing offers a loose coupling of application instances, one could think of the *creation of different synchronization levels* where a user can leave exact synchronization and perform some own operations before again joining the synchronization. This would enable a more flexible cooperation process. Also, new *compression methods* can be examined to be used in the event sharing or in video conferencing to further reduce the amount of data in a multimedia cooperation, at the same time also allowing users with low-capacity communication lines, e.g. mobile users, to take part in such a work session. For the application of Virtual Reality in the engineering process, a lot of research has to be done in the field of interaction. The main focus here is to create persistent knowledge from more or less volatile visualizations in immersive environments.

3.4 An Adaptive and Reactive Management System for Project Coordination

M. Heller, D. Jäger, C.-A. Krapp, M. Nagl, A. Schleicher, B. Westfechtel, and R. Wörzberger

Abstract. Design processes in chemical engineering are hard to support. In particular, this applies to conceptual design and basic engineering, in which the fundamental decisions concerning the plant design are performed. The design process is highly creative, many design alternatives are explored, and both unexpected and planned feedback occurs frequently. As a consequence, it is inherently difficult to manage design processes, i.e. to coordinate the effort of experts working on tasks such as creation of flowsheets, steady-state and dynamic simulations, etc. On the other hand, proper management is crucial because of the large economic impact of the performed design decisions.

We present a management system which takes the difficulties mentioned above into account by supporting the coordination of dynamic design processes. The management system equally covers products, activities, and resources, and their mutual relationships. In addition to local processes, interorganizational design processes are addressed by delegation of subprocesses to subcontractors. The management system may be adapted to an application domain by a process model which defines types of tasks, documents, etc. Furthermore, process evolution is supported with respect to both process model definitions and process model instances; changes may be propagated from definitions to instances and vice versa (round-trip process evolution).

3.4.1 Introduction and Overview

As *design processes* are highly creative, they can rarely be planned completely in advance. Rather, planning and execution may have to be interleaved seamlessly. In the course of the design process, many design alternatives are explored which are mutually dependent. Furthermore, design proceeds iteratively, starting from sketchy, coarse-level designs to detailed designs which are eventually needed for building the respective chemical plant. Iterations may cause feedback to earlier steps of the design process. It may also be necessary to revoke inadequate design decisions. Finally, design involves cooperation among team members from different disciplines and potentially multiple enterprises, causing additional difficulties concerning the coordination of the overall design process.

Technical tools such as flowsheet editors, simulators for steady-state and dynamic simulations, etc. are crucial aids for effectively and efficiently performing design tasks [354]. In addition, *managerial tools* are required which address the coordination of design processes. In fact, such tools are crucial for supporting *business decision making* [174]. In the course of the design process, many decisions have to be made concerning the steps of the chemical process, the relationships among these steps, the realization of chemical process steps by devices, etc. To perform these decisions, design alternatives have to be identified and elaborated, and the respective design tasks have to be coordinated regarding their mutual interfaces and dependencies. To support business decision making, managerial tools must provide chief designers with accurate views of the design process at an adequate level of granularity, offer tools for planning, controlling, and coordinating design tasks, thereby taking care of the dynamics of design processes.

The management system *AHEAD* (*A*daptable and *H*uman-Centered *E*nvironment for the M*A*nagement of *D*esign Processes [120, 161, 162, 207, 209, 212, 249, 355, 392, 474–476, 478, 488]) addresses the challenge of supporting dynamic engineering design processes. It has been developed in the context of the long-term research project IMPROVE [299, 343, 352] described in this volume which is concerned with models and tools for design processes in chemical engineering. The management tool AHEAD is primarily developed to support design teams in the industrial practice. In order to develop concepts and tools which can be transferred into practice, we have chosen to use a case study in the IMPROVE project as a reference scenario and a guideline for our tool design process ([17], and Sects. 1.1, 1.2). The case study refers to the conceptual design and basic engineering of a plant for the production of *Polyamide-6* (PA6). This approach has been greatly supported by the fruitful collaboration with our engineering partners in the IMPROVE project.

AHEAD equally covers products, activities, and resources and, therefore, offers more comprehensive support than project or workflow management systems. Moreover, AHEAD supports seamless interleaving of planning and execution – a crucial requirement which workflow management systems usually do not meet. Design processes are represented by *dynamic task nets*, which may evolve continuously throughout the execution of a design process [159, 160, 163, 242, 243, 472]. Dynamic task nets include modeling elements specifically introduced for design processes, e.g., feedback relationships for iterations in the design process which cannot be represented in project plans. This way, AHEAD improves business decision making since it offers a more natural, realistic, and adequate representation of design processes.

Initially, the AHEAD system focused on the management of design processes within one organization. In particular, we assumed that all management data are stored in a central database which can be accessed by all users. This assumption breaks down in case of *interorganizational design processes*. Each of the participating organizations requires a view on the overall design process which is tailored to its needs. In particular, it is crucial to account for information hiding such that sensitive data are not propagated outside the organization.

To support the management of interorganizational design processes, we have developed an approach which is based on *delegation* [30, 208]. A subprocess may be delegated to a subcontractor, passing only those data which are relevant for the contract. Both the contractor and the subcontractor use their own instances of the management system, which maintain their data in local databases. The management systems are coupled at runtime by exchanging state information.

We have further developed this initial cooperation approach and extended it to a *view-based cooperation model* for the AHEAD system supporting interorganizational development processes. Organizations can create dynamic process views onto their local processes and publish them to other organizations. We utilize process views to enable organizations to manage how their local processes are integrated with other processes. A broader spectrum of cooperation scenarios besides delegation is supported. Additionally, contracts between organizations can be explicitly modeled and configured according to individual cooperation needs.

The AHEAD system may be applied to processes in different domains – including not only chemical engineering, but also other engineering disciplines such as software, electrical, or mechanical engineering. In fact, the core functionality is domain-independent and relies on general notions such as task, control flow, etc. AHEAD may be *adapted* to a certain *application domain* by defining domain-specific knowledge. For example, in chemical engineering domain-specific task types for flowsheet design, steady-state simulations, dynamic simulations, etc. may be introduced.

Domain-specific knowledge is formalized by a *process model definition* (cf. Sect. 2.4) which constrains the *process model instances* to be maintained at project runtime. As a consequence, the manager may compose task nets from predefined types and relationships. The process model definition is represented in the Unified Modeling Language (*UML* [560]), a wide-spread standard notation for object-oriented modeling. A process model is defined on the type level by a class diagram which has been adapted to the underlying *process meta model* for dynamic task nets [388, 389].

The current version of AHEAD provides for *evolution* both on the definition and the *instance level*. Changes on the definition level may be propagated to instances during their execution. If required, process model instances may deviate from their definitions under the control of the project manager who may switch off consistency enforcement deliberately and selectively (i.e., in designated subprocesses of the overall process). Knowledge acquired on the instance level may be propagated to the definition level, resulting in improved versions of process model definitions. This way, AHEAD provides for *round-trip process evolution*.

AHEAD is a research prototype which cannot be applied immediately in industry in a production environment for various reasons. In addition to deficiencies with respect to stability, efficiency, and documentation – problems which are faced by many research prototypes –, an important prerequisite of industrial use constitutes the integration with other management tools which are used in industry. Therefore, we *integrated* AHEAD with *several commercial systems* for workflow, document, and project management. The ultimate goal of these research activities is *technology transfer* into industrial practice.

This section describes 10 years of research on management of design processes. It should be clearly pointed out that this research within IMPROVE was carried out in *close cooperation* with subproject A1 (see Sections 2.4 and 2.5) and I1 (see Section 5.1), but also with B1 (see Section 3.1). The latter subproject also supports processes, but on another level and with different support mechanisms.

The rest of this *section* is *organized* as follows: Subsect. 3.4.2 introduces the AHEAD core system, which supports integrated management of products, activities, and resources for dynamic design processes. In the core system, process model definitions were static, and management was constrained to local processes within one organization. The next subsections describe extensions of the core system, namely on one hand the adaptation capabilities of AHEAD as well as round-trip process evolution (Subsect. 3.4.3) and on the other hand interorganizational coordination of design processes (Subsect. 3.4.4 and 3.4.5). Subsection 3.4.6 is concerned with related work. A conclusion is given in Subsect. 3.4.7.

3.4.2 AHEAD Core System

Basic Notions

In general terms, *management* can be defined as "all the activities and tasks undertaken by one or more persons for the purpose of planning and controlling the activities of others in order to achieve an objective or complete an activity that could not be achieved by the others acting alone" [996]. This definition stresses coordination as the essential function of management.

More specifically, we focus on the management of *design* processes by *coordinating* the *technical work* of designers. We do not target senior managers who work at a strategic level and are not concerned with the details of enterprise operation. Rather, we intend to support project managers who collaborate closely with the designers performing the technical work. Such managers, who are deeply involved in the operational business, need to have not only managerial but also technical skills ("chief designers").

The distinction between *persons* and *roles* is essential: When referring to a "manager" or a "designer", we are denoting a role, i.e., a collection of authorities and responsibilities. However, there need not be a 1:1 mapping between roles and persons playing roles. In particular, each person may play multiple roles. For example, in chemical engineering it is quite common that the same person acts both as a manager coordinating the project and as a (chief) designer who is concerned with technical engineering tasks.

In order to support managers in their coordination tasks, design processes have to be dealt with at an appropriate level of detail. We may roughly distinguish between *three levels* of *granularity*:

- At a *coarse-grained level*, design processes are divided into phases (or working areas) according to some life cycle model.

- At a *medium-grained level*, design processes are decomposed further down to the level of documents or tasks, i.e., units of work distribution.
- At a *fine-grained level*, the specific details of design subprocesses are considered. For example, a simulation expert may build up a simulation model from mathematical equations.

Given our understanding of management as explained above, the coarse-grained level does not suffice. Rather, decomposition has to be extended to the medium-grained level. On the other hand, usually management is not interested in the technical details of how documents are structured or how the corresponding personal subprocess is performed. Thus, the *managerial level*, which defines how management views design processes, comprises both *coarse-* and *medium-grained* representations.

In order to support managers in their coordination tasks, they must be supplied with appropriate *views* (abstractions) of *design processes*. Such views must be comprehensive inasmuch as they include products, activities, and resources (and their mutual relationships, see Sect. 1.1):

- The term *product* denotes the results of design subprocesses (e.g., flowsheets, simulation models, simulation results, cost estimates, etc.). These may be organized into *documents*, i.e., logical units which are also used for work distribution or version control. Complete results are *subconfigurations*.
- The term *activity* denotes an action performing a certain function in a design process. At the managerial level, we are concerned with *tasks*, i.e., descriptions of activities assigned to *designers* by managers, but also complex tasks performed by subproject groups.
- Finally, the term *resource* denotes any asset needed by an activity to be performed. This comprises both human and computer resources (i.e., the designers and managers participating in the design process as well as the computers and the tools they are using). Please note that also resources might be atomic or composed.

Thus, an overall *management configuration* consists of multiple parts representing products, activities, and resources. An example is given in Fig. 3.61. Here, we refer to the Polyamide-6 design process introduced earlier. On the left, the figure displays the roles in the design team as well as the designers filling these roles. The top region on the right shows design activities connected by control and data flows. Finally, the (versioned) products of these activities are located in the bottom-right region.

Below, we give a more detailed description of Fig. 3.61:

- *Products.* The results of design processes such as process flowsheets, steady-state and dynamic simulations, etc. are represented by documents (ellipses). Documents are interdependent, e.g., a simulation model depends on the process flowsheet (PFD) to which it refers (arrows between ellipses).

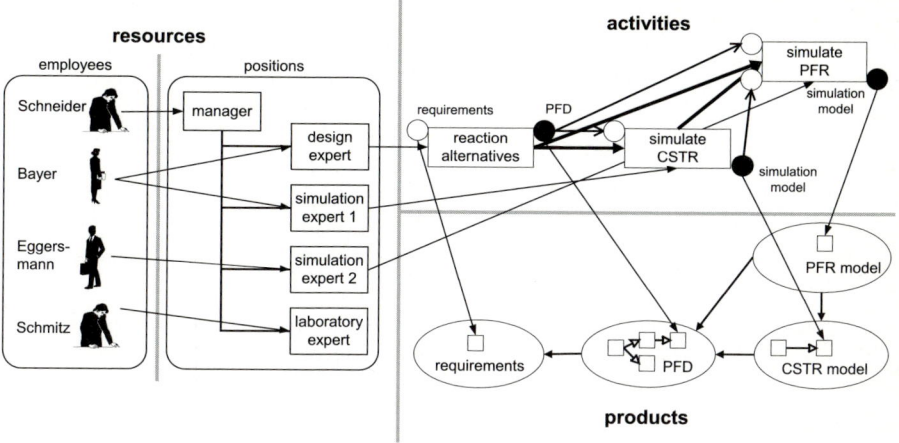

Fig. 3.61. Management configuration

The evolution of documents is captured by version control (each box within an ellipsis represents a version of some document).

- *Activities.* The overall design process is decomposed into tasks (rectangular boxes) which have inputs and outputs (white and black circles, respectively). The order of tasks is defined by control flows (thick arrows); e.g., reaction alternatives must have been inserted into the flowsheet before they can be simulated. Finally, data flows (arrows connecting circles) are used to transmit document versions from one task to the next.
- *Resources.* Employees (icons on the left) such as Schneider, Bayer, etc. are organized into project teams which are represented by organization charts. Each box represents a position, lines reflect the organizational hierarchy. Employees are assigned to positions (or roles). Within a project, an employee may play multiple roles. E.g., Mrs. Bayer acts both as a designer and as a simulation expert in the Polyamide-6 team.
- *Integration.* There are several relationships between products, activities, and resources. In particular, tasks are assigned to positions (and thus indirectly to employees). Furthermore, document versions are created as outputs and used as inputs of tasks.

It is crucial to *understand* the scope of the term "*management*" as it is used in this section. As already stated briefly above, management requires a certain amount of abstraction. This means that the details of the *technical level* are *not represented* at the managerial level. This is illustrated in Fig. 3.62, whose upper part shows a small cutout of the management configuration of Fig. 3.61. At the managerial level, the design process is decomposed into activities such as the creation of reaction alternatives and the simulation of these alternatives. Activities generate results which are stored in document versions. At the managerial level, these versions are basically considered black boxes, i.e.,

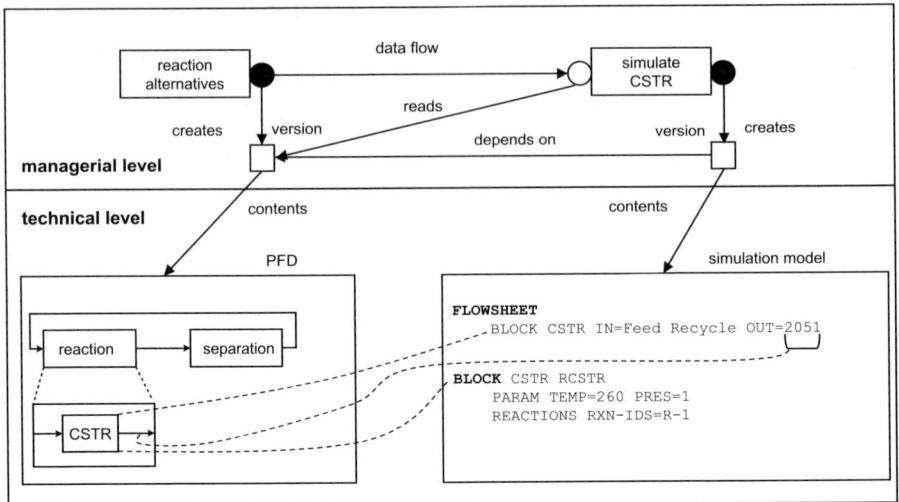

Fig. 3.62. Managerial and technical level

they are represented by a set of descriptive attributes (author, creation date, etc.) and by references to the actual contents, e.g., PFDs and simulation models. How a PFD or a simulation model is structured internally (and how their contents are related to each other), goes beyond the scope of the managerial level. Likewise, the managerial level is not concerned with the detailed personal process which is executed by some human to create a PFD, a simulation model, etc.

This does not imply that technical details are ignored. Rather, it must be ensured that the managerial level actually constitutes a correct *abstraction* of the *fine-grained information at the technical level* – and also controls technical activities. In fact, the management system described in this paper is part of an integrated environment for supporting design processes in chemical engineering. As such, it is integrated with tools providing fine-grained product and process support [21, 26]. The interplay of the tools of the overall environment is sketched only briefly in this section; see [348].

Particular attention has to be paid to the *dynamics* of design processes: The design process is not known in advance. Rather, it continuously evolves during execution. As a consequence, all *parts* of a *management configuration evolve* continuously:

- *Products.* The product structure is determined only during the design process. It depends on the flowsheets which is continuously extended and modified. Other documents such as simulation models and simulation results depend on the flowsheet. Moreover, different variants of the chemical process are elaborated, and selections among them are performed according to feedback gained by simulation and experiments.

- *Activities.* The activities to be performed depend on the product structure, feedback may require the re-execution of terminated activities, concurrent/simultaneous engineering calls for sophisticated coordination of related activities, etc.
- *Resources.* Resource evolution occurs likewise: New tools arrive, old tool versions are replaced with new ones, the project team may shrink due to budget constraints, or it may be extended to meet a crucial deadline, etc.

However, a management configuration should not evolve in arbitrary ways. There are *domain-specific constraints* which have to be met. In particular, activities can be classified into types such as requirements definition, design, simulation, etc. (likewise for products and resources). Furthermore, the way how activities are connected is constrained as well. For example, a flowsheet can be designed only after the requirements have been defined. Such domain-specific constraints should be taken into account such that they restrict the freedom of evolution.

Overview of the AHEAD System

Since current management systems suffer from several limitations (see introduction and section on related work), we have designed and implemented a new management system which addresses these limitations. This system is called AHEAD [212, 355]. AHEAD is *characterized* by the following *features*:

- *Medium-grained representation.* In contrast to project management systems, design processes are represented at a medium-grained level, allowing managers to effectively control the activities of designers. Management is not performed at the level of milestones, rather, it is concerned with individual tasks such as "simulate the CSTR reactor".
- *Coverage and integration at the managerial level.* AHEAD is based on an integrated management model which equally covers products, activities, and resources. In contrast, project and workflow management systems primarily focus on activities and resources, while product management systems are mainly concerned with the products of design processes.
- *Integration between managerial and technical level.* In contrast to project management systems, the AHEAD system also includes support tools for designers that supply them with the documents to work on, and the tools that they may use.
- *Support for the dynamics of design processes.* While many workflow management systems are too inflexible to allow for dynamic changes of workflows during execution, AHEAD supports evolving design processes, allowing for seamless integration of planning, execution, analysis, and monitoring.
- *Adaptability.* Both the structure of management configurations and the operations to manipulate them can be adapted by means of a domain-specific object-oriented model based on the UML [560].

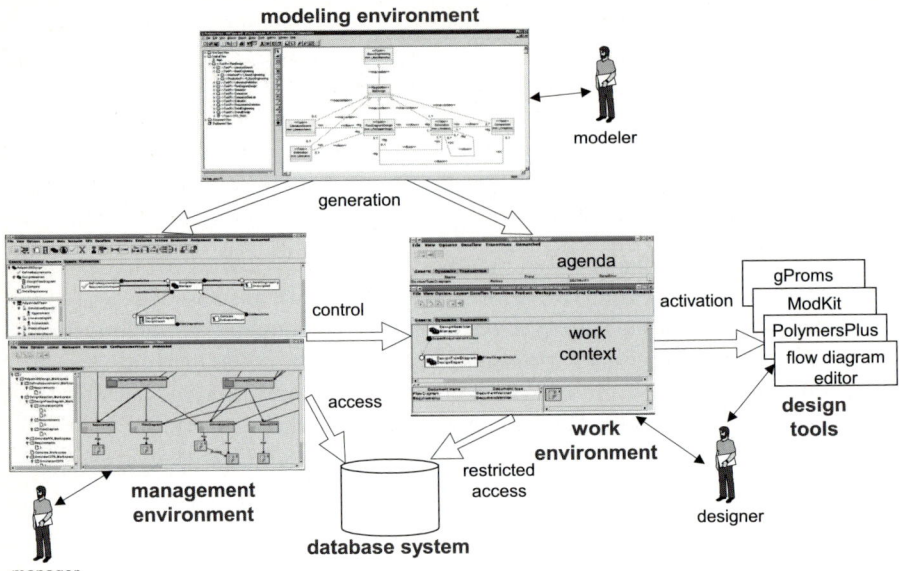

Fig. 3.63. Major components of the AHEAD system

Figure 3.63 gives an *overview* of the AHEAD *system*. AHEAD offers environments for different kinds of users, which are called modeler, manager, and designer. In the following, we will focus on the functionality that the AHEAD system provides to its users. Its technical realization will be discussed later.

The *management environment* supports project managers in planning, analyzing, monitoring, and controlling design processes. It provides graphical tools for operating on management configurations. These tools address the management of activities, products, and resources, respectively [244]:

- For *activity management*, AHEAD offers dynamic task nets which allow for seamless interleaving of planning, analyzing, monitoring, and controlling. A task net consists of tasks that are connected by control flow and data flow relationships. Furthermore, feedback in the design process is represented by feedback relationships. Tasks may be decomposed into subtasks, resulting in task hierarchies. The manager constructs task nets with the help of a graphical editor. He may modify task nets at any time while a design process is being executed.
- *Product management* is concerned with documents such as flowsheet, simulation models, cost estimations, etc. AHEAD offers version control for these documents with the help of version graphs. Relationships (e.g., dependencies) between documents are maintained as well. Versions of documents may be composed into configurations, thereby defining which versions are consistent with each other. The manager may view the version histories

and configurations with the help of a graphical tool. This way, he may keep track of the work results produced by the designers.
- *Resource management* deals with the organizational structure of the enterprise as far as it is relevant to design processes. AHEAD distinguishes between abstract resources (positions or roles) and concrete resources (employees). The manager may define a project team and then assign employees to the project positions.

Management of activities, products, and resources is *fully integrated*: Tasks are assigned to positions, inputs and outputs of tasks refer to document versions. Moreover, AHEAD manages task-specific workspaces of documents and supports invocation of design tools (see below).

AHEAD does not only support managers. In addition, it offers a *work environment* for *designers* which consists of two major components:

- The *agenda tool* displays the tasks assigned to a designer in a table containing information about state, deadline, expected duration, etc. The designer may perform operations such as starting, suspending, finishing, or aborting a task.
- The *work context tool* manages the documents and tools required for executing a certain task. The designer is supplied with a workspace of versioned documents. He may work on a document by starting a tool such as e.g. a flowsheet editor, a simulation tool, etc.

Please note that the scope of *support* provided by the *work environment* is limited. We do not intend to support design activities in detail at a technical level. Rather, the work environment is used to couple technical activities with management. There are other tools which support design activities at a fine-grained level. For example, a process-integrated flowsheet editor [21] may be activated from the work environment. "Process-integrated" means that the designer is supported by process fragments which correspond to frequently occurring command sequences, see Sect. 3.1. These process fragments encode the design knowledge which is available at the technical level. This goes beyond the scope of the AHEAD system, but it is covered by the overall environment for supporting design processes to which AHEAD belongs as a central component.

Both the management environment and the work environment access a common *management database*. However, they access it in different ways, i.e., they invoke different kinds of functions. The work environment is restricted to those functions which may be invoked by a designer. The management environment provides more comprehensive access to the database. For example, the manager may modify the structure of a task net, which is not allowed for a designer.

Before the AHEAD system may be used to carry out a certain design process, it must be adapted to the respective application domain [211]. AHEAD consists of a generic kernel which is domain-independent. Due to the generality of the underlying concepts, AHEAD may be applied in different domains such as software, mechanical, or chemical engineering. On the other hand, each domain has its specific constraints on design processes. The *modeling environment* is used to provide AHEAD with domain-specific knowledge, e.g., by defining task types for flow diagram design, steady-state and dynamic simulation, etc. From a domain-specific process model, code is generated for adapting the management and the work environment.

A Tour through the AHEAD System

In the following, we introduce the tool *support* provided by the AHEAD system with the help of a small *demo session*. The demo refers to the overall reference process of IMPROVE, namely the design of a chemical plant for producing Polyamide-6 (see Sect. 1.2). Here, we focus on the design of the reaction which proceeds as follows: After an initial PFD has been created which contains multiple design variants, each of these variants is explored by means of simulations and (if required) laboratory experiments. In a final step, these alternatives are compared against each other, and the most appropriate one is selected. Other parts of the reference process will be addressed in subsequent sections of this paper.

Modeling Design Processes

Before the AHEAD system is used to manage some actual design project, it is provided with a domain-specific *process model definition* (cf. Sect. 2.4) which should capture the conceptual design and basic engineering of arbitrary chemical design processes to meet our own demands in the IMPROVE project.

As we have discussed earlier, all parts of a management configuration evolve throughout the course of a design project. This kind of evolution is called *instance-level evolution*. While process model instances evolve continuously, we would like to constrain this evolution in such a way that a domain-specific design process is followed.

Object-oriented modeling is well suited to meet this requirement. The core of an object-oriented model is defined by a *class diagram* of the widely known Unified Modeling Language (UML). The UML class diagram declares types of tasks as classes and relationships between task types as associations which can be constrained by multiplicity restrictions. Object diagrams contain instances of classes (objects) and associations (links) which follow these restrictions. Object diagrams represent reusable task net patterns on the instance level. Both diagram types represent the structure of a process model. For behavioral aspects, state and collaboration diagrams are used.

Although the UML is a general object-oriented language, it has to be adapted to be suitable for process modeling. We use the *extension mechanism* of the UML to introduce new meta classes and meta attributes in the UML meta model [391].

An Adaptive and Reactive Management System for Project Coordination 311

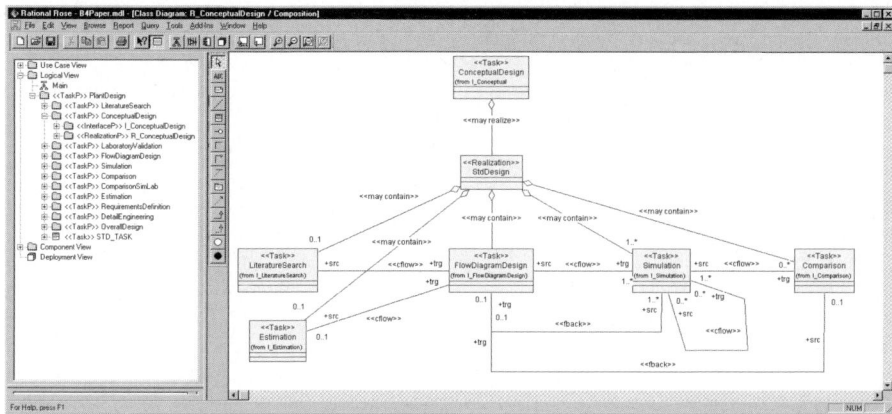

Fig. 3.64. UML model of a design process in chemical engineering (modeling environment)

In the following, we give a small *example* of *structural modeling* with the help of class diagrams; for more information, see [211, 390]. Furthermore, we will deal only with the modeling of activities, while a comprehensive process model must define products and resources, as well.

Figure 3.64 shows an *excerpt* of the *UML model* for design processes. The window on the left displays a hierarchy of packages, which are used to organize the overall model of the design process. On the right, a class diagram is presented which defines a part of the activities of the design process[32]. Here, *task classes* are modeled rather than specific instances. Instances are created dynamically at project runtime, implying that the topology of a task net is determined only at runtime. This object-oriented approach takes care of the dynamics of design processes and contrasts with the static workflows as defined in workflow management systems.

The *class diagram* introduces a complex task ConceptualDesign – dealing with the conceptual design of a certain part of the chemical process – and one of its possible realizations StdDesign. The realization contains multiple task classes, which is expressed by aggregations stereotyped with may contain. A contained task class may again be complex, resulting in a multi-level task hierarchy. For each contained task class, a cardinality is specified. E.g., 1..* means that at least one subtask is created at runtime[33]. Control flow associations (stereotype cflow) define the potential orders of task executions. E.g., a Comparison of design alternatives is performed at the very end of the design process.

---

[32] For the sake of simplicity, input and output parameters of tasks as well as data flows were removed from the diagram.
[33] The default cardinality is 1..1.

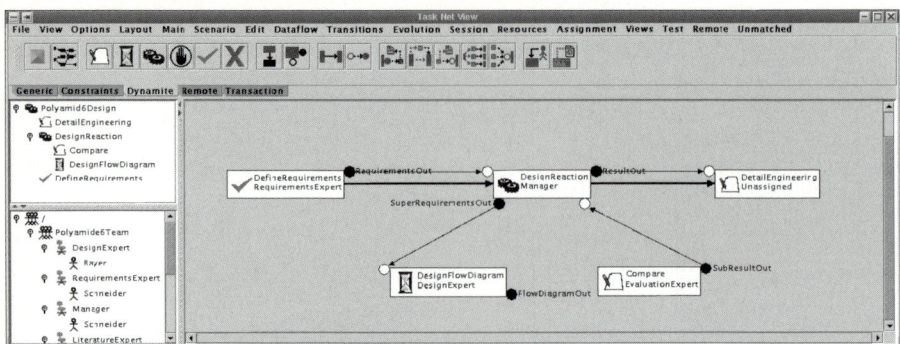

Fig. 3.65. Initial task net (management environment)

From the UML model, code is generated to *customize* the *functionality* provided by the AHEAD system. For example, the project manager may instantiate only the domain-specific classes and associations defined in the class diagrams. The core system as presented in this section enforces consistency with the process model definition. In this way, we can make sure that design proceeds according to the domain-specific model. A more flexible approach will be discussed in the next section (extending process evolution beyond consistency-preserving instance-level evolution).

Managing Design Processes

In this subsection, we illustrate the functionality of the AHEAD system provided at the instance level. This is performed with the help of a demo session which mainly focuses on the management environment, but also introduces the work environment.

Figure 3.65 presents a snapshot from the management environment taken in an early stage of the Polyamide-6 design process for reaction design. The upper region on the left displays a tree view of the task hierarchy. The lower left region offers a view onto the resources available for task assignments (see also Fig. 3.66). A part of the overall task net is shown in the graph view on the right-hand side. Each task is represented by a rectangle containing its name, the position to which the task has been assigned, and an icon representing its state (e.g., the gear-wheels represent the state **Active**, and the hour-glass stands for the state **Waiting**). Black and white circles represent outputs and inputs, respectively. These are connected by data flows (thin arrows). Furthermore, the ordering of task execution is constrained by control flows (thick arrows). Hierarchical task relations (decompositions) are represented by the graphical placement of the task boxes (from top to bottom) rather than by drawing arrows (which would clutter the diagram).

Please recall that the demo session deals only with the reaction part and particularly with its design (**DesignReaction** in Fig. 3.65). In this early stage, it is only known that initially some reaction alternatives have to be designed

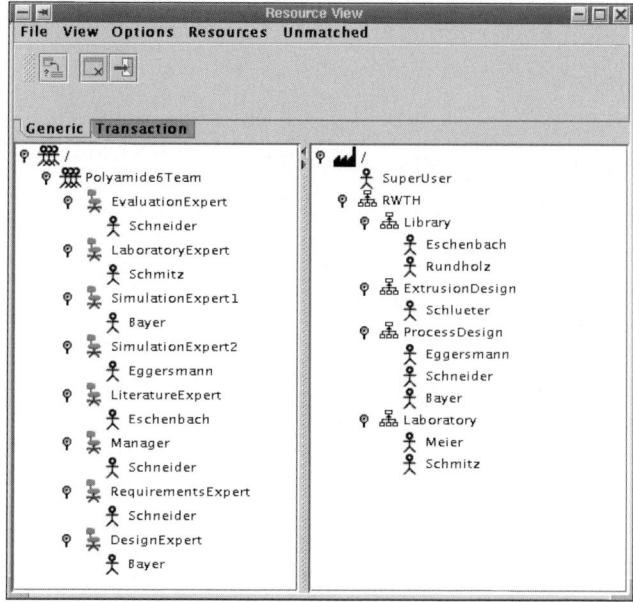

Fig. 3.66. Resource view (management environment)

(DesignFlowDiagram) and documented in a flowsheet. Furthermore, at the end these alternatives have to be compared (Compare), and a decision has to be performed. Other tasks of types defined in the class diagram of Fig. 3.64 are either left out (e.g. Estimation) or will be filled in later.

In addition to the initial task net, the manager has also used the resource management tool for building up his project team (Fig. 3.66). The region on the left displays the structure of the Polyamide-6 design team. Each position (represented by a chair icon) is assigned to a team member. Analogously, the region on the right shows the departments of the company. From these departments, the team members for a specific project are taken for a limited time span. Tasks are assigned to positions rather than to actual employees (see task boxes in Fig. 3.65). This way, assignment is decomposed into two steps. The manager may assign a task to a certain position even if this position has not been filled yet. Moreover, if a different employee is assigned to a position, the task assignments need not be changed: The tasks will be redirected to the new employee automatically.

The work environment is illustrated in Fig. 3.67. As a first step, the user logs into the system (not shown in the figure). Next, AHEAD displays an agenda of tasks assigned to the roles played by this user (top of Fig. 3.67). Since the user Bayer plays the role of the design expert, the agenda contains the task DesignFlowDiagram. After the user has selected a task from the agenda, the work context for this task is opened (bottom window). The work context graphically represents the task, its inputs and outputs, as well

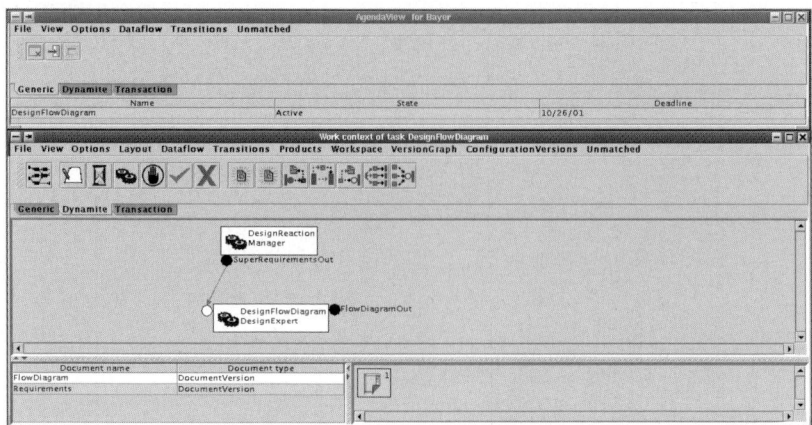

Fig. 3.67. Work environment

as its context in the task net (here, the context includes the parent task which defines the requirements to the flowsheet to be designed). Furthermore, it displays a list of all documents needed for executing this task. For some selected document, the version history is shown on the right (so far, there is only one version of the requirements definition which acts as input for the current task).

From the work context window, the user may activate design tools for operating on the documents contained in the workspace. Here, the user invokes a flowsheet editor [21] in order to insert reaction alternatives into the flowsheet for the Polyamide-6 process. The flowsheet editor, which was also developed in the IMPROVE project, is based on MS Visio, a commercial drawing tool, which was integrated with the PRIME process engine [371].

The resulting flowsheet is displayed in Fig. 3.68. The chemical process is decomposed into reaction, separation, and compounding. The reaction is refined into four variants. For our demo session, we assume that initially only two variants are investigated (namely a single CSTR and PFR on the left hand side of Fig. 3.68).

After the generation of the two variants, the manager extends the task net with tasks for investigating the alternatives that have been introduced so far (*product-dependent task net*, Fig. 3.69). Please note the control flow relation between the new tasks: The manager has decided that the CSTR should be investigated first so that experience from this alternative may be re-used when investigating the PFR. Furthermore, we would like to emphasize that the design task has not terminated yet. As to be demonstrated below, the designer waits for feedback from simulations in order to enrich the flowsheet with simulation data. Depending on these data, it may be necessary to investigate further alternatives.

Subsequently, the simulation expert creates a simulation model (using Polymers Plus) for the CSTR reactor and runs the corresponding simulations.

An Adaptive and Reactive Management System for Project Coordination   315

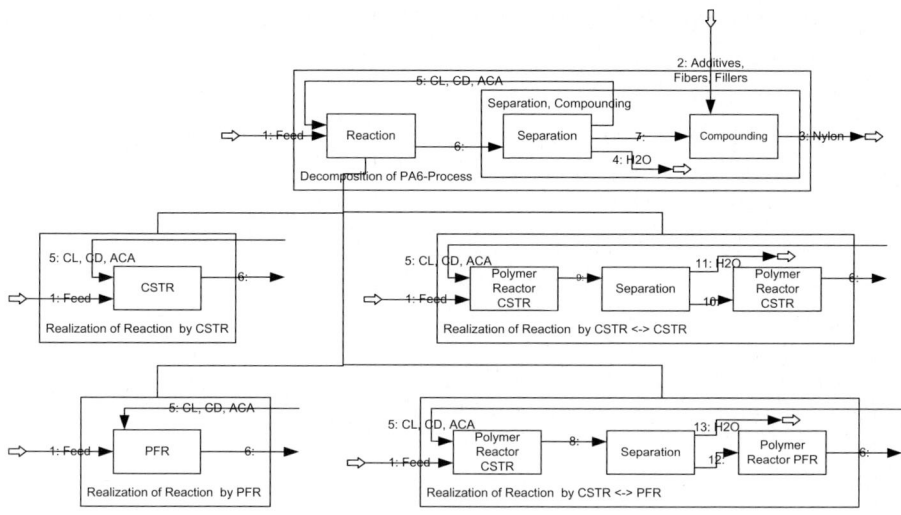

Fig. 3.68. Reaction alternatives in the process flowsheet

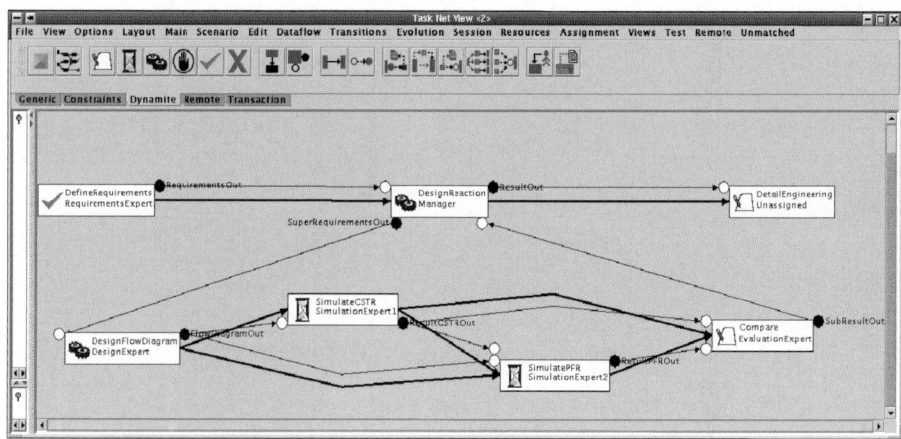

Fig. 3.69. Extended task net (management environment)

The simulation results are validated with the help of laboratory experiments. After these investigations have been completed, the flowsheet can be enriched with simulation data such as flow rates, pressures, temperatures, etc. To this end, a *feedback flow* – represented by a dashed arrow – is inserted into the task net (Fig. 3.70) [245, 246]. The feedback flow is refined by a data flow, along which the simulation data are propagated. Then, the simulation data are inserted into the flowsheet.

Please note that the semantics of the control flow from DesignFlowDiagram to SimulateCSTR is defined such that these tasks can be active simultaneously (*simultaneous engineering*) [576]. As a consequence, we cannot assume that

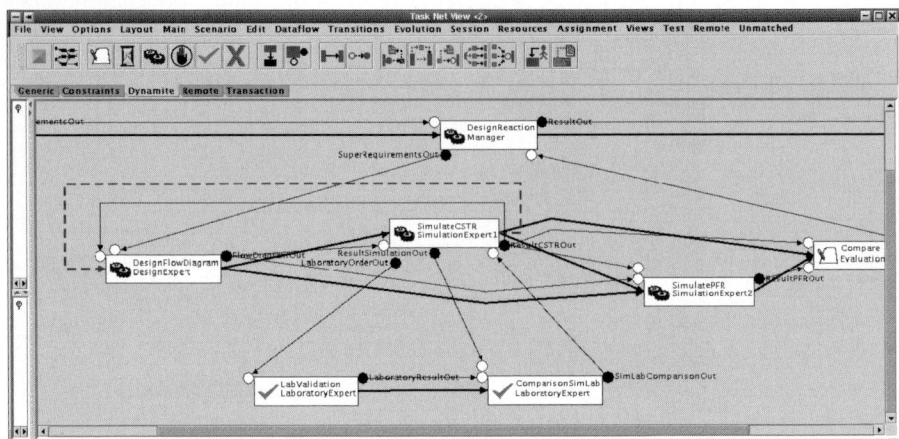

Fig. 3.70. Feedback and simultaneous engineering (management environment)

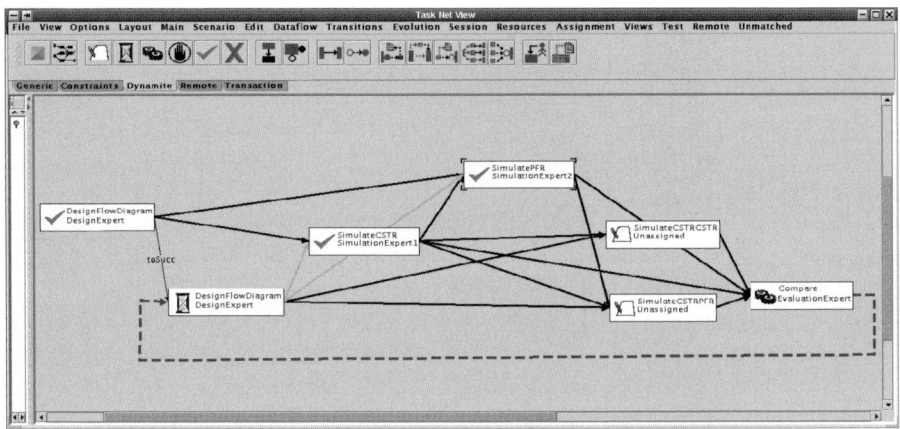

Fig. 3.71. Far-reaching feedback (management environment)

the work context of a task is stable with respect to its inputs. Rather, a predecessor task may deliver a new version that is relevant for its successors. This is taken care of by a sophisticated release policy built into the model underlying dynamic task nets [475].

After the alternatives **CSTR** and **PFR** have been elaborated, the evaluation expert compares all explored design alternatives. Since none of them performs satisfactorily, a far-reaching feedback is raised to the design task. Here, we assume that the designer has already terminated the design task. As a consequence, the design task has to be reactivated. Reactivation is handled by creating a new *task version*, which may or may not be assigned to the same designer as before. New design alternatives are created, namely a CSTR-CSTR and a CSTR-PFR cascade, respectively (see again Fig. 3.68).

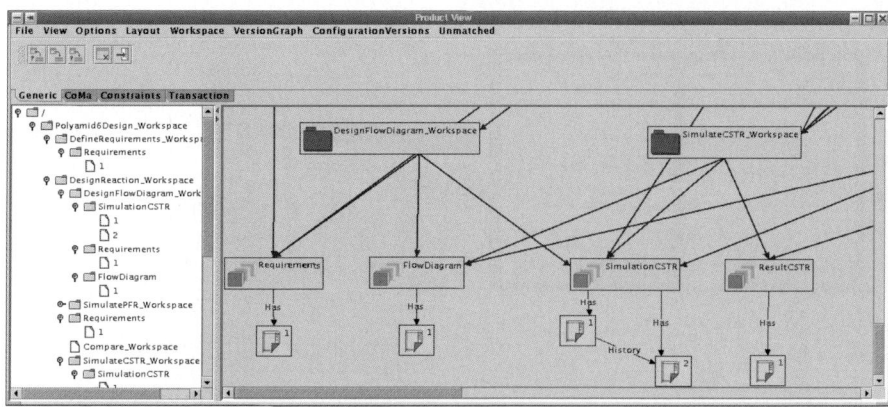

Fig. 3.72. Product view (management environment)

Furthermore, the task net is augmented with corresponding simulation tasks (Fig. 3.71[34]). After that, the new simulation tasks are assigned to simulation experts, and simulations are carried out accordingly. Eventually, the most suitable reactor alternative is selected.

So far, we have primarily considered the management of activities. Management of products, however, is covered as well. Figure 3.72 shows a tree view on the products of design processes on the left and a graph view on the right. Products are arranged into *workspaces* that are organized according to the task hierarchy. Workspaces contain sets of *versioned documents*.

Generally speaking, a *version* represents some state (or snapshot) of an evolving document. We distinguish between *revisions*, which denote temporal versions, and *variants*, which exist concurrently as alternative solutions to some design problem. Revisions are organized into sequences, variants result in branches. Versions are connected by *history relationships*. In Fig. 3.72, there is a history relationship between revisions 1 and 2 of SimulationCSTR, the simulation model for the CSTR reactor. In general, the version history of a document (flowsheet, simulation model, etc.) may evolve into an acyclic graph (not shown in the snapshot).

There is only one version of the flowsheet in Fig. 3.72. Here, we rely on the capabilities of the flowsheet editor to represent multiple variants. Still, the flowsheet could evolve into multiple versions at the managerial level (e.g., to record snapshots at different times). Moreover, in the case of a flowsheet editor with more limited capabilities (no variants), variants would be represented at the managerial level as parallel branches in the version graph.

Finally, it is worth noting that the support for process managers provided by the AHEAD system so far could easily be extended with respect to *process analysis* and *simulation* aspects. Within another sub project I2/I4 in

---

[34] Task parameters and data flows have been filtered out to avoid a cluttered diagram.

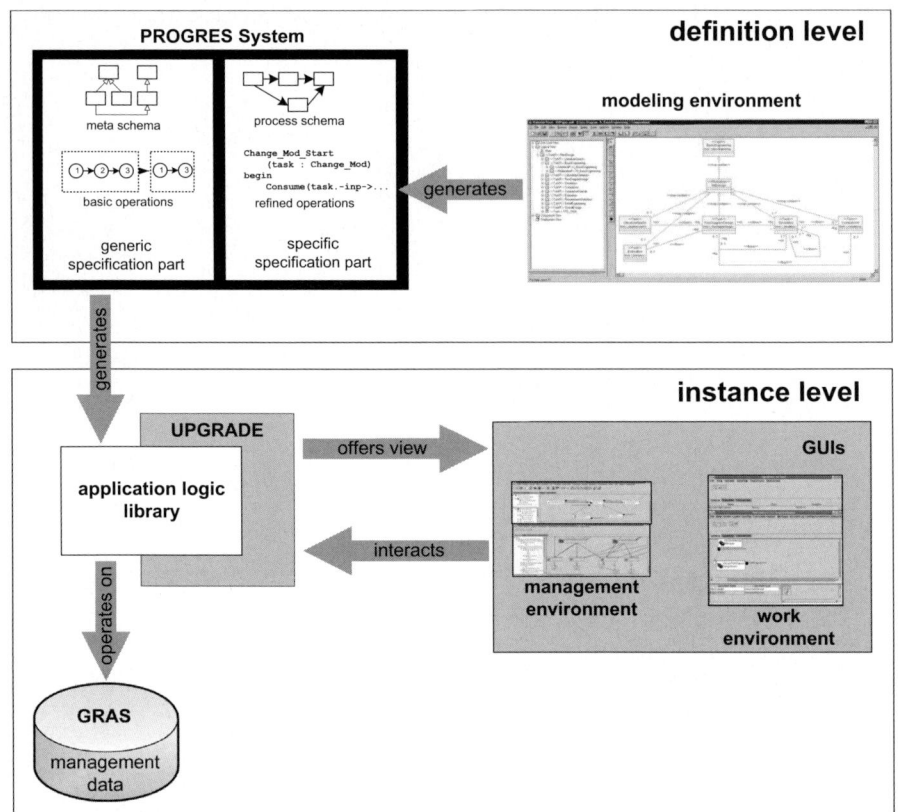

Fig. 3.73. Architecture of the AHEAD system

the CRC 476 IMPROVE, comprehensive support for detailed analysis of process activities, organizational structure, and information flow as well as the identification of weak spots within individual work processes has been developed. The chosen analysis and simulation approach and the research results are thoroughly described in Sect. 5.2 below. Although the focus in that work is on fine-grained work processes, a process manager could easily profit from similar analysis and simulation support on the medium-grained administrative management process layer above the work process layer. Further research in that direction is therefore needed.

Realization of the AHEAD Core System

Figure 3.73 displays the architecture of the AHEAD system. It also shows internal tools (left hand side) and thereby refines the overview given by Fig. 3.63.

Internally, AHEAD is based on a formal specification as a programmed graph rewriting system [206]. To this end, we use the specification language

PROGRES as well as its modeling environment, which offers a graphical editor, an analyzer, an interpreter and a code generator [414]. Both the process meta model and process model definitions are specified in PROGRES. The former was created once by the tool builders of AHEAD; the latter ones are generated automatically by the modeling environment (cf. Subsect. 3.4.3).

The overall specification, consisting of both the process meta model and the process model definition, is translated by the PROGRES compiler into C code. The generated code constitutes the application logic of the instance-level tools. The application logic library operates on the management data which are stored in the graph-based database management system *GRAS* [220]. The user interface of the management tools is implemented with *UPGRADE*, a framework for building graph-based interactive tools [49].

3.4.3 Process Evolution and Domain-Specific Parameterization

Motivation

As stated above, the AHEAD system may be applied to processes in different domains – including not only chemical engineering, but also other engineering disciplines such as software, electrical, or mechanical engineering. The core functionality is domain-independent and uses general notions such as task, control flow, etc. AHEAD may be *adapted* to a certain application domain by defining domain-specific knowledge [164]. For example, in chemical engineering domain-specific task types for flowsheet design, steady-state simulations, dynamic simulations, etc. may be introduced. Within a *process model definition*, domain-specific knowledge is defined which constrains the *process model instances* to be maintained at project runtime.

In the initial version of the AHEAD system, process model definitions were constrained to be static throughout the whole project lifecycle. Either no process model was defined at all, relying on a set of unconstrained *standard types*, or a process model had to be provided before the actual project execution could be started. Thus, evolution was constrained to the instance level (interleaving of planning and execution). However, gathering and fixing process knowledge beforehand turned out to be virtually infeasible for design processes, in particular in conceptual design and basic engineering of chemical plants. Therefore, support for *process evolution* was generalized considerably [171, 172].

While working on the Polyamide-6 reference process, it turned out that even process model definitions cannot be determined completely in advance. Therefore, we generalized evolution support to include all of the following features (cf. [171–173, 387, 390]):

- *Instance-level evolution.* Planning and enactment of dynamic task nets may be interleaved seamlessly (already part of the core system).

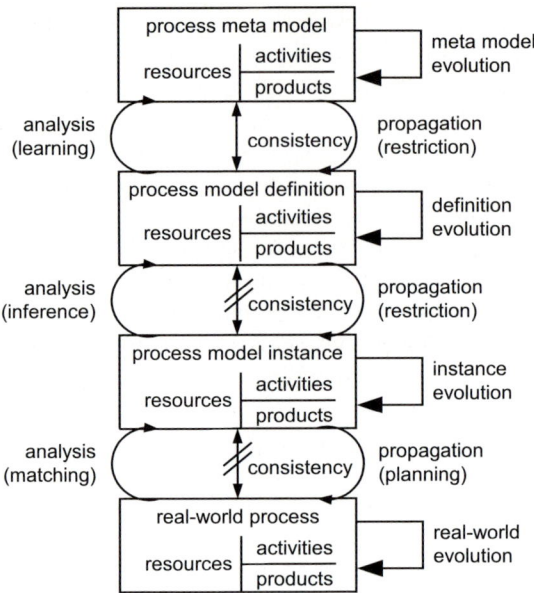

Fig. 3.74. Conceptual framework

- *Definition-level evolution.* At definition level, evolution is supported by version control at the granularity of packages (modular units of process definitions).
- *Bottom-up evolution.* By executing process instances, experience is acquired which gives rise to new process definitions. An inference algorithm supports the semi-automatic creation of a process model definition from a set of process model instances.
- *Top-down evolution.* A revised process model definition may be applied even to running process model instances by propagating the changes from the definition to the instance level.
- *Selective consistency control.* The project manager may allow for deviations of process model instances from their respective definitions resulting in inconsistencies. These deviations are reported to the project manager who may decide to reinforce consistency later on.

Conceptual Framework

Levels of Modeling

Our work is based on a conceptual framework which distinguishes four *levels of modeling* (Fig. 3.74). Each level deals with process entities such as products, activities, and resources. Here, we focus on activities, even though our framework equally applies to products and resources. Process evolution may occur on every level. Furthermore, adjacent levels are connected by propagation and analysis relationships. Propagation is performed top-down and constrains the operations that may be performed on the next lower level. Conversely, analysis works bottom-up and aims at providing feedback to the next upper level.

The *process meta model* introduces the language (or meta schema) for process model definitions. The meta model is based on dynamic task nets. It provides meta elements for structural (tasks, control and data flows etc.) and for behavioral aspects (e.g. state machines for tasks) of these task nets.

Process (model) definitions are created as instances of process meta models and are defined in the UML using class diagrams at the type level and collaboration diagrams for recurring patterns at the abstract instance level. Process definitions are organized into interface packages defining the interface of a task (in terms of its inputs and outputs) and realization packages (of a complex task) containing the class diagram and the collaboration diagrams of the respective subprocess. UML model elements are adapted to the process meta model with the help of extension mechanisms provided by the UML (stereotypes and tagged values).

Process (model) instances are instantiated from process model definitions. A process model definition represents reusable process knowledge at an abstract level whereas process model instances abstract from only one real world process. A process model instance is composed of task instances which are created from the task classes provided by the process model definition.

Finally, the *real-world process* consists of the steps that are actually performed by humans or tools. The process model is used to guide and control process participants, who conversely provide feedback which is used to update the process model instance.

Wide Spectrum Approach

In general, a *wide spectrum* of processes has to be modeled, ranging from ad hoc to highly structured. Moreover, different subprocesses may exhibit different degrees of structuring. This requirement is met by defining for each subprocess a corresponding package which contains a model at an adequate level of structuring. As illustrated in Fig. 3.75, we may distinguish four levels of process knowledge to be discussed below.

On the *untyped level* (Fig. 3.75a), there is no constraining process knowledge (ad-hoc process). The process manager may create and connect any number of tasks with the help of *unconstrained types* (which are constrained only by the underlying meta model).

On the *partially typed level*, the process modeler is capable of defining domain-specific types of tasks and relationships, but he also permits the use of unconstrained types on the instance level. In particular, this allows to leave out exceptions like feedbacks in the definition. When feedback does occur during enactment (e.g., a design error is detected during implementation), it can be handled by instantiating an unconstrained type of feedback flow without necessarily changing the process model definition.

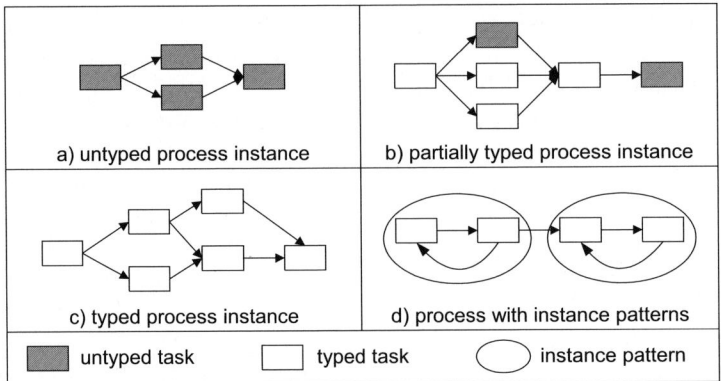

Fig. 3.75. Wide spectrum approach

The *completely typed level* requires complete typological knowledge of the respective subprocess and therefore permits only domain-specific types. Therefore, it excludes the use of unconstrained types and permits only domain-specific types. The only degree of freedom at the instance level is the cardinality of tasks which also can be constrained in the process model. For example, the cardinality [1:1] enforces exactly one instance of the respective task type.

The *instance pattern level* (Fig. 3.75d) deals with (abstract) instances rather than with types. Then, the process modeler may define an instance-level pattern which may be inserted in one step into a task net. An instance pattern for a whole subprocess is also called an *instance-level process definition*.

Consistency Control

Below, we discuss vertical *consistency* relationships between framework levels. As argued before, we assume consistency of process definitions w.r.t. the meta model.

Inconsistencies between process model instances and real-world processes are frequently caused by inadequate process models. The vast majority of process management systems demands consistency of the process model instance with the process model definition. As a consequence, the process model instance cannot be updated to represent the deviations taken by the process participants.

In our approach, we allow for inconsistencies between a process model instance and its definition. This way, the process model instance can match the real-world process as closely as possible. By default, a process model instance must be (strongly or weakly) consistent with its definition, but each subprocess can be set to allow for temporary inconsistencies (e.g., insertion of a task of some type that is not modeled in the respective process model). It is up to the process manager to decide whether these subprocesses containing *controlled*

Table 3.2. Potential consistency levels

	untyped	partially typed	typed
untyped	w	w	i
partially typed	i	w, i	i
typed	i	s, i	s, i

deviations finally have to be again consistent with their respective definitions or if they can be left inconsistent.

Like in object-oriented modeling, we distinguish between a structural model (defined by class and object diagrams) and a behavioral model (defined by state and collaboration diagrams). Accordingly, a process instance is *structurally (behaviorally) consistent* if it satisfies the constraints of the structural (behavioral) model.

We distinguish three levels of consistency ordered as follows: *inconsistent (i)* < *weakly consistent (w)* < *strongly consistent (s)*. We introduce the level of weak consistency to account for the use of unconstrained types (partial process knowledge). Table 3.2 summarizes potential consistency levels for combinations of instance- and type-level processes (rows and columns, respectively) depending on the respective degree of typing. For example, an untyped process instance is inconsistent with a typed process definition, which excludes the use of unconstrained types. Furthermore, a typed process instance is either strongly consistent or inconsistent with a typed process definition (weak consistency is only possible in the case of unconstrained types).

Process Evolution

In principle, *process evolution* may be considered at all levels of our conceptual framework though we assume a static meta model here to avoid frequent changes of the process management system itself.

Evolution of process instances is inherent to dynamic task nets. The process meta model is designed such that planning and enactment may be interleaved seamlessly. Evolution of process definitions is performed at the level of packages. To maintain traceability, packages are submitted to *version control*.

After new package versions containing improved process definitions have been created, *migration* of currently enacted process instances may be performed selectively. During migration, process instances are to be updated such that they are consistent with the improved process definition. It is crucial that temporary inconsistencies do not prevent migration because a consistent state is reached only eventually (if ever).

By permitting weakly consistent and inconsistent process instances, we may not only minimize the gap between process instance and real-world process. In addition, we support *bottom-up evolution* of process definitions by allowing for selective adaptation of incorrect or incomplete process models according to new experience expressed in the evolved process instance.

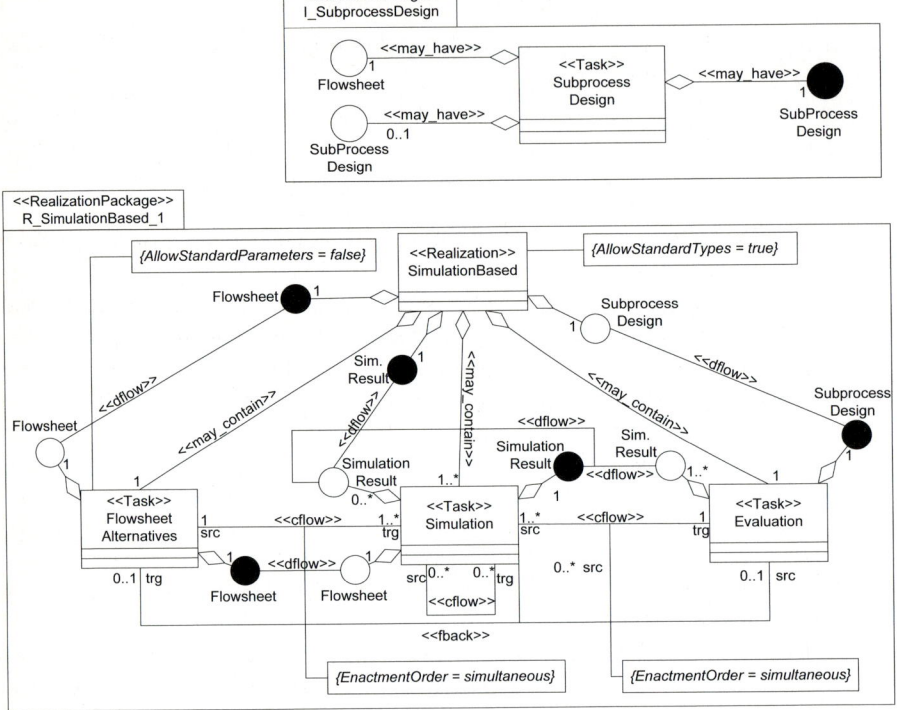

Fig. 3.76. Class diagram for a design subprocess

After having gained sufficient experience at the instance level, the process modeler may propagate these changes from the process definition to the instance level (*top-down evolution*). Changes may be propagated selectively, and inconsistencies may be tolerated either temporarily or permanently – depending on whether it pays off or it is considered necessary to reestablish consistency. Altogether, our approach supports *round-trip process evolution*.

Sample Process

The example below shows a *process evolution roundtrip*: During the execution of the design process, changes are performed which introduce inconsistencies with respect to the process definition. In response to this problem, an improved version of the process definition is created. Finally, the process instance is migrated to the new definition. In contrast to the previous section, we will deal with a different part of the overall reference process, namely the design of the separation (Sec. 1.2).

Figure 3.76 presents a process definition on the type level of a subprocess design as it can be used for any part of the overall chemical process (i.e., not only for the separation, but also for the reaction and the compounding). This version of the process definition will be replaced by an improved version later on.

The subprocess design is defined in two UML *packages* containing class diagrams for the interface and the realization, respectively. The class diagrams are adapted to the underlying process meta model by the use of either textual (<<Task>>) or graphical *stereotypes* (black/white circles for input/output parameters). Further meta data are represented by *tagged values* which are used to annotate model elements (shown as notes in the diagram).

The *interface* is defined in terms of inputs and outputs. A task of class **SubprocessDesign** receives exactly one flowsheet for the overall chemical process and (potentially) the design of a preceding subprocess. The output parameter denotes the result of the subprocess design, including the flowsheet for the subprocess, simulation models, and simulation results.

The *realization* is described by a class diagram containing a class for the respective task net as well as classes for the subtasks. Although multiple realizations may be defined, we discuss only a simulation based realization (**SimulationBased**[35]). The inputs and outputs attached to **SimulationBased** are *internal parameters* which are used for vertical communication with elements of the refining task net.

The refining task net[36] comprises several serial and parallel tasks and is defined in a similar way as in Fig. 3.64.

Modeling elements are decorated with *tagged values* which define both structural and behavioral constraints. A few examples are given in Fig. 3.76:

- *Structural constraints.* The tag **AllowStandardTypes** is used to distinguish between partially and fully typed processes (Fig. 3.75b and c, respectively). Likewise, **AllowStandardParameters** determines whether a task may have untyped parameters.
- *Behavioral constraints.* The behavior of control flows may be controlled by the tag **EnactmentOrder**. The value simultaneous allows for the simultaneous activation of tasks connected by a respective control flow.

In addition to the process definition given above, the composition of subprocesses (**PreStudy**, **ReactionDesign**, **SeparationDesign**, **Compounding** and **Decision**; see top part of Fig. 3.77) into an overall design process has to be defined. When the overall process definition is available, a process instance is created according to the process definition. In the sequel, we will focus exclusively on separation design.

Initially, separation design is decomposed into a task for designing **FlowsheetAlternatives** and a final **Evaluation** task for comparing these alternatives. Now, the following problem is recognized: In order to design the separation additional data on the reaction are required. Waiting for these data would

---

[35] The <<may_realize>> association to the corresponding task class (see Fig. 3.64) was omitted from the figure.

[36] Data flows along feedback flows were omitted to keep the figure legible.

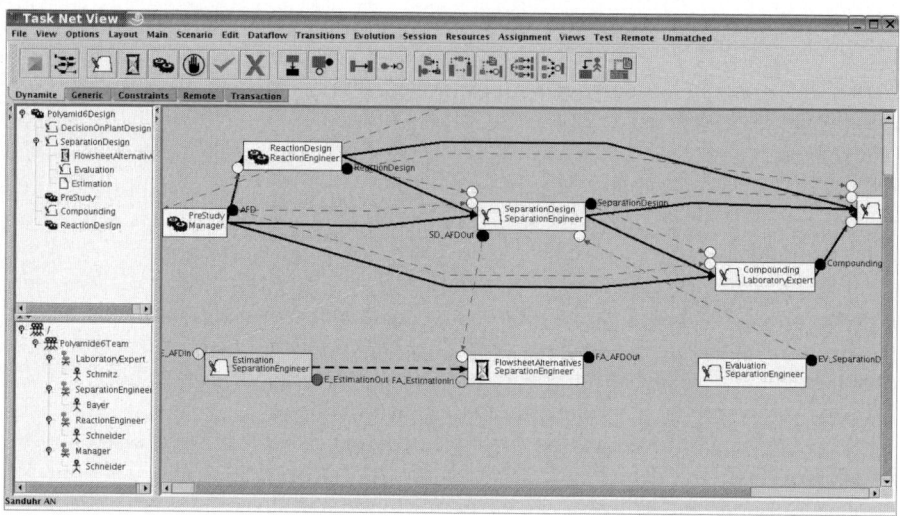

Fig. 3.77. Task net with inconsistencies

severely slow down the design process. Therefore, the manager of the separation design calls for and inserts an initial **Estimation** of these data so that design may proceed using initial estimations until more detailed data finally arrive.

Since the **Estimation** task is not defined in the class diagram, the manager makes use of the unconstrained type **Task** to insert it into the task net as an untyped task. This modification degrades the consistency level of the task net to weakly consistent. Unfortunately, inconsistencies are introduced, as well: The flowsheet design task has to be supplied with the estimation as input parameter. This was excluded in the process definition by the value false of the tag **AllowStandardParameters**. Therefore, the manager has to switch off consistency enforcement explicitly to make the modification feasible.

Figure 3.77 illustrates how weak (grey) and strong inconsistencies (red) are signaled to the manager[37].

Execution may continue even in the presence of inconsistencies. Further modification, like the insertion of parallel tasks **SimulationDistillation** and **SimulationExtraction** may cause additional inconsistencies.

At this stage, it is decided to clean up the process definition so that it includes the recent process improvements. Since the old definition may not be modified for the sake of traceability, a new version is created instead. Among others, the class diagram presented in Fig. 3.76 is revised (see Fig. 3.78, where the changes are emphasized in bold face).

---

[37] Unfortunately, there is hardly any difference between grey and red in grey-scale reproduction.

An Adaptive and Reactive Management System for Project Coordination 327

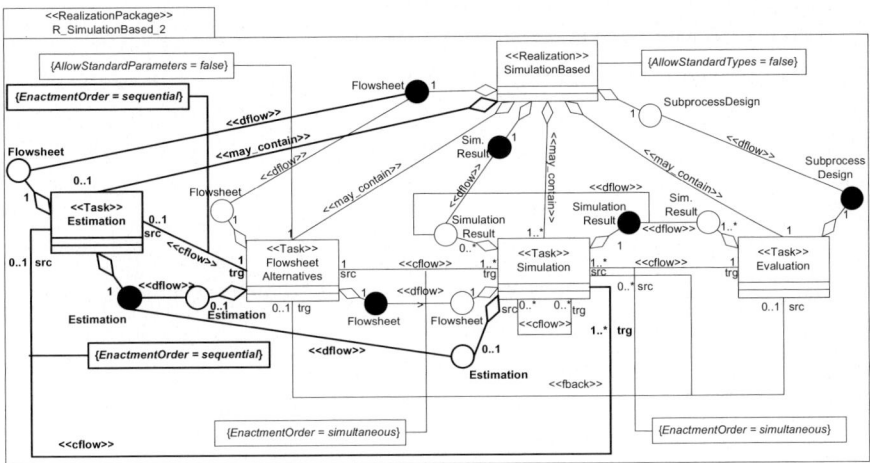

Fig. 3.78. Revised process definition

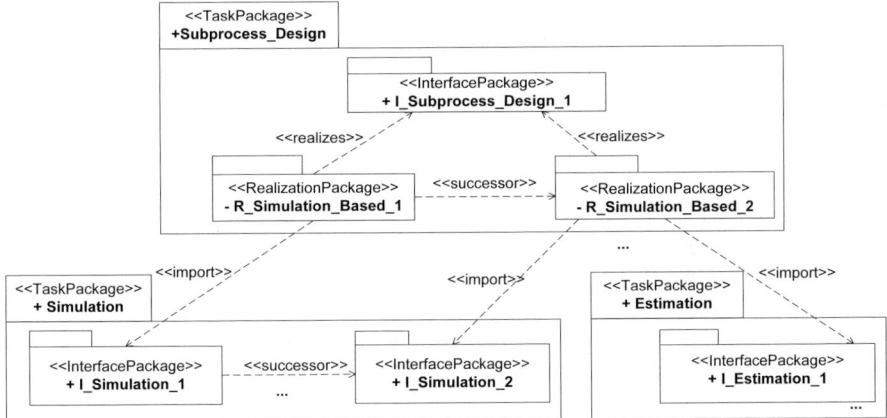

Fig. 3.79. Package versions

Figure 3.79 illustrates the evolution on the definition level by a *package diagram*. A task package serves as a container for interface and realization packages. The interface package for the subprocess design is not affected. For the realization, a new package version is derived from the old one. In addition, a new task package for the estimation is created. Finally, the interface package for the simulation task has to be revised such that the simulation task may be supplied with an estimate of a preceding subprocess.

The process evolution roundtrip is closed by propagating the changes at the definition level to the instance level. In general, migration has to be performed

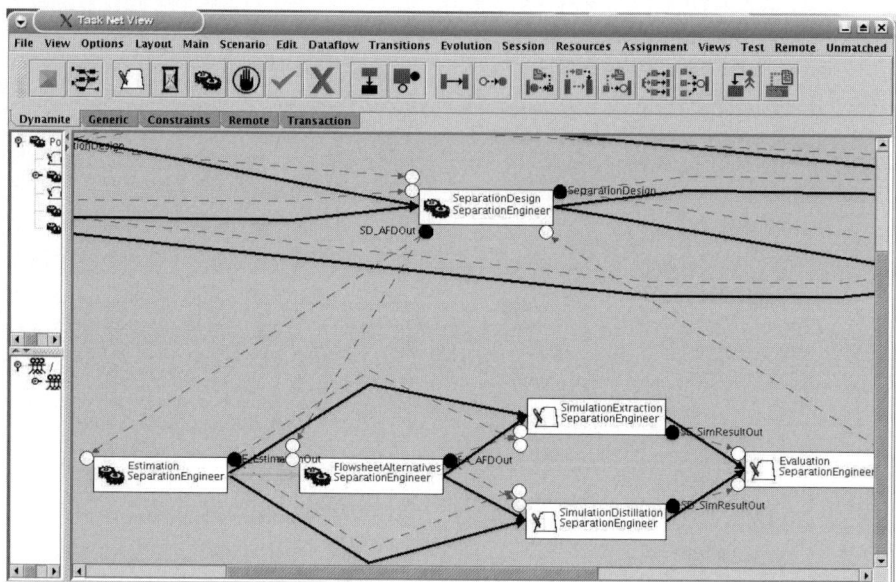

Fig. 3.80. Task net after migration

interactively since it is not always possible to uniquely determine e.g. the target type of migration.

All tasks whose types were already contained in the old definition can be migrated automatically to the new type version. In our example, this rule applies to the design task and the simulation tasks. In contrast, the estimation task's target type cannot be determined uniquely since it was introduced as an untyped task. After all objects have been migrated, the relationships can be migrated automatically. This is possible even for untyped relationships provided that there is only one matching relationship for each pair of object types.

The task net after migration is shown in Fig. 3.80. The control flow from the estimation task to the flowsheet design task is marked as behaviorally inconsistent (emphasized by red color). Both tasks are currently active, while the revised process definition prescribes a sequential control flow. This illustrates that migration does not necessarily result in a task net which does not contain inconsistencies. Migration can always be performed – even if inconsistencies persist.

Realization

Figure 3.81 displays the architecture of AHEAD extended by details of the modeling environment concerning process evolution (see upper right corner).

The process modeler uses a commercial CASE tool – Rational Rose – to create and modify process definitions in the UML. Rational Rose is adapted

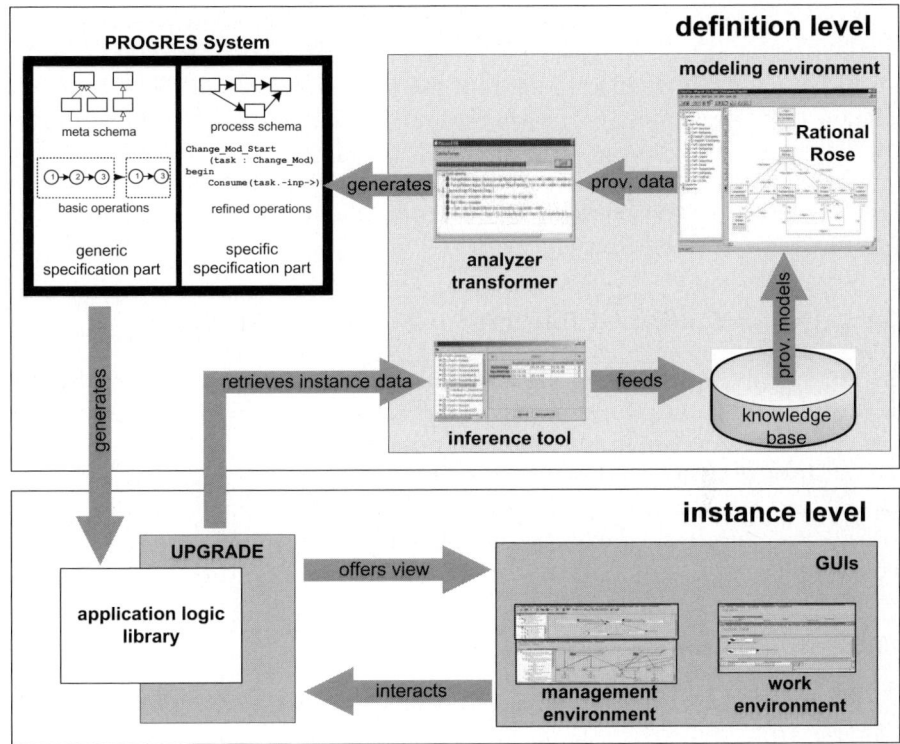

Fig. 3.81. Architecture of the AHEAD system with process evolution support

with the help of stereotypes which link the UML diagrams to the process meta model. A class diagram is represented as shown in Fig. 3.76. An **analyzer** checks process model definitions for consistency with the process **meta model**. The **analyzer** is coupled with a **transformer** which translates the UML model into an internal representation hidden from the process modeler [211].

Finally, the **inference tool** closes the loop by assisting in the inference of process definitions from process instances. The **inference tool** analyzes process instances and proposes definitions of task and relationship types. These definitions are stored in a **knowledge base** which may be loaded into Rational Rose. In this way, bottom-up evolution is supported. For a more detailed description of the inference tool, the reader is referred to [390].

To conclude this section, let us summarize how process evolution is supported by AHEAD. The sample process presented in the previous section assumes that a type-level process definition has already been created. For a while, the design process proceeds according to the definition. Planning and execution are interleaved seamlessly, the task net is extended gradually (instance evolution). Then, the manager detects the need for a deviation. Consistency enforcement is switched off in the task net for the separation design, and the estimation task is inserted. These steps are performed with the help of the management tool. Execution continues even in the presence of inconsistencies until it is decided to improve the process definition. To this end, the process modeler creates new package versions in Rational Rose. This results in an extension of the process definition, i.e., the old parts are still present. The extended definition is transformed into the PROGRES specification, which in turn is compiled into C code. Now the process manager may migrate the task net to the improved definition.

3.4.4 Delegation-Based Interorganizational Cooperation

So far, we have assumed tacitly that the overall design process is performed within one company. However, there are many examples of processes which are distributed over multiple organizations.

We have developed a delegation-based model for *cooperation between companies* and a generalization thereof. We first concentrate on the delegation-based cooperation and introduce a scenario for this kind of interorganizational cooperation.

Delegation of Subprocesses

Figure 3.82 illustrates the key components of the *distributed AHEAD system* [30, 166, 167, 169, 208]. The local systems are structured as before; for the sake of simplicity, the modeling environments are not shown. The extension of AHEAD to a distributed system is illustrated by the arrows connecting different instances of the AHEAD system.

AHEAD may be used to *delegate* a subprocess to a subcontractor. In general, a delegated subprocess consists of a connected set of subtasks; delegation is not confined to a single task. When the subcontractor accepts the delegation, a database is created which contains a copy of the delegated subprocess. Subsequently, execution of the subprocess is *monitored* such that the *contractor* may control the progress of work performed by the *subcontractor*.

The *delegation model* underlying the AHEAD system meets the following *requirements*:

- *Delegation of subprocesses.* A delegated subprocess consists of a connected set of subtasks. This way, the contractor may define *milestones* for controlling the work of the subcontractor.
- *Delegation as a contract.* The delegated subprocess serves as a contract between contractor and subcontractor. The contractor is obliged to provide the required inputs, based on which the subcontractor has to deliver the outputs fixed in the contract.
- *Autonomy of contractor and subcontractor.* The autonomy of both parties is retained as far as possible; it is restricted only to the extent required by the contract.

An Adaptive and Reactive Management System for Project Coordination   331

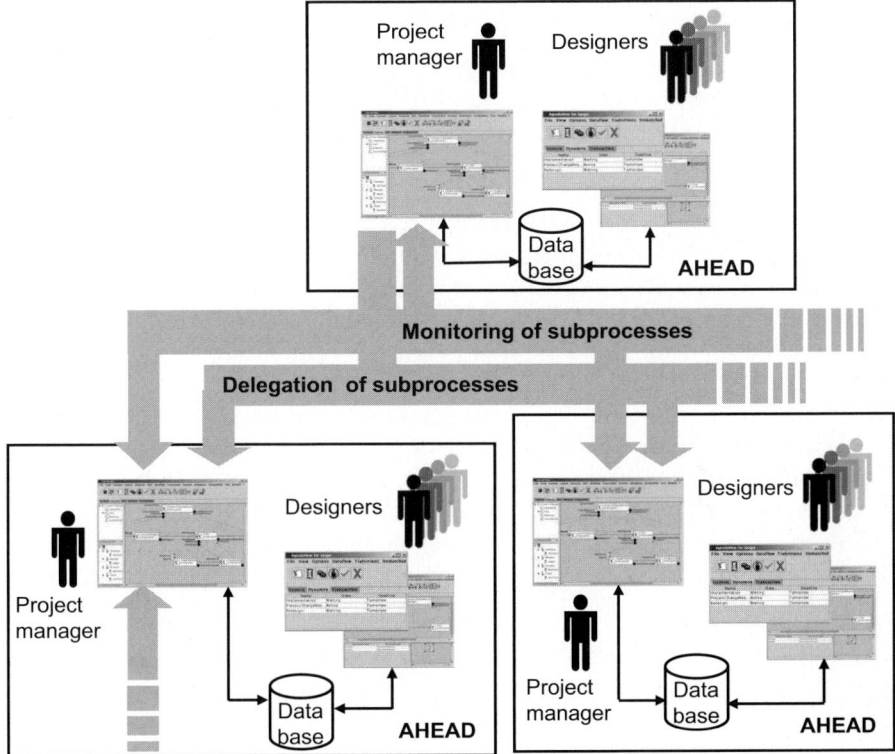

Fig. 3.82. Distributed AHEAD system

- *Need-to-know principle.* The parties engaged in a contract share only those data which are needed for the contract. This includes the respective subprocess as well as its context, i.e., its embedding into the overall process. Other parts of the process are hidden.
- *Refinement of delegated subprocesses.* The subcontractor may refine delegated subprocesses if this is required for managing the local work assignments. Since these refinements are not part of the contract, they are not visible to the contractor.
- *Monitoring of process execution.* The contractor is informed continuously about the state of execution of the subprocess delegated to the subcontractor. In this way, the contractor may monitor execution and control whether set deadlines are met.
- *Support of dynamic design processes.* Support for process dynamics is extended to interorganizational design processes. In particular, contracts can be changed dynamically. However, this requires conformance to a *change protocol* because cooperation among different enterprises requires precisely defined formal rules. The change protocol ensures that the contract may be changed only when both involved parties agree.

Delegation is *performed* in the following *steps*:

1. *Export.* The contractor exports the delegated subprocess into a file. A copy of the delegated subprocess is retained in the database of the contractor.
2. *Import.* The subcontractor imports the delegated subprocess, i.e., the file is read, and the local database is initialized with a copy of the delegated subprocess.
3. *Runtime coupling.* The AHEAD systems of contractor and subcontractor are coupled by exchanging events. Coupling is performed in both directions. This way, the contractor is informed about the progress achieved by the subcontractor. Vice versa, the subcontractor is informed about operations relevant for the delegation (e.g., creation of new versions of input documents).
4. *Changing the contract.* The contract established between contractor and subcontractor may be changed according to a pre-defined change protocol. The change is initiated by the contractor, who issues a change request. In a first step, the proposed change is propagated to the subcontractor. In a second step, the subcontractor either accepts the change – which makes the changes valid – or rejects it, implying that the propagated change is undone.

Please note that steps 1–3 are ordered sequentially. Step 4 may be executed at any time after the runtime coupling has been established.

Sample Process for Delegation-Based Cooperation

Scenario

When designing a chemical plant, expertise from multiple domains is required. For example, in the case of our Polyamide-6 reference process experts from *chemical engineering* and *plastics engineering* have to cooperate. Plastics engineering is needed to take care of the last step of the chemical process, namely compounding, which is performed with the help of an extruder.

The scenario to be discussed below involves two companies. The overall design of the chemical plant is performed in a chemical company. Compounding is addressed by an plastics engineering company. Designers of both companies have to cooperate closely with respect to the separation step of the chemical process since separation can be performed partly still in the extruder. In Fig. 3.83, a detailed extruder configuration including the polymer feeding, a mixing section, a degassing section followed by the fiber adding and the degassing of air is shown. The substances fed into the extruder still contain a small fraction of monomers which are fed back into the reaction step. Thus, a major design decision concerns the question to what extent separation can still be performed in the extruder.

Further on we will concentrate on the *delegation* of the activities from chemical engineering to plastics engineering. The compounding expert receives

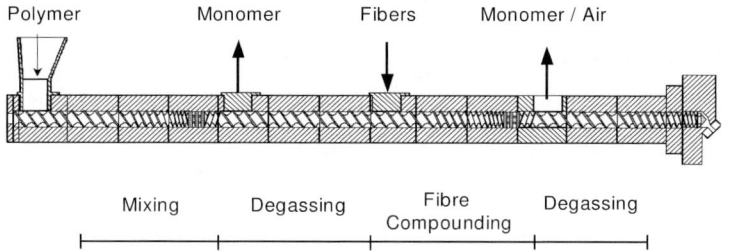

Fig. 3.83. Functional sections in a compounding extruder

the compounding steps and information about process boundary conditions such as mass flow, estimated viscosity, and thermophysical polymer properties (e.g., heat capacity, thermal conductivity). Afterwards he estimates compounding specific process parameters like the machine size, the extruders' rotational speed, the mass flow in every extruder, and the number of needed extruders.

Because the degassing process in the extruder can be quantified only with high experimental effort or by a simulation program [147], at first the degassing section is investigated by the compounding simulation expert. In a meeting, all necessary tasks are discussed and afterwards the compounding simulation expert starts a calculation to quantify the amount of degassed monomer while the compounding expert estimates the process behavior for the fibre adding section by his experience based knowledge. In the following meeting, first design results are discussed with the separation expert representing the chemical company. This collaboration for the design of the separation process is necessary, because the separation of volatile components like monomers and solvents from the polymer is possible both in e.g. the wiped film evaporator and the compounding extruder as mentioned above.

As a result of the interdisciplinary meeting, the members decide to make a detailed analysis of the homogenization processes in the mixing section by use of 3D-CFD tools (Computational Fluid Dynamics). Afterwards the results are discussed among the plastics engineers in a second meeting to prepare a report for the chemical engineering contractor.

The parallel activities in chemical and plastics engineering require powerful and smart management tools which can handle the highly dynamic concurrent processes. If any of the analyzed process steps turns out to be not feasible or not economically reasonable, various activities can be affected and a large part of the complete project has to be reorganized or in the worst case canceled.

Initial Situation

The example session described here deals only with the part of the overall design process which is related to the design of the extruder. The chemical company acts as a contractor and delegates the task of designing the extruder component to its subcontractor, the plastics engineering company.

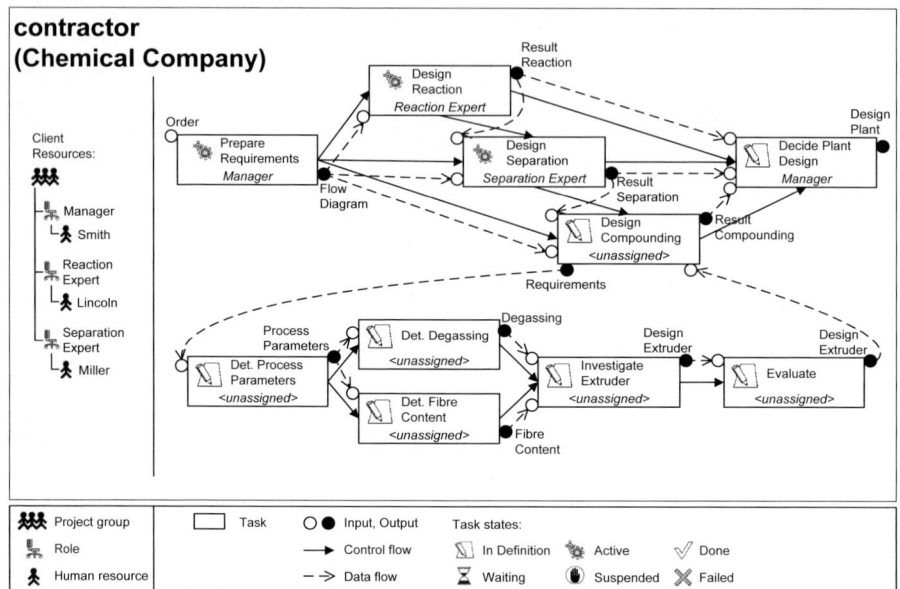

Fig. 3.84. Task net after refinement of task **Design Compounding**

The task net in Fig. 3.84 results after the manager of the chemical company, acting as the contractor, has refined the extruder design task by a subnet. The task **Design Compounding** and all subtasks will be delegated to the subcontractor. The task definition of **Design Compounding** can be seen as a *contract* between both companies, where the subcontractor has to produce a certain output (extruder design alternative) based on the inputs (requirements) which are provided by the contractor.

As stated before, in this subprocess an extruder is developed according to a set of desired product properties. The subtask **Determine Process Parameters** receives a product quality specification and the extruder's properties as input and produces rough estimates for the extruder's parameters. The content of fibres as well as the degassing of volatile components of the plastics are investigated in separate tasks. The subsequent investigation of the extruder's functional sections in task **Investigate Extruder** is based on the output of the three previous tasks. The results are evaluated and if the desired properties are met, the extruder design is propagated as a preliminary result to the parent task **Design Compounding**.

Establishing the Delegation

The delegated subprocess **Design Compounding** and its refining task net is are exported to a file. For further monitoring, all delegated tasks remain in the local data base and in the task net view on the contractor side but are assigned to a newly created resource **Remote: Plastics Engineering Company**. The

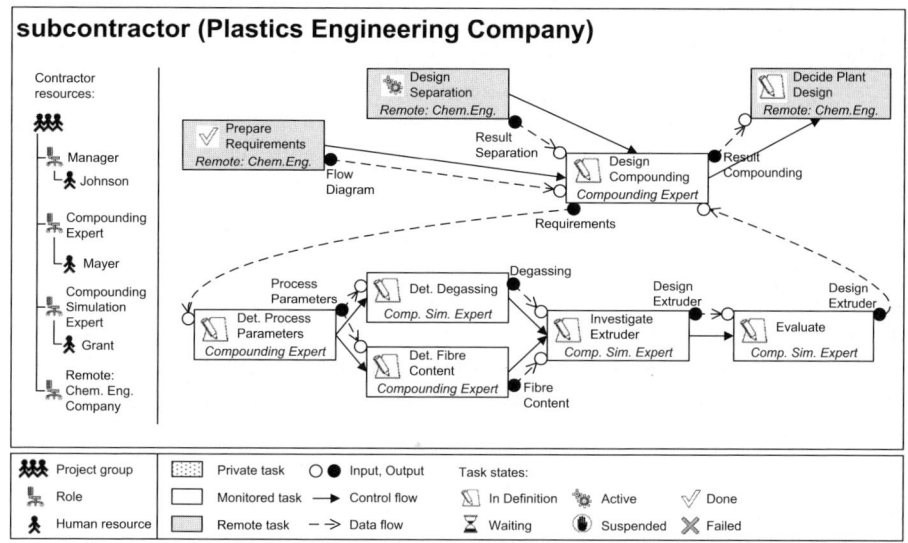

Fig. 3.85. Task net on subcontractor side after delegation and resource assignments

export file also contains contextual information about the delegated process, namely those tasks that are not delegated but connected via control flows to a delegated task (here: Prepare Requirements, Design Separation and Decide Plant Design). They can be *monitored* on the subcontractor side in contrast to *private tasks* that are not in the context (e.g. Design Reaction).

The plastics engineering company (subcontractor) imports the process description file into its AHEAD system. Figure 3.85 shows the corresponding task net and its context, which are instantiated in the local database on the subcontractor side. All delegated tasks are still in state In Definition, so that the manager on the subcontractor side can ask the contractor for a revised version if he does not agree with the contract consisting of delegated tasks, their parameters, control flows and data flows. If the subcontractor agrees to execute the delegated process, he may begin with the execution of the corresponding task net in his management environment, e.g. by assigning all delegated tasks to either the role Compounding Expert or Compounding Simulation Expert as shown in Fig. 3.85.

The management systems of contractor and subcontractor are loosely coupled together by exchanging events. The contractor is informed about changes of the delegated tasks' execution state which are considered *milestone* activities. Vice versa, the subcontractor is informed about changes of the context tasks which are executed on the contractor side. For instance, if Prepare Requirements is changed from Active to Done on the contractor side, a change event triggers the same change on the subcontractor side (cf. Fig. 3.85).

The delegated task Design Compounding is activated by a Compounding Expert. During execution of the delegated tasks, roles are assigned (e.g. Compounding Expert for Determine Process Parameters), task states are changed, and results are passed according to the defined data flows. All these updates of the delegated tasks by the subcontractor can be monitored by the contractor as well as the produced result, the first version of the extruder design.

Changing the Delegated Task Net Dynamically

In our example, the contractor and the subcontractor agree that the preliminary design alternative for the extruder could be optimized if the mixing quality of the materials in the extruder is investigated further. This is done by performing a three-dimensional simulation of the polymer flow in the extruder. The subcontractor agrees to carry out the additional simulation and the contract between contractor and subcontractor can be extended. Changing a delegated process after having started its enactment is not unusual in the design process of a chemical plant.

AHEAD supports dynamical changes of the contract between contractor and subcontractor. Changes are allowed only when both parties agree on them. Therefore, the delegated task net is changed according to a formal *change protocol*. The delegated task net is at every time in exactly one of the three *delegation states* Accepted, Change, and Evaluate. As described below, the transitions between these states define the commands which can be executed either on contractor and subcontractor side during the change process.

Initially, the subcontractor has issued the command Allow changes (from Accepted to Changed) to signal that he agrees to the change proposal of the contractor. After that, the contractor is able to modify the delegated process. The contractor adds a new task Determine Mixing Quality in the subnet of the Design Compounding and adds the appropriate control and data flow relationships from Determine Process Parameters and to Investigate Extruder. While the contractor changes the task net, all changes to the task net are propagated to the subcontractor. Eventually, the contractor may either discard his changes by using the command Reset Changes (Changed→Accepted) or he may signal that the structural changes are finished by using the command Changes Finished (Changed→Evaluate).

In our example, the subcontractor evaluates and accepts the proposed changes of the delegated process. Triggering the command Accept Changes (Evaluate→Accepted) yields an update of both processes on the contractor side and the subcontractor side according to these changes. As an alternative, the subcontractor may reject the change of the contract by use of the command Reject Changes (Evaluate→Changed). In this case, the changes are discarded and the contractor would be informed about the rejection. Both partners then would have to talk about the problem again before eventually the subcontractor would accept a proposal made by the contractor.

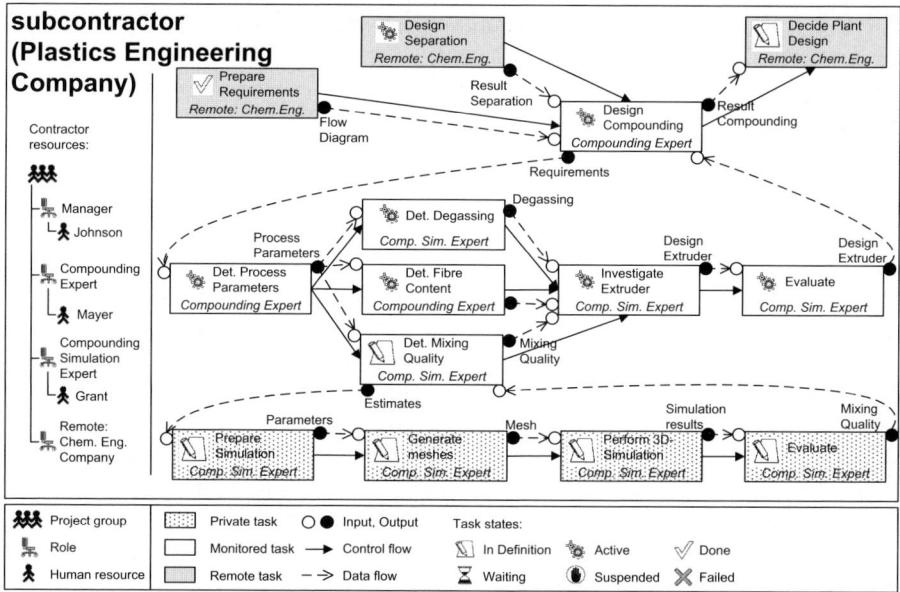

Fig. 3.86. Private refinements of the delegated process on the subcontractor side

Information Hiding Regarding Delegated Processes

The complex new task **Determine Mixing Quality** in the delegated process is refined by the manager on the subcontractor side by a private subnet to break it down into smaller working units and assign separate resources to each of the tasks. This refining task net comprises the tasks **Prepare Simulation**, **Generate Mesh**, **Perform 3D-Simulation** and **Evaluate** as shown in Fig. 3.86. The subnet is not part of the contract between contractor and subcontractor and can therefore be hidden from the contractor by means of private tasks.

Finishing the Delegation

After the process has been resumed, on the subcontractor side a second version of the extruder design has been finally produced and released to the task **Design Compounding**. This result should be taken as the final outcome of the delegated task. The subcontractor can signal this to the contractor with the command **Complete Delegation** stating that he wishes to complete the contracted delegation relationship. The contractor can confirm this with the command **Confirm** or reject it with the command **Reject**. If the result is accepted, the coupling of the two AHEAD systems of contractor and subcontractor is finished. In the other case, the rejection is signaled to the subcontractor and the coupling is maintained.

3.4.5 View-Based Interorganizational Cooperation

Motivation

In the previous subsection, a *delegation-based* relationship between cooperating organizations has been explained, where a contractor delegates a part of his process to a subcontractor organization. This model is now criticized. We want to *generalize* the *model* in order to support a broader spectrum of cooperation scenarios.

The previously discussed delegation-based relationship is restricted with respect to its flexibility and *adaptability* to different cooperation scenarios:

- The *visibility* of elements in the contractor process for the subcontractor can only be defined for *tasks* in the *direct neighborhood* of the delegated process. Thus, it is not possible to expose process parts without a delegation-relationship.
- AHEAD currently supports only a delegation-relationship for the cooperation between processes where both parties have *different* roles during the collaboration (namely contractor and subcontractor) implying different rights to define all cooperation aspects. However, other possible cooperation scenarios should also be possible. For example, the *same* rights and duties can be given to the partners of a *peer-to-peer* cooperation.
- Only *connected* parts of a process can be *delegated*. If multiple parts of a process are delegated, they are all regarded as independent new processes on the subcontractor's side. They cannot be composed into an overall process with a shared process context. Following this approach, the integration of pre-existing processes with each other is not possible.
- Cooperation can require *less formal* or *more formal* configurations regulating the procedures and mechanisms used by the organizations for defining, executing, and evaluating interorganizational processes. Therefore, flexible and configurable cooperative processes for interorganizational processes need to be supported by the AHEAD system. For example, not every delegation requires very strict and formal contracts about the agreements and procedures between the partners. Currently, the cooperation protocols for delegation within AHEAD are built-in and cannot be tailored to specific cooperation needs according to a higher or lower level of trust between the cooperating organizations.

Hence, we have identified two important *requirements* for flexible cooperation support in dynamic development processes: (1) An organization should be able to use powerful and flexible mechanisms for defining the *visibility* of process information shared with other organizations. (2) Interorganizational cooperation has to be supported insofar as the different processes of the cooperation partners can be *integrated* with each other not exclusively according to delegation-based relationships between them. A broad set of *customizable cooperation relationships* has to be supported instead.

Dynamic Process Views

Our approach to interorganizational cooperation in development processes builds on the definition of *dynamic process views* onto development processes as its foundation [175–177]. Dynamic process views support better visibility management for process elements carried out within an organization.

A dynamic process view is defined for a *process instance* (i.e. a dynamic task net) with its products and resources and it resembles a *subconfiguration* of the *process instance*. A process view constitutes a certain cut-out of its underlying process with products and resources which should be made visible to external parties.

A process view basically *contains* the following *elements*:

- A *view name* and a unique *view identifier*.
- A *subgraph* of a dynamic task net (partial abstraction): This subgraph represents a fragment of a dynamic task net which is structurally and behaviorally consistent with respect to the process meta-model of DYNA-MITE [243]. Zero or more tasks can be part of the process view and not all of a task's parameters need to be in the process view. Only a subset of the existing flow relationships between tasks needs to be represented in the process view.
- A *view product workspace* maintaining all view-related products and product versions which are contained in the underlying process and should be visible within the process view.
- A *view resource space* which contains all view-related resources, i.e., all abstract or concrete resources of the underlying process which should be made visible within the process view.
- *View definition rules*: A set of rules defines which elements of the underlying private process are also part of the process view. For instance, some specific model elements (i.e. tasks, products, resources, or flow relationships) can be assigned to the view, or all model elements of a specific type can be chosen instead.

The process view concept is *illustrated* by an *example* in Fig. 3.87. The top part of the figure shows a part of the Polyamide-6 process (used throughout the entire subsection) as it is seen from the perspective of the Chemical Company. In the middle part of Fig. 3.87, a process view definition named ReactionSimulationTasks for this process is shown which contains two simulation tasks Simulate CSTR and Simulate PFR with their input and output parameters from the overall process, while the control flow between both tasks is not included in the process view.

Process views can be used to provide *different perspectives* of the underlying private process. For instance, managers can use process views to gain overview with minimum technical process information. Technical experts can use process views containing all necessary process information with respect to

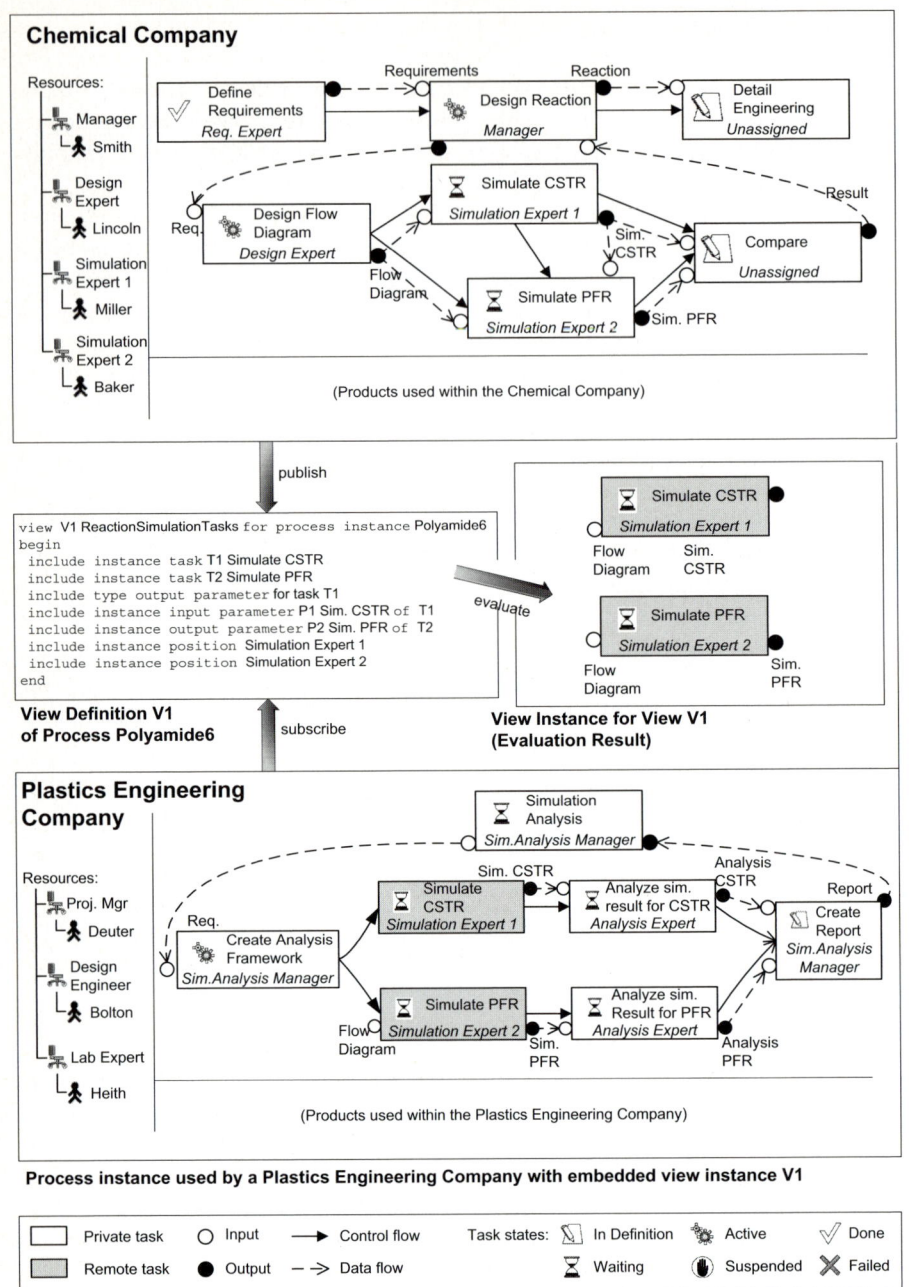

Fig. 3.87. Dynamic process view for the Polyamide-6 process

a specific information need. In our opinion, a view-based approach is a natural approach for managing the visibility of process elements to external parties.

Process view definitions are *published* by one organization (publisher) and they can subsequently be *subscribed* by other organizations (subscribers). The application of a view definition to a process results in a *view instance* containing all process elements which are visible according to the (automatic) evaluation of the view definition rules of the subscribed process view. The subgraph, view workspace, and view resource set of a process view exactly contain all process elements which are determined by the view definitions rule set.

Subsequently, the process elements of a view instance can be embedded into private process instances of the subscriber. Thereby, any new restrictions on the embedded elements, e.g. new incoming control flows, have to be negotiated between the subscriber and publisher before they can take effect.

Process view instances change, either when the underlying private process or when their corresponding view definition are modified. Therefore, view instances are *re-evaluated* whenever the underlying processes or the view definitions are changed, in order to update the contents of the process view. For example, the lower part of Fig. 3.87 shows the private process of the **Plastics Engineering Company** with the embedded view instance of **View 1**.

Private processes contain *local* as well as *remote tasks* embedded from other organizations within process views. The embedded view elements (here tasks **Simulate CSTR** and **Simulate PFR**) can be *interconnected* with tasks of the private process where the view is embedded by control flows, feedback flows or data flows to establish inter-process cooperation. This provides the basis for interorganizational cooperation as explained in the following paragraph.

View-Based Interorganizational Cooperation Model

We are now in the situation to introduce our *interorganizational cooperation model* which is based on dynamic process views [175–177]. The model is described according to *three* layers which are located on top of each other starting at the bottom of the layer stack (Fig. 3.88):

- The private processes are modeled on the *private process layer*, where the process manager of each organization defines, controls and monitors a task net instance reflecting the development process within the respective organization.
- Dynamic process views are located at the *process view layer* above. Parts of the overall process within each organization are made externally visible by the definition and publication of one or more process view definitions. These process view definitions are subscribed by other organizations, where the respective private processes are extended with the contents of the corresponding view instances. In our approach, the remote process view elements are directly embedded into the private task nets to allow

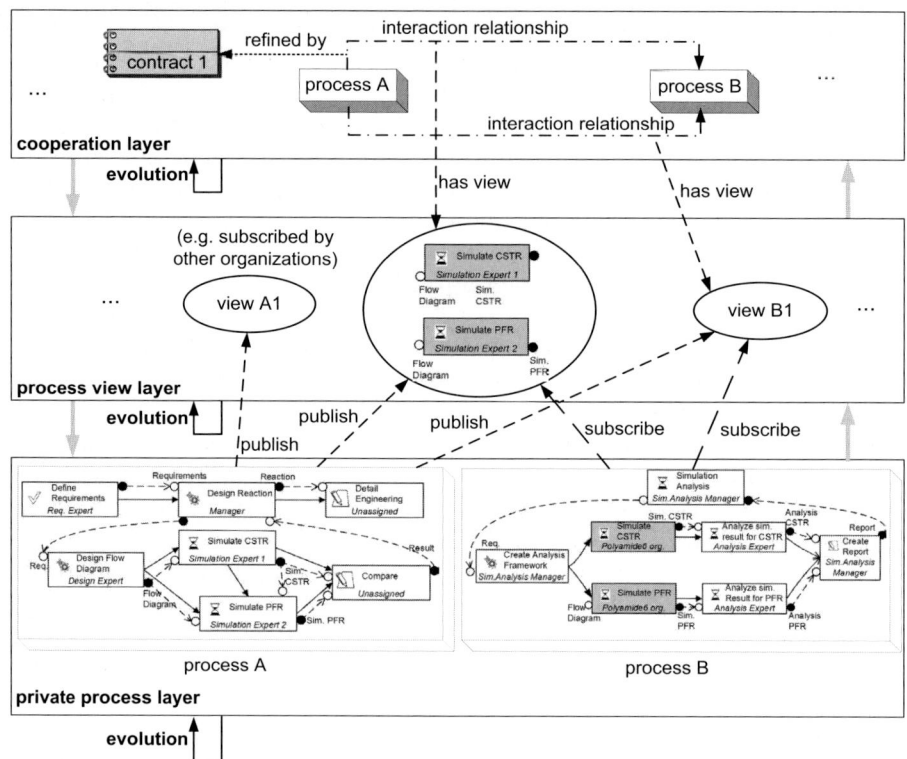

Fig. 3.88. Layers of the view-based interorganizational cooperation model

for a complete overview of all process elements together. Process views are used to enable inter-process coupling as described below.

- Details about the intended cooperation relationships between organizations are contained in the *cooperation layer* on top of the process view layer. Different kinds of *cooperation relationships*, e.g. outsourcing relationships, are introduced here (described later). Cooperation relationships model the interactions between project teams residing in different organizations and they prescribe how control and data can be transferred between the organizations.

Our approach to interorganizational view-based cooperation management comprises the following *cooperation phases*:

1. *Private task net planning.* Within each organization, the process manager plans its own private process instance.
2. *Process view publication.* The process manager creates process views to make certain cut-outs of his managed process instance externally visible. View definitions with view rules on the instance-level and the type-level

are created and subsequently published to other organizations. Selected tasks of a process view definition can be marked as *outsourced* in order to execute them within other organizations.
3. *Process view subscription.* The process manager of another organization subscribes the published process view definition to embed a corresponding view instance into his private process instance (dynamic task net).
4. *Process inter-connection.* By connecting private and remote process elements with each other, process instances are coupled across organizational borders.
5. *Cooperation policy definition.* The process manager can define the cooperation relationship on the cooperation layer. Additionally, he can assign selected process view definitions to the cooperation policy. Each relationship can be further refined with a contract, if needed.
6. *Process coupling.* The AHEAD systems of the cooperating organizations are coupled to exchange process update messages with each other. The different processes are executed locally in the organizations. All process views are updated upon changes of the underlying process instance. Whenever a process instance is modified locally, the respective AHEAD system computes all affected process views. Subsequently, it notifies all those remote AHEAD systems about the change, wherein the computed process views are embedded.
7. *Completion of process inter-connection.* The process manager of each organization decides autonomously when to terminate or cancel the process interconnection. Therefore, he marks a selected process view definition as completed. After that, the corresponding view instance is not updated any more and the process instances evolve independently from that moment on (although all embedded view elements remain in the private processes of all subscribers).

Layers and Components of the Cooperation Model in Detail

We now *describe* the different *layers* and components of our cooperation model in the following four subparagraphs in more detail.

Private Process Layer

Within the private process layer we allocate the process instances of each organization. Of course, all process aspects are visible within the organization. But due to a lack of trust, in most cases it is not suitable to expose private process details completely to other parties but only a certain fraction of the overall process. For this purpose, the process view concept is introduced.

Process View Layer

All process views are located on this layer above the private process layer. Process views are used to enable inter-process connection. The private processes can contain tasks which are executed locally as well as tasks which are embedded locally using subscribed process views from other organizations. A private process can contain local process elements as well as remote process elements. Therefore, process managers can oversee their local process together with all connections to process parts executed in other organizations within a single task net representation.

In order to achieve inter-process coupling, the process instances of the cooperating organizations can be connected by control flows, feedback flows, or data flows. We do not need an additional modeling language for modeling the coupling of processes. Instead, we re-use the known control flow, feedback flow, and data flow concepts of dynamic task nets. While other approaches favor to model intra-organizational and inter-organizational control and data flows differently, we aim at modeling both in the same way. From a manager's point of view, control flow is transferred between two tasks regardless if they both are locally executed or not. Of course, intra- and inter-organizational dependencies between tasks have to be handled differently, but they can be modeled the same way for ease of use. Intra-organizational and inter-organizational flows can be identified, because the source and target tasks of these flows are either both local tasks or not. We believe, that modeling intra-organizational as well as interorganizational cooperation in a uniform way is feasible and should be supported by a modeling approach that is simple to understand and to use by process managers.

Cooperation Layer

On the cooperation layer, we model *basic cooperation relationships* between processes. This model represents which connections to other organization's processes exist and how they relate to each other. We distinguish between monitoring relationships, interaction relationships, and outsourcing relationships.

A process view instance can be embedded into a private process of an organization in order to observe the progress of the process cut-out visible by that process view. Such *monitoring relationships* are the simplest form of cooperation because no direct inter-connection of process elements from different organizations is needed here. Using monitoring relationships helps process managers to oversee their own processes as well as interesting remote process parts in one uniform representation.

If local and remote process elements are connected by control flows, feedback flows, or data flows, we model *interaction relationships* between the coupled organizations on the cooperation layer. Interaction relationships resemble situations where control flow or data are transferred between processes. For example, local tasks can be restricted to start only after some remote tasks have terminated by inter-process control flows. This allows interweaving different processes in the sense that the processes are executed in parallel while they are loosely coupled at the same time.

We define *outsourcing relationships* to model cooperation in a customer-producer relation. In our approach, outsourcing means that an organization (termed as customer) can plan single process tasks or a task net fragment to be executed by another organization (termed as producer) within the process view definition. The outsourced tasks are then transferred to the other organization and regarded there as a local task in the future. The outsourcing organization will no longer be responsible for the outsourced tasks, because they are executed within the other organization. In an extreme scenario, cooperation can even happen without outsourcing (or delegation) at all. This represents a useful scenario when different organizations cooperate with each other with the goal to allow access to selected parts of the private processes while prohibiting any further process coupling. Interaction relationships and outsourcing relationships can complement each other and can exist between two organizations at the same time.

Contracts

The extent of trust is a key factor in cooperations and must be modelled appropriately. For instance, if an organization wants to delegate a process to another organization, *different cooperation relationships* are possible. If the contractor has not worked with the planned subcontractor before, a very strict and formal cooperation setting may be suitable. If the partners know each other well, or if they are engaged within a long-term relationship, it may be more appropriate to work together in a less formal relationship, without fixing all details fixed within a contract beforehand.

In our approach, *contracts* are used to tailor cooperation relationships to individual cooperation needs. For example, the object of discourse, the different partner roles, or other data are defined within the contract. Additionally, selected process views can be assigned to the contract if the task net structure of some process fragments shall be a part of the contract between the cooperation partners.

A broad spectrum of cooperation scenarios can be realized with different contract configurations. On the cooperation layer, all three basic cooperation relationships (monitoring, interaction, and outsourcing) between processes are orthogonal to contracts. They can optionally be refined by contracts. The basic idea is to implement a very light-weight default contract protocol for the interaction between partners and to provide contracts as a means to further *define* the *fine-grained structure of a cooperation* between partners if this is needed. Formerly, only one fixed contract between a contractor and a subcontractor was supported in the delegation-based management approach of AHEAD.

Sample Process for View-Based Cooperation

We now demonstrate the view-based approach to interorganizational cooperation. In the example described below, we focus on the part of the overall design process which is related to the design of the reaction and separation as well as the design of the extruder. The **Chemical Company** acts as a *customer*

346   M. Heller et al.

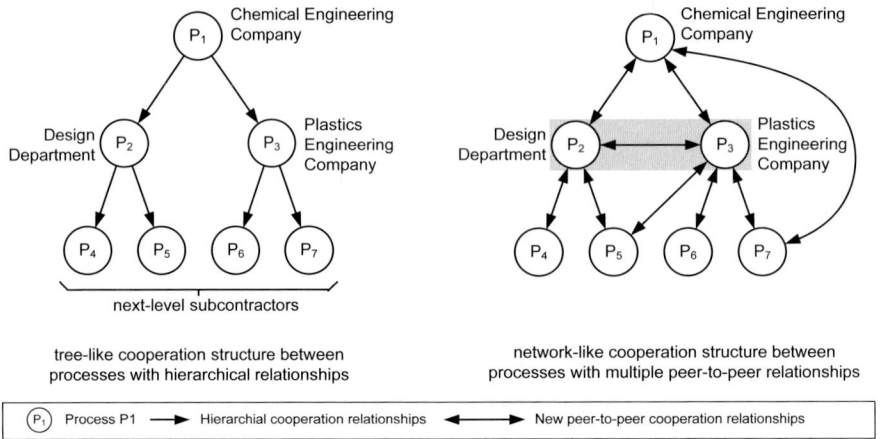

Fig. 3.89. Cooperation relationships in the example scenario

organization and *outsources* the task of *designing* the reaction and separation to another organization, the **Design Department**, having an own project manager who manages his own process, products, and resources independently. The task to design the *compounding* is outsourced to an **Plastics Engineering Company**.

The **Chemical Company** works together with its subcontractors, the **Design Department** and the **Plastics Engineering Company**, in outsourcing relationships. For the moment, we will deal with the situation after these outsourcing relationships have been established in order to show how a *direct cooperation* relationship between *both subcontractors* can be achieved with the process view concept. After that, we will explain how outsourcing relationships can be configured with process views.

This kind of direct cooperation between organizations resembles a *graph-like network cooperation* structure in a peer-to-peer mode which is not supported in the former delegation-based concept of AHEAD: Both subcontractors cannot cooperate with each other directly but only through their common contractor, the chemical company (shown in the left part of the Figure). In this way, only tree-like cooperation structures are possible. Although this delegation-based process decomposition approach is sufficient in many situations, often direct cooperation between all partners of a cooperative network of companies is needed as well.

Initial Situation

Fig. 3.90 shows the process part which has been delegated to the **Design Department** from the manager of the **Chemical Company**: Some tasks for the investigation of multiple reaction or separation alternatives will be carried out in the **Design Department**. The tasks **Define Reaction Alternatives** and **De-**

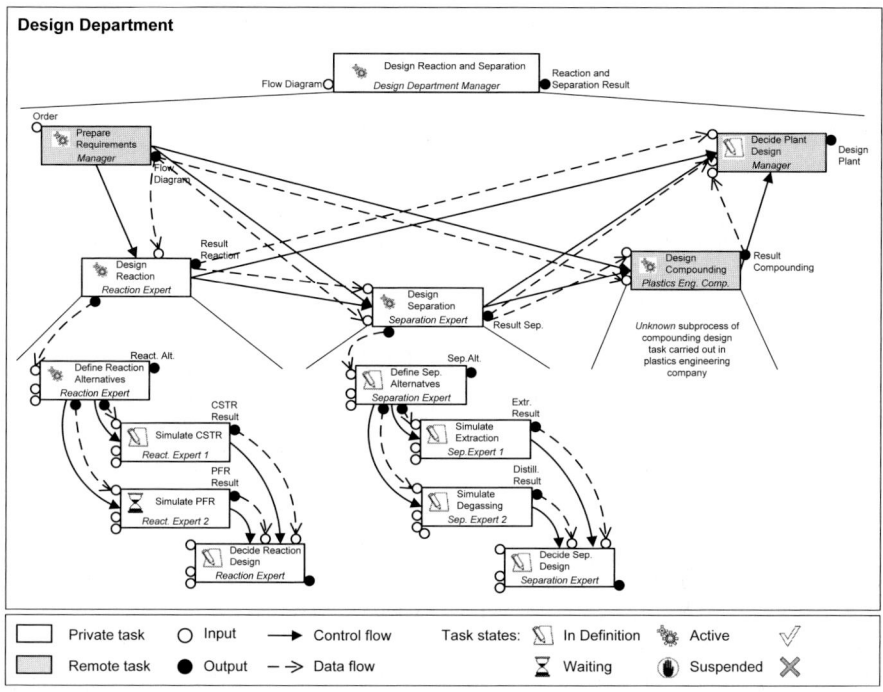

Fig. 3.90. Initial situation

fine Separation Alternatives have been refined, while four new tasks Simulate CSTR, Simulate PFR, Simulate Extraction, Simulate Degassing still need further refinement.

The subnet of the task Design Compounding contains several tasks (not shown in the figure): First, the expected output of the extruder will be roughly estimated in order to provide a starting point for the extruder design (task Determine Process Parameters). Second, separate tasks deal with the investigation of the fibers content and the degassing of volatile components of the plastics and all results will be used in the central task Investigate Extruder. Third, the extruder design is forwarded as a preliminary result to the parent task Design Compounding.

View Definition, Publication, and Subscription

The process manager of the Design Department can provide different views onto his private process (Fig. 3.91):

- A process view V1 provides information about the reaction part of the overall process. The tasks Define Reaction Alternatives and Simulate PFR for the simulation of the plug flow reactor is published for that purpose.

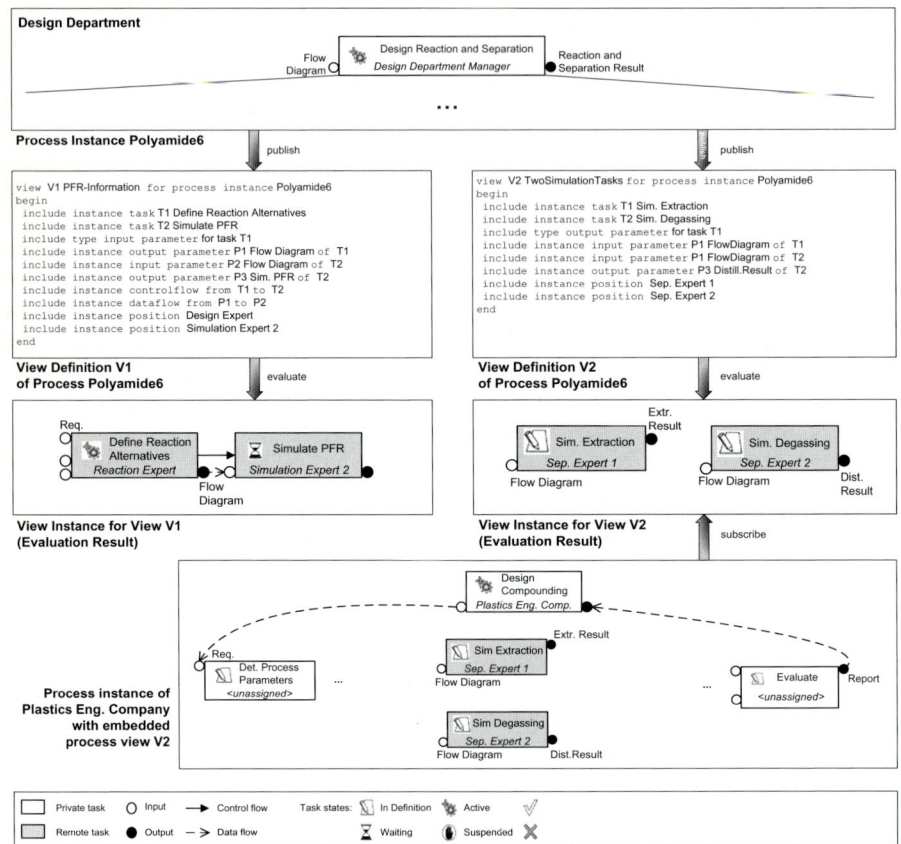

Fig. 3.91. Definition, publication and subscription of process views

- A process view V2 gives access to the simulation parts of the reaction design and contains the tasks Simulate Extraction and Simulate Degassing with some of their input or output parameters.

After both process views have been published, the manager of the Plastics Engineering Company can subscribe both process views and embed the corresponding process fragments into his own private process. This would result in the situation, that both processes are connected at two different locations (views V1 and V2) which can be planned and evolved independently. This demonstrates the advantage of our process view approach, where multiple cooperation contexts between both processes can be maintained simultaneously within logically separate process views. In the sequel, only the process view V2 is subscribed while the view V1 is neglected (lower part of Fig. 3.91).

Changes in the published process parts are transmitted particularly from the Design Department's AHEAD system to the Plastics Engineering Com-

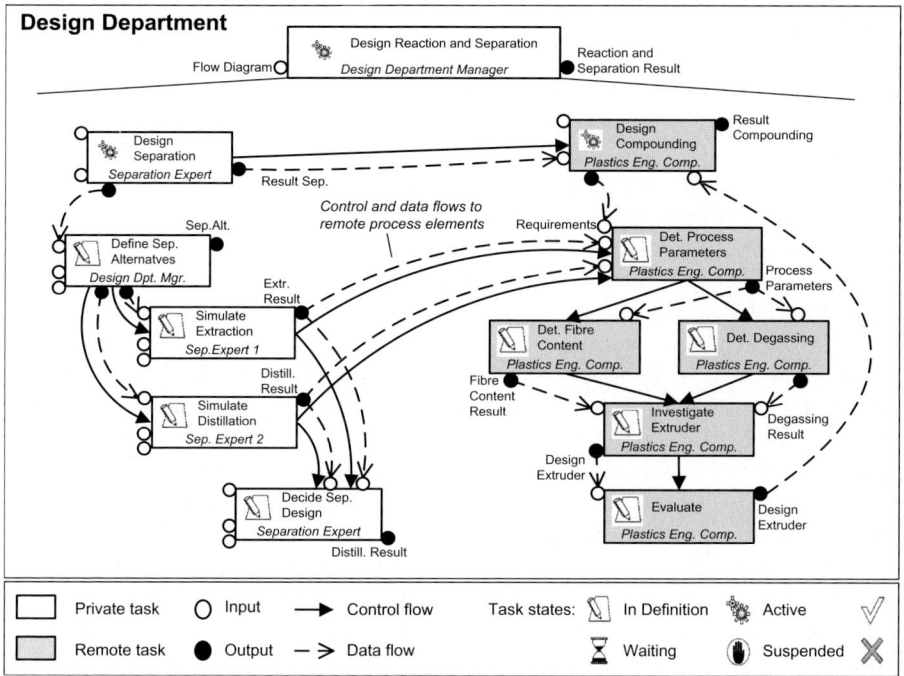

Fig. 3.92. Interconnections between Design Department process and Plastics Engineering process

pany's AHEAD system and displayed immediately. For example, the activation of task **Define Reaction Alternatives** on the side of the Design Department would be propagated to the AHEAD system of the Plastics Engineering Company through the process view V1.

Bottom-Up Process Composition

Both parallel processes in the Design Department and the Plastics Engineering Company can be inter-connected (Fig. 3.92). According to the plan of the process manager in the Design Department, the two simulation tasks Simulate Extraction and Simulate Distillation shall be synchronized with the task Determine Process Parameters of the other organization with respect to their execution states and documents shall be transferred between these tasks. This is an example of inter-organizational control and data flow.

The process manager of the Design Department creates new control flow and data flow dependencies between these tasks in his private process instance. It is important whether a flow dependency goes from a local task to a remote task or vice versa. *Locally relevant* inter-process dependencies between tasks (going out of a remote task into a local task) do not cause problems since they do not impose new behavioral restrictions on the remote tasks. But *remotely relevant* inter-process dependencies (going from a local task into a remote task) are problematic. In our example, both new control flows going into the remote task **Determine Process Parameters** are remotely relevant and the intended changes are only allowed if the manager of the **Plastics Engineering Company** agrees to them.

The process manager of the **Design Department** can either re-use an already existing process view definition or create a new process view definition. Here, he decides to re-use the process view definition **V2** and inserts all related tasks (**Simulate Extraction**, **Simulate Distillation**, and **Determine Process Parameters**) as well as all new control and data flows there. After that, he publishes the view (view definition evolution).

The manager of the **Plastics Engineering Company** subscribes the published view definition **V2** (if it is not already subscribed there). Then he inspects the changes in the view definition. If he accepts them, the changes become persistent in both systems. The **Design Department** manager could also make modifications to the changed task net fragment under discussion. He can even choose to discard the modifications if no consensus can be reached.

After the changes have been carried out in both management systems, the managed process instances remain coupled with each other. Both process instances evolve autonomously and they are only loosely coupled with each other through the two newly inserted control and data flows between elements of both processes.

Top-Down Process Decomposition with Outsourcing

We now demonstrate the outsourcing of a task from a customer organization for execution within a producer organization. The manager of the **Design Department** requests the **Plastics Engineering Company** to investigate the different alternatives for separation as soon as possible. In this way, possible design flaws within the separation alternatives or their interplay with other design details can be detected very early. This helps to reduce the risk of far-reaching process feedbacks in later project phases due to closer communication between the partners in the beginning. Then, the manager of the **Design Department** creates a new task **Investigate Distillation** and outsources it to the **Plastics Engineering Company** (Fig. 3.93).

In this situation, he refrains from re-using an existing process view and creates a new process view **V3a** instead with the new task and its control and data flows to the tasks **Define Separation Alternatives** and **Decide Separation Design**. He marks the task **Investigate Distillation** as outsourced in the view definition. After that, he calls a command to add a minimal *context* of the outsourced task in order to maintain consistency with the surrounding task net. In the example, the context comprises the predecessor task **Defines Separation Alternatives** and the parameter **Flow Diagram** as well as the control and data flows to task **Investigate Distillation**. The process view **V3a** is published by the manager of the **Design Department**.

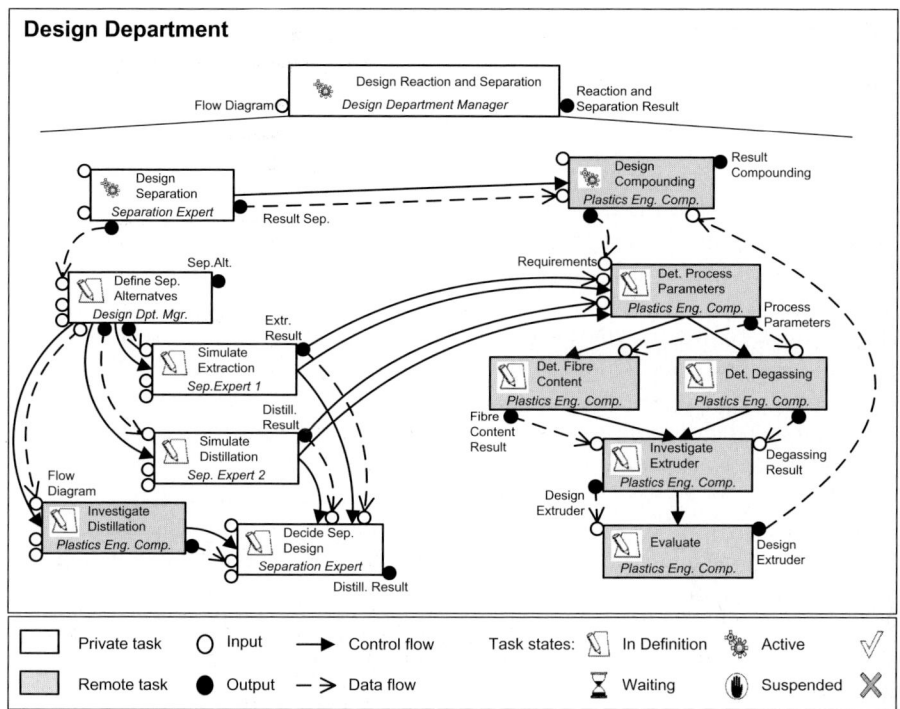

Fig. 3.93. Outsourcing of a task net fragment to external organization

Upon publication, the marking **outsourced** is detected and the system uses a special *outsourcing procedure* on both sides in the sequel. When the manager of the **Plastics Engineering Company** subscribes the process view **V3a**, he is asked to accept the process view as usual. When he accepts the view, he also accepts the announced task outsourcing therein. Then a new task instance **Investigate Distillation** is created in the private process of **Plastics Engineering Company**. This task is private upon creation and has to be published within a process view definition. Therefore, a new process view **V3b** (*back view*) is created and filled with the outsourced task and its context, where **V3a** and **V3b** are structurally the same, but the role of local tasks and remote tasks is reversed in the view. This new process view **V3b** is published to the **Design Department**. The manager there is also asked to accept this announced view. If he accepts also, then both managers have accepted the new cooperation situation and the outsourcing relationship between both processes is fully established. The outsourced task **Investigate Distillation** is executed by the **Plastics Engineering Company**. The other context tasks are not transferred between organizations so that they are executed by the **Design Department** as before. In this way, the delegation model presented in the previous section is simulated with the view model.

Integration of Workflow Processes in the Design Process

The view-based cooperation model presented so far addresses the interorganizational integration of design processes carried out in several organizations. In AHEAD, design processes are represented by *dynamic task nets*, which may evolve continuously throughout the execution of a design process. We now extend the cooperation model with an approach for the intraorganizational integration of processes executed within heterogeneous process management systems, for example workflow management systems.

Although the overall design process cannot be planned fully in advance and thus cannot be executed completely within workflow management systems, this may be possible for some fragments of the overall design process (e.g. the design of an apparatus may be predictable). If the structure of such *static fragments* of the design processes is well-defined and most of the needed planning information is available, then these fragments can be specified in advance on a fine-grained level as a workflow process. Although workflow management systems have originally been designed to support repetitive business processes (e.g., in banks or insurance companies), the use of workflow management systems for design process support is investigated in other research projects, as well (e.g., in [832]). In contrast to workflow management systems, AHEAD supports the seamless interleaving of planning and execution – a crucial requirement which workflow management systems usually do not meet [475].

We have developed an approach to integrate workflow processes into the overall design process and have realized a coupling of workflow management systems with AHEAD for use within an organization. Our approach is characterized by the following properties:

- Within an organization, *AHEAD* serves as the central instance for the planning of the overall process, e.g. it is used for its *global coordination*. The composition of process fragments into a coherent overall process is realized using dynamic task nets, so that the dynamic character of the design process is adequately supported by AHEAD.
- *Predefined parts* of the overall process are executed in *workflow management systems*. Existing process definitions can be reused (a-posteriori integration). This approach addresses the observation, that often in the beginning of a design process only part of it are understood well enough to support them using a workflow management system.
- Through a *view-based integration*, partial processes running in workflow management system can be represented in AHEAD as dynamic task net fragments within the overall design process. Thus the manager can monitor all parts of the process in AHEAD using only one adequate process representation regardless if or how they are executed by other management systems.
- In order to reduce the effort for the integration of multiple workflow management systems, we make use of the *neutral exchange format XDPL* from the Workflow Management Coalition [1059]. The transformation between

processes described in XPDL format to dynamic task nets does not need to preserve the full semantics of both formalisms, because this would lead to very rigid requirements for the systems to be integrated. Moreover, this is not necessary, since the workflow fragments are used in AHEAD for monitoring purposes only, and therefore it seems tolerable if some process information is lost during the generation of the workflow fragments.

There are several alternatives for the mapping of predefined partial workflow processes into dynamic task nets. On the one hand, the whole workflow process can be represented as a single task in the dynamic task net. Its activation reflects the start of the workflow instance within the workflow management system, while the termination of this task reflects the termination of the workflow instance. In this case, the fine-grained activity structure of the workflow process is hidden (*black-box approach*). This simple form of integration is sufficient in many situations, but because of the encapsulation it is impossible to monitor the progress of the activities in the workflow process within AHEAD.

On the other hand, all details of the workflow process can be mapped to a task net (*white-box approach*). This alternative suffers from the following disadvantages. First, this extreme form of transparency is often not feasible, if some details of the workflow should be hidden because of confidentiality reasons. Second, the language of dynamic task nets has to be capable of expressing all aspects and peculiarities of the modeling language used for the definition of the workflow process. Because the mapping is carried out to allow for the monitoring of these processes in AHEAD, we can afford to map only a filtered portion of all details of the workflow process, e.g. control structures. Third, both modeling languages are used on different levels and for different purposes. Workflow definition languages target at the automatic execution of the described workflows within the workflow management system. This requires describing a lot of necessary technical details on a very low abstraction level. In contrast, dynamic task nets describe processes with respect to the coordination of their tasks on a very high abstraction level.

Our mapping approach is in the middle of these two mapping alternatives. Only selected details of the workflow process, which are necessary to represent the coordination aspects of the activities in the workflow, are mapped into a dynamic task net (*gray-box approach*). For example, such workflow fragments do not contain workflow relevant process variables which are only needed internally by the workflow engine to automatically decide which activities to execute next upon the termination of workflow activity.

To illustrate our approach, we revisit our scenario on the design process of a plant for Polyamide-6 (PA6) carried out within a chemical company, which is used throughout this paper. There we can easily identify a static process fragment in the plastics engineering part of the process as a good example, namely the determination of the mixing quality within the extruder through a complex and expensive 3D-simulation (see Fig. 3.86). Because this process fragment is small, static and well-understood, it is feasible to model it as a

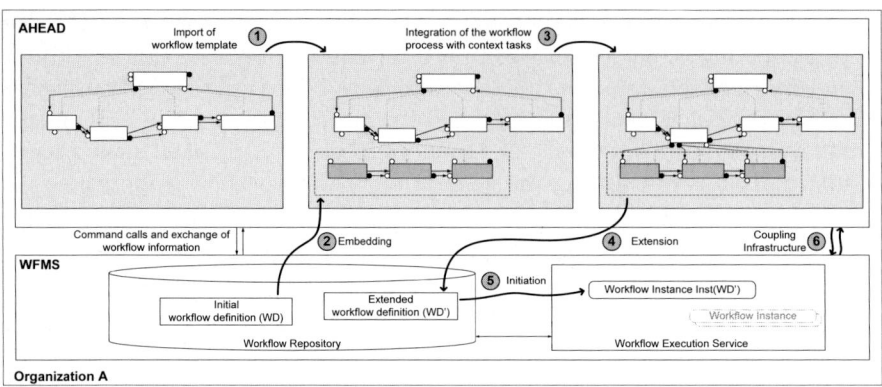

Fig. 3.94. Integration of workflow processes in AHEAD

workflow process and support its execution within a workflow management system.

The AHEAD system and the workflow management system SHARK from ENHYDRA [660] have been integrated with each other to support this scenario. The manager of the chemical company uses AHEAD to manage the overall design process, and a dedicated team of simulation experts is responsible for performing 3D-simulations. Because all 3D-simulations have to follow a best-of-breed practice, developed once within the company, a workflow process has been defined to enforce this simulation procedure. As illustrated in Fig. 3.94, we describe the different phases of workflow processing using an example session:

- *Workflow Embedding.* The manager decides within the compounding part of the process that a 3D-simulation is needed in order to analyze quality-affecting problems of the current parameterization of the extrusion process. He opens a browser displaying all available workflow processes, chooses the 3D-simulation workflow and imports its dynamic task net representation (called *workflow fragment* or *workflow template* below) into the compounding subnet (1). Then the workflow fragment is embedded within the subnet of the compounding task where other compounding subtasks also reside (alternatively it is possible to embed a new single task first and embed the workflow fragment as its subnet) (2). All formal parameters of the workflow regarding the input and output to it are located at the tasks which represent workflow activities processing these parameters. These are the first and last tasks in the workflow fragment.
- *Workflow Context Definition.* The manager connects the isolated workflow fragment with other tasks by adding control flows and data flows to at least the first and last tasks of the workflow fragment (3). After that, he sets the execution state of the workflow tasks to **Waiting** and subse-

quently all workflow tasks are set to this state, too. An extended workflow definition with additional workflow process data for the workflow context is generated (4). After that, the context definition phase is finished within AHEAD.
- *Workflow Instantiation.* AHEAD automatically contacts the workflow management system and requests the creation of an instance of the corresponding workflow process definition. A new instance of the workflow definition is created and a reference to the instance is handed back to AHEAD (5). After all defined input data is transferred and provided as actual parameters to the workflow instance, the workflow is finally started.
- *Workflow Monitoring.* Workflow activities are assigned to members of the 3D-simulation team. They can accept and start assigned activities, read and update workflow relevant data and finally commit workflow activities. The workflow management system automatically routes the control and data flow to the next workflow activities. All process changes which are relevant for the monitoring within AHEAD, like status changes or document processing, are forwarded to AHEAD via an event-based coupling infrastructure (6). The manager can thus monitor the progress of the workflow process within AHEAD.
- *Process Traceability.* After the termination of the last workflow activity, the workflow instance is terminated in the workflow management system automatically. In AHEAD, the workflow fragment is still visible with all terminated workflow tasks of this fragment. All documents produced during the course of the workflow remain accessible in AHEAD. This allows for traceability of the overall process regardless if a part has been executed in AHEAD or in a workflow management system.

With this approach workflow processes can be embedded within the overall dynamic process. The process manager can monitor and control the execution of workflow fragments within the AHEAD system. The coupling of workflow management systems and the AHEAD system is achieved by an event-based coupling infrastructure. Both systems generate events about relevant process changes and forward them to the coupled system via the coupling infrastructure [471].

Features of the View-Based Cooperation Model

The new view-based cooperation model can be *characterized* and summarized by the following *features*:
- *Dynamic process view model.* Managers can use dynamic process views to configure the access rights to selected parts of the private process by external organizations. Process views are highly flexible and support the provision of different perspectives onto a process fragment according to

individual cooperation needs. With the concept of process views, each process manager can manage autonomously which process parts shall remain private and which process parts shall be published to other organizations.

- *Process view evolution on definition-level and instance-level.* Planning and enactment of dynamic task nets may be interleaved seamlessly so that the private processes are constantly changed (process evolution). Consequently, our process view concept allows for the dynamic evolution of the process views as well. Process view instances are always kept consistent with their underlying process by incrementally updating the view contents according to the process view definition upon process changes.
- *Uniform modeling of processes and process inter-connection.* Intra- and interorganizational processes are uniformly modeled by re-using elements of dynamic task nets (e.g. tasks, parameters, control flows, data flows, and feedback flows). In this way, process managers do not need to use different modeling languages for the modeling of intra- and interorganizational process fragments.
- *Contract-based support for different cooperation scenarios.* The concept of process views allows to support different cooperation relationships between organizations. For instance, monitoring relationships, interaction relationships, or outsourcing relationships across organizations can be configured within the same process management system. Contracts can be established to fix all necessary agreements between the partners, like the different organizational roles with rights and duties, the involved process views, as well as different kinds of cooperation policies for changes of the contract or related process view definitions. Additional parameters can be stored in contracts as well (e.g. cost or time schedules).
- *Conformance monitoring and inconsistency toleration.* Another important feature of our approach (not presented here) is the monitoring and control of the inter-organizational cooperation. Upon modification, each process view is checked by the management system for conformance with the process meta-model of dynamic task nets. Detected violations of structural and behavioral constraints are reported to the process managers. They may either modify the process views in order to re-establish consistency or tolerate the violation.
- *Integration of workflow processes.* Workflow processes can be embedded into the overall dynamic task net in order to monitor and control their execution from within the AHEAD system. To achieve the desired integration, workflow processes are mapped to dynamic task nets and the resulting workflow fragments are subsequently integrated with the dynamic parts of the process. Therefore, all aspects of the design process within all of its static or dynamic parts are represented in a unique process modeling formalism. On the technical level, workflow management systems are coupled with the AHEAD-System using an event-based coupling infrastructure. At process runtime, both management systems exchange events to keep each other informed about relevant process changes.

System Architecture for Interorganizational Cooperation Support

Both the delegation-based and the view-based cooperation model are realized on the basis of an *event-based coupling mechanism* [208]. The *graph-based realization* of the coupling concept is described in Fig. 3.95. Two AHEAD systems are coupled together using a communication server.

Let us first concentrate on the AHEAD system on the left-hand side. Each AHEAD *system* consists of a graphical user *interface*, the AHEAD core (containing the application logic library and the UPGRADE framework) and the underlying graph database. The task net shown in the graphical user interface is created step by step by invoking special user interface *commands*, for example, to insert a new task or a new control flow relationship between two tasks. Each user interface command calls a graph *transaction* of the application logic in the AHEAD core. The execution of a graph transaction leads to the *manipulation* of the *graph data* stored in the graph database. In the example, at the graphical user interface a task T1 is displayed. Invoking a user interface command to activate task T1 leads to a *change* of one of the *attributes* of the corresponding graph node in the database. The database propagates all changes on the graph data back to the AHEAD core. According to these *change events* the current state of the graphical user interface is updated.

If one of the AHEAD systems is temporarily disconnected, the communication server stores the events for subsequent delivery. In the coupled system, corresponding graph *transactions* in the AHEAD core are called for each of these *change events*. Accordingly, the graph data stored in the graph database is manipulated and the graphical user interface is updated. Therefore, changes regarding the monitored task T1 are also displayed in the GUI on the right hand side. Every AHEAD system can at the same time act as a producer of change events regarding all elements which are monitored in coupled systems and as a consumer of change events regarding all elements which are executed elsewhere and only monitored locally.

The realization of the view-based cooperation model has required a number of *extensions* to this *coupling mechanism* with respect to the coupling of AHEAD systems and workflow management systems [129, 471]. Mainly, the application logic of AHEAD was substantively changed and extended in order to realize the new view-based concepts for process views, cooperation relationships, flexible configuration support, as well as the needed user interfaces for a view editor environment.

3.4.6 Related Work

AHEAD Core System

In the following, we will discuss the state of the art of *tool support* for managing design processes. From the previous discussion, we derive a set of crucial requirements for management tools for design processes:

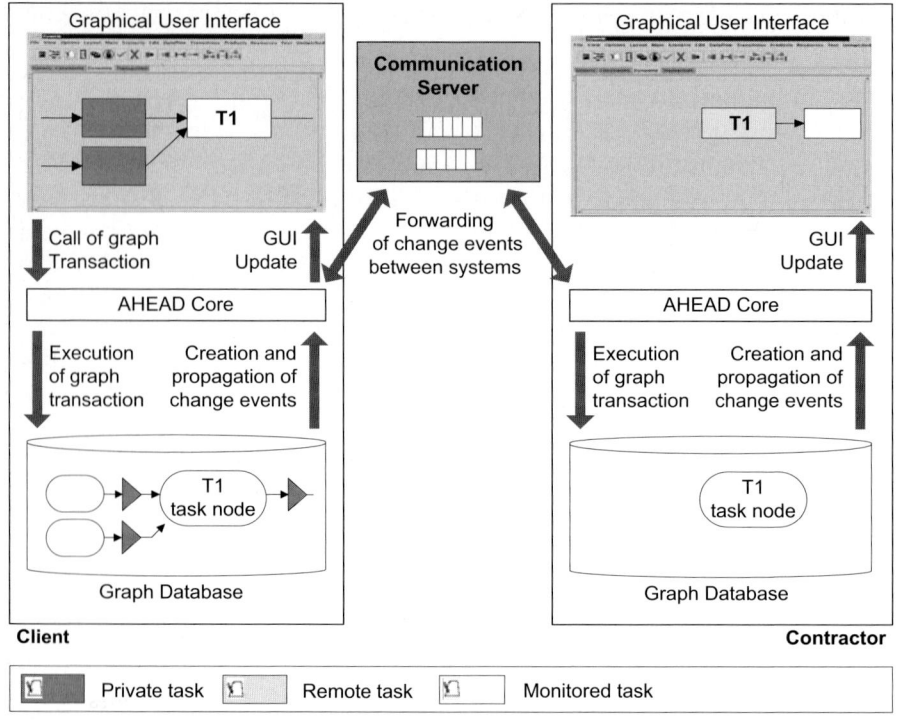

Fig. 3.95. Realization of the coupling of two AHEAD systems

- *Medium-grained representation.* The management of design processes has to be supported at an appropriate level of detail.
- *Coverage and integration at the managerial level.* Management tools have to deal equally with products, activities, resources and their relations.
- *Integration between managerial and technical level.* Managerial activities have to be coupled with technical activities: Designers have to be supplied with the documents they to be manipulated, as well as with the corresponding tools.
- *Dynamics of design processes.* Design processes evolve continuously during execution (product evolution, feedback, simultaneous/concurrent engineering).
- *Adaptability.* Management tools have to be adapted to a specific application domain and they must provide domain-specific operations to their users.

AHEAD meets all of these requirements. In industry, a variety of *commercial systems* is being used for the management of design processes, including systems for project management, workflow management, and product management, see below. All of these systems only partially *meet* the *requirements* stated above (Table 3.3):

Table 3.3. Comparison of AHEAD with commercial management systems

	AHEAD	project management systems	workflow management systems	product management systems
granularity of representation	medium-grained	coarse-grained	medium- and fine-grained	medium-grained
coverage at the managerial level	products, activities, resources	activities, resources	activities, resources	products, (activities)
integration with technical level	tool integration, document storage	not supported	tool integration	document storage
support for dynamic design processes	full support of process evolution	evolving project plans	limited (fixed workflows)	version control for documents
adaptability	UML models	not supported	workflow definitions	database schema

Project management systems [777] such as e.g. Microsoft Project support management functions such as planning, organizing, monitoring, and controlling. The project plan acts as the central document which may be represented in different ways, e.g., as a PERT or GANTT chart. It defines the milestones to be accomplished and provides the foundation for scheduling of resource utilization as well as for cost estimation and control. Project management systems are widely used in practice, but they still suffer from several limitations: project plans are often too coarse-grained, products (documents) are not considered, project plans are not integrated with the actual work performed by engineers, and there is no way to define domain-specific types of project plans.

Workflow management systems [763, 803], e.g., Staffware, FlowMark, or COSA, have been applied in banks, insurance companies, administrations, etc. A workflow management system manages the flow of work between participants, according to a defined procedure consisting of a number of tasks [836]. It coordinates user and system participants to achieve defined objectives by set deadlines. To this end, tasks and documents are passed from participant to participant in a correct order. Moreover, a workflow management system may offer an interface to invoke a tool on a document either interactively or automatically. Their most important restriction is limited support for the dynamics of design processes. Many workflow management systems assume a statically defined workflow that cannot be changed during execution. This way, dynamic design processes can be supported only to a limited extent (i.e., the statically known fractions can be handled by the workflow management system). Recently, this problem has been addressed in a few university prototypes (see e.g. [628, 688]).

In the context of this paper, the term *product management system* refers to all kinds of systems for storing, manipulating, and retrieving the results of design processes. Depending on the context in which they are employed, they are called engineering data management systems (EDM), product data management systems (PDM [722]), software configuration management systems (SCM [1000, 1049]), or document management systems. Documentum and Matrix One are examples of such systems which are used in chemical engineering. Documents such as flowsheet, steady-state and dynamic simulation models, cost estimations, etc. are stored in a database which records the evolution of documents (i.e., their versions) and aggregates them into configurations. In addition, product management systems may offer simple support for the management of activities (e.g., change request processes based on finite state machines), or they may include workflow components, which suffer from the restrictions already discussed above. Their primary focus still lies on the management of products; in particular, management of human resources is hardly considered.

The approaches cited above do not depend on a certain application domain. For instance, workflow management systems can be applied to business processes in different disciplines, and product management systems can be used in different engineering disciplines. Only a few approaches target the domain of chemical engineering directly. For example, KBDS [524] allows to manage different design alternative together with the change history; n-dim [1047] supports distributed and collaborative computer-aided process engineering. But these approaches do not really support the integrated management of processes, products, and resources. Moreover, they are restricted to their single application domain and cannot be used in different domains.

Process Evolution and Parametrization

The need for a *wide spectrum approach* to process management was recognized as a research challenge in [963]. It is explicitly addressed in GroupProcess [742], a project that has been launched recently, but does not seem to have produced technical results yet. In addition, this matter is addressed in some workflow management systems which originally focused on highly structured processes. For example, in Mobile [727] and FLOW.NET [773] the process modeler may define the control flow as restrictively as desired and may even introduce new control flow types. In addition, many commercial systems allow for deviations such as skipping, redoing or delegation of activities. Finally, exception handling [712] may be used to deal with errors and special cases. However, the main focus still lies on highly or moderately structured processes. In contrast, our approach covers the whole spectrum, including also ad hoc processes.

There are only a few other approaches to process management which are capable of dealing with *inconsistencies.* [616] and [861] both deal with inconsistencies between process definitions and process instances. In PROSYT [616], users may deviate from the process definition by enforcing operations violating preconditions and state invariants. However, all of these approaches do not deal with definition-level evolution, i.e., it is not addressed how inconsistencies can be resolved by migrating to an improved definition.

A key and unique feature of our approach consists in its support for *round-trip process evolution*. To realize this approach, we have to work both bottom-up and top-down: we learn from actual performance (bottom-up) and propagate changes to process definitions top-down. In contrast, most other approaches are confined to top-down evolution. For example, in [727, 764, 772, 792], the process definition has to be created beforehand, while we allow for executing partially known process definitions.

Modifications to process definitions may be performed in place, as in [598, 1044]. However, it seems more appropriate to create a new *version* of the definition in order to provide for traceability. Version control is applied at different levels of granularity such as class versioning [772, 792] and schema versioning [584]. Our approach is similar to class versioning (interface and realization packages for individual task types are submitted to version control).

Different *migration strategies* may be applied in order to propagate changes at the definition level to the instance level. A fairly comprehensive discussion of such strategies is given in [584]. We believe that the underlying base mechanisms must be as flexible as possible. For example, in [727, 772, 792], both structural and behavioral consistency must be maintained during migration. This is not required in our approach, which even tolerates persistent inconsistencies.

Finally, there are a few approaches which are confined to *instance-level evolution* (e.g., [526, 929]). A specific process instance is modified, taking the current execution state into account. However, there is no way to constrain the evolution (apart from constraints which are built into the underlying process meta model). In contrast, in AHEAD instances are evolved under the control of the process definition. Inconsistencies can be permitted selectively, if required.

Interorganizational Coordination

A lot of workflow management systems deal with *distributed processes*. However, a distributed process need not be interorganizational as addressed in this paper. The term "interorganizational" refers to cooperation between different enterprises, while the term "distributed" can be used to describe processes where tasks are distributed either within a single enterprise or across enterprises. For instance, the workflow management system *Mentor* [1055] supports distributed processes by providing multiple workflow servers. In this approach, work is distributed within a single enterprise among workflow servers, according to a sophisticated load balancing algorithm.

[1012] provides an overview of paradigms for *interorganizational processes*. Among others, the following paradigms are identified:

- *Process chaining.* From some process p, a process q is launched to continue the overall process. The only interaction between p and q occurs when q is started. Subsequently, p and q perform independently of each other.
- *Subcontracting.* A task t of the overall workflow is passed to a subcontractor, which executes t and passes the results back to the contractor. From the perspective of the contractor, t appears to be atomic. The contractor and the subcontractors interact both at the start and at the end of the execution of the subcontracted process.
- *Loosely coupled processes.* Processes are executed in parallel in different organizations. Occasionally, they interact at pre-defined communication and synchronization points.
- *Case transfer.* The workflow is seen as a case which has to be transferred among different organizations. Transferring the case includes transfer of documents and transfer of the current state of execution. Only one organization at a time may execute the case.

Some of these *aspects* are investigated in *literature*: The work of [1012] primarily focuses on case transfer and an extended variant thereof. In [1013], the same author discusses loosely coupled processes. The interaction paradigms process chaining and subcontracting are supported by the standards defined by the Workflow Management Coalition (WfMC [803]). In addition, subcontracting was introduced as early as 1987 by the Istar system [641] into the software engineering domain.

The *delegation* model of the AHEAD system adds a new paradigm to the classification scheme presented above. It differs from process chaining inasmuch as the contractor and the subcontractors do interact while the delegated subprocess is being executed. The delegation-based approach also differs from the case transfer model because both parties perform their parts of the overall process in parallel: The contractor is not suspended when a subprocess is delegated to a subcontractor. Delegation constitutes a significant extension of subcontracting because subprocesses rather than single tasks may be delegated in general. Like loosely coupled processes, contractor and subcontractor may interact during the execution of the delegated subprocess rather than merely at the start and the end, respectively. Delegation differs from loosely coupled processes because there is a hierarchical relationship between contractor and subcontractor (while loosely coupled processes are peer to peer in general). Finally, the delegation-based approach supports dynamic changes, while loosely coupled processes have been introduced for statically defined workflows.

Besides this work, we have extended AHEAD to provide additional support for the paradigm of *loosely coupled* process integration mentioned above and we introduced a new cooperation layer above the execution-oriented process view and private process layers. In the following, we restrict ourselves to highlighting related work addressing similar view-based approaches to support interorganizational processes.

Some other researchers like Finkelstein [669] use the concept of a view in different way than we do. These approaches focus on the consistent integration of these views in order to maintain a consistent and up-to-date representation of the whole development process by superimposition of all views. While these approaches focus on the problems of view-based process definition that arise with modifiable views, we use views which usually are not modified by anyone else than the view publisher, so we do not face problems of consistent integration to that extent. Because we do not use different modeling formalisms for all process views (we always use dynamic task nets in all process views), we do not face the problem that two views onto the same process part model different aspects of it in a conflicting way.

In our application domain of development processes in chemical engineering, we put more focus on the processes at the instance level rather than on the definition level when interorganizational cooperation is concerned. Using a view-based approach to process coupling, the views published for a process instance and the process instance can easily become inconsistent upon modifications because the processes evolve with the time. Process views are directly embedded into the private processes of other organizations (no integration process are used), where remote elements and local elements are connected with control flows, feedback flows, or data flows.

Several approaches target the modeling of the integration aspects between separate processes. To model the interconnection of existing workflow processes, the used workflow modeling language can be extended with additional modeling elements. For example, new modeling elements can be introduced to express the publication and interception of events which are exchanged between workflows processes of different organizations (like the approach described by Casati and Discenza [585]). Alternatively, explicit *synchronization points* can be modeled, as proposed by Perrin et al. [905]. In this case, the existing workflow modeling language is not extended and a separate modeling language is introduced. This approach allows to replace one of the two used modeling languages by another modeling language without affecting the other modeling language.

Van der Aalst [1011] focuses on independently running but loosely coupled interorganizational workflow processes, modeled in a language based on Petri-Nets. This approach is based on a predefined communication structure between the private partner processes which cannot be changed during runtime. Another approach is to split a workflow into several workflow fragments which are executed by the cooperation partners afterwards. Here, definition-time and run-time are strictly separated. In these two approaches, a top-down approach is used which is feasible if the overall process structure is known in advance. In our application domain of dynamic development processes, this is not feasible, since development processes cannot be planned fully in advance. New integration points between already existing partial processes of the partners should be creatable and modifiable whenever needed. So, a mixed top-down and bottom-up approach is more feasible. But at the same time it is important to ensure that all partial processes can be managed autonomously by the process managers of the cooperating organizations.

Three important view-based approaches of interorganizational workflows have been proposed by Liu and Shen [821], Chiu et al. [595], and Tata, Chebbi et al. [590, 993]. All three concepts provide support for routine business processes and they separate definition-time from run-time. Workflow definitions can be re-used as view definitions to model the public workflow parts which are accessible by other organizations. These view definitions cannot be changed after the overall workflow has been started. The workflow definitions usually do not need to be modified frequently, because the modeled business processes are not changed too often. For example, Liu and Shen use additional process definitions ("integration processes"), which contain the coupling of private workflow definitions and foreign view workflow definitions. This eases rapid composition of business processes from pre-existing processes as further goal of these approaches. In contrast, in our view-based approach the process views represent processes at the instance-level (not on the definition-level). Process views are directly embedded into private processes of other organizations (no integration processes are necessary). Furthermore, the other mentioned approaches do not focus on the interleaved definition and execution of process and views.

The view-based cooperation model in AHEAD also has related work in the research field of communication-oriented interorganizational cooperation. For example, Weigand and de Moor [1040] work on workflow modeling that considers both *customer relations* and *agency relations* to chart complex organizational communication situations. Here, "agency" means that a relation between a principal and some agents exists where both roles have different rights and duties. An agent acts for the benefit of someone, the beneficiary, and at the same time conducts an operation on behalf of someone else, the principal. The authors propose a modeling method with the following steps: (1) the process is defined and all process tasks can be decomposed into subtasks, (2) selected tasks can be delegated to intra-organizational resources for execution (introducing new *agency relations*), and (3) selected tasks can be outsourced to other organizations (introducing new *customer relations*). The authors present an extended workflow loop model to separate between the *workflow execution task* and the *control task*. This extended model is used for modeling both the agency and customer relations. In AHEAD, we deal with all three mentioned aspects of decomposition and composition of processes as well as intra-organizational and interorganizational cooperation relationships. Our new cooperation layer introduces three different cooperation relationships (monitoring, interaction, and outsourcing) as well as contracts for defining the formal guidelines structuring the cooperation.

3.4.7 Conclusion

In this section, we argued that design processes in chemical engineering are hard to support because they are highly creative, many design alternatives are explored, and both unexpected and planned feedback occurs frequently. These difficulties are taken into account by the *reactive* management system AHEAD which has been developed as the main contribution of the subproject B4 of IMPROVE. AHEAD addresses the management (or *coordination*) of complex and *dynamic design processes* in chemical engineering and supports the planning, execution and control of design processes, which continuously evolve during process execution. Design processes, e.g. for the design of a chemical plant, are represented as process model instances and process model definitions for the description of classes of design processes are created in order to adapt AHEAD to different application domains.

The system has a number of outstanding features which contrasts it from competing process management systems: First, AHEAD supports seamless *interleaving* of planning and execution which is a crucial requirement which traditional workflow management systems usually do not meet. Second, AHEAD *integrates* products, activities, and resources, and their mutual relationships on a medium-grained level. Third, process *evolution* is supported with respect to both process model definitions and process model instances; changes may be propagated from definitions to instances and vice versa (round-trip process evolution). Fourth, in addition to local processes, *interorganizational* design processes are addressed by providing flexible and configurable cooperation support. These contributions on the conceptual level have been demonstrated by several research prototypes. Summing up, the AHEAD system in its current state is the result of one habilitation project and four dissertation projects carried out by the members of the subproject B4.

Another important aspect of reactive management of design processes is the *incorporation* of process *knowledge* contained within *application* models which are developed by our partners in IMPROVE. We have not covered this topic explicitly here, because it is addressed in more detail in Sect. 6.4.

We have applied the AHEAD system successfully to the reference *scenario* studied in the IMPROVE project, which was elaborated in cooperation with industrial partners. But AHEAD is a research prototype which cannot be applied immediately in industry (i.e., in a production environment) for various reasons. In addition to deficiencies with respect to stability, efficiency, and documentation – problems which are faced by many research prototypes –, an important prerequisite of industrial use constitutes the integration with other management tools which are used in *industry*. Therefore, we have integrated AHEAD with several commercial systems for workflow, document, and project management in order to prepare the *technology transfer* into industrial practice as the ultimate goal of the research activities carried out within the subproject B4 of IMPROVE (see Sect. 7.7). Since we are convinced that the developed concepts and mechanisms in the AHEAD system can contribute significantly to the state-of-the-art of commercial process support tools, we will investigate in the future how dynamic processes can best be supported on the basis of existing management systems. Together with our partners from industry, this research is planned to be carried out within the transfer project.

4

Platform Functionality

Two subprojects of IMPROVE deal with *platform* problems. More precisely, they discuss the question, how the construction of new or the extension of given tools (see previous chapter) can be made *independent* of the underlying and used platform.

Each of these subprojects is represented by one section. In Sect. 4.1 *information* flow *management* and process *data warehousing* is discussed. It gives general support w.r.t. administration of heterogeneous data used in integrated environments. Especially, this subproject delivers specific support for its companion subproject on direct process support (see Sect. 3.1).

Section 4.2 deals with *management of platform services*, namely trading, load balancing, distribution, and similar questions. This, again, is a general question to be dealt in realizations of different and integrated tools in a distributed environment.

These two subprojects belong to the third layer of IMPROVE's project structure (see Fig. 1.27).

4.1 Goal-Oriented Information Flow Management in Development Processes

S.C. Brandt, O. Fritzen, M. Jarke, and T. List

Abstract. The research of the IMPROVE subproject C1 "Goal-Oriented Information Flow Management in Development Processes" aims at the development and evaluation of database-driven methods and tools to support and optimize the distributed storage and routing of information flows in cooperative design processes. The overall concept of a *Process Data Warehouse (PDW)* has been followed which collects, and selectively transforms and enriches required information from the engineering process. The PDW has been conceptually based on interrelated partial domain and integration models which are represented and applied inside a metadata repository. This allows to query and apply experience information based on semantic relationships and dependencies. Special attention has been paid to aspects of cross-organizational cooperation.

4.1.1 Introduction

The management of *organizational knowledge* is becoming a key requirement in many engineering organizations. In many cases, it is difficult to capture this knowledge directly, as it is hidden in the way-of-working followed by networks of highly qualified specialists. Moreover, much of this knowledge is strongly context-dependent, requiring the rule applications to be augmented by adequate situation analysis. Hardware and software tools used within the creative design processes in these organizations are strongly heterogeneous, involving significant effort of usage and very different kinds of data.

Structure of This Section

This section describes the research done as part of the subproject C1 of IMPROVE. During the nine years of its existence, several different aspects of *goal-oriented information flow management in development processes* have been approached, examined, and evaluated.

After an introduction into the topics of this section, Subsect. 4.1.2 will present some general issues of supporting creative processes by information science, including related research approaches. Subsection 4.1.3 will describe the prototype of the PDW designed to enable extended method guidance in cooperation with the PRIME process-integrated environment. The following subsection treats a different aspect, as it shows the management of heterogeneous multimedia trace information in plastics engineering. Subsection 4.1.5 introduces the *Core Ontology* and its extensibility as a new technological basis for the PDW. Additionally, some extensions are described, that integrate several of the models of IMPROVE. Afterwards, a concrete usage scenario of the PDW is described, which is extended in Subsect. 4.1.7 towards the problems of cross-organizational cooperation. The section closes with some conclusions and an outlook into further research problems.

Data Warehouses for Engineering Design

Data warehouses have established themselves in the information flow architectures of business organizations for two main reasons: firstly, as a *buffer* between operational and transactional tasks on the one hand, and analytical strategic tasks on the other; secondly, to capture the *history* of business transactions for the purpose of archiving, traceability, experience mining, and reuse.

The same basic arguments apply to *engineering* applications. In these applications, the *buffer function* of data warehousing may be even more important. Research results are often obtained by expensive simulations or even more expensive laboratory experiments, such that analytic processing on demand from information sources is only possible with exceptional effort.

Similarly, from the viewpoint of *history management*, many engineering organizations complain that simulations and experiments are repeated unnecessarily, or at least, that too few lessons for analogous cases concerning promising or useless simulation/experimentation are drawn beyond the experiences of individual engineers. Several organizations are therefore embarking on large-scale traceability or process-capture programs [920, 938]; other organizations pursue the introduction of large-scale document management systems [707] (e.g., the Documentum product [657]) that make at least a coarse-grained representation of products and processes available electronically.

This trend is particularly strong in the research-intensive and law-suit prone process industries (chemicals, oil, food, pharmaceuticals, biotechnology) where global competition with many mergers is largely decided by timely and cost-effective invention of novel products with high market potential. In these industries, data exchange standards, interoperation standards, web-based information distribution and portals, groupware and workflow are being developed within companies and on a scale of worldwide cooperation and competition. However, few coherent approaches have emerged.

Process Data Warehousing is proposed as a solution strategy for some of these issues. We define a Process Data Warehouse (PDW) as a data warehouse which stores histories of engineering processes and products for experience reuse, and provides situated process support. According to the authors' approach, a PDW synchronizes features from document management systems, engineering databases, and traceability tools through active repository technology. It is centered around a knowledge-based metadata repository which records and drives a heterogeneous engineering process, supported by selected materialized instance data. We follow a concept-centered approach expanding ideas from the European project "Foundations of Data Warehouse Quality" (DWQ, see Subsect. 1.3.4 and [769]). At least one major organization in the

German chemical and pharmaceutical industry has implemented a similar system, in part based on earlier research results of the authors [785].

In the context of IMPROVE, as described in this book, the PDW has been designed to offer design information from existing software environments, both to the users and the tools designed as part of this research. The primary purpose of the PDW is to offer an integrated service platform for model-based access of experience traces, to be used by the various tools and tool platforms of the CRC. By integrating and enriching information from the various sources of a design process, the PDW offers a uniform access structure onto this multitude of sources and resources. Additionally, experience reuse methods are applied to structure and analyze this data, enabling situation-based reuse as part of further development cycles. Constructed around a set of interrelated partial models, the conceptual core of the PDW is easily extensible for the application in various domains. Its flexibility also allows adjusting to the fluently changing requirements in creative domains like chemical engineering design.

While initially designed as a service platform to be used by other CRC subprojects, much of the functionality of the PDW has been validated by specialized information and experience management tools that have been developed as part of this subproject (C1). In the remainder of this section, this interplay between *service platform* and *application tools* will reappear regularly.

The PDW as a Service Platform

A short overview is given here about the various services offered by the subproject C1, and the experience repository of the Process Data Warehouse as its primary result. It is also described, in which way they are, or can be, used by other subprojects of the CRC. Offering the services described here has formed the driving force behind many of the C1 and PDW design decisions, to enable an integrated and homogenized access onto the most important artifacts of design processes.

By model-based integration of the data sources available in the design process, a comprehensive access layer has been developed, that allows to search for, access, and manipulate the instance data of different origins in a homogenized way. Thus, extended search functionality (e.g., semantic searching), and an integrated view onto the relations between the different sources has been established, especially by relating and combining the process and the product aspects. The services offered by this access layer have been used especially in conjunction with the PRIME environment of the subproject B1 (cf. Subsects. 3.1.3 and 3.1.5), and other integrated tools with PRIME (e.g., the AHEAD project management system of B4, cf. Subsect. 3.4.2).

In addition to relational databases, application tools, and other non-standardized databases and data sources, several special kinds of sources have been treated, as described in the following. Document management systems have been integrated, which allows to manage documents the same way as other artifacts. Documents and their contents can thus be linked with other artifacts of the supported processes, i.e., with the products of other tools. Unstructured documents can also be enriched with semantical annotations and categorization, based on the repository's integrated view.

A special approach has been researched to support the management and annotation of weakly structured multimedia artifacts, i.e., videos, resulting from technical simulations in plastics engineering. This approach was developed in tight cooperation with the subproject A3 (cf. Subsect. 5.4.2), and PRIME (cf. Subsect. 3.1.3).

Some of the research results have been extended onto the experience-based management of semantic information transfers in cross-organizational cooperative settings. This offers places for tight cooperation with aspects of interorganizational project management, as researched in subproject B4 (cf. Subsect. 3.4.4). Also, multimedia-supported direct communication can well be applied here, for remotely discussing delegation results (subproject B3, cf. Subsect. 3.3.2).

To demonstrate and evaluate the conceptual functionality offered by the PDW's services, several prototypical and specialized tools have also been developed as part of this subproject. In addition to preparing the ground for service offering, some of them also have been necessary for developing and administrating the PDW itself, e.g., for managing the conceptual models.

4.1.2 Information Management in Dynamic Design Processes

The inherent dynamics of the *work processes* pose one of the main problems in engineering design and development. In these processes, the requirements and other parameters change from one project to the next, and can also evolve during the lifetime of a single project. As no methods in the sense of "best practice" are known, the driving influence is the personal experience of the experts working on the project. In the following, these aspects will be treated in more detail, including the presentation of other approaches in this and related fields.

Issues

The software environment used within complex and creative design and development processes usually comprises tools from different disciplines, vendors, and usage paradigms. Each of these tools is normally based on its own proprietary model and contains only some generic import and/or export functions. A *unified access structure* is generally missing that would allow to access all the resources and results over the full lifetime of a project.

Cooperation, especially in *interdisciplinary* settings, requires to transfer information from one expert to another – and thus, from one tool to another. In many cases, the functions for data export/import, or even data integration, do not offer enough possibilities. Some information transfer still needs to be done "by hand", i.e. on paper, by verbal communication, or by re-entering data into a new tool.

All these work processes, including the correct and "optimal" usage of the tools themselves, are based on the experts' *experience* only. It is often necessary to transfer this tacit knowledge from one expert to another. This transfer normally requires a long-term process which prominently consists of more or less successful trials and errors.

Problems may also appear that cannot be solved by the experts inside the company itself which is currently designing a complex system. A certain part of the process needs to be *delegated* to another company. An external contractor (a party taking over a certain task) needs to be found by the delegating company (the contractee). After finding potential partners, business negotiations need to be conducted about the task(s) to be delegated, and about important constraints like timing and costs.

Because of all these problems, it would be helpful to offer *fine-grained* computer and information science support to the experts working on such non-deterministic processes. Here, some approaches for unified product data management, experience management and reuse, work process management, and cross-organizational cooperation will be introduced.

Related Work

During the last decade, many manufacturing enterprises have implemented Product Data Management (PDM) systems, and/or their extended successor of Product Life-cycle Management (PLM). Their aim is to integrate the manufacturing processes (usually CAM/CIM-based) with product design activities on the one hand, and Enterprise Resource Planning (ERP) processes on the other hand. Yet most of these existing systems still lack essential aspects needed for supporting phases of conceptual design, e.g. *knowledge* or *experience management*. Some more recent approaches exist to extend these systems by integrating concepts of artificial intelligence [780], or by using ontological models and tools [684]. Common to all these approaches is their placement in domains like automotive engineering where the design processes are relatively well-documented, strict and deterministic, allowing prescriptive definitions. On the other hand, highly dynamic design processes, as found in domains like computer-aided process engineering (CAPE), need far more flexible approaches for fine-grained process support. Otherwise, the unpredictability of the processes, and the complexity and size of the models, would leave them hardly manageable.

For some time it has been known that experience and understanding of one's own work is necessary to enable process evolution and improvement [741]. This insight has resulted in several approaches based on the basic concept of *experience reuse*. Many approaches to the problem of experience and knowledge management are based on the concept of *organizational memory*, as described in [606]. Another set of well-researched approaches is based on the definition and reuse of cases which represent knowledge, based on certain problem characterizations and the lessons applicable for reusing this knowledge. The possibilities offered by this Case-Based Reasoning (CBR) are described e.g. in [493].

The authors of the TAME project [535] propose a process model for supporting creative software development processes which is based on the their own experiences in software requirements engineering. This approach focuses strongly on quantitative and metrics-based method evaluation for the later steps of software engineering. In the Experience Factory approach, an independent logical organization is responsible for gathering the knowledge and core competencies of a development group and offering this information for reuse [534].

Some other research approaches exist in the area of *engineering design processes*. In [696], a knowledge-based approach for product design is examined which is based on integrating partial domain models and using patterns to represent the non-deterministic behavior of design processes. Another project is developing a process platform which supports the experience-based management and reuse of coarse-grained aspects of software development processes [854]. A different approach to reuse the experience of product development is shown in [1032]. This case stresses the manual processing of expert knowledge and its reuse by less experienced colleagues, by storing the knowledge and the contact information of experts who know more about it, in a corporation-wide experience portal.

An important research project to support the cooperation on the level of *business agreements* is "Negoisst" (see [405, 406]) which drives electronic negotiations between potential contract partners. It uses a three-phase state model to represent the different phases and steps of a contract negotiation. Semantic models and technologies are used to integrate the informal (textual) representation of a contract and its formal (exact) conceptualization. This allows web- or email-based discussions to be based on the exact concepts, attributes, and values. It also enables the tracing of the final results and their intermediate steps.

Early approaches on the technical level of *Enterprise Application Integration* (EAI) were usually based on proprietary formats and printed documents. Many developments have taken place since the advent of the semi-structured data model of XML. Current approaches are based on semantic models, technologies and languages like OWL (see [546]), together with standardized data formats and exchange protocols – electronic mail, XML, http, WebDAV. Semantic web services (OWL-S, see [831]) play a major role in this area as well. These technologies simplify, and in the first place enable, the integration between the application environments and business processes of different companies. Other, more informal systems – video conferencing, electronic mail, web server access, or web-based cooperative work support systems – may also be used for support on the technical side.

The results described in this section also need to be seen in relation to the *new tool functionalities* of the CRC, as presented in the previous chapter, especially with respect to the PRIME approach in Sect. 3.1. The PDW cooperates closely with the PRIME environment in two primary directions. The models, their realizations, and the extended semantic functionalities, as described in this section, offer a set of *services* to the fine-grained process support described there; on the other hand, the traceability mechanisms needed for process guidance provide part of the functionality required for the process data warehouse approach.

4.1.3 Approaching the Process Data Warehouse

Beyond the approaches described so far, there is the need of integrated technical support for the early phases of creative system design, specifically concerning complex technical systems. Therefore, the authors' research group has examined the possibilities offered by *recording* and *reusing* the traces of work processes in technical design. The research deals with the *product*-based view as well as the concepts of direct *process* support, already investigated in previous projects. It has led to supporting creative design and development processes by integrated method guidance [371]. These views have been extended and adapted to the domain of chemical engineering. Most of the ideas described in this subsection have been originally published in [193].

Process Tracing

The Process Data Warehouse (PDW) has been designed to capture and analyze the *traces* of design processes: products, process instantiations and their interdependencies. The artifacts (the technical system) to be designed and modified during the process are traced, and related to the processes which perform these modifications. From these semantically structured product and process traces, the relevant information can be *extracted* in an *analysis* step, and then *reused* in further process executions. This information can be presented to the experts as experience knowledge in order to solve the problems of later development cycles more easily, efficiently, and autonomously.

The central issue of this approach is that of supporting traceability. To enable traceability, first of all the conceptual relations between products, processes and their dependencies need to be examined. Therefore, in [376] the *traceability reference model* shown in Fig. 4.1 was abstracted from a large number of industrial case studies. This model distinguishes between product-oriented and process-oriented trace objects.

The *product-oriented* traces describe the properties and relationships of concrete design objects. A high-level object defines some goal or constraint that needs to be *satisfied* by a number of *product objects* on a more fine-grained level of modeling. This implies dependencies (*depends-on*) between

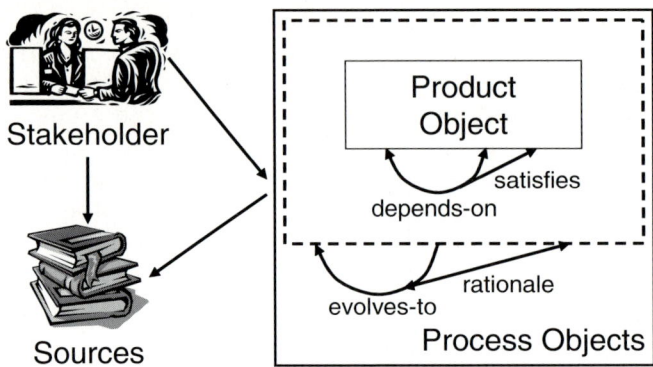

Fig. 4.1. Traceability reference model from [376]

these lower-level objects which also comprise the special cases of generalization and aggregation.

The *process-oriented* traces represent the history of actions that led to the creation of the product objects. Two link types exist between those process objects: *evolves-to* which describes the temporal evolution of a lower-level design object towards a higher level, and *rationale* which captures the reason for this evolution. The integrated presentation of the product and process traces in this "onion-shell" meta model symbolizes the fact that they cannot be reasonably separated as one strongly depends on the other.

As visible in the left part of Fig. 4.1, the role of the *stakeholder* during product creation or documentation is of importance as well. It is also necessary to record and connect the *sources* which contain and display the information.

The description of this reference model shows that recording the process traces needs to include all related influence factors, like the actual problem situation, the resulting artifacts, and the decisions that led to the final results.

From these traces the semantically relevant information can be extracted in an *analysis* step. Due to the complexity of the traces, automated analysis is impossible in most cases. When working on complex processes with only few repetitions and few concrete product instances, this analysis step can often be left out. The decision between the available information can be done in the moment of reuse. If there are too many data to be retraced this way, a so-called *method engineer* is responsible for extracting and explicitly modeling method fragments and situations, often supported by methods of data mining.

When an expert needs to solve a certain problem, the current process and product situation is analyzed by the PDW to find matching solutions from the recorded (and analyzed) traces. If an adequate method or product fragment is found, it can be offered to the expert for *reuse* through a *guidance* mechanism. Yet it is his own decision whether to adapt and use this information, to request more details, or to discard it. In many cases a small hint should suffice that the step currently enacted, conforms to the experience gathered up to now or, even more important, conflicts with it.

It also has to be recorded whether the problem was successfully solved, and how far the process support provided was appropriate, as a final *feedback* information. By using this information the system and the support it offers, can be evaluated and improved.

Extended Method Guidance by the PDW

To achieve the goals stated above, we derive from the concept of Data Warehousing as established for domains with fixedly structured process and product models, to enable the support for creative processes in chemical process engineering. The extensions necessary to support *process* data warehousing in this domain are mainly twofold. At the *conceptual level*, the "enterprise model" of the DWQ approach [769] has to be split into a set of loosely connected partial models which look at different facets of the chemical engineering process. At the *logical* and *physical level*, heterogeneity of the process engineering tools is far greater than traditionally considered in OLTP data sources [192]. Therefore, an intermediate standardization step is necessary, not only at the level of data but also at the level of services. This is due to the fact that often the data of engineering tools are not sensibly accessible directly but only via the tool services.

As a coherent conceptual model cannot be built, a multitude of *partial information models* have to be considered with poorly understood interconnections. In the IMPROVE context, these models have been systematically developed, resulting in the Conceptual Lifecycle Model CLiP and in its successor, the OntoCAPE ontology, as described in Subsect. 2.2.3. Empirical studies of chemical process design demonstrate that one family of closely related submodels, visualized through *flowsheets*, has a clearly dominating role in the communication between different designers. Coherence between the partial models is thus achieved through concepts contained in the process flowsheet. It forms the key access structure to heterogeneous information sources and documents.

These flowsheets may describe very complex processes and evolve in complex refinement structures, including operations such as enrichment of object definitions, decomposition of functions, specialization of choices, and realization of functions by (combinations of) device types. Additionally, process synthesis decisions are made under uncertainty of their impact. A complete analysis of all design choices is impractical due to the high effort in setting up simulations or laboratory experiments. However, this may result in backtracks in the engineering process. There also is a huge and continuously growing number of different devices, connections, and specialized functions that can be used in chemical engineering; estimates speak about roughly 50.000 types to be considered. The information models of a *meta database* for process data warehousing must therefore be easily extensible by new product and process knowledge, and cannot be mapped one-to-one in tool functionality.

The requirements of the process engineering domain strongly support the case of a *concept-driven* approach for data warehousing. They also require further refinements of metadata handling concerning information model integration, structural and behavioral refinement, the interplay of design and analysis, and the extensibility with a growing body of knowledge.

The diversity of the information models at the conceptual level is exacerbated by the diversity of *data formats* and *service accessibility* at the technical level of engineering tools and databases. Current process engineering environments often hide this problem through monolithic software architectures with fixed means of access. These make it close to impossible to include company-specific knowledge or home-grown specialist tools.

The European process industries have therefore embarked on the CAPE-OPEN initiative [71] in order to accomplish a *standardization* of simulation interfaces, such that a component-based approach can be followed. This standard has been defined at the *conceptual* level through UML models. At the *middleware* implementation level, the standard is both defined in DCOM [847] and CORBA [877]. In IMPROVE, the CORBA version is used.

CAPE-OPEN has identified the following *standard components* of a process simulator from the conceptual point of view [997]:

- *Unit Operation Modules*, often merely termed *units*, represent the behavior of physical process steps (e.g. a mixer or a reactor).
- *Physical Properties (Thermodynamics) Packages*: An important functionality of a process simulator is its ability to calculate thermodynamic and physical properties of materials (e.g. density or boiling point).
- *Numerical Solvers*: The mathematical process models of a unit operation or a complete plant are large and highly non-linear. As analytical solutions are impossible, iterative, numerical approaches are used to solve the equations.
- *Simulator Executive*: This is the simulator's core which controls the set-up and execution of the simulation, i.e. analyzing the flowsheet and calculating the units. Furthermore, it is responsible for a consistent flowsheet set-up and error checking.

The Process Data Warehouse, as described in this subsection, has been implemented using the deductive object-base *ConceptBase* [204]. The lifecycle model CLiP was realized in ConceptBase (see Subsects. 1.3.4, 2.1.3, and 2.2.3), while the tool environment uses the interfaces offered by the CAPE-OPEN initiative to integrate physical properties, mathematical models, and simulators. Subsequently, it will be shown how this setup is used in an example scenario. In this scenario, the services offered by the *PDW Query Assistant* are used by the PRIME process integrated environment (cf. Subsect. 3.1.3) for *extended situation analysis*.

The PDW Query Assistant

As an example, it will be presented here how the standard interfaces from CAPE-OPEN supplement the process data warehouse in the chemical engineering domain. The prototype *combines* techniques for integrating the highly *heterogeneous information* sources in the application domain with the *standard interfaces* for unit operations and physical properties packages. The process data warehouse client operates on this prototype, called "cross-tool situation analysis". It uses the domain knowledge captured in the meta database to give guidance to the chemical engineer via the process-integrated flowsheet editing tool of the PRIME environment (see Subsect. 3.1.3 and [194]).

To this aim, the current development situation is analyzed via the product state of several source databases or tools. The scenario shown here is a simplification of a part of the design scenario described in Subsect. 1.2.2. Here, the selection and adaption of a mathematical model is treated, to simulate the separation of the final product Polyamide-6 from the residue monomer Caprolactam.

In the early conceptual design stage a chemical engineer draws a flowsheet of the plant. The blocks in the flowsheet represent unit operations, such as mixing, reaction or separation. These functional units can be further refined and realized in concrete apparatuses, the behavior of which can be simulated by mathematical models.

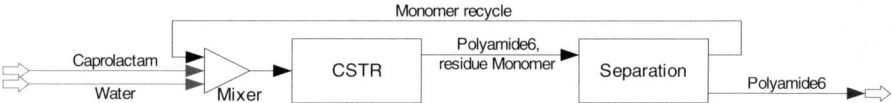

Fig. 4.2. A simple example flowsheet

As a (very simple) example, we consider the flowsheet in Fig. 4.2. The flowsheet has been designed in the process-integrated flowsheet editor that forms part of the PRIME environment (cf. Subsect. 3.1.3). The simulation model for the reactor device (CSTR – continuous stirred tank reactor) is already given. The developer's task is to find a suitable model to represent the separation. The designer can choose between several separation models, including complex combinations with recycle streams (backflows).

The detailed setting is: Two input streams feed an initial mixer. The substances Caprolactam (monomer) and water are fed through the streams into a mixing device, then into a reactor modeled by a CSTR. The result is the Polyamide-6 product and the residual non-reacted monomer. These substances have to be separated, as the monomer has to be fed back into the reactor. Now the task of the chemical engineer is to find a useful model for the separation unit.

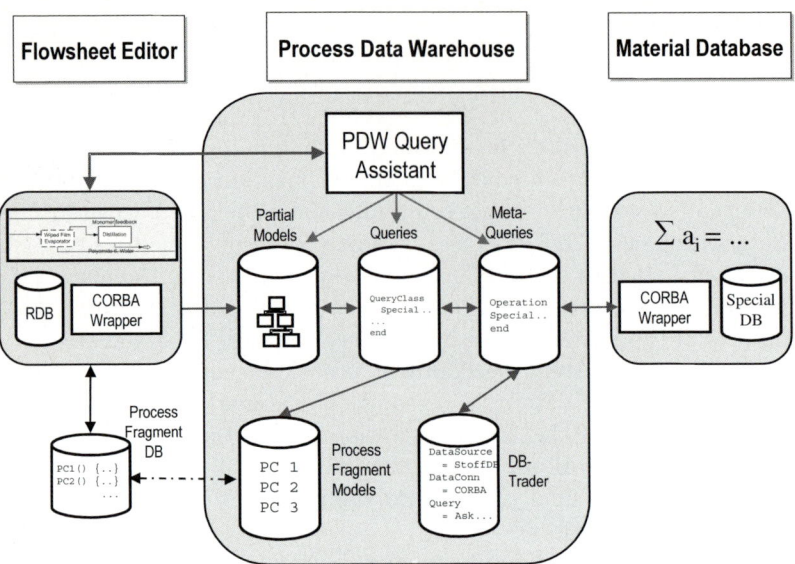

Fig. 4.3. The PDW Query Assistant

The *extended situation analysis* function of the process data warehouse, as described in [193], is able to provide some hints: which models should be considered for this task if some of the properties of the reactor's output stream (containing Polyamide-6 and Caprolactam) are known, e.g., temperature and pressure of the mixture in the stream, the fraction of each substance in the stream, and their boiling temperatures.

Figure 4.3 describes the functionality of the *cross-tool situation analyzer* of the warehouse. A client tool (here: the flowsheet editor) calls the the service offered by the PDW Query Assistant, a Java-based control program. The call contains an identifier for the selected flowsheet element (the separation device) and an operation to be executed on the device (realize). This information is available in the process-integrated flowsheet editor as part of the current *situation, intention* and the currently running *process fragment* (cf. Subsect. 3.1.3). This process fragment will also be responsible for handling the information returned by the Query Assistant later on.

The warehouse itself contains several *sub-meta databases* that are queried during the processing of this request for extended situation analysis:

- *Call back patterns* are used to determine which additional information is needed to answer the request.
- The *DB trader* contains information from which tool or database and how this information can be accessed. The calling client will be one of the tools that are accessed. In the example, a larger part of the flowsheet is needed to classify the context of the call.

- The *mediator patterns for materialization* are then used to materialize the additional data into the data warehouse such that they become instances of the *partial models* of the data warehouse.
- The *analysis and result patterns* are now applied to calculate useful results for the original query.
- The presentation of the results is highly dependent on the client tool that initiated the query. The *client model* is used to transform the result into a suitable form. The special process integrated features of the flowsheet tool can be used to directly insert a proper refinement of the separation into the flowsheet. Details on the flowsheet tool and its interaction with the process data warehouse can be found in [194].

These data sources are accessed a nested way so that only those information sources are accessed, that are needed to answer the specific query, instead of gathering all information in advance.

The call back queries used in these steps are *not purely queries to source databases*. For example, the needed simulation results of the reactor are results of an aggregation function. In this sense the results are the results of a (highly complex) query on the data warehouse store. As simulating is a time consuming and expensive task we also store the results in the data warehouse for reuse. To gain access to the units the DB trader contains meta information about the CAPE-OPEN components. As a result of the usage of the CAPE-OPEN compliant units we do not need to handle very different simulators such as Aspen Plus, Pro/II or gPROMS, but only have to create the CAPE-OPEN objects used by the units. This especially concerns the material object for each substance contained in the input ports of the unit. The process data warehouse produces these CORBA objects and is then able to start the simulation of the unit.

As final step, the results of the analysis, i.e. one or more models appropriate for representing the simulation, are delivered back to the tool that requested the extended situation analysis. In the scenario described here, the PRIME process fragment is then responsible to offer the model alternatives to the user, and to allow him or her to integrate them into the flowsheet as refinements of the separation block.

4.1.4 Heterogeneous Trace Management in Plastics Engineering

In this subsection, a special case of experience traces will be addressed. In contrast to the structural models described in the last subsection, multimedia information usually does not provide any possibility of extracting meaningful semantic information. As a concrete application scenario, the visualizations resulting from three-dimensional plastics engineering simulations can be stored as short *video* clips, and then *structured* and *annotated* according to an appropriate domain model. This allows the *retrieval* and thus the *reuse* of these complex simulation results, together with the domain experts' interpretations.

Thus, a service is offered that combines the domain models and application cases from the subproject A3 (cf. Subsect. 5.4.2) with the experience-based functionality of B1 (PRIME, cf. Subsect. 3.1.3). This service is then used by the application tool *TRAMP* which will be described in the second part of this subsection. The research results presented here have been originally published in [188], [197] and [198].

Simulation Analysis in Plastics Engineering

As part of the polymerization scenario described in Subsect. 1.2.2, the compounding extruder is to be designed by a company specializing on extruder design and construction, supported by simulations. As the necessary knowledge is often not present in chemical companies, aspects of *cross-organizational cooperation* come into play, which will be treated later in Subsect. 4.1.7.

The *compounding* of thermoplastic polymers usually employs closely intermeshing, co-rotating twin screw extruders. Based on a modular concept, the screw geometry must be designed most precisely to realize the desired mixing of fillers and other material modifications, in a maximally effective and economical way. A detailed analysis of the flow effects inside a compounding extruder must be resolved in three dimensions. For complex flow channels, this is only possible at high numerical effort using the Finite Element Method (FEM) or the Boundary Element Method (BEM). Currently, these methods can only be used on fully filled screw sections, but the modeling of flows with free surfaces is under development.

A more detailed description of the simulation and design tasks can be found in Subsect. 5.4.2 (subproject A3). For the possibilities of experience-based support for extruder design, refer also to the FZExplorer tool in Subsect. 3.1.5. Here, we will concentrate on structuring the domain model for annotation of the videos resulting from simulation.

In the example scenario, the BEM-based program BEMFlow is used because of its efficient way to generate the simulation meshes. BEMView is a special postprocessor for the visualization of BEM-calculation results in videos. The visualization by means of stream lines facilitates the investigation of flow phenomena of interest such as local spots with low residence time or vortices [145].

The main purpose of these process analysis activities is to evaluate goal achievement in terms of polymer product properties. For example, the residence time as well as the deformation history of a single particle is an indication for the polymers' thermal and mechanical degradation.

Figure 4.4 shows how the results of simulation calculations can be categorized in three *goal categories*. Primary and secondary effects are direct results from the conservation equations or can easily be calculated. By means of these values the user can assess the process behavior, but cannot quantify abstract phenomena like the deformation or the melting of polymer. For this case, ex-

Primary Effects	Secondary Effects	Tertiary Effects
Directly from the Conservation Equations	Calculated by differentiation and integration from primary effects	Abstract phenomena with an empirical definition
• Pressure • Velocity • Strain • Viscosity • Density	• Acceleration • Gradients' Tensor • Volume Flow • Force	• Melting • Deformation • Mixing • Residence Time Distribution

Fig. 4.4. Layers of effects/goals in polymer flows

plicit modeling of processing goals is necessary which in plastics engineering are called tertiary effects [676].

It is obvious that, besides other goals, the consideration of cost is very important. Sections or zones of the extruder that need high capital investment for their realization have to be further investigated. The same applies to zones that can cause side effects to other components of the plant, or cause high running production costs.

As a result of both, 1D and 3D simulation, the screw-configuration of the extruder will be modified in an iterative design and analysis cycle for each functional zone. The results of the simulations have to be interpreted by the users who then have to asses product quality based on their experience [145].

TRAMP: Linking Goals, Domain Ontologies, and Multimedia Scenarios Efficiently

The multimedia (mostly video) visualizations of the flow through an extruder are a side effect of running a 3D simulation. This side effect is extremely important, as only this visualization enables the experienced plastics engineer to evaluate a proposed decision alternative with respect to the actual goals and obstacles relevant to the next stage in the supply chain. Additionally, it is possible to visualize the primary vector-oriented simulation results directly as interactive media.

However, extruder design usually requires many simulations before a satisfactory solution can be found. Comparing all the videos with respect to multiple goals can take a long time even for a single design step. The problem (but also the opportunities to establish a good solution and avoid a waste of time) grows when the reuse of similar situations is enabled, so even more videos must be viewed and compared. Simple linkage of video clips to goals, as studied by [723], is not sufficient here.

Drawing on the analogy of the well-known phenomenon of *"zapping"* rapidly across TV channels to find interesting ones, we have therefore developed a system which allows *"semantic zapping"* among multimedia sce-

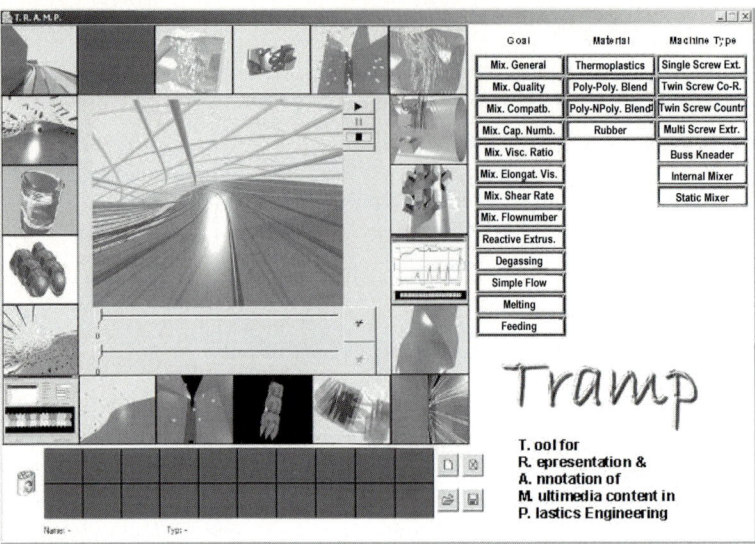

Fig. 4.5. Tool for Representation & Annotation of Multimedia content in Plastics Engineering (TRAMP)

narios according to both goals and domain ontologies. This system allows to *organize* and *annotate* the information both according to the MPEG-7 multimedia metadata standard [419], and according to the goal hierarchies and domain models of the Process Data Warehouse. Additionally, the PRIME trace database (see Subsect. 3.1.3) was extended with semi-structured interfaces and storage mechanisms based on XML, to allow the integration of the PRIME process and decision traces with the the semantic multimedia annotations of the PDW. For demonstrating and evaluating the functionality and usage of the system, a tool called *TRAMP* (Tool for Representation and Annotation of Multimedia content in Plastics engineering) has been developed, as described in the following.

Figure 4.5 shows a screenshot of TRAMP. The three columns of buttons on the right were generated from the domain models. They represent three different *dimensions* of characterizations. The left column contains a list of relevant goals (tertiary effects) to be achieved, whereas the other columns refer to the domain categories of materials and extruder types.

The following dimensions, or *domain categorizations* can be selected:

Goal: mixing (general), mixing (quality), mixing (compatibility), mixing (viscosity ratio), mixing (elongation viscosity), mixing (shear viscosity), mixing (flow number), reactive extrusion, degassing, simple flow effects, melting, feeding.
Material: thermoplastics, polymer-polymer blend, polymer-non-polymer blend, rubber.

Machine Type: single screw extruder, twin screw extruder (co-rotating), twin screw extruder (counter-rotating), multi screw extruder, buss co-kneader, internal mixer, static mixer.

By selecting a combination of buttons (multiple choices are possible in each column), the thumbnail gallery gets filled with visualizations of 1D and 3D simulation results relevant to the indicated combination of goals, materials, and device types. By dragging one of these thumbnails into the center, the corresponding multimedia objects gets enlarged and – if it is a video – played, thus enabling human judgement, but also rapid zapping to another thumbnail candidate. When changing the center object, the context also shifts, so new similar objects can appear in the thumbnail gallery, old ones can vanish, and slowly the context of goal, materials, and device metadata can shift as well.

Alternatives the engineer finds particularly interesting (in the positive or negative sense) can be drawn into the personal collection at the bottom left of the tool, and annotated with *arguments* linking them to the goal hierarchy, or to choices in a decision editing tool. This allows semi-automated construction of decisions to be taken, and, later on, to document the *decision* itself, with all relevant alternatives and arguments.

As a result, these process and decision documentation traces can be reused by tools that access the underlying services of the PDW. For example, the TRAMP tool is directly integrated with the decision editing functionality of the PRIME environment (subproject B1, cf. Subsect. 3.1.3). Other tools can also access these services, to track and reproduce the simulation steps and their rationale, and possibly to intialize necessary modifications.

4.1.5 The Ontologies of the PDW

On the path from the conceptual prototype described in Subsect. 4.1.3 towards a flexible and extensible service base and application framework, a decision was taken to use the *ontology languages* from the Semantic Web approach for modeling the Process Data Warehouse (PDW). This decision went together with the porting the simple class layer of the conceptual lifecycle mode CLiP into the ontology-based version OntoCAPE (see Subsect. 2.2.4). As ontologies do only support one modeling and one instance (token) level, it was necessary to abandon the powerful features offered by meta modeling.

To achieve the projected functionality, this new realization of the PDW was developed around the so-called *Core Ontology*. This central conceptual model was originally a result of the IMPROVE subproject C1 and the research described in this section. Some aspects of the Core Ontology have already been introduced and described in Subsect. 2.2.4. Based on these ontological models, the PDW has been designed to offer an integrated access environment onto all the product and process artifacts of the supported design processes. In the following, it will be described how these models and services are exemplarily applied and used by the application tools of the PDW. In the end of the subsection, it will be shown how the various subprojects of the CRC can, or do, use these services for their own integrated purposes. The conception and application of the Core Ontology was originally published in [64], [65] and [66]; its application in chemical engineering is also described in [62] and [63].

The Core Ontology

The various partial models of the Process Data Warehouse are interconnected through the Core Ontology which consists of four primary areas of conceptualization: *products, processes, descriptions,* and *storage*. This central model comprises the concepts of process modeling and enactment, of products and documents, dependencies, decision support and documentation, for the description of content and categorizations, and other integration models.

Around these fundamental and domain-independent models, *extension points* are placed that can be used to add the models of a specific application domain or other specializations. The concrete data are then stored as instances of the appropriate ontological concepts. This allows modifications and extensions of the partial models used, even during project execution.

For reasons of interoperability, the *Ontology Web Language* (OWL) [546] standard, as already used for OntoCAPE (Subsect. 2.2.4), would have been the first choice for the representation of ontologies. However, current OWL-based ontology repositories do not offer an efficiently searchable storage in (relational) databases, nor do they easily support client/server-based architectures to, e.g., facilitate update synchronizations.

The *KAON system* [871] was chosen instead. It is based on the Resource Description Framework (RDF) that also forms the base of OWL. KAON enables semantic queries directly on the backend repository (stored in a relational database) by transforming the query into SQL, at the cost of loosing some of the expressiveness of OWL. Translation of OWL ontologies, such as OntoCAPE, into the KAON system has been realized, based on their common RDFS characteristics.

The Core Ontology is formed by the aforementioned four areas of conceptualization, arranged around the object as the abstract central concept. It is shown in Fig. 4.6.

- The *product area* (top) contains basic models for the artifacts created or modified during the design processes – documents, document versions and their structural composition.
- The *descriptive area* (left) contains basic concepts for describing the content or role of documents and products on a high semantic level. This includes content descriptions, sources, and categorizations which are grouped into categorization schemes. Type definitions are also placed here that characterize, e.g., a large number of products that have been created from the same template.

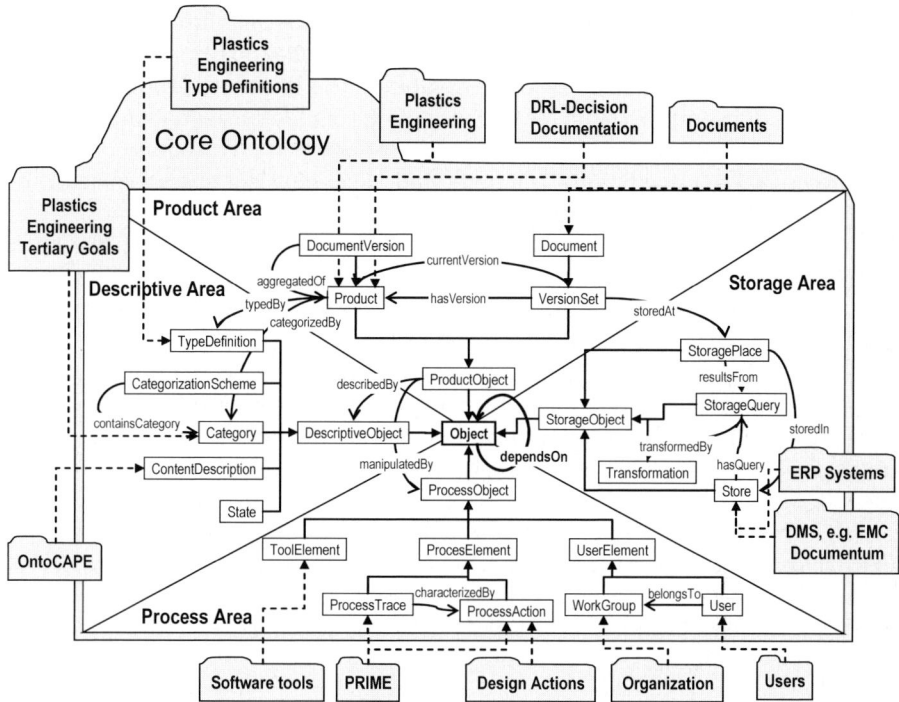

Fig. 4.6. The Core Ontology with some embedded extension models

- The *process area* (bottom) contains the concepts needed to describe the process steps which modify the artifacts. This comprises process definitions or actions which can be enacted (method fragments), process traces resulting from enactment, and users who guide the enactment.
- In the *storage area* (right), external stores and repositories are integrated into the PDW. This applies to document management systems, databases, external tools, and others. Rules for the execution of queries on the storage backends, and the necessary transformations of the results (and other document contents) are also represented here.

Dependencies have been introduced as a global concept to enable specialized relations between elements independent of their concrete relationships. They are also described in the traceability reference model in Subsect. 4.1.3. This hierarchy of relationship types is modeled inside an additional area which can be seen as orthogonal to the four areas.

Around the core ontology, several *extensions* can be found. The most elaborate of these is OntoCAPE, the already mentioned large-scale ontology for the description of the process engineering domain (Subsect. 2.2.4, [489]), which covers fields like physicochemical properties, process equipment, and mathematical modeling. Here, it extends the descriptive area by refining the *Content Description* concept (refinement is indicated by dashed arrows in Fig. 4.6).

As another example, the storage area offers the basic models for file storage inside a document management system, relating file-based documents with their conceptual representation inside the Process Data Warehouse. This allows accessing the documents' contents and their physical storage places, including the visualization or modification inside appropriate tools.

To apply these concepts, the EMC Documentum system [657] has been integrated by extending the concepts, and implementing specialized functionality. The integration of ERP systems has exemplarily been realized for the SAP R/3 system [947].

Based on the ontology concepts of *StorageQuery* and *Transformation* (see Storage Area in Fig. 4.6), a flexible mechanism for integrating external data sources or stores has been developed. Specific connectors for the various kinds of stores – relational databases, tools, ERP systems, etc. – are able to save the data resulting from specified *Storage Queries* in an intermediate XML format. After transforming these XML documents into a generic XML or OWL format, a generic importer is responsible for importing the data into the PDW, creating instances of the appropriate concepts. This transformation, e.g. by stylesheet transformations (XSLT, [602]), is organized by related instances of the ontology's *Transformation* concept.

Product Models for Plastics Engineering

As an example, it will be described here how the concepts and ontologies for *plastics engineering* are embedded around, and into, the Core Ontology. The structural composition of compounding extruders by a set of functional zones like *Polymer Stream, Boiling Degassing* or *Reactive Extrusion*, and their realization by conveying elements and kneading blocks, is described in further detail in [147].

Several important regions of this model can be identified in Fig. 4.7.

- The extruders and their structural composition, as created by the MOREX tool (see Subsect. 5.4.2), are extensions of the *Product* and *Document* concepts. Each *Document* has a set of *Document Versions* that in turn contain the *Extruder* realizations, and other *Plastics Engineering (PE) Devices*.
- The various types of *Functional Zones* are added as specializations of this concept, such as *Polymer Stream* or *Boiling Degassing*.
- The *Screw Elements* themselves are split up into two different kinds, the *Kneading Blocks* and the *Conveying Elements*. Many different types of these elements can exist, each with different default attribute values (e.g., *Length* or *Diameter* as shown in Fig. 4.7). It was decided to use a *Screw Element Type* concept, and two specializations, as part of the descriptive area. Instances of these concepts define the possible screw element

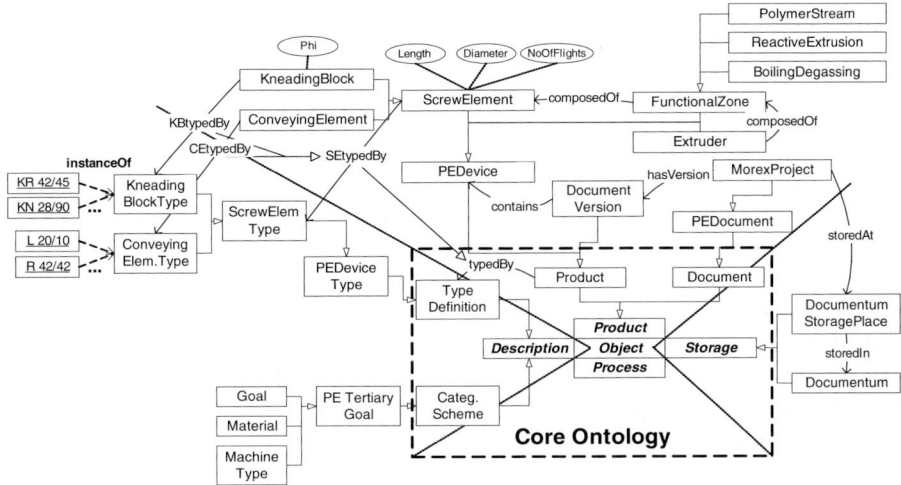

Fig. 4.7. Embedding the plastics engineering domain

types. Each concrete kneading block or conveying element is *typed by* one of the derived screw element type instances: kneading block type or conveying element type, respectively. The instantiations shown in the figure (KR 42/45, R 42/42) are, of course, not complete (indicated by "..").

- The *tertiary goals* of extruder simulation as introduced in Subsect. 4.1.4, and some other categorizations, are modeled as categorization schemes in the descriptive area. Each of the schemes is composed of a set of categories (which are not shown in Fig. 4.7). This allows to integrate the TRAMP functionality of annotating weakly structured (multimedia) documents.

- As the MOREX tool stores its projects in simple unversioned XML files, the document management system (DMS) Documentum [657] has been integrated into the PDW for file storage. This allows the *versioning* of the project files on the one hand, and the extruder realizations stored inside the PDW on the other hand. Also, when checking in a new version into the DMS, the PDW is automatically notified of this new version. This allows to import the information, or to annotate it using the available categorization schemes. A simple function has been implemented to import the MOREX project files into the PDW repository; this function still has to be integrated into the storage transformation concept as described above.

Process Models

Process models form a very important part of IMPROVE. In diverse subprojects, they are modeled in different levels of granularity and with emphasis on different aspects. The PDW Core Ontology was designed to allow the integration of these models around one central core. One important aspect is the separation of *Process Actions*, i.e. task, activity or method definitions that can be enacted or executed by users and/or tools, and *Process Traces* that result from enactment. More about this characterization can be found in Subsect. 3.1.2.

For the construction of the PRIME environment (Subsect. 3.1.5) on top of the PDW, the NATURE process meta model [201] has also been embedded as an extension of the Core Ontology. The concepts of *Contexts* (EC, PC, CC) have been derived from the Process Action, while the process tracing concepts are children of Process Trace. Organizational elements like users, work groups, and companies have also been extended from the appropriate concepts.

Similar work is being done to integrate the administrative and workflow level concepts of IMPROVE into the Core Ontology. This concerns the C3 modeling formalism and method as described in Subsect. 2.4.4, and the AHEAD task management concepts from Subsect. 3.4.4.

Ontological Search and Access Services

As part of the subsection closing here, it has been described how the models of various other CRC subprojects have been, or could be, integrated into the experience repository of the PDW, including the instance data based on these models. This allows to offer a homogenized access, search, manipulation, and analysis service to all interesting parties. Especially the data integration across the various models, layers, and application domains offers a unified view onto the traces that is not available otherwise. This includes the following:

B1: the process integration and support models of PRIME, based on the NATURE process meta model;
A3: the models for extruder design, and for the annotation of simulation results;
B4: the process models for coarse-grained intra- and inter-organizational task management support; and
I2: the C3 process modeling formalism.

As another example for using this service, the next subsection will describe how the application tools of the PDW itself can be used to work with, browse through, and apply the information of the repository as part of the experience-based support of work processes.

4.1.6 The Process Data Warehouse for Engineering Design

To demonstrate the conception and realization of the Process Data Warehouse, we will describe how a concrete *application scenario* is supported by the experience reuse functionality of the PDW. The scenario of designing and analyzing the compounding extruder by one-dimensional simulation is part of the complete scenario described in Subsect. 1.2.2.

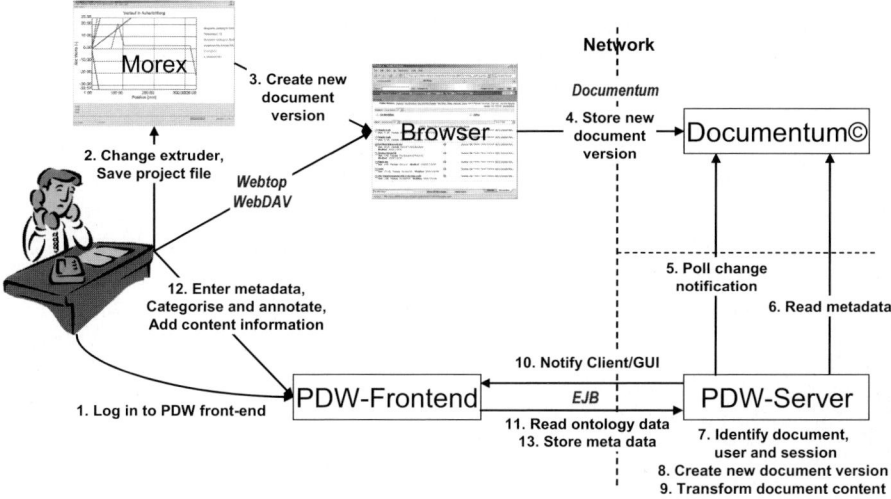

Fig. 4.8. Scenario: creating a new document version

Capturing Experience Traces

Figure 4.8 describes the activities which result in the *capturing* of the product and process *traces*. This includes the interactions between the user (a compounding or plastics engineering expert), the MOREX design tool (as described in Subsect. 5.4.2), the document management system Documentum [657], and the Process Data Warehouse. The continued scenario which shows the *reuse* of this recorded experience, will be described afterwards.

The following steps are shown here:

1) The experts starts the PDW front-end and logs into the PDW Server.
2) The expert works with the MOREX tool to design and simulate the functional zones and their screw element realizations.
3) When reaching an important stage in the design, the experts checks in a new version of the project file into Documentum. This can be done via the Documentum Webtop browser plugin.
4) The new version is stored in Documentum.
5) The PDW polls for change notifications via the Documentum audit mechanism. The creation of the new version is thus recognized.
6) The metadata of the new document version is read by the PDW: document name, version and ID, user and session information, etc.
7) The PDW uses this metadata to identify the correct *StoragePlace* and *Document* concept instances inside the repository.
8) A new *Document Version* instance is created in the repository and related to the *Document* instance as its new and current version.

9) The document content is fetched and analyzed, and the structural composition is then transformed into the conceptual representation of the PDW (*Extruder, Functional Zone, Screw Element* and their specializations in Subsect. 4.1.5).
10) The PDW front-end is notified of the change.
11) The front-end reads the new information from the PDW server.
12) The expert can now annotate and enrich the document and the structural information of the new extruder realization. The categorization schemes can be used for this purpose.
13) The annotation information is stored in the PDW repository.

After these activities, the project file is stored in the DMS, while its *structural content* (the realization of the extruder) is stored inside the PDW repository, together with additional information like domain model annotations and enrichments. Importing of information into the PDW is done via an XML format that is directly based on, and thus convertible into, the core ontology concepts. External resources need to be transformed into this format, based on the rules defined by derivations of the Core Ontology's *Transformation* concept, and implemented e.g. by XML stylesheet transformations (see Subsect. 4.1.5).

The same flow of activities can be used with any different tool and document information. Depending on the degree of integration, the document content may be converted into the ontological format of the PDW (see also Subsect. 3.2.1 about fine-grained product relationships and conversions), or enriched and annotated based on a coarse-grained categorization model. While the former has been applied in the scenario described here, the latter concept is used in the TRAMP tool as described in Subsect. 4.1.4.

Reusing Experience Traces

In a different design project within the case study described here, possibly an altered organizational context, an expert reuses the recorded information by searching the PDW via its front-end. Older, or recurring traces can be found with the help of a query language that is based on the semantic relationships of the PDW's integrated partial models. Similar to the concepts of Case-Based Reasoning (CBR, see [493]; see also Traces-Based Reasoning, [848]), the comparison and adaptation of the retrieved cases enables their direct reuse.

This results in an *advisory system* where the retrieved cases are manually adapted to the current context by the user. The case reuse and the adaption are both traced for later review and repetition. To simplify the search for matching cases, the integrated tools can directly query the experience repository based on their current situation or context. Another approach applied here is the manual building of semantic example-based queries (Query-By-Example, QBE).

The system also offers the possibility of analyzing the data to *detect recurring fragments*. Fine-grained product and process support can then be offered to the user by guiding him or her through these fragments, or even enacting them in the process engine of the process integrated environment (see PRIME in Subsect. 3.1.3).

The *experience reuse framework* consists of the Process Data Warehouse, the process-integrated development environment PRIME, the Documentum repository, and a set of integrated tools, e.g., MOREX. This framework is able to specify the current problem situation based on the integrated rules, and tries to find a matching process trace or a recurring method fragment in the experience base of the PDW.

In the scenario described here, the expert needs to find sensible realizations of the functional zones that form the compounding extruder. In MOREX, he or she selects the current functional zone (a "Filler Adding", in the case described here), and "asks" for a set of alternative realizations. The current problem situation is analyzed by the environment, and a *semantic query* is composed that can be executed on the experience repository. This problem definition consists of elements of several of the four areas introduced in Subsect. 4.1.5.

Each of the elements is the concrete instance of one of the semantic concepts contained in the PDW's ontologies.

- The *product* part of the situation is composed of the design elements displayed in MOREX, i.e. the extruder, the functional zones, and already finished screw element realizations. The materials being compounded, and attributes like temperature or pressure, also need to be taken into consideration. For some of the elements, their user interface state is also important, as the currently selected zone is to be used as the "central" aspect of the query.
- Some of the relations and dependencies reach into the *description* area, as chemical components, states, and categorizations are found here.
- The most important *process* element is the user's intention in this situation, i.e. "Realize the current functional zone". This is normally determined by a user interface element being activated, e.g. a menu item being clicked. The current activity type of "Create extruder realizations" – synthesizing a model, in contrast to analyzing a number of alternatives or deciding on one of them – is also part of the situation.

This *situation definition* can then be passed on to the Process Data Warehouse to search for matching experience information. In the example scenario, several different realizations are found and returned. This information is then presented to the expert in the PDW client front-end, as visible in Fig. 4.9. Two different visualizations can be applied here. A generic representation shows the concept instances, their attributes and relations in UML instance notation, while the specific representation in this case shows a graphical snapshot of the screw elements.

Now the expert needs to decide which realization to use. He may decide on *reusing* one of the alternatives offered. Then a method fragment inside PRIME is activated that directly executes the steps in the flowsheet editor which are

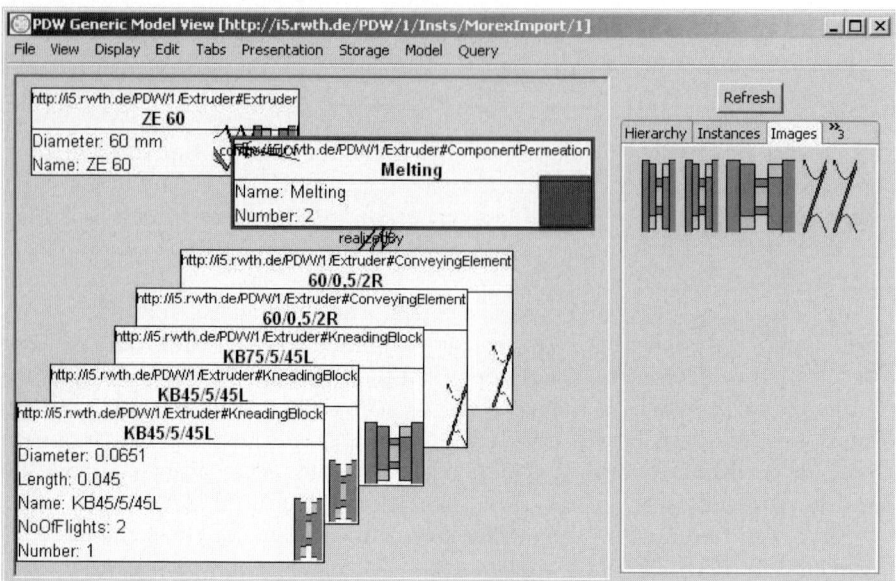

Fig. 4.9. Generic (left) and specific (right) visualizations of one returned realization alternative

needed to add the new refinement and to insert the selected alternative. If he or she wants to *adapt* the realization, or to create a new one, this can be done using the normal functionality of MOREX.

In any case, the solution is recorded in the Process Data Warehouse. Apart from the product information (the selected realization), the intermediate process steps and situations are traced and related to the decision that led to this alternative, including some additional arguments the expert may want to enter manually.

This is only a preliminary decision, as the chosen realization still has to be analyzed. The PDW can also serve as the *integration point* for simulation runs in the appropriate tool(s). After the simulation, the same cycle will have to be repeated for other, alternative refinements. In the end, the expert can decide – and document – which of the alternatives should be kept for further steps in the development process. The arguments entered earlier and the simulation results related to them will be needed for supporting the decision process here. This information is only kept in the PDW.

Choosing an extruder realization, furthermore simulating the realization, documenting the results and entering some notes or arguments, forms a repeated cycle that concludes with the final decision-making. This may be recognized as a single recurring *method fragment*, through analyzing the process traces. In this analysis phase, the method engineer may decide to add a loosely modeled method fragment into the experience base. It can then be activated by the expert when facing this or a similar problem situation again, to guide him or her more efficiently through these tasks.

Graphical User Interface

As already partially visible in Fig. 4.9, several ways of *visualizing* the content of the PDW experience repository exist.

- The *generic* visualization is based on displaying the concepts, instances, relationship types and relation links, and the attribute types and attribute values in UML static structure notation. This is also called the semantic repository browser (see next subsection). It is the primary interface of the PDW front-end, mainly aimed at the domain specialists responsible for creating and structuring the domain models (also called the *Knowledge Engineer*). The generic visualization also offers the possibility to add icons and images to all elements (instances and concepts), to allow the domain end user to recognize the usual symbols connected with certain elements of his or her work, as the class and instance notation in itself may be too hard to understand. Examples can be seen in Fig. 4.9 and Fig. 4.11.
- The *semi-specific* visualization uses additional information about certain concepts, instances and their relationships to achieve a more conclusive presentation. As an example, the *ordered composition hierarchy* (tree) in Fig. 4.10 displays the composition and realization hierarchy of an extruder. All information necessary for this presentation is stored as meta-level annotations of the appropriate concepts (*Extruder*, *Functional Zone* and *Screw Element*), and their respective relations. Thus, the same presentation can be easily applied to any other model fragment that conforms to the same or a similar structure. Another possible application is to present a flowsheet as a network of devices and streams, by using a different kind of annotations.

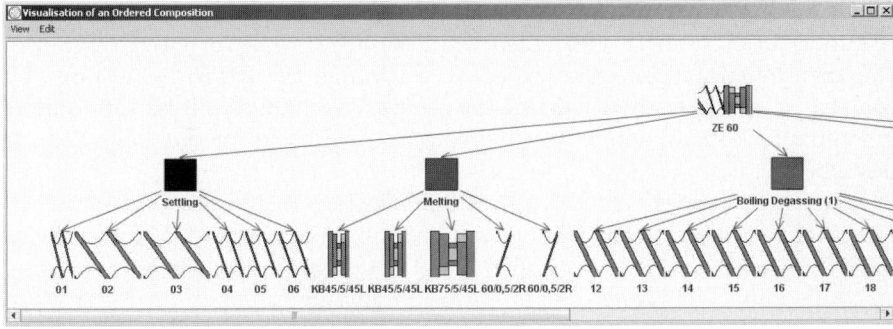

Fig. 4.10. Semi-specific presentation of an ordered composition hierarchy: the realization of an extruder

- For most application domains, *specialized* user interfaces for the PDW will have to be developed. At the moment, work is under way for a graphical interface to give production line foremen a concrete view of the production history of their line, e.g. process faults and countermeasures taken by the operators.

4.1.7 Cross-Organizational Cooperation

In the interdisciplinary setting of an engineering design process, knowledge and tools needed for a certain step are sometimes not found inside the company that is working on the design. Other companies then need to be integrated into the design process by delegating part of the process to them. While the AHEAD administration system in Subsect. 3.4.4 treats the administrative issues of passing part of the process to an external contractor, here we will concentrate on *experience-based guidance* of the *information transfer* itself between contractee and contractor.

In such a *cross-organizational* cooperation process, information needs to be passed based on strict rules of intellectual property and need-to-know. Besides the problems of business agreements, communications, and the technologies for information transfer, other aspects need to be addressed in this case.

Only very specific and selected information may be passed across organizational boundaries. The cooperation partner may only be allowed a restricted and well-controlled *view* onto the repository. To define such a view on the side of the contractee (the delegating party), cooperative discussions among the experts are necessary. The reuse of these traces in later cooperation cycles is enabled by recording the discussion results, decisions taken and their arguments, and the data to be transferred, into the PDW.

While the task is being solved at the cooperation partner's site, information may be recognized there to be missing. It has to be inquired from the contractee. The PDW allows to find this data and, possibly, find the decision that led to its exclusion. This decision may now either be revised (allowing the data to be sent), or confirmed. As a last part of the cooperation process, the information returned from the contractor needs to be reviewed, discussed, and then integrated into the project flow through the PDW's repository. Of course, the design of the PDW is kept generical enough to mirror the support functionality, if it is applied by the contractor, instead of the contractee as described here.

The three-dimensional BEM-based simulations in plastics engineering as described in Subsect. 4.1.4 will be used as a concrete example. The knowledge for this kind of analysis is usually found inside an extruder manufacturing company only. The initial design of a compounding extruder, as described in the IMPROVE scenario in Subsects. 1.2.2 and 4.1.6, is placed in a chemical company. Thus, the integration of the two scenarios needs the two companies to cooperate.

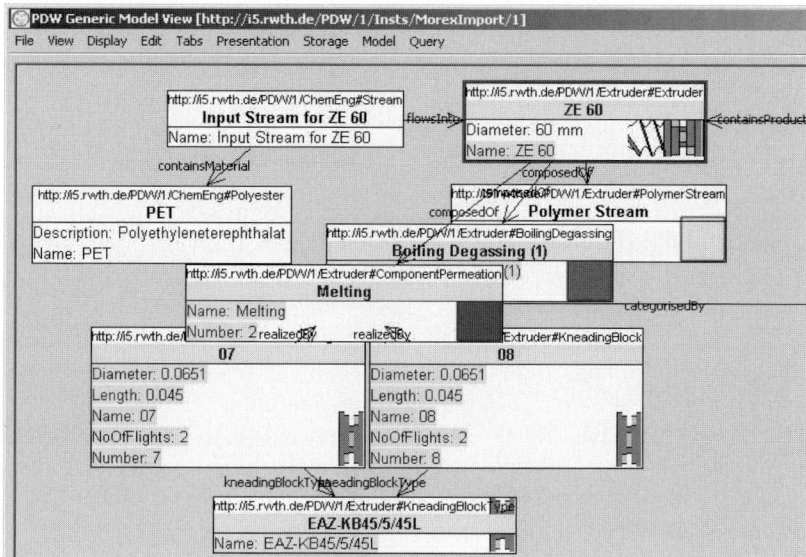

Fig. 4.11. The semantic repository browser with the extruder and related objects

The Process Data Warehouse (PDW) can be used here to help finding the necessary prerequisites. The designed realization of the extruder needs to be analyzed in reference to earlier projects. If an identical realization was already examined in an earlier simulation, it may be possible to reuse the results. If the examination of a similar realization was delegated before, and the results of it were acceptable, the same delegation may be sensibly used again. In the scenario at hand, it is decided to delegate this simulation to a certain external contractor.

The task and its requirements need to be discussed, and aspects like expenses and delivery dates need to be agreed on. Here, the aforementioned Negoisst system may be used [405]. The results of such a negotiation – the terms agreed on – might then be used as input information to the PDW, for tracing the fulfillment phase of the agreement.

On the contractee's side, the PDW as *data mediator* and integrated data storage should already contain all necessary information in an explicit form. Of course, it has to be examined and decided which information is needed for the delegated task, and which information may not be transferred, as it contains intellectual property. Using the conceptual object of the extruder as stored in the PDW, the semantic repository browser allows the user to browse and navigate through its database. Starting from this object, all other connected instances can be browsed, searched for, and displayed. The different types of relationships and dependencies, including generalization and specialization, can be followed. In Fig. 4.11, the instances, their relationships and their attributes can be seen in the UML static structure notation of the PDW front-end.

Based on the results of this semantic navigation, a concrete view can then be defined by "*marking*" some of the displayed instances, including certain attribute values, and certain relationships and relationship targets. As shown in Fig. 4.11, the extruder and some related instances are marked, including some of their attribute values. This marking is represented visually by coloring and highlighting the appropriate text labels. The extruder, two of its functional zones, the screw element realizations of these two zones, and their respective screw element types, are marked here. The information about the other functional zones and about the input stream is not marked, including the chemical component being extruded (PA6); neither are some instances from the organizational context like project, document or user.

The navigation path and the processes that led to this view, and the view itself, are recorded in the PDW as traces. This includes decisions about why to include or exclude certain data. The materialized view is then exported into an ontological (i.e. semantic) format like OWL.

This allows to transfer the data to the cooperation partner. If he supports the same – or a matching – domain model, the information can directly be used there. Otherwise, a transformation into some proprietary format, a generic or specific XML representation, or a structured textual form (e.g. HTML) can be done. The information itself can be transferred via (secured) electronic mail, or it can be accessed via the inter-organizational interfaces of the PDW which are based on web services and web pages.

As OWL supports the access and inclusion of ontology files via `http://` web addresses, the web interface of the PDW is able to offer the appropriate files for direct opening in OWL tools such as Protégé [979]. *Semantic Web Services* [831] as the coming standard for Enterprise Application Integration (EAI) may also be used here. By extending the syntactical interfaces definitions with their semantics, the semantic models can be directly integrated with the transfer mechanism.

Instead of navigating through the concepts and relationships to determine the data for transfer, an older view (possibly from a different project) can also be searched for in the experience base. The related information – the type of project, the cooperation partner, the extruder type and its realization – are used to graphically formulate a search query. This query is then executed on the experience base.

The resulting set of information views passed to contractors in earlier projects, are displayed in the PDW client. This allows to reuse them by "exchanging" the central concept instance (the extruder). The project, the user, and some other parameters also need to be bound according to the current situation which enables the PDW to automatically find the related data. The resulting view can be adapted, transferred, and recorded as described above. A view definition can also be derived from the view in a Query-by-Example manner, stored persistently, and possibly transformed into the semantic query language.

The scenario ends with the discussion of the information returned from the contractor. Video conferencing tools, as developed by the subproject B3, can well be applied here to allow direct communication between the experts of chemical and plastics engineering (cf. Subsect. 3.3.2). After entering the final decision into the PDW, earlier design steps might need to be revisited if a central assumption of the extruder design had to be changed, e.g., the rotational speed. In cooperation with the AHEAD administration system, PRIME and the PDW can follow the traces and determine the necessary steps (see Subsect. 3.2.6).

The scenario of cross-organizational cooperation has been described, as supported by the application tools of the PDW. Of course, many of the steps are designed to be used by, or in cooperation with, existing and newly designed domain applications, and thus as services offered by the PDW. This primarily includes the tools of the CRC subprojects themselves. As described in the beginning of this subsection, the AHEAD administration system realizes delegation- and view-based support for inter-organizational cooperation processes on the task management level (cf. Subsect. 3.4.4). The services offered by the PDW can be used to integrate the coarse-grained task management (*"What should the contractor do?"*, *"Which documents does he need?"*) with the semantic information access (*"What information does the document need to contain, so that the contractor can achieve the given task(s)?"*). On the fine-grained process support side, there is a tight connection between the PDW and the PRIME environment, especially concerning the recording of process traces, as described above for the side of the contractee. Last but not least, an important task of the PDW in this cross-organizational setting is the ontology-based information exchange, offering interfaces that are to be accessed by any kind of tool available at the cooperation partner's site, e.g., via the web or web service interfaces.

4.1.8 Summary and Conclusions

The conceptualization and implementation of the authors' Process Data Warehousing approach has been illustrated, as realized as part of the research done in the subproject C1. Aiming at *goal-driven information flow management*, three different prototypes have been described in their application and evaluation for engineering design process support.

The *Query Assistant* of the PDW has been designed to aid the PRIME process-integrated environment by extended situation analysis and method guidance. It has been demonstrated in the context of a process engineering design scenario, guiding the expert through the realization of a separation unit as part of a polymerization plant design. For the enrichment and annotation of weakly structured traces, the *TRAMP tool* has been described with its support functionality for plastics engineering simulations. The multimedia artifacts (videos and images) resulting from three-dimensional BEM simulations of compounding extruders, can thus be recorded and reused as experience traces.

The *Core Ontology* has been introduced and exemplarily extended with the domain model(s) of extruder design. The current prototype of the PDW, as realized on top of this Core Ontology, has been demonstrated in an example scenario from plastics engineering. It has been shown how the process steps and product artifacts are captured, structured, and later on offered for reuse in the scenario. Extensions for cross-organizational engineering have been introduced and described.

The research described here resulted in the Process Data Warehouse (PDW) as a process and product *tracing* and *reuse platform* for engineering design processes. Tools have been realized and integrated to directly apply this reuse functionality. Additionally, this service platform is offered to the other projects of IMPROVE. Using an integration-by-concept approach, a document management system was connected to the PDW to enable distributed and versioned file storage.

By integrating the partial models of several of the other subprojects – A2, A3, B1, B4, and the C3 modeling formalism – some support has already been realized, while other application scenarios are still in design or planning. Using the integrated models and the services offered by the PDW environment, the tools of these subprojects are provided with extended functionality; even basing their data directly on the models and storage functionality of the repository is possible.

For future research, several open problems need to be examined. The aspects of *cross-organizational cooperation* have only been treated in an initial approach that needs to be extended further. Also, there are plans to extend the traceability approach onto the full life cycle of a chemical plant, integrating the design traces with operation, modification, and reengineering processes. Work also has to be done in the transition between *process* and *control engineering*, as integrated support for passing and converting design information between those phases is mostly missing.

For the plant operation phase, the integration-by-concept scheme is being extended to enable the integration of *ERP systems* (Enterprise Resource Planning, especially SAP) on the one hand, and the shopfloor MES (Manufacturing Execution Systems) and other operation data recording systems on the other hand. The latter systems are usually even more heterogeneous than those used during design. Often, no integrated capturing of the data is done at all. A project is in preparation to transfer the results of the research presented here, into the design and management of rubber recipes, and onto production control systems of the extrusion processes based on these recipes (see Sect. 7.5).

4.2 Service Management for Development Tools

Y. Babich, O. Spaniol, and D. Thißen

Abstract. Tools used in development processes in chemical engineering and plastics processing are often highly *heterogeneous* with respect to the necessary software and hardware. In the previous chapter, several functionalities for improving a development process in different aspects was presented. But one problem still remains: when coupling such heterogeneous tools, many technical details are to be considered.

This section describes a *service management platform* that aims at *hiding such technical details* from new tools as well as from developers, and at ensuring a *performant execution* of all tools and services. The integrative support presented in this section is located at a lower level. Here, the focus is on the provision of a transparent, efficient, and fault-tolerant access to services within the prototype developed in IMPROVE. A framework was developed that allows efficient communication support, and the management of both, integrated tools and external services. The framework also allows an a-posteriori integration of existing development tools into the management platform. Framework and management are applied at platform level in this section.

4.2.1 Introduction and Problem Area

Tools and methods described in Chap. 3 implicitly rely on *infrastructure components* at a lower level of abstraction. These are hardware, networks, operating systems, and distribution platforms as opposed to the higher-level logic of direct support for developers [264]. This section deals with a service management platform that offers common interfaces for *communication and cooperation* to other tools and internally maps physical resources to logical ones (see Subsect. 4.2.3). This *frees* the other projects from the need to consider *technical details* of platforms, tools, services, and documents [443].

The *solution* presented here is *based on CORBA* [877], a wide-spread and mature middleware platform. CORBA provides direct support for a multitude of service management aspects in distributed systems [262]. Furthermore, there are many stable implementations. Thus, a first step was the selection of an appropriate CORBA implementation, based on an *evaluation of the runtime characteristics* (see Subsect. 4.2.3). For an a-posteriori integration of non-CORBA-based tools (e.g. those which were designed for Microsoft Windows) with the CORBA infrastructure, Microsoft's *COM/DCOM* [847] interface is used and *wrappers* were developed to integrated those technologies (see Subsect. 4.2.4).

At the same time, by introducing a service management platform which separates tools, documents, etc. from physical resources, *management of efficient and fault-tolerant execution* of development tools can be integrated transparently [422]. The service platform includes various functionality to do so. A basic functionality is the integration of a *service trader* which can find services and tools that can perform certain tasks with a certain quality. These tasks include e.g. simulations and direct communication. Here, response time is most important, thus the *service selection* optimizes the response time (see Subsect. 4.2.5).

Furthermore, the tools and services must be *configured* according to the needs specified in the higher layers of the environment. During the execution, applications must be *observed*. In case of a failure they must be either *recovered*, or the execution must be *transferred* to another instance of the same (or a compatible) application (see Subsect. 4.2.6). In addition, *Web Services* technology is supported as more and more such services are available on the Internet (see Subsect. 4.2.7).

4.2.2 Overview of Related Work

A comparison with some closely related publications is provided in the following subsections. Here, only a brief *overview* of related work in the area of service management and trading in distributed systems is given.

While service management systems for specific application domains are generally disregarded, definitive progress has been made in standardizing *application domains as part of middleware platforms* [878]. The *virtual enterprise* concept is similar to the scenario discussed in this book. General approaches to implement a framework for the management of virtual enterprises are suggested, for instance, in [893].

Trying to optimize the runtime performance of the tools in the development environment, requires an adequate *Quality of Service (QoS)*. QoS in the network can be implemented with IntServ, DiffServ or MPLS [1068]. However, in the scenarios considered here, a host's performance usually plays a greater role for the overall performance. Moreover, no QoS management is available in the application domain considered. There are, however, other, more flexible approaches, e.g. the application of mobile agent technology to network and system management, evaluated for example in [557] and [1001].

Service trading is a mechanism that supports the selection of one out of a set of similar services offered by different providers. An early work to consider *performance aspects in service selection* was [1056], which proposed the usage of a "social" service selection mechanism. Such a mechanism does not guarantee an optimal selection for each client, but tries to optimize the global behavior of a system by using load balancing. *Load balancing* in distributed systems was a successor of techniques such as scheduling of batch processing in time sharing systems [965]. Papers such as [953] describe so-called relative load metrics for the load balancing process.

4.2.3 Architecture of the Management System

This subsection considers two topics: On one hand it gives an overview for our service management system by sketching its architecture. On the other hand it describes the analyzes we have carried out before deciding for CORBA as the underlying implementation of our system.

The Architecture

The *architecture*, depicted in Fig. 4.12, reflects the tasks that are performed by the *management platform*. Task distribution, performed primarily by the service trader and load balancer (see Subsect. 4.2.5), plays a central role here. Configuration, performance, fault, and security management are separate entities and are described in Subsect. 4.2.6. The problem of the integration of legacy tools is described in Subsect. 4.2.4.

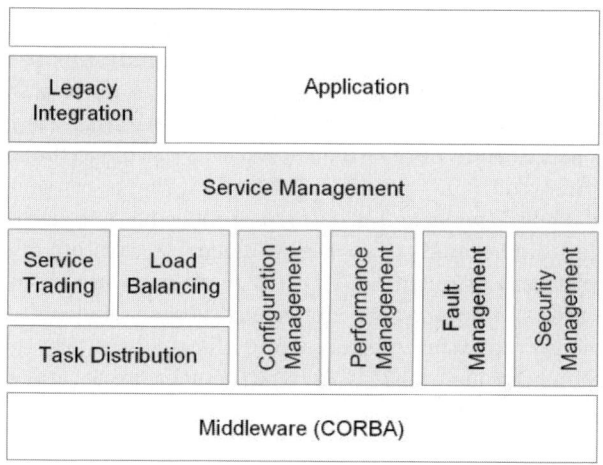

Fig. 4.12. Architecture of the management platform

The *management system relies* on the *Common Object Request Broker Architecture* (CORBA) to deal with the problem of heterogeneity. CORBA is a prominent example of a distribution platform [877]. The CORBA architecture consists of *five functional areas* [1024]:

- *Object Request Broker* (ORB), which is responsible for transparent communication between objects, i.e. clients and servers,
- *CORBAservices*, that perform basic tasks for implementing and using objects in a distributed environment,
- *CORBAfacilities*, a set of generic services,
- *Domain Interfaces*, oriented towards a specific application domain, and
- *Application Services*, developed by third party companies.

The *data transfer* in CORBA is realized by the ORB which transmits requests and results between clients and servers. This mechanism is based on the *Remote Procedure Call* (RPC). Data structures that should be transmitted must be defined in the CORBA *Interface Definition Language* (IDL). All objects must be described as IDL data types. Prior to sending, the ORB serializes (marshals) these objects and transfers them as strings of octets.

As all communication was envisaged to be based on CORBA, the first step was an *investigation of data transfer methods provided by CORBA*, to find out if the communication mechanisms are suitable for the application domain.

Evaluation of Data Transfer in CORBA

The CORBA implementation *Orbix* from IONA Technologies [750] was used in the project. Yet, before implementing all project software on top of Orbix, it was *examined* with regard to *performance in data transmission* [447]. Orbix is a fairly efficient implementation. It implements large parts of the CORBA specifications, and has a very broad installation base in companies. Another important criterion was Orbix' provision of extended features that support object and communication management. Although it had a number of deficiencies at the time of decision making, it was clear that it was rapidly developed further [421].

Studies on Orbix (version 2.3; later a transition to a newer version was made) consisted of the analysis of the influence of different hardware architectures on communication in CORBA, the overhead produced by communication operations, a comparison between different modes of data transfer, the applicability of updating methods, and the realization of multicast communication. Synchronous, oneway, deferred synchronous, and the use of the CORBA Event Service for asynchronous communication were investigated. The data transferred in each experiment were strings, due to the lack of container objects in CORBA.

The following *results* were obtained:

- Relatively *low data transfer rates* of some Kbyte/s reduce the usefulness of CORBA. Even if concrete values depend on server and network infrastructure, the general conclusion is that CORBA should not be used for transmission of continuous data streams like e.g. video or audio.
- For *direct transmission*, the synchronous data transfer mode is more suitable than the asynchronous mode. If *multicast* is required or frequent disruptions occur, the asynchronous data transfer mode is preferable.
- For all transfer modes, the transmission of data within a small number of *large packets* is the most efficient method.

Furthermore, it is important how *servers obtain continuous data* (e.g., load data from monitors, see the next subsections). We considered two strategies called *caching* and *polling*. Caching involves a 'passive' server that regularly collects data sent by other system components, whether or not the information is actually needed. In case of polling the update process is initiated upon request from the server.

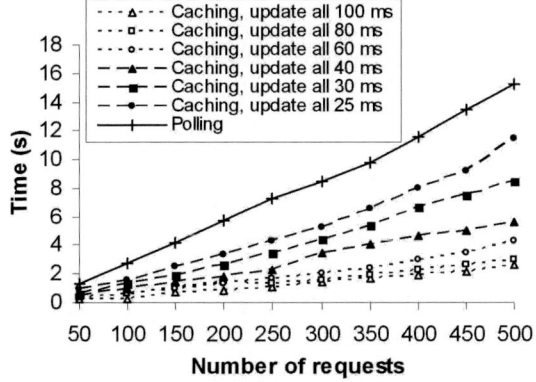

Fig. 4.13. Comparison of caching and polling performance in Orbix

To *compare* these *strategies* the following *scenario* was set up. A client requests a service from a server, which in turn needs information from three other system components to provide this service. This information is transmitted as a sequence of strings, using polling and caching.

Figure 4.13 shows the *results* obtained for this scenario. It can be seen that the *requests* are processed *faster* when *caching* is used rather than polling, even if the server is under high load through a large number of requests and small update intervals. Thus, caching seems to be the more suitable method. However, the *network load* caused by caching must not be ignored. In case of 500 client requests, polling achieves a transfer rate of about 0,1 Kbyte/s. The caching method with one update each 25 ms needs to transfer more than 60 Kbyte/s. This value is independent of the number of requests, whereas the transfer rate in the case of polling decreases with a reduced number of requests. Scalability is another difference between caching and polling. Because of the transmission process structure, *polling* has no *scalability* problems. In contrast, there are limits for the caching mechanism, see Fig. 4.14.

These results had shown that it is not recommended to do all data transfer via CORBA. For certain data types, e.g. video and audio data, other transfer mechanisms had to be integrated. Nevertheless, for small amounts of data, especially control information and for monitoring purposes, it was decided to use CORBA because of its ability to handle heterogeneity.

4.2.4 A-posteriori Integration

A coordinated interworking of tools is necessary since failure or blockage of one component must not defer the entire development process. Hence, bottlenecks and breakdowns have to be found and eliminated in the fastest possible way. Consequently, a service management *system* for managing the *execution* and *interaction* of the development *tools* and the supporting services is needed

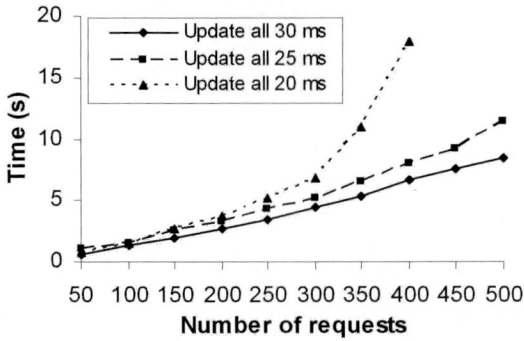

Fig. 4.14. Scalability of caching

to guarantee high availability, fault tolerance, efficiency, and reliability. For the reasons discussed above, CORBA was chosen as basis for this system to handle heterogeneity and to enable the cooperation of the development tools.

Unfortunately, existing tools usually do not provide an off-the-shelf CORBA interface out-of-the-box, which would allow for control operations to be performed. Often, however, the tools can be controlled through other, operating-system-specific or proprietary interfaces. In order to integrate tools into our CORBA-based environment, *wrappers* were developed that perform the communication between the CORBA service layer and the tools. Thus, the tools are represented by CORBA interfaces. Wrappers were enriched with tool management functionality and constitute so-called *management proxies*.

Access to Legacy Applications

There are several potential ways of *integrating* the *management extensions' functionality* into systems like those developed in IMPROVE.

- *Direct integration*: This is a very specific approach, where the management code is directly embedded in the original source code. It allows full control over the application and provides all information necessary for the management evaluation. Such an embedded approach, however, is not advisable, as it is application-specific and very difficult to maintain and to modify. This is often even impossible because the source code is not available.
- *Interface extension*: Object interfaces in CORBA are defined in IDL, in order to hide implementation details and heterogeneity in distributed systems. Extending the IDL specification with management functions, yields a modular approach. The management functionality provided by the object can be accessed by management components via the standardized interface and thus be modified without having to reconstruct the entire

system. However, although the process of adding the code can be automated as part of the IDL compilation process, it still requires the explicit modification of source code.
- *Linking*: Adding management functions to an object at a later point can be done through the linking process. Pre-configured libraries containing the management part can be linked to the original application.

Although the linking approach does not require an explicit modification of an object's source code, it still requires the code to be available for re-compilation. However, in the case of management of the development tools, problems arise from the fact that an *a-posteriori integration* has to be made. As the source code of most development tools was not available, the approaches mentioned above cannot be used. Instead, a new approach was needed which is capable of *adding* a *uniform management functionality* to each of the development tools.

Unfortunately, the tools offer *different capabilities* for adding management functionality. Some provide *OLE interfaces*, which enable access over a CORBA-COM interface. Although the general approach is the same, no uniform access functions are given, because the OLE interfaces of the applications are different. Some applications offer *CORBA bridges*, which usually provide limited access operations. Some applications, however, offer neither.

In order to provide a uniform management functionality which allows managers to access these applications via standardized interfaces and thus avoids proprietary and application-specific management solutions, a *new approach* was chosen, which is presented in the following.

Management Proxies

For the integration of legacy applications into a CORBA-based management system, different approaches are possible: The most simple approach is to *use only the management information* which can be *provided by the ORB*. This could be the number of requests to a server or the response time of the server. This can for example be achieved via so-called interceptors, see the CORBA specification [877]. Yet, this approach merely gives minimal access to servers, and only a limited set of management information can be used. Furthermore, this approach requires applications to provide a CORBA interface.

A more promising approach deploys the concept of management wrappers, or *management proxies* [275], see Fig. 4.15.

Such a proxy *hides* the *details* of an application by encapsulating its interface and offering a *new CORBA interface* to clients and management applications. The requests made by clients are simply passed to the original interface of the application. This way, for a client the functionality of the application remains the same. Internally, the wrapper contains additional functionality, which is made available to management applications (as well static management components as mobile agents) through a uniform CORBA interface.

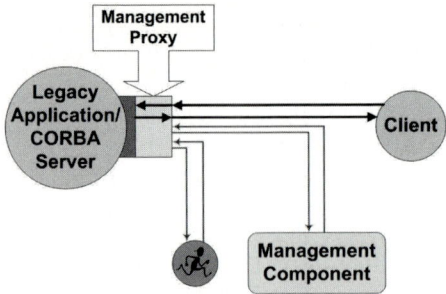

Fig. 4.15. Management proxies

Thus, the service manager itself does not need to adapt to each managed application; it has one *general management interface*, which is mapped by the proxy internally to specific application interfaces. This concept allows a manager to obtain more detailed management information and to perform more complex management actions by using the specific interface of an application and additional features of the wrapper, for example measurements, surveying of thresholds, and statistics. Thus, this concept is *suitable as an a-posteriori approach*.

Moreover, the uniform interfaces allow *new management approaches* to be easily deployed. For instance, a mobile agent locally executing a management task can access the applications management data via the same methods as a central management component. Based on the needs of the aspired environment, this structure of the wrapper's interface was determined. It must be taken into consideration that the introduction of the management wrappers *adds an overhead* to the underlying system, causing *increased execution times*.

4.2.5 Trader and Load Balancer

To save cost and time, the *tools and services* in a development process should be *executed* as *efficiently* as possible, i.e., if there are several instances of a service that could be integrated, the cheapest one or the one with the shortest processing time should be used.

Thus, a method for assessing and *choosing tools and services* with respect to their *quality* was developed [265]. The term quality, or *Quality of Service*, refers to non-functional properties of a service, for instance cost, performance, or availability. Suitable services must be found by the management layer upon request. It is of further substantial benefit if applications and services, such as simulators or communication tools, can be selected from a set of interchangeable instances. This is very likely to be the case in collaborative distributed scenarios.

The choice of a service that meets the needs of the caller is performed by a component called *service trader*. This is a variation of a name server, where searching for services is done in terms of service type (the kind of service) and non-functional service properties. The CORBA specification includes a *standard trading service*, which has, in our view, a major *drawback* (e.g., for multimedia applications): It only performs a hard match (yes/no) of the service parameters onto the request parameters, and does not allow for selection of services that do not exactly fulfill the requirements but have almost the desired properties.

Trader

Our approach [265, 440, 445] *enhances* the simple selection mechanism of the CORBA trader to *consider QoS*. Here, an importer (which is the client using a trader to search for a service) describes a service by service type and service properties as before, but it is possible to *specify roles for several values* of the same property, namely a target value for the property, a lower bound (minimum acceptable quality) and an upper bound (maximum useful quality). This enhancement allows a client to formulate wishes and limits on service properties, see Fig. 4.16. Additionally, preferences for specifying the importance of properties can be expressed. These values, describing different aspects for the same quality criterion, are called **Service Request Property**. The importer can specify a whole vector of such properties to express its complete requirements on a service through different QoS aspects. In a similar way, a server offering a service can describe its limitations and capabilities by expressing upper and lower bounds for the quality aspects it can deliver.

When a client searches for a service, it now formulates the **Service Request Property Vector** and sends it, together with the searched service type, to the trader. The trader matches it against all **Service Offer Property Vectors** of services with the same service type, and selects the best match. In this context, matching means that a *quality score* is computed for each service offer, and the service with the *lowest value* is chosen as the *best service*.

Service Request Property Vector
$\mathbf{r} = (r_1, ..., r_n)$ with

Service Request Property
$r_i = \{$ Target t,
Upper Bound u,
Lower Bound l,
Preference p
$\}$

Service Offer Property Vector
$\mathbf{o} = (o_1, ..., o_m)$ with

Service Offer Property
$o_j = \{$ Upper Bound u,
Lower Bound l
$\}$

Fig. 4.16. Service properties in the request and offer descriptions

The matching is done in two steps. First, a so-called *distance* is computed between the single elements of **Request** and **Offer Vectors**. For that purpose, a set of rules was defined [265] to consider all roles which can be specified for one element. The result of this computation is a vector of differences between the *client's demands* on certain quality aspects and a *server's capabilities*. A number of distance functions on vectors are well-known in analysis, especially the maximum, Euclidean, and Manhattan metrics. Those were integrated to map the difference vectors to scalar values, assessing the usefulness of a service for the client's demands. *Minimization* over the values for all available service instances returns the best fitting service, and the client can start using this service directly.

Load Balancer

Trading is a valuable concept to support binding between clients and servers in large open systems. Still, individual servers can be overloaded. To avoid such an overload, *load balancing* can be used. A load balancer tries to realize a perfect distribution of the clients' requests to the available servers. Yet, it can only select one server in a particular group; in large open systems, the load balancer would have to know exactly the type of the service to select a server. Additionally, the load balancer can only select a server based on its load, not on its service properties.

Thus, a *combination of trader and load balancer* seems to be a suitable solution for a load-oriented assignment of servers to clients in a distributed environment [439, 441, 442]. A *simple approach* would be the usage of load values as dynamic service properties within the trader. The trader could perform a load distribution based on these attributes. Yet, it could be hard or even impossible for a service provider to offer an additional interface where the trader could request the information about dynamic attributes, especially in cases where legacy applications are used. Furthermore, this concept is inflexible, as a more differentiated interpretation of the load value would be hard. Therefore, this simple solution is *not feasible* in our environment.

For a better *enhancement* of a *trader* with a *load balancing mechanism* some *design issues* had to be kept in mind: It must be possible to use the trader with or without load distribution. The load distribution process has to be transparent for the user, but it should offer the option to influence the process, e.g. by defining special service properties. Such properties could refer to the information whether or not load balancing should be performed, or which influence the load parameter should have compared to the service quality. The load balancer should be integrated into the trader to achieve a synergistic effect by exchanging knowledge between trader and load balancer [448, 449]. Furthermore, the load balancer should be flexible to enable the use of several load balancing strategies and load definitions.

The *architecture* of the *enhanced trading system* is shown in Fig. 4.17. A monitor is installed on each server host to observe all local servers. The

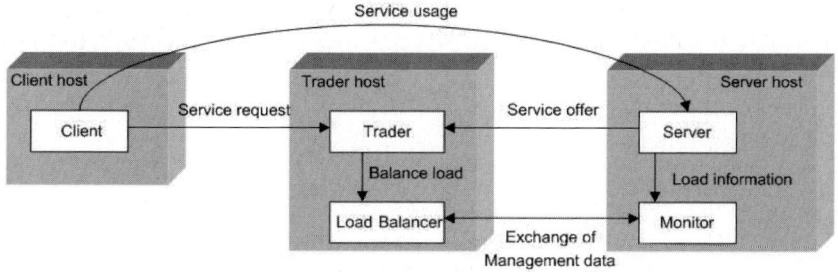

Fig. 4.17. Architecture of a trader combined with a load balancer

monitor is connected to the load balancer, which is located on the trader host. A client imports a service from the trader and uses the selected server.

Service usage can be determined using a variety of *metrics*, e.g. the CPU load, the network load, or the load caused by I/O operations. To determine the CPU load, the servers' queue length, the service time, and the request arrival rate can be used. Each participating server is equipped with a sensor, which collects this information and sends it to a monitor. As most applications in our scenario are legacy applications, the management wrappers were used to enhance an application by the necessary functionality. Load information includes the *service time*, the *process time of the service*, the *available CPU capacity*, and the *queue length*. The load information is passed to the monitor. This *monitor* manages a local management information base of *load information* and enables the load balancer to access it. It has a list containing all hosted servers and their respective load. As different load metrics should be possible, all load information - which is transmitted by a sensor - are stored.

The monitor not only stores the received load values, but also *calculates* additional, more *'intelligent'* values. This includes the computation of a floating average value for the load values as well as an estimation of the time required to process all requests in a server's queue. This estimation uses the mean service time of the past service usages and the time for the current request to estimate the time the server will need to process all requests.

As no outdated load information should be used by the load balancer, a monitor uses a *caching strategy* to *update* the *load balancer's information* at the end of each service usage. Some values, e.g. the queue length and an estimation of the time to work, are also sent to the load balancer upon each start of a service usage. Based on the load values' access and change rates, a *dynamic switch between caching and polling* can be performed. This mechanism is shared by load balancer and monitor. In case of the polling strategy the monitor knows about access and change rates, thus it can switch to the caching mechanism. On the other hand, if caching is used the load balancer has this information and can switch to the polling mechanism if necessary.

Based on a client's service specification the trader *searches its service directory*. Services meeting the specification are stored in a result list. The *sorting* of the service offers is done according to the *degree of meeting the client's quality demands*.

For the integration of a load balancer this sorting is not sufficient, as the *servers' load* must also have an *influence on this order*. Thus, we had to introduce some modifications to our trader. When a new entry is added to the result list, the trader informs the load balancer about the corresponding server. As the load balancer only knows about load aspects, it cannot do the sorting according to the client's constraints. To enable the consideration of both the trader's sorting and load aspects, the trader must assign a *quality score* characterizing the degree to which the client's requirements are met by each service offer.

After searching the whole service directory, the trader *calls the load balancer* to *evaluate the most suitable service* offer instead of sorting the result list relating to the quality scores. To influence the evaluation, information about the client's weights regarding quality score and load are also passed to the load balancer as well as the metric to combine both values. The load balancer returns an index identifying a service offer in its list. The object reference belonging to this offer is returned to the client. In addition to the load balancer's mechanisms the trader implements a random strategy to determine an order for the services found. This can be seen as a static load balancing strategy which can be used for evaluating the gain achieved by using the load balancer's strategies.

The *load balancer manages two tables* (Fig. 4.18). One contains the management information about the known servers (*ServerMonitorTable*). In this table, each server in the system is listed together with the monitor responsible for measuring the load, and the load itself. The other table, *ScoreTable*, is created when the trader receives a service request. Each service offer found by the trader for this request is recorded in the table together with the quality score computed by the trader.

After the trader has searched the whole service directory, the *load balancing process* begins. The approach chosen here consists of two steps. First, the load for all recorded servers is obtained from the *ServerMonitorTable* and inserted into the *ScoreTable*. Getting the load for all service offers at this time implies that no old load information is used. The 'load' field in the *ServerMonitorTable* does not contain a single value, but a set of load values for all different load balancing strategies.

Currently, *three strategies* which try to *minimize the system's load* regarding to a particular load metric are implemented:

- *Usage_Count* (UC) only counts the past number of requests mediated by the trader.
- *Queuelength* (QL) considers the current number of requests in a server's queue.

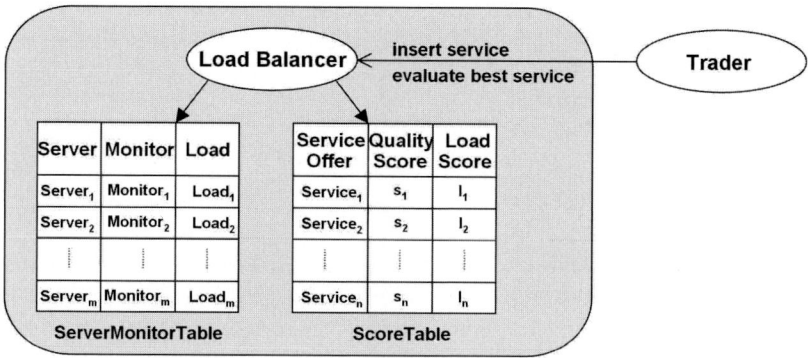

Fig. 4.18. Internal data used by the load balancer

- *Estimated_Time_to_Work* (ETTW) calculates the estimated time a server will have to work on the requests currently in its queue.

The *load value* represents a *score* for a server. That is, the server with the lowest score has the lowest load. The load values corresponding to the chosen load balancing strategy are copied into the *ScoreTable*. The second step combines the score obtained by the load balancer with the quality score calculated by the trader. Metrics are used to calculate an overall score for each service offer.

Measurement results indicate that the random strategy yielded the worst request distribution with respect to mean response time, which was to be expected (Fig. 4.19). UC is second worst in high load situations, as it does not consider the service times for the incoming requests. For lower system loads, UC is more suitable than ETTW. As the service times vary heavily, errors may occur in the estimation of ETTW, which then cause a wrong decision for the next request distributions. Only for high load situations this error is smaller than the unfavorable distribution caused by UC. The error in estimation is also the reason for ETTW performing poorer than QL. QL only counts the number of unserved requests; in this case, a distribution without more information about the requests is better than using potentially wrong information.

The optimization of service selection with respect to the servers' respective load is a *worthwhile enhancement* of the trader. The *response times* of servers offering a service which is available in several places can be *significantly reduced*. The cost for this advantage is an increased service mediation time, but this overhead is very small. The usage of trader-internal knowledge, like a server's number of mediations, is of substantial advantage only in idealized scenarios. In a heterogeneous environment, it does not help to significantly improve the load situation. For such environments, *dynamic strategies* are more suitable. A simple strategy, e.g., considering the servers' queue length, proves best for most situations. The weighting of the load influence against the ser-

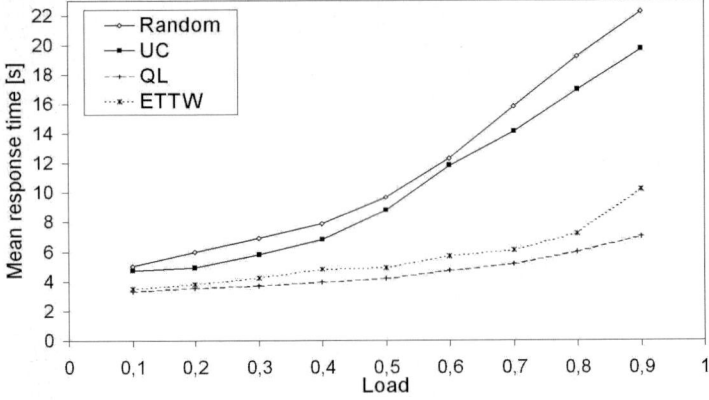

Fig. 4.19. Gain of load balancing strategies depending on the system load

vice quality should be the user's choice, but some experiments indicated that it is best to give them equal weight. Network transfer quality was a topic also considered, but it had been shown that much more effort is necessary than justified [114].

4.2.6 Service Management Components

A service trader is only used to support the binding between a client and a server; after the binding process, the trader is no longer involved. That is, the trader cannot guarantee that the quality is kept during the whole service usage process. Therefore, a *management system* is needed to *control* the execution of all *applications* and their runtime behavior.

In classical network management, such functionality is subdivided into five categories: *configuration management, fault and performance management* (which we treat as one category), *security management*, and *accounting management* (not relevant in the given context). The realization of the functionality is done separately for each category.

Furthermore, the management system is split into *management proxies* that reside on the same hosts as the managed applications (see Subsect. 4.2.4) and the *management components* that are responsible for particular tasks (fault, security, etc.) and can reside somewhere in the network or can be implemented as mobile agents (see below). A *hierarchical approach* to create a structure of management components for scenarios with cross-enterprise cooperation is discussed in [438]. In the following, only the management components for the different functional areas are described.

Dynamic Configuration Management

In network and system management *configuration* refers to the *placement, adjustment, and interworking of hardware and software components* in a compound of applications. Providers of developer tools can suggest configurations in order to provide better performance or better reliability. Additionally, a system administrator can define optimality criteria. The decision about configuration parameters can depend on the current usage pattern, so dynamic reconfiguration may become necessary.

In our system, the configuration manager *supplies control data* (such as parameters that are to be monitored) to the management proxies and monitors at run time. It is based on the *Common Information Model (CIM)* [636], which is a standardized model for representing various aspects of management of applications, systems, and networks.

CIM is subdivided into *three layers*: the core model, common models, and extension schemas. The core model consists of a small number of classes for a unique description of all components. Common models enhance this basic information for special domains like database, application, or networks. Extension schemas can be defined for adopting the descriptions of the common models to specific products.

To use CIM, several *enhancements* had to be made. A *Restriction* class was introduced to describe situations when a product licence is limited to certain network addresses, or to describe information necessary to enable high-performance execution of tools. As CIM lacked a user concept, we extended it by a *User* class [438].

We subdivided the system description into the *static structural model* view and the *dynamic configuration view*. The current configuration of the development environment based on the information model is provided by the configuration manager. It holds all static environment information. The configuration manager is dynamic, i.e. it can reconfigure the environment at runtime. To ease the use of the service in the distributed environment, it is implemented according to the CORBAservice paradigm.

Run-time information about integrated tools is obtained with the help of *management proxies*, as described in Subsect. 4.2.4. Additionally, *host monitors* (Subsect. 4.2.5) provide application-independent performance parameters of systems and networks.

From the specification of a service given by the information model *policies* are generated which define rules for monitoring and controlling the service. Management proxies and host monitors obtain policies from the configuration manager. They are *instructed* about the parameters which need to be monitored during service execution.

The *system administrator* can use the configuration manager to *formulate restrictions* for applications, to *install* new software, or to make *reconfigurations*. Additionally, the administrator can obtain a dynamic view from the monitors via the performance and fault service. In case of a performance problem or a fault, the dynamic view provided by monitors, and the static view provided by the configuration service, can be used by the system administrator to *detect the problem's source*, choose a reconfiguration method, and compute the new configuration. One approach for partly automating this error handling, involving mobile agents, is described in the next subsection.

Distributed Management with Mobile Agents

For effective performance and fault management *mobile agents* were employed [269]. Mobile agents are autonomous pieces of application code that can move from host to host carrying their code, data, and status with them. This leads to local instead of remote communication, which reduces network load and response times.

There are special *platforms for mobile agents*, but inter-agent communication can also be based on CORBA [266, 267, 273], which integrates them better into the management system and allows to use the agents for managing CORBA objects.

The general assumption that mobile agent based solutions perform better than those relying on remote communication is not correct, as it ignores the impact of the *overhead* introduced by *mobile agent migration* [263]. Rather, the benefit of migrating agents depends on many factors such as mobile agent size, size and number of communication requests, parameters of the underlying network, etc.

Using mobility only in certain situations can help to improve performance; thus the concept of *strategic mobility* was developed in which an agent decides whether or not it makes sense to migrate or if a remote call is to be preferred [270–272, 274]. To develop a decision algorithm for mobile agents which enables them to *decide* if *migration or remote communication via RPC* performs better in a given scenario, we first had a look at the basic properties of both alternatives. We evaluated the respective execution times in different scenarios to provide the agents with decision rules based on observations of the number of interactions, the size of requests, and the status of the network. The strategically mobile agents can decide freely whether or not to migrate prior to processing a task.

To *evaluate* this *new mobility concept*, the *expected response times* have been calculated. The same has been done for static agents, which always communicate remotely, and for mobile agents which exclusively communicate locally, i.e. always migrate prior to communication. At the same time, several ratios of correct and erroneous decisions of strategically mobile agents were compared to examine best and worst cases of agent decisions.

The *results* are shown in Fig. 4.20 (left). The more a scenario is in favor of RPC communication, the better RPCs will perform and the lower the expected response time will be. For mobility, an analogous situation can be observed.

Figure 4.20 also shows that the *performance of strategically mobile agents* can *differ widely*, depending on the quality of the decision process. The two extreme cases of a strategically mobile agent are also shown in the left graph of Fig. 4.20: A strategically mobile agent always making the right decision on the one hand vs. one always making the wrong decision. Between these extreme cases there is a wide range of possible outcomes of strategic mobility, depending on the decision algorithm of an agent.

The right graph in Fig. 4.20 reveals a *high potential of strategic mobility* with regard to performance. Compared to pure RPC and migration based task processing, the strategic mobile agents can achieve a high performance improvement. However, it is important to keep in mind that inadequate decisions can also considerably deteriorate the performance.

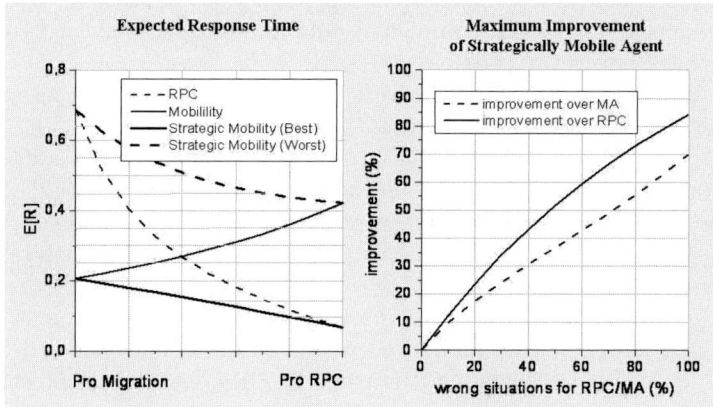

Fig. 4.20. Evaluation of strategic mobility

The usefulness of migration depends on the interaction pattern. We *abstained* from using *mobile agents* for the collection of dynamic service properties required by the *load balancer* [362], because the interaction pattern suggests to use the PRC.

Nevertheless, *mobile agents* were used for *detecting network and system problems* which can alter the service properties as considered within the trading process [361, 364].

Error Detection with Alarm Correlation

Frequently, there are *complex dependencies* between services and some events that occur during their execution. In such cases it is *not easy* to find the *reason* for a *particular error* which occurs during interaction between several tools and services.

In common network management, network or system components (which in this context are treated as *managed objects*) can notice an exception, i.e. an error or a bottleneck. In this case they send an *alarm message* to the *responsible management component*. Similar to this standard approach, the *tools* in the application area are *integrated as managed objects* by the management proxies. The management proxies can sent out alarms if necessary.

To provide a management that is transparent to the upper layers, these *alarms* have to be *collected and analyzed*, to pinpoint and solve the problem. Often, only a *specific combination* of alarms is *meaningful*. For instance, when simulation data is sent from one tool to another and the data flow stops, information on, e.g., the source host, the source application, the network, the network subsystem of the target host, or some combination of them can be the source of the problem. The process of finding a meaningful combination of alarms is called *alarm correlation*. Usually, a set of rules is defined that describes dependencies in a complex system and allows for locating the component which causes the problem.

In a first step, a *system for alarm correlation* to collect data and react on alarms was developed purely *based on mobile agents* [268]. The system's drawback was *poor scalability*. For large systems with a multitude of alarm types and complex correlations, the logic of mobile agents would grow and make the migration overhead unacceptable. This could be solved by using only simple correlation mechanisms, but then the error recognition rate would deteriorate considerably.

For this reason, an *alarm correlation service* was designed as a static component, only supported by mobile agents [272]. It should be noted that this service is implemented like a CORBAservice, thus there could be several cooperating instances of the service distributed over a large system, responsible for individual administrative subsystems.

The *structure* of an *alarm correlation service* is shown in Fig. 4.21. The three main functionalities of the correlation service are

- collection of static and dynamic information and suitable representation,
- collection of alarms, and
- correlation of alarms and providing results to mobile agents.

Each correlation service is composed of *three modules*, which correspond to those three functionalities. The *information module* collects static and dynamic information on all system components, builds a class hierarchy, and constructs a correlation graph. This graph is used as input to the *correlation module*, which converts this information into a system matrix, holding all information about correlations of alarms and sources of errors. A mobile agent does not need to make a correlation itself, it can simply contact the correlation module, which constructs a codebook according to the correlation requests of the mobile agents, containing only the information relevant for the mobile agent. The *monitoring module* takes care of the collection of alarms and reports these alarms to the correlation module.

The correlation services *communicate on two levels*. On the correlation module level, they exchange subscriptions of events for cross-domain correla-

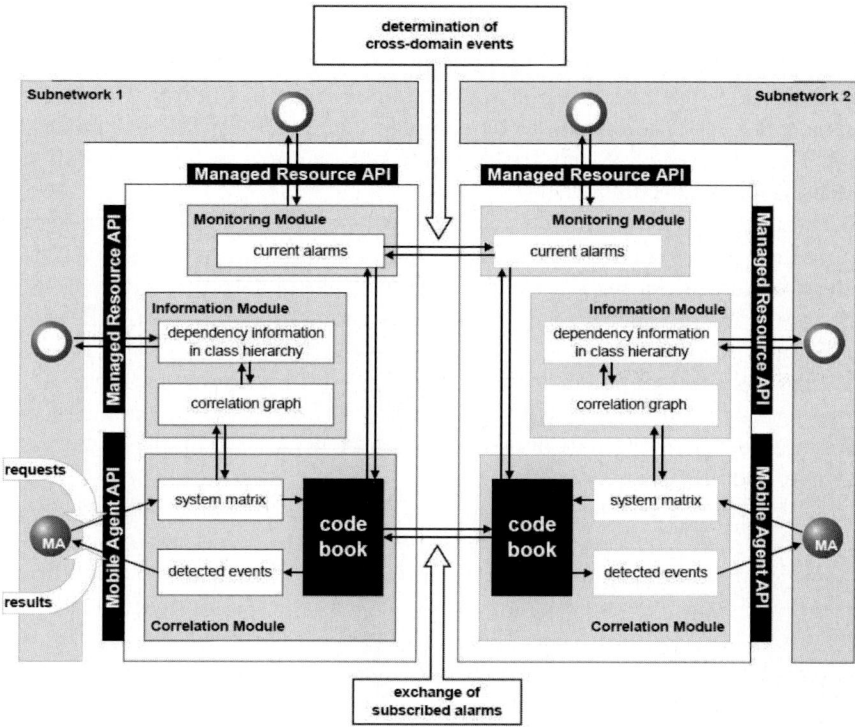

Fig. 4.21. Structure of the alarm correlation service

tions. On the monitoring module level, they exchange collected events according to these subscriptions.

The *role of the agents* is *restricted to error handling* and does not involve error detection. In contrast to the attempt to realize alarm correlation only with mobile agents, this implementation shows much better scalability. Unfortunately, an automated reaction to identified problems often is not that easy to realize, as it usually needs a system administrator to react (e.g. in case of hardware failure). Thus, for some faults, the correlation service only sends a notification to the administrators.

Security Management

Data transmitted between computers may be subject to corruption and theft. Industrial espionage can aim at obtaining simulation results and information about the workflow, or an adversary may want to manipulate important data, which would lead to incorrect results. Thus, *security* management was an *important* part of the project.

Main Concepts of Security Management

One part of the solution is *encryption,* which is a must when transmitting sensitive data over the Internet, but less important in physically closed environments. Encryption needs to be complemented by user and site *authentication.* Here, it is crucial that the users do not compromise the authentication mechanisms by circumventing it.

Inside a *company's network,* the *cornerstone security measures* are control of physical access, prevention of social engineering, and systems for intrusion detection and response. The former two are more of an organizational nature; here, we briefly present an approach to intrusion detection, for details we refer to [78] and [75].

Intrusion detection allows to secure the system against attacks by recognizing suspicious activity. Intrusion detection techniques to monitor users, processes and communication protocols can be generally categorized into misuse and anomaly detection [823].

Misuse detection is *based* on the specification of the *undesirable or negative behavior* of users, processes and protocols. It tries to detect patterns of known attacks within the data stream of a system, i.e. it identifies attacks directly. In order to do so it explicitly specifies attack patterns and monitors the data stream for any occurrences of these patterns. The problem is that the set of possible attack signatures is usually not completely known. Therefore, it is difficult to make a clear statement about the limits and the coverage of a misuse detection technique. In monitoring of a critical infrastructure, misuse detection techniques alone cannot be used for a reliable state determination.

A dual approach is the specification of the desired or *positive behavior* of users, processes and protocols. Based on this normative specification of positive behavior attacks are identified by observing deviations from the norm. Therefore, this technique is called *anomaly detection.* The positive behavior can be specified through learning or specification.

The approach of *specification-based anomaly detection* was first proposed in [789] and is based on the formal description of positive behavior. Specification-based anomaly detection techniques do not rely on attack signatures as they compare the actual behavior of a protocol or a process with the expected behavior given by a specification. Therefore, the limits and the coverage of a specification-based anomaly detection technique are clearly defined, although not every deviation necessarily constitutes an attack.

However, from the monitoring and control perspective, even *harmless anomalies* are usually of interest to a network administrator, as they are a potential *indication of error conditions* or *misconfiguration.* Hence, specification-based anomaly detection can have an advantage over misuse detection if the specification itself is known.

Specifications of communication protocols and processes are either given (e.g. a protocol specification) or can be derived. Thus, *specification-based anomaly detection* can easily be used for a *reliable state determination.* However, the general approach does *not allow* for further *differentiation of detected anomalies*, which would be essential for the initiation of effective countermeasures.

The concept of *transaction-based anomaly detection* [76] provides the necessary classification of anomalies.

Transaction-Based Anomaly Detection

The *transaction* concept is a major cornerstone of database theory. Transactions are used to describe atomic operations, i.e. operations which are free from interference with operations being performed on behalf of concurrent clients; and either an operation must be completed successfully or it must have no effect at all.

Transactions can be characterized by the *ACID properties*:

1. *Atomicity*: All operations of a transaction must be completed, i.e. a transaction is treated as an indivisible unit.
2. *Consistency*: A transaction takes the system from one consistent state to another.
3. *Isolation*: Each transaction must be performed without interference with other transactions.
4. *Durability*: After a transaction has successfully been completed, all its results are saved in permanent storage.

For database transactions, the properties of atomicity, consistency and isolation guarantee the avoidance of database anomalies (e.g. inconsistencies, phantom updates etc.). But the *ACID properties* are also suited to *classify anomalies and related attacks* in critical communication infrastructure.

Information exchange processes in a communication infrastructure can be modeled as transactions that have to fulfill the ACID properties. If a transaction does not properly proceed and finish, the *ACID properties provide a direct categorization* of the related anomaly. Based on this categorization, appropriate and effective countermeasures can be applied. A direct violation of the atomicity property, for example, corresponds to a denial-of-service attack, as the transaction is not completed and therefore the requested service is not provided. A buffer overflow represents a violation of consistency, and a race condition a violation of isolation. Other attacks can be classified accordingly. The corresponding anomalies can be detected by comparing protocol and process runs with the given specifications, which are represented by extended finite state machines.

A component-based *prototype for intrusion detection and response* has been realized. It is based on Microsoft's component architecture, the Component Object Model (COM) and its extension, the Distributed Component Object Model (DCOM).

The *architecture* of the prototype follows the *layered approach* of the TCP/IP protocol stack. For each layer and protocol a corresponding analyzer is provided. Currently, the stateless analysis of Ethernet, IP, ICMP, and

UDP packets and the stateful analysis of fragmented IP and TCP packets is supported.

Mathematical analysis and performance measurements have shown that it is possible to monitor a 100 Mbit/s network with standard hardware and packets being analyzed on three layers (data link, network, transport layer). However, for the monitoring of additional layers or higher bandwidths special hardware or a combination of several intrusion detection systems has to be used.

Based on the *reported anomalies* and information from other sources such as network management, firewalls, etc. *countermeasures* can be initiated. To initiate and control these countermeasures an appropriate mechanism is required. Continuing the analogy between intrusion detection and database theory, the theory of *active databases* (see e.g. [899]) is applied. The theory of active databases is built around the concept of *active rules*. An active rule has the following *ECA* form:

```
on event        E
if condition    C
do action       A
```

Based on active rules and the different information sources an *active response system* can be defined as a tuple (S, E, R), where $S = \{s_1, s_2, \ldots s_l\}$ denotes the set of monitored system components, $E = \{e_1, e_2, \ldots e_m\}$ denotes the set of related events, and $R = \{r_1, r_2, \ldots r_n\}$ denotes the set of ECA rules defined for S and E.

Figure 4.22 shows a *simple rule set* for monitoring and control of a critical infrastructure. The rule set is layered according to the criteria given above, with r_i denoting a rule, S_i denoting a layer, and P_i denoting the priority of layer S_i. The first three layers belong to the class of *corrective actions*. Layer 5 gives an example for a *preventive* and layer 6 for a *forensic action*.

For each layer a corresponding *post condition* is defined. The first three layers ensure that each transaction obeys the A, C, and I of the ACID properties. The activation of a rule in a layer (e.g. r_{22} in S_2) can result in the activation of other rules within other layers (e.g. r_{51} in S_5 and r_{61} in S_6). For the different layers different priorities are defined. Rules belonging to a layer with higher priority can trigger rules at a layer with lower priority, but not vice versa.

An implementation showed that employment of several intrusion detection systems in parallel not only *increases the fault tolerance* but is also a *scalability* parameter, allowing for adoption to changing network load; for further results see [77].

4.2.7 Management of External Services

While the project started with choosing CORBA as communication infrastructure and all applications and services were seen as objects [182], *Web Ser-*

Layer 1 ($P_1 = 1$)		
r_{11}:	ON	e_{11}: (IDS-ANOMALY-EVENT, Atomicity violation, SYN-FLOOD);
	DO	a_{11}: (terminate_connection) $\wedge$ raise_event(e_5) $\wedge$ raise_event(e_6) within (t_{11} ms)
Post condition:		$\forall x_i H(x_i)$ with x_i denoting the transactions and $H(x_i) \triangleq$ Transaction is atomic.
Layer 2 ($P_2 = 1$)		
r_{21}:	ON	e_{21}: (IDS-ANOMALY-EVENT, Consistency violation, BUFFER-OVERFLOW);
	DO	a_{21}: (restart process) within (t_{21} ms)
r_{22}:	ON	(Consistency violation, TCP-FIN scan);
	DO	a_{22}: (terminate_connection) $\wedge$ raise_event(e_5) $\wedge$ raise_event(e_6) within (t_{22} ms)
Post condition:		$\forall x_i H(x_i)$ with x_i denoting the transactions and $H(x_i) \triangleq$ Transaction is consistent.
Layer 3 ($P_3 = 1$)		
r_{31}:	ON	e_{31}: (IDS-ANOMALY-EVENT, Isolation violation, RACE-CONDITION);
	DO	a_{31}: (install_backup) within (t_{31} ms)
Post condition:		$\forall x_i H(x_i)$ with x_i denoting the transactions and $H(x_i) \triangleq$ Transaction is isolated.
Layer 4 ($P_4 = 1$)		
r_{41}:	ON	e_{41}: (IDS-SYS-EVENT, Acceptable load exceeded);
	DO	a_{41}: (reconfigure_IDS) within (t_{41} ms)
Post condition:		$\forall x_i H(x_i.\text{load} < X)$ with x_i denoting the monitored components and $H(x_i.\text{load} < X) \triangleq$ The load of the monitored components is acceptable.
Layer 5 ($P_5 = 2$)		
r_{51}:	ON	e_{51}: (IP address blocking);
	DO	a_{51}: (block_IP_address) within (t_{51} ms)
Post condition:		$\forall x_i H(x_i)$ with x_i denoting the connections and $H(x_i) \triangleq$ All connections are regular.
Layer 6 ($P_6 = 2$)		
r_{61}:	ON	e_61: (Increase_log_level);
	DO	a_61: (Reconfigure_log_level) within (t_{61} ms)
Post condition:		$\forall x_i H(x_i)$ with x_i denoting the monitored components and $H(x_i.\text{Log\_Level}) \triangleq$ The security level of the monitored components corresponds to the current log level.

Fig. 4.22. An example rule set for intrusion detection and response

vices [1061] are becoming increasingly popular for loosely coupled distributed applications that can be used across enterprise boundaries. They are not as efficient as CORBA and similar middleware technologies, and can hardly compete in terms of management flexibility. In fact, Web Services are generally used as front-ends to some legacy systems. Yet, while keeping CORBA as the basis for the management layer we also want to offer *managed access to external services* (not integrated into our environment) and provide a prototype implementation for Web Services.

Ensuring Quality of Service for external services is difficult or even impossible but necessary when they are used in conjunction with integrated tools. We extended our architecture to allow for *plugging external Web Services into the CORBA based system* with quality aware service selection. The monitoring subsystem was extended to observe the behavior of remote Web Services and to forecast the possible values of the relevant QoS parameters.

Communication with a Web Service (using the XML-based Simple Object Access Protocol, SOAP [974]) is normally comprised of uncompressed text messages. The invocation of a Web Service is performed by a specially formed request to a web server or an application server. No absolute guarantees can be made either regarding the reachability of a particular Web Service or its temporal characteristics.

Most research on *QoS* with respect to Web Services (e.g. [523, 625]) makes use of *Service Level Agreements* (SLA) between the companies involved. This is, however, not always possible, and new Web Services cannot be used immediately.

In these approaches the *provider* is primarily *in charge* of offering the required *QoS*. There is no possibility to use Web Services offered from a less intelligent infrastructure, and if the provider fails to observe the SLA there are no recovery mechanisms foreseen for the client.

In contrast, *our focus* is on communication with *potentially unreliable services* and *consumer side measures* to handle the failures and to optimize the execution quality. There are several points in the execution path of a Web Service that need to be addressed in order to give at least *statistical QoS guarantees*. They include: the network; the server side; the registry; and the client side. We focus on the client-side implementation, having no control over the remote Web Services and the network.

Integration of Web Services into the Environment

There are many tools that help encapsulating existing CORBA applications and offering them as Web Services, but here we face a complementary task. We assume that we have *no control over the network and the Web Service* itself, and focus on the *QoS relevant measures on the consumer side*.

Several *integration scenarios* of different levels of integration have been considered:

1. The framework's applications can directly handle Web Services.
2. A registry is coupled with the trader.
3. The applications do not communicate directly with Web Services.
4. Applications are not necessarily aware of the Web Service protocols.

In *our environment* applications and services communicate in a very specific way. Translating CORBA requests to SOAP cannot be performed by a generic gateway without the complete knowledge of the interfaces. *Solution 3* is therefore *most feasible*, allowing for QoS monitoring and control.

All requests are sent as CORBA requests to a *Web Service gateway* that translates them into SOAP requests and forwards them to the corresponding Web Service. In this case, the *run-time behavior* of the Web Service can be *observed* and *information for load balancing* can be gathered. The major drawback here is that a single gateway could become a bottleneck if there

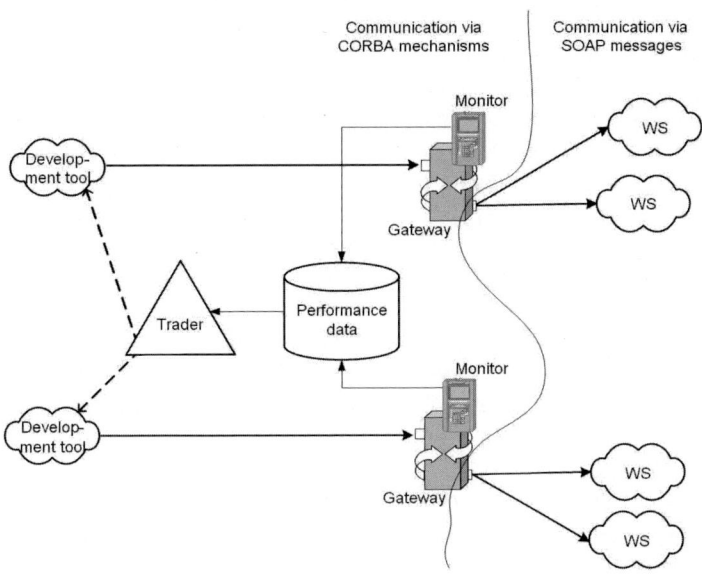

Fig. 4.23. Access to Web Services (WS) in the management layer

are too many requests, or too large ones. This can be solved, for example, by creating a dedicated gateway instance for each Web Service.

Figure 4.23 shows the *extensions for access to Web Services* that have been developed. They do not affect the principal architecture as described in Subsect. 4.2.3.

Finding Web Services that can be used from within our environment is not trivial. Using UDDI would be the most natural approach but there is no common, world-wide UDDI registry. Instead, many companies maintain their own registries. The difficulty is that potentially interchangeable services may have different syntax or, conversely, services with syntactically identical interface may have different semantics. Hence, in the absence of a universal ontology that would allow for automatic checking of semantical equality we have decided to manually load the descriptions of Web Services of interest into a *separate repository*, and to not use the UDDI features directly. The manual or semi-manual selection process will typically include searching and checking for formal metadata (primarily syntax), informal annotations (usually semantic) and potentially test requests and the inspection of results if the description is not exhaustive.

Service Assessment and Selection

The objective of the *selection process* is to improve the execution characteristics of those Web Services that are used quite often. We describe a *statistical approach* [6], which provides the more reliable results the more intensive the usage of the observed Web Service is.

We primarily consider the following aspects of QoS: *availability and reliability* on the one hand, as well as *response time* on the other hand. Clearly, we can only observe the cumulative behavior of the Web Service itself and the network. Execution behavior is logged and the most recent information is used for comparison of Web Services potentially to be selected. We measure the rate of successfully answered requests to address both availability and reliability.

With respect to *response time* we want to address both *static*, long-term service properties and *dynamic*, short-term characteristics. Static properties are usually determined by the complexity of the task performed by the service, the capacity of the server and the network between client and server. They are usually accounted for by calculating a mean value, in this case the average response time.

Dynamic properties depend on the current server load and congestions in the network, as well as on short recoverable breakdowns of a server or the network. As future server and networks states are not known in advance, a *prediction* has to be made. Such a prediction is not very reliable if it is only based on the most recent information. Rather, fluctuation patterns, periodicity, and deviations from the average also need to be considered in service selection.

The *selection process*, however, must not suffer from too complex algorithms. Thus, we need a reasonable *compromise* between *accuracy and speed*. Therefore, we ignore potential patterns that may change dynamically anyway. We can thus concentrate on the recent average, the trend, and recent deviations.

In the prototype implementation, the *access to the Web Services* is realized by *gateways*, placed between the applications that want to call a Web Service and the Web Services themselves. The gateways *monitor* execution success and execution times of the services that were called, and store this information in a database. This information is used for selection of services as follows.

First, a *probability of successful completion* of the next request is estimated for each service. Next, a *prediction of the response time* under the condition of successful completion is done. Finally, both values are added with some weighting factor, and the *service* with the *best value* is *selected*.

Since the calls are discrete events we also choose a *discrete time scale*, as in Lamport's clocks [801]. This is motivated by the impossibility to assign meaningful values to real points in time when there were no prior invocations of the service. Moreover, we assume that the time span between two subsequent calls is of no great importance for the evaluation and selection algorithm, or that at least the relevant effects are accounted for indirectly, e.g. by preserving the daytime fluctuations in the measurement series.

To provide a smooth user interaction on the client side, the *maximum response time* should be selected for optimization rather than the average.

According to the Extreme Value Theory [587] the maximum values of a random process can be approximated by a *Gumbel distribution*. We selected this distribution for characterization and selection of the Web Services, and took an approach similar to [917], applied to end-to-end service parameters instead of network traffic.

The *Gumbel distribution for maxima* is given by

$$F(x) = \exp\left[-\exp\left(-\frac{x-\alpha}{\beta}\right)\right]$$

with location parameter α and shape parameter β.

To *find the parameters* α and β of the approximating Gumbel distribution it is sufficient to calculate the expected value and the variance. This is done as follows.

Let $y(\tau)$ be the *total execution time* of a Web Service (including the round trip time of the communication) measured at the current time τ, or another varying parameter of Web Services that we use for selection. A monitor logs the values $y(\tau_0)$, $y(\tau_0+\delta)$, $y(\tau_0+2\delta)$, etc., where τ_0 is the first measured value and δ is an elementary time slot, which is always equal to one in case of a logical clock. Subsequently, envelopes for the measured values (e.g. maximum execution time envelopes) are computed for the last M blocks of duration $T\delta$. They describe the dynamics from *short-term fluctuations* ($^iY_1$) to *long-term trends* (^iY_T) where $i = 1 \ldots M$ and $k = 1 \ldots T$:

$$^iY_k = \frac{1}{k\delta} \cdot \max_{\tau-(T-k)\leq t\leq \tau} \sum_{u=t-k+1}^{t} y(u)$$

Now the empirically expected value $\overline{Y}$ and the variance σ^2 can be computed:

$$\overline{Y_k} = \frac{1}{M}\sum_{i=1}^{M} {}^iY_k$$

$$\sigma_k^2 = \frac{1}{M-1}\sum_{i=1}^{M} \left({}^iY_k - \overline{Y_k}\right)^2$$

and a set of Gumbel parameters can be determined as:

$$\beta_k = \sqrt{\frac{1}{1.644934}\sigma_k^2}$$

$$\alpha_k = \overline{Y_k} - 0.577216\beta_k$$

With this set of Gumbel distributions a set of predicted maxima $F_k(\tau+\delta)$ for the next call can be calculated, but for *selection of the best service* it is more practical to *consider the confidence interval* and choose the service with the smallest value of

$$\max_{k=1\ldots T} \left(\overline{Y_k} + \epsilon \sqrt{\sigma_k^2} \right)$$

with some confidence parameter ϵ.

The algorithm thus has *three optimization parameters* to achieve a reasonable trade-off between accuracy and efficiency: T, M, and ϵ.

Apart from the temporal behavior, *reliability* is also taken into account. *Penalties* are assigned to Web Services that fail to send a response, so that these services can be excluded from being selected for some period of time.

Evaluation of Web Service Integration

The *selection* aims at *minimizing* the probability of forwarding the request to a server that has a high *response latency* or a high *failure-to-success ratio*. The execution times of Web Services that were tested was within the range of few seconds if they had been reachable and not under heavy load. The delay can vary heavily depending on the time of day. This can be mainly attributed to the server load but also to network congestions. Under these conditions the overall performance of the extended CORBA environment does not noticeably degrade compared with services integrated into the CORBA environment. Delay variations and occasional service failures confirm that a flexible, automatically adapting selection algorithm can considerably improve the Quality of Service.

The *computational overhead* for the selection algorithm is *quite low* and negligible for the interaction with a human developer. For delay sensitive tasks that can be found, for instance, in visualizations in a video conference, remote calls to Web Services should be avoided altogether and full integration should be preferred.

4.2.8 Conclusion and Outlook

Support of collaborative engineering design processes involves many applications, services, and platforms. To provide smooth operation, a management system is necessary. We introduced a *service management layer* that is based on the CORBA middleware technology and prototypically implements the core functions. We suggested an architecture that takes into account the *heterogeneity*, the *need for integration of legacy tools*, and that can manage the *Quality of Service*. Many of its concepts are useful not only for highly specialized environments, like support of engineering design processes, but also in *open service markets* [444].

CORBA as the *basis* for the management layer offers good performance, and extensibility. Even when integrating other technologies such as Web Services, it is still a very reasonable choice for the core infrastructure.

A-posteriori integration of existing applications is not trivial and can hardly be fully automated, see also Sect. 5.7. However, the trading process and course grain management operations could be kept quite generic, so that little manual work is necessary to integrate new tools into the service management layer.

Not all services can be integrated into the collaborative environment. We investigated *loose management of Web Services* offered by *third parties*. The basic approach towards their integration into the service management platform is the same, but they have to be handled differently: since direct management is not possible, only statistical quality can be dealt with. As Web Services have become an important concept in industry, we aim at a closer integration of Web Services within our environment. *First steps* have been done to *transfer* our concepts *from CORBA to Web Service* technology and to consider automated *composition of services* [446].

Developers with *light-weight mobile devices* have not been considered explicitly in our scenarios. Their role is likely to increase in the next future, so that a management layer like ours should be extended to support nomadic clients using mobile middleware.

Security is crucial to the acceptance of support systems as described in this book. We investigated various aspects, yet a real operational system must be more comprehensive and integrate security policies and tools of the institutions and companies involved. This is also part of our current work.

5

Integration Aspects

This chapter deals with *integration* in a further sense. Thereby, integration has *different meanings*, each meaning being represented in one section of the following chapter.

Whereas Chap. 2 discussed integration on application domain models, Chap. 3 in the sense of bridging insufficient tool support by new cooperative functionality, and Chap. 4 by giving a uniform platform integrating various existing ones, we now deal with interesting *further aspects* of integration.

In Sect. 5.1, we mainly discuss integration with respect to work processes in *industry*. A tool for denoting work processes is introduced as well as a procedure for work process modeling, the practical use of both being demonstrated in case studies.

Section 5.2 gives the first links to labor research. It demonstrates a *simulative method* by which design processes can be analyzed in order to find out suitable organizational forms.

Section 5.3 gives information about a tool able to *integrate* different and *heterogeneous* unit simulations into one plant-wide simulation.

In 5.4 it is shown, how *plastics processing* is integrated in the chemical engineering development process. Furthermore, specific flow simulation approaches together with their virtual reality presentation are studied.

Section 5.5 integrates the novel informatics concepts of Chap. 3, by showing their *synergistic* mutual *application* (two-level integration). Three examples for this synergistic integration are given.

Section 5.6 is the second link to labor research. It discusses the *ergonomic evaluation* of our findings in IMPROVE.

Finally, 5.7 studies *software engineering* and *architecture* modeling *aspects* common to all tool construction or extension processes within IMPROVE. Wrapper techniques and a specific architecture tool are presented.

5 Integration Aspects

This chapter corresponds to the *project area "integration"* of Fig. 1.27. However, it also discusses further results not delivered by the subprojects of that area.

5.1 Scenario-Based Analysis of Industrial Work Processes

M. Theißen, R. Hai, J. Morbach, R. Schneider, and W. Marquardt

Abstract. In this section, the modeling procedure for design processes introduced in Subsect. 2.4.2 is discussed from a more application-oriented point of view. The Workflow Modeling System WOMS, which has been developed for the easy modeling of complex industrial design processes, is described. Many case studies have been performed during the elaboration and validation of the modeling methodology and the tool, several of them in different industrial settings. In this contribution, some case studies are described in more detail. Two of them address different types of design processes. A third case study, demonstrating the generalizability of our results, deals with the work processes during the operation of a chemical plant.

5.1.1 Introduction

As emphasized in Sect. 2.4, empirical studies are a prerequisite for capturing the knowledge required for analyzing and improving design processes. Whereas in Sect. 2.4 the focus is on the models used for representing design processes, this section treats the implementation of the procedure for modeling, analyzing, and improving work processes, which has been sketched in Subsect. 2.4.2. We first discuss some general issues related to the practical application of the procedure in industry (Subsect. 5.1.2). The implementation of the procedure requires an easy-to-use modeling tool for creating and processing work process models. To this end, the Workflow Modeling System WOMS has been developed by the authors. The tool and its functions are described in Subsect. 5.1.3. WOMS supports the semi-formal C3 notation (cf. Subsect. 2.4.4), which has been developed taking into account the characteristics of creative design processes (cf. Subsect. 2.4.3). In this contribution, three case studies are described in more detail (Subsect. 5.1.4). Two of them address design processes (process and product design in an international chemical company; basic and detail engineering at an engineering consultant). The third case study deals with the operational processes of a chemical plant.

5.1.2 A Procedure for Work Process Modeling and Improvement in the Chemical Industries

In Subsect. 2.4.2, a procedure for modeling, analyzing, and improving design processes in chemical engineering is presented. In the following, we give a more detailed presentation of the practical issues arising when the procedure is applied in an industrial setting. This subsection gives a summary of our experiences gained in several case studies with industrial partners. As a large assortment of literature from these communities is available, we focus on those issues that result from the peculiarities of creative design processes. The numbering of the steps in the following paragraphs refers to the C3 model of the modeling procedure (cf. Fig. 2.12).

Step 1: Defining the Modeling Scope

Defining the Goal

Before a work process is modeled, the *questions to be answered* and the *issues to be addressed* by means of the model have to be identified. They depend on the goals of a particular case study. In general, the overall goal is to improve a real work process. Nevertheless, as there is usually at least some consciousness about the problems in the real process, this goal can often be specified more precisely in order to focus on the relevant issues and limit the complexity of the models to be created. Some examples of specific goals are

- to reduce the cycle time for executing the work process,
- to balance the workload of the available human or technical resources, or
- to identify the requirements for an improved IT system supporting the work process.

As far as a-priori knowledge about the work process permits, these goals should be specified; as discussed below, they affect the process aspects to be captured in the model and the choice of an appropriate *as-is process*.

Setting the Modeling Focus

The *work process models* to be created during the modeling sessions must be *sufficiently rich* to provide the information relevant for the goals. For instance, if one of the goals were to reduce cycle times, then information like the number of actors performing a certain activity and the total working time spent for the activity would have to be captured. In contrast, such aspects could be neglected in a case study addressing an improved information flow between the experts involved in the work process.

Choosing an As-Is Process

The first model to be created when performing a case study should reflect the design process as it is currently performed (*as-is process*). That way, the *advantages and shortcomings of the current practice* can be identified (and later on considered when an improved *to-be process* is defined).

The adequate *level of generality* of the *as-is process* depends on the complexity of the process in consideration (see also Subsect. 2.4.5). For typical *business processes*, such as procurement processes for standard plant equipment, it is often possible to start with a generalized model, whereas generalized models of *design processes* are usually too complex to be created in a single step. Instead, models of one or several representative *concrete processes* should be created first; after their validation, they can support the creation of more complex generalized models.

Like the modeling focus, the choice of appropriate as-is processes depends on the goals. For example, in a case study aiming at the definition of best practices, a successfully completed project can serve as a first step towards the to-be process. If time allows, it is also useful to model the critical parts of less effective projects. If the reasons for the problems which came up during the project can be identified, the to-be process can be amended to avoid similar problems in future projects.

Planning the Modeling Sessions

In particular for complex work processes with many actors, it is helpful to create a first coarse-grained model with a manager and possibly the leaders of the involved groups. This model can be used to *identify the required participants for the subsequent modeling sessions*, which will focus on different parts of the process in order to create a more detailed model. The number of participants in the sessions should be limited (not more than 5-7 people) so that all of them can participate actively in the creation of the model. Finally, a schedule listing the participants and topics of each session should be created and distributed to ensure the availability of the involved persons.

Step 2: Recording the As-Is Process

When our case studies were performed, the concrete processes had already been completed. Thus, no records explicitly destined for our modeling activities were available. However, during design projects a plethora of records is produced which can serve as additional sources of information complementary to the memories of the actors, including the products of the design process (such as process flow diagrams and equipment specifications), written communication between team members, minutes of project meetings, and possibly the information stored in a project management system. In our case studies, it has proven helpful to guarantee easy access to such information during the modeling sessions.

Steps 3 and 4: Modeling and Validating the As-Is Process

The modeling of an as-is process is essential for further elicitation of the pending issues and weaknesses in a work process. Before an improved to-be process can be created, all participants must be aware of the problems in the current work processes.

The modeling of an as-is process is done cooperatively by the staff involved in the work process and one or preferably two moderators. During a modeling session, the participants describe their activities during the work process.

The moderators *guide the sessions* in an *interview-like style*. They pose questions, structure the evolving process knowledge, and clarify issues. Simultaneously, the moderators create a work process model by means of WOMS, a modeling tool for work processes (cf. Subsect. 5.1.3). If possible, the model is projected onto a screen that is visible for all participants. This immediate visualization of the work process supports a first validation of the model, in particular the identification of contradictory statements about the work process and the correction of misunderstandings between the participants.

We have observed that the *modeling sessions encourage the participants* to reflect, discuss, and scrutinize their work processes. Often, improvements for shortcomings are suggested. They are recorded in the minutes for later reuse, but are not modeled explicitly. Before any profound assessment of alternatives can be made, an overall view of the work process is needed. Spontaneous proposals might overcome local difficulties in one part of the work process, but they are likely to cause problems in another part. Given the complexity of industrial work processes, such issues cannot be detected before a complete process model is available.

Step 5: Analyzing the As-Is Process

The analysis aims at the *identification of shortcomings* in the as-is process. Typical questions that are posed comprise:

- Are all activities necessary with respect to the purpose of the process? Does an appropriate performer handle each activity? Does each activity occur at the best point in the process?
- Are appropriate media used for information exchange? Is the information provided to the recipient in a comprehensible form? Is the recipient notified in case information is updated?
- Are the project meetings efficient? Do they have a well-defined goal? Do their different professional backgrounds impede the communication between the participants?

The shortcomings are inserted into the model of the as-is process (by means of the *shortcoming element* of C3, cf. Subsect. 2.4.4). The result is a comprehensive model correlating the activities with the associated shortcomings as well as with their position in the overall process. Thus, the hasty changes mentioned above can be avoided when the to-be process is defined.

Step 6: Defining the To-Be Process

Based on the results of the precedent analysis, an *improved work process* is created. Changes of the as-is case may affect all elements of the work process model. For instance, superfluous activities may be canceled, the order of activities may be rearranged to better comply with the objectives for the overall process, and further activities may be introduced. Inappropriate data formats are replaced by alternatives accounting for the needs of the information consumer.

The to-be process undergoes the same analysis as the as-is process in order to detect further shortcomings before the process is put into practice.

Step 7: Formalizing the Process Model

Formal work process models are a prerequisite for automating well-understood parts of work processes. So far, our industrial case studies have focused on semi-formal modeling. Besides other issues, the *formalization of models of industrial work processes* will be addressed in the transfer project described in Sect. 7.3.

Step 8: Implementing the Improved Process

The final model of a to-be process can serve the implementation of an improved process in different ways.

In the first instance, it is an *excellent guideline* for the people actually performing the work process. They become aware of their contribution to the overall process. Even in situations that are not covered by the work process model, the model supports the choice of an appropriate reaction. For instance, assume that in a design process a chemist discovers an error in a physical property data sheet that has been forwarded to a group of chemical engineers. The information flow depicted in the work process model allows to determine those people and groups that have based their work on the incorrect information. As no time must be wasted for a costly and long search, the concerned people can be informed immediately.

The work process model also provides valuable information about the *requirements for an integration of the software tools* used by the actors. For example, it can be decided whether a dynamic coupling of two software tools would be necessary due to a continuous information exchange, or whether a simple file conversion would be sufficient.

Furthermore, the model contains the required process knowledge in case some *workflow management system* should be installed for supporting repetitive sub-processes of the overall design process. An efficient usage of these systems is often impeded by insufficient and imprecise definitions of the workflows to be supported.

5.1.3 Workflow Modeling System WOMS

During the first empirical studies within IMPROVE, semi-formal work process models were *created manually*. Sometimes small paper charts posted on a white board to represent modeling elements like activities or tools were

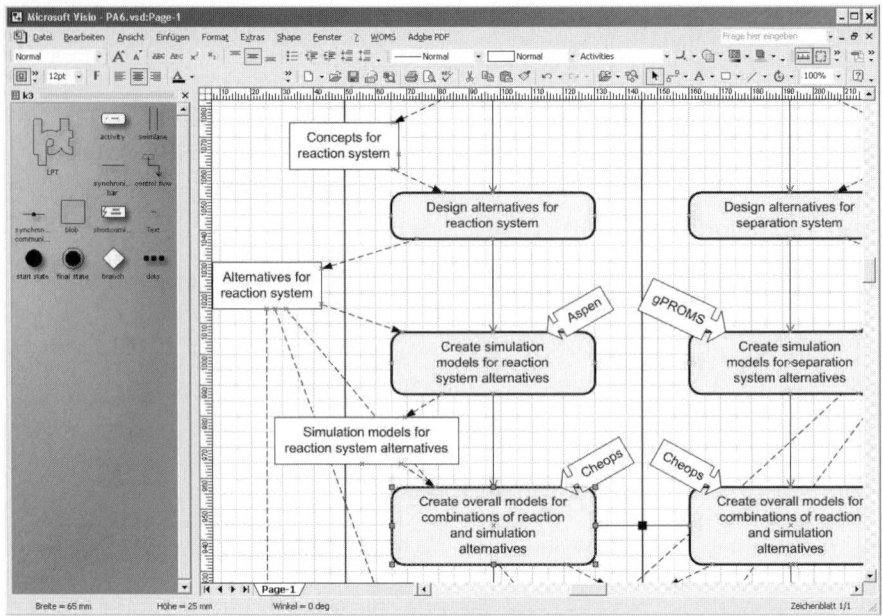

Fig. 5.1. WOMS user interface

used, which could be rearranged easily. This way, the involved designers could participate actively in the modeling process. However, further processing of such models was difficult and time consuming, which impeded their analysis and storage with software tools and their exchange between geographically distributed team members.

Soon, the *need for a modeling tool* emerged, which should allow to implement the modeling procedure with acceptable time and effort for all participants. As argued in Subsect. 2.4.3, conventional modeling languages such as *activity diagrams* of UML [560], the *Business Process Modeling Notation* [873], or *Event Driven Process Chains* [949] do not meet the requirements imposed by the characteristics of creative design processes. Hence, none of the existing modeling tools for these languages fitted our needs. Instead, we had to create a modeling tool for C3, the modeling language created during IMPROVE for the semi-formal representation of design processes in chemical engineering (cf. Subsect. 2.4.4).

This tool, the *Workflow Modeling System WOMS*[38], was developed in parallel with the further elaboration of the modeling procedure and the C3 language. WOMS is based on the commercial drawing tool Visio [845]. The

---

[38] The term *workflow* in the name of the tool is due to historical reasons; according to the terminology defined in Subsect. 2.1.2, *Work Process Modeling System* would be more appropriate.

Scenario-Based Analysis of Industrial Work Processes   439

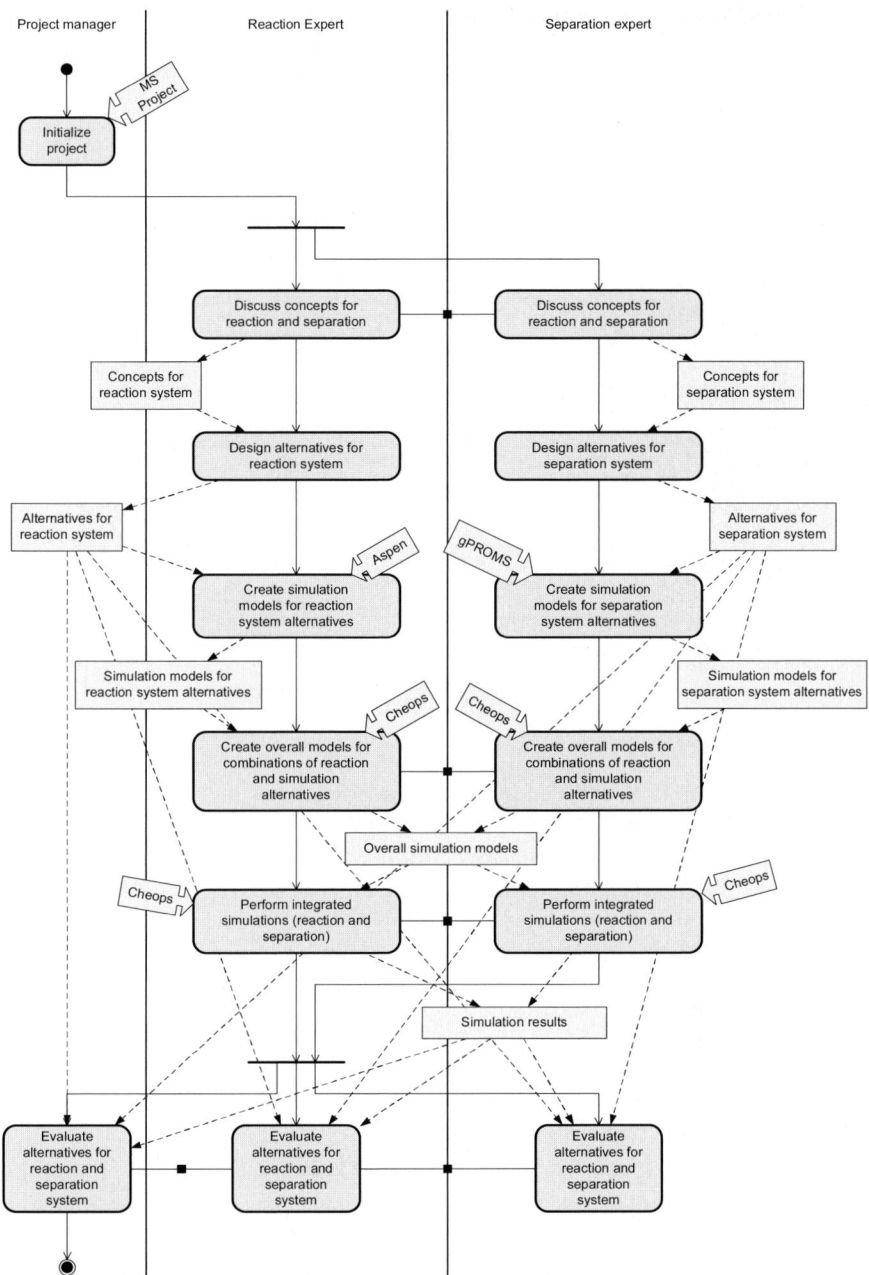

Fig. 5.2. Simple WOMS model of a PA6 design process

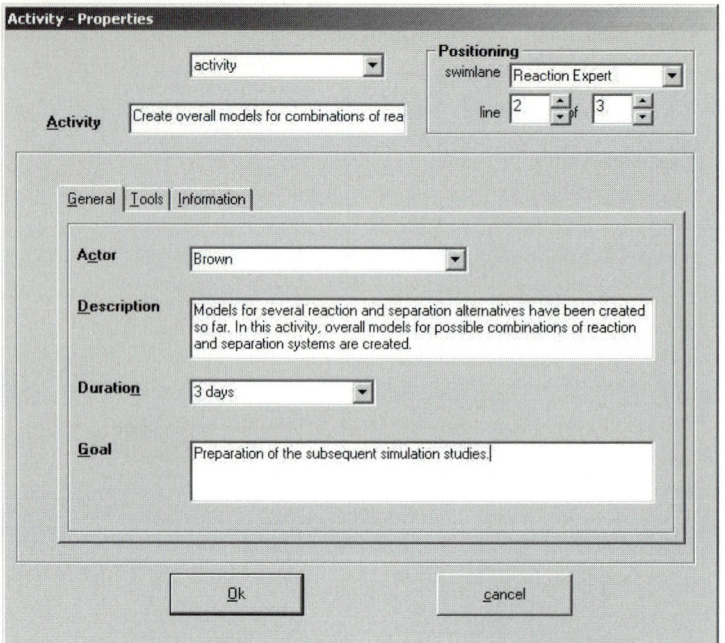

Fig. 5.3. Pop-up window of an activity

first version of the tool, created in 2003, was not more than a stencil providing shapes for C3 elements (such as activities and roles), which can be placed in the modeling window by simple drag-and-drop operations. Intensive usage of WOMS in different areas motivated its continuous extension with additional functions for the creation, exchange, analysis, and execution of work process models. In the following paragraphs, these functions are described. It should be noted that an important part of the functionality of WOMS is based on the integration of existing software tools and technologies, such that the implementation effort could be kept low.

Support for Creating Work Process Models

The *principal elements of the WOMS user interface* (see Fig. 5.1) are the stencil with C3 modeling elements (on the left) and the modeling window itself (on the right). The modeling window in the figure shows part of a simplified model of the PA6 design process from the IMPROVE reference scenario (cf. Subsect. 1.2.2). In the remainder of the subsection, this model will serve to illustrate some functions of the tool; for reference the entire model is depicted in Fig. 5.2.

For some modeling elements such as roles, activities, tools, and information items, attributes can be specified in *pop-up windows*. To give an example,

Fig. 5.3 shows one page of the pop-up window of an activity. Several text fields permit to give the role the activity is assigned to (called *swimlane* in WOMS), the actor who has performed or should perform an activity, a more extensive textual description of the activity complementing the short text in the graphical model, and so on. Existing information items and tools are shown in a list inside the activity window and can be assigned to the activity. This prevents multiple definitions of the same object with different names and ensures the syntactical correctness of WOMS models.

WOMS models of complex work processes consist of a large number of activities and information items, and consequently they comprise a multitude of control and information flows, often spanning large parts of the model; this is in particular true for information flows, because information produced in one activity is typically used in several subsequent activities (for instance, see the multiple information items entering the evaluation activities at the bottom of Fig. 5.2). As a result, graphical representations of complex work processes can become unclear or even incomprehensible. In order to simplify the handling of complex models, certain *types of modeling elements can be hidden*. Thus, users of the tool can focus their attention on those aspects of a process which are relevant in a certain context, without being distracted by less important elements. For example, when the order of activities is edited, only activities and control flows need to be shown, whereas tools and information items can be hidden.

WOMS provides a *hyperlink function*, which enables a user to link a resource file or a directory to the corresponding information item in the model. For instance, assume that the simulation models for different alternatives for the reaction system (cf. Fig. 5.2) are stored in a single directory; this directory could then be linked to the information item named *Simulation models for reaction system alternatives*. Once such links are established, users can access the resources from WOMS by clicking on the information elements. This function is in particular useful for documenting concrete processes, because the model can then be used as an access structure to the documents created and used during the work process. This way, an integrated documentation of the complete history of the design process and the documents produced can be created.

Support for the Exchange of Work Process Models

To enable the *exchange of work process models* between WOMS and other software tools, two tool-independent output formats, a HTML format and a XML format, are supported.

HTML Export

The HTML export creates a *vendor-neutral representation* of the model that can be viewed and explored with any web browser. The resulting HTML page

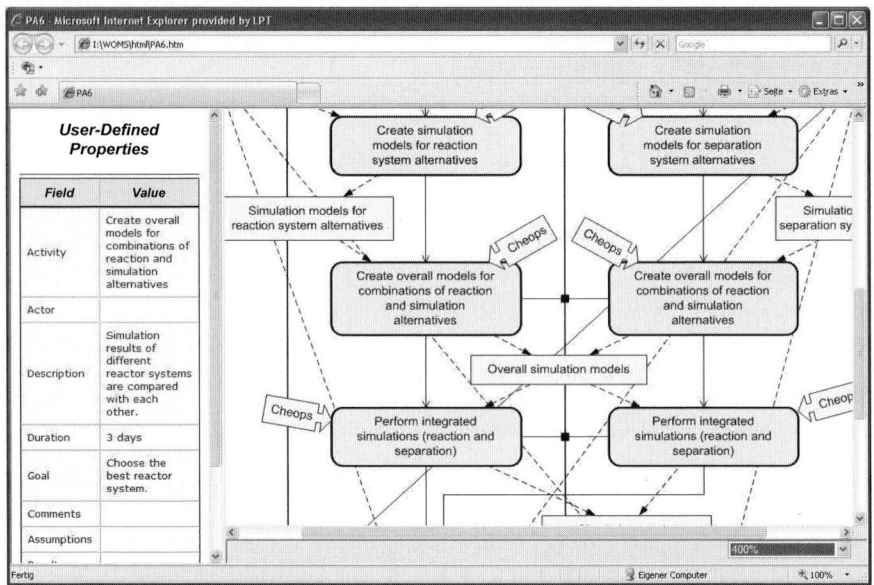

Fig. 5.4. HTML page showing the WOMS model from Fig. 5.2

contains two frames, the right one showing the graphical WOMS model, and the left one the attributes of a selected element (see Fig. 5.4). By choosing a suitable zoom factor, users can obtain an overview of the entire work process model or focus on the details they are interested in. This output format facilitates the publication of a WOMS model in the internet or intranet. This way, a given model of a work process can be shared at any time with a selected group of users, even if they do not make use of the WOMS tool. This feature is especially useful if a design process is carried out by an organizationally or geographically distributed team.

XML Export

Using the XML export function, a work process model can be stored in a *structured standard format*, which is a universal exchange format for a number of software tools. As XML files can be easily rearranged or altered into a user-defined format by means of XSLT (*Extensible Stylesheet Language Transformations*, [602]), the XML export is useful when WOMS models are to be used by other software applications. Some examples are given in the following subsections.

Support for the Analysis of Work Process Models

Explicit *graphical representations* of work processes (i.e., in C3 notation like in the modeling tool or in the HTML output) enable a first process analysis

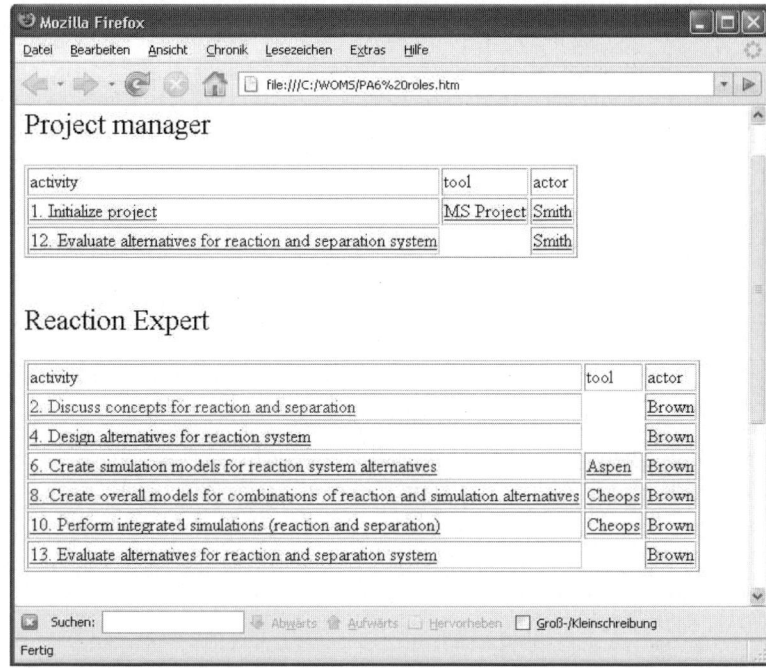

Fig. 5.5. Part of the role centric view of the WOMS model from Fig. 5.2

by an individual or by a group of people in order to detect deficiencies of the work processes modeled.

In addition, specific *views* can be generated by transforming the XML output of WOMS by means of template rules defined in *XSLT*. We briefly discuss the different views in the following.

Role-Centric View

The role-centric view is displayed in a table, which lists the activities, actors, and tools that are assigned to each role. The role-centric view of a design process model provides answers to the following questions:

- Which roles are involved in the design process?
- Which tasks are assigned to a certain role?
- Who performs the activities of a certain role?
- Which tools are used by a certain role?

The expertise needed for certain roles can be easily determined with the role-centric view. All roles are arranged alphabetically. Figure 5.5 shows a part of the role-centric view of the sample design process shown in Fig. 5.2.

Actor-Centric View

The actor-centric view gives information about the expertise and the work load of individual designers. It enables to answer the following questions:

- Which actors are involved in the design process?
- What are the roles played by a certain actor?
- Which tasks are assigned to an actor?
- Which tools are used to perform the tasks?

Tool-Centric View

The tool-centric view can be used to support the management of technical resources. This view allows to answer the following questions:

- Which tools are required for the entire design process?
- Which tool is required to perform a certain activity?
- Which role uses a certain tool?
- Which actor uses a certain tool?

Activity-Centric View

The activity-centric view contains comprehensive information about the whole design processes; it is displayed as a table listing all activities. For each activity, the following information is depicted:

- the input and output information,
- the predecessor and successor of the activity considered,
- the role and actor associated with the activity,
- the tools needed to perform this activity,
- its duration, and
- a textual description.

Information-Centric View

Finally, an information-centric view lists all information items and gives the activities that create or use it. This view is important for the rearrangement of a work process, because the logical dependencies between different activities can be determined with this view.

The major items in each view, except the activity-centric view, are arranged alphabetically. In the activity-centric view, the activities are arranged chronologically. In addition,, hyperlinks between different views are also supported to facilitate the analysis of large and complex design processes.

Table 5.1. Application of WOMS

Guidelines

- *Namur Arbeitsblatt 35: Abwicklung von PLT-Projekten* [863] (handling PCT projects, see also [397])
- *VDI guideline 3633: simulation of systems in materials handling, logistics and production* [1020]

University

- IMPROVE reference scenario (see also Subsect. 1.2.2)
- self-observation during modeling and simulation
- business processes at Fakultätentag für Maschinenbau und Verfahrenstechnik

Industry

- optimization projects
- product and process development process (see also Subsect. 5.1.4)
- design processes in basic and detail engineering (see also Subsect. 5.1.4)
- operational processes of a chemical plant (see also Subsect. 5.1.4)

Support for the Automation of Design Processes

In case a routine design process is well-understood, it can be completely specified in advance on a fine-grained level for execution in various project contexts. The *execution* of such *generalized process templates* can easily be automated by means of workflow engines; the prescriptive work process models required can be created by WOMS and further enriched and transferred into a formal XPDL model [1059] by means of XSLT [154]. The established formal design process model can be processed directly by a workflow engine such as Enhydra Shark [660].

During the execution of the design process, the workflow engine manages the flow of work between participants and passes tasks from one participant to another in the predefined order. To allow the designers to quickly identify their current tasks, each designer gets a list with the assigned tasks. This list is updated automatically when the designer completes one task or receives a new task. Work process monitoring is also supported by the workflow engine Enhydra Shark to provide information on the current state of the design process, e.g., which tasks are completed or which are still in progress.

Applications of WOMS

WOMS has been applied in several case studies carried out by the authors and their colleagues (see Table 5.1). It should be noted that not all of these studies address design processes. The flexibility of WOMS and the underlying C3 notation made it possible to apply the tool also for more or less predetermined work processes such as business processes. In the following subsections, some of the case studies in cooperation with industrial partners are discussed in detail. Here, we give a brief description of the different applications of WOMS, which were – to a varying degree – relevant for the successful completion of the different case studies.

- *Documentation.* During its execution, a work process can be modeled and thus documented by the actors themselves. The resulting model is a concise representation of the work process as it has actually been carried out. Information items can be linked to documents produced during the process to provide an easy-to-use access structure.
- *Planning.* Before or during a work process, the activities to be performed can be scheduled. In case of a complex process involving several actors, discussions between team members about the schedule are simplified because a common understanding of the tasks to be performed by each actor and their interdependencies can easily be established this way.
- *Education.* Well-understood chunks of routine design processes can be represented by WOMS as intuitive graphical models to document best practices in a corporation. This model can serve as a guideline for a novice to learn how certain tasks ought to be preferably performed. The model can also be used as a template for the planning or for the documentation of to-be or as-is processes after copying and modifying the template to the desired work context.
- *Analysis.* Models of design processes can be analyzed to achieve different goals. An analysis of an as-is process typically aims to identify weak points and possible improvements of the design process. It should aim at conclusions how design processes of the same or a similar type can be reengineered to be more efficient in future projects. An analysis of the information flow can help to identify the types and number of documents exchanged between software tools. Such findings form the starting point of requirements definition for tool integration. In case an inter-organizational (as-is or to-be) process model is analyzed, the character of the collaboration can be easily assessed. In particular, the interface between the organizations, which is implemented mainly through the documents exchanged between activities performed in different organizations, can be identified. Such analysis techniques are effectively supported by the various views on the work process models introduced above.
- *Communication.* A common understanding of complex design processes is necessary for an effective cooperation between the designers involved. Graphical models of design processes can serve as a common basis for a discussion of the features of a particular design process.

5.1.4 Case Studies

In this section, three industrial case studies are described which demonstrate the application of the modeling procedure and of WOMS in different settings. The first two case studies address different types of *design processes*. The third one deals with the start-up of a chemical plant, i.e., with an *operational process*.

Product and Process Development at a Large Multi-National Company

Product and process development processes in multi-national companies are characterized by an enormous complexity; actors in dozens of roles perform hundreds of interdependent activities, often located in geographically distributed teams. This case study, still in progress, deals with the product and process development processes for specialty chemicals at Air Products and Chemicals, Inc., Allentown, PA, USA. The overall goal is to reduce cycle times, as time-to-market is essential for the economic success of innovative chemicals.

Two approaches are pursued in the project. The first one is to provide a detailed compilation of the best practices in product and process development as performed at Air Products. Preliminary discussions with Air Products have shown that an improved communication between different departments is desirable. In order to understand and improve information exchange between different departments, the interactions between different departments need to be captured explicitly in a concise model of the development process. Such a model, combined with contact information about the people involved in a particular project, allows to determine all actors and roles who are affected by a particular situation, thus streamlining the work process.

The second approach is to apply simulation studies in order to identify bottlenecks in the process and to examine the influence of additional human or technical resources on the cycle time.

As both approaches require a generalized, detailed, and (in case of the simulation approach) formal model of the work process, they can only be realized in the long term. Several iterations of the modeling procedure are required to construct the required models. So far, models of two concrete projects have been created on a medium level of detail. The first model deals with a rather small project; it was planned as a demonstration of our methodology and was used by Air Products to assess the suitability of the methodology for their needs. The second model describes a rather complex design process with more than a dozen roles, several dozens actors, and more than one hundred activities. The project is continued as part of the transfer project described in Sect. 7.3.

Basic and Detail Engineering at an Engineering Consultant

A further case study has been performed at an engineering consultant company in Germany which offers services like conceptual, basic, and detail engineering of chemical plants as well as the supervision of their construction and start-up. The subject of the case study were the work processes during basic and detail engineering. The case study was motivated by the insufficient integration of some standard software tools used at the consulting company, including spreadsheets, databases, and CAD tools. The problem as stated by our project partner was that data had to be entered several times in different tools. The solution to this issue as envisioned by the consulting company was a better integration of the different software tools.

However, we could convince the management that an integration of the software tools would be infeasible without a precise definition of the requirements for their integration. We proposed to first analyze the work processes during basic and detail engineering in order to better define these requirements. We also assumed that part of the integration problem could be solved by reengineering the work processes, for instance by replacing some software tools whose usage was not mandatory.

Thus, it was planned to model a design project which had suffered from the insufficient tool integration. This model was elaborated in several interview and modeling sessions, equivalent to a total time effort of approximately 15 hours. During the sessions, it turned out that several deficiencies of the work process did not result from the integration problem. In the following, we give two examples.

- As the responsibilities for some routine activities were not clearly defined, such tasks were often not done by the most appropriate actor. For instance, one of the engineers participating in the modeling sessions stated that he always did the calculations for heat exchangers himself. Another engineer pointed out that such calculations could also be done by technicians with a considerably lower wage rate. The first engineer was not aware that also technicians had the required qualifications.
- In some cases, inappropriate procedures and documents had been established for the information transfer between engineers and technicians. Engineers used some simple forms, implemented in a word processor, in which they entered basic data for equipment such as reactors. These forms were electronically sent to technicians, who had to add further data. Though, the simple forms did not provide the entry fields required by the technicians; instead, they used other forms, implemented in a spreadsheet application, in which they had to reenter the data delivered by the engineers. During the modeling sessions it emerged that the two types of forms were simple electronic versions of some paper forms which had been used several years ago and which may have been justified at that time. Nevertheless, there was no reason why the more detailed forms should not be used by the engineers. This way, reentering of data could be avoided in a very simple way.

Thus, some deficiencies of work processes during basic and detail could be remedied by rather simple measures such as assigning activities to roles and introducing uniform data forms. Unfortunately, this cooperation was suspended when the engineer who had initiated the project changed to another employer. Nevertheless, the case study has shown that participative modeling of work processes makes actors reflect and discuss their daily practices. In particular, shortcomings could be detected and feasible solutions could be found without performing an explicit analysis step.

Operation of a Chemical Plant

In contrast to the case studies described before, which deal with design processes, the case study described in this subsection addresses an operational process: In cooperation with Bayer Technology Services (BTS), the start-up process of a semi-batch column has been investigated [436].

Before we started the project, the start-up process was specified by means of several check lists, each of them providing detailed instructions for subprocesses like *preparation* or *inertization*. These check lists were represented in a simple textual form. A first version of the check lists had been created after the construction of the plant, and occasionally they had been modified to take into account the experiences made by the operators during their daily work. However, complete revisions of the check lists had never been done, such that no explicit record of the valuable know-how of the operators existed. Furthermore, the effects of some modifications of the control system of the plant had never been incorporated in the lists, which therefore contained some out-dated information. In consequence, not all instructions given in the check lists were followed by the different operators, who rather performed individual start-up procedures, resulting in a considerable variance of the start-up time and the quality of the chemical product.

The main goal of this cooperation was to ensure better start-up processes. In particular,

- the time required for the start-up should be reduced,
- failures during start-up (and also during operation) should be minimized, and
- the reproducibility of the start-up should be improved.

In order to reach these goals, a procedure comprising three partially overlapping steps was applied: (1) the collection of the available knowledge of the start-up process, (2) the definition of an improved up-to-date specification of the process, and finally (3) the implementation of the improved specification, i.e., measures to ensure the execution of the process according to the specification.

(1) *Collection of the available knowledge.* The existing knowledge about the start-up process was distributed among the different members of the operating personnel and other experts, in particular the responsible plant

engineer. In addition, the old check lists as well as process and plant data were available. In order to capture this knowledge, a WOMS model of the process was created and incrementally enriched and modified to include information from different sources.

As the existing check-lists already provided a detailed and structured representation of the process, they were used to create a first version of the WOMS model. Subsequently, discussions with the operating personnel and the plant engineer addressed the shortcomings of the process as specified in the old check lists and the variants actually followed during plant operation.

(2) *Definition of an improved up-to-date specification of the start-up process.* Several deficiencies of the original process could be remedied. For instance, in some cases the location of equipment, hand valves, etc. hat not been considered when the original check lists had been created. Thus, a staff member strictly following these sequences would have to cover a considerable distance within the plant during a start-up. By rearranging some of the sequences, such distances could be reduced.

(3) *Implementation of the improved start-up process.* As the plant will further on be operated manually, the implementation of the improved start-up process had to ensure that the operating personnel will follow the procedures defined there. To this end, the relevant information must be passed to the staff. As the representation of the start-up process by means of check lists is widely established, it has been decided to transform the final WOMS model into a set of check lists.

5.1.5 Conclusion

We have discussed several issues concerning the application of the generic modeling procedure for design processes introduced in Subsect. 2.4.2. The Workflow Modeling System WOMS is a prerequisite for the successful implementation of the modeling procedure. Several case studies in both academic and industrial settings have proven the practicability of the modeling procedure.

So far, our cooperation projects with industrial partners have focused on the actual *modeling* of design processes (and, in the explorative case study described in Subsect. 5.1.4, the modeling of an operational process). In the transfer project sketched in Sect. 7.3, the methodology will be further elaborated. First, more attention will be given to the *formalization* of work process models, as it offers promising advantages for both process analysis and implementation. Secondly, the methodology will be extended to cover *different types of work processes*, including different types of *design processes*, but also *operational processes*.

5.2 Integrative Simulation of Work Processes

B. Kausch, N. Schneider, S. Tackenberg, C. Schlick, and H. Luczak

Abstract. The design and optimization of creative and communicative work processes requires a detailed analysis of necessary activities, organizational structure, and information flow as well as the identification of weak spots. These requirements are met by the C3 modeling technique, which was specifically developed for design processes in chemical engineering (cf. Subsect. 2.4.4). C3 is also the foundation of the simulation-based quantitative organizational study described in this section. Therefore, a transformation technique from semi-formal models of work organizational dependencies into formal workflow models has been developed and implemented. The verified results of test-runs show the various fields of application of this technique, including its benefits for the reduction of cycle times, for the optimization of the operating grade of the employees, and for the capacity utilization of tools and resources.

5.2.1 Introduction

Only 13 percent of work in projects in Germany is actually value-adding, resulting in a total "loss" of approximately 150 billion Euros [700]. The reasons for these deficits are wrong decisions during project selection and also the insufficient definition of goals. While these problems affect the project environment in the context of business philosophy, there is another area that affects the project structure itself. This area covers the development and continued use of findings and information in projects. So the project planning at its very early stage, along with the accurate implementation of employee competence and availability as well as resources, must be improved. Many existing tools for workflow planning do not provide sufficient functionalities to take all important factors in consideration at the same time and to get a quantitative comparisons between different project structures. In addition, the high amount of different influencing factors makes it impossible for the planner to select the best project structure according to different criteria like project duration, project costs, or human resource allocation. Simulation has been proven to be an effective tool for planning and improving complex systems according to multicriteria optimization problems. However, at present simulation is mainly applied for technical problems. The approach described in the following shows its application in the field of project planning.

The C3 technique for recording and modeling cooperative work processes (cf. Subsect. 2.4.4) is the foundation of the simulation model presented here. In order to allow for a dynamic simulation of different organizational designs including measures of technical work support, however, the C3-model must be extended by elements and attributes, e.g., according to time and frequency distributions and combinatorics. These should be systematically combined in various constituent models that roughly cover the areas of human, technology, and organization. Furthermore, "organizational calculation rules" that permit a simulation of the relations in complex organizations must be developed first.

In the sense of a sensitivity analysis, simulation is not restricted to the identification of time-critical activities; it also allows the human-oriented evaluation of organizational alternatives such that an integrative work process design and assessment becomes possible. So far, the organizational analysis and optimization of industrial design processes are restricted to sound considerations of the management, either ex post based on some documentation of a process, or ex ante based on a draft of the target process. A simulation-based instrument with the necessary functionalities is not at hand. Such an instrument must be suited for the participatory design of design processes in chemical engineering. This can be reached by a graphical modeling language whose elements and relations are easy to use.

Our research aims at the creation of an organizational simulation model, which is adequate for the participatory design of complex workflows and project structures (see Sect. 3.4) in the process industries. The research approach is based on the C3 notation. The notation's plain graphical basic elements and the consideration of an advanced methodical approach, e.g., for communication analysis, is an adequate foundation for simulation studies by means of high level Petri nets.

In other areas (e.g., production processes [630, 1071], design processes in mechanical engineering [793, 925]), simulation is often used for analyzing well defined work processes. Similarly, the simulation of activities in design processes and their interrelations can be used as a foundation for the prospective analysis and design of a design process to be implemented. Thus, variants of a work process (e.g., a work process with additional work tools) can be examined. Also, the influence of humans – which in the scope of this sub-project have a particular role in the sense of socio-technological system design – is a major difference to conventional simulation approaches.

According to VDI guideline 3633 [1020], there are four basic categories with respect to the level of detail when human behavior in production processes is considered: (1) material flow simulation, (2) person-integrated and person-oriented simulation, (3) anthropometric simulation and (4) biological simulation. However, only the first and the second category are more closely examined in the following, as they are the only ones related to the activities in IMPROVE.

5.2.2 Workflow Simulation Model

A workflow simulation model of the PA6 design process has been developed in subproject I4 of IMPROVE. Although steady state and dynamic simulations are well known aids for the effective and efficient design of engineering artifacts [354], project planning and coordination are typically not supported by simulation in a satisfactory manner. In consequence, managerial tools are required that address the coordination of design processes. Although several supporting tools exist (cf. Subsect. 3.4.1), so far the support of planning processes in project management by means of simulation is unusual.

With respect to the human-centered simulation approach, it is necessary to consider the two different representations of humans in the simulation environment. In guideline 3633, VDI distinguishes between person-integrated models (person as reactive action model) and person-oriented models (consideration of various additional traits possessed by person) [1020, 1072].

Furthermore, two basic forms of model logic can be found in simulation models of design and development processes [1018]:

1. In the case of actor-oriented simulation models, system dynamics are produced by actors (persons or organizational units) based on specific activities [261, 426, 599, 604, 771, 813].
2. In process-oriented simulation models, system dynamics are produced by activities through the usage of resources (persons, tools) [571, 596, 597, 691, 925].

According to this terminology, the model presented below is person-oriented and process-oriented.

5.2.3 State of Research

In the field of design and development processes, only very few high fidelity simulation models can be found.

The Virtual Design Team (VDT) is an actor-oriented model for the simulation of product development projects that was created at Stanford University. Early versions of VDT were already able to model actors and activities as well as the information flow between them [599, 604]. In subsequent versions, the different goals of actors, the construction of exceptions, and exception handling [771, 812, 813] were accounted for. A process engineering context was not considered in this model. The methodology does not support the participative creation of simulation models.

Independent of VDT, Steidel [426] developed a further detailed actor-oriented simulation model for product development processes at Berlin University of Technology. This model also ignored the particularities of process engineering. Likewise, the participative creation of the simulation model or the optimization of workflow management were also not supported by the methodology. Additionally, bottlenecks caused by the insufficient availability of resources could not be represented. In chemical process design, such bottlenecks can be caused by the restricted capacities of laboratories, for example.

Also at Berlin University of Technology, Raupach and Krause [925] have examined a process-oriented approach for the simulation of product development processes such that consistency can be asserted in different design solutions. In this approach, the product structure is accounted for in great detail. Neither a process engineering context, nor the participative creation of the process model, nor the optimization of workflows are addressed. Interdependencies between project success criteria and factors influenceable by technical planning have not been examined.

At Massachusetts Institute of Technology, several process-oriented simulation models have been developed. We briefly discuss two examples: Browning [570, 571] used a design structure matrix which makes the modeling of complex projects very uncomfortable and unnecessarily complex. His simulation model was based on the assumption that an unlimited supply of resources (in this case, employees) exists. In consequence the simulation results of this model are limited in their predicative and predictive power. Cho's [596, 597] simulation model does account for the limitation of resources available in design and development project, but a corresponding processing of multiple activities is not yet possible. There is no relation to chemical engineering; also, participative modeling is not intended. Interdependencies between project success criteria and factors influenceable by technical planning can hardly be considered.

A process-oriented model for the simulation of a factory planning project was developed by a research group at the University of California at Berkeley [691]. This model accounted for the effects of altered requirements on the planning process and on the duration of construction projects. Particularly, so-called postponement strategies are examined, in which the start of a succeeding operation is delayed on purpose in order to increase the quality of the work results of the preceding operation. Similarly to other approaches, the simulation model assumes an unlimited supply of resources. However, participative modeling and the optimization of work processes are not dealt with. Interdependencies between the technical planning of influenceable factors and project success criteria are not sufficiently taken into consideration.

Like the approach followed in IMPROVE, which is described in the following subsections, the method by Krause [794] uses colored Petri nets in combination with stochastic procedures in order to sufficiently depict these decisions during simulation. First, the planner roughly models the activities of a design or development process; these activities are then further specified during the simulation run by means of a library. This dynamic calculation of the model structure adequately depicts the uncertainty-afflicted character of planning processes. However, participative modeling and optimization of the processes according to defined restrictions and target criteria are missing.

Recently, a further person-centered simulation model was developed by Licht [261] at RWTH Aachen University. It offers an – according to our requirements – more suitable approach for analyzing design and development processes in the chemical industries. The model includes many different specific aspects of the process, such as the type and the complexity of products, the characteristics of employees, tools, and organizational structure, etc. Due to the person-oriented approach, the model also serves as a realistic method for employee management by addressing the employees' behavior. A negative consequence, however, is that the model is very complex and therefore quite difficult to apply.

5.2.4 Implementation of the Workflow Simulation Model

The simulation model presented here offers a suitable technique for project planners in order to compare several alternative ways of project organization with respect to the number of involved persons, tools, time budget, and further decision variables at an early stage. With its close connection to C3, the model enables a transparent, very concise, understandable, and well-applicable representation of project organization. The goal of this simulation model is to combine the advantages of C3 (inherent simplicity and intuitive applicability) and the advantages of the simulation (the possibility of analyzing, planning, and rearranging the design or development process based on mathematical constraints). In addition, the model offers the chance to optimize the process with respect to duration and costs.

To maintain the distinctiveness of the C3 language, the simulation model was implemented using a person-oriented and process-oriented approach. For the formal representation and implementation of the simulation model, Timed Stochastic Colored Petri Nets were used. The development process was mapped to a directed graph consisting of places, transitions, arcs, and additionally its markings. A great advantage of this approach is that a stepwise simulation can easily identify weak points. In this case, Petri net tokens as representatives for active work activities indicate the status as well as the progress of the design process; they indicate possible weak points and bottlenecks resulting from this status.

The simulation model was implemented using the Java-based high level Petri net simulator Renew [796, 797]. Renew is a tool for the development and execution of object-oriented Petri nets. It provides synchronous channels and seamless Java integration for easy modeling.

The entire Petri net model, according to the description of the PA6 design process, is composed of different sub-networks (*constituent models*). The core element is the *Activity Network*, which connects i) the activities, ii) the employees, iii) the work tools or resources, and iv) the information to be processed. These four elements and their behavior are represented in more detail in four sub-networks that use different smaller networks for additional functionalities, such as the import and export of data or the visualization of the simulation progress.

In the following, we present the Activity Network and its constituent models.

The Activity Network

The design and development of a new or the modification of an existing chemical process or plant usually take place in process respective plant development projects. The complexity concerning the organizational structure as well as the workflow dynamics of such projects should become apparent and – as far as possible – it should be reduced. The model concept of the Activity Network – first generated by using the C3 notation and later on transformed in an executable Petri net – describes the work processes within the project. The individual phases of the project are divided in activities. This decomposition into constituent models is done to enable the planner to rearrange the different activities and their sequence without touching the detailed constituent models of the different activities, actors, employees, tools, resources, and information items. Predecessor-successor-relationships between the activities, which specify their logical execution order, are represented in the Activity Network. The Activity Network is visualized using the C3 modeling language (cf. Subsect. 2.4.4). The work activities of the Activity Network are assigned to organizational units for execution. Apart from the chronological sequence of activities, the assignment of work equipment to the activities is also displayed in the Activity Network. Thus, necessary input and output of the activities is represented in the Activity Network itself, but administrated in more detail in an adequate partial model (*Information Model*, see below).

An example of a simple Activity Network of a process with nine activities is displayed in Fig. 5.6 in C3 notation. This model has been created using the workflow modeling system WOMS, a software tool for the participative creation of C3 models (cf. Subsect. 5.1.3).

To develop the project simulation model, we consider – as an example – the PA6 design process (cf. Subsect. 1.2.2), which was used as a reference scenario in IMPROVE. This model describes the different activities for the conceptual design of a production process for Polyamide-6; it consists of 79 activities organized in eight organizational units (swim lanes) in the work organizational model. The Activity Network describes the structure of the design process. The predecessor-successor relationships between individual activities are defined in this Activity Network and in its corresponding Petri net.

In C3, there are several possible relations between activities such as sequential activities, concurrent synchronized activities, and activities that model the communication between different organizational units or persons. Within the scope of the activity-driven approach, activities can be represented by places in a Timed Colored Petri Net; thus, the tokens visualize the control flow, which again determines the activity sequence.

In order to simulate sequential activities, the termination of a predecessor activity has to be checked by means of a transition. In addition, the required tools and information as well as at least one person who is able to execute the activity must be available. The example process in Fig. 5.6 contains several sequential activities. Activity 1 and Activity 4 of swim lane 1, for example, are sequential activities that need one tool (catalogue) and some information for their processing. The corresponding activities in the simulation model are presented in Fig. 5.7 on the upper right side.

The simulation of synchronous activities is carried out by checking the termination of both synchronized activities before executing the subsequent

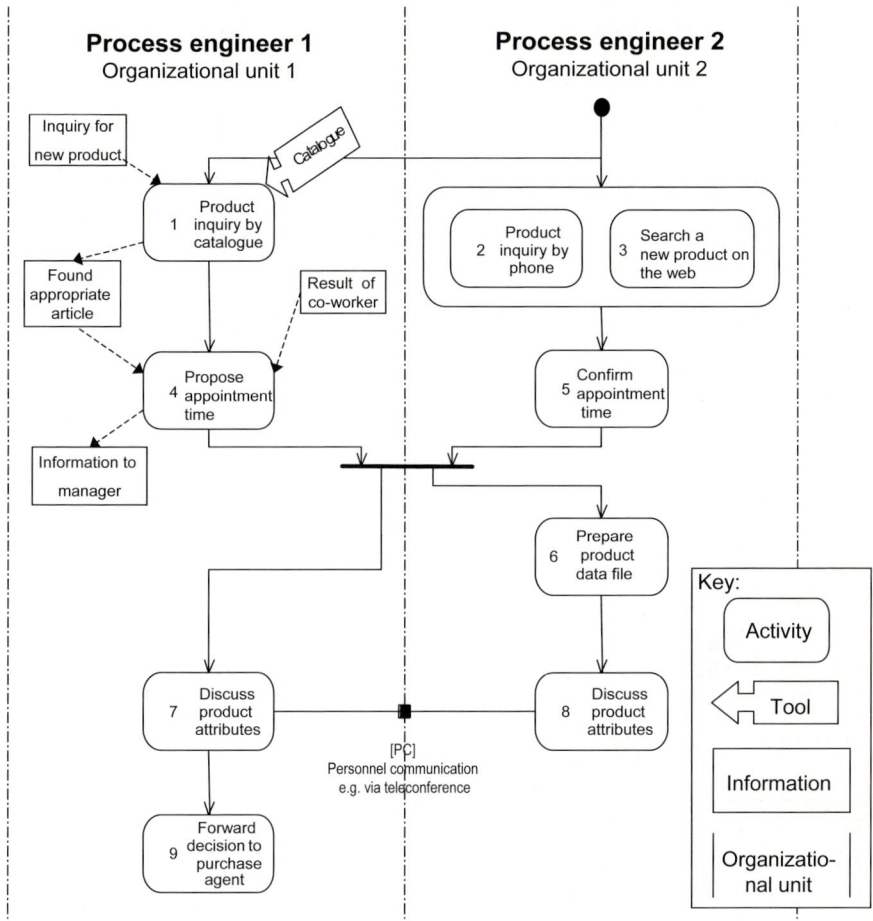

Fig. 5.6. An example of an Activity Network in C3 notation

activities. Furthermore, the execution of the activities relies on the availability of the required tools and information. A sufficient number of persons allocated to the activities can result in the parallel execution of synchronized activities. For instance, Activity 4 and Activity 5 in Fig. 5.6 are synchronized activities situated in different swim lanes. These two activities are transformed into two places of the Petri net, which are, according to the C3 model, graphically arranged in two different swim lanes (cf. Fig. 5.7). The successor activities, "prepare product data file" and "discuss product attributes", have to wait until both activities are terminated.

Communication can begin as soon as both developers arrive at the communication activity. The communication between different organizational units is a parallel activity that starts and ends in both swim lanes at the same

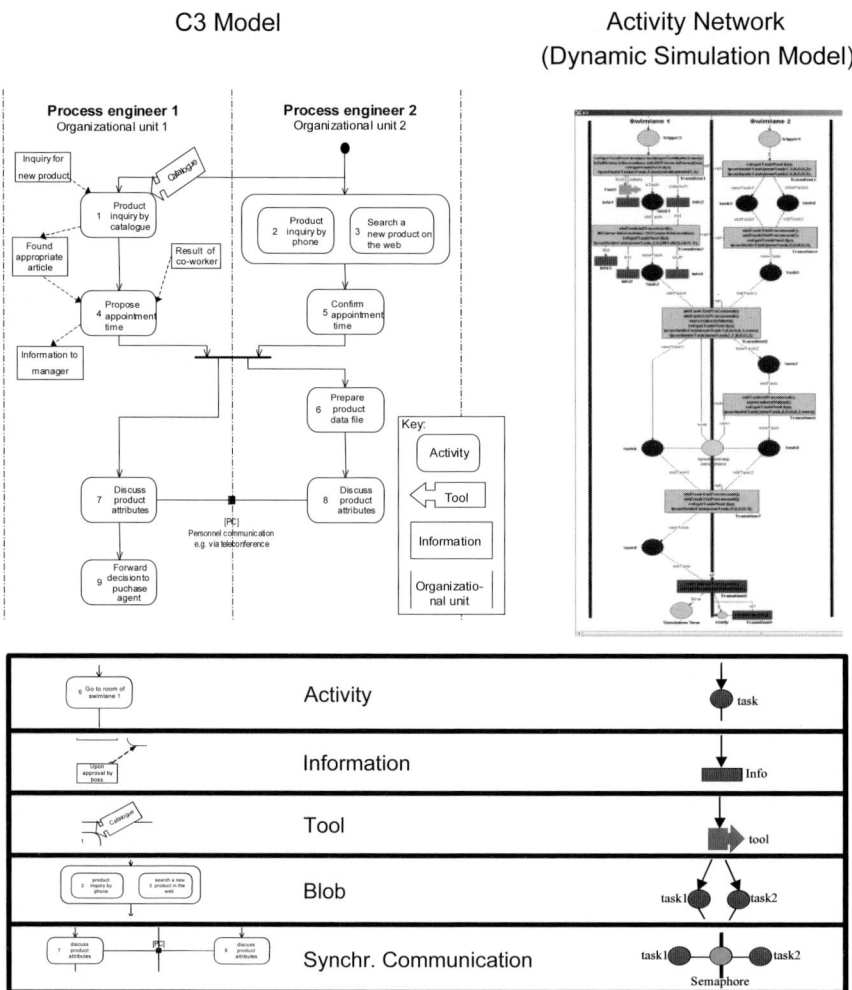

Fig. 5.7. Transformation rules from C3 to the Activity Network (Petri net)

time. The simulation of a single communication event can be realized by a semaphore variable that guarantees a synchronous processing. The discussion of product attributes between two persons or organizational units (Activities 7 and 8 in Fig. 5.6) is such a kind of activity. Both activities have to start and end simultaneously.

Figure 5.7 shows the C3 example model on the left and the transformed Petri net (implemented in Renew), which represents the Activity Network, on the right. The underlying basic transformation rules are given at the bottom of the figure. For illustration, Fig. 5.8 shows a section of the Activity Network of the PA6 design process.

Based on the process-oriented approach, the Activity Network the individual employees, the tools or resources, and the information items. Rough correlations, such as possible alternatives, the coordination of work processes, and resources required for the individual activities, are kept in the Activity Network. However, the exact processing of activities is represented in the *Activity Net*.

The Activity Net

Whereas the Activity Network consists of different activities of the design process, the behavior of each individual activity is described in the *Activity Net*.

For each activity, there is information about the subject matter to be processed, a necessary work tool, a skill profile of possible persons to execute the activity, input and output information of the activity, and a duration distribution. The distribution of the duration includes an expected value and a variance value. The distribution of the time consumption (e.g., Gaussian, right- or left-skewed β-distribution) may also be used for the calculation of possible buffer times. For the execution of an activity, a qualified person and, if necessary, adequate tools are selected to achieve the goal of the activity. As a result, the net for the representation of the execution of a single activity builds the link between the partial models of the work tool and the employee (Person Net and Tool Net, see below).

For each person, a value is determined that reflects his or her qualification for a certain activity. This value is calculated from the weighted sum of the person's assigned characteristics (cf. Person Net). The weighting and the different attributes are not constant and can vary depending on the area of application. The most highly qualified person will we chosen for the activity. If several persons are qualified for an activity, the person with the best efficiency, i.e., the best anticipated quality of the working results within the shortest time, will execute the activity. This efficiency is affected by the quality level Q_L, which is calculated as follows:

$$Q_L = \alpha \cdot P + \beta \cdot Q_w + \gamma \cdot Q_t.$$

The weights α, β and γ determine give the influence of an attribute on the quality level of a person. According to the model concept of the person, the attributes productivity (P), qualification based on the field of work (Q_w), and the ability and qualification to handle a work tool (Q_t) are viewed as independent variables.

With respect to these constraints, the Activity Net aims to reserve and provide the person who will execute the activity. Similarly, the resources to be used for the activity (such as tools or laboratories) are chosen and reserved.

Occasionally, it may happen that the basic skills needed for a certain activity are not possessed by anyone available. In this case, the activity cannot

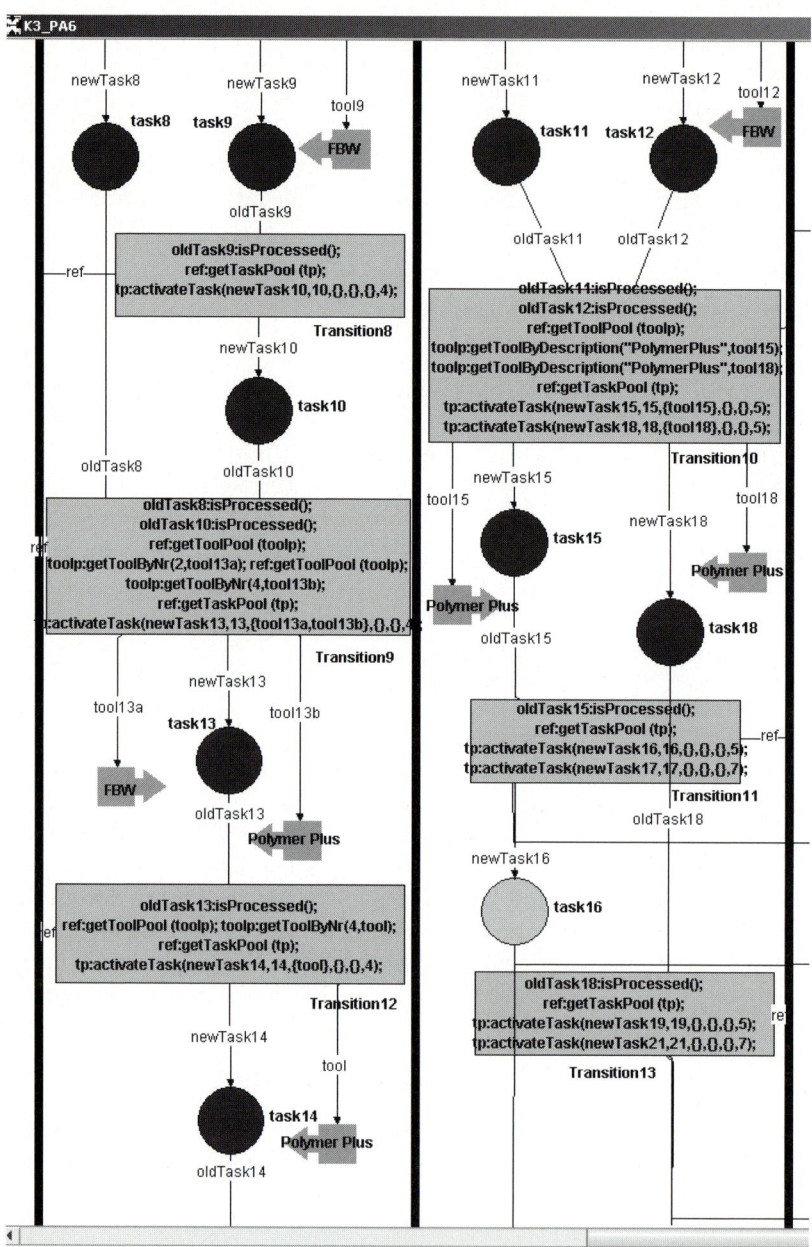

Fig. 5.8. Section of the Activity Network of the PA6 design process

be completed until someone qualified for the activity is available. The activity can only be executed if the adequate employee and essential tools and resources are available. The duration depends on the underlying distribution (Gaussian or Beta), on the qualification for the necessary tool, on the field of work, and on the productivity of the person who is employed for the activity.

As mentioned above, the duration of an activity is another variable determined by the Activity Net. Effort and duration for the processing of an activity depend on the estimated average execution time and on the qualification and proficiency level of the specific employee. The choice of work tools used along with the procurement of additional information can also have an effect on the duration and processing of an activity. In order to estimate the execution time of an activity, the concept of a probability distribution is employed. The first step is the estimation of a mean processing time. The variance around this mean value is represented by a Gaussian distribution. To allow for the realistic trend of activities to take longer than expected, the Gaussian distribution can be replaced by a right-skewed β-distribution. A normal distribution with relative variance between 10 and 30 percent of the mean was used for the runs of the simulation model.

The Activity Pool is an auxiliary network for the administration of the activities in a process. This comprises the import of the activities and their attributes from a database as well as the formatting into a new format adequate for the further processing in the Petri net simulator. All activities are initialized, imported, and managed by the Activity Pool.

The Person Net

According to the person-oriented simulation approach, the definition of the characteristics of employees is of particular importance. At the same time, an attempt is made to model persons as realistically as possible. This includes the employees' characteristics and abilities that have an influence on the allocation of persons to the various activities, the execution times for the activities, and the work quality achieved in the activities. Therefore, the employees involved in the process, along with their behavior, characteristics, and capabilities, are implemented in the Person Net.

The described attributes of an employee are summarized in the following:

- *Productivity of an employee:* To each person, a numerical value is assigned that describes the productivity of the person. This value allows to select the most qualified employee for an activity. It also has an influence on the execution time of an activity.
- *Qualification in terms of a particular area of work:* The activities of the process are arranged in swim lanes in accordance with C3 modeling. These swim lanes describe areas of work, such as the simulation of chemical processes (done by simulation experts) or the design of separation units (done by separation experts). The persons possess abilities and skills that qualify

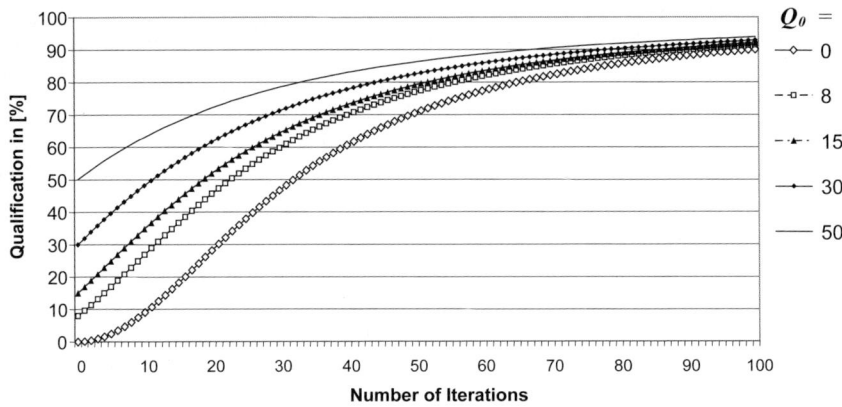

Fig. 5.9. Learning curve with different start qualifications

them for the execution of activities in certain areas of work, but they may make them unsuitable for others. According to these qualifications, the appropriate persons can be assigned to the activities to be executed.
- *Ability to deal with particular work tools:* Several activities require a work tool such as a software tool or a machine. The persons possess abilities and qualifications that describe how well they can handle certain tools. This means that a person must not only have the appropriate qualifications to execute an activity, but must also have the ability to carry out the activity with the necessary tool.
- *Learning aptitude:* An employee begins his career with certain basic qualifications, i.e., knowledge, skills, and abilities that have been acquired during education or training, and further inherent characteristics like retentivity or technical comprehension. During the course of a career, however, a person's abilities can change. Due to routine activities and new methods and expertise, certain qualifications can actually be improved. Conversely, abilities not used over a long period of time can also decrease. This capacity to learn and unlearn is represented by means of a learning curve that is attributed to each person. Activity-specific abilities of a person are improved, thereby increasing the corresponding attributes of the person when an activity is executed. This learning ability of the employee follows the characteristics shown in Fig. 5.9.

It was not possible to deal with the learning ability of every single employee; instead, a general function was used. One aim of further research activities is to individualize the learning and unlearning rate by employee-specific factors based on empirical studies and results attained in the field of psychology [647] and education science.

Integrative Simulation of Work Processes

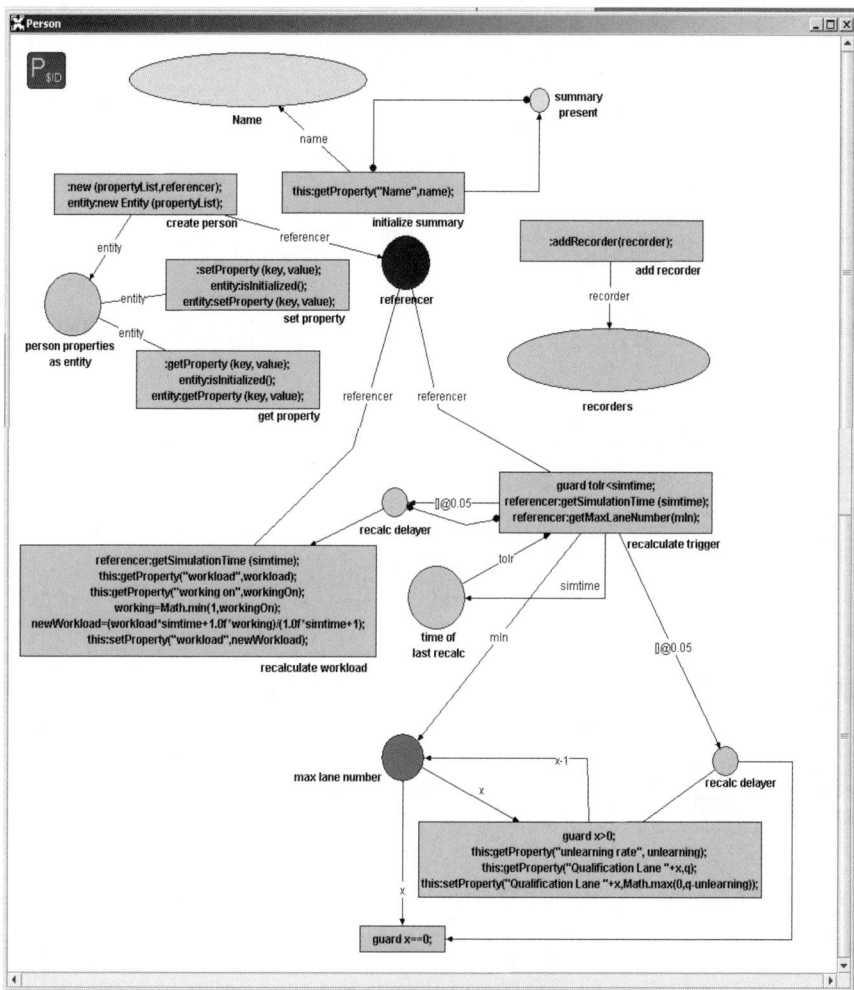

Fig. 5.10. Person Net

Personal qualifications are taken into account in the model concept to recognize that each person is able to carry out a variety of activities. This portfolio of possible activities can be directed at specific job descriptions that are representative for the different organizational units and work means related to the process. In Fig. 5.10, a section of the Person Net is shown.

The management of employees is organized in an auxiliary net, the so-called Person Pool. Here, the current number of available persons as well as their current status – "currently in processing" or "free for the next available

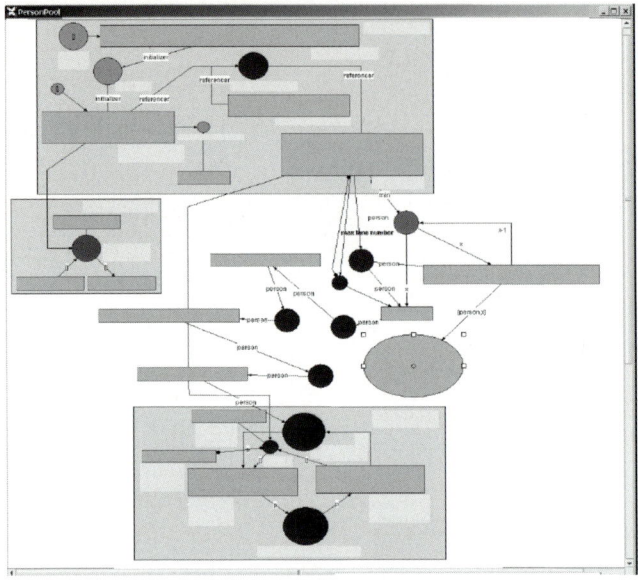

Fig. 5.11. Person Pool

activity" – is deposited. Before an activity is executed, an adequate employee is searched in the Person Pool. The Person Pool net is shown in Fig. 5.11.

The Tool Net

The influence of work tools on the execution of activities is represented by means of the *Tool Net*. As mentioned above, the assigned work tools can have an influence on the completion of individual activities. The allocation of work tools for activities results from the work organization of the project. Information about possible assignments of tools to activities is already included in the Activity Network.

Due to their scarcity, work tools must be reserved prior to their use. Also, a tool can be used by only one employee at a time, though more than one tool can be used for a specific activity. The amount of possible work tools in the project cannot be exhaustively declared since the amount of possible activities in need of completion, detached from individual case examples, cannot be given a priori.

Thus, similar to the Activity Network and the work organization, a list of work tools must be created. This list is specific for the design or development project to be simulated. The level of detail is also to be specified individually for each case. For instance, it may suffice to differentiate between drawing boards and CAD systems as exemplary tools for the creation of technical drawings; in other projects, a distinction between different operating systems for the CAD tools may be necessary.

The work tools available for the process are administered in the Tool Net and its corresponding auxiliary net, the Tool Pool. In the model presented here, a name and a distinct identifier are sufficient to characterize a work tool. Similar to the Person Pool for employees, the Tool Pool implements the maintenance of work tools; the current number as well as status of available work tools is represented here. A tool from the Tool Pool is allotted to a person needing the tool. However, in case another person is already using this specific tool, waiting times must be accepted.

Information Model

Information is already assigned to activities in the Activity Network and has an influence on the duration of the project. Information elements can be grouped into input and output information.

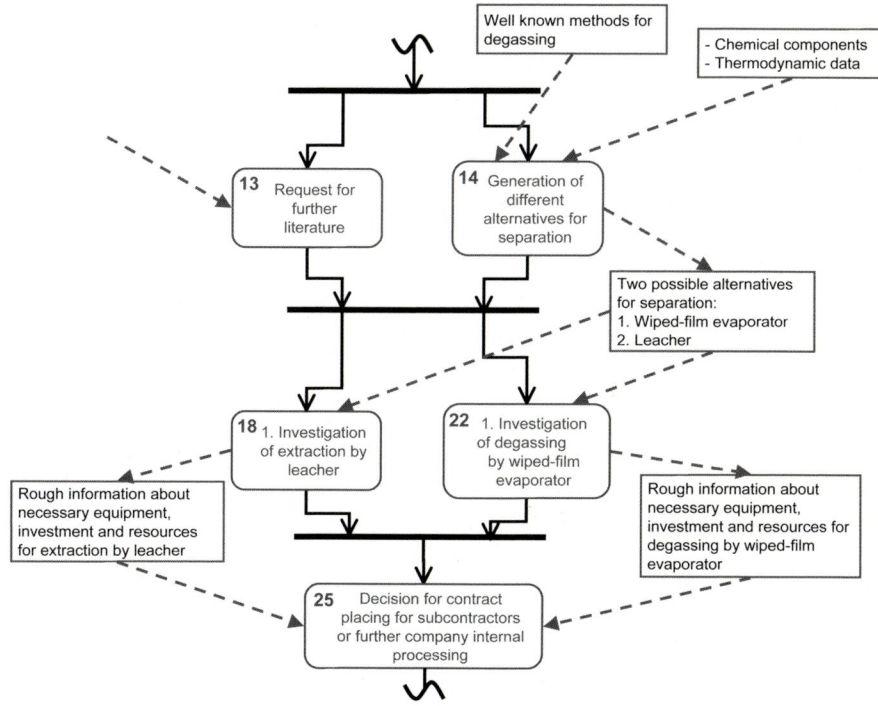

Fig. 5.12. Section of the PA6 design process

Input information such as files or documents is essential for the execution of an activity. Depending on the quality or type of the information, an activity cannot start or it must be executed differently if information is missing. For example, Activity 14 in Fig. 5.12 requires some information about degassing methods as well as information about the chemical components involved and their thermodynamic properties. Based on this information, an expert generates alternatives for the separation such as a wiped-film evaporator or a leacher. Both methods are analyzed in more detail (Activities 18 and 22).

Networks for Additional Functions

In addition to the constituent models described so far, a supporting model composed of further auxiliary networks exists. In this supporting model, functions such as the initialization of the model and the output of simulation results are implemented. It acts as a link between the various nets.

The input data of the simulation model (the description of the activities in the process, the required personnel, the necessary resources, the number and characteristics of the employees involved, and work tools available) is organized in tables. A user interface has been designed and implemented for the modification of these tables. To give an example, Fig. 5.13 shows the form for the specification of the attributes of a person such as the abilities to use the different tools or to perform special activities depending on the organizational unit. These parameters can be varied by changing the values in the boxes. Alternative values that have to be calculated separately must be divided by semicolons; such values are highlighted in green color.
In a second step, the number of possible combinations can be restricted. The interface provides the possibility to reduce and combine possible variables in test scenarios. These scenarios are administrated in a clearly arranged tree structure as shown in Fig. 5.14.

The test scenarios are saved in a special data structure and can be viewed with the help of the initialization network. This data structure contains tables that describe the parameters of the simulation model that have an influence on the duration and on the resource utilization.

Additional functions, e.g., the calculation of the normal distribution of the execution time or the printout of simulation results, are implemented in independent Java classes, whose functions are invoked and performed in the corresponding parts of the network.

5.2.5 Validation of the Simulation Model

Validation means to prove that a developed system (product, program, protocol, ...) meets some goals that have been specified a-priori. Thus, before models can be used to identify causes and effects [1009], it must be checked whether they are valid representations of the systems in consideration. To meet these requirements for the simulation model of the PA6 design process,

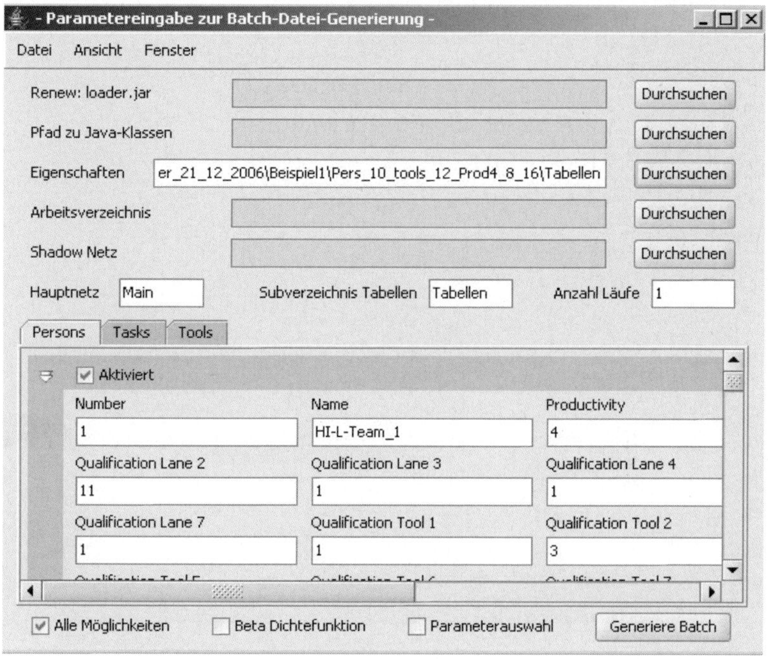

Fig. 5.13. User Interface for the variation of parameters

the model was developed in close cooperation between process experts from industry and academia. In addition, VDI 3363 [1020] suggests the comparison of real data to simulation results. Therefore further projects will be recorded and transferred into a simulation model in the future. A further step is to compare the real project structure, the project duration, and the operating grade of the employees with the simulation results.

Concerning the structural validation of the simulation model, the adjustment of the numerous parameters is particularly critical; these parameters include the number of actors (more than 20 attributes for each actor), the number of tools (five attributes for each tool), the dispersion and the variation of the activity durations, the probability of occurrence of certain activities, and many more. The values of these parameters can result in extremely complex system dynamics. Therefore, in the first runs of the organizational simulation, the number of persons was varied, and afterwards it was set to the optimal number. Subsequently, the number of tools was also varied. The other factors were not examined in the first test runs. Then, the influence of the number of actors and tools on the simulation results for the total time of project duration was examined in order to judge the internal validity of the simulation model. To do so, the expected durations of the individual activities were established in multiple expert workshops. As described in the following, these test runs

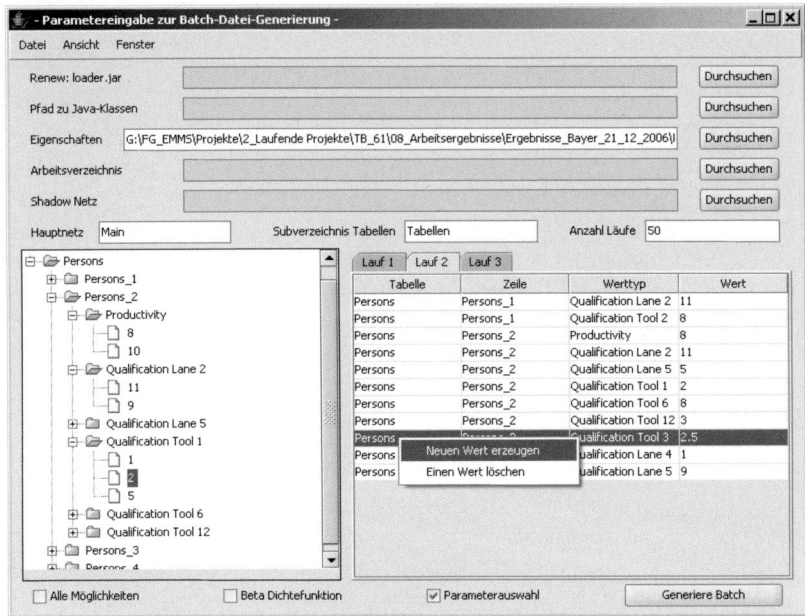

Fig. 5.14. User Interface for the configuration of test scenarios

showed consistent behavior of the simulation model with respect to the variation of the total time of project duration, to the organizational structure of the activities, and to the operating grade of the employees when the input variables (number of persons, number of tools) were varied.

Dependence of the Total Duration on the Number of Employees

The relationship between the total duration of the project and the number of organizations involved – in the present case identical to the number of persons – was analyzed in simulation runs. The analysis was restricted to the simple case that a single activity is executed by a single person. Further extensions of the simulation model will allow for the synchronous execution of single activities by several persons.

Hypotheses

The following three *null hypotheses* were formulated for a comparative assessment of the simulation results related to the PA6 design process:

- H_{01}: "The dependent variable 'total time of project duration' (TT_{pd}) is not influenced by the independent variable 'number of persons involved' (N_{ip})."

- H_{02}: "There is no effect of the independent variable 'variance of the activity duration' (V_{td}) on the dependent variable 'total time of project duration' (TT_{pd})."
- H_{03}: "The dependent variable 'total time of project duration' (TT_{pd}) is not influenced by the independent variable 'total number of tools' (TN_{ot})."

The Petri net simulation was used to investigate these different comparative hypotheses. Independent variables were the number of persons, the variance of the duration of each single activity, and the number of available work tools. The main dependent variable was the total time of project duration TT_{pd}. These results can conversely give the indication of the optimal project constellation of persons and work tools.

Analysis of Input Data

In practice, all simulation models are stochastic models, i.e., both input and output variables are random variables. In a simulation run, only one specific constellation of possible random variables can be generated, and only the corresponding simulation results can be analyzed. In the present case, the actual time consumption of each individual activity is calculated from the input duration and the attributes of the activity, the tools, and the persons. This input duration disperses between freely definable limits, normally distributed around a predicted mean value. The determination of this variation is acquired with random numbers and ranges to 99 percent between freely definable limits of ± 10, 20, or 30 percent. The random numbers are between zero and one; they were tested for autocorrelations smaller than 0.005 for a sample of 1000 random variables ($u_1, ..., u_{1000}$). By means of the Box-Müller Method [855], the equally distributed random numbers were converted into random numbers ($z_1, ..., z_{1000}$) with a normal distribution ($\mu = 0$, $\sigma = 1$):

$$z_1 = \sqrt{-2\ln(1-u_1)} \cdot \cos(2\pi u_2)$$

and

$$z_2 = \sqrt{-2\ln(1-u_1)} \cdot \sin(2\pi u_2)$$

It is assumed that the predicted time consumption is normally distributed within the given range.

Pre-test Conditions

To compare different constellations of input variables in this project, the total number of involved persons was varied between one and eleven. For more than five persons, no significant reduction of the total time of project duration could be detected. To ensure that 5 is the optimum number of persons for this project, the number of persons was increased up to eleven. For six or more persons, no significant effects could be detected (satiation). The variance of each expected activity duration, estimated by experts, was regarded as an independent variable. This variable was changed in the simulation experiment in three steps (10, 20, and 30 percent of the mean) such that for this pretest $n = 10$ runs were performed for each of the possible 33 combinations of variables (the number of combinations results from of the number of persons (11), multiplied with the number of variances (3)). The sample size of ten runs is suggested by Goldsman and Nelson [695] as adequate for determining the optimal number of final runs. As each run took roughly two hours, the total time for the simulation runs was approximately one month. This extremely low system performance is due to the insufficient appropriateness of Petri nets for reading, sorting, and writing tables, a functionality that is used very often for selecting and evaluating the different attributes in this simulation concept. Another reason is the online processing and the visual presentation of the simulation progress and its results. Finally the high amount of interfaces in this modular simulation system causes many approval processes between the modules of the model.

The corresponding hypothesis states that the duration of the design project decreases for each additional employee. Experts had predicted that the influence of the number of employees would have the most significant impact on the variable 'total time of project duration'. The experts also had given another reason to analyze this independent variable: personnel expenditures affect more than 80% of the total costs of development projects. Therefore, the total duration and the operating grade of the employees were to be analyzed.

Simulation Results

First, ten simulation runs were performed with some selected combinations for the number of persons ($0 < N_{ip} < 12$) and for the variance of the individual activity durations (V_{td}=10%, 20%, 30%). For these results, a two-way analysis of variance (ANOVA) was performed. The factorial ANOVA is typically used when the experimenter wants to study the effects of two or more treatment variables [945]. This method allows to test multiple variables at the same time rather than having to run several different experiments. By means of this method, interaction effects between variables can be detected.

A highly significant ($F_{10;297} = 1226,015; p < 0.0001$) dependence of the total time of project duration on the number of persons (independent variable) was discovered. The variance of the predicted processing times, however, was not significant.

Therefore, the null hypothesis H_{01} must be rejected. Instead, its negation H_1 ("The dependent variable 'total time of project duration' (TT_{pd}) is influenced by the independent variable 'number of persons involved' (N_{ip})") is confirmed. Furthermore, the expectation that a proper estimation of the duration of each activity can abbreviate TT_{pd} is *not* fulfilled; instead, the null hypothesis H_{02} is confirmed.

This result holds independently of the predicted duration; it describes a balancing effect on the variance of a large number of activities ($a = 79$).

Table 5.2. Post-hoc test

Variation/Change of project duration by variation of number of persons(*)

Number of persons	Mean (in sim.units)	Test runs	Std.deviation (in sim.units)	Changes of proj.duration in sim.units	Changes of proj.duration in %	Sig.
1	453,5	10	4,7			
2	269,9	10	6,5	-183,6	-40,49	0,000
3	220,4	10	8,3	-49,5	-18,34	0,000
4	191,8	10	5,1	-28,6	-2,98	0,000
5	182,9	10	3,8	-8,9	-4,64	0,038
6	182,6	10	7,6	-0,3	-0,16	1,000
7	182,4	10	5,1	-0,2	-0,11	1,000
8	182,6	10	5,0	0,2	0,11	1,000
9	182,5	10	5,5	-0,1	-0,05	1,000
10	181,4	10	3,9	-1,1	-0,60	1,000
11	183,5	10	7,1	2,1	1,16	0,999
Total	219,41	110	78,9			

(* with boundless number of tools and 10% deviation from expected task duration)

Experience shows, however, that projects usually do encounter delays, which is why the variance in the redesign of the simulation should be replaced by a right-skewed β-distribution. This measure shifts the mean of the random activity durations to the right side of the distribution. Hence, in contrast to a symmetrical Gaussian distribution, for more than 50% of the activities a longer execution time will be used in the simulation than estimated by the experts. This way, the quality of planning and the response to unexpected activity delays can be improved (longer buffer times, better risk management).

According to Eimer [653], the measure of effect (o^2) of the individual variables occurs as follows: 97.4% of the variance is due to the number of persons, whereas the variance of the individual activity durations, the interaction factor (number of persons, multiplied with the variance), and the errors are not significant. Thus, independent of the variance of the individual activities, there are no significant differences within the individual groups. Next, the total time of project duration was related to the different values for the number of persons. Starting at one person, the number of persons involved was successively increased by one. In addition, the variable "variance of activity duration" (V_{td}) was held constant at 10%.

The Tukey test was used post-hoc to compare the individual groups consisting of a constant number of persons to each other (see Table 5.2): up to 5 persons, increasing the number of persons resulted in a significant decrease of the total time of project duration (TT_{pd}). Beyond 5 persons, no significant reduction of the total time of project duration can be attained (cf. Fig. 5.15).

Furthermore, the simulation has shown that the duration can be reduced by approximately 60 percent by employing five persons instead of a single person. Though, six or more persons do not result in a significant additional

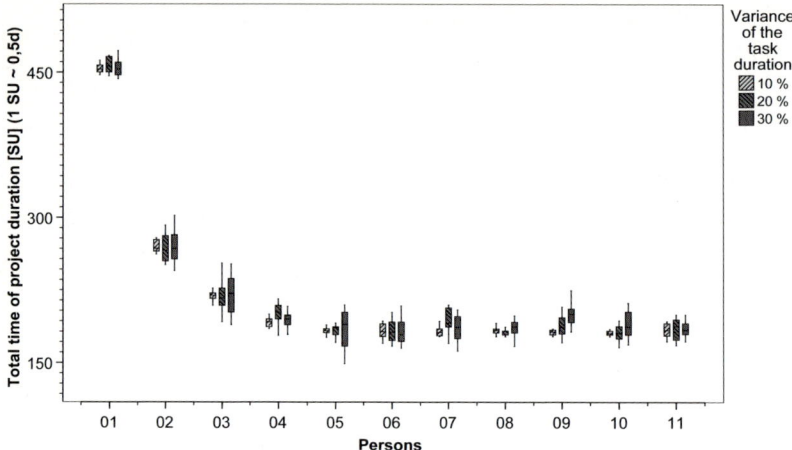

Fig. 5.15. Effect of the number of persons on the simulated total time of project duration and on the variance of the expected value (average values shown)

reduction. This is due to the structure of the activity network for the project: there are never more than five activities which can be executed at the same time. As we have assumed that an activity is performed by a single person, additional persons do not have any effect. If several persons can be employed simultaneously for particular activities, these circumstances will change. This motivates further research to implement the time-shared execution of a single activity by several actors. This will require additional attributes for the activities. For instance, the activity duration depends on the number of actors available. There are activities that must be executed by more than one person at the same time. For other activities, there is a maximum number of actors. This approach will be examined in further studies.

There is a further influence of the synchronous communications on the project duration. Synchronous communications between the activities occupy the required persons of the participating organizational units. Employees are picked from the activity network and "scheduled" for the discussion by the simulation model. These employees cannot execute other activities during this time. Such communication relationships are a characteristic for design processes, and therefore their effect should be examined more carefully in future.

The arrangement of the activities and the workload of the employees can be analyzed based on the graphical representation of the simulation results.

For instance, Fig. 5.16 the shows the results of a simulation with 2 persons (actors), whereas Fig. 5.17 gives the results for 5 persons. In the second simulation, the activities are parallelized as far as possible. Nevertheless, the decrease of the total project duration is not large (180 simulation units [SU]

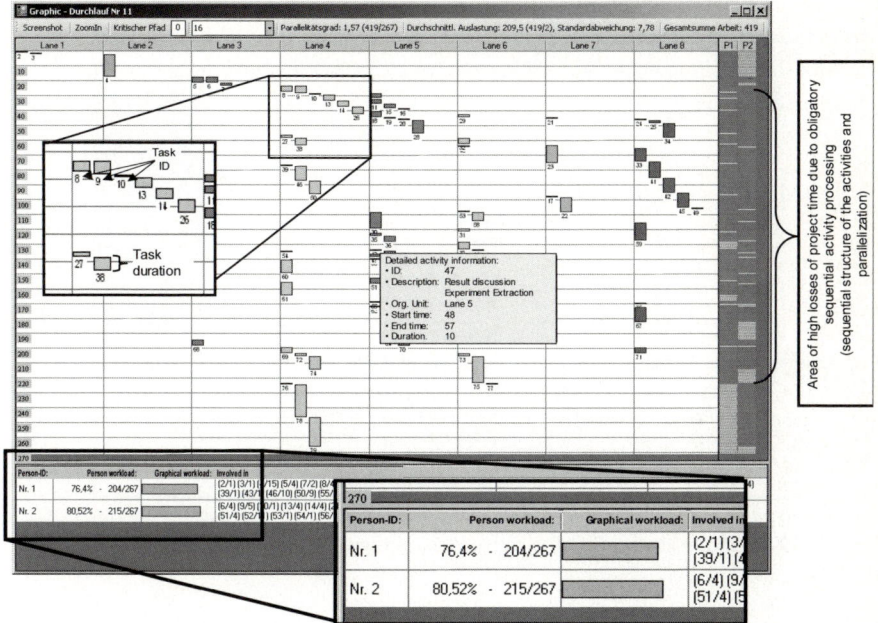

Fig. 5.16. Simulation results for two actors visualized as bar charts

in contrast to 267 SU for the first simulation). The calculated project duration ranges from 127 to 142 days for the first simulation (two persons), and from 88 to 110 days for the second simulation (five persons). This effect results from synchronous activities such as communication, which impede any re-arrangement of the activities. A further effect are different average workloads: The average workload of the two persons in the first simulation is approximately 79% (employee A: 80%; employee B: 78%), whereas the average workload for the five persons in the second simulation is approximately 40 percent (individual values between 16% and 58%).

Dependence of the Total Project Duration on the Number of Work Tools

The influence of the number of tools on the total simulation time was examined in further simulation runs. According to the results of the first examination, the parameter 'number of persons' was set to the optimum value (5 persons, cf. Fig. 5.15).

As the total project duration does not significantly depend on the variance of the expected process duration (as described above, variances of 10%, 20%, and 30% have been examined; see also Fig. 5.15), a variance of 30% was chosen in an arbitrary manner. That way, the different activity durations mentioned

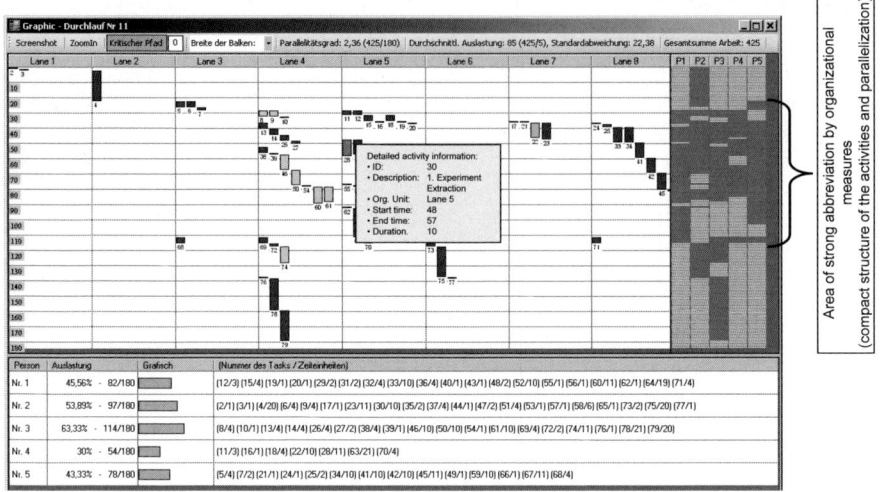

Fig. 5.17. Simulation results for five actors showing a different project organization

by the experts can be considered. Ten simulation runs were conducted for each combination of the parameters.

Hypothesis

The total quantity of arbitrary work tools (total number of tools, TN_{ot}) does not play a crucial role; in contrast, the number of very specific work tools, depending on the structure of the process, is important.

Therefore, a fourth hypothesis is introduced:

- H_{3a}: "There is no significant difference between
  - increasing the *total number* of available tools in a project and
  - increasing the number of *project-specific* tools

  on the total time of project duration."

For the PA6 design project, the minimum total number of tools is 9. This means that each work tool must be available at least once; otherwise the project cannot be carried out. If some work tools are available more than once, the effect on the total project duration is specific to the work tool. Moreover, several work tools are needed only once or only in a work area with sequential activities; in these cases, additional work tools of the same type have no positive effect on the duration of the process.

Through simulation, it is thus possible to identify those work tools which have a significant effect in bottleneck situations in the overall process.

To substantiate this fact, two groups of simulation runs were performed. In the first group (a), the total number of tools was varied (9, 18, 27, and 36 work

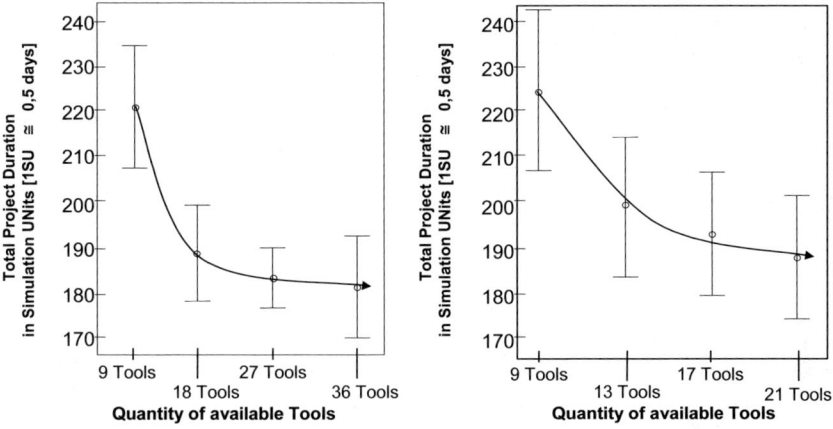

Fig. 5.18. Connections between the total time of project duration and the number of tools under procedures a) and b)

tools), whereas in the second group (b), only the number of four selected work tools was changed (4, 8, 12, and 16 instances of these four tools). These four tools were identified to be the reason for bottlenecks. Increasing the number of these tools could also have an influence on the duration of the example process due to its structure. This procedure is used to identify the optimal number of specific tools. In industrial practice, it allows to reduce the capital investment for such tools.

Simulation Results

The results for both groups of simulation runs are displayed in Fig. 5.18. They confirm the hypothesis that the simulated time is more dependent on the number and combination of specific tools than on the total number of work tools.

In the second group, only four specific tools were added in each case. A Levene test proved that the variances within the two groups did not differ significantly. Thus, the results for the two groups could be compared. A significant ($F_{6;69} = 110,081$; $p < 0.05$) reduction of the total time of project duration (TT_{pd}) between 9 and 13 work tools could be shown using a one-way ANOVA. Between 13 and 17 tools and beyond, the reduction of time consumption is not significant.

Thus, the "naive" duplication of *all* work tools – resulting in a total number of 18 – has the same result as duplicating only *4 selected* tools – corresponding to 13 tools in total. Analogous statements hold for total numbers of 27 and 17, as well as 36 and 21. In addition, the sensitivity analysis showed that increasing the number of persons has a stronger influence on TT_{pd} than increasing the number of work tools. As the measure of effect o^2 for manipulating the number of work tools on the total time of project duration is only 48.98% [653], there is only a weak effect if the number of work tools is varied.

A slight regressive tendency can be seen when the number of tools increases. This can be explained through the structure of the process; there is no situation in which more than three identical work tools are needed simultaneously.

5.2.6 Summary and Outlook

The simulation model described here consists of five constituent models. It allows project planners to study and optimize the structure of design and development projects and to allocate human resources effectively. The approach allows to apply simulation studies for planning design and development projects both before project start and during the projects.

The simulation model offers a graphical representation of the process due to its close connection to the C3 modeling language and the Renew simulation tool.

The influence of the quantity of persons and tools were investigated in the first simulation runs. These experiments produced satisfactory results. Additional analyzes are planned for the further validation of the simulation model.

We have identified two extensions to be studied in the future in more detail. These are the inspection of a right-skewed β-distribution as well as the execution of a single activity by several persons.

Additionally, investigations are planned concerning the influence of staff qualification on cycle times. The employment of highly specialized experts in comparison to the employment of workers with broad qualifications (generalists) should be observed. Also, the variation in the weighting of different employee characteristics is to be surveyed.

The application-oriented enhancement of the simulation model presented here is planned in the transfer project "Simulation-supported Workflow Optimization in Process Engineering" (cf. Sect. 7.4). The long term goal is the all-around support for project engineering through advanced project simulation in order to increase validity and to optimize time and resource planning. Thus, improved risk management in daily project planning is also allowed for.

5.3 An Integrated Environment for Heterogeneous Process Modeling and Simulation

L. von Wedel, V. Kulikov, and W. Marquardt

Abstract. The development of chemical processes requires the consideration of different areas such as reaction, separation, or product conditioning from several perspectives such as the economic efficiency of the steady-state process or the performance of the control system during operation. Mathematical modeling and computer simulation have become vital tools in order to perform such studies. However, various applications are supported by a number of tools with different strengths and weaknesses. Unfortunately, the formulation of the mathematical model and the data structures underlying the implementations of these tools are incompatible. Their integration at runtime for chemical process development poses technical problems to the engineer who carries out the work. This section describes an environment that accounts for the differences of mathematical modeling and simulation tools, and facilitates modeling and simulation across the boundaries of incompatible tools. The use of this environment is illustrated in the context of the IMPROVE reference scenario, addressing the production of Polyamide-6 from ϵ-caprolactam (cf. Subsect. 1.2.2); the simulation tools used in this case study are currently used in industrial design processes.

5.3.1 Introduction

Mathematical modeling of chemical processes has become a convenient means for process engineers in order to plan, evaluate, and assess design alternatives for chemical plants. Continuous improvement of modeling and simulation tools over a long period of time has resulted in the mature commercial systems available today. These tools provide domain-specific *modeling languages* for representing a chemical process through mathematical abstractions (such as variables and equations). Further, they provide *algorithms* to perform a numerical analysis of the models. Such an analysis, carried out in the sense of a virtual experiment inside a computer, is subsequently termed *simulation*. Simulation permits to study aspects of a chemical process that are expensive, difficult, or even impossible to study in real life. Such aspects can relate to economical, environmental, or safety aspects. For example, the effect of different controller settings on product quality and quantity, or the behavior of a process in hazardous situations may be safely evaluated.

In this section, the term *model* refers to a mathematical model which consists of equations describing the physical and chemical behavior of the process under consideration. Modeling and simulation are also employed in other areas. For example, a simulation study of the work processes during the conceptual design of a chemical process is presented in Sect. 5.2. However, the concepts used in languages for workflow analysis (such as *activities* or *actors*) differ strongly from the mathematical models and numerical methods employed in the simulation of chemical processes, which is the topic of this section.

Modeling and Simulation in the Chemical Industries

Several decades of research in simulation technology have led to a high level of maturity of modeling languages and numerical algorithms to analyze chemical process systems. More recently, a *lifecycle modeling* approach has been identified to be an important research topic in model-based process engineering [303]. This approach is characterized by an integrated view of the work processes conducted during the design and the operation of a chemical plant as well as of the information created and consumed by the activities during these work processes. The ultimate goal is to understand process design as a continuous work process in which all information required is invented (and entered) only once and is then enriched along the work process towards the final goal, i.e., the complete specification of a chemical plant.

However, current tools are often too narrow in focus, and many of them are used by different specialists when designing or analyzing chemical processes; in the following, we give some representative examples:

- Simple economical considerations are based on linear mass balance models, which are often evaluated using a spreadsheet application such as Microsoft Excel.
- The performance assessment of process flowsheets is often supported by steady-state process modeling tools (also known as flowsheeting packages) such as Aspen Plus [516].
- Important process units such as a reactor or a non-standard separation unit (e.g., a crystallizer) are often modeled by means of special-purpose packages.
- Tasks like controller design or the assessment of start-up procedures require dynamic simulation studies, which are supported by dynamic simulators such as gPROMS [916].

Hence, chunks of the overall design process are well supported by diverse software tools. However, the desired integration of these chunks into a coherent work process requires also an *integration of the respective software tools* to overcome the solution islands. A *manual integration* requires (at least) the re-entering of information to specify the simulation of the various tools. Manual integration could be overcome, if an omnipotent simulation tool were available which covers all aspects of the design process. Considering the efforts that have been invested into the various process modeling and simulation packages, it becomes quickly obvious that the development of such an omnipotent software tool would be a formidable task. Instead, a promising pragmatic approach seems to be the *a-posteriori integration* of the existing simulation tools at run-time.

Integration of Heterogeneous Models

The information system described in this contribution integrates process modeling tools to be employed during the process design lifecycle such that the engineer may freely choose among available tools for modeling and simulation on an arbitrary level of granularity. When developing new models, it shall be possible to employ the best-suited modeling tool for different parts of the chemical process without bothering about conceptual or technical aspects of their aggregation in advance. The reuse of existing models and their aggregation to new ones should be supported, regardless of the modeling paradigms or software technologies that were used to develop and implement the models.

Hence, *model integration* must be supported to permit model development based on existing simulators representing parts of the complete chemical process. This integration is particularly difficult because the existing models have been developed using different process modeling tools, which are incompatible with respect to the underlying modeling concepts and languages as well as simulation algorithms. Hence, an a-posteriori integration approach is needed to permit an efficient reuse of these models.

Model integration comes in two flavors [303]. *Horizontal model integration* covers process parts of comparable granularity, but of different kind to be combined to a model of the whole process. For example, a CFD model (*computational fluid dynamics*) of a reactor might be integrated with a neural net model of a melt degassing unit and with lumped mass and energy balance models of the remaining process units to form the model of a complete polymerization process. In contrast, *vertical model integration* refers to the case where submodels are introduced in a supermodel in order to increase the degree of detail of the description of the physical and chemical phenomena occurring. For example, the reaction rates of a polymerization model may be considered in detail to reflect the complete kinetics covering all the reaction steps rather than lumping all physico-chemical effects with an overall reaction rate.

The technical implementation of model integration is determined to a large extent by the way the models are represented [303]. We distinguish procedural and declarative representations, respectively. *Declarative model representations* are coded in a generic modeling language and are completely independent of the solvers applied. The term *solver* refers to any numerical algorithm applied to a model to accomplish a simulation or an optimization. The integration of declarative models requires the aggregation of submodels on the level of the modeling language and is therefore similar to the concept of data integration in software engineering [1038].

In contrast, *procedural model representations* are intimately intertwined with a solver and are coded in some programming language. The integration of procedural models requires some concept of control integration [1038] to execute the partial models according to the way they are interconnected to form the complete model.

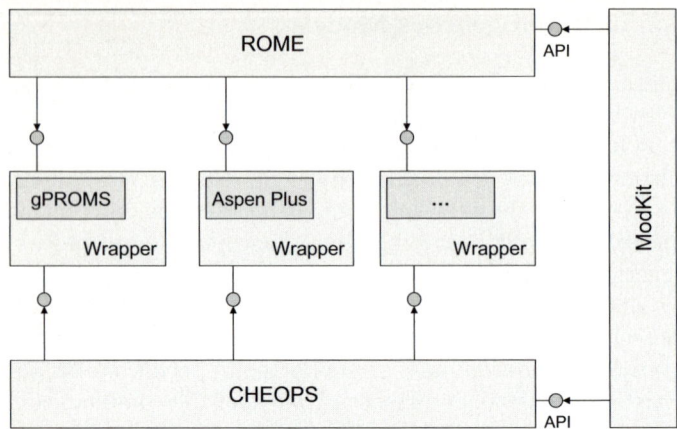

Fig. 5.19. Conceptual overview of the integrated modeling environment REALMS

The distinction of declarative and procedural representations is directly related to the classification of the models from the point of view of an external solver [303]. *Open-form model representations* provide interfaces to access the full equation system of the model, for example, in form of a CAPE-OPEN equation set object [894]. Alternatively, the *closed-form model representation* provides interfaces which only enable to set inputs and to retrieve outputs of the model. Typically, declarative representations require an external solver and use an open-form interface, whereas procedural representations come with an integrated solver and usually have an interface of the closed-form type.

An Environment to Support Heterogeneous Process Modeling

A modeling environment supporting the reuse of *heterogeneous software module* of very different nature as described above roughly requires the support of three important aspects of model integration [459]. The modeling environment REALMS developed in the research group of the authors essentially consists of three software tools, each of them covering one of these aspects (see Fig. 5.19):

- *Model maintenance and archival* along the lifecycle of process design must be supported. The model repository ROME [24] provides functionality for model management and archival. ROME accesses proprietary modeling tools in order to extract information about the models being maintained. ROME stores only metadata about the models being archived and can therefore easily reference models in a variety of representations, open-form or closed-form as well as declarative or procedural models.
- A *symbolic perspective* of model integration permits the aggregation of model building blocks that originate from heterogeneous process modeling

tools. The modeling toolkit ModKit [52, 54] uses model building blocks made available by the model repository ROME to support model development across the boundaries of existing modeling languages and tools. Since ModKit permits aggregation as well as refinement of models, horizontal and vertical integration of model building blocks is supported.
- The execution of *heterogeneous simulation experiments* is enabled by the simulation environment CHEOPS [409]. It supports *run-time integration*, i.e., heterogeneous models can be used together to simulate a complete plant. CHEOPS builds on the native solving capabilities of the proprietary simulators to ensure that the models are computed with exactly those algorithms that have been used during their development. This applies to declarative as well as procedural models. In case open-form and closed-form models are used in a mixed setting, suitable wrappers are supplied to make the different forms interoperable.

Furthermore, in order to simplify the technical realization and maintenance of the environment, abstract interfaces for *tool wrappers* have been introduced to render models from external tools, including gPROMS and Aspen Plus (cf. Fig. 5.19). These abstract interfaces permit the development of generic functionality in the environment. Furthermore, future changes in the simulation tools only influence the corresponding wrapper implementation; they do not need to be accounted for in the implementation of the functional modules of the overall environment.

Before discussing the individual elements of the heterogeneous modeling environment in detail, we present a scenario to illustrate the issues addressed above (cf. Subsect. 5.3.2). The model repository ROME is discussed next in Subsect. 5.3.3, followed by an explanation of ModKit in Subsect. 5.3.4. More details on CHEOPS are given in Subsect. 5.3.5. Subsection 5.3.6 presents the work process for solving the scenario problem introduced in Subsect. 5.3.2 by means of the heterogeneous modeling environment. Concluding remarks and a summary are finally given in Subsect. 5.3.7.

5.3.2 Production of Polyamide-6 – An Illustrative Scenario

The development of a Polyamide-6 (Nylon 6) production process as described in Sect. 1.2 is employed to illustrate the issues discussed in the previous subsection. Besides being a process of industrial relevance, it has certain properties which stress the importance of a neutral model integration platform. First, the behavior of polymer materials is more difficult to describe than that of ordinary fluids which are handled quite well by most state-of-the-art simulation packages. Further, non-standard pieces of equipment are used to realize the Polyamide-6 process in a technically and economically efficient manner. In addition, the complete process is supposed to be analyzed including the downstream extrusion of the material. This extrusion step is not only required to formulate the polymer product into a particulate material, but it also could

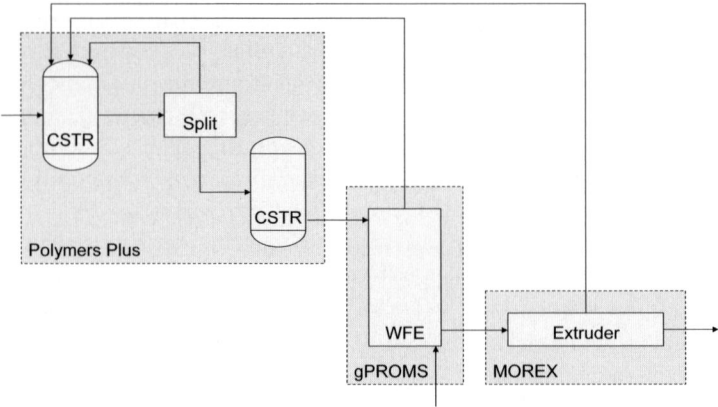

Fig. 5.20. Model flowsheet of the process; *CSTR*, *Split*, *WFE*, and *Extruder* refer to model types, while *Polymers Plus*, *gPROMS*, and *MOREX* denote available simulation tools

be used as an additional separation step to recycle unconverted monomer to the reaction and separation sections of the process [99].

The monomer feed is converted into Polyamide-6 by polycondensation and polyaddition reactions [930]. This reaction step can be realized by a complex reactor which can be modeled as a sequence of stirred tank and plug-flow reactors. An exemplary model flowsheet comprising two reactors (*CSTR*) with an intermediate water separation (*Split*) is shown in Fig. 5.20. Such a model of the reaction section can be analyzed by means of Polymers Plus, an extension of Aspen Plus for handling polymer materials [513].

The reaction section is followed by a separation section which separates unconverted monomer from the effluent of the reaction section. Two alternative realizations using either a leacher or, as shown in Fig. 5.20, a wiped-film evaporator (*WFE*) are described in [99]. Models of an appropriate level of detail are neither available for the leacher nor for the wiped-film evaporator in standard libraries of process modeling tools. Therefore, customized models have been developed for both apparatuses by means of the gPROMS modeling environment.

Finally, the polymer is processed in an *extruder*, where polymer properties (such as the chain length distribution) are adjusted by means of a sequence of different functional zones of the extruder. Further modifications of the polymer properties can be achieved by adding additives to the polymer melt. The extrusion step can be simulated by a special purpose tool called MOREX [146, 394]. Given geometry data of the extruder, it calculates the required energy demand and the properties of the resulting polymer. In addition, it can be used to calculate the vapor stream of ε-caprolactam which can be stripped of if a considerable amount of ε-caprolactam is present in the polymer melt.

As shown in Fig. 5.20, there are two (potential) recycles across the system boundaries covered by the simulation tools. In order to analyze their economic benefits and to determine to what extent the extruder should perform a separation function, an integrated treatment of the simulation problem is required which ideally reuses the separate submodels implemented in different tools.

Since these submodels are incompatible, their run-time integration for the simulation of the overall process is difficult. The integration is complicated by the fact that several experts are responsible for modeling different parts of the process. A unified platform for model storage and archival facilitates an overview on the mathematical models that have been developed for a certain sub-process. Second, the models are represented in a combination of declarative and procedural models. The reaction model is represented in a procedural manner in Aspen Plus. Evaluation of the closed-form model is possible through the automation interface of Aspen Plus. The separation model exists as a declarative representation in gPROMS and can be evaluated using the CAPE-OPEN equation set object interface. The extruder model is coded as a set of procedures within the tool MOREX, which permits evaluation through an automation interface.

5.3.3 ROME – Management and Archival of Heterogeneous Process Models

In the current situation, modeling in the chemical industries is characterized by a number of modelers working independently. The modelers are often not aware of models which have already been developed by others previously, and chances are low that models are reused when their developers are no longer members of the process design team. In the Polyamide-6 scenario presented above, the different models may have been developed in different projects (probably even for different plants) so that there is probably no single person that knows about the existence of all models that are applicable in a certain context.

Model management can ensure that models developed in early stages of the design phase can be reused properly in later phases. The model repository ROME has been proposed to provide model management functionality and a central storage for models across projects [463]. The repository stores models in a neutral format and integrates different applications by providing import and export capabilities from and to existing applications [552]. In the environment presented, ROME acts in the sense of a model server [288, 293, 303, 532], supplying models and model-based services for a variety of activities in the area of process design or operation.

In order to achieve this objective, ROME stores existing *model implementations* (cf. Fig. 5.21) in their native representation. Such model implementations for proprietary modeling tools comprise text files such as the ASCII files of gPROMS or Aspen Plus, but also binary components such as dynamic link libraries containing foreign objects for gPROMS.

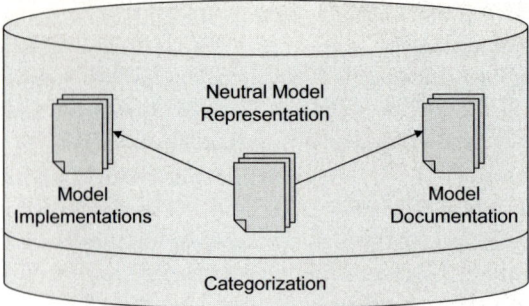

Fig. 5.21. Conceptual view of model repository contents

It has proven useful to abstract these proprietary models into a *neutral model representation* (cf. Fig. 5.21) to allow functionality to be developed without the need to consider the specific tools with which these models have been built. This neutral model representation is used as a substitute for the incompatible native model implementations in the sense of metadata. Such metadata are described by an object model and cover the structure of the model (e.g., blocks and their connectivity) as well as its behavior (e.g., variables denoting process properties and equations representing relations among properties). It should be noted, that this neutral representation is an abstraction and not a complete translation. The actual model development process as well as the step of computing a model is still based on the original implementation of the model rather than on its abstraction stored in ROME. Otherwise, it would not be possible to reuse the evaluation or solution functionality of the original model that is provided by the respective process modeling environment.

Required metadata are extracted automatically when a model is imported into the repository. For the scenario presented above the following actions are performed to import the elementary models: In case of Aspen Plus, the automation interface is used to extract blocks and their connecting streams from the model definition. For gPROMS, a fragment of XML is added to the model input file as a special comment. This XML fragment is extracted and interpreted at import. A fully automated solution would be feasible but has not been realized due to the effort associated with parsing the gPROMS model definition. For MOREX models [146, 394], a fixed set of connecting ports is specified in the implementation of ROME. For other tools to be connected to ROME, the degree of possible automation for import may vary according to documented interfaces or file formats that are required to access information about the model to be imported.

Functionality such as searching or browsing for models within ROME is performed without actually using the incompatible model implementations. Instead, the metadata abstraction stored in the model repository is traversed for this purpose. In addition to simple search strategies, models can be arranged into hierarchically organized *categories* (cf. Fig. 5.21) to build up a library of process models which is easy to navigate for the user.

As a further advantage, *model documentation* (cf. Fig. 5.21) can be attached to the neutral model representation; it is maintained independently from the tool in which the model was developed. Besides storage and organization of model implementations, the model abstraction in ROME is also able to represent structured, hierarchically decomposed models. This property is used by ModKit (cf. Subsect. 5.3.4) to aggregate heterogeneous process models. Further services like configuration and version management can also be based on the uniform model representation in future developments.

ROME has been implemented using C++ [985] and the object-oriented database management system VERSANT [1022]. It provides an API for external services through the CORBA [877, 1024] middleware standard. This communication layer has been implemented using omniORB [701].

More recently, the AixCAPE consortium ([496], see also Subsect. 7.1.3) and LPT [278] have continued work in the area of model management in cooperation with industrial partners (Shell, Lanxess, BASF) and developed a model management system similar to ROME in scope and functionality. The metamodel has been largely retained, but is focused on industrial requirements such as storage of chemical components and their relations with process models. In addition, metadata about property models employed (e.g., NRTL, UNIFAC, etc.) and reaction stoichiometries have been added. As opposed to ROME, this extension permits detailed queries about the availability of models considering certain mixtures of substances in combination with particular reactions and property models. Importing capabilities have also been improved further so that users can now add models to the system through a web interface. Currently, automated import is possible for Aspen Plus, Pro/II [749], and ChemaSim, the in-house simulator of BASF. As a technical basis for the system the web-based application server ZOPE [1073] together with the content management system Plone [911] have been used to realize the web-based application entitled *MOVE* (*Model Organization and Versioning Environment*, [1029]). At the time of writing, first installations have been deployed on-site at industrial project partners, who use MOVE routinely as part of their business. Further industrial partners are evaluating its use and consider extensions for importing from other modeling tools as well.

5.3.4 ModKit – Model Construction with Heterogeneous Process Models

The modeling toolkit ModKit aims at simplifying the model development process by providing reusable model building blocks [52]. Further, ModKit provides interactive support for the user during the assembly phase of the model building blocks.

A graphical editor is used to define the structure of the process model. The editor manipulates structural modeling objects representing compartments (devices) of the process and flows (connections) between them. These objects have attribute tables which store the phenomena occurring in the structural entities of the process. With this structured information the modeler can easily use the behavior editor to provide the equations and variables, called behavioral modeling objects, describing the behavior of the process. As an alternative, reusable model building blocks on various levels of granularity (from phases to complete pieces of equipment) may be reused from a palette by cloning and tailoring them to the modeling context. A documentation editor allows to add informal documentation to the structural modeling objects. This information is organized in hypertext nodes, allowing to reference other modeling objects and documentation thereof. A decision made during the process of setting up a model can be documented using an implementation of the IBIS model [608].

A first development phase of ModKit [52, 54] is based on the G2 expert system shell [720] together with additional functionality to analyze models and simulate them on different simulators. Code generators for target languages such as gPROMS [916] or SpeedUp [512] are available in order to translate the modeling concepts into the language compatible with the desired modeling tool. This code generation capability permits the use of different numerical solution algorithms without additional effort for reformulating the model in different modeling languages. A screenshot of the implementation is given in Fig. 5.22.

More recently, a re-implementation of ModKit (named *ModKit+*, [151]) was undertaken in order to overcome shortcomings of the G2-based prototype in the areas of persistence, modularization, and licensing cost associated with the expert system shell. This new development phase aimed at an integration of ModKit+ with ROME in order to permit model building blocks maintained in ROME to be used for model development in ModKit+. In addition, models natively developed using ModKit+ are also stored in ROME so that aggregation of process models from elementary pieces can make use of heterogeneous process models imported from simulation tools as well as process models developed on a first-principles basis using ModKit+.

Whereas ROME addresses the management of information about heterogeneous models, ModKit+ provides domain specific logic about model building and uses elementary services provided by ROME. The result of these activities is a process model in which the equations and variables are augmented with information about the underlying physico-chemical concepts they are describing. This complements the ability of the model repository ROME to store and manage legacy models in the often proprietary modeling languages of simulation tools [293]. These languages usually include only information on equations and variables and their structure. They do not provide means to associate phenomenological concepts or extensive documentation with the mathematical representation of models.

ModKit+ is implemented in Python [824], a scripting language which permits tight iteration cycles within the software development process; that way,

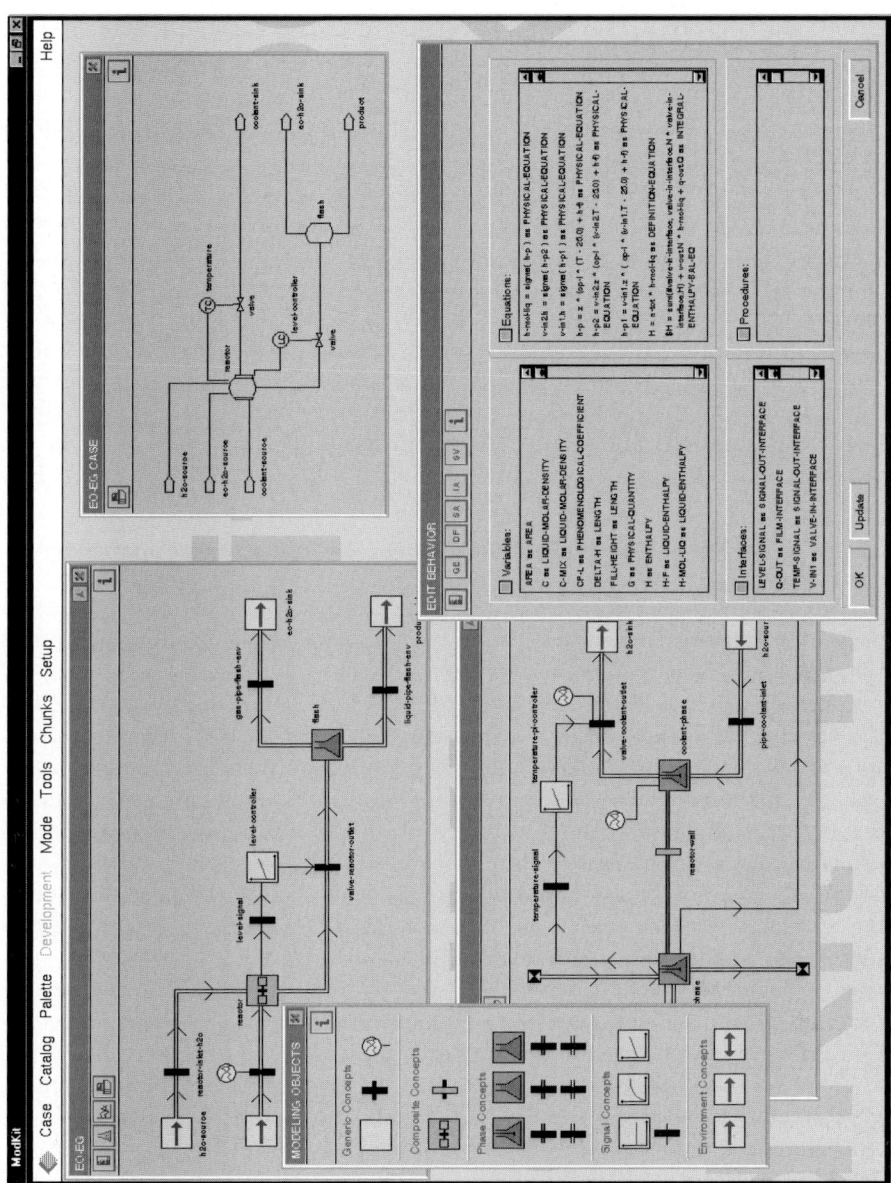

Fig. 5.22. Screenshot of ModKit

experiences concerning the usability of the tool could easily be accounted for during the development process. Communication with the model repository ROME is established through the CORBA communication middleware and has successfully been tested in a network of computers with mixed Windows and UNIX operating systems.

In order to realize the Polyamide-6 scenario problem presented in Subsect. 5.3.2, a process description is defined using ModKit+. The elementary models for the reaction section, the separator, and the extruder, already imported into ROME, are added as submodels of the overall process. Further, a mixer is defined in order to combine feed and recycle streams; corresponding mass balances are added to the elementary mixer model. After this modeling activity the model repository ROME contains all necessary models for the overall Polyamide-6 process. The model behavior is partly described by equations (for the mixer) and partly described by model implementations in the form of input files for the modeling tools Aspen Plus, gPROMS, and MOREX.

5.3.5 CHEOPS – Integrated Simulation of Heterogeneous Process Models

In the environment presented in this contribution, the modeling process is supported by ModKit+, which allows for model integration on the basis of conceptually consistent model building blocks with common interfaces. When it comes to simulation, integration of process models faces a technical problem due to different implementation of the models in the respective simulation tools. An attempt to recode these models would be tedious and often impossible without the loss of model quality or validity. Instead, reusing the implementation of the models within specialized tools (e.g., the fluid dynamics model of a reactor in a CFD tool) is the desired option. Hence, a simulation environment is needed that simulates complex models built from several submodels, which are possibly of different types.

Such a simulation environment should support technical solutions to enable communication with various external simulation tools and an internal mechanism for the integration of existing model implementations where the overall problem has been formulated in ModKit+. To support lifecycle management, the problems to be solved by such an environment should involve steady-state and dynamic simulation, parameter identification, and optimization.

These concepts are implemented in the integration platform CHEOPS (*Component-Based Hierarchical Explorative Open Process Simulator*) [252, 409, 462]. The platform provides generic component prototypes and interfaces for the integration of models, solvers, and tools. The generic components are instantiated at run-time by concrete software components and classes representing actual unit operation models, solvers, etc. That way, arbitrary components from the list of available components can be used in the simulation. The list of model and solver components can easily be extended with the components that comply with the abstract structure and interface definitions.

The *flowsheet* component represents the structure of the full problem and contains references to *unit operation* components and *couplings*. Couplings describe the exchange of information between the units and determine the topology of the flowsheet. Unit operations are the containers to store the model for a single unit of the flowsheet.

The abstraction of the model in CHEOPS is a *CHEOPS model representation* component, which defines the inputs, outputs, states, and parameters of the model available for CHEOPS. The CHEOPS model representation is associated with the *model source*, which is a tool-specific implementation of the model. CHEOPS distinguishes the open-form representation of the model, which provides access to all states and the equation system, and the closed-form representation, which provides access only to model inputs and outputs. Separate CHEOPS model representation classes are defined for each type of the model, *OpenFormModelRepresentation* and *ClosedFormModelRepresentation*, respectively.

The type of model representation determines the type of the solver components used in the simulation. CHEOPS supports the equation-oriented and the modular simulation approach. The *equation-oriented simulation approach* can be used if all models are formulated in open form. A joint model is derived by retrieving and concatenating the model equations of the individual models and adding extra identity relations for coupling variables. Such a model can be solved in CHEOPS by general numerical solvers, which are available from a library of numerical algorithms (such as *LptNumerics* [408]) containing numerical codes from a number of sources. This is an implementation of the concept of horizontal model integration.

Alternatively, closed-form model representations require a *modular simulation approach*, where each closed-form model is computed using the internal solver of the software tool the model is implemented in. The algorithm sets the model inputs, performs control over the simulation, and retrieves the outputs of each model through the commonly defined interface of the closed-form model representation, independently of the specific implementation. These outputs are propagated to the inputs of downstream units, and the simulation continues until all the units are computed. If the flowsheet contains recycles, an iterative strategy is performed until convergence of the flowsheet variables in tear streams is achieved.

Both the equation-oriented and the modular simulation approaches are implemented in CHEOPS for steady-state and for dynamic models.

The tool-specific part of the implementation is done in the dedicated *tool wrapper*, which is an instance of the closed-form model representation component derived for the specific tool. Its function is the translation of commands and data between the tools' specific interfaces and the generic interface of CHEOPS. This enables CHEOPS to communicate with the tools as if they were its own components and provides the technical and algorithmic solution for the tool integration. Currently, CHEOPS internally supports gPROMS [916] and MODELICA Equation Set Objects [655], and contains tool wrappers for Aspen Plus [516], Hysys [519], Parsival [601], and FLUENT [507]. Further tools can be integrated into this framework with a moderate effort to develop appropriate tool wrappers components.

Besides steady-state and dynamic simulation, CHEOPS supports a number of further applications like solving steady-state parameter identification and optimization problems for open-form models. It also supports a 'hybrid' model formulation, where only one model is explicitly formulated in an open form, while the other one is represented in a closed form, and the full problem requires transformation to a single representation. The framework supports the addition of further application components.

CHEOPS is developed using C++ first under Unix, then under Microsoft Windows. However, certain components, in particular the tool wrappers, can be developed using other programming languages (e.g., Python, Fortran) even under different platforms. An inherent feature of CHEOPS is the support of inter-platform communication between the components using CORBA middleware. For instance, the tool and the corresponding wrapper component can run on a software platform other than Windows.

The specification of a simulation problem in CHEOPS is done by means of setup files in XML format which describe the structure of the flowsheet to be solved as well as variables of various types (scalars, vectors, time profiles, and distributions). The variables are classified into inputs, outputs, parameters, and states. Inputs and parameters should be specified by the user. The setup files define references to the models, their types and associated tools, and the type of simulation with a respective set of simulation options.

These setup files can also be exported from an existing model specification in ModKit+. In this case ModKit+ acts as a mediator between the user, the model repository ROME, and the simulation framework CHEOPS. ModKit+ then provides specifications of the problem, the simulation setup, and correct references to the models which must be imported as files (cf. Subsect. 5.3.3).

A different method to construct the setup files for a simulation in CHEOPS is realized in the IMPROVE demonstration platform described in Sect. 1.2. There, an *integrator* composes a simulation model by querying the flowsheet editor for the components to be simulated and retrieves the corresponding simulation documents for these individual components from the AHEAD system (cf. Sect. 3.4).

With the features discussed, CHEOPS constitutes a powerful platform for heterogeneous process simulation, which can collaboratively work with ROME and ModKit+ at various stages of the lifecycle of the process. Current applications of CHEOPS include the simulation of the Polyamide-6 process, the simulation of a process for ethylene glycol production described by semi-empirical models [409], the modular dynamic simulation of a pentaerythritol crystallization process [252], and the coupled simulation of crystallization and fluid dynamics problems [253].

5.3.6 Solving the Scenario Problem

The integrated simulation of the Polyamide-6 process involves models originating from the modeling tools Aspen Plus, gPROMS, and MOREX as presented in Subsect. 5.3.2. The first step is to import these existing models into the model repository ROME. This import should be done by the various modelers as soon as their models have reached a useful and stable state.

The input files for all models are imported by a tool that is invoked via the command line of the Windows operating system shell. Based on the file extension of the model input file, this tool decides which wrapper to run. The wrapper will then inspect the given file and determine the metadata to be transferred to the model repository. Most notably, this concerns the connection points of the model itself and the variables describing these connection points. This information is at least necessary in order to permit connection to other models in the modeling tool ModKit+.

As a second step, ModKit+ is used to define the flowsheet model of the ε-caprolactam process, using the imported models from Aspen Plus, gPROMS, and MOREX as building blocks. The flowsheet topology is defined according to Fig. 5.20. The next step concerns the specification of a simulation experiment to be executed by the CHEOPS simulation framework. ModKit+ supports this step by generating a template of an input file for CHEOPS, which contains the necessary information except for the actual values of the feed streams and the parameters of the simulation. These have to be filled in by the user before he finally launches the simulation in CHEOPS.

CHEOPS obtains this setup file in XML format from ModKit+. Tool wrappers are started according to this XML file. The input files required for the modeling tools Aspen Plus and gPROMS are obtained from the model repository ROME. CHEOPS applies a sequential-modular simulation strategy implemented as a solver component because all tool wrappers are able to provide closed-form model representations. The iterative solution process invokes the model evaluation functionality of each model representation, which refers to the underlying tool wrapper to invoke the native computation in the modeling tool the model originated from. Finally, the results of all stream variables are written to a Microsoft Excel table when the simulation has terminated.

The resulting heterogenous process model facilitates to study the complete process based on detailed models of all process sections including recycles. Without a tool such as CHEOPS, the manual effort for setting up a tailored integration of the tools described would be far too high to be economical. Rather, simplified models would be used for some sub-processes (e.g., the separation equipment) as opposed to the detailed models developed in gPROMS and MOREX. These models would then be fitted against results obtained from stand-alone simulation experiments with these tools. However, this approach would not permit to study the influences between separation performance and the recycle, for example. In addition, the integrated simulation using CHEOPS ensures that the results obtained are indeed computed consistently, whereas the manual transfer of results through communication media like phone or email may lead to transmission errors or a wrong set of values being employed to specify a simulation.

5.3.7 Conclusions

The analysis of the current situation of tool support in chemical process modeling reveals that a significant potential is left to be exploited by considering modeling from a lifecycle perspective. In particular, the use of a neutral model definition and the integrated consideration of models and the work processes for their creation have been identified as important issues. Further, the integration of existing tools and models into a software environment to support modeling in the process design lifecycle must be ensured. An advanced software architecture has been presented as an initial step to address these problems. An overview of individual subsystems of the architecture has been presented and their relevance with respect to the idea of modeling support along the process design lifecycle has been emphasized.

We are confident that the solution sketched in this section does actually contribute to a better and more widespread use of models within process design. However, the heterogeneity of the model representations used by different modeling tools is hard to overcome by a single research initiative given the required robustness for such a software environment. Standardization activities in the area of process modeling and simulation such as MODELICA [655] or CAPE-OPEN [997] are therefore important steps towards a fully interoperable world of process modeling and simulation software. They are, however, only an enabling factor but cannot deliver complete solutions for the problems posed. Hence, future work must further develop open standards and promote their use within an environment as the one outlined above before industrial solutions can finally be achieved and offered to the end-users in the process industries.

5.4 Design Support of Reaction and Compounding Extruders

M. Schlüter, J. Stewering, E. Haberstroh, I. Assenmacher, and T. Kuhlen

Abstract. This section describes the different dimensions of integration inside the plastics processing domain as well as cross-organizational integration and collaboration issues. The presented results range from work process modeling up to technical process analysis for the design of compounding extruders in the chemical engineering context. Standard practices for the design of polymer compounding extruders were analyzed and afterwards formalized in cooperation with subproject I1 using methods and tools developed and used in the CRC 476. Fragments of these workflows were redesigned using innovative informatics functionality provided by the CRC's B-projects which provided the novel tool functionality. Exemplarily, the extruder simulation tool MOREX was integrated with the process-integrated modeling environment PRIME and coupled with BEMflow. The distributed analysis of 3D simulation results using KOMPAKT and TRAMP was developed, and a scenario showing the integration of the project management system AHEAD with the plastics engineering design tools was designed participatively. Another focus was set on an integrated visualization environment using Virtual Reality technology for different data from a number of simulation tools.

5.4.1 Introduction

Within the IMPROVE project, the *plastics processing and engineering* domain is characterized by some significant differences when compared to the chemical process engineering domain. In the course of the relatively short history of industrial plastics processing since the mid 20th century, plastics processing methods and machinery have been particularly developed based on experience knowledge and experimental efforts. New machine concepts and technologies were required especially for conveying, manipulating, and mixing of these fluids which are characterized by very high viscosity and shear thinning flow behavior.

In particular the new and complex machines and processing methods were developed and optimized by separate engineering groups in different companies. The *compounding extruders* which transform the polymer into a ready-to-use material in the form of pellets, belong to the domain of plastics processing which is disjunct from the domain of chemical engineering. Within the design and development of industrial chemical plants (see Sect. 1.2), from the workflow modelling view the design of compounding concepts and machines normally is a completely separated work process (see Fig. 5.23).

But also inside the plastics engineering domain, the knowledge of the experts was gradually complemented by *explicit process models* and documented, *formalized* knowledge which was elaborated by the scientific community, e.g. within research projects. Therefore, with the help of experiments and computer simulation, abstract and validated process models were developed for

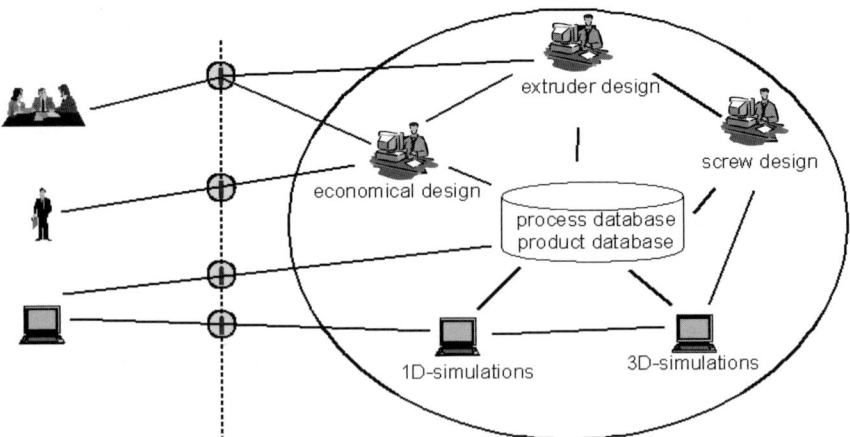

Fig. 5.23. Information barriers in current chemical process design processes

nearly every processing technology. Based on such process models, *simulation programs* are available today also for the qualitative and quantitative analysis of polymer compounding processes. These methods can especially be used in conceptual process design phases to asses and analyze several process alternatives objectively (see Sect. 1.2).

Thus, a common basis for the methodic integration in comprehensive design processes is available. In this subproject the *integration* has been investigated from the *conceptual* level up to the concrete *implementation* and *validation* for the first time. The aspects and dimensions of integration presented and discussed here contain interfaces to all participating project partners inside the CRC 476.

In the following sections, the integration aspects are shown along *three different sub-scenarios* which are all based on the demonstration scenario presented in Subsect. 1.2. The special characteristics of theses use cases are identified as:

- The interaction and collaboration with the chemical engineering expert for separation processes (Fig. 1.10 in Subsect. 1.2.2), where the focus is on the exchange of process and product data within the domains.
- The interface between the compounding expert and the compounding simulation expert, where inter-domain optimizations are developed and applied.
- The interplay with the work processes at the 3D simulation expert workplace, who is considered to be located in an external company. On the administrative level this use case demonstrates the delegation based cooperation (see Subsect. 3.4.4). On the operational level these work processes were analyzed and improved by application of virtual reality-based methods and tools.

5.4.2 Domain Characteristics and Scientific Challenges

Polymer compounding is practiced on different levels of machine size and mass throughput. Polymer production plants with very high capacity of a single polymer and use the biggest compounding extruders, whereas plants for the production of a variety of special polymer materials, produced in smaller quantities, use much smaller compounding equipment and have to be flexible in their production program. In the first case, the big compounding extruders, e.g., for polyolefines, can have screw diameters up to 380 mm and a length of about 10 m. Mass throughputs come up to 70 t/h and even more. As *large-scale* and *mid-scale extruders*, which are integrated in chemical plants, cause the highest investment costs, we will focus on this class of machines in the following. Here the ratio of investment to design costs is very high.

Polymer Compounding

The tasks of *polymer compounding* can be the mixing of the polymer with additives and fillers (such as glass fibers), and the degassing of volatile components such as monomers. It can even mean to carry out chemical reactions in the extruder. Usually extruders with two intermeshing screws, so called co-rotating twin screw extruders, are used.

As a result of the compounding process, the raw polymer is transformed into plastic material with a well defined property profile. The specific properties of the product are influenced by the melt flow and mixing processes in the extruder.

Co-rotating twin screw extruders are the dominant type of compounding machines. A broad variety of compounding tasks can be executed efficiently [790]. The main advantages of this type of machine are the following [730]:

- The design structure of the extruder is extremely flexible due to the modular screw and barrel concept.
- The volatile components can be removed (degassing).
- Residence time distribution and mixing capability can be influenced by the screw design.

As outlined above, compounding extruders usually contain several different *functional sections* which are combined and highly integrated like the devices of a chemical process plant. The extruder's functional sections are all coupled along the axis by the fact that all screw elements rotate at the same rate. Along the screw and barrel system, additives or fillers can be added to the polymer stream by auxiliary sidefeeders or hoppers (cf. Fig. 5.24, second barrel element from the left). Also volatile components can be removed by degassing notches (cf. Fig. 5.24, second barrel element from the right). Due to the modular screw and barrel concept the extruder can be configured and specialized in nearly any way to match specific process requirements. Additionally, two or more extruders can be combined to form an extruder cascade.

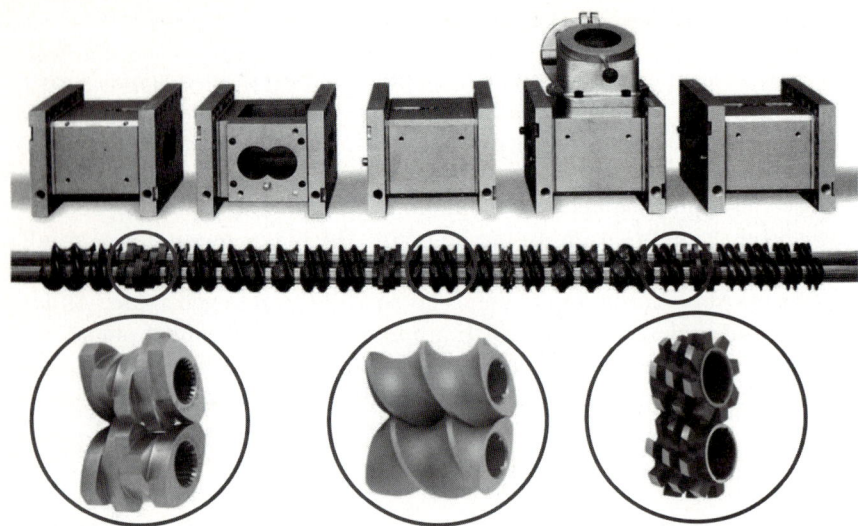

Fig. 5.24. Modular machine concept of co-rotating twin screw extruders

Design of Compounding Processes

The process of designing a compounding *process* has always to be seen in tight relation with the design of the *extruder* it is to be executed on. Both the initial *design* of such a process and its *optimization*, consist of the follwing tasks:

- the pre-estimation of process characteristics and process costs,
- the selection of appropriate extruder size,
- the design of screw and barrel configuration,
- the selection of auxiliary units (sidefeeders, vacuum pumps etc.), and
- the definition of process parameters (e.g. temperature, screw speed, residence time etc.).

Nowadays, the extruder design itself and the specification of process parameters are usually done on an *empirical basis*. This is due to the fact that a high number of different screw and barrel elements exist which allow for a huge number of combinations. Some special features of the compounding process, for example the degassing capability, can not be assessed adequately by empirical methods.

The high integration density of the many complex partial processes, their interaction and interdependencies in one single extruder require a high level of knowledge about these processes in order to combine them in an optimal way. Therefore, the use of *simulation* software for the *enhancement* of *process understanding* and knowledge is helpful and can significantly reduce experimental effort and costs.

Optimal mixing capability of a compounding extruder can only be achieved by intricate forms of melt flow, especially in the intermeshing sections of the screws, and in special kneading elements. For the melt flow analysis in these sections, it is necessary to use simulation software based on 3D models using the Finite Element Analysis (FEA) or the Boundary Element Method (BEM).

Improvement and Impact by Integration and Innovative Methods

The work presented here is focused on the conceptual (re)design and integration of design and development processes. The simulation and visualization of *melt flows* in co-rotating twin-screw extruders as part of the design processes will be discussed as follows, as well as supporting these processes by experience knowledge and innovative multimedia tools:

- integration of domain simulation software with the PRIME framework [371] for reusing experience knowledge for extruder design tasks;
- management of organization-spanning administration processes in cooperation with the AHEAD system (Fig. 3.85);
- using multi-media video conference system for supporting communication in distributed design and analysis processes;
- managing and retrieving multi-media data from simulation processes by the TRAMP tool;
- using methods of virtual reality for exploration and analysis of 3D simulation results.

5.4.3 Computer-Aided Analysis of Extrusion Processes

For the coarse estimation of extruder size and screw speed, simple mass and energy balances based on a fixed output rate can be used. For the more detailed design of a twin-screw extruder configuration it is necessary to *combine* implicit experience knowledge with simulation techniques. Theses simulation techniques cover a broad range from specialized programs based on very simple models up to detailed Computational Fluid Dynamics (CFD) driven by Finite Element Analysis (FEA) or Boundary Element Method (BEM).

The complex geometry of the compounding extruder's flow channels, particularly the intermeshing zone of the screws, *complicates* the modeling of the polymer flow (see Fig. 5.25). For example, even in a highly simplified flow channel model, substantial effects, such as high velocity gradients in the intermeshing zone, cannot be completely modeled. As the flow description is not complete, detailed analysis of flow effects can not be done with this type of software. Therefore the conservation equations must be solved in three dimensions. For the considered complex flow channels this is only possible with a high numerical effort (FEA or BEM).

Both methods imply *major drawbacks* especially for the analysis of compounding processes. On the one hand, the BEM lacks a mature implementation for the *non-Newtonian* (shear-thinning) material behavior. On the other hand the FEA needs a high manual and computational effort for the *transient*

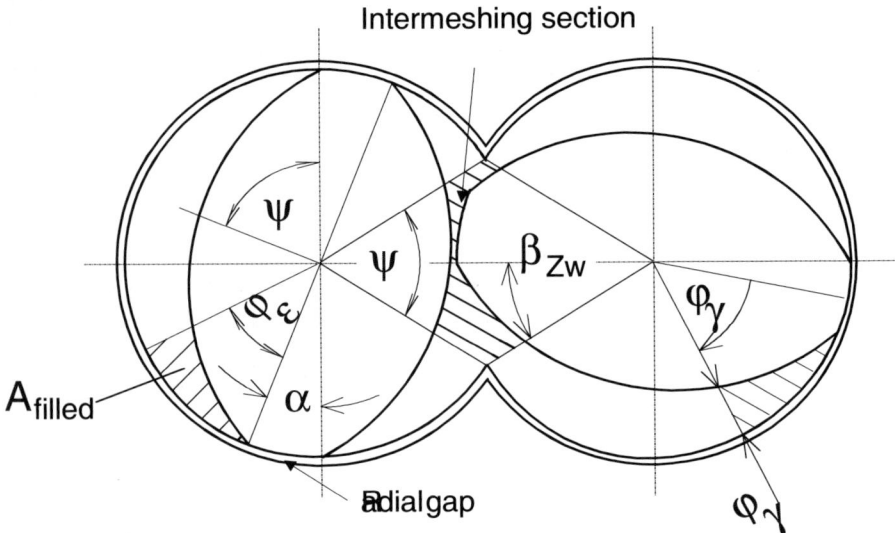

Fig. 5.25. Geometry cross section for closely intermeshing co-rotating twin screw extruders

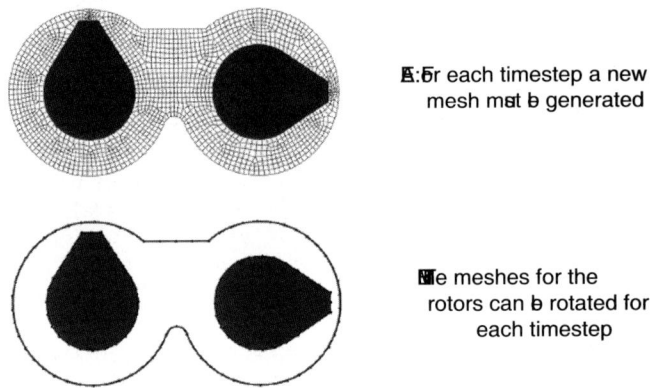

A: For each timestep a new mesh must be generated

The meshes for the rotors can be rotated for each timestep

Fig. 5.26. Comparison of FEA and BEM meshes for transient flow channels

flow channel geometry in twin screw extruders (see Fig. 5.26). Currently, both methods can only be applied for *fully filled* screw sections, however the modeling of flows with free surfaces is under development [1008].

As highly specialized simulation software, the FEA and the BEM can be seen *complementarily* to each other. Thus, all outlined methods were chosen for the conceptual redesign of integrated work processes in the CRC scenario to combine their advantages. This will be discussed shortly in the following subsections.

Extended Analytical 1D Process Models

The MOREX simulation software is based on a *physical process model*, which regards aspects of flow and heat transfer in closely intermeshing, co-rotating twin screw extruder [699]. It contains the description of the screw geometry for the conveying elements and kneading blocks (Fig. 5.25). The conservation equations are solved in an extended 1D model with partial models for the flow in the intermeshing section and in the radial gap in a vertical cross-section of the extruder. In this model, the extruder's flow channel is divided into *slices*, for each of which an energy and mass balance is formulated. For example, integral pressure and temperature values along the screw axis are calculated for the flow channel cross sections with a geometry as shown in Fig. 5.25. Also the beginning and the end of the completely filled sections is determined automatically.

Conveying elements (bottom mid of Fig. 5.24) differ due to their *pitch* and the *number of flights*. The model which is essential for the description of the flow in these elements, considers besides the above parameters the leakage flows in the radial gap as well as the particular geometric proportions in the intermeshing zone. The latter are important for the calculation of the maximum drag volume flow.

The cross-sectional areas and the pitch of a main conveying and rear conveying channel are calculated for the description of kneading blocks depending on the *off-set angle* of the single discs of a kneading block [699]. With these parameters the total throughput rate of a kneading element can be determined. For the calculation of flow processes, the special geometry has to be taken into account. The high energy input effects a *rapid melting* of remaining non melted pellets. The heat transfer coefficient between the polymer melt and the barrel surface must be set for the calculations.

FEA Simulation Characteristics

The most common way for three-dimensional simulations is the application of the FEA. The 3D analysis of flow processes in co-rotating twin-screw extruders is very time-consuming, because for transient geometrical adjustment in the intermeshing zone a new volume mesh must be generated for each new screw position (see top of Fig. 5.26).

To reduce this effort, the software Polyflow (Fluent, Lebanon, USA) contains a special module to avoid the remeshing of the flow channel for every single timestep. This is called the "Mesh Superposition Technique", where the inner barrel and the screw are meshed separatly. The discrete meshes are *overlayed* to create one system where the surfaces of the screw define the channel boundary. A major issue with this method is that the flow channel volume varies as the intersection of the surface elements leads to unequal sums over all elements. This is compensated by a *compression factor* on which the simulation results react very sensitively.

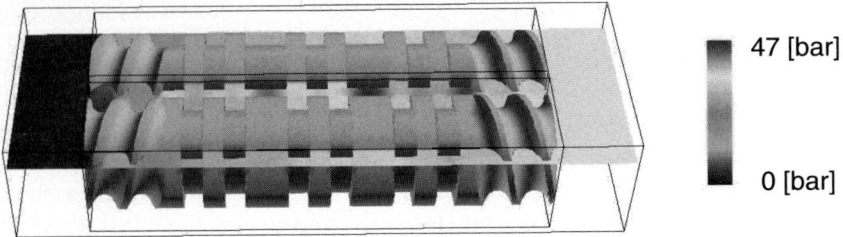

Fig. 5.27. FEM result: pressure distribution within an extruder mixing zone

The calculation commonly starts with a simple FEA model and then incrementally adds model refinements. Examples for model refinements are

- time-independent → time-dependent;
- isothermal → non-isothermal;
- Newtonian → non-Newtonian.

After every simulation pass, the results are tested for plausibility. Furthermore, the independence of the simulation results of the fineness of the mesh is tested.

Figure 5.27 shows the pressure distribution in a mixing zone of a twin-screw extruder as an example for FEA simulation results [149]. The pressure is visualized on a cutting plane through the flow area. The flow direction is from the right to the left.

BEM Simulation Characteristics

As one alternative in this subproject, the complex flow conditions in co-rotating twin screw extruders are analyzed by means of the simulation software BEMflow [889], which is based on the Boundary-Element Method (BEM). For this method *just the boundaries* of the examined region have to be meshed. The particular surfaces can be moved relatively to each other.

As an alternative to the FEA, which is based on volume integrals, the conditional equations can be formulated as surface integrals following the divergence theorem. It relates the flow of a vector field through a surface to the behaviour of the vector field inside the surface. For polymer flow this means, that all phenomenea and effect *inside* the volume are completely determinated by the *conditions* on the *volumes boundary*. This is also called the Boundary Element Method (BEM) and can be used for 3D analysis analogous to the FEA [890]. For practical use, this approach differs in some points from established commercial FEA systems. First of all, the result of the solved integrals are only availible on the volumes or flow channel surface. That means, that for a detailed analysis and optimization inside the flow one has to explicitly calculate the flow state for every single point which has to be analyzed inside the volume. As the approch is relativly new and not yet widely used, no specialized pre- and postprocessing tools for the workflow exist. This is compensated by converting the data to standard interface formats and then using generic commercial CAD or visualization tools.

From the mathematical point of view the complexity is reduced because the system of equations which has to be solved is a function defined on the *two-dimensional manifold* of the control volumes boundary and leads to a dimension reduction. Practically the discretisation of the boundary usually is more simple than the meshing of complex three dimensional volumes. Especially this pertains to the transient flow channel geometry in co-rotating twin screw extruders. The surface meshes for the screws can independently be rotated inside the screw and barrel mesh analogous to the batchwise working internal mixer (Banbury Mixer) shown in the bottom part of Fig. 5.26.

To determine the flow status inside the volume, so called 'Internal Points' are calculated in the BEMflow software. From these points on, *streamlines* are calculated for a number of specified timesteps. For steady state flows, streamlines and pathlines are identical, while for transient flow channels as in the extruders the pathlines have to be calculated from the results for every single timestep respectively every relative screw position. This leads to an additional task for the flow analysis [145].

5.4.4 Integrated Workflows in Domain Spanning Design Processes

Based on the aforementioned basics of polymer compounding and process analysis the focus will now be on the *formal integration issues* within the CRC and its scenario. All the presented use cases were improved by the requirements from the application domains and vice versa. For example the discussion and formalization of the interfaces between plastics and chemical engineering lead to simulation tools enhancements and new features which themselves showed effects on the work processes within the domain and to external partners.

In order to conceptually design and integrate the discussed issues efficiently within the chemical engineering design processes, a comprehensive *analysis* and *formalization* of the overall work process was done. Only in this way dependencies and similarities could be found and documented completely. The identified relations were used for a conceptual redesign of the engineering tasks, particularly the integrated demonstration scenario (see Sect. 1.2). In Fig. 5.28 the adopted methods from the CRC partners are shown, ordered by their formalization grade and granularity. The modeling started at the upper right by defining simple use cases (using UML notation) as a basis for the C3 models for the demonstration scenario which was elaborated *participatory* together with the project partners (see Subsect. 1.2.1). In an iterative design process the models were refined, validated and redesigned several times in the course of the subproject.

One major result of these requirements engineering tasks is the integration of MOREX and PRIME to gain fine granulated, *experience based user support* while using the software for simulation calculations [198]. The use and

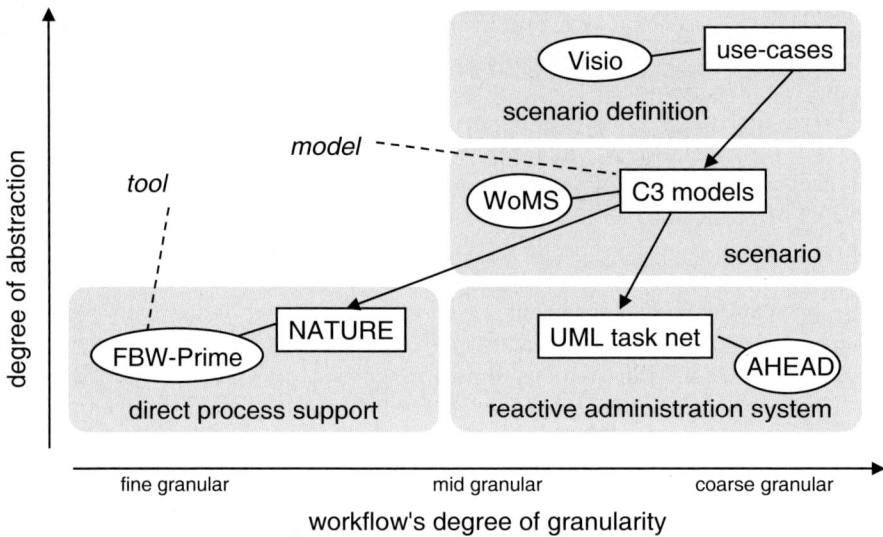

Fig. 5.28. Granularity and formalization grade of the used methods

integration of MOREX in the scenarios marked the initial point on the integrated simulation environment and other innovative technologies discussed in the following subsections.

A precondition for all the integration was the availability of a domain spanning product model for the polymers material description. Therefore, simultaneously to the process modelling of the demonstration scenario a partial model for the produced material based on CLiP (see Subsect. 2.2.3) was adopted in the plastics engineering domain. This will be discussed later on.

Contribution to the Demonstration Scenario

The following *plastics engineering* specifications for the required polymer and for the polymerization process were fixed in the early design phases:

- residual content of caprolactame ≤ 0.1 %
- residual content of cyclic dimer ≤ 0.04 %
- residual content of water ≤ 0.01 %
- relative viscosity in m-Cresol at 2.7 (determined using a 1 % by weight solution of the polyamide in m-cresol at 25° C)
- cylindric pellets with 2 mm length and 2 mm diameter
- propionic acid as reaction controller
- complete recycling of caprolactam and oligomer
- flame resistance of minimum UL-94 V-0
- tensile strength of 180 MPa minimum

In Fig. 1.10, one can see the domain spanning scenario with three roles from the plastics engineering domain. In the following it will be discussed, how the workflow has been analyzed and optimized in detail. The activities (cf. Fig. 2.13) are further decomposed and refined in Fig. 5.29 and Fig. 5.31. It can be seen that an *intricate interplay* between the involved roles and the external party from chemical process engineering takes place.

The result of this exemplary partial process is the C3 model for the delegation of the computational process anlysis to an external partner in the scenario. Complementary to these interdisciplinary work processes also collaboration aspects *within* the plastics engineering domain were examined. Therefore the tasks of process analysis by integrated 1D and 3D simulation as well as the interactive exploration of huge simulation result data in a virtual 3D space were improved by innovative tools and methods driven by real world scenarios.

The development and elaboration of the demonstration scenario, the requirements for the processing process and the polymer were carried out in subproject I1 of the CRC 476 [355]. The process chain and a flow sheet of a chemical processing plant for the production of polyamide 6 were already shown in Fig. 1.2 and Fig. 1.3. The *plastics processing* related aspects of the partial scenarios from compounding extruder design, the integration of the activities of plastic and process engineering to make use of synergies in the process design and application of concurrent and simultaneous engineering to reduce the development times are discussed in more detail below.

A generalized coarse granular sequence of the demo scenarios aggregated activities is shown in the C3 model Fig. 1.10 as an overview and framework for the detailed work procedures. On the left side, the chemical engineering and on the right side the plastics engineering development and design processes are depicted. The single activities before the start of development in manager level (left) should not be further deepened. The process simulation results have to be *validated* by laboratory scale experiments. More detailed information about the demonstration scenario can be found in [124].

In the following the focus is on the activities of the roles of the compounding technology and their interfaces to other activities (see Fig. 1.10). The compounding *concept* is in this scenario developed by the compounding expert, who delegates the simulation part to the simulation experts and the 3D simulation experts. To estimate the *process quality*, together with the simulation experts a simple 1D simulation is made to evaluate the compounding concept.

As for the mixing capabilities of the process no reliable information is available, an extensive 3D flow analysis is planned and disposed. Simultaneosly, the compounding expert together with the separation expert from chemical process engineering are working on the degassing concept (4). *Independently* inside the plastics engineering domain the compounding expert and the 3D simulation expert are discussing and reviewing the simulation results (5). To evaluate the *degassing concept* with the monomer feedback, an integrated overall simulation (reactor, separation and extruder) is planned (6). In a meeting in which all roles are present (not shown in Fig. 1.10) the plant's process concept and first equipment specifications are fixed.

Based on the diagram in Fig. 1.10 two alternative workflows for this scenario in the compounding domain were further elaborated for the aforementioned and described activities. For the first alternative (A, Fig. 5.31) the 3D simulation will be operated in an *external company*. In this case, additional requirements to almost every used tool are necessary. The second alternative (B, Fig. 5.32) models the case, that all simulation tasks are located in one company. In reality this is mainly the case for big raw material producers which have own resources for detailed CFD studies.

Common Workflow for Both Alternatives

At first, in the details of the demonstration scenario discussed above, the activities are *identical* as far as the 3D simulation. At the beginning of the design process, it is the task of the process engineering, to analyze and formalize the requirements for the chemical process with a literature research and based on personal experience, and to design an integrated compounding concept for the whole chemical plants including the compounding extruder.

The integrated polymerization and compounding concept for *energy efficiency* is designed in a way that the raw polymer which leaves the separation device is not to be cooled down and melted again for compounding. So in this scenario the extruder is fed with polyamide melt and does not need a feeding, compacting or melting section like for the compounding of plastics resin. The specified residual content of caprolactame of 0.1 % requires a two stage degassing concept for the extruder processing concept. Only by experimenal or simulation based analysis the degassing capabilities can be quantified.

Another requirement (see above) refers to the polymer's *flammability*. To meet the UL 94 specification, the addition of chemical additives is necessary. Here a concentration of 8 % of ethylenebistetrabromophthalimide and 8 % antimonytrioxide is chosen to gain the class V-0 [680, 1005]. A special mixing section including kneading blocks and left handed conveying elements enables the homogenization of the compound.

The required *tensile strength* of 180 MPa is realized by adding approx. 40 % of glass fibers [581, 591]. The air which is brought in together with the fibers must be removed downstream in another degassing section.

Considering these requirements, the compounding expert designs a *machine concept* and an initial screw configuration (cf. Fig. 5.30). To get a first guess of the process quality, e.g. temperature profile, shear energy and pressure profile, the simulation expert configures and executes a first simulation with MOREX (see Subsect. 1.2.1). By this simulation the length and position of the completely filled screw sections, where 3D simulations can be applied, is determined. A *quantitative analysis* of mixing capabilities is not possible with MOREX because of the simplified process models such that 3D simulation has to be arranged with the 3D simulation expert.

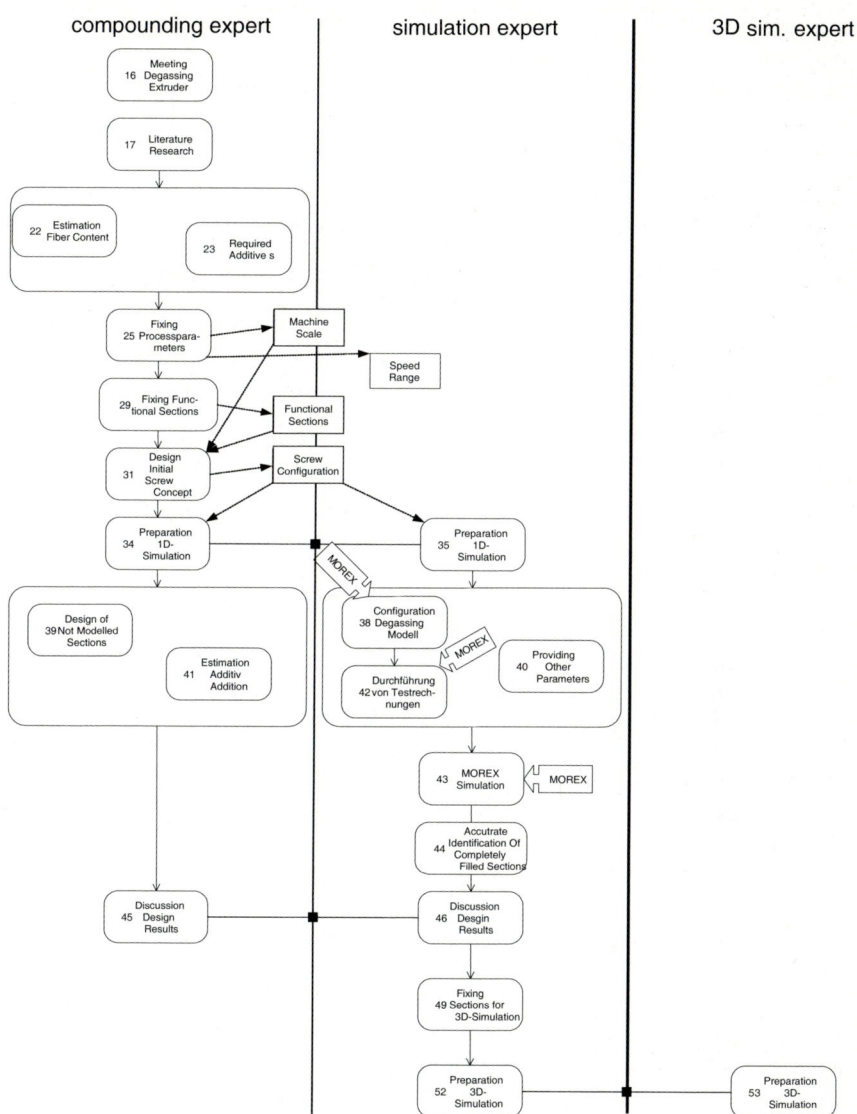

Fig. 5.29. Decomposition of the polymer compounding activities of the demonstration scenario (I)

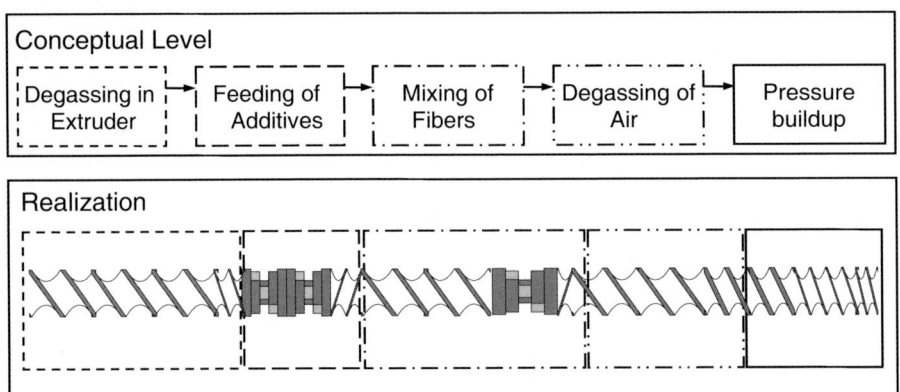

Fig. 5.30. Compounding concept and draft of screw configuration

As the following tasks differ due to the above discussed forms of organisation within the company, two alternatives have to be described.

Alternative (A)

In this variant, the 3D simulation to predict the mixing quality, is done by an *external service provider*, as not all raw material producers have an own employee for that. For this case, the extruder manufacturing company offers the knowledge and competencies for the 3D simulation. The communication within the design team with external participants is realized by phone, video conferences and email. Hence for *nondisclosure issues* the interfaces to the external partner are non- or semiformal, which means, that just a few process parameters and anonymized data will be exchanged but no complete specifications. This is called the 'need to know' principle [1051].

The C3 model for this part of the scenario can be seen in Fig. 5.31. As mentioned before, information like process, material and geometry parameters aren't exchanged automatically by service oriented applications in terms of product data models but as single values in e-mails or by fax. For *3D simulation preprocessing* firstly the screw geometry is modeled using a commercial CAD system followed by meshing the surfaces and then converting the meshes to the simulation programs format. After this being done, the 3D simulation expert starts the BEMflow calculation and analyzes the results.

As it seems that the dispersive mixing of the additives could be improved by increasing the melt shear rate, in the *video conference* the 3D simulation expert proposes a modification of the screw configuration as well as the screw speed. This is accepted by the team leader, a new MOREX simulation is done and the compounding extruder configuration is fixed.

This workflow is supported by several tools developed by the CRC project partners (e.g. KOMPAKT, AHEAD, BEMView) and is further described in

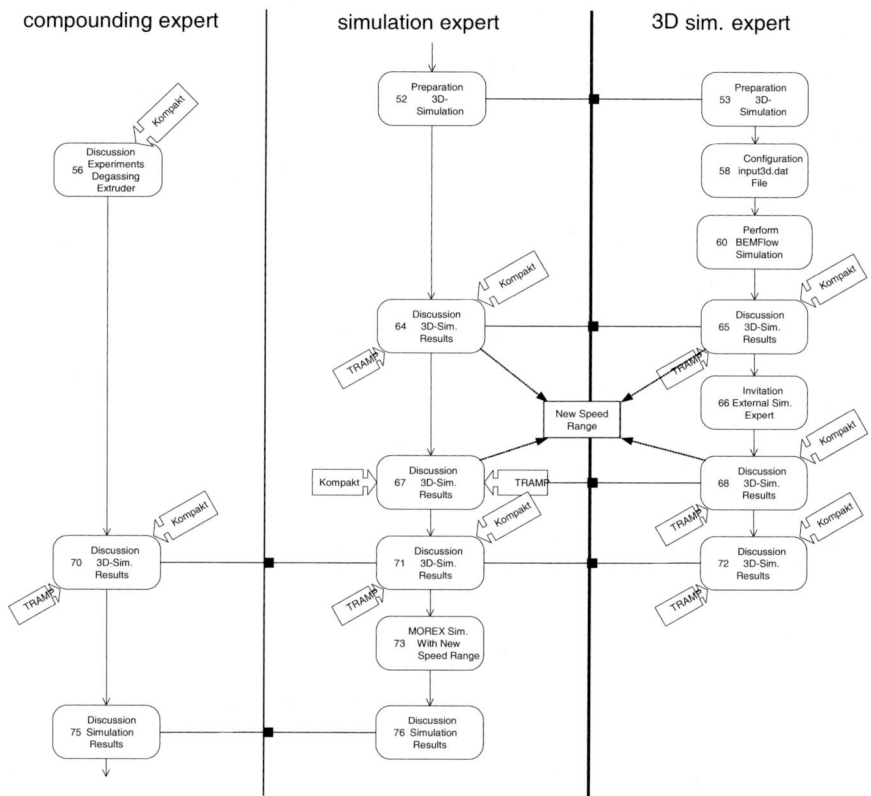

Fig. 5.31. Decomposition of the polymer compounding activities of the demonstration scenario (II)

Subsect. 3.3.2. Especially the distributed analysis of 3D results with a visualization tool shows the impact and improvement using *event sharing* (see Subsect. 3.3.3) for synchronizing locally running BEMView instances as well as the role of the dynamic and reactive administration tool AHEAD for advising the workflow fragment.

Simultaneously to the 3D simulation issues another team consisting of the separation expert, compounding expert and simulation expert is working on the *degassing concept validation*. Therefore an integrated overall CHEOPS simulation is prepared and executed to quantify the degassing capacities of every single system component. This is described in Subsect. 5.3.5.

Alternative (B)

Within the scenario's second alternative all design and simulation tasks for compounding processes are done in one company. This means that a *tight coupling* and *integration* of the computer based tools is possible and can be

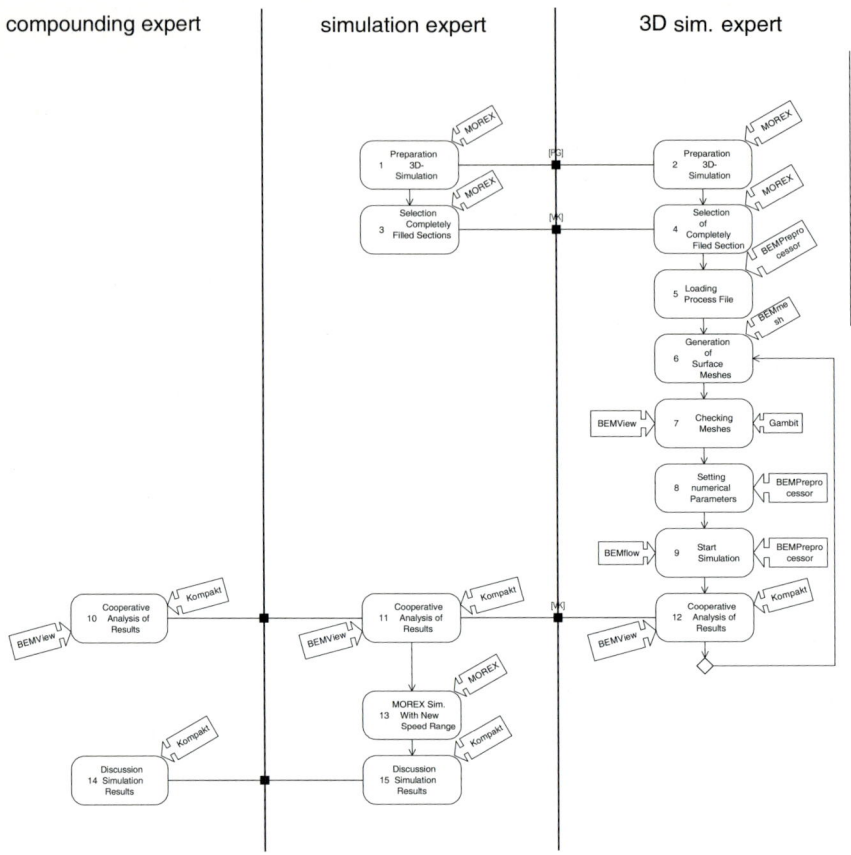

Fig. 5.32. Decomposition of the polymer compounding activities of the demonstration scenario (III)

automated. Consistent data models and services can be used for information exchange, communication, support and administration. Such a framework based on generalized data models and with the a posteriori integration of tools was one major challenge and goal of the CRC 476 (see Sect. 1.1).

In this second alternative the use of standardized interfaces based on integrated product and process models enables the design and seamless interplay of several modules e.g. for the generation of simulation input data. The restructured scenario builds the basis for the *integrated 1D and 3D simulation* where MOREX and BEMFlow were coupled (see Subsect. 5.4.5).

Already before the 3D simulation the *media database* TRAMP offers access to multi-media clips containing documentation of earlier simulations. Screenshots as well as animations can be searched for annotations and browsed based on a category scheme [196]. After a short study of recent simulations and resulting annotations, the design process starts with configuring the extruder screw in MOREX.

The completely filled sections are determinated by a MOREX simulation analogous to alternative (A). But now the user is supported by an integrated *surface mesh module* where the time consuming process of manual mesh generation is automated by special algorithms. Additionally, all simulation parameters can be configured inside the MOREX module, such that no further preprocessing in other tools is necessary before starting the calculation.

The TRAMP database content (see Subsect. 4.1.4) is generated by the systems users during the analysis of flow effects with commonly used postprocessors. After identifying significant and characteristic effects, screenshots and animations are recorded by a capture program and then stored with annotations into the database by using the *category scheme*. Similar as in alternative (A) after discussing and reviewing the 3D simulation results, another MOREX simulation with the modified process parameters is done to check the impact on the other partial processes inside the extruder.

Common Sequence of Both Alternatives

For both scenario alternatives the degassing of monomer (caprolactame) and other volatile components is an important aspect for the integrated chemical process and compounding extruder design. If these interrelationships can be considered and quantified both at chemical process engineering and plastics engineering, *synergies* can be developed for dimensioning the different degassing modules. It has to be taken in consideration that the degassed monomer has to be fed back to the reactor. This way, the degassing in the extruder can contribute to an improved and efficient production process.

The C3 models development from analyzing the existing processes until the optimization can be found in [398]. As described in detail before, the following aspects were improved by the redesign:

- Improvement and integration of the workflow by simultaneous engineering and coherent data models.
- Distributed analysis of 3D simulation results using the event sharing mechanism.
- Management and adminstration of domain spanning work processes between the simulation expert and the external 3D simulation expert.
- Integration of MOREX to CHEOPS simulation environment to enable an integrated process simulation (especially for degassing issues).
- Integration of MOREX to PRIME and to the flowsheet editor to enable experienced based user support to the simulation expert.

The identified potential for optimization by the use of consistent process and product models for data exchange already in the *early design phases* enables broad understanding and quantitative knowledge about process alternatives.

Dependent tasks can be initiated automatically when the needed information is available. Independent tasks can be started directly with preliminary presumptions if the results can help to assess other tasks within a first analysis.

5.4.5 Models, Methods, and Tools Supporting the Integration

Complementary to the scenario fragments described above, concrete *integration work* was done by defining requirements as well as developing, implementing and validating product and process models. This was done by realizing the demonstration scenario from the compounding extruder's design viewpoint in the CRC 476 context together with the project partners. Thereby the A3 subproject contributes to the integrated process and product models and provides a real world background story.

The other way around the A3 subproject participated from new and innovative methods and models as discussed above. In the following subsection *selected integration aspects* are discussed with focus on the application inside our domain. Informatics details and innovations are presented by the other project partners' contributions in this book. Especially the storyboard for the delegation to external service providers for the AHEAD system should be mentioned here (see Subsect. 3.4.4 and [169]), as well as the multi-media communication supported by KOMPAKT within the demonstration scenario (see Subsect. 3.3.2).

Refined and Extended Material Product Model

To enable coherent storage and information exchange inside the domain [146] and to collaboration partners, an adequate data model for *material properties* and *behavior* was necessary. Therefore, a concrete refinement and extension of the CLiP data model (see Subsect. 2.2.3) towards an implementation model has been done for the support of design processes in thermoplastic polymer processing.

In plastics engineering, especially in the domain of extrusion lines design and compounding plants design, a comprehensive, unified method for managing material data has not been addressed in previous work. There are basically two aspects in material data management in this domain. One is about the *efficient exchange* of data between applications as well as that between these applications and others from chemical engineering used for example during the design of the polymer production plant. The second aspect is the convenient *storage* of data, especially parameters of mathematical models in files or databases. Apparently, these two aspects need to be supported by a unified material data model.

In this application example, a unified data model for plastics engineering has been developed based on the *generic conceptual model*. With the support of the material management component, simulation tools can be used in a stand-alone version without connections to databases. In the following, the

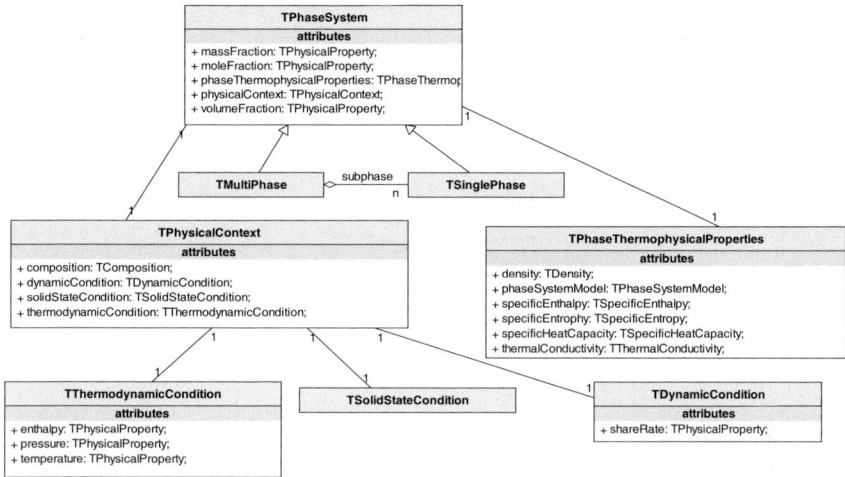

Fig. 5.33. Implementation of CLiP phase systems partial model based on the composite design pattern

data model is explained with a focus on the phase systems' dynamic material behavior. It should be mentioned that, as determined by the domain of this application (i.e. plastics processing), it is essential to model the non-equilibrium properties and behavior of material.

In most of the existing work in plastics engineering, polymer flow is modeled only considering single phases. However, often *multiple phase systems* like gas and melt phase or two different polymers phases in blending processes are processed. Thus, a multiple phase model for describing the polymer system and its *dynamic behavior* is required. To meet this requirement, the CLiP concept of multiple phase system in the generic conceptual data model is adapted for the implementation model shown in Fig. 5.33. Here, a phase system has a list of phase system properties containing the thermophysical and the rheological properties.

The *physical context* concept in the conceptual model is extended to describe the behavior of plastics in the form of pellets through the class solid state condition which encapsulates properties such as pellet type. This part of the implementation model concerns the mathematical modeling of some of the properties of polymers, which correspond to their dynamic or flow behavior. A class for a concrete mathematical model not only holds *declarative information* such as the list of parameters, but also provides a method for *calculating* the value of the property modeled. This method requires an implementation which is usually different from the one for another mathematical model. Therefore, mathematical models are organized in this application through further classification.

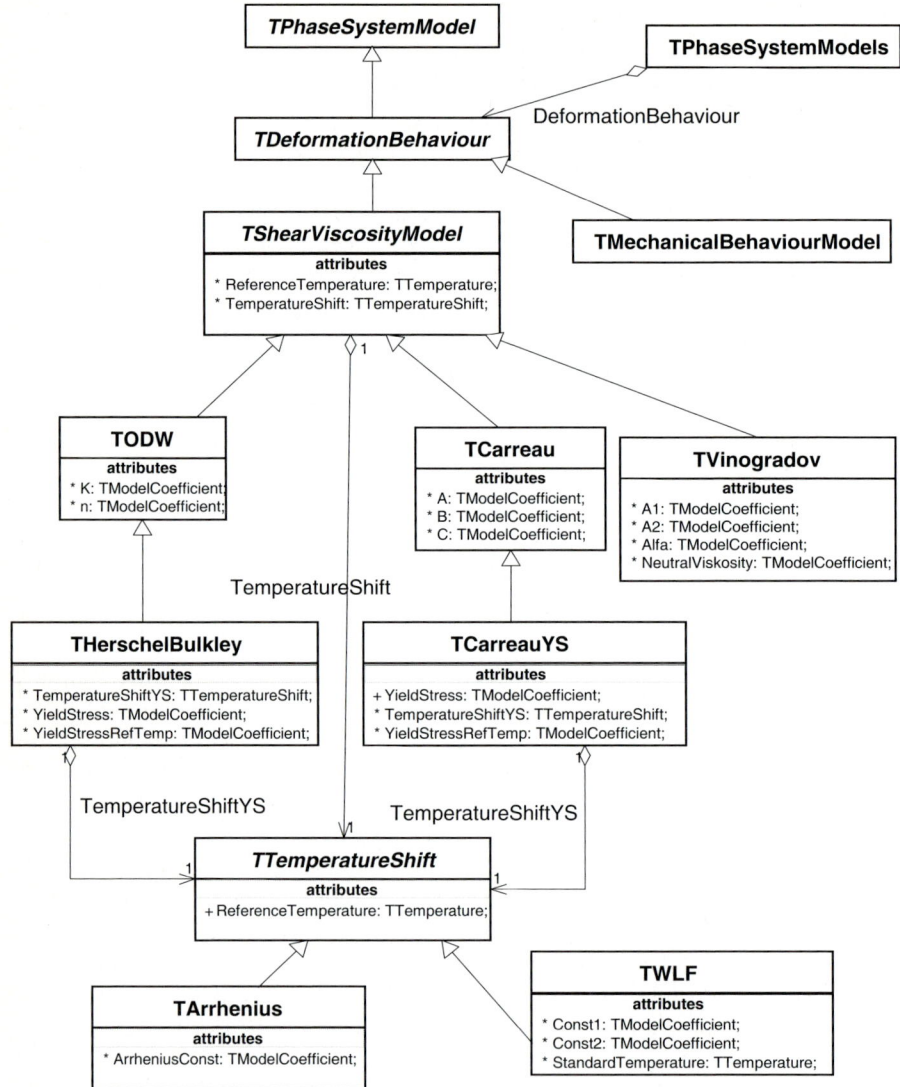

Fig. 5.34. Extended partial model *dynamic material behavior*

The hierarchy of *mathematical model* classes is developed as follows. First, the concept phase system model in the conceptual model is set as the super class of all concrete phase system models holding common attributes such as assumptions and validity range. The rheological behavior model, as a subclass of phase system model, represents a type of mathematical models that is of major concern of this application. Especially, the derivation of the subclasses of shear viscosity model has been given most consideration up to now. As a major issue in modeling plastics melt flow, the typical viscoelastic material behavior has to be considered by solving the conservation equations [829]. Extrusion dies, single- and twin-screw extruders, and injection molds contain flow channels for the polymer flow. Usually, the dependency of the shear-viscosity on the shear rate is described by models like the power law or the Carreau-equation. This has lead to the classes ODW (for the power law), and Carreau, respectively (see Fig. 5.34). If a temperature shift (defined as a subclass of phase system model) of the yield stress has to be considered, the approaches can be extended with an Arrhenius or WLF model. In the data model they are called HerschelBulkley and CarreauYS. Material parameters for simulation models are often measured by companies (e.g. the raw material producers) and research institutes with their own methods and instruments. It is even possible that new parameters are introduced into a certain model. Therefore, the data model is subject to extensions in order to represent new mathematical models.

MOREX Integration to PRIME and the Flowsheet Editor

In addition to the usual machine oriented modeling of compounding extruders the functional view on these machines will be introduced. This is a specialization of the CLiP partial model **ProcessingSubsystem** and was inspired by the CLiP modeling approach. It represents a *logical abstraction* of the extruder, which is composed of screw and barrel elements, onto the functional sections for material processing. This method shows an analogy to the abstraction in chemical process engineering flowsheets, where the single components initially are represented by simple blocks connected by material and engergy streams and are refined and specialized step by step [119].

Concretly the *arrangement* of functional sections in the compounding extruder can be compared to chemical unit operations in an abstract flow diagram. By connecting several functional sections, a compounding extruder can be configured like a chemical process including feedback connectors. For all functional sections the input and output streams are constant except for sidefeeder or degassing sections. In Fig. 5.35 a schematic diagram from the abstract to the realization level by enrichment and decomposition is shown.

The functional sections in co-rotating twin screw extruders are sequentially arranged and usually have no back coupling *upstream* due to the partially filled sections. Thus material modifications or flow effects have no influence and are decoupled one another unlike in single-screw extruders, for example.

Based on the concepts described above MOREX has been integrated to PRIME (PRocess Integrated Modeling Environments, see Sect. 3.1). This integration provides *context-dependent assistance* to the user and is not explicitly shown in the design process fragment in Subsect. 1.2.2. For example, PRIME shows different decision possibilities that have proven to be reasonable in the past. The first step was to develop a product model of the MOREX interface.

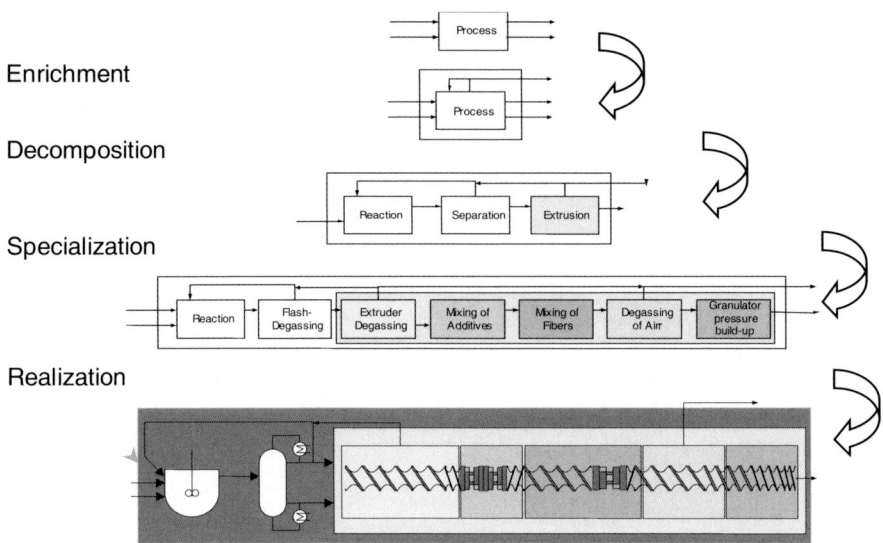

Fig. 5.35. Concept of functional zones in the flowsheet editor analogous to chemical unit operations

With this context model, the workflow of the user is traced. If simulation parameters are changed (e.g., the number of revolutions of the extruder), this information is transferred to PRIME.

Integration of 1D and 3D Simulation

The above described scenario of integrated MOREX and BEMFlow calculations is based on models and modules described in the following. The material model (see subsection *Refined and Extended Material Product Model*) in MOREX is already used to store the material parameters, thus a transfer of these between other programs and the management of the parameters in a central database is possible.

The model for the *geometry description* in MOREX contains a complete three dimensional, parametrized description of conveying- and kneading elements. Based on this model a surface mesh can be exported to the BEM-software. For the structure of these meshes the cross section can be seen in Fig. 5.36. Additionally the visualization of the screws in MOREX is based on these meshes. The boundary conditions for the numerical methods as well as the velocity profile at the flow channel inflow and the viscosity can be given in a specified module in MOREX, resp. are overtaken from a previous MOREX calculation.

Screw elements such as specialized mixing-elements, which are not supported for automatic mesh generation in MOREX, can be modeled with 3D-CAD system and meshed afterwards. Thereby it is possible, to convert meshes

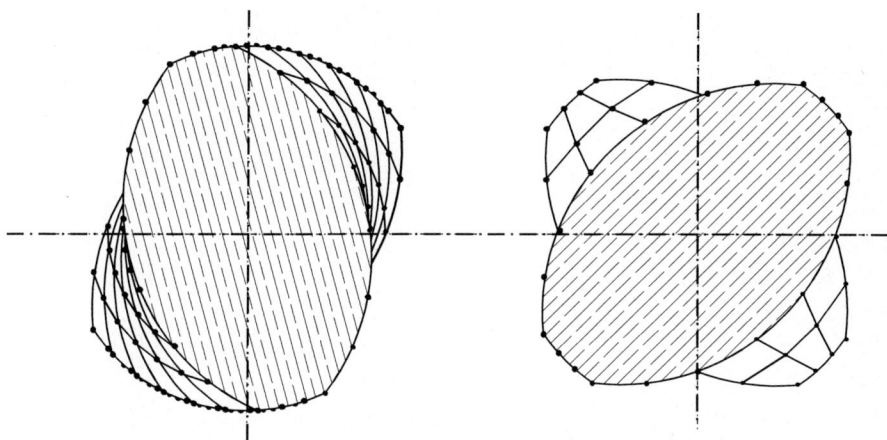

Fig. 5.36. Principle mesh structure for conveying elements and kneading blocks

which are generated with I-DEAS with a special tool delivered with BEMflow. For MOREX an additional converter for meshes which are generated with the FLUENT pre-processor GAMBIT has been developed.

To visualize and explore BEM results, the postprocessor *BEMView* has been developed [144, 145]. One functionality of BEMView is to visualize the BEM mesh, such that the user can check it before starting the simulation. With BEMView, the boundary conditions can be visualized as vectors and the screws can be animated. BEMView can be seen as a part of the demonstration scenario in Subsect. 1.2.2 and is further described in Subsect. 1.2.4. It reads the BEM simulation results file and provides a quick visualization of BEM results like particle streamlines already on a desktop computer. Another possibility for results visualization is the visualization within a Virtual Reality (VR) environment.

5.4.6 Interactive Exploration and Visualization of Simulation Result Data

Furthermore, a workflow combining 1D- and 3D-simulation tools has been studied. The VR prototype TECK has been developed to offer visualization and *interactive exploration* of simulation results [149, 315, 428, 429]. The VR interface allows an interactive *configuration* of extruders and an online visualization of 1D-simulations. The tool can be used for the configuration process as well as the knowledge transfer between different users, as simulation and documentation data are displayed in real-time in a single environment.

In order to configure extruders and to *visualize* 1D-simulation results in a VR environment, the software tool TECK (Twin-screw Extruder Configuration Kit) has been developed. For extruder configuration, TECK offers the following functions:

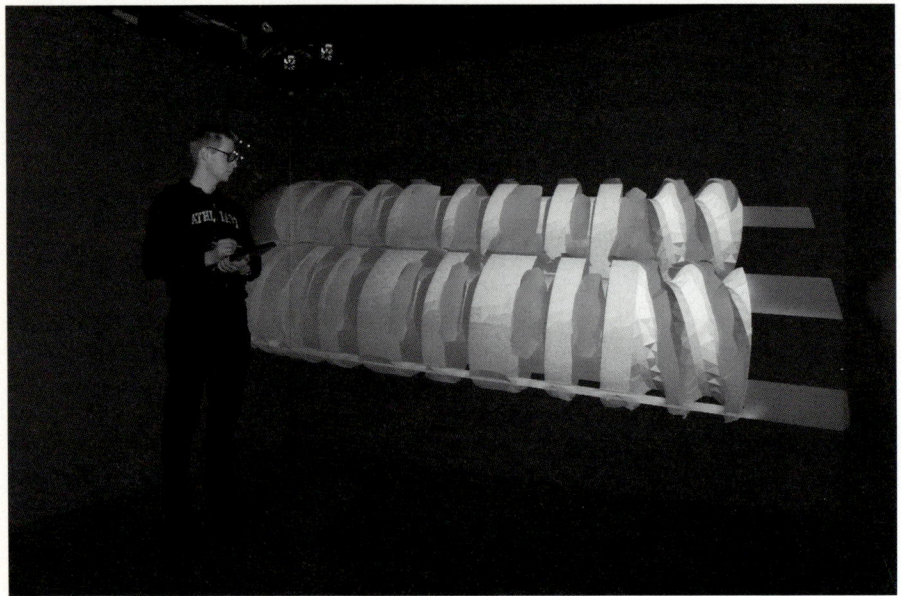

Fig. 5.37. Interactive exploration of simulation results in the CAVE

- Selection of screw elements from an element catalogue.
- Positioning of an element in the screw configuration.
- Variation of screw element parameters.
- Visualization and interaction with simulation results.

For the visualization of simulation results in a Virtual Reality environment, it is necessary to *convert* the data into a file format that is appropriate for Virtual Reality. Consequently, several converters have been developed in order to convert the simulation data (MOREX, BEM, FEM) into the VTK (Visualization Tool Kit) file format [5].

Different extraction functions, for example cutting-, clipping- or iso-planes for interactive data exploration, are implemented in TECK. They are all *filtering functions* for the simulation data sets that help to reduce the overall visual complexity. Fig. 5.37 shows a user who interacts with simulation data in the CAVE. He is using the iso-surface and cutting-plane functions.

ViSTA-FlowLib is a software framework developed at the Institute for Scientific Computing of RWTH Aachen University [386]. It comprises algorithms for the interactive visualization of data sets produced by Computational Fluid Dynamics (CFD). Special attention is paid to unsteady and large-scale data sets.

The technique of virtual tubelets has been developed for the visualization of *particle streamlines*. The streamlines are visualized as geometrical tubes. This allows a more intuitive tracing of particles in a three-dimensional environ-

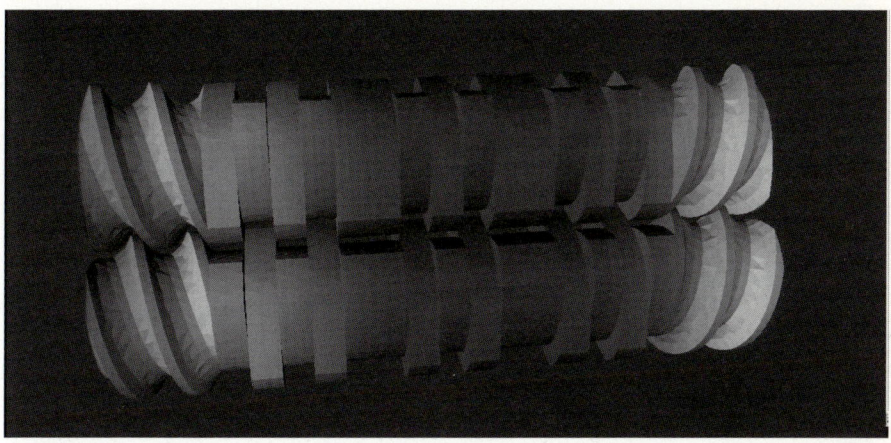

Fig. 5.38. Visualizations of MOREX simulation results are depicted as colored surfaces, using per vertex coloring (left) and a texture mapping based per pixel coloring (right). TECK uses texture mapping in order to realize a per pixel coloring that avoids artefacts resulting from per vertex coloring schemes.

ment. An example of the virtual tubelets is shown in Fig. 3.54. The tubelets are realized with polygons that are oriented to the observer (billboard-technique). The colors of the streamlines demonstrate the calculated values of the Ottino mixing criterion. This criterion refers to the rate of surface enlargement of a fluid particle in the flow; it is an example for a tertiary flow parameter.

TECK uses MOREX as its primary source for simulation data by *direct run-time coupling*. In particular, when a user changes parameters in TECK, a MOREX 1D-simulation is triggered as a remote procedure, and the simulation results are returned and visualized. The *instant visual feedback* allows the frequent variation of simulation parameters by the engineer. Information about the pressure and temperature distribution is obtained at an early stage of the simulation and configuration process, without using 3D-simulations. Once the configuration shows desired attributes in the 1D-simulations, it can be considered for the more time consuming 3D-simulations as depicted above.

The VR environment helps in the overall process, as it uses real-time rendering algorithms and allows the usage of large-screen displays that better match the human physical resolution of the visual sense. 1D-simulation results are displayed with high frequencies using texture mapping and advanced graphics hardware technology. This visualization enables the precise mapping of simulation results to the extruder geometry as exemplarily shown in Fig. 5.38.

5.4.7 Conclusion

In this section, it has been shown that the research done by the subproject A3 had a huge impact on the *issues of integration* between the domains of plastics engineering, chemical engineering, and computer science. Before the start of the CRC, the plastics engineering domain was characterized by very detailed and precise knowledge about its production processes on the one hand, but nearly no expertise with regard to abstract data and workflow modeling or domain spanning integration issues, on the other.

The integration endeavor of *linking* work and design processes between the chemical process engineering and the plastics engineering domain was achieved by abstract data modeling as well as by the mediation of domain specific engineering knowledge. Within the elaboration of this basics, we have formalized work processes and build domain specific data models together with the CRC partners.

The huge impact of this work was especially visible in the transfer of *concrete results* within the MErKoFer and the VESTEX projects. In MErKoFer (see Sect. 7.5), the process-integration approach of PRIME (cf. Subsect. 3.1.3) was extended and adapted for the support of machine operators in rubber extrusion processes. This approach was supported by *extensive data mining*, and by modeling the domain knowledge based on the Core Ontology of the subproject C1, and validated directly in *industrial* work and production processes. In the VESTEX project, the use of *virtual reality* to support and improve work processes in the plastics engineering domain was investigated, particularly concerning the design of compounding extruders.

For the future, several activities are planned to continue the *successful interdisciplinary work* between engineering and informatics. For example, the MErKoFer project results will be further developed and commercialized, and the CRC scenarios extended to other applications or domains. Thereby, the created data models, tools and methods are to be enhanced and augmented further. As most of today's and tomorrow's innovations will be based on the optimization of current technology through software and computer science, the subproject A3 shows a *broad potential* to exploit prospective synergies.

5.5 Synergy by Integrating New Functionality

S. Becker, M. Heller, M. Jarke, W. Marquardt, M. Nagl, O. Spaniol, and D. Thißen

Abstract. The novel informatics concepts presented in Sects. 3.1 to 3.4 can be integrated again to fully exploit their synergistic potential. This section shows three interesting examples of such synergistic integration. The examples bridge different roles or companies in the design process. They also bridge between the efforts of different research groups within IMPROVE.

5.5.1 Introduction

In Sect. 1.1, we introduced novel informatics concepts for supporting design processes. This section is intended to discuss concepts and implementation of their *synergistic integration*.

The section gives three examples of synergistic integration (see Fig. 5.39). In Subsect. 5.5.2 the process-centered *process flow diagram* tool, called FBW[39] (subproject B1, see Sect. 3.1), is connected to the *reactive management* system AHEAD (subproject B4, see Sect. 3.4) via an integrator (subproject B2,

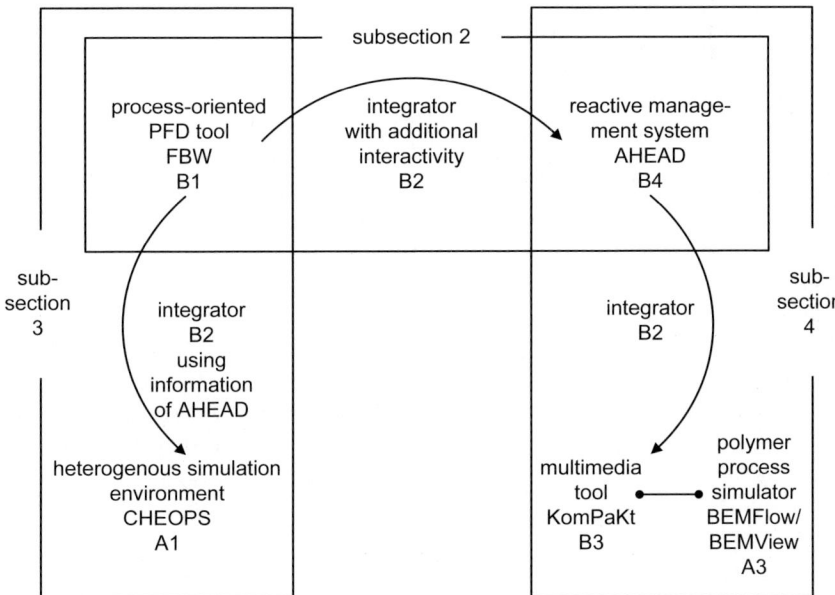

Fig. 5.39. Synergistic integration between different IMPROVE software tools

---

[39] The terms flowsheet and process flow diagram are used as synonyms, they are called Fließbilder in German. FBW is a corresponding flowsheet tool.

see Sect. 3.2). The *integrator* combines two master documents and offers additional functionality in comparison to the integrators of Sect. 3.2.

The following Subsect. 5.5.3 combines *FBW* with the *heterogenous simulation environment* CHEOPS (subproject A1, see Sect. 5.3), again by an *integrator*. The necessary information is available in the PFD and in the integration documents linking PFD and existing simulation models.

The last subsection of this paper describes how a *multimedia tool* (subproject B3) can be connected to *AHEAD*, also by an *integrator*. In particular, the multimedia tool integrates BEMFlow and BEMView (subproject A3, see Sect. 5.4). Furthermore, the integration spans across organizational units.

Of course, many further examples of synergistic integration could have been presented.

5.5.2 Integration between Technical Documents and Process Management

In most design processes, consistency management support is offered only for the results of technical activities like process flow diagrams or simulation models. Changes of technical data may also have consequences for the future course of the design process on the managerial level. As the PFD is a *master document* serving as a reference for the chemical process to be designed, it can also serve as an interface to the management of the design process. Therefore, any modifications of the flowsheet will likely have *consequences* on the *management* of the design process.

In this subsection, the integration of the technical and the managerial level of the design process is shown. It is achieved by *integrating* the technical data contained in the *process flow diagram* with the *process* and *product* data on the *managerial* level. The main goal of this integration is to provide advanced functionality to the chief designer and design process manager as users of the process flow diagram tool FBW and AHEAD, respectively. The aim is to interactively determine the consequences of major technical changes and, especially, their impact on the design process.

To achieve this ambitious goal, the functionalities of FBW and AHEAD are synergistically combined, to offer additional benefit to the end users which cannot be provided by any of the tools. The integration itself is performed by a *FBW-AHEAD integrator* to link both tools. This integrator combines *two master views* of the design process, namely process flow diagrams as the technical master documents and the task net as the master document for developers' coordination.

Our *approach* for integrating the data of FBW and AHEAD is illustrated in the left part of Fig. 5.40:

- FBW is used to synthesize and to maintain all the process flow diagrams of the plant. The PFD contain all essential plant units on the technical level together with their mutual interdependencies. FBW allows to organize

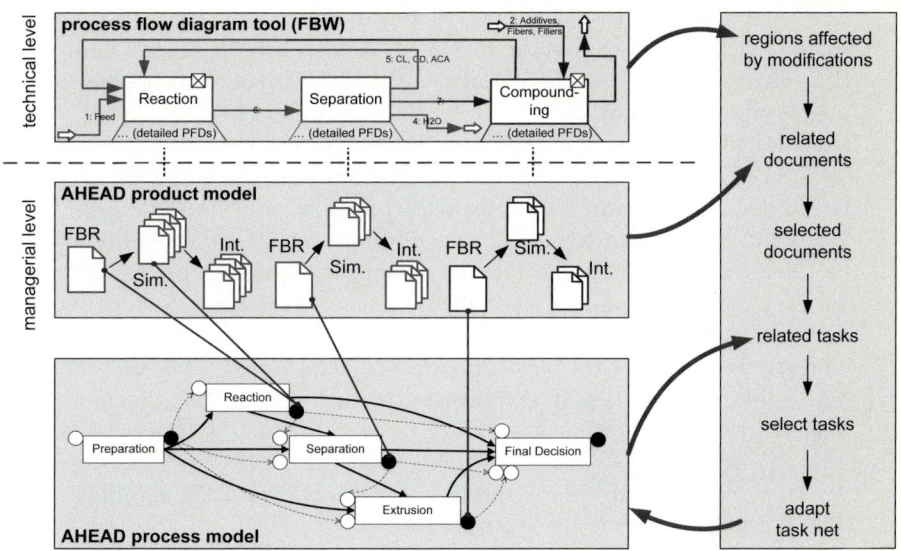

Fig. 5.40. Synergy between the process-centered PFD tool FBW and the reactive management system AHEAD

process flow diagrams hierarchically, such that any flow diagram can be further refined with detailed ones.
- Exploiting the dependencies between flow diagrams, the user of FBW can determine all *regions* in the overall process flow diagram which are *affected* by a possible change in the product data.
- AHEAD stores more coarse-grained data about design documents which are needed for management purposes. Technical process flow diagrams are represented within the product model of AHEAD as product *versions* which are interconnected by *dependency relationships*.
- AHEAD also maintains a model of the overall *design process* with all design *tasks* and their *interdependencies*. Within the process model of AHEAD, the input and output documents of the design tasks are defined and related to product versions of the product model (indicated as links between product model elements and process model elements).
- Integration of process flow diagrams and the corresponding products or tasks in AHEAD requires an *integration document*. This document contains the interdependencies between elements of the process flow diagram on the one and design product as well as process elements of AHEAD on the other hand. This integration document is *created* and *updated* by the *integrator* between FBW and AHEAD.

We now briefly sketch how the *integration procedure* of the FBW-AHEAD-integrator (right part of Fig. 5.40) works:

1. The chief designer starts the process flow diagram tool (FBW) and *selects* some devices which have to be modified. He can use the tool to *determine* the *consequences* of the modifications for other parts of the design in the process flow diagram.
2. Additionally, the chief designer can get a list of all *documents* stored in AHEAD which are *related to* the affected *regions*. For example, all simulation documents related to the affected devices are returned.
3. Next, the chief designer *marks* those *documents* that are probably *affected* by the initial modifications. The selection discussed in 1. above occurs only within the FBW. However, the selection also covers other documents such as those of the simulations.
4. The affected *tasks* of the work process are *determined*. The chief designer gets a list of all tasks of the process model of AHEAD which use the affected documents either as input or output documents. These tasks process the affected documents and therefore they have *to be checked*, e.g. whether they need to be revised in order to cope with the modifications.
5. The chief designer finally *selects* those *tasks* which are affected by the modification according to his knowledge as a technical expert.
6. Now the design process manager receives the list of all tasks which might be affected by the modifications. He starts the integrator which *suggests modifications* to the tasks in the task net. For all affected tasks, it is estimated whether new feedback flows need to be created for running tasks, already terminated tasks need to be restarted, or whether tasks remain unaffected with respect to the modifications.
7. The list of change *proposals* produced by the integrator can be reviewed by the process manager and each change can be interactively *accepted*, *rejected*, or *modified*. After the list of all proposed task modifications has been processed, the process manager can trigger the necessary modifications in the task net while the process is running.
8. The process manager now *interactively changes* the management data using AHEAD functionality. As a result, the process manager gets an adapted task net that reflects the necessary re-work needed to answer the initial process flowsheet modifications performed by the chief designer using FBW.

The *modification* have to come in *cycles* since the consequences of a change can only be determined locally. Corresponding changes are managed. Then, they are carried out by engineers with different roles. When doing the changes, further consequences come up. They, again, have to be managed and carried out. This procedure is continued iteratively.

It can even get worse in case a change is due to a *backtracking* step. When carrying out the corresponding changes iteratively, further backtracking steps might occur. This example shows that managing the changes within a design process is an activity affecting the whole process and, therefore, is a difficult task.

This *integrator* between the FBW and AHEAD *differs* from other integrators mentioned in Sect. 3.2. It offers specific user interfaces for both the chief designer and the design process manager roles. As a result, the integrator has been implemented manually without using the integrator framework introduced in Sect. 3.2. Nevertheless, some concepts of the original integrator approach have been applied, resulting in a clean architecture and a straightforward integration algorithm.

This integrator clearly shows the *benefit* of the *synergistic combination* of new tool functionality which has been developed within IMPROVE. It combines the process-centered FBW (subproject B1) with the reactive management system AHEAD (subproject B4) using an integrator (subproject B2).

5.5.3 Support for Configuring CHEOPS Simulation Models

In the CRC scenario (cf. Sect. 1.2), *CHEOPS* [409], described in Subsect. 5.3.5, is used for the *simulation* of the *overall* PA6 production *process*. CHEOPS uses different existing simulation tools to carry out the overall process simulation by an a-posteriori runtime integration approach. The task of CHEOPS is to perform all partial simulations, each with the appropriate tool, to exchange simulation results between them, and to converge recycle streams which may occur in the flowsheet.

CHEOPS is controlled by an XML-file containing references to the partial simulation models and their parameters and initial guesses. In our scenario, this *file* is *generated* with the help of an *integrator*.

The generation of the CHEOPS control file is again an example for the *synergy* resulting from combining the advanced functionalities and underlying concepts created in the IMPROVE project. Only this combination allows to *derive* the control file with *minimal user interaction*.

Generating the CHEOPS control file involves *data* from *several* IMPROVE *tools* (cf. left part of Fig. 5.41):

- FBW (see Sect. 3.1) contains the overall structure of the plant as process flow diagrams consisting of plant regions and their interconnections. Thus, it can also provide the *structure* of the corresponding *overall simulation* model. Furthermore, it contains detailed flow diagrams for all regions. Only a tool with a clean concept of hierarchical flow diagrams like FBW can provide the necessary *information* about the *relationships* between flowsheets of varying degree of detail.
- AHEAD provides a medium-grained model of all the products of the design process. For the generation of the CHEOPS control files, we employ information about detailed process flow diagrams existing for a given plant *region*, *corresponding* simulation *models* created in which simulator, and integration documents containing the *interrelations* between the two. Again, these data are only provided because of the advanced model structure of AHEAD: There are partial models for the design process, the resulting products, and the employed resources.

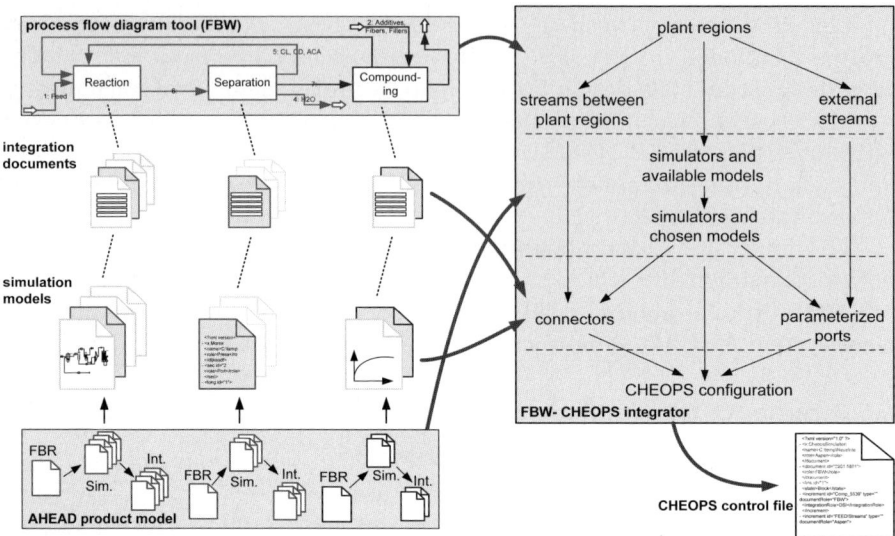

Fig. 5.41. Synergy between flowsheet tool FBW, management system AHEAD and integrators: creation of the CHEOPS simulation file

- *Integration documents* referring to detailed process flow diagrams and the corresponding simulation models provide the *relationships* between *external streams* of each plant region and the *corresponding streams* in the simulation model. They are created by integrators during the elaboration of flow diagrams and simulation models. This information is needed by CHEOPS to *connect* the simulation *models* provided by the simulators at runtime (see below). Without using integrators, collecting this information would be difficult.

The main *activities performed* by the integrator and their dependencies are depicted in the right part of Fig. 5.41. We will explain each activity in more detail subsequently to illustrate how the data sources are used to create the CHEOPS control file.

First, the plant *regions* to be contained in the simulation are *marked* by the engineer and the integrator is started using FBW. A PRIME process fragment (see Sect. 3.1) ensures that the identifiers of the selected *regions* are *passed* to the integrator as parameters. Next, the integrator *identifies streams* coded in a particular simulator connecting plant regions and external streams acting as inputs to the overall simulation by reading the abstract process flow diagram.

By querying the AHEAD product model, all *simulation models* available for the plant regions to be simulated are *determined*. Then, user interaction is performed to *select* one *model* for each plant region.

Now, for each stream connecting a source and a target plant region, a CHEOPS *connector* has to be *set up*. A connector is used by CHEOPS at runtime to get values from the simulation results of a certain output stream of the source simulation model and to provide these results as input parameters for a certain input stream of the target simulation model. Up to now, only the identifiers of the abstract and detailed process flow diagram streams are available to the integrator. Thus, the integrator uses the integration documents relating detailed flow diagrams to the corresponding simulation models to *locate* the related *streams* in source and target *simulation models*. Their identifiers are then used for the definition of the connectors.

For each external input stream, a set of *input values* has to be provided to CHEOPS. Using the same approach as for internal streams, the corresponding simulation model *stream* is *determined*. Then, the simulation model is queried for all *parameters* that have to be set for this stream. Now, the user is able to supply the required values.

After all necessary data have been collected, the CHEOPS *configuration* is *saved* in an XML file and the CHEOPS *simulator* is *activated* using a PRIME process fragment.

The *integrator* sketched here is *different* of those described in Sect. 3.2: It integrates more than two documents, it works unidirectional, there are no conflicting rules, and the user interaction is very specific (see also Subsect. 3.2.6). Thus, this integrator was implemented manually, only partially making use of the integrator framework introduced in Sect. 3.2. All integration rules are static and have thus been hard-coded.

It is only possible to use this simple integrator implementation here, as all information needed is already available in a structured way. This demonstrates how the novel tools do not only perform the tasks they have been designed for but also provide the basis for realizing even *more advanced support*.

5.5.4 Synergy with Multimedia Communication

As discussed in Sect. 3.3, supporting designers' communication by new forms of media can improve a design process. A problem with introducing novel media and communication forms, however, is their *acceptance* by the users. As a first step towards a communication *platform*, KomPaKt (see Subsect. 3.3.2) provides a single interface for all new communication forms,.

These new communication forms, for example, were used to *support* the *cooperation* of *plastics processing engineers*: The tool BEMView for 3D presentations of extruder simulation results was enhanced with an event sharing mechanism, allowing for cooperative work on simulations by geographically and possibly organizationally distributed users. Event sharing on one hand enables the synchronized presentation of the 3D simulations on several computers. This is extensively discussed in Sect. 3.3. On the other hand, it is also possible to switch over to a mode of loose synchronization, in which different engineers can view the simulation results from different perspectives. This way, two new forms of cooperation can be realized. In addition, it is also possible to capture simulation runs and the associated discussions in the design team as a video sequence enriched by respective annotations.

To further improve acceptance, an integration with AHEAD (Sect. 3.4) is accomplished. AHEAD manages the resources of a design process, which also include the involved project members. By implementing a connection to AHEAD, *information* about *designers* and their contact data could be transparently gathered by KomPaKt. Additionally, the *documents* used in some part of the design process can be determined and used in a conferencing session.

KomPaKt is managed by AHEAD, as all other tools used in the design process. Thus, for example, *planned meetings* for discussions and decisions about intermediate results can be modeled as own tasks of work process management by AHEAD. Additionally, KomPaKt can be used for *spontaneous* conferencing, if an engineer wants to discuss with team members in case of problems occurring unexpectedly. In this case, KomPaKt uses the project team management information of AHEAD to determine the possible participants of the conference. Last but not least, *results* of a multimedia conference (e.g. protocols, video recordings, annotated documents, etc) can be *stored* by using the document management part of AHEAD.

For an integration with AHEAD, the communication process during a cooperation is to be seen independent of the process models of the other IMPROVE subprojects in the first step. The communication process extends the C3 model of communication and cooperation possibilities. To do so, detailed information about *synchronous communication* for certain tasks is provided. This is achieved by using a *layered model* (see Sect. 3.3) of communication relationships. This model is connected to the model of AHEAD to offer the possibility to model multimedia conferences as separate management tasks, for enabling KomPaKt to get configuration information from AHEAD etc.

Advanced functionality cannot only be realized by the integration of KomPaKt with AHEAD but also with the *other tools*. In case of integration with PRIME, multimedia cooperation can become a part of a *process fragment* to store the multimedial outcomes of informal discussions when a process fragment supports inter-organizational cooperation. Also, synergies can be realized between KomPaKt and the *integrators* if the interactive parts of an integration process are supported with new cooperation forms.

5.5.5 Summary

This section has introduced three *examples* of *synergistic* integration of novel *support* concepts and tools. It is shown that this integration results in more than the sum of the corresponding functionalities. It also shows that this synergistic functionality is, essentially, a consequence of *connecting interesting engineering results* (either technical or managerial) during the design process.

There are many further possibilities for the realization of synergistic functionality. The selected examples are especially interesting, as they also *bridge* different *roles* or *organizational units* within the design process. They also bridge between different *groups* within IMPROVE.

5.6 Usability Engineering

C. Foltz, N. Schneider, B. Kausch, M. Wolf, C. Schlick, and H. Luczak

Abstract. The number of employees working exclusively with computers increased almost 20 percent within the last four years. In technical offices and in the field of research and development the largest amount, 94 percent, can be found. Hence the focal point of the research project was composed of the development and prototypical implementation of software tools for the support of the work of process development engineers. This field of application provides a large range of innovative computer support because of its high amount of creative work processes hardly to support by strictly structured software tools. Additionally the exploration of many design alternatives and weakly structured constraints between different activities with unexpected or planned iterations can scarcely supported in a conventional way. For this research objective several specialized applications were developed and the ergonomics of their user interfaces were evaluated. These application specific evaluation methods form the groundwork for the redesign based on the design recommendations.

The following section describes the foundations of software ergonomics with corresponding international norms. It will also be shown, that software ergonomics knowledge alone, however, is not sufficient in order to be able to assess the quality of the software support. So the well known ergonomic requirements were supplemented by methods for work analysis. In general, the support of creative and communicative work processes requires a detailed analysis of necessary activities, organizational structure, and information flow including the identification of organizational bottlenecks. The basis for the conducted evaluation of work thus constitute three different models of work analysis which are introduced and discussed.

A suitable procedure was presented, not only for the development of evaluation criteria, but also for application scenarios for the following evaluation of subjects. The scenarios will be briefly described, and the results of the various evaluations of subjects will be introduced. These findings show the quality of the developed prototypes in terms of their area of application and give recommendations for possible improvements.

5.6.1 Introduction

The *number of people using computers* increased continuously in recent years. According to the German Federal Statistical Office, in May 2000 52 percent of employees in Germany worked at PC workstations; in March 2004 the number was already increased to 59 percent. The amount of workers that use PCs for their jobs varies considerably depending on the type of occupation. The largest amount of PC users, 94 percent, are the group of those who are employed in (technical) offices and in the field of research and development and are therefore in the main focus of software-ergonomic improvements. Empirical studies in recent years describe the change of the content requirements that can be observed through the introduction of computer-operated work means in offices and PC workstations [891, 1028]. Along with constantly increasing demands for qualification, employees that work primarily with software systems must strive for continuous learning. In addition, the meaningful improvement of social, communicative and methodical competencies is necessary. A significant strategic advantage results if these requirements can already be supported by the software tools through a user-centered system design.

5.6.2 Basic Principles of Software Ergonomics and Work Analysis

Software ergonomics or usability engineering can be regarded as the intersection between the research areas of ergonomics and computer science [926]. Experts in this field (software ergonomists or usability engineers) analyze, design and evaluate the use of interactive computer systems to optimize the interaction of all components that determine the work situation: human, task and organizational frame [710, 825, 864, 941]. When using a computer, the worker is confronted with two kinds of problems [645]:

- *task-related*, i. e., to solve the work task on the basis of his professional knowledge, skills, and abilities;
- *interaction-related*, i. e., to use the software system utilizing the graphical interface, keyboard, etc.

To cover these aspects when designing a software system, methods of work and more specific task analysis must be applied and ergonomic requirements have to be met. Furthermore, to validate if both task-related and interaction-related issues have been considered, an evaluation is necessary.

Ergonomic Requirements

The legislation of minimum *requirements for workplace design*, particularly in the design of workstations with monitors, is set through legislation in the context of occupational health and safety. The EU visual display unit guideline 1 [502] is currently the most important regulation in the area of workstations with displays. The German Occupational Health and Safety Act [577] ensures that EU guidelines are put into national legislation, accident prevention and further regulations.

All aspects of working with computers are covered in the multi-part standard ISO 9241 [632]. Originally titled "Ergonomic requirements for office work with visual display terminals (VDTs)", it has been retitled by ISO to "Ergonomics of human-system interaction" in 2006.

An overview of the goals and the other parts of the standards is given in Part 1. Guidance on task requirements is given in Part 2. Parts 3 to 9 address the physical characteristics of computer equipment and environment, for example work desk, display, keyboard etc.

Parts 10 to 17 deal with usability engineering aspects of software, also referred to as software ergonomics. Two parts are of particular importance.

Part 10 – renamed Part 110 in the 2006 edition – titled "Dialogue principles" presents a general set of principles for the design of different types of dialogue. Part 11 "Guidance on usability" gives a general guidance on the specification and measurement of usability.

The other parts deal with the presentation of information (Part 12), user guidance (Part 13), menu dialogues (Part 14), command dialogues (Part 15), direct manipulation dialogues (Part 16) and form filling dialogues (Part 17). In the 2006 edition of ISO 9241 new parts such as "Guidance on World Wide Web user interfaces" (Part 151) and "Guidance on software accessibility" (Part 171) have been added.

Aside from this standard, guidelines and so-called "style guides" of software manufacturers like Microsoft or Apple exist to ensure a uniform look and feel [886, 964]. They also contribute to a standard interface design. Therefore, new applications, such as those developed in the scope of IMPROVE, try to adapt to these guidelines and to use standardized symbols and functions. In this manner, the software system should be familiar to the user. Furthermore, the programming effort can be reduced.

As mentioned above, the principles of dialogue design described in Part 10 of ISO 9241 are of particular importance. These seven design and evaluation criteria are described briefly in the following.

Suitability for the Task

Suitability for the task describes the *characteristics of a user interface* to effectively support the user in doing his work. This means that the dialogue only shows information which is related to the actual problem and therefore important for solving the task. To avoid confusing the user and to ensure efficient work, additional or unnecessary commands should not be provided.

Self-Description

A system has the ability of self-description if every dialogue step is *directly understandable* by feedback signal, or if it can be made understandable to the user on his demand. Not every user has the same knowledge of computer or software systems, so it is very important that the feedback signals are adapted to the situation and to the vocabulary of the user.

Controllability

To ensure that an interactive system is controllable, the user should be able to *regulate the speed of the dialogue procedure*. Additionally, the user should be enabled to decide on type and extent of the input and output as well as on the order of the dialogue steps. Controllability of a system also includes the ability to undo the last dialogue steps.

Conformity with User Expectations

Conformity with user expectations describes the *uniformity and consistency of a dialogue system*. Consistency means the rule observance of the design of the user interface and the dialogue steps, so that similar user inputs lead to similar reactions of the system in related situations. Conformity with user expectations also includes the adaptation of the dialogue system to the user's characteristics. It includes understanding the state of knowledge of the user in his field of work, his education and experience, as well as the generally accepted conventions.

Error Tolerance

A system is called error tolerant if the intended task result can be achieved without or with *minimum effort of correction* by the user despite incorrect input. The dialogue system should support the user in finding and avoiding input mistakes, and it should secure that under no circumstances an incorrect user input leads to data loss or even a system break down.

Suitability for Individualization

A dialogue is suitable for individualization if the dialogue system is *adaptable to the work task* as well as to the *preference and skills* of the user. On the one hand, the system should be able to conform to the speech and cultural singularity of the user. On the other hand, the system's speed of reaction should adapt to the user's needs; it should also be possible to individualize the extent and availability of commands of a work situation.

Suitability for Learning

A dialogue is called suitable for learning if it *supports the user in learning* the dialogue system. The dialogue system should enable the user to familiarize himself with the basic design principles and operational concepts. In doing so, the user can visualize a picture or model of the function of the system.

However, ergonomic requirements alone are not sufficient. They are closely connected to the user's task: "The dialogue principles in ISO 9241-10 cannot be used for design or evaluation without first identifying the context of use. Some of the dialogue principles are closely related to specific aspects of the context of use" [554].

Therefore, ISO 13407 "Human-centered design processes for interactive systems" [631] describes a generic approach to a systematic consideration of context or task-related issues.

Work Analysis

While ergonomic requirements address the interaction-related issues of working with computers, task analysis [280, 782] or work analysis are used to *solve the task-related problems*. The existing techniques for work analysis can be divided into normative, descriptive and formative approaches [118, 923, 1023].

First, *normative models* ("The one best way?") prescribe how a work system should behave. Normative models can be found in textbooks about systems engineering [901], or more specifically, in monographs, e.g. about chemical process design [556, 559, 957]. The emphasis is on identifying what workers (or engineers) should be doing to get their job done. However, the relatively well-ordered transformation from an abstract problem formulation to an exact equipment illustration does not describe in sufficient detail what engineers are really doing. Design is iterative and depends on the experience, the skills and competence of the engineer [989]: "Furthermore, no two designers design a complex process following exactly the same steps" [957]. In particular, normative models describe novice rather than expert performance [924]. To avoid misunderstandings, normative models are important for curriculum purposes but not sufficient to derive implications for the design of computer support-based systems.

Second, *descriptive models* ("What workers really do") seek to understand how workers (or engineers) actually behave in practice. This goal is accomplished by conducting field studies. However, as far as chemical engineering is concerned these studies are very rare [16, 125, 1047]. As an exception, the well-known participant observation of an engineering design project involving the design of a high-pressure, high-temperature system for testing materials in a simulated coal gasification environment can be mentioned [713]. Moreover, a huge effort is necessary to gain insight into a complete development process from the cradle to the grave. Additionally, it would only be a unique example. Even a development process with a similar problem formulation may differ significantly from the other. Furthermore, results from those studies are limited in at least two ways. Current practice is, on the one hand, always tied to existing technology, i.e. it is device-dependent and contains work-around activities that are caused by inappropriate computer support [1023]. On the other hand, engineers are adaptive, so the introduction of a new design results in new work practices [279]. This interdependence is known as task-artifact cycle in the human-computer interaction literature [583, 697]. In summary, descriptive techniques are important and useful in order to understand what engineers really do and what they would like to do. Nevertheless, there are serious limitations in extracting design implications from descriptive models.

Third, *formative models* ("Workers finish the design") focus on identifying requirements, both technological and organizational, that need to be satisfied if a device is going to support work effectively. The workers will be given some responsibility "to finish the design" locally as a function of the situated context. Formative approaches overcome the difficulties which occur when normative or descriptive approaches for system design are used [1023]. Currently, two formative approaches are described in the literature: contextual design [555] and cognitive work analysis [1023], based on the work of Rasmussen [924]. The latter one, cognitive work analysis (CWA) comprises five steps or concepts and will be discussed in more detail.

The first concept, *work domain*, represents the system being controlled, independent of any particular worker, automation, task, goal, or interface, i.e. the work domain shows the possibilities for action. The goal of this design procedure is to identify the requirements which have to be realized to guarantee effective work of a supporting system. Only the tool-independent description of work activity permits an analysis which can lead to the development of designing methods enabling the user to accomplish his work in an adequate way, as well as making it possible to design new ways of operation [924, 1023]. The work domain will be represented by a two-dimensional abstraction-decomposition space (ADS). The first dimension, the abstraction hierarchy (AH) or functional means-ends dimension, supports knowledge-based reasoning and decision making in terms of functional relationships among the information objects. To form a two-dimensional problem space, orthogonal to the AH a decomposition dimension specifies the part-whole relationship of a system. Each of these levels represents a different level of granularity. Therewith, two main strategies for problem solving - abstraction and aggregation - can be covered [922, 924, 1023]. Certainly, very often a change of the decomposition level is coupled with a change in the abstraction level. Nevertheless, these two dimensions are conceptually separate.

The second concept, *control task*, are the goals that need to be achieved, independently of how they are achieved or by whom. In other words, the focus is on identifying what needs to be done, independent of the strategy (how) or actor (who). The third concept, *strategies*, are the generative mechanisms by which particular control tasks can be achieved, independent of who is executing them. They describe how control task goals can be effectively achieved, independent of any particular actors. The fourth concept, *social organization and cooperation*, deals with the relationship between actors, whether they be human workers or software tools. This representation describes how responsibility for different areas of the work domain may be allocated among actors, how control tasks may be allocated among actors, and how strategies may be distributed across actors. Finally, the fifth concept, *worker competencies*, represents the set of constraints associated with the workers themselves. Different jobs require different competencies. Thus, it is important to identify the knowledge, rules, and skills that workers should have to fulfil particular roles in the organization effectively [924, 1023].

Information Representation for Computer-Supported Chemical Process Engineering

The computer-aided design of work processes in process engineering has, as in other branches, contributed to a multiplication of information volume and an increase in modification iterations, thereby it let to an increasing complexity and intransparency of work processes. In fact, in the last decade systems were developed that make the entire storage of all process development-related information possible [559, 1047].

However, chemical process design is not supported appropriately and continuously by software tools [195, 335, 401, 834]. Furthermore, often these tools usually can only be used by experts instead of "normal" chemical engineers [834]. In other words, software support in chemical engineering is concerned with task-related and interaction-related problems.

As far as the task-related issues are concerned, the following can be stated. Chemical process design is characterized as a complex, iterative, and creative activity typically starting as an ill-defined problem [1047]. In order to create new processes and plants or to retrofit existing ones, an interdisciplinary team develops and uses different models [17, 401]. Decisions in chemical process design arise from goals and constraints incorporated in these models, constituting knowledge about the process and the plant, respectively.

In other words, in a complex network of causal and functional dependencies, the chemical process engineer should be able to *solve problems and make decisions more quickly* with the aid of a computer-based support system. Therefore, it is necessary that these systems not only support the method knowledge of the developer, but that they also correspond to the way of thinking the developer has become accustomed to, i.e. the task-specific mental model. The smaller the gap between the computer-aided representation of knowledge structures and the developer's mental model to be carried out, the more efficiently the developer can fulfill his tasks [924, 1023]. The support of the special abilities of the human problem-solver through structuring and information visualization is therefore a major concern rather than computer-supported automation of problem solving. Therewith, this knowledge describes the work domain the process designer is acting on.

To develop a representation of the work domain of chemical process design the introduced *abstraction-decomposition space* has been developed [117–119]. The announced abstraction hierarchy or functional means-ends dimension supports knowledge-based reasoning and decision making in terms of functional relationships among the information objects. Each level in the hierarchy represents the goals or ends for the functions of the level below and potential resources or means for the level above. In other words, the AH spans the gap between purpose and material form.

For chemical process design the part-whole dimension contains the following five levels: system (e.g., process), subsystem (e.g., separation process), functional unit (e.g., distillation column), subassembly (e.g., valve tray), and component (e.g., float valve).

Similarly, the functional means-ends dimension discriminates five levels (Fig. 5.42):

- *Functional purpose:* textual description of what is desired like "production of Polyamide-6, residue of water less 0.01% ..." for the system or "allow part load" for a float valve on the component level;

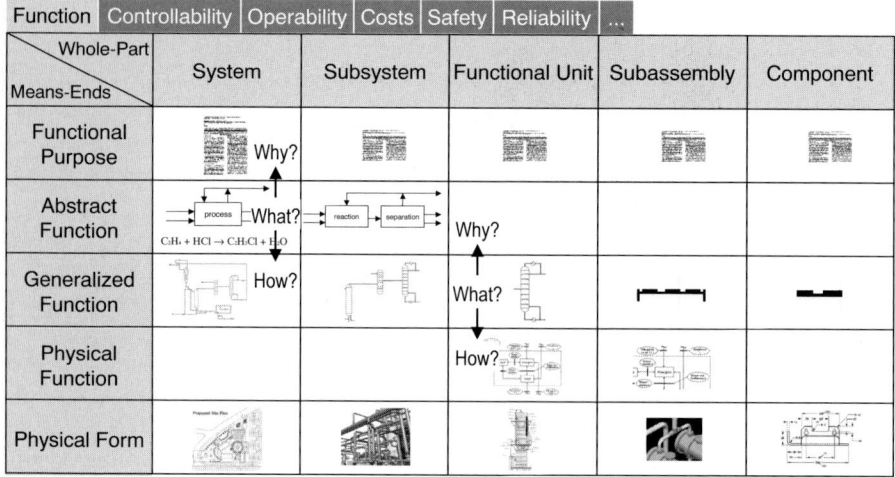

Fig. 5.42. Abstraction-decomposition space (ADS) for chemical process design [117–119]

- *Abstract function:* chemical reaction paths, basic functions (react, separate, etc.), physical property data;
- *Generalized function:* mostly unit operations like continuous stirred tank reactor or plug flow reactor for "react" and distillation column or evaporator for "separate" and also new combined operations; assumptions are necessary due to lack of some data in advance; calculations with linear mass- and energy balances; short-cut methods;
- *Physical function:* rigorous process models in different refinements; physico-chemical phenomena included; complex mass, momentum, and energy balances; assumptions about data (e.g., recycling rate, physical properties) are replaced;
- *Physical form:* 2D-drawings and 3D-models of all equipment, plant layout.

Indeed, there are further important aspects such as controllability, operability, costs, safety, reliability, etc. which impose constraints on process design. However, they are secondary objectives always related to the functional representation. Hence, they can be characterized as supplementary layers linked to the proposed problem space (tabs at the top of Fig. 5.42).

Evaluation

A *usability evaluation* is any analysis or empirical study of the usability of a prototype or system. The goal of the evaluation is to provide feedback in software development, supporting an iterative development process [886, 941].

In general, two types of evaluation can be distinguished: summative and formative. The latter takes place during the design process to identify aspects of the design that can be improved and to provide guidance in how to make changes to a design. The first type of evaluation is done to assess a design result and is most likely to happen at the end of a design process. Consequently, evaluation methods can be separated in two different classes, analytic and empirical, respectively.

Analytic evaluation methods can be used early in the development process, well before there are users or prototypes available for empirical tests. Furthermore, it is often less expensive than making studies with users. Examples of analytic methods are heuristic evaluation, cognitive walkthroughs, usability-expert reviews, group design reviews [864, 1025]. A hazard of analytic evaluation is that system developers or software designers may feel that they are being evaluated [941].

Empirical evaluation methods involve actual or designated users. The methods can be relatively informal, such as observing people while they explore a prototype, or they can be quite formal and systematic, such as a tightly controlled laboratory study of performance times and errors or a comprehensive survey of many users [941, 943]. Independent of this differentiation, in general, qualitative and quantitative methods of both data collection and data analysis can be distinguished. While *quantitative research* focuses on how to operationalize or quantify the attributes to be measured, *qualitative research* interprets verbal or non-numerical data [563].

The great variety of quantitative and qualitative approaches may provoke to neglect of one important quality factor: validity. Cook and Campbell [610, 611] suggest to use the concepts validity and invalidity to refer to the best available approximation to the truth or falsity of propositions. The concept validity is subdivided into four types:

- *Statistical conclusion validity.* Statistical conclusion validity refers to the validity of conclusions about whether the observed covariation between variables is due to change. In other words, it refers to the confidence with which one can say that there is a real difference in Y scores between X cases and X' cases.
- *Internal validity.* Internal validity is concerned with whether covariation implies cause. In other words, it deals with the logical question, how to rule out alternative explanations such as that Y caused X or that both X and Y stemmed from unmeasured factor Z.
- *Construct validity.* Construct validity refers to the validity with which cause and effect operations are labeled in theory-relevant or generalizable terms. In other words, a study needs to have clearly defined theoretical concepts and conceptual relations as well as clearly specified mappings of those concepts into empirical operations.
- *External validity.* External validity refers to the validity with which a causal relationship can be generalized to various populations of persons, settings and times.

It will probably be apparent that the methods used to increase internal validity and statistical conclusion validity and the techniques to gain precision will threaten the external validity of that particular set of data, but the relation is not a symmetrical one. Things that aid external validity, e.g. large and varied samples, may either hinder or help internal validity or have no effect on it. Moreover, it is certainly not the case that things that decrease internal validity will somehow increase external validity.

In summary, a multiplicity of methods exists. However, an empirical method on its own is neither right nor wrong. All methods mentioned above offer both opportunities not available with other methods and limitations inherent in the use of those particular methods [563, 838]. Hence, it is not possible to simultaneously maximize three major goals of research evidence [838]:

(A) the *generalizability* of the evidence over populations of actors;
(B) the *precision* of measurement of the behavior and precision of control over extraneous facets or variables that are not being studied;
(C) the *realism* of the situation or context in relation to the contents to which you want your evidence refer to.

To increase one of these criteria means reducing one or both of the other two. In brief, field studies gain realism (C) at the price of low generalizability (A) and lack of precision (B). Laboratory experiments maximize precision of measurement and control of variables (B), at the price of lack of realism (C) and low generalizability (A). Surveys have high generalizability (A) but get it by giving up much realism (C) and much precision (B). The nature of this strategic dilemma is shown in Fig. 5.43, which shows a set of eight alternative research strategies or settings in relation to one another. The strategies are arranged in four sections:

  I: setting in natural systems;
 II: contrived and created settings;
III: behavior not setting dependent;
 IV: no observation of behavior required.

In the figure, three points with different weighting of the major research goals can be identified: A is the point of maximum concern with generality over actors, B is the point of maximum concern with precision measurement of behavior, and C is the point of maximum concern with system character of context.

Despite naming and classifying the research methods slightly differently from the terms used above, it is well suited for our purposes. Any research strategy is both limited in what it can do and flawed, although different strategies have different flaws.

Once again, there is no single best strategy because each is inherently limited. However, they are all potentially useful. In considering any set of

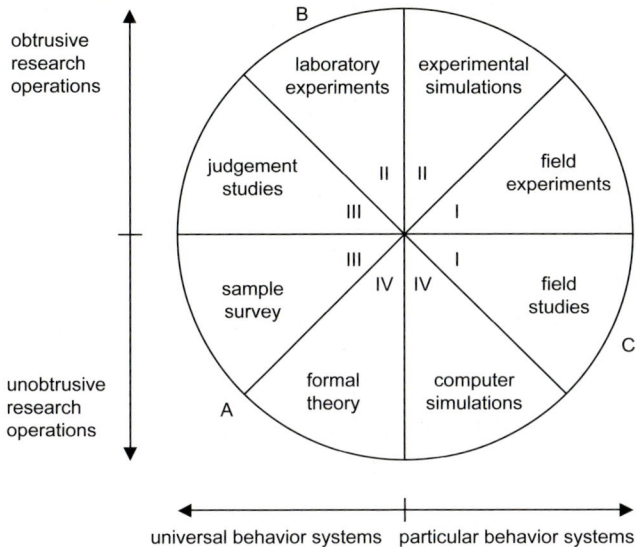

Fig. 5.43. Research strategies [838]

evidence, one should take into account what strategies were used in obtaining various parts of it, hence the strength and limitations of that evidence at the strategic level. Any study needs a plan for what data will be gathered, how that data will be aggregated and partitioned, and what comparisons will be made within. As evident from the preceding section, the choice of one or another of the various strategies will limit the kinds of design one can use [122].

Consequently, the software tools developed as part of IMPROVE were examined both analytically and empirically. Depending on the state of development of each software, one or more evaluation techniques were chosen, and by means of concrete examples of designing user interfaces, suggestions for a better configuration were acquired. In the following, we discuss the analysis and evaluation of the design support system EVA, of the flowsheet editor FBW (see also Subsect. 3.1.3), of the administration system AHEAD (see also Subsect. 3.4.2), and finally of the communication platform KomPaKt (see also Subsect. 3.3.2).

5.6.3 EVA – Design Support System

The design of chemical plants – in particular during the early stages – is characterized by an intensive co-operation in a large project team. The project team usually consists of members with widespread scientific backgrounds and expert knowledge. The team members mainly execute creative and modestly structured tasks including frequent demand-driven and spontaneous communication processes required for the co-ordination process. Due to the specific characteristics and typical shortcomings of such a co-operation, it should be possible to improve this process by the use of tailored information- or groupware-systems. For example, experts are often integrated into ongoing chemical design projects at short notice and on a short-term basis in order to solve acute problems, e.g. modeling of the reaction kinetics. The frequent modification of the group structure leads not only to difficulties in information storage and exchange but also to a different understanding of ideas and goals, because internal rules are unknown or have changed. The development of the groupware-system *EVA* (design support system for chemical engineering, in German: *E*ntwicklungsunterstützung *V*erfahrenstechnischer *A*nlagen) was intended to provide appropriate support to overcome such problems.

Requirements Engineering and System Design

As a first step of the development of a groupware-system, communication and co-operation processes must be investigated from different points of view to identify the requirements of potential users [612]. Therefore, it is necessary to investigate characteristics of their communication and co-operation and to identify possibilities of how to improve these processes. Furthermore, theoretical models of communication and co-operation should be analyzed to find innovative ways of supporting co-operation. During the development of the groupware-system EVA a requirements analysis was conducted by (a) a field study to analyze the characteristics of co-operation in the field of chemical engineering, (b) screening of the theoretical concepts of co-operation and workspace models to develop the basis for new functionalities [486]

The groupware system EVA consists primarily of a shared work space that all users of a chemical design project can access (Fig. 5.44). In this workspace all of the documents created during a project are inserted and displayed in a structured way. As a result, the work area presents the center for the project-internal and primarily asynchronous information exchange. It also gives a structured overview of the relevant information and procedures throughout a development process in chemical engineering. The structure of the work area was derived from the abstraction-decomposition space introduced in Fig. 5.42.

Therefore, a document placed into the workspace must be enriched with context information, so that process designers are able to comprehend the contextual meaning of a document and immediately identify the main relations between documents.

Context information is represented by the following four design elements:

1. *Document relations:* Colored lines between two documents explain whether the following document is a decomposition (red line) or a variant (or version) of the previous document (blue line). In case of decomposition, the interface between these subranges can be specified within an interface

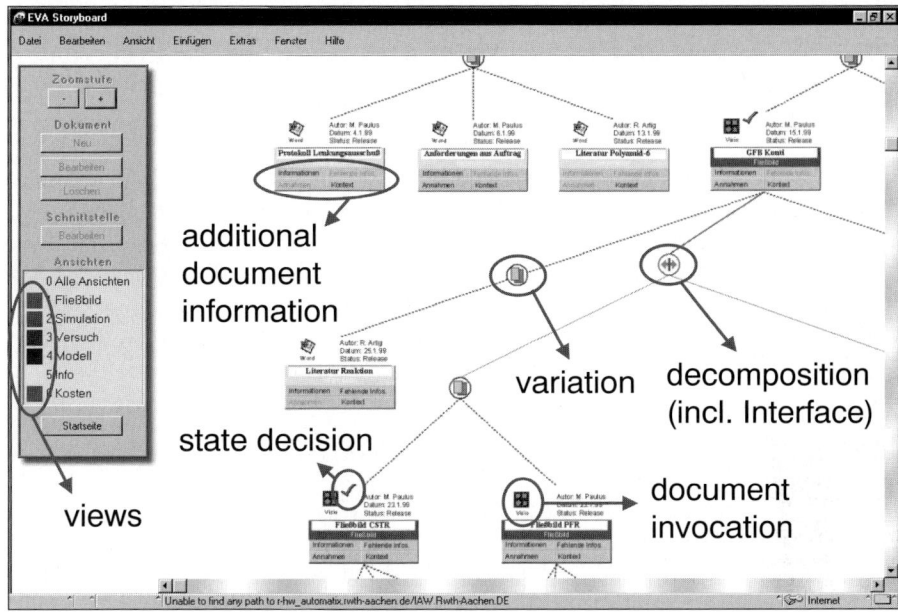

Fig. 5.44. Screenshot of EVA [483, 486]

symbol (e.g. mass flow). Thus, propagation mechanisms inform all authors of subrange documents connected to this interface about changes in the interface specification (e.g. a recycle flow has been computed).
2. *Views:* Documents are categorized and displayed in six different views (flow diagram, simulation, model, experiment, costs, information). These views were derived from corresponding classes of a product model for chemical plants. Due to the fact that the process flow diagrams are of essential importance, the presentation of all documents is oriented to their structure.
3. *Additional document information:* Substantial information is documented in text boxes close to the document icons in form of a short overview with the following attributes:
   - Important information concerning the creation of the model (simulation results, physical property, etc.).
   - During a chemical design process, designers often make assumptions to be able to continue their work (e.g. boiling point of an unknown element in a simulation of a reactor). Usually, these assumptions are verified in a later stage.
   - If no assumptions can be made, this missing information causes a task to remain unfinished. By explicitly naming this lack of information, unfinished tasks are marked for further adaptation.

- Additional (contextual) information about the thematic and spatial environment of developers, tools used, scheduling, etc. This context information is intended to generate awareness about the informal group structure in a project. Users can adjust the amount of information displayed.
4. *Decision state:* The document status (green hook and red cross) displays whether an alternative was selected or rejected and provide an indicator for further handling of a document. In this way, all decisions in the course of a project can be captured. Beyond that, the rational behind the decision is documented by an application of the Issue Based Information System, IBIS [798]. The benefit of documenting design rationale is pointed out e.g. by Conklin and Yakemovic [607] (see also Sect. 2.5).

Inserting a document in EVA can be accomplished while working with another engineering tool (e.g. process flow diagram tool). By retaining the option to store a document in the engineering tool it is possible to store the document in EVA, too. For this reason, the position in the structure provided by EVA and the appropriate relationship with the previous document (blue or red line, see above) must be determined. Additionally, EVA queries further information about the inserted document.

Evaluation and Results

For the evaluation of EVA two main goals were targeted:

- The effects of the four design elements embedded in EVA were to be analyzed empirically with regard to the usage behavior of process designers.
- The evaluation should discover the chemical engineers' method of information reception and representation according to the use of EVA. It is expected that the revealed mental structures will lead to more detailed design requirements for EVA, concerning both the graphical design of the user interface and the design of the cognitive concept of EVA.

To achieve these goals an evaluation with three subsets was conducted. The first analysis focussed on effects in the use of the four design elements. With the second analysis the information reception during the use of the design elements was investigated. The third analysis targeted the mental compatibility of the structures used in EVA [484, 485].

Basis of the laboratory examination was a prototypical realization of the user interface, which provided the required possibilities of interaction to manage the different tasks. All explanations and questionnaires were provided by documents. During the experiment, the actions of the participants were stored in log files and recorded on videotapes. To investigate the influence of the design elements of EVA on the behavior of chemical developers, a four-group test design was selected. Therefore, the EVA prototype was tested in a realistic scenario, the "development of a Polyamide-6 scenario" [124]. Each group was provided with a different number of design elements within its EVA prototype. The EVA prototype was implemented on a Lotus Notes 5 Platform to ensure flexible navigation functions and efficient navigation between the 14 different types of views of the scenario for all four groups.

Altogether, 20 chemical design experts (average age: 30.5 years; average professional experience: 2.9 years) and 17 students of chemical process engineering (average age: 26.29 years); average professional experience (including practical training): 0.68 years) participated in the experiment. The tests took 1 to 1.5 hours per person. Four pre-tests and six tests with the EVA prototype were executed. The pre-tests served the purpose of acquiring the personal data of the participants and ascertaining possible interference variables such as logical reasoning or the possibility of recollection. The other six tests examined the software-ergonomic criteria task adequacy, ability of self description, conformance of expectation and transparency, as well as the influence of a different number of design elements. During these tests, the participants had to name different objects and their functions, and they were asked to arrange the structure of the objects and their relations used in EVA. This was performed by the structure-laying technique according to Groeben and Scheele [702].

Test persons were divided into four test groups based on the personality traits determined in the pre-tests. Each group had a different design model from EVA at its disposal. The number of design elements was thus increased successively so that group zero worked with a version of EVA that did not have any design criteria and group three had three design criteria available.

The four groups had to solve a task with different versions of EVA comprising different numbers of design elements, as well as complete a multiple-choice questionnaire. In addition, the subjective estimation of the functional structure of each participant was secured to enable the development of design requirements.

The test of the software-ergonomic criteria showed that most of the objects and functions were named correctly, which indicated an adequate use of symbols in the user interface of EVA. An exception constituted the graphical symbol "user interface", which was named correctly by only 43 percent, and whose function was named correctly by only 19 percent of the participants.

Eventually, the data of all participants were analyzed statistically using SPSS. The results are visualized in Fig. 5.45 and Fig. 5.46 using box plots. The median is illustrated as a horizontal, thick line. Minimum ($\perp$), maximum ($\top$), extreme values (*), and outliers (o) are also shown. The box stretches from the lower hinge – defined as the 25th percentile – to the upper hinge – the 75th percentile – and therefore contains the middle half of the scores in the distribution. Figure 5.47 shows the medians, quartiles, extreme values and outliers of the groups, with reference to the number of correct answers in the multiple-choice test. The evaluation of the results of the four groups signified that the number of correct answers of the multiple-choice test increased with the number of design elements available for each participant. The average values of the four groups (for both experts and students) differed significantly

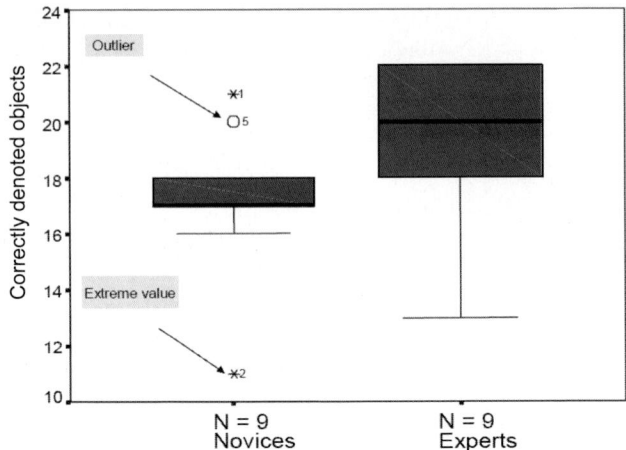

Fig. 5.45. Correctly denoted number of objects (max. 28)

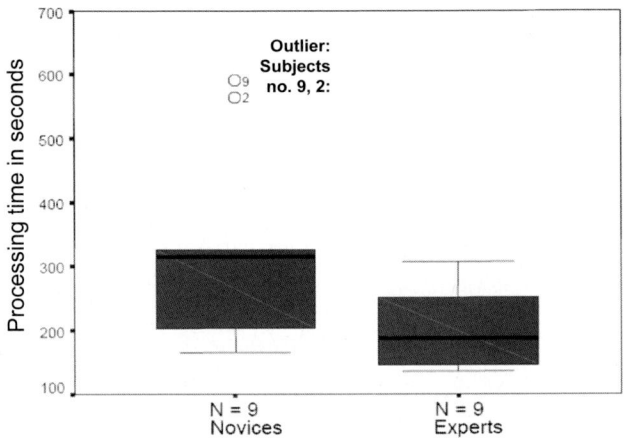

Fig. 5.46. Time on task

between group 0 (without design elements) and group 3 (three design elements). Additionally, the expert group showed significant differences between group 0 and group 1 (one design element).

After significant differences between the results of groups in general were found by a one-way analysis of variance ($\alpha = 0.01$), significant differences of group medians were tested by the Newmann-Keuls test ($\alpha = 0.01$).

The results of the simultaneous comparison of group medians is shown in Table 5.3. The variable k indicates the distance between the compared groups. In the second column, the difference between the medians of the two

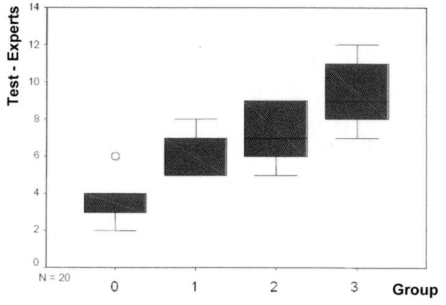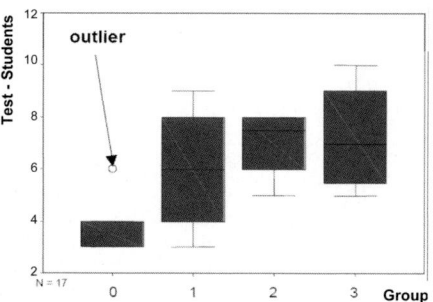

Fig. 5.47. Correct answers in the multiple-choice test

k=4	Group 3 - Group 0	Total	4,8 > 2,07	significant
		Experts	5,6 > 4,44	significant
		Students	4,0 > 3,04	significant
k=3	Group 2 - Group 0	Total	3,2 > 2,11	significant
		Experts	3,4 > 2,45	significant
		Students	3,0 < 4,81	not significant
	Group 3 - Group 1	Total	2,4 > 2,11	significant
		Experts	2,8 > 2,45	significant
		Students	2,0 < 4,81	not significant
k=2	Group 3 - Group 2	Total	1,6 > 1,30	significant
		Experts	2,2 < 2,60	not significant
		Students	- -	-
	Group 2 - Group 1	Total	0,8 < 1,30	not significant
		Experts	0,6 < 2,60	not significant
		Students	- -	-
	Group 1 - Group 0	Total	2,4 > 1,30	significant
		Experts	2,8 > 2,60	significant
		Students	- -	-

Table 5.3. Results of the Newmann-Keuls-Test

investigated groups is compared with q. The parameter q is a function of the quartiles of the t-distribution, the degree of freedom, and the variables α and k.

The results also indicated that the reception and comprehension of the semantic structure of project transitions depended on relations between documents, which were indicated with colored lines and which generated a structure of decompositions and variations. A positive influence of the use of views in EVA could not be found.

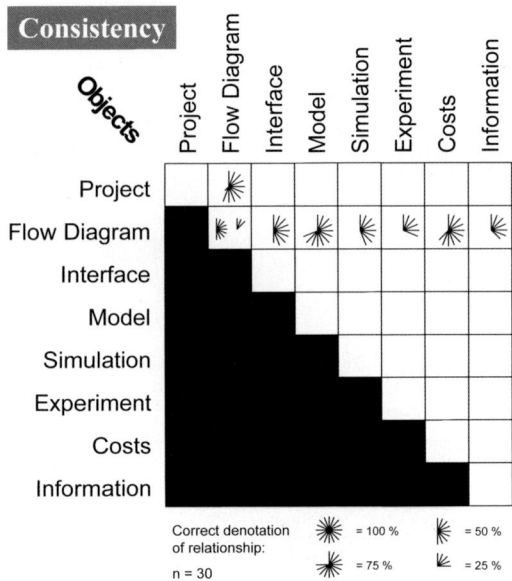

Fig. 5.48. Consistency in the test of transparency

In the scope of the analysis of information reception, the participants had to arrange the structure of the objects used in EVA. The outlined structure can be interpreted as the information pattern. As a result, all correctly denoted relations as well as all incorrectly denoted or missing relations were documented.

The matrix in Fig. 5.48 shows the frequency of correct denominations of relation concerning the transparency experiment. The values obtained a range between 43 percent and 68 percent, with one exception: the second relation between process flow diagrams (process flow diagram - "is part of" - process flow diagram), in the sense of a decomposition, was denoted correctly by only 13 percent of the participants.

The matrix in Fig. 5.49 shows the number of divergent denotations. Most frequent divergent denotation information objects were assigned to the object "project" (instead of the object "flowsheet"); simulation objects were assigned to the object "model" (instead of the object "process flow diagram"); and test objects were assigned to the object "simulation" or "model" (instead of the object "process flow diagram"). Because this divergent assignment corresponds to the usual methodology in chemical design projects, it is assumed that the conceptual structure of EVA does not correspond to patterns of process designers.

These results indicated that the underlying concept of EVA is sufficiently suitable to fill a pattern with necessary declarative and procedural information

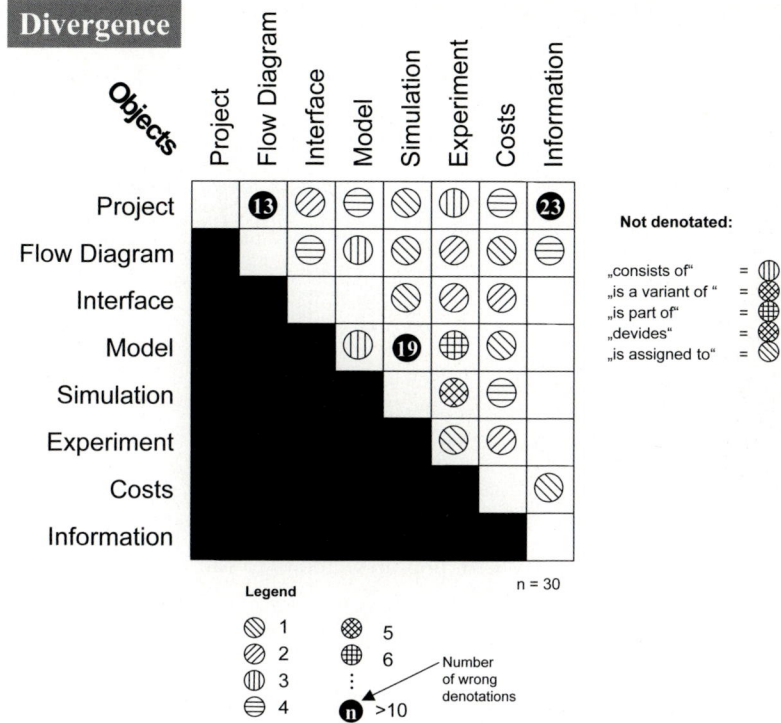

Fig. 5.49. Divergence in the test of transparency

(considering the general condition, that a participant had approximately one hour to recognize the informative structure of EVA).

Overall, the evaluation helped to find further design suggestions for EVA. Among other things, the presentation of the interface between parts of decomposition had to be adapted regarding symbolism and embedding in the concept of EVA. The graphical representation of decomposition and variation relations between documents brought such significant advantages that an extension of the concept was needed. These hints are relevant for the extended arrangement of the user interface for developers of the administration system in project B4 of IMPROVE.

5.6.4 FBW – Flowsheet Editor

The process-integrated Flowsheet Editor (in German: **F**ließ**b**ild**w**erkzeug, FBW) was developed in IMPROVE project B1 (cf. Subsect. 3.1.3). It is based on the insight that process flow diagrams play a central role in the design process [21, 470]. However, block flow diagram (BFD) – also called abstract flow

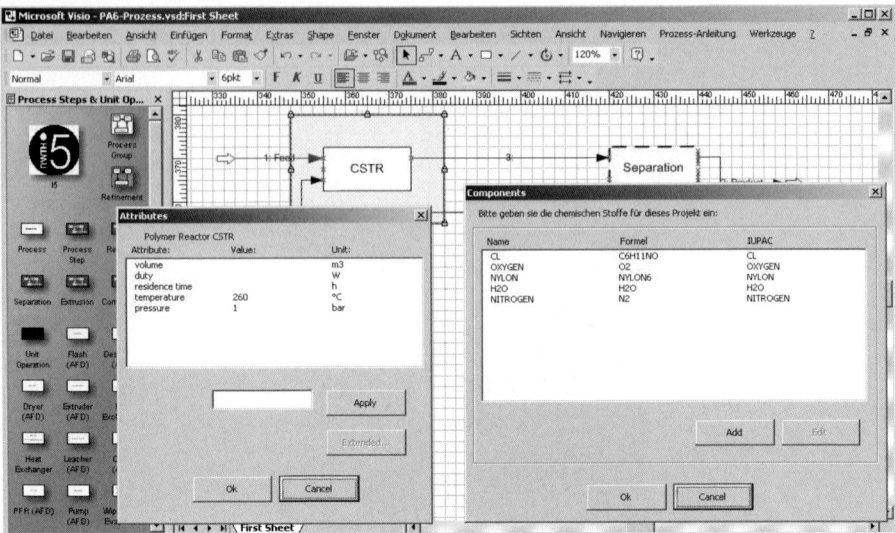

Fig. 5.50. Screenshot of FBW with pop-up windows to edit the chemicals and their material data (cf. [196])

diagram (AFD) – process flow diagram (PFD) and piping and instrumentation diagram (PID) are totally separated as far as tool usage is concerned.

But for the process engineer there is a clear cognitive connection between the different diagrams and the information about the process to be designed, such as textual information about constraints and requirements, reaction path etc. as shown in Sect. 5.6.2, see Fig. 5.42, page 534. Moreover, different variants of a flowsheet may share identical parts. So far, different variants of a flowsheet are stored in different files, i.e. diagrams, so that these identical parts cannot be recognized easily.

Therefore, the process-integrated flowsheet editor has been built, presenting different variants of a flowsheet in one file. Furthermore, a generic mechanism for the a-posteriori process integration of existing tools has been used to integrate the flowsheet editor with other domain-specific (simulation) tools (with AHEAD, cf. Subsect. 5.5.2, and with Cheops, cf. Subsect. 5.5.3), complemented by generic tools for the documentation of design rationale and visualization of traces.

To build a flowsheet the chemical process engineer can use different templates for blocks like "reaction" or "separation", see Fig. 5.50, left side. Later on he can concretize those blocks, e.g. defining the "separation" as "liquid separation". Similar to EVA (Subsect. 5.6.3), alternatives and refinements are indicated through different colors and lines [21].

The data model of FBW is shown in Fig. 5.51. This data model and the abstraction-decomposition space (Fig. 5.42) describing the work domain of a

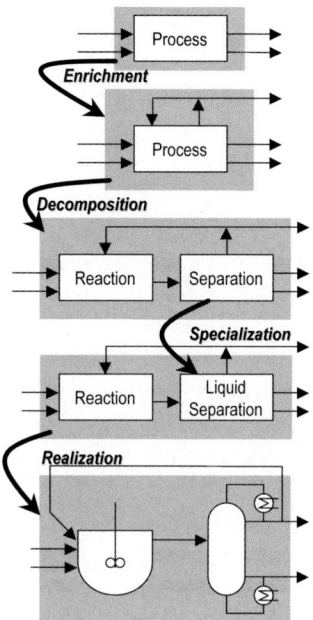

Fig. 5.51. Data model of FBW [470]

chemical process engineer have been developed in parallel. As a result, some differences exist. Nevertheless, every step in the one-dimensional data model can be mapped to a vertical or horizontal step in the two dimensional ADS.

Altogether this approach is completely different from the existing ones in chemical process design. Thus, an engineer utilizing this tool will have to work in a way different than usual. In terms of usability, this is the task-related part (cf. Sect. 5.6.2). Consequently, the FBW must be tested in a realistic scenario with experienced chemical engineers to evaluate the impact on effectiveness and efficiency on the design process. To minimize the effects of interaction-related problems, the software prototype *must* fulfill all software ergonomic requirements, particulary the principles *self-descriptive* and *controllability*. Otherwise, a developer will struggle against the user interface instead of getting an insight into the new concept of the flowsheet editor.

Therefore, the first step was an expert review [864, 964] with emphasis on the three areas:

- design and labeling of menus,
- scenario *navigation and information search*,
- scenario *creation and modification of a flowsheet*.

This analytical evaluation was performed by two experts using the Polyamide6 scenario as a basis. The results and suggestions for a better design have been documented in a technical report [115]. Some important findings are the following:[40]

- A function for "Undo" is missing (error tolerance, controllability).
- Context sensitive menu entries cannot be recognized or are indicated in different ways, e.g. grey or with {} (conformity with user expectations, self-descriptiveness).
- The indication for context sensitive entries is missing in all dialogs accessible with the right mouse button (self-descriptiveness, controllability).
- Pop-up window for "Show all streams" cannot be closed (controllability).
- Existing chemicals cannot be edited in the "components"-window, though showing a corresponding button (self-descriptiveness, controllability).
- Navigation between two variants is only possible when moving up – on the less concrete level – and then moving down again (controllability).

In the end, we decided that the FBW should not be tested with chemical process engineers in a realistic scenario until an improved user interface is available.

The second step was an analytical evaluation using the ADS and the other concepts of Cognitive Work Analysis [924, 1023]. Therewith, some additional design suggestions, e.g. direct representation of reaction path and material data to relieve the developers working memory, and constraints on usage, e.g. if the process design is done by more than one user, have been derived.

All in all, the flowsheet editor FBW is a promising new approach to chemical process design in the early stages because it offers new, meaningful functions and integrates different tools. However, without improving the interaction-related usability issues, the task-related new concept cannot be evaluated in a realistic scenario with experienced chemical process engineers.

5.6.5 AHEAD – Administration System

The AHEAD system – Adaptable and Human-Centered Environment for the Administration of Development Processes – developed in IMPROVE by project B4, supports the management of design processes in chemical engineering as well as in mechanical and software engineering.

The system provides four environments supporting different kinds of users: the work environment assists developers, the management environment supports project managers in analyzing, planning and controlling, the progress environment assists specification experts, and the modeling environment provides support for domain experts in engineering predefined task sequences and task types. AHEAD manages products, activities and resources in an integrated way. Furthermore, evolving design processes are supported by seamless interleaving of planning and execution. The process management system

---

[40] The violated software ergonomic principles are shown in brackets.

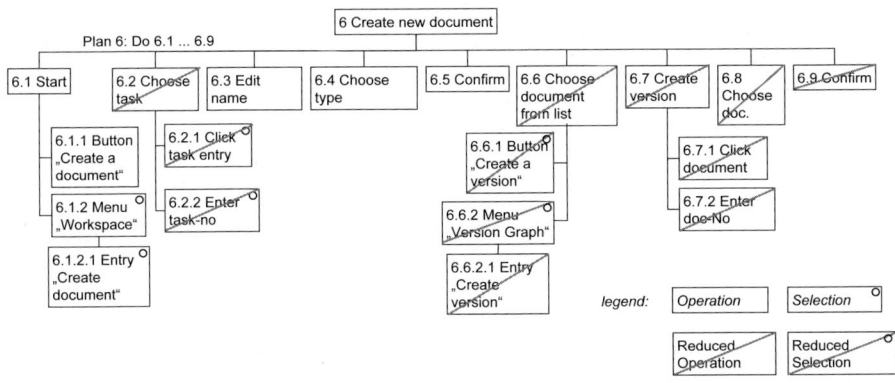

Fig. 5.52. HTA diagram visualizing some mayor differences between the original and the alternative design (according to [120])

can be adjusted to an application field such as chemical engineering by defining specific types of products, processes and resources. AHEAD is based on graph transformations and employs the wide-spread object oriented modeling language UML for acquiring process knowledge from domain experts.

The aim of the evaluation of the Linux-based administration system AHEAD was the user-centered design of the developer user interface [120], which supports both coordination and processing of design activities in process engineering. The diagram-notation of hierarchical task analysis [633, 782] was used to analyze the activities, operations and movements (cf. [522, 711]) a user has to execute for different work activities in the original user interface. In addition, a heuristic evaluation [864, 964] was performed to expose further shortcomings of the system. The analysis showed that operations could be simplified by combining several sequences of activities (cf. [774]). Additionally, the notation and arrangement of functions had to be improved to fulfill ergonomic criteria. Based on these results, an alternative user interface (cf. Fig. 5.52) was designed and implemented as a horizontal prototype [739, 864] with Borland's Delphi in MS Windows.

This interface is a mock-up with which all relevant activities can be performed. However, the mock-up cannot be connected and used with the full-functional AHEAD system. As can be seen from Fig. 5.52 some menu items were eliminated by the software-ergonomic review and redesign. The scenario, developed for evaluation and improvement of the user interface aimed to the creation of a new document. Divided into subtasks for the work analysis, this goal is achieved by successively going through nine subtasks (6.1 to 6.9), which are represented in Fig. 5.52. The decision if the creation is started by menu or button represents the initial activity (6.1.1 or 6.1.2). After that, one task must be chosen by entering the task-number or clicking on the task. This step is dispensable because of the already opened work-context and the thus preselected activity. The next three steps, namely the naming of the new document (6.3), the selection of the document format (6.4) and finally the confirmation (6.5) are required and irreducible. That is different with the following four tasks. The automated creation of an initial document version avoids unnecessary user interaction and can shorten the time consumption with concurrent error reduction (cf. [111]).

Finally, a comparative empirical study of the original and the alternative user interface of the AHEAD system was conducted. As there is a detailed description of the examination in [111, 120], the following only refers to the most important aspects.

A notebook with both operating systems, Windows 98 and SuSE Linux 7.3, was used for the analysis of both user interfaces of AHEAD. A questionnaire helped to acquire personal data of the participants. The overall duration of the examination process was recorded on video and the processing time of each participant was measured using a stopwatch.

Due to the formative character of the evaluation, only ten male participants aged between 27 and 34 years with a mean of 28.8 years were recruited for the empirical study. Since it was not possible to recruit experienced chemical engineers, people with experience in weakly structured engineering processes such as software development were chosen. They were programmers, system administrators, students and graduates of the branches of mechanical engineering, civil engineering and computer science. The professional experience averaged 2.5 years, and eight of the ten men worked for the IT-sector. Ninety percent of the participants used a computer several times a day, and all participants were familiar with at least one version of the Windows operating system. One half was familiar with one of the Unix operating systems including Linux.

The ten participants were split into two groups (EG1 and EG2) of five people for the execution of the test. The participants were selected at random for the two groups. Each participant used both interfaces whereas one half, EG1, started with the original interface and proceeded with the alternative one. The other group used the interfaces vice versa.

The analysis of the original and alternative user interfaces lasted between 75 and 120 minutes and was structured in a preliminary interview, an introduction and the examination. During the preliminary interview and the introduction, personal data of the participants were acquired and the interviewees were told about the goal as well as the process of the analysis. The interviewees also read a users' guide of the administration system AHEAD. The examination of the original and alternative user interface was executed alternately, whereas the procedure was identical. First, the interviewee had to name the objects and functions shown on a screenshot of the agenda. Afterwards, two different conditions according to the screenshot had to be produced. Again, the participant had to identify several objects, after which he had to create another condition, namely the fetching of documents. Subsequently, three additional tasks had to be solved, each within three minutes. If

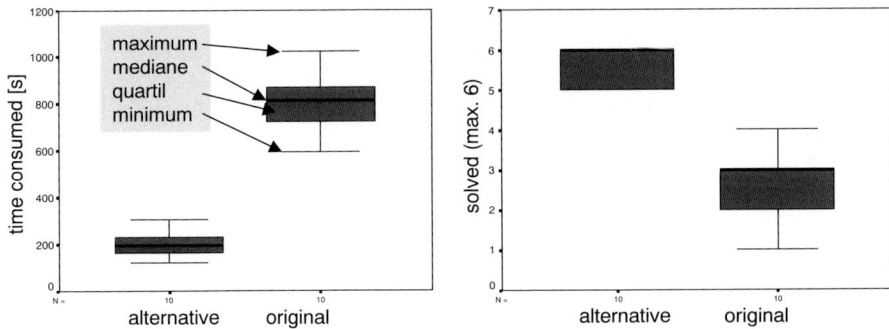

Fig. 5.53. Time consumed (solved states and tasks only) and amount of solved states and tasks using original and alternative user interfaces of AHEAD's developer environment

it took longer, the task was interrupted and the superintendent explained the procedure. Thereafter, the subjective usability estimation of each participant was captured using the IsoMetrics-questionnaire [687, 1054].

Following the analysis of the user interface, the results were statistically evaluated. In order to compare two sample averages, t-tests for dependent samples were used. Figure 5.53 exemplarily displays the box plots for the time consumed to solve states and tasks on the left side (t = -11.485; p < 0.01). On the right side the box plots for the number of solved states and solved tasks are presented (t = 8.333; p < 0.01).

There are great differences in the values for both effectiveness and efficiency measures. Even the best value for the original interface is far away from the worst value for the alternative interface.

Overall, the analysis and evaluation showed that the accomplishment of the participant using the alternative user interface differed significantly from the accomplishment using the original user interface. Moreover, it was proved experimentally that the results are independent from the order of presenting the software. Consequently, the order of presenting has no influence on the accomplishment of the participant.

Furthermore, looking at the current literature of research about team efficiency [213], the administration system AHEAD meets several requirements that are made for software supporting Concurrent Engineering [214]. It must be critically noted, however, that the Tayloristic form of division of work – the project manager delegates tasks directly to the developer – is contrary to psychological assessments of work [711, 1003] as well as to the results of research mentioned above [214]. Concerning AHEAD, though, it does not mean that the basic technical functions provided are of little use in supporting development processes. The presented functions should be divided in a different manner between "manager" and "developer", to allow a higher autonomy in choosing their work tasks.

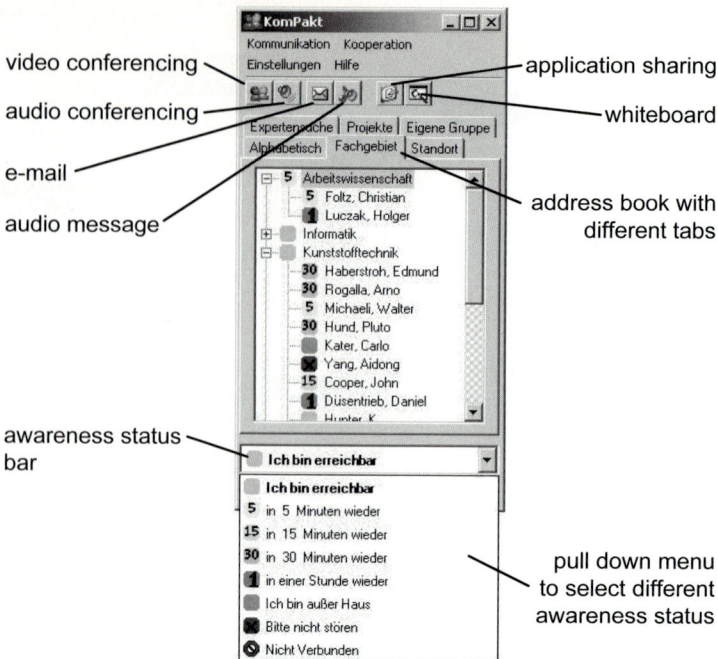

Fig. 5.54. KomPaKt User Interface [380]

5.6.6 KomPaKt – Communication Platform

The communication platform KomPaKt (Subsect. 3.3.2) was developed by IMPROVE project B3 and integrates both synchronous and asynchronous communication tools under a unique user interface. The aim is to satisfy the different cooperation and communication needs of a developer. Therefore, it allows access to the user information saved in the administration system AHEAD.

Synchronous communication comprises services such as audio-visual real-time data exchange, which can be used for conferences between locally distributed design team members. Asynchronous tools such as email or audio message allow communication if the team members are not available simultaneously.

The interface concept of KomPaKt is based on results of awareness research (e.g. [483, 703, 709]), knowledge management (e.g. [309, 866]), case studies in cooperation and communication research (e.g. [214, 715, 833]) and experience in the field of tele-cooperatively supported automotive development [121, 282].

Figure 5.54 presents the user interface and gives some basic information about KomPaKt's functionalities. A detailed description of the design process and the evaluation of the user interface is given in [380].

The comparative empirical evaluation of KomPaKt was executed using a conventional desktop system with the operating system Windows 2000. The entire experiment was recorded on video, and the processing times of the participants were noted using a stopwatch. The 18 participants in the study consisted of nine novices and nine computer experts using the same interface. Comparing the performance of both of these groups should ensure that the results of the evaluation can be generalized to many potential users.

The novices were all female students of social sciences aged 19 to 26 years. Three of the nine novices used a computer daily, the others weekly or monthly. The novices' average use of the operating system Windows was 3.8 years. Seven out of nine were not familiar with any instant messaging and presence awareness programs such as ICQ, AOL Instant Messenger, MSN or Lotus Sametime; only three had used ICQ or MSN before, but there was no permanent lasting usage.

The experts, five men and four women, were between the ages of 23 and 34 and studied or graduated in computer science, civil engineering and mechanical engineering. All of them worked at the computer daily and used the communication tools mentioned above. The experts had worked with Windows and Unix operating systems for 7.5 years on average.

The laboratory study of the communication system KomPaKt was guided by a questionnaire and divided into three parts: a preliminary interview, an introduction and a test.

In the preliminary interview personal data such as gender, age, computer experience etc. were gathered. Afterwards, information of the aims and functionalities of KomPaKt were given. If necessary, the interviewer answered questions from the interviewee before finishing the introduction.

The test started with the presentation of a screenshot of KomPaKt, similar to the one in Fig. 5.54. Then, the participant was asked to name the objects shown, such as buttons for different communication and cooperation services, to test the self-descriptiveness. Second, three screenshots showing different states of the software were shown. The interviewee briefly described their perception and then tried to reproduce these states on his/her own. Third, the participant had to solve three simple tasks, e.g. "Please connect to the network". Here, the time consumed was recorded, whereas in the first and second part of the test just a right or wrong was noticed. Finally, the IsoMetrics-questionnaire [687, 1054] was used to capture the subjective usability estimation of each participant.

The difference between the objects denoted correctly by novices (17 objects) and experts (19 objects) is not statistically significant. In addition, both novices and experts solved the same quantity of tasks (5.8). In contrast, the time consumed to solve tasks differs significantly ($p < 0.046$) between novices (326.3 s) and experts (205.7 s). This difference can be ascribed to the knowledge and routine of computer experts, whereas the novices needed time to orientate within the software.

5.6.7 Summary

The software tools developed to improve chemical design processes started from different origins and attained a different stage of maturity. Therefore, specific methods for both design and evaluation were required.

In this section, several *suitable techniques for usability engineering* have been presented and selected according to the development status of the particular software.

With the help of scenarios, *innovative prototypes of user interfaces* [774, 921, 964] have been designed. Suggestions for an improved design were compiled using analytical and experimental methods.

The usability results show that the developed prototypes *offer adequate methods to support work processes in chemical process design*. However, for an application in an industrial context, further research and development activities are necessary.

5.7 Software Integration and Framework Development

Th. Haase, P. Klein, and M. Nagl

Abstract. The a-posteriori integration of heterogeneous engineering tools, where tools are supplied by different vendors, constitutes a challenging task. In particular, this applies to an integration approach where existing engineering tools are extended by new functionality which, again, can be integrated synergistically. Responding to these challenges, an approach to tool integration is described which puts strong emphasis on software architecture and model-driven development.

Starting from an abstract description for an integration software architecture, this architecture is gradually refined down to the implementation level. To integrate heterogeneous engineering tools, wrappers are constructed, abstracting from technical details and providing homogenized data access. This approach to tool integration is supported by a collection of tools for software architecture design and model-driven wrapper development, all based on formal graph models and transformation rules. This collection of tools considerably leverages the problem of composing a tightly integrated development environment from a set of heterogeneous engineering tools. So, we give specific architecture tools for the problem of tool integration following an a-posteriori approach.

5.7.1 Introduction: Tool Integration

Concerning tool support for development processes in engineering disciplines [342, 352, 353], the typical situation can be described as follows (cf. Sect. 1.1): *Various tools* are available, each of which supports a certain and *specific part* of the development process. In chemical engineering for example, there is a tool to compile the design of a chemical process by means of flowsheets and data sheets and another tool for the simulation of the chemical process. However, the *overall design process*, i.e. the dependencies and consistencies between single activities and their resulting products, use of experience or direct communication between members in the design team, or the integration of technical and management activities are *not considered* by the existing tools.

Moreover, as tools are provided by different vendors, they are based on different system platforms, proprietary document formats, and conceptual models of the application domain. So, there is a heterogeneous landscape of *existing tools*. These tools constitute a *proven solution* for carrying out a certain activity. Engineers are familiar with these tools; their use can be best practice in the application domain. Maintenance as well as further development is guaranteed due to established vendors, wide deployment, and an actively pursued dialogue between vendors and their clients. For there reasons, it is economically not feasible [973] to replace these tools by newly built ones.

Preparation for A-posteriori Integration

To fulfill the requirements of comprehensive support but using existing tools we favor an *a-posteriori* or bottom-up integration approach [135] in order to improve the computer-based support for design processes in chemical engineering [225, 343–345]. The term "a-posteriori" refers to the fact that existing systems have to be *prepared* afterwards according to the needs of integration. Here are some *system-technical aspects*, e.g.

- the possibility of sharing and exchanging common data or documents among tools,
- the ability to activate tools or parts of the tools' functionality under the control of another tool,
- the notification about certain events and the invocation of corresponding actions.

Furthermore, these preparations have to be realized across distributed and heterogeneous operating system platforms. Further issues of preparing a tool for integration deal with *reengineering* tasks on the level of a *tool's functionality* [31, 210, 295]. Some parts of the given functionality of an existing tool may contain hard-wired subprocesses affecting different design domains. This could be a mix of engineering activities, e.g. a tool for process flowsheet design as well as for process simulation, but also an intersection of engineering and administrative activities, e.g. a tool with a built-in version control. These functions have to be separated in order to have a clear separation of concerns [346]. This makes it possible to substitute functionality by a more general and suitable tool.

Wrapping

The reorganization of a tool's internal structure requires access to of the tool's source code. In the case of commercial tools this code is usually not available. Therefore, an internal reengineering of the tool is not possible. The existing tool can only be encapsulated by a software component called *wrapper* [972] in order *to adapt* it with respect to the needs of integration.

Wrappers have several facets [136]: One is (a) to abstract from *technical details* like the programming languages used for implementing the tools' programming interfaces, or the middleware used for interprocess communication. Another task of a wrapper is (b) to offer *a view* on the data of a tool. Thereby, (b1) views of one tool have to be "semantically homogeneous". Analogously, (b2) views of different tools should be homogeneous. This way, the proprietary data structures are hidden and the building of further data models on top is facilitated. The third purpose of a wrapper is (c) to accomplish a clear *functional interface* to existing tools. Depending on the tool, it can be (c1) a coarse-grained interface to start/stop the tool or to load a document into the tool (black-box view). Alternatively, (c2) it can be a fine-grained interface to invoke tool specific commands (white-box view). Again, the functionality has to be homogeneous with regard to the granularity of the commands. This is achieved by composing low-level commands to a new and higher one. Finally, (d) the wrapper has to *prevent access* to undesirable commands in the sense of the above discussion about functionality reengineering.

New Integration Functionality and Synergistic Cooperation

So far, *wrapping* addresses the classical, well-known integration dimensions found in literature [950, 1038], namely (1) data integration, (2) function integration, (3) control integration, (4) platform integration, and (5) user interface integration (not discussed above). These dimensions only cover a limited range of integration problems and mainly focus on the *technical perspective* by linking existing tools together.

A tool integration solution that delivers a real *added value* for the engineer and bridges the gaps sketched in Sect. 1.1 of this book is not achieved by simple data exchange, batch-wise command chains, or a uniform user interface. The *tools* have to be *extended* by additional integration functionality as described in Sect. 1.1.

To shortly summarize the *concepts* for *tool extension*[41]:

- *Fine-grained process-integration* support engineers by process fragments, based on the engineers' experience, which partly automate execution of command sequences (Sect. 3.1).
- *Incremental integration tools* assists in keeping interdependent documents consistent with each other (Sect. 3.2).
- *Multi-media communication tools* give engineers at different locations the opportunity to discuss and resolve design problems (Sect. 3.3).
- *Reactive administration tools* allow for integrated and dynamic management of products, activities, and resources of a development process (Sect. 3.4).

The extended tools may again be integrated to cooperate *synergistically* [348]. Scenarios [17] for the synergistic cooperation were presented in Sects. 1.2 and 5.5, e.g. the integration tool between the process flowsheet editor Comos PT and the management system AHEAD. This example will be picked up again below to illustrate how architecture modeling and transformation as well as wrappers can be used for tool integration.

Integrated Engineering Development Environment

Existing engineering tools together with their extensions and synergistic cooperation form an *integrated engineering tool environment* to significant improve the computer-based support of design processes [295–299, 343–346].

---

[41] In the sequel, the term "extended tool" refers to a tool with an extension which realizes one part of the new integration functionality.

Figure 5.55 again illustrates such an engineering design environment on a *coarse-grained architectural level*. The bottom layer of Fig. 5.55 consists of independent existing engineering tools: A chemical process flowsheet editor (Comos PT [745]), three chemical process simulation tools (Aspen Plus [516], CHEOPS [409], and MOREX [147]), and a visualization tool for 3-D-simulation results (BEMView [197]). Based on a common platform infrastructure (cf. Chap. 4) these tools are integrated with the extensions shown at the top layer of Fig. 5.55: Fine-grained process-integration (PRIME, cf. Sect. 3.1), integration tools (cf. Sect. 3.2), multi-media communication tools (KomPaKt, cf. Sect. 3.3), and reactive administration tools (AHEAD, cf. Sect. 3.4).

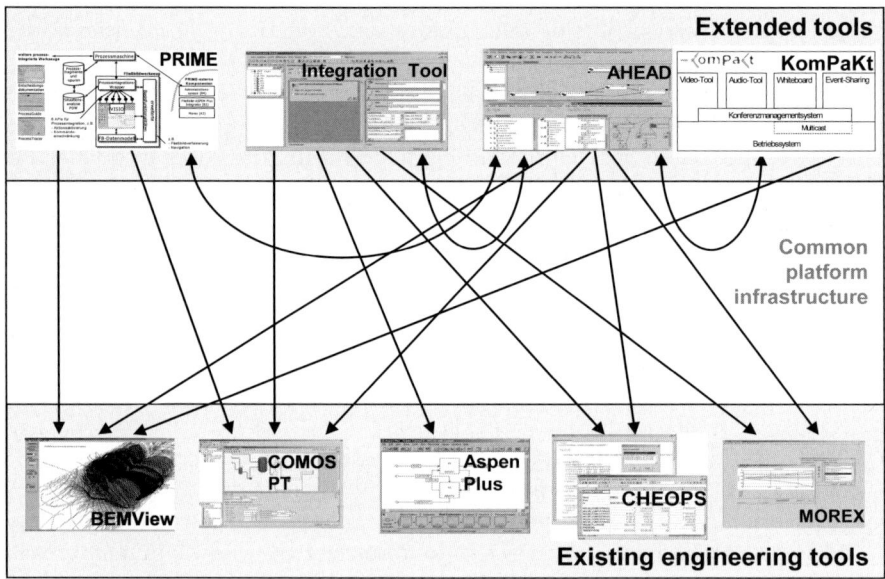

Fig. 5.55. Integrated engineering development environment

An arrow between two tools A and B denotes a «Uses» relationship indicating that tool A uses data and/or functionality from tool B. A horizontal relationship refers to a synergistic cooperation between two extended tools that can be handled in the (accessible) code of the involved extended tools, whereas a vertical relationship between an extended and an existing tool implies the use of an external wrapper for an a-posteriori integration of the existing tool.[42]

---

[42] In principle, a horizontal relationship can also be realized by a wrapper. In some cases wrapping requires less effort than modifying the code of historically grown software systems.

Software Development Process within IMPROVE

Building an integrated tool environment constitutes a large (software) development process. To ensure that such a process is carried out in a coordinated and systematic way, *integration* has to be regarded on the *architectural level*. This is done by defining the essential components and subsystems necessary for performing the integration.

The construction of a precise and formal architectural description facilitates the detection of general system parts to build up a framework of reusable software components [43, 44], e.g. for tool wrapping (*software product reuse*). Also, *software process reuse* is fostered in the sense that specific components can be generated automatically from specifications.

Realizing integration on an architectural level means to *describe* the "*glueing parts*" necessary for performing the integration. The architecture of the integrated overall environment defines its modules regarding the kinds of interfaces the tools to be integrated have to offer, the interfaces to be wrapped in order to be homogeneous, the tools and wrappers to be distributed, the interfaces to be accessed via certain middleware techniques and so on. It does not consider the internal structure of the tools to be integrated. The architecture also describes the functionality to be realized by extending parts of the overall environment, in particular, how tool extensions are built and how synergistic cooperation is achieved.

Following these ideas, the tool integration process, consequently, has to put strong emphasis on *software architecture* and *model-driven development*. This should not only be regarded on the conceptual level. Rather, a corresponding collection of supporting tools facilitating integration has to be realized. The following subsections sketch this architecture-centered and model-driven development approach to tool integration[43] [26]:

1. *Concepts for architecture modeling* (Subsect. 5.7.2). Firstly, in order to define a suitable vocabulary for architecture modeling an adequate architecture description language is required. Such a language [331] defines the basic modeling idioms, like modules, subsystems, components, interfaces, and relationships, from which an architecture is built.

   Furthermore, various perspectives of an architecture have to be distinguished [224] (*logical/concrete* or *static/dynamic* perspective). They are described by corresponding subsets of the modeling language. To avoid a loose collection of graphical notations for modeling such views, an interconnection semantics of the different views has to be defined. Altogether, these modeling elements and concepts form the *conceptual frame* for the architecture design process.

---

[43] The following tool integration approach is obviously not restricted to a certain application domain. The discussion of this aspect will be taken up in Sect. 7.8.

2. *Pattern-based architecture modeling and refinement* (Subsect. 5.7.3). The software architecture of the integrated design environment is modeled initially on a high level of abstraction (see, for example, Fig. 5.55). The *initial* architecture is then gradually refined by means of architectural transformations which take care of technical details and introduce technical components such as tool wrappers required to achieve integration. The transformation process results in a *concrete* architecture consisting of all the components which have to be implemented (either manually or automatically). The architecture refinement process is not performed in an ad-hoc manner. Rather, it is controlled by domain-specific knowledge about a-posteriori software integration. This knowledge is expressed by *architecture transformation patterns*.
3. *Model-driven wrapper development* (Subsect. 5.7.4). In case of a-posteriori integration, tools supplied by different vendors using different data management systems etc. have to be handled. To make use of these tools *wrappers* have to be provided to render well-defined tool interfaces which are suitable for integration. Wrapper development can be decomposed into two levels. *Technical wrappers* are responsible for hiding technical details of the interfaces provided by the tools. In contrast, *homogenizing wrappers* located on top of technical wrappers realize the necessary data and functional abstraction. To *reduce* the development effort for building a wrapper, a visual *model* of the wrapper is specified. Based on this model, the executable code for the wrapper is *generated*.

5.7.2 Conceptual Framework for Architecture Modeling

The aim of this subsection is to provide a *conceptual framework* for architecture modeling which establishes the basis for the following subsections. This includes the definition of the underlying modeling elements and principles from which a system's architecture is built, as well as the distinction of different perspectives of an architecture [224, 331]. The latter constitute the main focus as they serve as a guideline for the architecture modeling process. Figure 5.56 gives an overview on the conceptual framework.

Basic Architectural Terms

The following definitions focus on the contents (not the purpose) of an architecture specification.

Interface: A collection of exported resources like operations, types, constants etc. constitute an interface. Basically, the term is used as in programming languages providing a module construct [336]. Frequently, some clients of a module need more transparency on the realization of the interface than others. This situation is covered by introducing a separation between *public*, *protected*, and *private* resources.

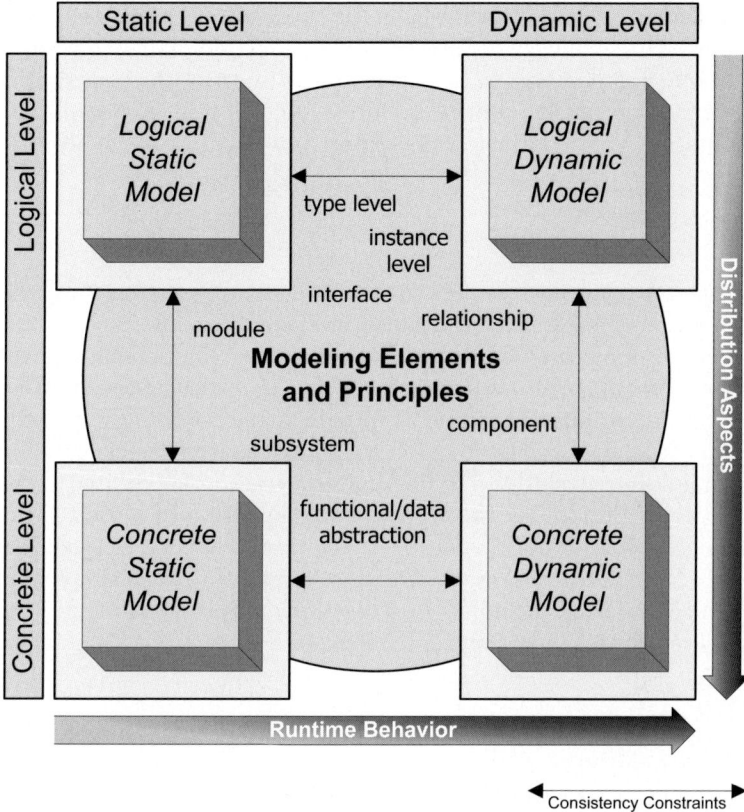

Fig. 5.56. Conceptual framework for architecture modeling

Module: A module is a logical unit of a software system with a clearly defined purpose in a given context. It consists of an export interface defining which resources (data and/or operations) the module offers to the rest of the system, an import interface defining which resources from other modules the module may use to realize its export interface, and an implementation in some programming language. The internals of the module (the implementation or body) are encapsulated. Therefore, a module can be viewed as an *abstraction*: the interface provides access to "abstract" resources; the module abstracts from the realization of these resources. The term "interface" of a module, without further qualification, refers to the export interface. Furthermore, a module is the atomic architectural unit of reuse. Modules can be used in a context or system different from the one in which they were developed.

Subsystem: A subsystem constitutes a collection of components (see below). Subsystems have interfaces, like modules. The import interface of a subsystem is the union of all import interfaces of its internal components, minus the resources defined by components within the subsystem. The export interface of a subsystem is an explicitly defined subset of the union of all export interfaces of the internal components. In all other aspects, most of the characterizations given for modules can be applied to subsystems as well. Especially, internal components not contributing to the export interface are hidden outside the subsystem.

Component: A component is either a module or a subsystem.

Relationship: A relationship refers to the dependency between components in the sense that some resource contained in the export interface of one component (the resource provider) is usable by another component (the resource client) where it appear in its import interface. Considering the static view of an architecture, an «Uses» relationship indicates a *potential* use of some resource. An actual use is to be found in the source code of the client's body. On the dynamic level, a relationship designates a use to be executed at runtime of the program. Besides the static/dynamic distinction of a use relationship, certain *structural* relationships between modules also exist: (1) *local containment* and (2) *specialization/generalization*. Usability relationships are found in three specific forms including *local usability* within local structures, *general usability* offering layering, and *inheritance usability* within inheritance structures.

Architecture: An Architecture refers to the structural layout for a software system. It defines all of the system's components and their relationships by means of their import and export interfaces, but not by their implementations. The term includes different facets to be explained below.

Any module provides an abstraction of how the interface resources are implemented in a programming language. In particular, *two* dimensions of (module) *abstractions* can be distinguished:

Functional vs. data abstraction: Functional abstraction refers to the case where a module has some kind of transformation/coordination character. Hence, an interface resource transforms some kind of input data into corresponding output data, or the component coordinates the resources of lower components. Functional abstraction facilitates the hiding of algorithmic details of this transformation/coordination. In contrast, data abstraction is present if the module encapsulates the access to some kind of "memory" or "state". Then, the module hides the realization of the data representation. The module's interface only shows how the data can be used, not how it is mapped onto the underlying storage.

Type vs. instance abstraction: This distinction stems from the necessity to distinguish those modules encapsulating a single state or a concrete control flow as well as those offering a template/type to dynamically create a state or control flow at runtime.

Architectural Views

One of the basic ideas is the distinction of *two directions* with respect to what is modeled and how it is modeled (cf. Fig. 5.56). This results in different views concerning on one hand the static/dynamic (horizontal) and on the other hand the logical/concrete properties (vertical) dimension.

In the top-left corner of Fig. 5.56, a system's *static* structure is defined with its components, their (import and export) interfaces and their relationships. To describe the *dynamic* behavior of the system one or more interaction or collaboration diagrams (top-right corner of Fig. 5.56) may be used for example. Both specifications are restricted to the *logical level*, i.e. they strictly adhere to the concepts of modularity and encapsulation.

However, there are many reasons why an architecture cannot be implemented exactly the way it is specified on the logical level. One of them is due to the desire to specify further *information* in addition to the logical structure. Some examples are

- the annotation of *concurrency* properties of components to distinguish components which comprise a process for example;
- the introduction of components to handle *distribution*, e.g. for parameter marshaling, finding a service provider etc.;
- the *extension* or *adaption* of the architecture in order to integrate components with a different architectural structure in case components of external libraries or of external tools are used;
- the specification of the *implementation* of *usability* relationships, e.g. via (remote) procedures calls (RPC), exceptions, interrupts, event-triggering, or other forms of callback mechanisms.

All these activities require modifications of the logical architecture of a system if interfaces change, new components are introduced, or implementation details are added. The resulting architecture, therefore, has a different quality than the logical architecture: It does not aim at the best abstract structure with respect to maintainability etc.; rather, it describes the concrete implementation of a system. Therefore, it is called a *concrete architecture*.

It should be noted that *different concrete architectures* for one logical architecture exist in general. These may reflect a sequence of possibly different interdependent decisions of the above list, or different realization variants. For example, one concrete architecture may be equipped with a RPC and another one with a CORBA (Common Object Request Broker Architecture [877]) implementation of interprocess communication.

In the context of *reverse-* and *reengineering*, an existing system is first analyzed [81–83, 88, 179, 286, 287]. Some concrete architecture, derived from the source code or other documentation, describes the actual situation. Then, the logical architecture can be distilled from this concrete architecture which, in turn, will probably form the basis for the restructuring of the system and the respective new concrete architecture. To some extent, such an iterative reverse- or reengineering process is presented in Subsect. 5.7.4 by constructing a wrapper for a given tool.

Not only the different logical and concrete architectures, but also the *transformations* leading from one architecture to another contain important *design knowledge*. Both, the original design decisions as well as the how and why of later modifications are necessary to understand a system's structure and to facilitate reuse. An explicit transformation step, for example, offers a convenient place to document differences between the logical and the concrete levels.

Furthermore, if a specific transformation occurs frequently, it can be formalized and tool support can be provided for its application [45, 85–87, 89]. For example, given the knowledge on how the architecture changes if some relationship between two modules is implemented as a remote method call using CORBA (i.e. which components are added, how existing components are modified [374, 375]), a tool can provide a command to apply this transformation. In this sense, the design knowledge of how to *modify* an architecture to *meet* some *purpose* can be specified, communicated, reused, and *supported* by *tools*. This is the topic of the following subsection.

5.7.3 Pattern-Based Architecture Modeling and Refinement

The *term software architecture* is defined as a description of "the structure of the components of a program/system (and) their interrelationships" [331, 686]. This description serves different purposes, e.g. for *analyzing* certain software *qualities*, such as adaptability, maintainability, or portability, or managing the software development process [536].

This simple definition disregards, that *more than one structural perspective* together *with* the corresponding *dependencies* will be necessary. Structural perspectives, for example, include a conceptual, a development, and a process view [603]. Therefore, "high-level" diagrams such as Fig. 5.55 are helpful to get a first impression of the overall system's structure, but are not an adequate description of a software system to serve as a blueprint for building the system.

The intended *architecture transformation tool* – described in the following – guides the software engineer in gradually refining an abstract (logical) architecture to ultimately result in a corresponding concrete one. In contrast to other architecture design tools, we aim to offer *specific support* for software architecture design. "Specific" has two aspects: for one, it refers to the specific *domain*, here the construction of integration solutions as a special field of systems' programming, and two, the tool is devoted to a specific *task*, namely the a-posteriori integration of given applications.

The first step of the integration process as described in the first subsection, namely pattern-based architecture modeling and refinement, is concerned with *refining* the *relationships between* the (extended and existing) applications, i.e. the *tools* of the design environment depicted in Fig. 5.55. Refining the architecture focuses on just these integration aspects. The internal architecture of the tools to be integrated will not be considered in detail. Rather, the *"glueing parts"* needed for performing the integration are investigated [137].

In fact, it turns out that the refinement results in a set of fairly sophisticated subsystems which are designed systematically by applying *architectural transformations*. These transformations represent *patterns* [580, 674, 682] expressing the domain-specific knowledge about a-posteriori integration, in particular, the possible alternatives for wrapping a tool.

According to the distinction of diverse architectural views (cf. last subsection), the *refinement* process *distinguishes* between a *logical* architecture abstracting from technical details and a *concrete* architecture which realizes the logical architecture. Starting from a high-level simple architecture, wrappers are introduced and decomposed to result in a refined logical architecture (Fig. 5.57). Subsequently, this logical architecture is further refined into a concrete architecture which eventually takes care of all the details of the underlying technical infrastructure (Fig. 5.58).

Gradual Refinement of Architecture

This subsection demonstrates how the *coarse-grained* "architecture" of the *design environment* (cf. Fig. 5.55) is gradually *refined* towards a detailed architecture description, considering aspects like (i) decomposing components, (ii) introducing wrappers, and (iii) distributing components via certain middleware techniques.

As an *example* we look at the integration tool between the process flowsheet editor Comos PT and the management system AHEAD (cf. Sect. 5.5). The *functionality* of this tool is shortly summarized as follows. It supports the chief designer and the project manager to analyze the impact of changes in the process flowsheet (made by the chief designer) with respect to corresponding changes in the task net (maintained by the project manager). For this purpose, the tool reads the modified parts of the process flowsheet, determines the affected document revisions controlled by the product configuration component of AHEAD and suggests to the project manager, which tasks that work on these revisions have to be (re-)activated. The project manager may accept this suggestion and the integration tool changes the task net accordingly.

Refinement of the Logical Architecture

The system description sketched in Fig. 5.55 serves as a starting point for the architecture refinement process. The coarse logical structure for the example mentioned above is depicted in box 1 of Fig. 5.57[44].

As the *first* refinement *step* the *access* to the *application* to be integrated by the `Integrator Tool` is defined. This can be done either by accessing the

---

[44] The explanations focus on the left «Integrates» relation between the `Integrator Tool` and `COMOS PT`. The right «Integrates» relation can be refined analogously.

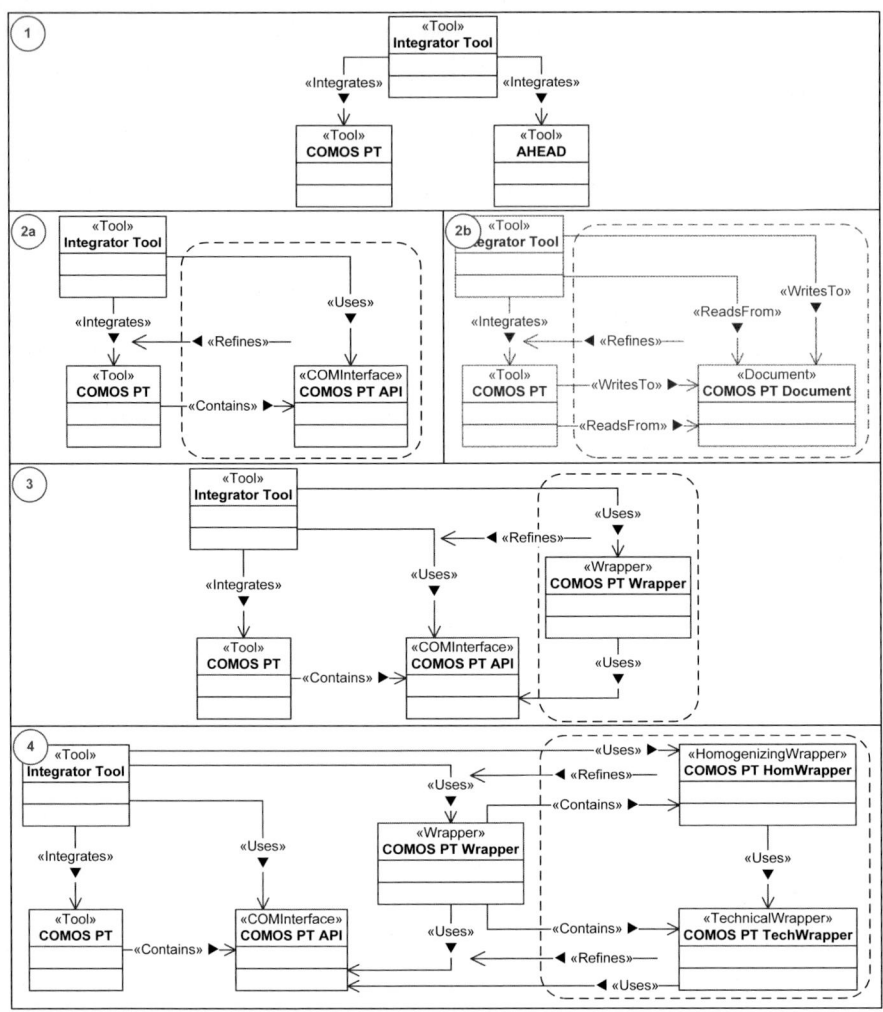

Fig. 5.57. Refinement of logical architecture

application via an API (application programming interface) (cf. box 2a of Fig. 5.57) or, in case no API is offered by the application, via the documents produced by the application (cf. box 2b of Fig. 5.57). The latter refinement alternative is applicable if the application is equipped with an XML import and export function, and if only data integration is intended. Mixtures of alternatives 2a and 2b are possible as well (not shown in Fig. 5.57): If the API, for example, is a read-only interface, the read access is realized via the API, while for the write access the document solution is used.

Choosing alternative 2a leads to the model shown in box 2a of Fig. 5.57: The «Tool» COMOS PT is extended with an additional «COMInterface» COMOS PT API representing the API that is used by the Integrator Tool.

This «Uses» relation between the Integrator Tool and the COMOS PT API is subsequently refined in following two steps: A «Wrapper» COMOS PT Wrapper is introduced (cf. box 3 of Fig. 5.57) which is subdivided into a so-called *homogenizing wrapper* (COMOS PT HomWrapper) and a *technical wrapper* (COMOS PT TechWrapper) (cf. box 4 of Fig. 5.57). This is done for the following reasons. The proprietary data model provided by the tool's API has to be transformed by the homogenizing wrapper into a data model expected by the Integrator Tool. In this context, the technical wrapper offers the homogenizing wrapper a location- and implementation-independent access to the tool's API. How the homogenizing and the technical wrapper can be further refined, will be explained in Subsect. 5.7.4.

The refinement steps shown in box 2a (or alternatively in box 2b) require *user interactions*. It is the software engineer's knowledge to decide how the «Tool» COMOS PT is accessed by the Integrator Tool and, in case of an API, which technique is used to realize the API. After determining this interactively, the transformations shown in box 3 and box 4 can be performed by an appropriate architecture modeling tool automatically. When, for example, the software engineer decides later that no homogenizing wrapper is necessary, he can delete this component manually.

Refinement of the Concrete Architecture

So far the logical architecture of the system is specified. The next step is to define the concrete architecture. Therefore, the logical structure is transformed into a concrete one (cf. box 5 of Fig. 5.58): Instances of the components «Tool» and «Wrapper» are *transformed* into instances of component «Process». They represent a process in the sense of a operating system. While the «Contains» relations are kept, the «Uses» *relations* are transformed into equivalent «MethodInvocation» or «InterprocessCall» relations.

A «MethodInvocation» represents a *local communication*, while an «InterprocessCall» represents a *distributed* one. Therefore, a «MethodInvocation» relation is only feasible between components that are contained within the same «Process», whereas an «InterprocessCall» relation is only allowed accordingly between components of different processes. An architecture modeling tool can again carry out these transformations automatically.

Specifying how the «InterprocessCall» *relations* will be *implemented* are the final steps of architecture refinement. As the COMOS PT API is implemented by the tool's vendor using COM (Component Object Model [846]), the «InterprocessCall» relation between the COMOS PT TechWrapper and the COMOS PT API is simply refined into a «COMCall»(cf. box 5 of Fig. 5.58). In case of the «InterprocessCall» relation between the Integrator Tool and the COMOS PT HomWrapper, different *alternatives* are possible.

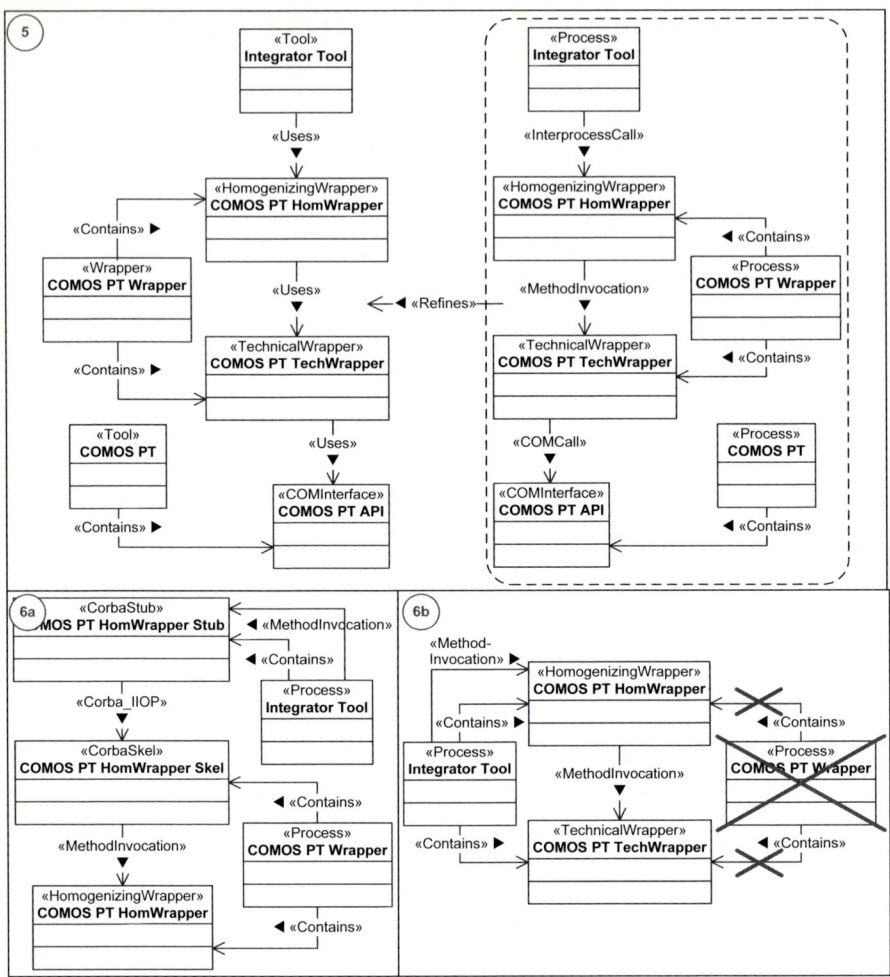

Fig. 5.58. Refinement of concrete architecture

Realizing the `COMOS PT Wrapper` as an *independent* operating system *process* is one alternative.[45] The interprocess communication between the `Integrator Tool` and the `COMOS PT HomWrapper` can then be implemented e.g. *using CORBA*. This alternative offers the opportunity to distribute the `Integrator Tool` and the tool to be integrated (`COMOS PT`) over various nodes in a computing network. If the software engineer decides so, the architecture is refined as shown in box 6a of Fig. 5.58: The `Integrator Tool` and the `COMOS PT Wrapper` are extended by a corresponding stub and skeleton («CorbaStub» and «CorbaSkel») and a «Corba_IIOP» (Internet Inter-ORB Pro-

---

[45] This alternative was already suggested by the initial transformation of the logical into the concrete architecture.

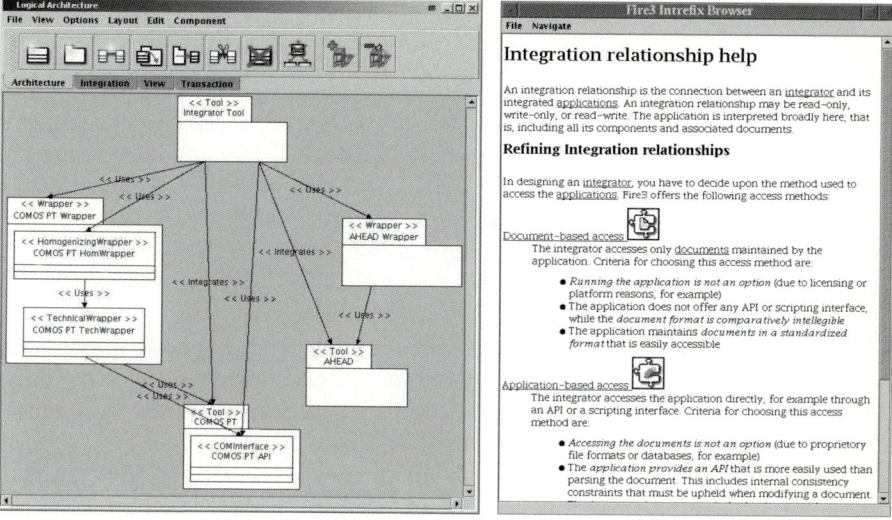

Fig. 5.59. Sample screenshots of `Fire3`: logical architecture (left) and help texts (right)

tocol) relation is established between them. Furthermore, a «MethodInvocation» relation is established between the `Integrator Tool` and the `COMOS PT HomWrapper Stub` and between the `COMOS PT HomWrapper Skel` and the `COMOS PT HomWrapper`, respectively.

If *no distributed solution* is desired, the independent «Process» `COMOS PT Wrapper` is resolved, i.e. the `COMOS PT Wrapper` is deleted, the `COMOS PT HomWrapper` and the `COMOS PT TechWrapper` are realized as *local components* of the `Integrator Tool`, and the interprocess communication between the `Integrator Tool` and the `COMOS PT HomWrapper` is substituted by a «MethodInvocation» relation as well (cf. box 6b of Fig. 5.58).

Tool Support for Architecture Modeling and Refinement

The creation of an architecture design tool is not new. We have studied *architecture design* languages [334] and accompanying *tools* for some years [73, 224, 260]. Most of these tools claim to be usable for any context of software development. The operations offered to the software engineer are thus common to all domains and do not give specific support.

With the "Friendly Integration Refinement Environment", abbreviated as `Fire3`, an architecture design tool was developed [141, 142] which is *specific* to the task of *a-posteriori application integration*. In particular, the tool is specific for dealing with the transformation from logical architectures to concrete architectures. Figure 5.59 gives an impression of the tool: The left screenshot of the user interface shows the integration scenario displayed in Fig. 5.57.

Knowledge about *integration architectures* is captured in `Fire3` in multiple ways:

- *Transformations*: Changes in the architecture often affect multiple and different parts simultaneously. When, for example, a single component of an integrated application is addressed, a particular wrapper is needed.
- *Stereotypes*: The tool uses a variant of UML (Unified Modeling Language [560, 880]) to display its architecture diagrams. As the tool further classifies types of classes, packages, and components, UML-stereotypes were introduced. They convey this more complex model to the software engineer immediately. All diagrams contain such stereotypes.
- *Analyzes*: Using particular types of components from a specific application domain, the use and arrangement of these components is restricted. While specific transformations prohibit erroneous conditions in the first place, analyzes point to problems such as incomplete specifications.
- *Help texts*: To inform the software engineer about the options he may choose from and to give him arguments for his decisions, help texts offer the declarative information he needs. They help to alleviate identified problems and facilitate direct activation of the necessary repair actions.
- *Illustration*: As the resulting architecture can get very complex even for a simple integration scenario, typical uses of the architecture can be illustrated by means of animated collaboration diagrams to present the interaction of the various components.

`Fire3` *covers* the multiple *refinement steps* of architecture design as discussed in the example above. For each of the refinement steps, there exist specific operations to support the software engineer in defining the results on that stage or in refining them to get to the next step. Additional help texts (cf. Fig. 5.59) explain the usage of the refinement operations. Different views allow the software engineer to focus on the context specific to the refinement step he is working with.

Implementation of Tool Support

We start with a *summary* of the *implementation* of the architecture tool. The internal application logic of `Fire3`, i.e. the data model and the corresponding operations to manipulate the data model, are *formally specified* by a programmed *graph rewriting system* using the PROGRES language and environment [412, 414]. The code is generated from this specification and the code is put into a framework for visual tools. Hence, a rather elaborated tool construction process is put in place.

In the following, some basic *aspects* of the tool's underlying *graph schema*, exemplarily shown in Fig. 5.60[46], will be *discussed*. This graph schema, which

---

[46] The figure uses an UML-like notation. The stereotype «`NodeClass`» indicates an *abstract class*, thus no instances are allowed, whereas a node tagged with the stereotype «`NodeType`» refers to an instantiable one.

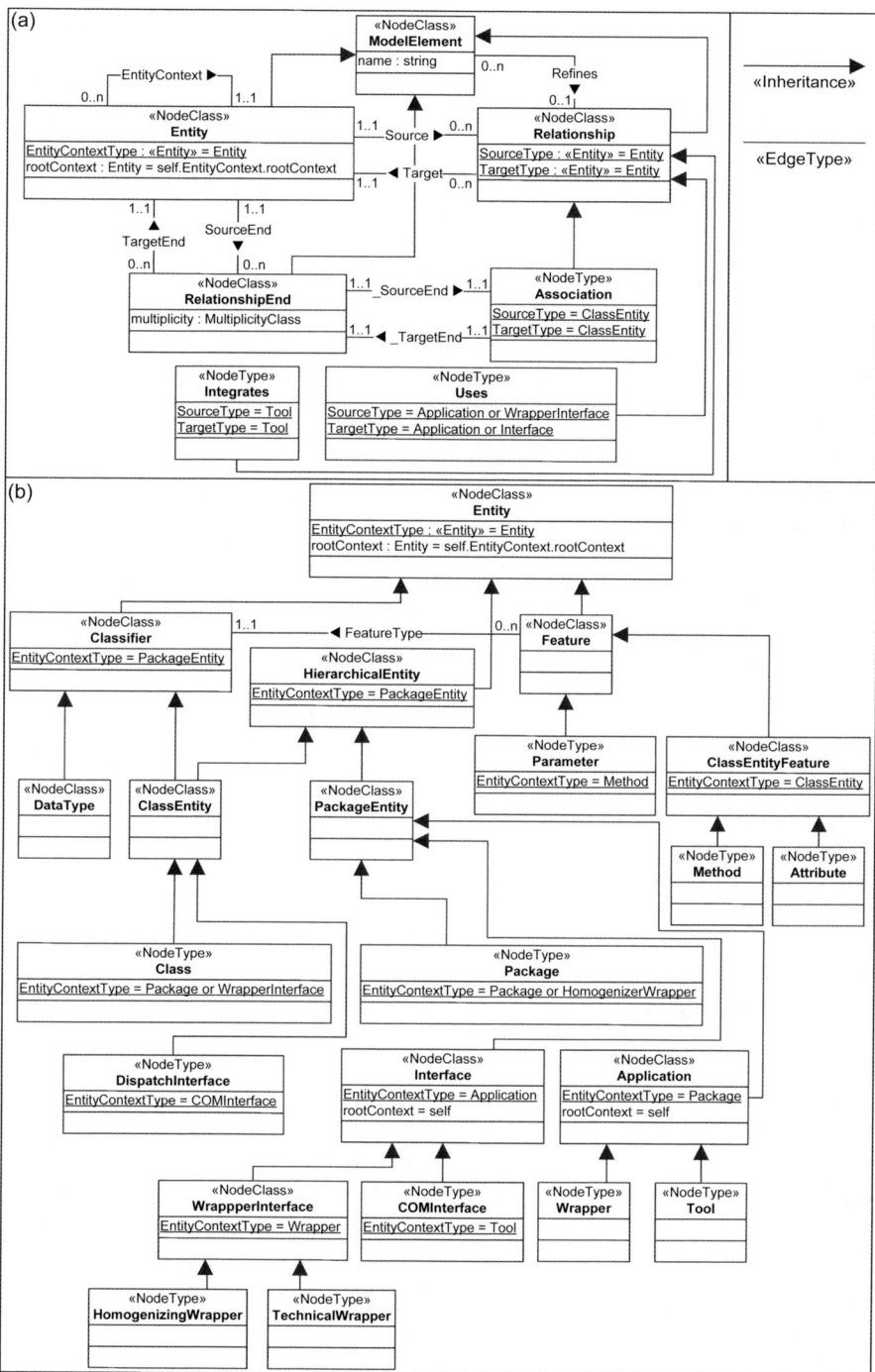

Fig. 5.60. Graph schema for the logical static architecture view (cutout)

can be regarded as the *meta model* with respect to the models of Figs. 5.57 and 5.58, formally defines the relevant modeling concepts, both for general architecture modeling, as introduced in Subsect. 5.7.2, and for the domain-specific area of architecture modeling for a-posteriori application integration. It is a modified and extended variant of the UML meta model [880].

The *root* of that graph schema is the *node class* `ModelElement` which carries common attributes for all modeling concepts like a `name`. This basic node class is specialized on the next inheritance level into the node class `Entity` and the node class `Relationship` (cf. part (a) of Fig. 5.60).

A *relation* is *modeled* as a node in order to enable attributed links between entities. Furthermore, relations are binary and directed. They are connected with their source and target entity via a corresponding `Source` and `Target` edge. Further specializations of the node class `Relationship` are e.g. the node types `Association`, `Integrates`, and `Uses`. Supplementary, an `Association` is extended with a `RelationshipEnd` for the source as well as for the target entity to allow the annotation of multiplicities for that relation. As a `RelationshipEnd` is also derived from the node class `ModelElement`, its `name` attribute can be used to enrich the source and the target entity with a role identifier.

For each `Relationship` type, i.e. the node class `Relationship` and all its subnodes (both node classes and node types), exist certain *restricting constraints* relating to the set of valid types for the source as well as for the target entity. For example, an `Integrates` relation can only be established between two instances of the node type `Tool`. These constraints hold for all instances of a certain relationship and are defined through the meta attributes `SourceType` and `TargetType`[47].

Furthermore, a `Relationship` can be *refined* by zero or more *instances* of a `ModelElement`[48]. The `Refines` edge allows to keep track of the various refinement steps shown in Figs. 5.57 and 5.58.

Additionally, entities are linked by an implied *containment relation*, i.e. this relation is not modeled by an explicit subnode of the node class `Relationship` but by the edge `EntityContext` between two entities. Each `Entity`, the source entity of that edge, has to be contained by exactly one other `Entity`, the target entity of the edge.

Restricting *constraints* also exist for *containment* relations. These constraints describe which entities are allowed to be contained by another. For example, an `Interface` always belongs to an `Application`. The constraints are defined by the meta attribute `EntityContextType` of the node class `Entity` and, therefore, they are limited to the *type* level.

---

[47] In Fig. 5.60 the type «Entity» for the meta attributes `SourceType` and `TargetType` denotes the power set of the set of all node types directly or indirectly derived from the node class `Entity`.

[48] Instances of a node type are also considered as a instance of all direct or indirect supernodes of that node type.

```
transformation + NewEntity
( entityName : string [1:1]; entityType : type in Entity [1:1];
  context : Entity [1:1]; out newEntity : entityType [1:1] )
[0:1] =
```

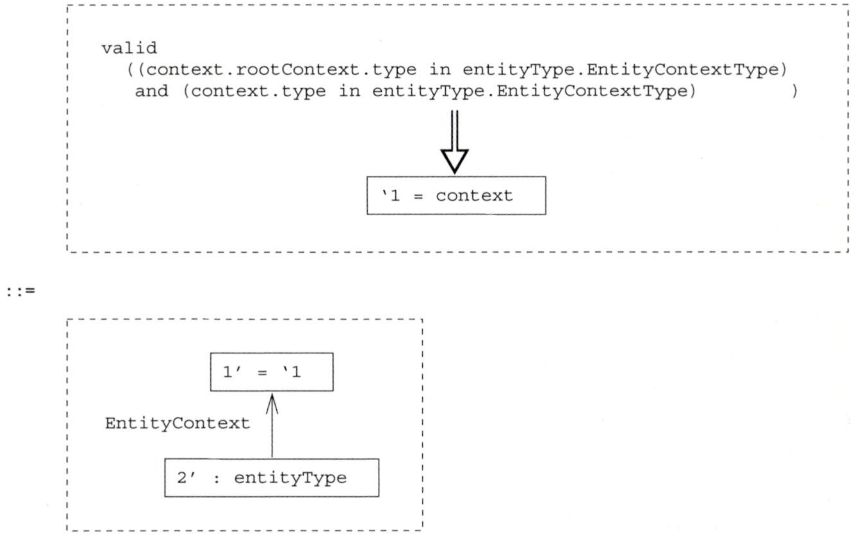

```
┌─────────────────────────────────────────────────────────────┐
│   valid                                                     │
│     ((context.rootContext.type in entityType.EntityContextType)
│      and (context.type in entityType.EntityContextType)    )│
│                             ⇓                               │
│                      '1 = context                           │
└─────────────────────────────────────────────────────────────┘
```

::=

```
┌─────────────────────────────────────────────────────────────┐
│                      1' = '1                                │
│                        ↑                                    │
│      EntityContext                                          │
│                   2' : entityType                           │
└─────────────────────────────────────────────────────────────┘
```

```
transfer 2'.name := entityName;
return newEntity := 2';
end;
```

Fig. 5.61. Generic graph transformation to instantiate a new entity node

In addition, *constraints* on the *instance level* have to be defined regarding the transitive closure of the containment relation. For example, the constraints defined by the meta attribute `EntityContextType` allow a `Wrapper` instance to contain a `HomogenizingWrapper` instance, the `HomogenizingWrapper` instance can contain a `Package` instance, and the `Package` instance can contain a `Tool` instance. To avoid this undesirable situation, as a `Tool` refers to an external application possibly having an `Interface` that is used but not contained by a `Wrapper`, the *derived attribute* `rootContext` of the node class `Entity` calculates for each `Entity` node n its root node r with regard to the transitive closure of its containment within other entities. The type of r has to be as well an element of the type set defined by the meta attribute `EntityContextType` of n.

All of these constraints are ensured by the *generic* graph *transformation* shown in Fig. 5.61 to instantiate a new node for a given subtype of the node class `Entity`. Relationships between entities are instantiated analogously.

Finally, part (b) of Fig. 5.60 illustrates how the *node class* `Entity` is further *specialized* to general modeling concepts, e.g. `ClassEntity` or

PackageEntity[49], and domain-specific modeling concepts such as Wrapper or Application.

The *implementation* of Fire3 makes use of the *model-driven* development idea. As already stated, the internal application logic of Fire3 is implemented by a declarative specification using the PROGRES language and its environment [412, 414]. Based on this specification, C-code is generated, which is embedded into UPGRADE [48, 49, 206], a JAVA-based framework for building user interfaces for graph-based applications. UPGRADE offers different default user interfaces for visualizing and editing. This way, a first executable prototype can be realized without any further manual implementation.

However, the default user interface is rudimentary and only suitable for testing. Further development effort is needed for *adapting* the framework to provide a more *appropriate user interface*. Besides the configuration of built-in filter mechanisms, e.g. to exclude certain "help" edges and nodes from visualization[50], this includes the implementation of different node representations for different node types, e.g. a class or a package view (cf. the screenshot in Fig. 5.59). A corresponding layout algorithm has to be implemented for this purpose.

5.7.4 Model-Driven Wrapper Development

A wrapper acts as an *adapter* "convert(ing) the interface (of a given tool) ... into another interface clients expect" [682]. Therefore, the application of wrappers *enables* the *reuse* of existing software in a new context [972] as it realizes a transparent access to existing interfaces.

Different Tasks for Wrappers

The development of a wrapper includes several tasks. Syntax and semantics of the given interface, the *source interface*, to be wrapped as well as of the interface required by the client, the *target* interface, have to be *specified*. Furthermore, the *transformation* of the source into the target interface has to be defined. Consequently, a wrapper is not a monolithic component, it is rather a subsystem consisting of several subcomponents [136].

In the previous subsection, the architecture was refined such that the problem of wrapper construction is decomposed into *two* levels (cf. Fig. 5.57) [135]: (a) *Technical wrappers*, realizing the access to the source interface, are responsible for hiding technical details of interfaces provided by existing tools. For

---

[49] With respect to the definitions given in Subsect. 5.7.2, a ClassEntity can be compared to a *module* whereas a PackageEntity realizes the concept of a *subsystem*.

[50] For example, in Fig. 5.60 the node RelationshipEnd and its associated edges are visualized by an edge-node-edge filter as two direct edges between the nodes Entity and Association. The attributes of the node RelationshipEnd are visualized by a specific edge representation as attributes of these edges.

example, clients of technical wrappers are shielded from the underlying communication infrastructure such as COM or CORBA. Furthermore, the operations provided by the tools are mapped semantically 1:1 onto the interface of the technical wrapper. (b) In contrast, the *homogenizing wrapper* located on top of the technical wrapper realizes the required data and functional abstraction.

Besides location and implementation transparency, a given interface is *normalized* by a technical wrapper as it abstracts from specific operation names. In addition, the technical wrapper aggregates sequences of primitive operation calls to new operations in order to normalize the granularity of operations. The normalization on this level only covers syntactical aspects and does not include any kind of semantical adjustment. This is the task of the *homogenizing* wrapper, which adapts the data model of the tool to be wrapped to a uniform semantical level.

Moreover, both the technical as well as the homogenizing wrapper's interface can be further divided into (i) a *coarse-grained* interface to start/stop a wrapped tool or to load a document into a wrapped tool (black-box view) and (ii) a *fine-grained* interface to invoke tool specific operations, e.g. accessing the data maintained by the wrapped tool (white-box view).

Wrappers are used on *different levels* of the integrated engineering design environment (cf. Fig. 5.55):

- *Existing* engineering *tools* are integrated a-posteriori into the environment with the help of wrappers.
- *Extended* new *tools* on top of existing tools were developed using different methodologies and corresponding implementation frameworks. To integrate such extended tools, wrapping causes in some cases less development effort than reengineering.
- Finally, all tools, existing as well as new ones, have to be embedded into the common *platform* infrastructure, which is again achieved by wrappers [275].

Therefore, a wide range of wrappers with distinct characteristics are necessary. The second major aim of subproject I3 was to substitute manual ad-hoc implementation of wrappers by a well-understood *development process* for *wrapper construction* [26], which can be supported by tools and fosters reuse on the process as well as on the product level.

Methodology for Wrapper Development

The development of a set of concrete wrappers in the domain of tool integration resulted in a better understanding of the principles of wrapper design and implementation. This emerging knowledge facilitates the definition of a methodology for wrapper development. Next, the methodology was refined, so that wrappers can be described through a declarative specification from which, finally, the executable program code for the wrapper can be generated.

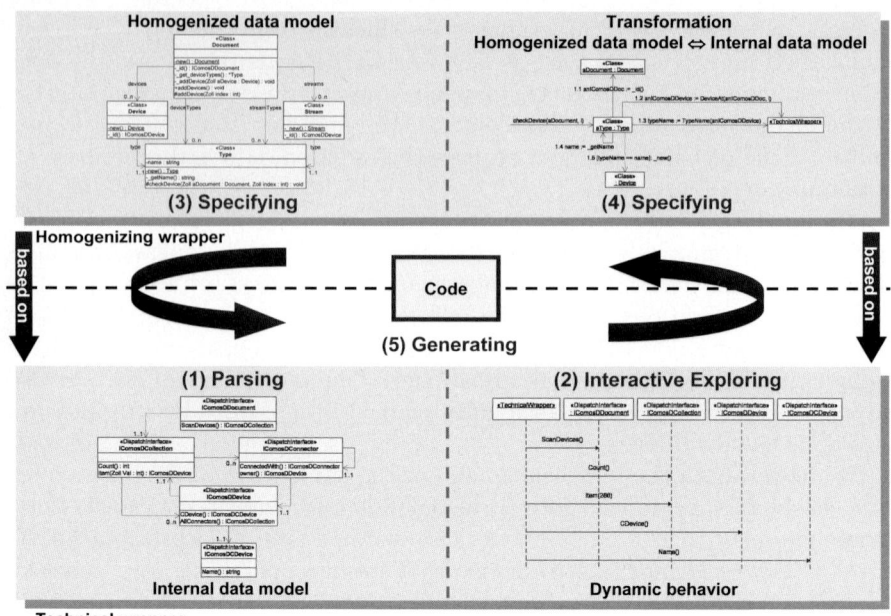

Fig. 5.62. Model-driven wrapper development process (overview)

For each phase of the development process, the software engineer, designing a wrapper, is guided by an appropriate suite of *tools* supporting its activities in *specifying wrappers*. In the following, the different phases of wrapper development and corresponding tool support will be presented in detail.

The methodology for wrapper development is illustrated in Fig. 5.62. It is subdivided into four *construction* phases (phase 1–4), which are followed by a *generation* phase (phase 5). In each construction phase specific models are defined, describing both static (phase 1 and 3) as well as dynamic aspects (phase 2 and 4) of the wrapper. The models are either generated (semi-)automatically (phase 1 and 2) or specified manually (phase 3 and 4).

The methodology is *applicable* both for *technical* as well as for *homogenizing wrappers*. The lower part of Fig. 5.62 deals with technical wrappers abbreviated by TW while the upper part refers to homogenizing wrappers abbreviated by HW.

The semantics of the models to be built in phases 1–4 are formally defined by a meta model using, again, the PROGRES language[51]. The models in their entirety constitute a complete, abstract, and *formal specification* of a wrapper, which is independent from a programming language, such that executable program code can be generated in phase 5.

---

[51] The graph schema shown in Fig. 5.60 comprises static aspects like `ClassEntity`, `Attribute`, `Method`, `Parameter`, or `Association`. Modeling of dynamic behavior is enabled by an extension of this graph schema as depicted in Fig. 5.67.

In detail, the following *models* are specified in the four different *phases*:

1. In the context of IMPROVE, commercial and third-party tools had to be integrated into the engineering design environment. They were mostly equipped with a *COM interface* (e.g. Aspen Plus, Comos PT, or Documentum). Hence, our approach concentrated on COM (Component Object Model [846])[52]. COM follows the object-oriented paradigm, i.e. COM components represent subsystems, consisting of a set of classes including attribute and method definitions, relationships between classes, and a set of interfaces to access the subsystem. For every COM interface, fixed by standardization, a textual description is available in form of a so-called *type library* or a *dynamic link library*, respectively.

   The first step of the wrapper development process is to understand the given COM interface of a tool to be wrapped on the syntactical level. Therefore, its type library is *parsed* and *transformed* into a *language-independent graph model*, according to the schema in Fig. 5.60. Language independence is achieved, for example, by mapping specific COM data types to general ones.

   Furthermore, the derived model is semantically enriched by the parser, in comparison with the type library description, as *relationships* are *explicitly modeled* by edges between `ClassEntity` nodes (cf. Fig. 5.60). Additional semantical information offers the opportunity for further analyzes, e.g. the calculation of object metrics, like weighted methods per class (WMC) or coupling between object classes (CBO) [594]. This can provide first hints concerning the complexity of a tool's interface.

   So far, the model generated automatically by the parser represents the *import interface* of the *technical wrapper* that is used to access the wrapped tool. In the bottom-left quadrant of Fig. 5.62, the model is visualized as an object-oriented class diagram.

2. Second, the *export interface* of the technical wrapper has to be defined. It consists of *two types of operations*: (i) atomic operations of the given COM interface (e.g. to insert, modify, or delete a primitive data element), which can be renamed, if wanted, or (ii) composed operations, each of them built-up from a sequence of atomic operations (e.g. to insert, modify, or delete a complex data structure).

   Composing atomic operations requires, besides syntactical knowledge about the given interface, their semantical interpretation with respect to the *dynamic behavior* of the tool to be wrapped. A simple, frequently occurring example is as follows. It cannot be derived from the syntactical description of an interface, whether indexing a collection begins with zero or one. Normally, this information is received from additional documentation offered by the tool's vendor. As experience shows, documentations

---

[52] Nevertheless, the approach is general enough to be also used with other middleware techniques, such as CORBA (Common Object Request Broker Architecture [877]) or EJB (Enterprise JavaBeans [504]) (cf. Sect. 7.8).

are often incomplete, erroneous, and, therefore, do not include the desired information.

For this reasons, tool support was developed to assist the software engineer in specifying composed operations. It consists of a *test environment* to *interactively explore* a COM interface and its underlying tool at runtime. User's interactions are traced by the supporting tool and serve as foundation for composing operations. Based upon traces, which describe atomic operation sequences, a *composed operation* is defined by *selecting* a subsequence from the *trace*. The tool automatically infers the signature for a subsequence, i.e. required input parameters, the output parameter, and corresponding data types. Furthermore, the generalization for a subsequence is possible by substituting constants by variables, which are added to the signature as new input parameters.

The *composed operations*, specified interactively by the software engineer assisted by the tool, constitute the *export interface* of the technical wrapper[53]. A composed operation is represented in the bottom-right quadrant of Fig. 5.62 as a sequence diagram.

3. After determining import and export interface of the technical wrapper, the *homogenizing wrapper*, realizing an intended view on an encapsulated tool, is specified. This happens by modeling the *static data* structure, *materializing* the *view* via an object-oriented class diagram (cf. top-left quadrant of Fig. 5.62). The set of all public methods of classes defined in the class diagram represents the export interface of the homogenizing wrapper.

Again, the software engineer is supported by a corresponding tool to construct the model. The tool realizes, among other things, a *modeling environment* for building object-oriented class diagrams. Main *concepts* offered for modeling are (i) packages, (ii) classes, (iii) attributes, (iv) methods, and (v) associations between classes (cf. Fig. 5.60). Moreover, attributes and methods (a) can be classified with regard to belonging to the instance or the class level and (b) their visibility can be determined (public, protected, or private). Associations are refined by accessory multiplicities. We believe that these modeling concepts are sufficient for specifying the homogenizing wrapper.

Models are *analyzed* by the tool with regard to *correctness* and *completeness*, e.g. naming conflicts are detected or missing methods, attribute, or parameter types are indicated[54]. Correctness and completeness of the models are necessary to allow generation of executable code in phase 5.

---

[53] In particular, a subsequence, selected from a trace, can consist of a single atomic operation call. This way, also atomic operations of the given interface can be added to the export interface of the technical wrapper.

[54] The type of a method is the type of its return value. If no value is returned, the type is set to `void`.

Furthermore, the tool *automatically extends* the *models*. In particular, when a class or an attribute is defined, corresponding methods are added to instantiate the class or for getting and setting the attribute. Analogously, methods for navigating and iterating are appended to associations. Enriching the model by these additional methods prepares for the following phase 4. The methods are declared as private and their semantics is fixed by the implementation of the code generator.

4. Finally, a *mapping* between the data *model* of the *homogenizing* wrapper, defined in previous phase 3, and that of the *tool* to be wrapped has to be *specified*. Invocation of the homogenizing wrapper by an external client at runtime leads to various changes of the materialized view of the homogenizing wrapper. For example, objects are instantiated or deleted, associations between objects are added or removed, or attribute values are changed. To keep the materialized view of the homogenizing wrapper consistent with its underlying tool, corresponding modifications have to be applied on the wrapped tool's data structure.

For this, the *internal behavior* of public and protected *methods*[55] is modeled by *collaboration diagrams* (cf. top-right quadrant of Fig. 5.62). Such collaboration diagrams describe on the one hand, how the homogenizing wrapper's data structure is modified. Therefore, the private methods (see phase 3), e.g. for instantiating an object, are used. On the other hand, the methods of the technical wrapper's export interface (see phase 2) are used to manipulate the wrapped tool accordingly.

The *abstract syntax* of collaboration diagrams is *defined* by the graph schema shown in Fig. 5.67, which will be explained later. A formal definition of the abstract syntax, again, enables the modeling environment to guarantee (syntactical) correctness and completeness with respect to context-free as well as context-sensitive conditions, such as type conformity of actual and formal parameters to facilitate the generation of executable code.

The models constructed in phases 1 to 4 are sufficient to *generate* executable program code for the specified wrapper. Up to now, a PROGRES/UPGRADE prototype is created that realizes a *test environment* for the wrapper. Like the tool for exploring COM interfaces interactively, the generated test environment facilitates *interactive* testing of the wrapper.

Wrapper Development Example

We will now *demonstrate* the above introduced wrapper development *process* by an *example*, where a wrapper for the flowsheet editor Comos PT is specified [27].

---

[55] While public methods are accessible by external clients of the homogenizing wrapper, protected methods are used for internal realization of public methods, but not visible for clients.

Phase 1: Getting the Syntactical TW Interface by a Parser

A cutout of the *internal data model* of Comos PT, described by its type library, is illustrated on the left side of Fig. 5.63[56]. It is represented as a UML-like class diagram that has been *generated by* the *parser* in phase 1.

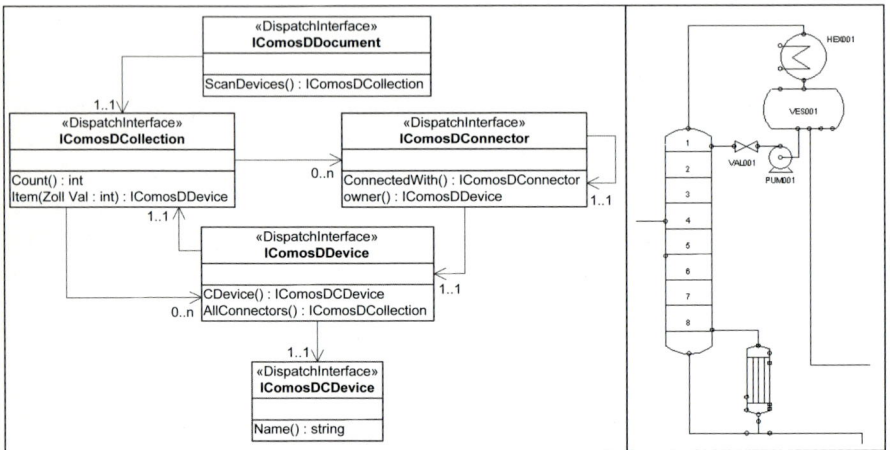

Fig. 5.63. Data model (cutout, left side) and screenshot (right side) of Comos PT (phase 1)

The right side of Fig. 5.63 shows a *sample* screenshot of a Comos PT *flowsheet* (class `IComosDDocument`), which includes a collection (class `IComosDCollection`) of devices (class `IComosDDevice`), i.e. a distillation column, two heat exchangers, a vessel, a pump, and a valve. These devices are connected via certain streams, symbolized as edges between the devices. Internally, a stream is also represented as an instance of class `IComosDDevice`.

`IComosDDocument` offers the method `ScanDevices()` to *access* the collection of *all devices* within a flowsheet. For *iterating* through a collection, the interface of `IComosDCollection` consists of two methods: (i) `Count()`, which returns the size of the collection, and (ii) `Item(Val:int)` to access a specific element within the collection by its index.

To *distinguish* different kinds of *devices*, each `IComosDDevice` is connected with exactly one `IComosDCDevice` denoting its type. Accessing a device's type is enabled by the method `CDevice()`. The name of the type can be queried by the `Name()` method of class `IComosDCDevice`.

Moreover, the connection points of devices (cf. left side of Fig. 5.63) are explicitly modeled by connectors (class `IComosDConnector`). Therefore, each

---

[56] In total, the data model of Comos PT consists of 99 classes with altogether 1131 attributes and 5758 methods.

`IComosDDevice` refers to a *collection* (class `IComosDCollection`) of its *connectors* (method `AllConnectors()`). The methods of class `IComosDConnector` are used to navigate (i) to the corresponding connector of the connected device (method `ConnectedWith()`), and (ii) to its own device (method `owner()`).

Phase 2: Interactively Exploring the TW Semantics

The wrapper's task is to transform the given data model of Comos PT into that of the homogenizing wrapper and vice versa[57]. The technical wrapper's interface used to access Comos PT is determined in phase 2 by *exploring* Comos PT *interactively* with the help of tool support.

After parsing the type library of the COM interface of Comos PT (phase 1), the exploring tool starts the underlying application as an operating system process via a generic start operation that every COM interface implements. The return value of this operation is a reference to the initial object of the COM interface. The exploring tool determines dynamically the object's class and, based upon the static information parsed out of the type library (cf. Fig. 5.63), a *GUI* (graphical user interface) for the given COM object is generated by the exploring tool. Using this GUI, the software engineer can *query* the *values* of object attributes or can *invoke* the object's *methods*.

An attribute value or a return value of a method can be either an atomic value, like a `string` or an `int`, or a reference to another COM object. In this case, another GUI, according to the referenced object, is generated allowing to *inspect* the *referenced object*. This way, the software engineer explores the COM interface of an application interactively.

Furthermore, the *user's interactions* are *traced* by the exploring tool. Figure 5.64 shows a sample trace captured during exploring. Assuming that a reference to an instance of class `IComosDDocument` was already explored, the trace, visualized as an UML sequence diagram, demonstrates (1) how to access the collection of all devices within a flowsheet, (2) how to determine the number of included devices, (3) how to receive a specific device, (4) how to navigate to a device's type representation, and (5) how to gain the type representation's name (cf. box 1 of Fig. 5.64).

Based upon this trace, three *methods*, constituting the technical wrapper's *export interface*, are *specified* by *selecting* specific *subsequences* out of the trace:

1. The first method `NumberOfDevices` (cf. box 2 of Fig. 5.64) consists of method calls (1) and (2) and returns an `int` value indicating the number of devices within a flowsheet given by the input parameter `anIComosDDoc`. The method's signature is inferred by the exploring tool automatically, i.e. `IComosDDocument` is added as input parameter.
2. Analogously, method `TypeName` is defined (method calls (4) and (5)), returning a `string` value denoting the type's name for the given input parameter `anIComosDDevice` (cf. box 3 of Fig. 5.64).

---

[57] For simplification, connections between devices are not considered.

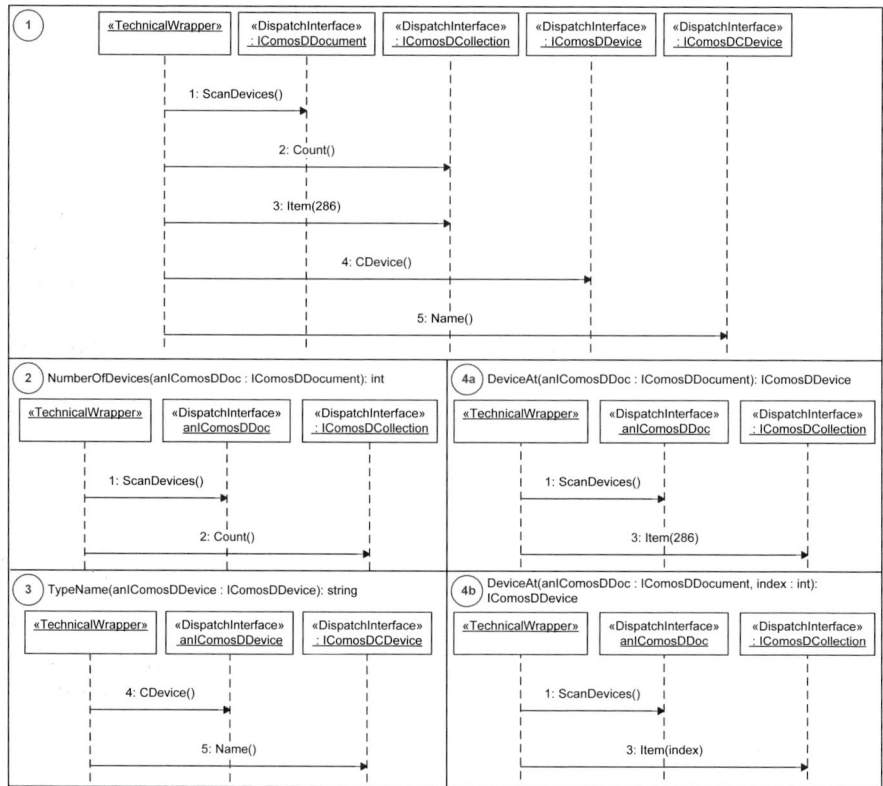

Fig. 5.64. Interactive exploring (sample sequence trace) and aggregation of subsequences (phase 2)

3. Method `DeviceAt` illustrates another feature of the exploring tool: After specifying the method (method calls (1) and (3)), which returns the specific `IComosDDevice` with index 286 included within the flowsheet given by the input parameter `anIComosDDoc` (cf. box 4a of Fig. 5.64), it is generalized, i.e. the `int` constant 286 is substituted by an input parameter `index`, so that any device within a flowsheet can be received (cf. box 4b of Fig. 5.64). Therefore, the signature of method `DeviceAt` is updated by the exploring tool.

Phase 3: Designing the HW Data Model

The *intended homogenized data model*, realized by the wrapper to be designed, is addressed in Fig. 5.65. The class `Document` represents a Comos PT flowsheet, indicated by the return value of method `_id()`, which returns a reference to an instance of class `IComosDDocument`.

In contrast to the original data model of Comos PT, in the homogenized data model, *streams* (class `Stream`) are separated from other flowsheet devices (class `Device`), as they are explicitly *modeled* by an independent *class*. For both classes, the corresponding Comos PT classes are again identified by the return value of method `_id()` (class `IComosDDevice` in both cases). To recognize different device types, *types* are modeled by *instances* of class `Type`. Its attribute `name` designates a certain type. Instances of class `Type` are split into two *disjunct collections* (edges `deviceTypes` and `streamTypes`, respectively, between class `Document` and `Type`).

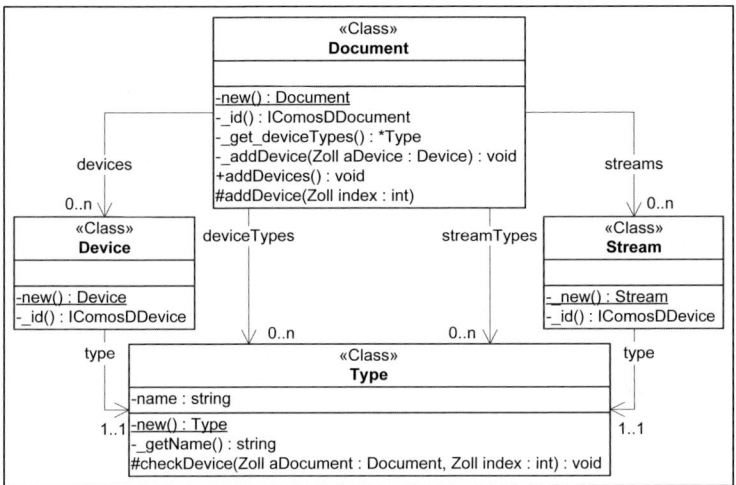

Fig. 5.65. Homogenized data model (phase 3)

Each `Device` and `Stream` is *associated* with its according `Type` via an edge type. In this case, a condition holds true that instances of class `Device` can only be connected with an instance of class `Type` included in the `deviceTypes` collection. The same is to be considered for instances of class `Stream`.

Phase 4: Defining the Mapping from HW to TW

Figure 5.66 demonstrates the final step of wrapper development, namely the specification of the *mapping* between the *given data model* (cf. Fig. 5.63) and that of the *homogenizing wrapper* (cf. Fig. 5.65)[58]. Therefore, the methods of the technical wrapper's export interface (cf. Fig. 5.64) and the private methods of the homogenizing wrapper's classes (cf. Fig. 5.65) are applied.

In our scenario, we require that *methods* are already specified to *create* an initial *instance* of class `Document` and to *connect* it to an according instance of class `IComosDDocument`. Furthermore, the `deviceTypes` and `streamTypes` collections of the instance of class `Document` are defined.

---

[58] UML-like collaboration diagrams are used in Fig. 5.66.

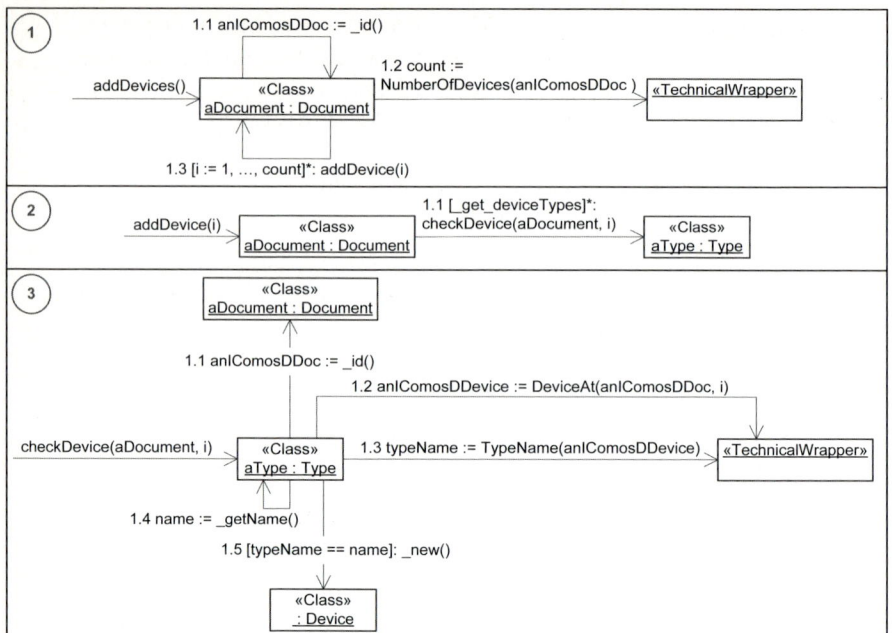

Fig. 5.66. Modeling the transformation between the homogenizing wrapper and the given interface (phase 4)

In the following, we will show, how a method `addDevices()` of class `Document` is modeled. The method (i) reads all devices out of the flowsheet, (ii) determines whether it is a device according to the homogenized data model, and, if applicable, (iii) creates a new instance of class `Device`. The *modeling* of the method takes place in three *steps*:

1. Firstly, within method specification `addDevices()` (cf. box 1 of Fig. 5.66) the reference to the instance of class `IComosDDocument`, representing the flowsheet, is queried (1.1) and the technical wrapper is asked for the number of devices within this flowsheet (1.2). Next, a loop is defined (1.3), bounded by the return value of (1.2), calling method `addDevice()` of class `IComosDDocument` in each iteration. The loop variable, indicating the index of the currently handled device, serves as input parameter for method `addDevice()`.
2. The specification of method `addDevice()` consists of a single method call (cf. box 2 of Fig. 5.66): For each element in the `deviceTypes` collection of the instance of class `Document`, the method `checkDevice()` is called with the instance of class `Document` and the index i as actual parameters.
3. Finally, method `checkDevice()` of class `Type` has to be modeled (cf. box 3 of Fig. 5.66). Again the reference to the instance of class `IComosDDocument` is queried (1.1) and the technical wrapper is asked for a reference to the

actual device (1.2) as well as for the name of the device's type (1.3). If the device's type name is equal to the name of the current `Type` instance, queried in (1.4), a new instance of class `Device` is created (1.5), as the flowsheet device represents a `Device` according to the homogenized data model.

Code Generation for the Mapping

From the detailed explanations of phases 1 to 4, it should be clear that the models of Figs. 5.63 to 5.66 fulfill (syntactical) correctness and completeness, guaranteed by the modeling environment. Hence, executable code can be generated. Up to now, again, an UPGRADE prototype, i.e. *PROGRES* and *JAVA code*, is *generated* allowing the interactive testing of the specified wrapper.

Implementation of the Wrapper Development Tools

While the tools supporting phase 1 and 2 were implemented with conventional object-oriented programming languages (C++ and SMALLTALK), the modeling environment `Fire3`, supporting phase 3 and 4, was developed using the PROGRES language. Besides the implementation, the *architecture* and the technical *infrastructure* of the entire tool suite is sketched.

Modeling Environment

Concerning the *implementation* aspects of `Fire3`, we will concentrate on modeling the *dynamic behavior*. The graph schema shown in Fig. 5.67 defines the abstract syntax for the collaboration diagrams used in phase 4. It is an extension of the formerly introduced graph schema of Fig. 5.60, which serves as meta model for defining the static data model in phase 3.

Extensions refer to the formal definition of modeling concepts for (i) *message sequences* (node type `MessageSequence`), defining the behavior of a method (edge type `Defines`) at runtime. A message sequence consists of (ii) single method calls (node class `MethodCall`) with (iii) accessory actual parameters (node class `ClassifierInstance`). Furthermore, a method call can be restricted by (iv) a condition (node class `Control` and its subnodes, respectively).

To ensure correct code generation, the following context-sensitive *conditions* have to be *fulfilled*[59]:

1. $\forall$ mc $\in I_{\texttt{MethodCall}}$:
    mc.Calls.EntityContext = mc.Callee.InstanceOf

---

[59] I_N denotes the set of all instances of node class or node type N.

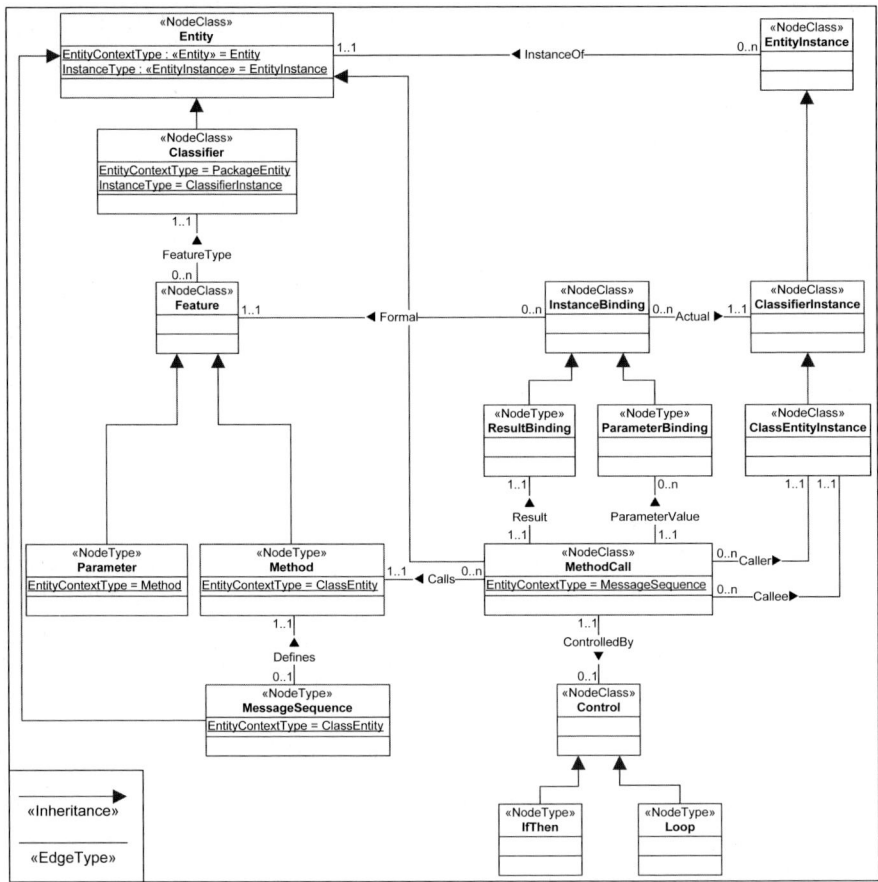

Fig. 5.67. Graph schema for modeling the dynamic behavior (cutout)

2. $(\forall\ mc \in I_{\texttt{MethodCall}}) \land (\forall\ p \in I_{\texttt{Parameter}})$:
    mc.Calls = p.EntityContext
    $\Rightarrow \exists\ ib_1 \in$ mc.ParameterValue:
      $(ib_1$.Formal = p$) \land$
      $(\nexists\ ib_2 \in$ mc.ParameterValue: $ib_1 \neq ib_2 \land ib_2$.Formal = p$)$

3. $\forall\ mc \in I_{\texttt{MethodCall}}$:
    mc.Calls = mc.Result.Formal

4. $\forall\ ib \in I_{\texttt{InstanceBinding}}$:
    ib.Formal.FeatureType = ib.Actual.InstanceOf

The first condition guarantees that the called method belongs to the callee, while the second and third condition assure that exactly one actual parameter

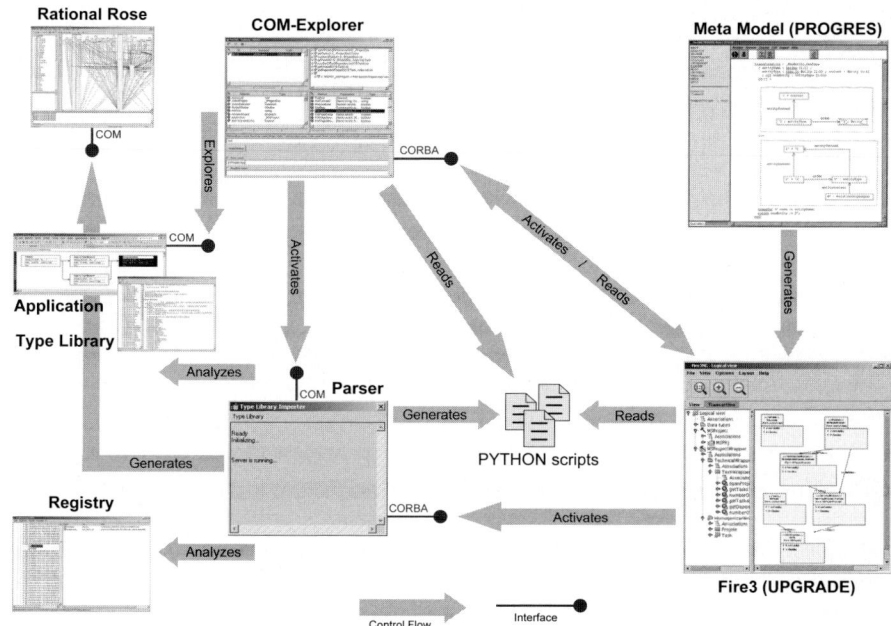

Fig. 5.68. System architecture of tool support for wrapper development (overview)

is assigned to every formal parameter of the method's signature. Type conformity between formal and actual parameters is denoted by the fourth condition. To meet these conditions, corresponding graph transformations exist.

Architecture and Technical Infrastructure

Just as the integrated engineering design environment introduced at the beginning of this section (cf. Fig. 5.55), the tool environment for wrapper development itself is also built on top of *different* operating system *platforms* (LINUX and WINDOWS), *middleware* techniques (CORBA and COM), and programming *languages* (PROGRES, JAVA, C++ and SMALLTALK). This diversity is due to the application of established and suitable programming infrastructure. PROGRES/UPGRADE, for example, which is up to now bound to a LINUX platform, has been used for implementing the modeling environment Fire3. In contrast, the tools to be wrapped demand a WINDOWS platform as we focus on tools with a COM interface[60].

Figure 5.68 gives an overview of the *system architecture*. Based upon the PROGRES specification (cf. Figs. 5.60 and 5.67) the UPGRADE prototype Fire3 was generated. At runtime, the parser for analyzing the COM interface of an application (phase 1) is activated via CORBA. To determine the type library's location, the parser analyzes the WINDOWS system registry. After

---

[60] This was a requirement by other subprojects within IMPROVE.

parsing, the results are stored as PYTHON scripts, which can be executed by Fire3.[61] This way, the knowledge about the internal data model of the tool to be wrapped is brought to Fire3. According to phase 2, the exploring tool (COM-Explorer) is activated, again via CORBA. It makes also use of the PYTHON scripts to get the knowledge about the given data model. Next, the tool to be wrapped can be explored via its COM interface. The results are given back to Fire3 and modeling of phase 3 and 4 can begin.

5.7.5 Related Work

The *discussion* of *related work* is *structured* according to the preceding subsections, i.e. for each subsection there is a corresponding part addressing related work.

Modeling and Refinement of Software Architectures

The observation that the structure of a software system can be specified as coupled units with precise interfaces is a major contribution of software engineering. It is almost as old as software engineering itself [897]. Due to the definition of a *software architecture*, as given in Subsect. 5.7.3, it is not surprising that *graph transformations* were identified as a simple and natural way to model software architectures. Consequently, the rules and constraints for the dynamic evolution of the architecture, e.g. adding or removing components and relations between them, can be defined as graph transformations [260]. Following this idea we use PROGRES [412, 414] to describe both of these aspects in an unified way.

Several related approaches are described in literature: Le Métayer [804] uses graph grammars to specify the *static* structure of a system. However, the dynamic *evolution* of an architecture has to be defined independently by a so-called coordinator. A uniform description language based on graph rewriting covering both aspects is presented by Hirsch et al. [731]. In contrast to PROGRES, this approach is limited to the use of context-free rules for specifying dynamic aspects. Similar to our approach, Fahmy and Holt [663] also apply PROGRES to specify software architecture transformations.

These and other approaches for architecture modeling [224] claim to be usable to specify architectures *independent* from the *domain* and do not consider the needs for *domain-specific* architectures [843]. Therefore, PsiGene [934] allows to combine design patterns as presented in [580] and to apply them to class diagrams. A technique to specify patterns in the area of distributed applications and to combine them to a suitable software architecture is shown in [374].

---

[61] Furthermore, the parsing result can be visualized using the commercial tool Rational Rose [743].

While these approaches offer solutions for architectural patterns on a technical level, e.g. distributing components and defining patterns for their communication, they do not address the problem of *semantic heterogeneity*. Numerous standardization efforts deal with that problem to define domain-specific interfaces based on corresponding architectural frameworks, e.g. OMG (Object Management Group) domain specifications [881], ebXML (electronic business using eXtensible Markup Language [646]), or OAGIS (Open Applications Group [885]). However, legacy systems can only be adapted to such standards if they are wrapped. In this section, we have shown how wrapping can be performed systematically on the architectural level.

Interactive Modeling and Construction of Wrappers

An *architecture-based* approach for developing *wrappers*, similar to ours, is described by Gannod et al. [683]: Interfaces to command line tools are specified as architectural components by using ACME [685], a generic architecture description language. Subsequently, based upon the specification, the wrapper source code for the interface is synthesized. In comparison with our methodology, Gannod et al. only cover the construction of the technical wrapper. Any kind of data homogenization is not considered.

To *enrich* the expressiveness of a given *interface* to be wrapped, Jacobsen and Krämer modified CORBA IDL (interface definition language) [877] to add specifications of semantic properties, to facilitate a wrapper's source code to be extended by additional semantic checks automatically [765]. For wrapping tools in an a-priori manner, i.e. in these cases where the semantics of the tool's interface is well-known, such descriptions are applicable for synthesizing the wrapper. Unfortunately, in the context of a-posteriori integration the semantic properties to be specified for generating the wrapper are unknown. This was one reason for developing our interface exploration tool.

Other attempts to automatically discover the structure and behavior of a software system come from the field of *software reengineering*. Cimitile et al. [600] describe an approach that involves the use of data flow analysis in order to determine various properties of the source code to be wrapped. A necessary prerequisite for this – and most of the other techniques in the area of software reengineering – is the availability of the source code that is to be analyzed. Again, a-posteriori integration, as presented in this section, is not constrained by this requirements.

The solution we have chosen is an application of the *programming by example principle* [620, 816]. Several approaches for wrapping semi-structured data sources, such as web pages, following this principle can be found in literature. Turquoise [849] is a prototype of an intelligent web browser creating scripts to combine data from different web pages. The scripts are derived form the user's browsing and editing actions, which Turquoise traces and generalizes into a program. Similarly, NoDoSE [495] combines automatic analysis with user input to specify grammars for unstructured text documents. An automation of the generalization step, necessary in every programming by example approach, is presented in [800]. For a set of web pages, single wrappers are specified manually. Then, an automatic learning algorithm generates a generalized wrapper by induction.

While these programming by example approaches concentrate on data integration, we are even more interested in *functional* and *event integration*, e.g. for offering the integration tool between the process flowsheet editor Comos PT and the management system AHEAD a visual browsing functionality.

5.7.6 Summary and Open Issues

In this section, we have presented an architecture-based and model-driven approach to the *a-posteriori integration* of engineering tools. This approach has been realized within subproject I3 of IMPROVE (cf. Subsect. 5.7.1).

Based on the underlying conceptual framework for software architecture modeling (cf. Subsect. 5.7.2), an integrated *architecture* development *environment* was elaborated using gradual refinement from a coarse-grained logical to a fine-grained concrete architecture (cf. Subsect. 5.7.3). We illustrated, how architecture refinement can be formalized by appropriate domain-specific patterns such that the development of suitable tool support is possible.

Furthermore, we have demonstrated that tight integration can be achieved even in the case of a-posteriori integration of heterogeneous tools (cf. Subsect. 5.7.4). Wrappers are used for this purpose. As the *wrapper development process* is strongly architecture- and model-driven, the process can be performed at a fairly high level of abstraction with considerably reduced effort.

To evaluate the practical relevance of the approach and its generality, the results of subproject I3 will be applied and extended within a DFG *transfer project* in the area of business applications (cf. Sect. 7.8).

Besides its evaluation in industry, current and future work will address some *extensions* to the approach presented. For instance, for modeling a wrapper, some concepts including inheritance are missing. The use of inheritance would facilitate modeling and would simplify the resulting models (cf. Fig. 5.65). Another open problem addresses bidirectional consistency between the materialized view, realized by the homogenizing wrapper, and the data model of the underlying tool. Up to now, changes of the materialized view are propagated to the wrapped tool but not vice versa. For the latter, the events thrown by a tool, if existing, have to be regarded within all four phases of the wrapper development process. This requires extensions to the technical infrastructure as well as of the modeling formalism on the conceptual level. Both aspects will be also addressed in the transfer project.

6

Steps Towards a Formal Process/Product Model

This chapter discusses *results* towards a formal *process/product model* (abbr. PPM) we have achieved so far. The results are of a preliminary nature. Hence, there is plenty of room for further research.

We have sketched the problem of developing a formal PPM in Sect. 1.1. Section 2.6 has discussed, how the different submodels of the application domain model are interrelated to each other. We have given details in Chaps. 3 to 5 on how tools are systematically constructed in a model-driven way. We are summarizing our findings in a dedicated chapter, since the formal PPM is the most challenging problem we addressed within IMPROVE.

The *chapter* is structured as *follows*: Sect. 6.1 gives an overview of a PPM in the context of model-driven tool construction. Sections 6.2 to 6.4 present results for the PPM we have achieved during methodological tool construction for novel cooperative support (Sects. 3.1, 3.2, and 3.4). These sections cover case studies, where the model-driven aspect of tool construction was studied more extensively than elsewhere. The functionality of these tools has to meet the specifications made on the application domain model layer. Finally, Sect. 6.5 gives a summary of what we have achieved as well as a long list of open problems.

The objective of this chapter is three-fold: Firstly, it again gives a motivation for working on the interesting problem of developing a formal PPM. Secondly, it presents preliminary findings achieved within IMPROVE. Thirdly, it is an invitation for other groups to share our endeavor towards a formal PPM, covering all aspects from application domain modeling down to details of tool construction.

6.1 From Application Domain Models to Tools: The Sketch of a Layered Process/Product Model

M. Nagl

Abstract. This section gives an introduction on the relation of the process/product model to tool development. We concentrate on tools offering novel functionality (experience-based, consistency, reactive, see Sect. 1.1 and Chap. 3). We motivate again the challenging task of developing such a comprehensive model. We give a summary how application domain models are structured (for details see Sect. 2.6), and we present how the corresponding information fits the tool construction process. A categorization of open problems related to our preliminary results for a coherent, comprehensive, and uniform model is given at the end of this section. The problems, themselves, are presented in Sect. 6.5. The section also serves as an overview for this chapter.

6.1.1 Introduction

In Sect. 1.1, we already stated that the development of a comprehensive and formal process/product model (abbr. PPM) is the *scientific key problem* of IMPROVE. This chapter discusses what we have achieved so far w.r.t. this problem.

PPM Characteristics

We already discussed that the PPM should be structured into different layers, from application domain models to platform models (cf. Fig. 1.6). On every layer, there is a complete description of the product, i.e. the result of the design process (*overall configuration*). Trivially, we find *hierarchies* on every layer. In top-down direction, from layer to layer, the number of details is growing. The model is transformed from layer to layer, from an explicit model to more implicit descriptions (code). The product is complemented by the associated *design process* on every layer. Like the product, the process is hierarchically structured. An abstract process model is transformed into tool commands and corresponding code.

In Sect. 1.1, it was also discussed that one characteristic of IMPROVE is that we *started* with *application problems* and *models*. In particular, we did not build tools which are evaluated afterwards and are either appreciated or rejected. Instead, we studied industrial work processes first to see what kind of support is necessary. Thus, we first fixed application domain models and the required functionality of tools, before building these tools.

It was also discussed in Sect. 1.1 that IMPROVE follows an *a-posteriori approach* in the sense that existing tools were further used, if available and useful. This principle makes the problem of deriving a suitable PPM even harder, since it has to incorporate the functionality of any kind of given tools.

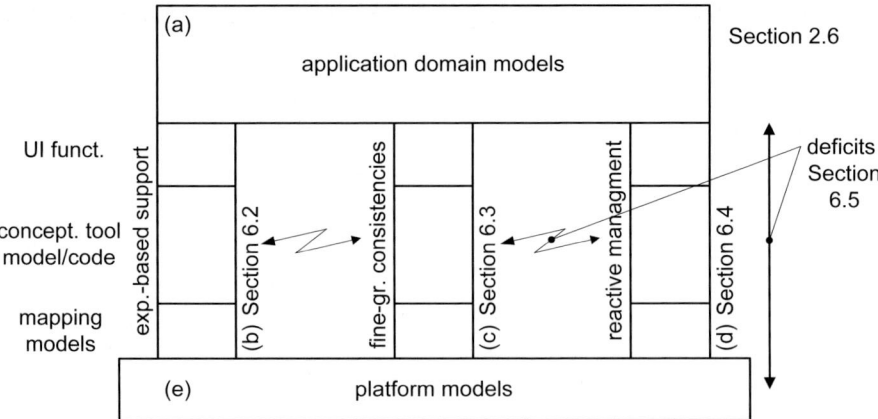

Fig. 6.1. Layered PPM: overview of results and chapter

PPM Survey

Figure 6.1 gives a *survey* of the results to be discussed in this section and, later on, in this chapter. We see the *layered PPM* according to Fig. 1.6, ranging from application domain models (a) down to platform models (e). In between, we see three columns, which are related to novel tool functionality, as introduced in Sect. 1.1 and discussed in detail in Sects. 3.1, 3.2, and 3.4, namely experience-based process support (b), consistency handling by integrators (c), and reactive project management (d). These tree columns are only examples for the relation of the PPM to the tool construction process. In the corresponding projects this relation has been studied more carefully than in others.

Column (b) has been a *cooperation* of subprojects A1/B1, (c) of A2/B2 and, finally, (d) of I1/B4. For subproject names and roles, see Fig. 1.27. Hence, the development of these columns, as well as their relation to application domain models, was a joint effort between *engineering* and *informatics*.

For any of the three tool development case studies we will present a *methodology* how to get to the tools, beginning with the corresponding information of the application model. We do not have a coherent and uniform methodology for all three cases yet.

Overview

This *section* is structured as *follows* (see again Fig. 6.1): We start by summarizing the application domain models in the next subsection (part (a) of Fig. 6.1). Especially, we explain that the different parts of the application model fit together (for details, see Sect. 2.6), each giving a specific perspective. Then, we sketch which part of the application model is relevant for which tool functionality, i.e. we explain the relation between application model and the three tool models (cf. column (b), (c), and (d) of Fig. 6.1). Finally, we sketch those deficits of the PPM which are already evident from the coarse perspective given in this section. This section ends with some concluding remarks.

The whole *chapter* is correspondingly structured as *follows* (see again Fig. 6.1): There are three following sections according to the three columns ((b), (c), and (d) in Fig. 6.1). These three sections show that tools – according to the three novel concepts – can be developed in a well-understood construction process. In the last section of this chapter, we give a long list of open problems according to the PPM development.

6.1.2 Application Models: Domain and Organizational Knowledge

Application knowledge – also to be used for tool construction – should be divided into two different parts (cf. Fig. 6.2): *domain knowledge* which is not dependent on the organization, where chemical processes are designed, and *organizational knowledge*, which depends on the organization. Both parts are (implicitly) contained in the application models as described in Chap. 2. In this section, we want to explicitly distinguish between domain and organizational knowledge. We first give a discussion for the product part of the PPM and later of the process part. This distinction between domain knowledge on one side and organizational knowledge on the other, or products and processes, is mainly used in Sect. 6.5 to define further and open problems.

Domain Knowledge Models

Domain knowledge consists of subdomain models to divide the overall design process into coarse clusters of activity (see Fig. 1.1), which we also called *working areas*. Examples are synthesis, analysis, economic investigations etc. These working areas usually occur in any design process and are, therefore, independent of the organization. Coarse-grained information laid down about working areas and their relation have been called *partial models* in Chapter 2. What might be specific for an organization is, whether corresponding activities occur and, if, how they are structured internally.

Domain knowledge also includes product data models (cf. Chap. 2). They describe which entity types appear in the domain (in our case chemical engineering) and how they can be grouped according to inheritance, aggregation etc. Ontologies should be organization-independent, i.e. not to go to details which are specific for an organization.

Also, there are *dependencies* between *entities* from one or from different subdomains, which say that these entities are semantically related to each other. More precisely, these relations connect entity types. The relations do not depend on in which documents the corresponding entities appear.

Summing up, domain knowledge consists of a coarse-grained division of the work process into working areas (subdomains) in form of partial models,

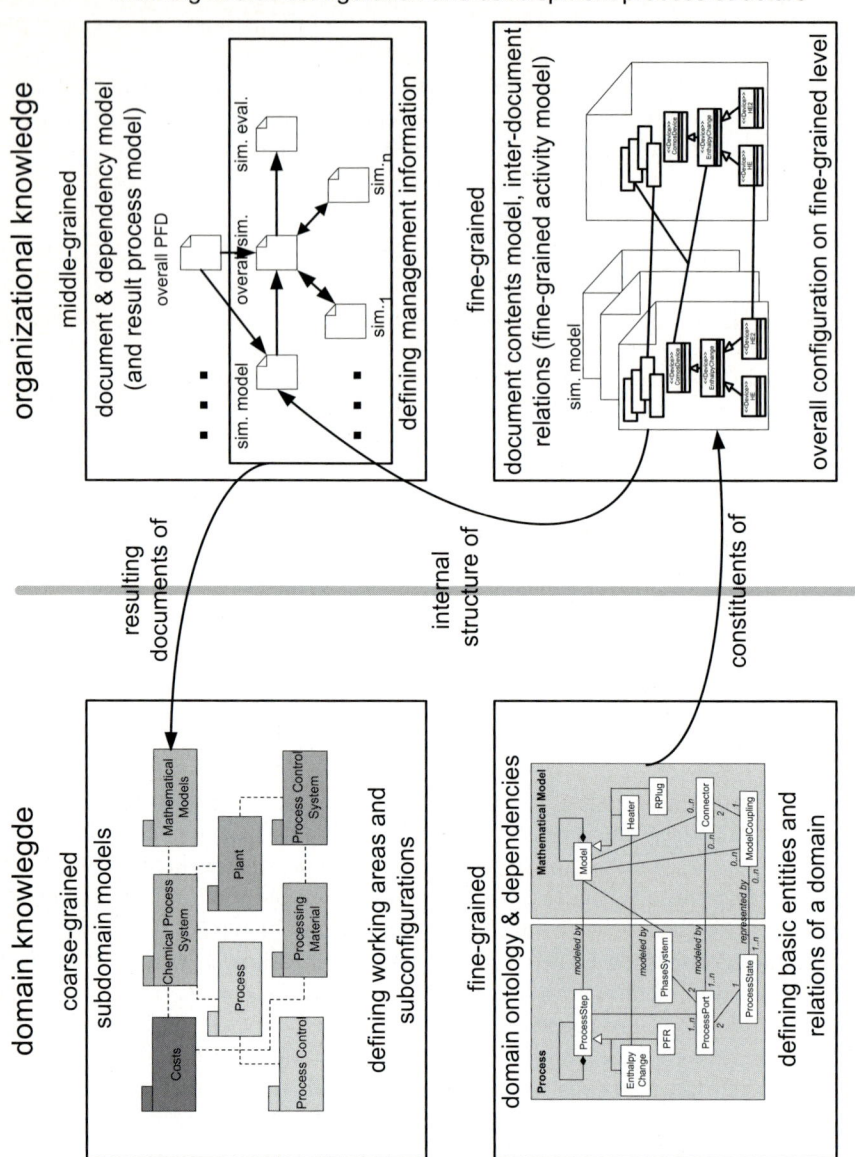

Fig. 6.2. Application model containing domain and organizational knowledge (the figure concentrates on the product perspective and mostly on type information)

of entity types referring to the products of the design process in a domain, and dependencies between entity types in form of an ontology.

As we have seen in Chap. 2, such domain knowledge models can be *layered*, transitions between layers express e.g. class-instance relations (see CLiP in Sect. 2.2). There may also be *clusters* of related items, which are concatenated in partial models (again to be found in CLiP or OntoCAPE).

Organization Knowledge Models

The activities within a working area and the corresponding design results (documents, subconfigurations) and the contents of these results depend on *organizational know-how*. This knowledge *differs* from organization to organization. Let us start with the design products and later discuss the design process.

If we structure a plant it is e.g. dependent from the company how a process flow structure is looking like. One company might prefer one single PFD covering the whole plant. Another might use a coarse PFD surveying the important parts of the plant, and use further PFDs to detail these parts. In general, this implies different result structures: Which documents describe a result, how do these documents reflect hierarchies, and which dependency relations exits between documents. We call the corresponding determination the *document and dependency model* (DDM).

The reader should note that dependency here only means, one result B depends on another result A, in the sense that if A is changing, B has potentially to be changed as well. The reader should also note that the DDM is also on *type level*. So, it defines document types and their relations, and not concrete documents and links between them. The overall configuration, as introduced in Sect. 1.1, depends on the DDM used in a company.

As the product structure on a medium-grained level depends on the organization, the internal structure of these documents is dependent as well: Which details appear within a document, how are these arranged, how are cross-references handled etc.? We call the corresponding part of the organizational model the *document contents model* (DCM). This model is again on the type level.

As already discussed, many fine-grained relations exist between increments of different documents, usually developed by different engineers. If now the documents have a specific internal structure, so do the mutual relations between increments of different documents. We call the model part, responsible for these fine-grained relations, *interdocument relation models*. These relations have to be consistent with the above mentioned entity type dependency relations (cf. Sect. 6.3), defined in domain knowledge. Again, the interdocument relation models are defined as relations between different types of increments of different documents.

The above models DDM, DCM, and the models for interdocument relations are on type level. A *concrete overall configuration* (cf. Sect. 1.1) is on *object level.* (Trivially, it also belongs to organizational knowledge.) It contains the occurring documents and their relations (being consistent with the DDM), the contents of these documents (being consistent with the DCM), and the interdocument relations, being consistent with the interdocument relation models. Theses structures on object level are not to be seen in Fig. 6.2.

We see that on the *domain knowledge* side, we are mostly on *type* level. On the *organizational* side we have *type* and *object* information.

Process Models Have to Be Divided as Well

Above, we mainly argued on the product view. We are now switching to the process view (coarsely shown in Fig. 6.2). The working area model was domain-specific, but enterprise-independent (top of left side). One level down in detail we get into the dynamics problem (cf. Sect. 3.4, but also 2.4 and 5.1). The corresponding activities thereby belong to complete documents. As the documents and dependency model was depending on the organization, so does the corresponding middle-grained process model. Let us call it *result process model* (top of right side). For any resulting document, there has to be a corresponding activity delivering the result. There, we find corresponding roles and team members. For any dependency relation between documents there is a corresponding change activity which changes a document B, if dependent on A, and if A has been altered. The result process model describes the task types and relations between them, so it is a model and not a concrete net. The corresponding task nets on object level have to be consistent with the result process model.

One further level down we have the process which builds up or maintains a document. The process determines how this is done. We call this *fine-grained activity model* (cf. Sects. 3.1, 2.4, and bottom of right side of Fig. 6.2). Of course, this level is again organization-dependent, as the structure to be built/changed is dependent. On this level we also find processes changing the internal structure of one document, thereby answering changes of another document. Again, we have concrete and fine-grained process structures on object level, being consistent to the activity model.

Domain versus Organization Knowledge

We see that going one level down of working areas, we get *organization-dependent knowledge* (documents, dependencies between documents, subprocesses and their relations). Even more, if we regard the contents of documents or their fine-grained work processes, we are on the organization-dependent side. The types of entities and of their relations, however, have to be consistent to the organization-independent *domain knowledge*.

So, the types of possible entities in form of ontologies are independent of the organization. The aggregation of the corresponding *entities* within *documents* and their fine-grained relations within and between documents, are organization-dependent. Hence, the configuration of documents and their contents are organized differently in each company. The same argumentation holds for the process side. This organization, however, has to *obey* the *determinations* given in the domain knowledge models or in the organizational models.

If, however, in a domain we have an *agreement* in form of a *standard*, which document types occur and which contents are to be found within these documents, this organizational knowledge becomes organization-independent. Thus, it can be counted as domain knowledge. Such a standardization often exists in engineering disciplines, in most cases this standardization is not complete. We see, that the division between domain and organizational knowledge can vary and that it depends on the amount of agreement in form of standards.

Application Domain Models: Products, Processes, Tools

Above, we have discussed the application model layer and we have divided it into domain knowledge and organizational knowledge. We could also have divided this layer into *products*, *processes*, and *further aspects*. This is to be shortly discussed here. Thereby, we only discuss information on type level.

The *product view* of the application model consists of a division into subdomains and of a domain terminology (ontology) together with further type dependencies. Both form a basic level of notions for the product side. The product view further contains the document and dependency model, the document contents model, and the interdocument relation model.

On the *process side* we have the working area model, being a part of the domain knowledge. Furthermore, we have the organization-dependent result processes model, and the fine-grained activity model.

A further aspect is, how to make a *connection* to process or product *models* of *existing tools* in the sense of a-posteriori integration. This is not an easy question, as models of given tools have to fit the application model layer in some way. Furthermore, we extend their functionality and, in some cases, we have built new tools as no suitable ones were available. This corresponds to documents and increments accessible and modifiable by tools, as well as to the commands manipulating these units. Please note that we speak of application *models* and not of *detailed* user *interface* determinations. So, if there is a tool to structure one big PFD, this tool can be used to create a document (document model) and offers certain commands for units and their composition to appear in this document (contents model).

It also should be remembered that results of the development process are on the *structural side* (how a plant is built up, of which part a simulation consists of), but also on the *value side* (what quantity has the output, how costly is the chemical process etc.). In the application models, *both aspects* have to be regarded, together with their mutual connections.

6.1.3 Application Models as Input for Constructing Novel Tools

In Chap. 3, we have discussed novel tools for supporting chemical engineering design processes. The *development process* of these tools essentially *contributes to* the *PPM*, because any tool construction process has to use the application layer of the PPM model. It has to extend it on lower levels, to fulfill the needs of tool implementation. This is the main message of this chapter which is discussed in the following three sections.

Application Models as Input

We shortly demonstrate here that the tool construction process needs *information* from the *application model* as *input*. This nicely shows the integrative approach of IMPROVE between engineers and computer scientists. We will explain the connection between application models (horizontal part (a) of Fig. 6.1) and tool construction (vertical columns (b), (c), and (d) of this figure). We do this in a sketchy way, as details are given in the three following Sects. 6.2, 6.3, and 6.4, explicitly discussing these columns.

Process-centered, fine-grained *technical* tool *support* (see Sects. 3.1 and 2.4) needs the knowledge contained in the *fine-grained activity* model. Indirectly, also, the middle-grained result process model is necessary, in order to see in which or between which documents the fine-grained support is to take place. As fine-grained operations also take the state of documents into account, the *document contents* model is needed as well.

Fine-grained interdocument *consistency handling* by integrators (see Sects. 3.2 and 2.2, 2.3) is on the product side. So, it does not need process knowledge. However, it needs the *document* and *dependency* model, the *document contents* and *interrelation* model, and the underlying *ontology* containing corresponding type relations.

Finally, middle-grained *reactive project management* (see Sects. 3.4 and 2.4) does not need fine-grained models at all. However, process as well as product knowledge is necessary. So, the *document* and *dependency* model is needed as well as the *result process* model.

Not All Tool Information on Application Model Layer

As we shall see, a significant amount of *information needed* for the tool construction process is not available on application model layer. This is in order as engineers responsible for the application model need and should not care about tool construction details. However, we also miss application-dependent information, which is too detailed to be found in the application model.

Let us explain the difference between information on the application model side on one hand and the necessary information for the tool construction process by taking reactive management as an example: *Further submodels* are necessary as those for variants, versions, or subconfigurations, or the relation between processes and products. They are application-dependent but regarded to be too detailed to appear on the application model layer. The same holds true for technical and *semantical details*, so as to specify when development tasks can start depending on the state of previous tasks. Thus, for the tool construction process, there is missing information, belonging to the application layer of Fig. 6.1, which is too specific and detailed to be handled there.

Next, *UI behavior* has to be *conceptually determined*, as entities to be accessed/manipulated by the tool have to be found as well as their relation to entities of the application model. This belongs to the upper parts of the three columns (b)–(d) of Fig. 6.1. Please note that detailed UI considerations are not to be handled in the PPM.

This leads to *principle problems* concerning the *connection* of *application models* to *lower levels*:

(a) How much technical and semantical knowledge is to be found on application layer? Who is delivering the missing and necessary application-dependent knowledge? If this knowledge is not found on the application layer, on which layer is it found?
(b) Clearly, the application model engineer should not become a UI specialist. However, the connections of the conceptual UI model to the application model have to be made clear. How and in which cooperation form is that done?
(c) There is also knowledge on the level of conceptual internal models for tools (middle layer of the columns) which is necessary and which has to be determined in connection with the application engineer. Again, how is this done?

This problem will be discussed in more detail in Sect. 6.4 exemplarily for column (d) of Fig. 6.1, though it also appears for columns (b) and (c) as well. Especially, these questions will be addressed in Sect. 6.5, dealing with the status and open problems of the PPM.

6.1.4 Categories of Open Problems for the Model

This subsection aims at *sketching open problems* which have to be addressed to achieve a comprehensive *PPM*. We only give an explanation on the level of problem categories rather than on that of individual problems.

Problem Categories for PPM Development

The *problem categories* for PPM development are the following (cf. again Fig. 6.1):

1. *Horizontal model integration*: The three columns define exemplary occurrences of PPMs beyond the application layer. They are related to three

sample tool construction/extension processes. For each specific tool construction process, the column works out the layers below the application model. The PPM in these three columns have not been investigated w.r.t. similarity or uniformity. So, it is not regarded in general up to now, how the three vertical columns (b)–(d) do fit together. Horizontal integration is considered yet only on application layer (see Sect. 2.6 and the last subsection). It is also regarded on platform layer, where the models are rather technical and implicit.

2. *Vertical integration and coherence*: The results to be described in this chapter are each specific for a group of IMPROVE using a specific modeling approach and a specific tool development machinery. So far, we have not identified which determinations occur on which layer. For example, it should be further elaborated how the UI model and other specific details mediate between application models and further layers of the PPM.

3. *Synergy*: In Sect. 5.5, we have described that the novel support concepts can be synergistically integrated (for example combining direct process support with integrators). This synergy crosses the columns in Fig. 6.1, as the novel concepts were realized by different groups. So, at the moment synergy is just programmed and not explicitly dealt within the PPM. This synergy model relating models of different columns is still missing.

4. *General and specific models*: As we have seen on application layer level, there are general parts (domain knowledge) and specific parts (organizational knowledge). Other divisions of "general" and "specific" can also be regarded, as common structures of different documents on conceptual tool models (see Chap. 3). In an ideal form, the PPM clearly separates different levels of generality (e.g. general aspects of development process structures, development processes in chemical engineering, for a specific subdomain, within a certain company, certain habits, etc.).

5. *Parameterization*: When having separated general from specific aspects, the parameterization aspect can be addressed. How to adapt a model to a specific context (company), to a specific chemical process, to a specific subdomain of chemical engineering etc.? Parameterization is handled within the three columns but not by a general mechanism.

6. *Adaptation to emerging knowledge*: We have learned (e.g. in Sect. 3.4) that process and product knowledge is emerging. It even can emerge during project execution. This not only requires model adaptations through all hierarchies within one PPM layer but also between different layers. This has been regarded only within one column. Can this adaptation, either between projects or during a project, be formulated uniformly within the whole PPM?

7. *Deviation to allow distributed models*: We have also learned that in the case of distributed development, different companies have to share some models but also need local deviations in order to get along with their contextual development cultures. Again, there are first models in column (d). However, this is a topic which has to be handled on all layers, also for

the transition between these layers, and uniformly between the different columns.
8. *Application to other domains*: The question arises, how to transfer our modeling and tool construction knowledge to other domains beyond chemical engineering. The problems are essentially the same in all design and development processes. The application model parts of above (but also the missing details) are the parameters to be changed, when switching the application domain, e.g. to software development.

Notations, Methodologies, and Tools for Modeling

Even more, there are open problems corresponding to the notations, methods, tools, and use of all of them (cf. Sect. 1.1).

Notations (languages) for describing a PPM on all different layers are not available and have to be invented. They have to cover all aspects of the model and all the different information being contained in layers. A *methodology* to use these notations building up a PPM is also not available.

Furthermore, *tools* to support notations or a methodology, when developing a PPM, are needed as well. Finally, notation, methodology, tool development, and their use are intertwined with each other. So, there are not only open problems w.r.t. modeling but also w.r.t. an infrastructure to support modeling.

With respect to notations, methods, and tools we also have some *partial results* in this book. For example, we find the C3 notation for work processes, notations for representing application domain models, or PROGRES as type- and instance-based notation for conceptual modeling of tools. Furthermore, we find methodologies for using these languages. We also find tools and tool adaptations to support product and process modeling.

However, we do not find a *modeling suite* for the layered PPM, where adaptation of modeling for different domains and experience knowledge, methodological use of modeling, transitions between layered models, etc. are uniformly supported.

6.1.5 Conclusion

Summary: In the first subsection, we introduced specific instances of a PPM resulting from three sample tool development processes. We gave a categorization of application models in the second subsection. This categorization distinguished domain-independent and -dependent models as well as the process and product view. We need this categorization essential for the open problem Section 6.5. In the third subsection, we have sketched which parts of the application model are needed for the construction of tools providing novel design process support. We also discussed that the application model is incomplete and has to be enriched during tool construction. In the last subsection, we have sketched problem categories to be addressed, in order to get a comprehensive, formal, and uniform PPM of design and development processes.

This section mainly serves as an introduction to the chapter. Section 2.6 has demonstrated that horizontal integration has been largely achieved on the application model layer. In contrast, such horizontal integration is still an open question on the layers below. The next three sections describe contributions to the PPM below the application layer gained in three specific tool construction processes (see (b), (c), (d) of Fig. 6.1). As we will show, vertical integration of the PPM for specific tools has been accomplished, constituting a good starting point for horizontal integration of the PPM beyond individual tool construction processes. The last section of this chapter will elaborate on a long list of open problems using the categories introduced in this introductory section.

6.2 Work Processes and Process-Centered Models and Tools

M. Miatidis, M. Theißen, M. Jarke, and W. Marquardt

Abstract. The first vertical column of the layered process/product model (PPM) addresses the direct, experience-based support at the technical workplaces of designers. More specifically, we demonstrate the transition from application domain models to executable tool models, focussing on the process perspective of the PPM. This vertical column is jointly realized by the A1 subproject, providing the fine-grained application domain models, and the B1 subproject, dealing with their conversion to executable tool models to be used by process-integrated tools. In this contribution, we provide an outline of the cooperation results.

6.2.1 Introduction

Direct, experience-based process support at the technical workplaces of designers requires to present crystallized knowledge from previous design processes to a designer while he is interacting with heterogeneous software tools. Thus, on the one hand, a fine-grained formal representation of both the design activities and design products is needed. On the other hand, the various software tools involved in the design process must be able to enact the process without interfering with the designer's creativity.

In this section, we describe the transition from application domain models to executable tools. This transition forms the first vertical column (b) of the layered process/product model (PPM) shown in Fig. 6.1. Two subprojects of IMPROVE are involved: The A1 subproject addresses the analysis of the design processes in the IMPROVE reference scenario (cf. Sect. 1.2.2) and the provision of fine-grained work process and decision models (Sects. 2.4 and 2.5). These models serve as a starting point for the development of tool support employing the process integration mechanism developed by the B1 subproject (Sect. 3.1).

The section is structured as follows: First, we give an overview of the application domain models of A1 and the executable tool models of B1, respectively (Subsect. 6.2.2). We successively describe the transition from application domain to tool models in Subsect. 6.2.3. Then, in Subsect. 6.2.4, we *relate application and tool models to the overall layered PPM*. Finally, we conclude with a summary and a discussion of still open issues (Subsect. 6.2.5).

6.2.2 Application Domain and Tool Models

Each of the subprojects A1 and B1 uses modeling approaches that are well-suited for the representation of the relevant aspects addressed by the subproject. The A1 models represent the perspective of the application domain, i.e.,

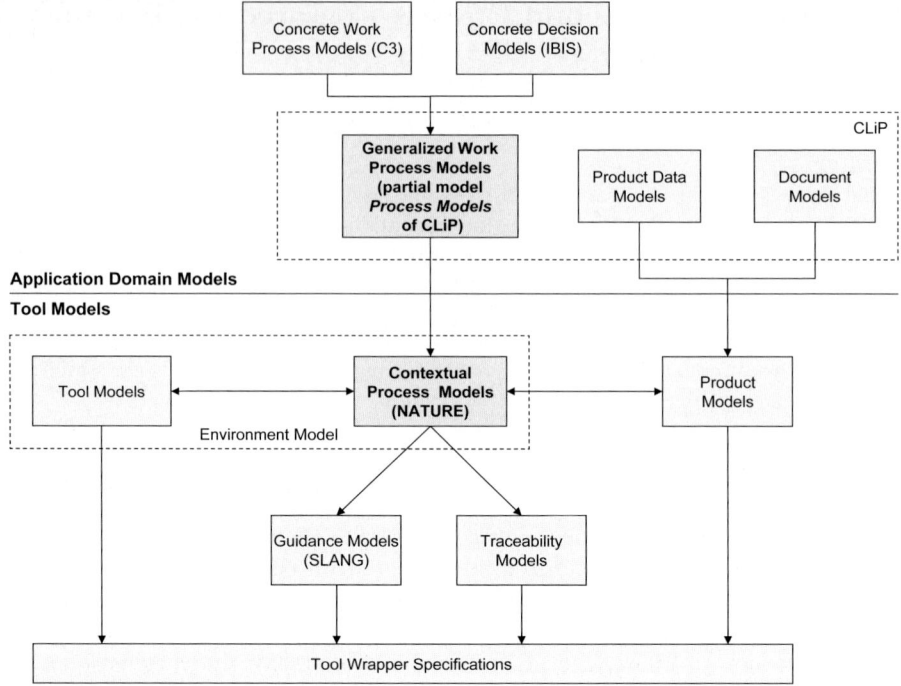

Fig. 6.3. Integration of work process models with process-centered tool models

design processes in chemical engineering, whereas the tool builder's perspective (B1) strives for enactable models for process-integrated tools.

We first discuss the application domain models of the vertical column (see upper part of Fig. 6.3):

- *Work processes* performed during a *concrete design project* are represented using the semi-formal C3 notation (see Subsect. 2.4.4). In particular, C3 provides modeling concepts for the activities in design processes, the information items (products) created and employed during the activities, and the complex interdependencies between different activities (such as synchronous communication between several actors) and between activities and information items (information flow).
- Notations like IBIS [798] and DRL [808] (see also Sect. 2.5) can be used for modeling *design decisions and the underlying rationale in concrete design projects*.
- For the formal representation of *generalized work process models*, the partial model *Process Models* of CLiP (Subsect. 2.4.6) can be used. Generalized work processes serve as templates for similar design projects in the future. As *Process Models* inherits some concepts from IBIS, it also provides a formal representation of rationale-related aspects.

- In addition to *Process Models*, CLiP contains several partial models addressing the representation of the *products* and *documents* involved in design processes (see Subsect. 2.2.3).

The lower part of Fig. 6.3 shows the five submodels employed during tool construction:

- A *contextual process model* plays a central role for the modeling of method guidance provided by process-integrated tools. Its representation is based on the NATURE process metamodel [201] that is able to capture the context of the designer while a design activity is enacted. A context is associated with a given situation made up of product states and with the decision which can be taken in the situation. Different granularity levels of contexts can be handled, including the direct execution of elementary actions, systematic plans composed of sub-contexts, and choices among alternative contexts (see also Fig. 3.11). The alternatives of a choice context can be annotated with positions and arguments that catch the underlying design rationale.
- A *tool model* describes the services and the GUI capabilities of a process-integrated tool. Such a description is mandatory for the external triggering of tool services and for adapting the tool GUI according to process definitions in order to provide integrated method guidance from inside the tool. Process and tool models are integrated in so-called *environment models* that uniformly describe the automation, guidance, and enactment services which establish the foundation for process integration.
- A *product model* extends the existing data models of process-integrated tools with further domain-specific product elements and their relationships.
- A *guidance model* enriches NATURE descriptions of method fragments with additional semantics in order to make them interpretable by a process engine. The PRIME framework is open and flexible enough to support various process-modeling languages for the interpretation of method fragments. For the IMPROVE case study, two basic interpretable languages are employed: An extended version of the SLANG language (based on Petri nets, [527]) is used for method fragments involving multiple data dependencies. Method fragments that simply shift control from one context to another are represented by UML state charts [880].
- A *traceability model* extends a NATURE process model in order to enable the almost automatic capture of extended traceability information concerning the design history inside process-integrated tools. It is able to represent the process steps executed, the products involved, the major decisions taken, and their possible interrelations.

6.2.3 From Application Domain Models to Tool Models

The first vertical column of the layered PPM addresses the refinement of application domain models to tool models, the latter serving as a specification for a process-integrated environment based on process-integrated software tools. In the sequel, we describe the basic steps of this transition.

To a large extent, these steps are accomplished manually; they require a tight cooperation between two specialized groups of experts: application domain experts and tool builders. The first group comprises experts from the chemical engineering domain, who create application domain models on a semi-formal or formal level, whereas members of the latter group are responsible for their transformation to fine-grained tool models and the implementation of tool wrappers according to the tool models.

Initially, application domain experts define a domain ontology that captures the knowledge of the fine-grained cooperative work among designers. The aim of this task is the identification and formalization of design knowledge based on the recording of real design processes by means of the C3 notation. Thanks to its ease of use, C3 allows the active participation of experienced designers (i.e., chemical engineers and technicians who are typically not familiar with more elaborate, but also more complex process modeling techniques) in the creation of design process models. These models of concrete design processes must be generalized in order to make them applicable for similar design projects in the future, and they must be formalized to eliminate any ambiguity. The transition from semi-formal C3 models of concrete design processes to formal CLiP models of generalized processes is described in detail in Subsects. 2.4.2 and 2.4.5.

From the software engineering perspective, work process models in CLiP are evaluated by method engineers in cooperation with chemical engineers, and they are translated into method definitions represented as NATURE contextual models. Thus, the interpretation is primarily based on the mapping of CLiP activities to the three types of NATURE contexts.

The transition can be supported at the metamodel level as follows:

- *Decision activities*, modeling either the selection or the evaluation of several alternatives, can be directly mapped to *choice contexts*. Then, each alternative activity is also mapped to an alternative context of the corresponding type. Further, IBIS decision models can be directly represented by *arguments* for or against each alternative in the choice context. Support of DRL models is an issue of future work.
- *Synthesis* and *analysis activities*, containing other subactivities, can be represented as *plan contexts* that can be provided as alternatives of a choice context. Then, each subactivity is modeled by its corresponding context type and inserted in the control flow of the parent plan context.
- Elementary activities that are directly executed by tool actions and, thus, cannot be further decomposed, can be represented as *executable contexts*.

- *Input information* of a CLiP activity can be directly represented as a collection of *products*.
- For the representation of *output information*, we can distinguish two cases: When referring to elementary actions that cannot be further refined, output information can be modeled as product modified by the applied action of the corresponding executable context. In case of a composite activity, output information cannot be directly modeled and attached to a plan context. Yet, it is subsumed by the output information of the executable contexts transitively contained in them.

Once the NATURE contextual model has been established, the *tool wrapper builder*, responsible for the process integration of software tools, plays a dominant role in the remainder of the transition process. In cooperation with the method engineer, he evaluates the degree to which a relevant software tool (or even a set of tools) conforms to the requirements for process integration as detailed in Sect. 3.1. Based on this evaluation, it is decided whether a full or partial process integration of the tool in consideration is possible. Subsequently, executable contexts are coupled with existing tool services, plan context definitions are represented by means of interpretable formalisms (i.e., SLANG or UML state charts), and, if a full process integration is possible and desired, new command elements are introduced, which represent the intentions of choice context alternatives.

The final step of the transition is the implementation of tool wrappers according to the generic tool wrapper architecture of PRIME, whose role is the dispatching of process requests to tool-specific ones, according to the environment model definitions. Inside the tool wrapper, a product model is implemented that must be consistent to the product and document models in CLiP. This model is realized using a tool-specific or XML-based format.

6.2.4 Relation to the Overall Process/Product Model

In this subsection, we relate the application and tool models presented so far to the overall layered PPM (Fig. 6.1).

The overall PPM is organized along the classification dimension of four layers: application model layer, external model layer, internal model layer and basic model layer. In a hierarchical top-down decomposition, each layer uses concepts from its direct upper layer and enriches them with additional aspects.

The vertical column shown in Fig. 6.3 can be directly mapped to the layered architecture illustrated in Fig. 6.4. In the following, we outline the basic concepts and modeling abstractions employed at each layer.

Layer 1 consists of all high-level application models presented in this section. Application models can be divided in two categories: the semi-formal and the formal ones. Semi-formal models are described using the C3 notation to capture design processes at a coarse-grained level as well as DRL or IBIS to represent the rationale of the major decisions made by the developer.

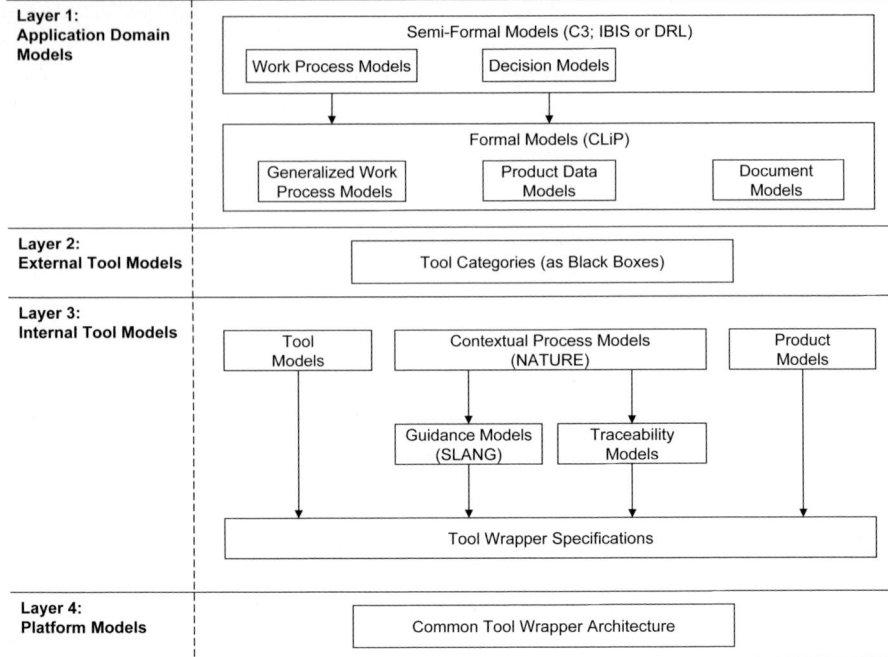

Fig. 6.4. Relation of the models for experience-based support to the layers of the overall PPM

Reusable process and decision patterns are represented as generalized templates in CLiP. CLiP also provides concepts for detailing the data model of the products which are transformed by the design activities.

Layer 2 provides the external representation of all engaged tools. Until now, there is no possibility to describe all details of the participating tools (e.g., the sequences of the commands enacted inside tools are not represented). Tools are just seen as 'black boxes' that support the execution of specific process steps.

The internal tool models needed for tool construction reside at layer 3. Such tool models comprise the contextual process, tool and product models that are based on the disseminated application models. The first two flavors use generic modeling formalisms for their representation (i.e., elements from the environment metamodel), whereas the latter depends on the tool. At this layer, tools are further refined to the services and GUI capabilities to be considered for their process-conformed behavior. Contextual models are further extended to formulate guidance and traceability models inside the tool wrapper.

Finally, layer 4 focuses on platform-related aspects. Its responsibility is to hide the platform-specific details from layer 3 and ensure the unobstructed use of process-integrated tools across many platforms. This purpose is largely attained through the construction of process integration wrappers according to the common tool wrapper architecture (Sect. 3.1). These wrappers use cross-platform protocols for communicating with tools.

6.2.5 Summary and Outlook

In this section, we have presented the first vertical column of the overall PPM addressing aspects of experience-based support. We have presented the corresponding application domain and tool models, described the transition from the first to the latter, and shown how these models relate to the four layers of the overall PPM.

In the following, we outline two open issues that still remain to be solved:

- The transition from application domain models to tool models is performed manually. In consequence, its results are prone to inconsistencies and errors. The implementation of converters for automating (at least parts of) this transition could overcome these difficulties.
- The transition as described above relies on a representation of process templates using the partial model *Process Models* of CLiP. Substituting *Process Models* with its recently developed successor, the Process Ontology described in Subsect. 2.4.6, is straightforward. Similarly, the other partial models of CLiP can be replaced with OntoCAPE (cf. Subsect. 2.2.4). In contrast, the integration of the Decision Ontology (see Subsect. 2.5), which replaces the simple IBIS-based rationale model inside *Process Models* with a more expressive variant of DRL, is still an open issue.

6.3 Model Dependencies, Fine-Grained Relations, and Integrator Tools

S. Becker, W. Marquardt, J. Morbach, and M. Nagl

Abstract. The models developed within subprojects A2 and B2 together form one of the vertical columns of the process/product model. The application domain models of A2 are refined to tool models of B2 such that integrator tools can be realized. The process of building integrators is rather well understood in general, as is the process of refining the application domain models of A2 to tool models of B2. Nevertheless, important parts are missing for a concise and layered process/product model.

6.3.1 Introduction

In this section, we present the transition from *application domain models* to *executable integrator tools* (cf. Sect. 3.2). The transition represents a part of the overall process/product model (PPM), namely the vertical column from subproject A2, supplying the application domain models, to subproject B2, importing these models and refining them to executable specifications for integrator tools. The main focus is on the product perspective of the process/product model, as integrators deal with products of the development process.

This *section* is structured as *follows*: In the next subsection, we give a short summary of the product perspective of the application domain models necessary for integrators and developed within A2, and the tool models of B2. Both are represented from the PPM perspective. Then, we discuss the transition from application domain models to tool models in Subsect. 6.3.3. After that, we relate these steps to the PPM layer structure presented in Sect. 6.1. The last subsection discusses open issues of our approach.

6.3.2 Application Domain and Tool Models

In the following, the different submodels of the application domain and the tool models, as presented in Chap. 2 and Sect. 3.2, will be summarized. Here, we focus on the submodels being relevant for an integrator development.

Summary of Application Domain Models

In the application domain models, *four submodels* relevant for the realization of integrators can be identified (see upper part of Fig. 6.5 and Sect. 6.1):

- *Working areas* (called partial models in CLiP) define a coarse-grained model. Working areas represent activities as well as their results. The

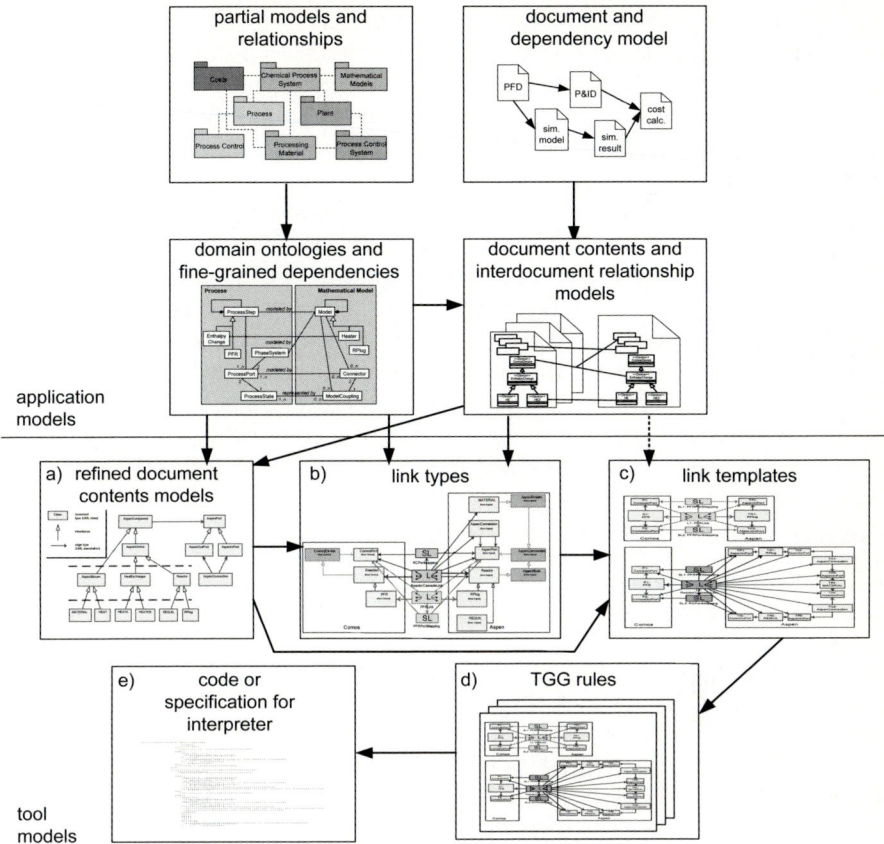

Fig. 6.5. From application models to tools

working area model contains the definition of all relevant clusters of activity together with their coarse-grained *relations*. The activities and their results are not further structured.
- For each working area, one or more *ontology modules* defining its basic entities are specified. *Fine-grained relationships* between entities of the *same* working area are introduced as well. All definitions are provided as multi-layered class hierarchies specifying inheritance, aggregation, instantiation, and association relationships. For example, different types of reactor models are introduced in the area **mathematical model** and aggregation relationships define that each reactor model can aggregate an arbitrary number of external **connectors**. Furthermore, also fine-grained relationships between entities of *different* partial models are declared. For instance, a relationship **modeled by** between a reactor in the **behavior** partial model and a reactor model in the **process** model is given.

- A middle-grained model is used to declare types of *documents* and their *dependency* relations. Just the existence of these documents is declared and not their contents. For instance, the document types PFD (denoting process flow diagrams) and SimulationSpec (specifying mathematical models for process simulation) are introduced as well as the dependency relationship between them (cf. Subsect. 2.3.2).
- For each document type, a *document contents model* determines which entities from the partial models' ontologies can be contained in a corresponding document instance. For instance, the PFD document can contain arbitrary process steps defined in the partial model process as well as substances defined in the partial model chemical process material. Fine-grained interdocument relationships that are not already included in the ontologies are added. Furthermore, the document structure can be refined, indicating the order or the type of structural entities (such as heading, text block, or table) that have to be contained in the document (cf. Subsect. 2.3.4).

Summary of Tool Models

Five submodels of the tool model are used, when following the approach sketched in Sect. 3.2 (see lower part of Fig. 6.5):

- *Formal and refined document contents models* are needed to supply all information which is later used for the definition of integration rules. Currently, only type hierarchies are used, defining all entities and relationships that are to be considered during integration. Future work will deal with adding more structural information. These models are similar to the document content models of the application layer which, at the moment, are not elaborated. However, they are much more detailed and have to be formal. Also, further information is needed here that is of no interest on the application domain layer.
- *Link types* are defined to relate certain entity types of one document to entity types of another. Arbitrary many-to-many relationships are supported. For instance, it can be expressed that a combination of ideal reactor models (CSTR, PFR) and interconnecting streams as well as the aggregated connectors within a SimulationSpec document correspond to a single reactor and its ports represented in a PFD document. Link types are used for two purposes: First, they provide a formal notation for a part of the organizational knowledge. Second, they constrain link templates that are defined in the next model.
- *Link templates* define fine-grained corresponding patterns in different documents that can be related to each other. Each link template is an instance of a link type. For example, a SimulationSpec pattern consisting of a cascade of two ideal reactors connected by a stream could be related to a PFD pattern consisting of a single reactor. This link template would be an instance of the link type used as example above. Link templates are not type but instance patterns.

- Link templates are purely declarative. That is, they do not provide any operational semantics. Instead, they have to be further refined to *operational* forward, backward, or consistency checking *rules* using the *triple graph grammar* approach (TGG) [413].
- Triple graph grammar rules cannot be executed directly. Instead, they have to be translated into appropriate graph rewriting operations which, in turn, are translated to executable *code*. Alternatively, triple rules are translated to textual *rule specifications* that are executed by the integrator framework.

The application domain models presented so far (see again the upper part of Fig. 6.5) are not necessarily of a sufficient degree of fine-granularity and formality to be directly translated to the according tool submodels. There are some steps to be performed manually, delivering further and missing information. Additionally, some modifications have to be done from one tool model to the other. In the following subsection, we will indicate which particular information has to be added for each type of a tool model, and where (i.e., tool builder, domain expert, or tool) the corresponding information needs to be delivered from.

6.3.3 From Application Domain Models to Tools

Using the middle-grained document and dependency model, all the *locations* can be *identified*, which can be *supported* by using an *integrator*. For the realization of each integrator, all tool submodels have to be provided.

In the following, we give an overview from where the required information for the different tool submodels can be obtained from. Some of the information can be *collected* from application domain submodels while others have to be *provided* by the tool builder.

How to Get to Tool Models?

First, *document contents models* for any of the documents to be integrated have to be refined and *formalized*, providing enough details for the tool building process (see a) of Fig. 6.5). A basis for that are the application document contents models (see upper part of Fig. 6.5) that specify which entities of partial models are relevant for which document. The definition of the corresponding entities and some of their relationships can be extracted from the ontologies.

In case these application submodels are completely formal and contain all the necessary information, no transformation needs to be done. Otherwise, the tool builder has to *translate* semi-formal application models into a *formal* notation, which is a UML class diagram in our approach.

The refined document contents model needs *not* contain *all definitions* given in the application document contents models. Just those entities and relations have to be specified that are to be considered during integration. On the other hand, some information required for tool building is not included in the application domain models, as it is of no interest for the application model builder. For instance, *technical* document entities, necessary to build efficient tools, have to be *defined additionally*. Links have to be introduced to access increments efficiently, to abbreviate access paths, and the like.

Next, for each integrator, a set of *link types* must be defined. These are based on the document contents models and the additional relationships defined in the document contents and the inter-document relationship model. Inter-document relationships that are already defined in the application domain models (within ontologies as well as document contents and relationship models) can be transformed into link types. This can be done *automatically* if the application models are formal. Otherwise, a *manual translation* has to be performed.

It is very *unlikely* that the interdocument *relationships* of the application layer are *complete* in the sense that no additional ones are needed. This is due to two reasons: Firstly, many of the languages commonly used for application domain modeling do not support the definition of complex many-to-many relationships. Secondly, from the application point of view, it is sufficient to depict the most important relationships. A complete set of inter-document relationships is only needed when actually building an integrator.

As a result, in most cases a lot of *link types* have to be *defined manually* by the tool builder (cf. b) of Fig. 6.5). This has to be performed in close cooperation with an application domain expert to ensure the definition of correct link types. The reader should note that relations between entity types on one hand come from the ontologies of the application domain, saying that an entity type is semantically related to another one. On the other hand, link types are related to organization knowledge, as increments of certain types are within documents, the relations between which have to be obeyed by an integrator. Organizational knowledge determines, which documents exist and which structures appear within them.

Link templates are defined as instances of link types using UML object diagrams (see c) of Fig. 6.5). As link templates use patterns to define possible fine-grained relationships between the documents to be integrated, they are much more precise than link types. All pattern nodes and edges have to be instances of entities and associations defined in the refined document contents model. Furthermore, the link that relates the patterns has to be an instance of a link type. Thus, all link templates can be checked for consistency with respect to link types and the underlying document content models. So, link templates belong to organizational knowledge, as the patterns on both sides are parts of document structures, which are part of this knowledge.

Although patterns are restricted, there are still a lot of options for the design of a certain pair of patterns. Thus, the definition of patterns is currently carried out completely and manually by the *tool builder*. Some related patterns can be (manually) *derived* from the refined document contents models, especially, if they belong to a generic document structure (e.g., a hierarchy). Others can only be *defined* in close cooperation with an application domain expert.

There is an inherent *problem* with patterns. On the one hand, the definition of related fine-grained patterns is often not pursued during application domain modeling, because it is a very complex modeling task, especially, if interdependencies of different patterns are concerned. Furthermore, it is typically not in the focus of the application domain expert. The tool builder, on the other hand, lacks the necessary domain knowledge to deliver the information. One option to solve this issue is to jointly extend the interdocument relationship model at the beginning of the tool construction process. Then, a more precise definition of necessary information for tool construction is at hand.

After link templates have been defined, operational forward, backward, and correspondence analysis *rules* can be *derived* following the TGG approach (see d) of Fig. 6.5). If the link templates are restricted to using only context and non-context nodes connected by edges, this derivation can be performed *automatically*. If further graph transformation language constructs (such as paths, set-valued nodes, etc.) are used, not all operational rules are deterministic. As non-determinism is not supported by our rule execution approach, in this case the TGG rules have to be postprocessed *manually*. This is done by the tool builder, as only few domain knowledge is required. Another manual task is necessary, if attribute assignments have been defined that cannot be inverted to match the derived rule's direction.

We support two different approaches for the execution of integration rules (cf. Subsect. 3.2.5 and e) of Fig. 6.5). *Code* can be generated automatically that executes the integration rules. Alternatively, *rule specifications* can be stored as XML files that are read and interpreted by the integrator framework. Some additional coding has to be done (i) to provide a friendly, domain-specific user interface, and (ii) to connect to the existing applications providing the documents to be integrated. This is done by the tool builder following the requirements of the application domain.

Application Models vs. Tool Models

In summary, the *relations* between *application* and *tool models* are as follows:

- Not all information of document contents models of the application domain layer is needed for the process of integrator development. Only the parts describing relations between documents need to be regarded.
- However, further technical details are needed for tool models, e.g. to access increments, to navigate efficiently, etc. These details have to be added by the tool builder.
- Link types usually have to be elaborated manually. They represent types of increments to be connected by an integrator. Today, the interdocument relationship models on the application domain layer are not of enough detail.

The following information is not part of application models, but *needs application knowledge*:

- Link templates are on the object level. As the complete application model layer describes type information, these templates have to be added by the tool builder.
- Rules can be derived from link templates only in simple forms. Usually, they are also developed by the tool builder.
- The rest, either code or rule specifications, are derived automatically from rules.

6.3.4 Relation to the Overall Process/Product Model

In this subsection, we *relate* the different submodels used for the definition and realization of integrators to the overall structure of the layered *process/product model*, as sketched in Fig. 6.1.

For fine-grained integrators, mainly the *product perspective* of the PPM is relevant. Nevertheless, there are dependencies between development processes and the requirements for the integrators to be used within these processes. This aspect is not further discussed here. Here we focus on the product perspective.

The PPM, as explained in Sect. 6.1, is organized top-down in *five layers*, each refining the next upper layer by adding specific aspects. This is also true for the integrator column (b) of Fig. 6.1. Figure 6.6 gives details for this column, which comprises an overview of the layers and an assignment of each of the integrator-relevant submodels to one of the layers, as discussed in the previous subsections.

We will now discuss the *relevance* of each layer for integrator realization. Furthermore, we will identify *requirements* for the *PPM* from the perspective of integrator realization. It will also be shown that the way from application models to the realization of integrators, though being well understood, still does not comply entirely with the structure of the PPM.

On *layer 1*, all application domain submodels, sketched so far, are available. *Layer 3* consists of all tool models, explained so far. The latter also comprises the source code of / specification for the integrator.

Layer 2 contains models describing the *external presentation* (UI) of all relevant tools. For integrators, it comprises the user interface, the tools we integrated, the external presentation of the links administrated by the integrator, and the behavior of the integrator.

Currently, all these UI aspects are *not explicitly defined*. The determinations of the external tool properties are partially contained in the refined document contents model on layer 3. The link presentation is fixed for all integrators. The behavior of the integrator can be adapted within a certain range. This feature is currently realized by the tool builder during the implementation of an integrator. Hence, all these aspects are only implicitly handled by the tool builder.

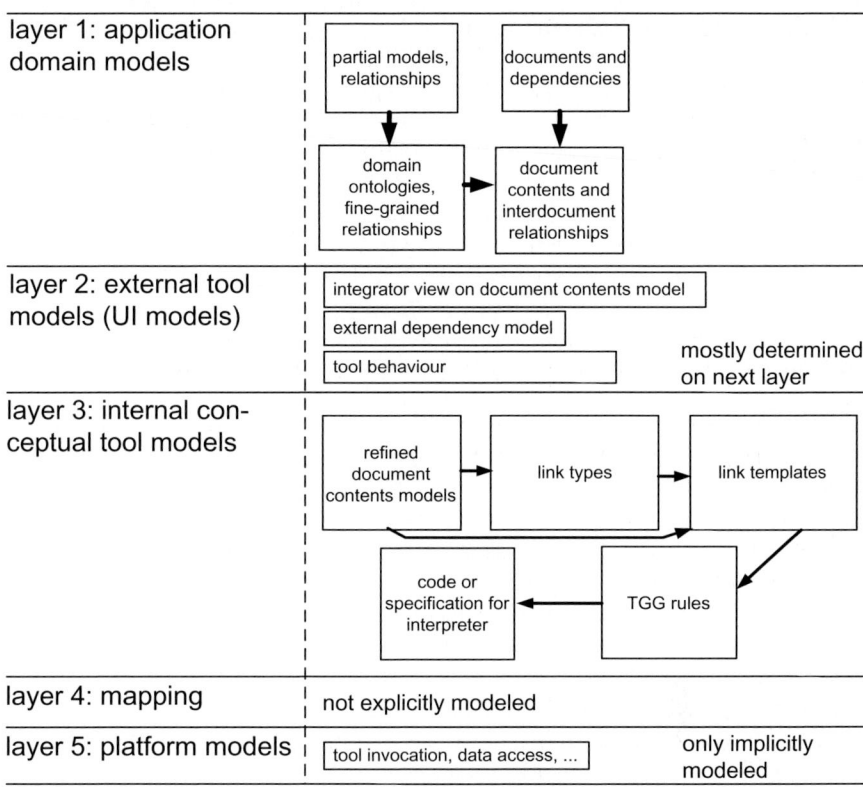

Fig. 6.6. Layers of the process/product model and their contents, here for integrator construction

On layer 3, we find *internal tool models* – either as abstract rewriting rule specifications being executable by an interpreter, or as equivalent source code. Furthermore, we find the models belonging to the *methodological construction* of integrators (link types, link templates, TGG rules), which were discussed above. The document contents models at the moment are also found on this level, they should be a part of the application domain model.

The tools on layer 3 should be kept independent of the underlying platform. The platform-related aspects are to be added on *layer 4* (mapping) and on *layer 5* (platform models). Again, this is only done *implicitly* for the integrator realization. All integrators are limited to specific platforms. Nevertheless, some independence for the tools to be integrated is provided. As the access to existing tools is handled by wrappers, different communication methods can be used to connect the integrator to tools running on *arbitrary platforms*. These wrappers make use of the platform functionality provided by the platform subprojects of IMPROVE. Altogether, these aspects of mapping and platform models are implicitly coded into the wrappers and not rigorously modeled.

6.3.5 Summary and Open Issues

In this section, an overview of the *different application* domain and *tool models* needed for the realization of integrators has been given and their mutual relationships have been explained. Additionally, the *integration* of these models into the *PPM framework* has been discussed.

There is an *advanced methodology* for integrator construction. This methodology explicitly handles abstractions on 2 of 5 layers of the PPM. There are a lot of topics, where models are explicitly handled and refined top-down across the layers.

Nevertheless, for a *complete* process/product model, there are still some abstractions missing and, therefore, some *open problems* remain:

- The transitions from application models to tool models are not formalized. Currently, the transitions are performed manually or are assisted by simple converters that still require manual postprocessing.
- The specification of fine-grained interdocument relationships in the application domain model is not sufficient to derive integration rules. Currently, the tool builder has to perform the definition of related patterns manually with the help of a domain expert. Whether it is possible to extend the application domain model without making it too complex and too tool-related remains an open issue.
- Even, if there is a formal document contents model as part of the application domain model, it needs to be extended to provide advanced tool support for defining related patterns.
- The approach sketched so far is tailored to the kind of integrators being in our research focus. It should be adaptable to other types of integrators or other integration approaches as well.
- In the previous subsection, we pointed out that layer 2 (external presentation, UI) and layer 4 (platform mapping) of the PPM are not yet represented in our current integrator methodology. They are only handled implicitly up to now.
- The distinction between domain- and company-specific knowledge has to be elaborated and clarified within the integrator construction process. This is especially important for integrators, as many of them are to be found in a development scenario. Also, the adaptation to a specific context is a matter.

6.4 Administration Models and Management Tools

R. Hai, T. Heer, M. Heller, M. Nagl, R. Schneider, B. Westfechtel, and R. Wörzberger

Abstract. One of the vertical columns in the overall process/product model deals with the cooperation of subprojects I1 and B4. Both study the support for reactive process management in dynamic development processes. In this section we highlight the transition from application models developed in subproject I1 to tool models for reactive management of subproject B4. We summarize our findings w.r.t. the development of management tool models as well as their connections to application models. The section focuses on a process-oriented viewpoint. Products and resources of development processes can be discussed analogously. We identify the missing parts which need to be further investigated in order to get a comprehensive and integrated process/product model, here for reactive management.

6.4.1 Introduction

In this section, we describe the transition *from application models to executable tools*. The vertical column of the PPM from subproject I1 to subproject B4 is regarded, dealing with reactive process management, see Fig. 6.1. The application models capture process-related aspects of work processes within development processes in subproject I1, and subproject B4 has developed executable tool models to derive *process management* tools supporting development process managers and process engineers. So, this section is to describe and evaluate the *contribution* of both subprojects to the dominating problem of developing a PPM.

The section is *structured* as follows: First, we summarize the application models of subproject I1 and the executable tool models of subproject B4 in Subsect. 6.4.2. The main part of this section deals with describing the transition from application models to executable tool models (Subsect. 6.4.3), and its relation to the formal process/product model (Subsect. 6.4.4). We finish the section by giving open problems and a conclusion.

6.4.2 Application Models and Tool Models

In this subsection we describe the *models* developed within the *subprojects* I1 and B4. The integration of these models is addressed in the next subsection.

Summary of Application Models

An *overview* over the *application models* and the tool builder's models relevant for development process *management* is given in Fig. 6.7. In the upper part of the figure, the IMPROVE application domain models in chemical engineering are shown:

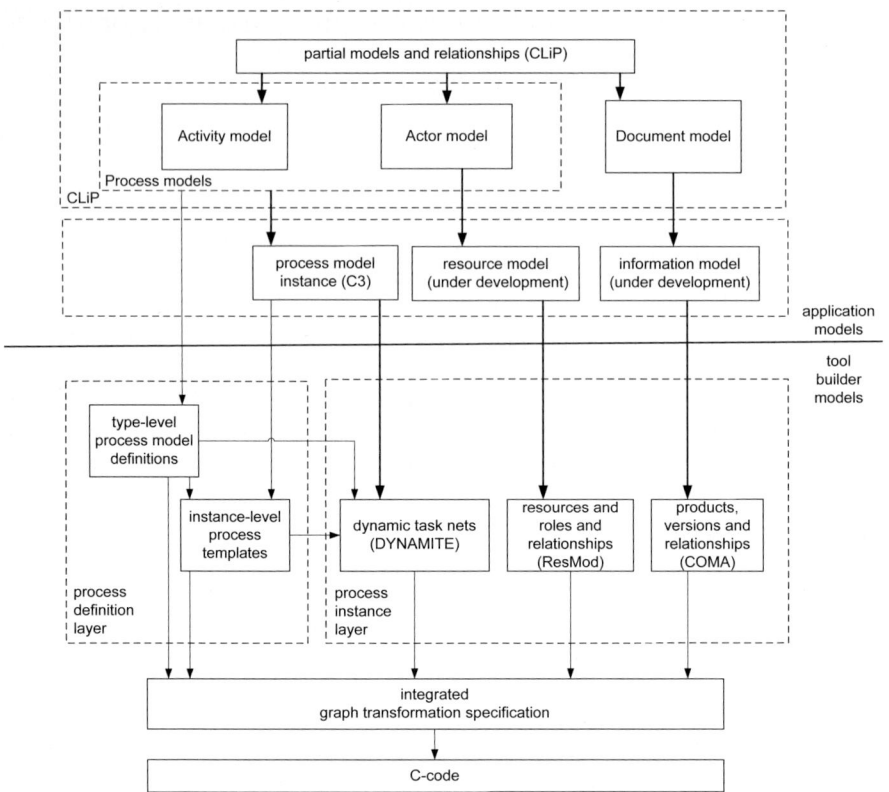

Fig. 6.7. Application and tool models for management

1. The object-oriented data model *CLiP (Conceptual Lifecycle Process Model)* [14, 19] for product data of the design process and the corresponding work process, as described in Sects. 2.2 and 2.4, defines partial models structuring the engineering domain into several working areas. The relationships between the partial models are also contained in CLiP. For instance, there is a partial model `Process Models` (details below). Within `Process Models`, the model `Activity` and the model `Actor` are connected by the relationship `skill`.

   In this section, we mainly *focus* on *process-related modeling* and therefore we highlight the partial model `Process Models` (cf. Chap. 2). This submodel defines several modeling concepts to express important structures for work processes like `activity class`, `input and output information`, `tool`, `goal` and their relationships. For example, the `Simulation reactor` activity creates the information `Reactor simulation result`. This information is defined in the partial model `Document model`, which contains the information-related modeling concepts. The resource-related modeling concepts can be found in the CLiP partial model `Actor model`.

Roles are only implicitly defined, as a skill-oriented approach is used in order to define the work capabilities required for a certain role. For example, an activity `simulation reactor` may need actors with a skill `Reaction kinetic knowledge`.

2. Work processes on the *instance level* are modeled by the C3 process modeling notation [116]. On the instance level, complete work processes can be modeled and analyzed, e.g., to identify potentials as to shorten development time. For example, all important aspects of the Polyamide-6 case study used in IMPROVE have been modeled using the C3 process modeling language (cf. Sect. 1.2). For this purpose, C3 contains elements to model process-, information-, and resource-related aspects of work processes. The basic elements of C3 are `activity`, `role`, and `information` with relationships like `control flow` and `information flow`. Currently, the clear focus is on the process side while the information models and role models only cover basic aspects and are currently under development.

Summary of Tool Models

The *tool models* are displayed in the lower part of Fig. 6.7. While the semi-formal application models describe the application domain, the tool models have to be formal in order to be used for building process management tools.

1. In the center the process model for dynamic *task* nets (DYNAMITE), the *product* model (COMA), and the *resource* model (ResMod) are shown. These models are tightly integrated with each other and form an integrated management model allowing to describe development processes on type level (knowledge) as well as on instance level (concrete projects or templates).
   For *example*, a *dynamic task net* resembles a running development process with all tasks, resources, and all documents which are created during the process. Elements of dynamic task nets are `task`, `input parameter`, `output parameter`, `control` or `feedback flow`, `data flow`, etc.
   As stated above, in order to be able to build tools, the *execution semantics* have to be formally specified. The models are generic in the sense that no application-specific information is contained. They form the core of the process management system AHEAD which supports the interleaved planning and execution of development processes.
2. The generic models of AHEAD can easily be *adapted* to an application *domain* or to specific project *constraints* by defining standard activity types, standard workflows, and workflow templates. We have used the wide spread modeling language UML to model such constraints and parametrization aspects in an object-oriented way on the class level.
   Class diagrams and *object* diagrams are used to create process model definitions or process templates, respectively. For example, standard types for activities can be defined by introducing new classes like `Design Reactor`

or `Simulate Reactor` within in a class diagram together with relationships between classes, like `sequential control flow`.
3. The three generic models DYNAMITE, COMA, and ResMod are formalized in *graph transformation* specifications. Specific adaptations of these models defined in class diagrams and object diagrams are transformed to specific graph grammar specifications.
4. Both *generic* and *specific* graph *specifications* are *combined* and executable C code is generated from them. This C-code is embedded into a tool building framework. In our case, the configurable user interface and the core application logic of AHEAD can be generated semi-automatically as explained in detail in Sect. 3.4.

The *transition* between *application* models to executable *tool* models cannot be easily realized. For example, some information which is necessary in a tool model might not be explicitly modeled on the application layer, as different aspects are targeted at both layers. In the following section, we describe how all necessary information is obtained by either adding additional information or by deriving the information from existing application models.

6.4.3 From Application Models to Tools

In order to offer a domain-specific support system for the management of development processes, the information of all *tool models* is needed to build the AHEAD system (subproject B4). *Additionally*, the *application* models developed in subproject I1 are exploited to provide the missing context information, needed to adapt the AHEAD system to a specific context [154].

Instance Level Application Models

On the instance level, dynamic task nets, products, and resources used within a specific development process are modeled. On the application model side, similar process information is foremost contained in C3 nets. Application modeling experts create *process templates* in the form of C3 nets to define best-practice work processes. These C3 nets can be transferred structurally into *dynamic task nets*. Currently, a set of structural restrictions has to apply for the C3 nets used. Dynamic task nets can be generated on the tool side as the surrogates of the process templates modeled as C3 nets. We have realized an *integrator* for the mapping of C3 nets into dynamic task nets (see Sect. 3.2).

Product-related *information* can also be contained in C3 nets and be transferred into dynamic task nets or the product model of AHEAD, respectively. Currently, we do not extract these data with the integrator. In the C3 net, it can be captured that input or output documents of activities require a specific document type. *Resource*-related information like actors or actor roles required to perform an activity are, however, carried over into a dynamic task net.

Class Level Tool Modeling: Connection to Application Models

On the class level, we are dealing on the tool side with process model definitions and process templates to define structural and behavioral knowledge about processes. This information is located at the class or *type level*, e.g. task types, document types etc. and their relationships for a specific context can be defined.

The CLiP models and domain ontologies can be searched for *standard types* of work process activities or document types. The partial models usually contain such elements which have been identified to be of broader relevance to the respective application domain of the partial model. For example, the activity type `Design Reactor` might be contained in a UML class diagram according to the domain ontology of the partial model `Work Processes` where the activity concept is located. Essentially, the same or similar modeling concepts are used in CLiP and in the metamodel underlying the process model definitions and process templates of AHEAD. Currently, the information found in CLiP models has to be integrated manually into UML class diagrams used in the AHEAD approach. For example, for the activity type `Design Reactor` in CLiP, a new class `Design Reactor Task` is created in the class diagrams. In this case, the AHEAD system will allow for the instantiation of tasks with this specific type in dynamic task nets.

Relationships defined in CLiP models can be reflected in class diagrams by introducing new associations between classes. Currently, this is not possible in our approach, as only a limited set of default associations has been implemented to connect classes, like associations denoting control flow relationships or data flow relationships. We use UML stereotypes to define the type of an association link between two classes. Up to now, other relationships, e.g., those with a more semantical character, are neglected, although their integration is easily possible by simply using additional UML stereotypes as annotations for associations.

Application Modeling for Tool Adaptations

We use the broadly distributed UML notation for the explicit purpose that *application* modeling *experts* can create process model definitions and process templates for AHEAD. Nevertheless, application domain experts and *tool building experts* can work together to arrive at such models more quickly until the application expert has gained enough experience in adopting our specific use of the UML for process modeling purposes. Specific tasks, like introducing new unforeseen dependencies between classes or new requirements leading to necessary technical modifications of AHEAD, can be discussed in a short-circuit mode of cooperation. By the way, this approach of bringing experts of different domains closer together, was often followed in the IMPROVE project.

If process model definitions or process templates have been defined to introduce specific adaptations of the otherwise unrestricted generic AHEAD model, some further steps are needed. These steps are only necessary because of the specific tool generation approach followed in the B4 project, which is based on the semi-automatic generation of user interface prototypes from graph transformation specifications: (a) First, the tool builder uses a transformation tool to transform the UML class or object diagrams into graph transformation specifications, which contain the specific parametrization data for AHEAD. If no process model definitions are defined, AHEAD uses built-in default types for tasks and documents etc. (b) Second, the tool builder combines the newly generated specific graph specification with the generic graph transformation part containing the AHEAD management model. From this overall specification, executable C-Code is generated in an automatic step. (c) Finally, the tool builder has to integrate this C-code with the pre-configured user interfaces to form the overall AHEAD system.

6.4.4 Relation to the Overall Process/Product Model

In this subsection, we look at the application and tool models in order to present their *relations* to the layered overall *process/product model* of Fig. 6.1.

The overall process/product model is decomposed into five layers: application model layer, external model layer, internal model layer, mapping layer, and basic model layer. Going from top to bottom, each layer adds specific aspects which have not been covered at the layers above. We now discuss the *vertical column (d)* regarding reactive management (cf. Fig. 6.1).

The way from application models to tools for reactive management does not match smoothly with the idea of the overall process/product model developed so far. We are now going to identify which aspects of each layer are relevant for reactive management and highlight some open problems.

Layer 1 of the overall process/product model deals with all *application domain models* for the process and its products mentioned in this section. Among them, we can identify domain knowledge models to structure the application domain and organizational knowledge models which contain knowledge how processes are carried out in different subdomains or companies. For example, work processes modeled by application domain experts on a medium-grained level can be found on this layer. Similarly, medium-grained product models and resource models also belong to layer 1.

Layer 2 contains the *external models* of tools for different users. The user interface notations used for modeling different process or product aspects of tools should be found on this layer. Likewise, the representation of complex commands of the tools is located here, because they offer application-oriented functionalities for the user. Currently, we do not have such explicitly modeled external process or product models within IMPROVE. External models are indirectly introduced on the next layer.

On layer 3, *internal models* of tools are located. They are best represented by formal and executable models which are immediately usable to derive tools. All tool models belong to this layer, such as the AHEAD management model

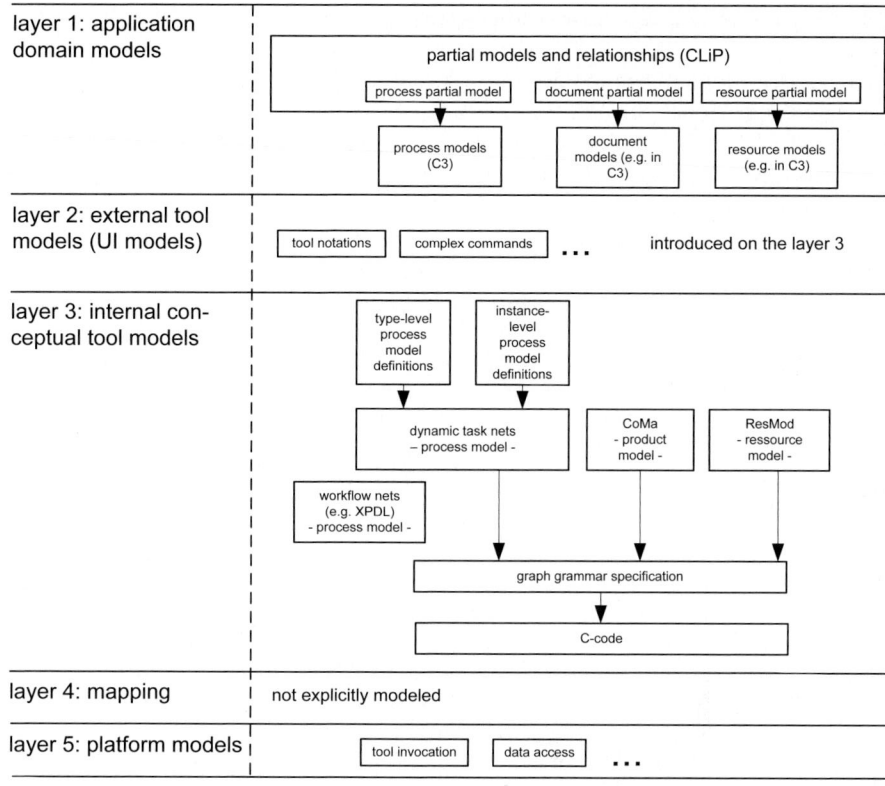

Fig. 6.8. Application models and tools models

(DYNAMITE, COMA, and ResMod). These models have to ensure that the user interface and functionality described on layer 2 is fulfilled by the tools. To be more specific: UI details are found here, in particular there are specs for complex commands. How the UI is built is not specified but introduced in the tool construction process.

Finally, layers 4 and 5 contain models containing all *platform aspects* and a *mapping* from conceptual models to platforms. These models are only relevant in order to implement the tools independently of specific execution platforms. As previously mentioned, the realization of the complex tool components on the basis of operating system processes is handled on this layer. Currently, the tools for reactive management developed within IMPROVE are limited to specific operating systems and programming languages. However, AHEAD is coupled with other tools of IMPROVE using platform services such as access to document repositories.

6.4.5 Open Problems and Conclusions

In this section, we discussed for the vertical column (d) reactive management how the *transition* from *application models to tools* takes place. We focused on the process-side and explained, how application models and tool models are related to each other. Finally, we discussed the relation of achieved results to the process/product model.

Although we have acquired a good understanding of application models and tool models corresponding to management support on a medium-grained level, some *open problems* still remain:

1. It has to be investigated if some of the *application models* could be *extended* in order to match the tool models more closely, or vice versa. Currently, these model gaps are bridged manually and for specific cases only.
2. As we stated above, a tool-independent process/product model on the *external model* layer is still *missing*.
3. We have achieved some results regarding the transformation of application models to tool-specific models when specific tools are chosen [154]. For that purpose, we have already described how a framework for the definition, analysis, improvement, and management of inter-organizational design processes involving different modeling formalisms and heterogeneous workflow tools could look like. This framework combines models, methodologies, and tools from the IMPROVE project. Its key features are to bridge the gap between modeling and execution of inter-organizational design processes and the seamless execution support for dynamic and static parts of the overall process, both by appropriate process management systems. These *results* need to be *generalized*. For example, a methodology could be developed for the extraction of common modeling concepts and the harmonization of the process models into a single uniform model which could serve as a mediation model between specific process models. The same applies for product models.

6.5 Process/Product Model: Status and Open Problems

M. Nagl

Abstract. In this section we give a summary and evaluation of what we have achieved w.r.t. the PPM. Especially, we discuss the open problems still to be solved. The message is that there are a lot of nice results, but there has also a lot to be done in order to get a uniform and comprehensive PPM.

6.5.1 Introduction

Sections 6.2, 6.3, and 6.4 described some *promising results* about how to get tool functionality by well-defined tool construction processes starting from elaborate application domain models, as described in Sects. 2.6 and 6.1. Furthermore, these sections also discussed how the information accumulated during these processes should be organized in a layered PPM.

Nevertheless, there are *more open problems* than solutions. We have already sketched these problems in the sense of a categorization in Sect. 6.1. We have detailed the problems identified during our work on the tool construction case studies in Sects. 6.2 to 6.4.

This *section* will give a *discussion* of these *open problems*: They are completely enumerated, ordered, and ranked according to their degree of severity. When discussing the problems, we are also discussing the results we have achieved so far.

The *section* is structured as *follows*: We take up the categorization of modeling problems of Sect. 6.1 and give a detailed explanation. Then, we address the corresponding modeling infrastructure problem. In total, we come up with 17 big problems to be solved yet. In the conclusion we give some arguments that the PPM problem was too hard to be solved within one project and/or the available time.

The *evaluation discussion* of this chapter is further elaborated in Chap. 8, which gives a résumé of IMPROVE's main results from other perspectives.

6.5.2 Status and Modeling Problems

Horizontal Integration Problems

(A) The case study represented in Fig. 6.1 by column (b) uses one and the studies of column (c) and (d) use another specification methodology and tool implementation machinery, namely NATURE (b), see Sects. 3.1 and 6.2, and Graph Transformation for (c) and (d), see Sects. 3.2, 6.3 and 3.4, 6.4. Hence, these *conceptual realization models differ* tremendously. In the Graph Transformation approach (columns (c) and (d)), we specify tool behavior by corresponding before/after states of a document's internal form. Accordingly, the description is an abstract implementation.

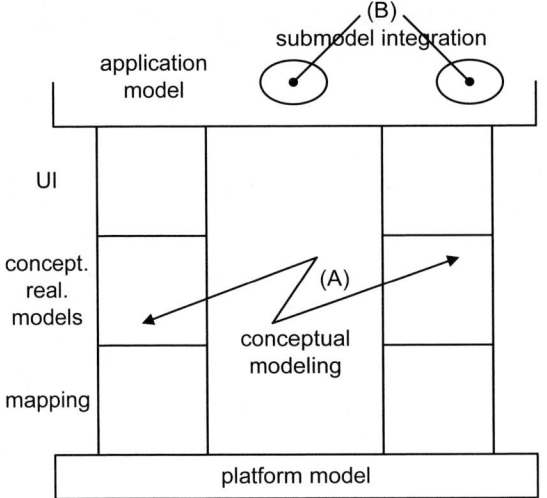

Fig. 6.9. Horizontal model integration

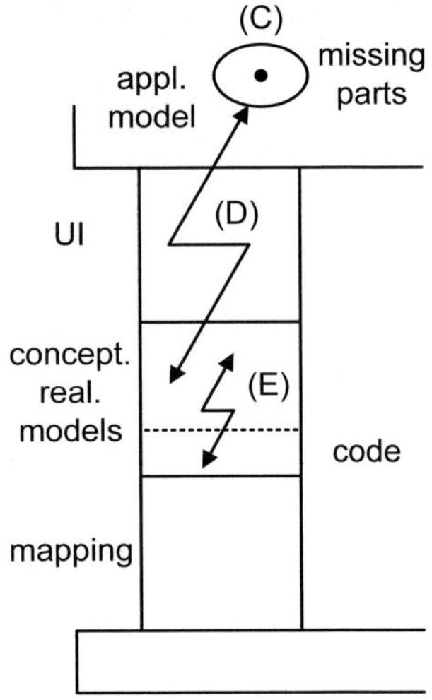

Fig. 6.10. Vertical model integration

In the NATURE approach (see column (b)), the implementation is done manually. The metamodel description determines some frame to be followed.

On the application domain layer (a) of Fig. 6.1, there is only one model, however being composed of different submodels, see Sects. 2.6 and 6.1. The same is true on the platform layer (e). Correspondingly, the mapping layer is less critical. The UI layer is handled implicitly, as has been argued. Summing up, the *main problem* is on the *conceptual realization* layer (see Fig. 6.9).

So, what is missing is either a new general modeling formalism for conceptual realization models which can replace NATURE and Graph Transformations (or any other suitable approach) as a superset. However, this idea is not very realistic. What is more realistic is that some *"glue model"* can be given, by which it is possible to define the connections between suitable formalisms. Here, more specifically, that means that a NATURE spec and a Graph Transformation spec can be unified in one spec.

(B) On the *application* layer, also some *problems* remain, as it was shown in Sect. 2.6:

The challenge in application domain modeling is to deal with the complexity of the domain, here of chemical engineering. Application domain modeling should aim at a general purpose *model* which can be *used* in *various contexts*. However, since modeling has to account to some extent for the purpose of the desired model, there is an inherent conflict. A design of the model architecture for *extensibility* is the key to success. Though various structuring means have been incorporated in the model to account for further extensions, there is room for improvement. In particular, some experimental validation in case studies is required to justify the design decisions.

A more specific aspect is the *integration* of product data, documents, work processes, and decisions into *one model*. The approach taken so far seems promising but still needs more work to refine and validate the resulting model.

Last but not least, there is a need for specific features in the model architecture which allows the *tailoring* of the model to an *organization* in an easy way to reflect the specific organizational details. This is particularly true for documents and work processes which reflect to a large extent the culture of an organization. Inter-organizational issues like the integration of information models in different organizations need to be considered, too.

Vertical Integration and Coherence

Each of the three case studies covers a demonstration of a coherent methodology for tool construction and for the corresponding modeling steps. The following three problems regarding vertical integration and coherence remain (cf. Fig. 6.10):

(C) Which *parts* of the *model* are required for tool construction and are *not covered* by the *application domain* model? Examples include missing model parts (e.g. for version/variant/configuration control) or semantic details of models and tools (e.g. developer task synchronization). They are typically not introduced on the application layer. If all these details were included, this layer would not only give application domain models, but ultimately specify tools and their behavior.

(D) *Structural* and *behavioral UI* details are also to be specified before conceptual tool construction can start on layer 3. UI specification should not be understood as, for example, layout of masks or similar details. However, the UI functionality has to be made precise in the sense of specifying requirements for tools. There are *two possibilities*: Either there is a model layer 2 for UI and tool functionality, which, realistically, has to be developed by both domain experts and computer scientists. Or, since UI specifications contribute to the application domain model and also to the conceptual tool realization model, we can regard the second layer as a modeling activity, the results of which are being delivered on the models of the first and the third layer. The results of this joint activity then are given by the domain experts on the first, and by computer scientists on the third layer.

(E) Another topic is how to proceed on layer 3 when *transforming conceptual realization models to code*, which is also considered to be a model, but of an implicit form. In the three Sects. 6.2, 6.3, and 6.4 all possible transformations have been practiced: methodological but manual coding, direct interpretation of a specification model, or generating equivalent code from a specification model. Hence, there is only the problem left to understand the differences between these transformations and to apply these transformation approaches uniformly to get a suitable implementation.

Synergy

Synergistic integration (see Sect. 5.5) is not handled in the PPM up to now. Hence, there is the question, *where* to place this *part* of the *model*. The corresponding specification should not to be included in the application domain model, because the definition of synergy refers to tool functionality. Also, synergy is not easy to be determined on a conceptual realization layer, since different specification formalisms are used in the tool realization case studies (i.e. column (b) on the one and columns (c) and (d) on the other hand).

(F) It is unclear at the moment, what kind of information has to be modeled *on the application layer* to cover synergy between extended tools. If, for example, we look at the case study of Sect. 6.3, we see that models of the application layer do not determine integrators, but only prerequisites which integrators have to obey. In the same way, synergy would have to be specified. But what are these prerequisites? Specifications about the UI are also needed, the question only is where and how it is provided (see vertical integration discussion of above). Specification on the conceptual realization layer is difficult, as different formalisms are involved (see horizontal integration of above). For columns (c) and (d) it is, in principal, possible to write one graph transformation specification to express a synergistic tool integration effect, as it has been done e.g. for the integrator between PFD and the management configuration in Sect. 5.5. However, the generated code has to be executed on one computer. Recently, we have studied distributed graph transformation specifications [377], where the generated code can be executed on different machines. This, however, would only solve the synergistic integration of column (c) and (d), but not the integration with (b).

General and Specific Models

There are only *partial answers* to the problem how to distinguish between general and specific models. This problem has to be addressed on the application layer, as well as on the conceptual tool models layer below.

Within the *application* domain models, we should *distinguish* between general *domain* knowledge and specific *organizational* knowledge. This distinction has not yet been addressed explicitly (see Chap. 2) and it is not easy to make (see remarks in Sect. 6.1).

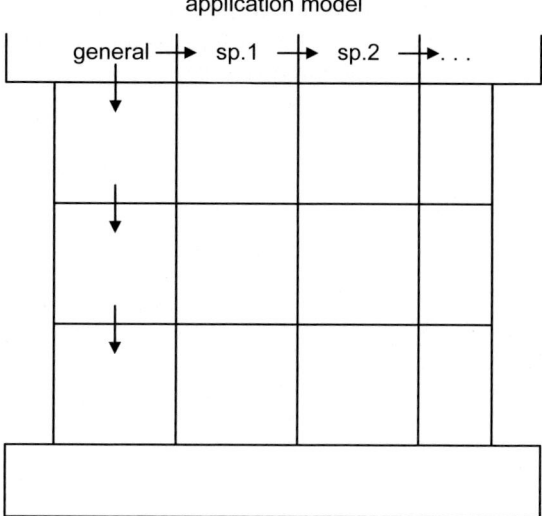

Fig. 6.11. Different degrees of specific determinations to a general one on one layer and between layers

In the case study reported in Sect. 6.2 (column (b) in Fig. 6.1), tool-specific models are determined individually. The NATURE *context model* only defines what is common to all direct process support tools.

In the case study on integrator modeling (see Sect. 6.3, column (c) in Fig. 6.1) we find *general* specifications, either as a basic layer of the conceptual realization model in the form of graph transformation *rules*, or in a coded form as a *part* of the integrator framework. *Specific* models are introduced to represent link types, link templates, and rules of TGGs. Thereby, different forms of determinations for "specific" are introduced in one step.

Models in the case study of Sect. 6.4 (column (d) in Fig. 6.1), on the conceptual realization layer, have an explicit basic specification, which is *independent* of the application domain. The *characteristics* of an application domain are specified interactively by the so-called parameterization environment (types of tasks, documents etc.). A graph transformation specification is generated from these interactive specifications, which extends the basic and domain-independent specification. Again, using this parameterization environment, different kinds of determinations for "specific" could have been introduced.

This parameterization environment could have been used to introduce *different degrees* of *specific determinations*, one after the other (see Fig. 6.11): A determination sp.1 could introduce a specific area within chemical engineering as, for example, Polyamide-6 development. A following one sp.2 introduces the specifics of a company, sp.3 those of a specific methodology used in department of a company, etc.

(G) A general and uniform modeling solution would result in a *model* on every layer to distinguish "general" from different forms of "specific" (cf. again Fig. 6.11). The specific details have to be determined on application model layer, e.g. by interactive determinations, for the application model or by determinations to be defined on lower layers. A modeling approach *distinguishing uniformly between "general" models and different degrees of "specifics"* on any of the different modeling layers, from application layer down to tool code, is still missing. A prerequisite would be a unified model to combine specifications for different types of tools (see argumentation on horizontal integration of above). Also, the distinction between the general part of the model and different forms of specialization parts should be handled uniformly, when mapping from one layer down to the next (see again Fig. 6.11).

Parameterization

The question of parameterization is closely related to the distinction between general and specific models of the last subsection. Again, we have partial results on the application layer as well as in the tool construction case studies. Parametrization means not only to distinguish general and specific determinations. It, even more, means to have suitable *mechanisms* to easily *formulate* the *specific determinations* and to *add* them *to the general part* of a model.

On the application layer ((a) in Fig. 6.1) we should parameterize on model construction time between *domain* knowledge and specific *organizational* knowledge.

Parameterization in the case study of Sect. 6.2 (column (b) in Fig. 6.1) is done as follows: Process *fragments* can be exchanged, they are expressed directly in a process modeling language. The underlying metamodel remains unchanged.

In the integrator case study of Sect. 6.3 (column (c) in Fig. 6.1), on conceptual realization layer, *dependency* relations on type level can be specified which serve for parameterization in the sense of parameterizing the tool to corresponding entity type situations. Object *patterns* can be defined to parameterize it to concrete aggregations of objects. Both belong to organizational knowledge, as both demand for determining which types or which entities occur in which documents. Finally, parameterization in the sense of tool *behavior* is done by transformation rules. So, altogether, there are different parameterization mechanisms, all of them being applied at tool construction time.

Parameterization in the case study of reactive management (see Sect. 6.3, column (d) of Fig. 6.1) has to account for domain and organizational knowledge, but also for tool behavior. It is done interactively using the *modeling environment*. These interactive specifications are translated to graph transformation specifications, extending the general graph transformation specification, which is independent of all parameterization details. Hence, all parameterization mechanisms are employed at tool realization time.

(H) As stated above in (G), there are results on the application layer as well as in the tool models of the case studies, but there is *no* general and *uniform solution* at hand *for parameterization*, which distinguishes between the "general" model and "specific" parts on every layer and serves for the transformations between layers (see Fig. 6.11). This is due to the fact that different modeling techniques and different forms of specification mechanisms for general and specific determinations are used at different locations of the PPM.

Adaptation to Emerging Knowledge

The adaptation of the PPM to emerging knowledge can be discussed in brevity, because the argumentation is essentially the same as in the case of parameterization. There is no difference to parameterization, if adaptation due to evolution is only between tool development projects. Then, a new tool construction and model adaptation process can be carried out. If, however, adaptation is supposed to happen while tools are in use, then some round-trip engineering has to take place [235]. In this subsection, we only discuss *adaptation* while *tools are in use*. This kind of adaptation has been exemplarily employed in the integrator (see Sects. 3.2 and 6.3) and in the reactive administration (Sects. 3.4 and 6.4) case studies.

(I) There are a lot of specific solutions for adapting to emerging knowledge at tool construction or at tool use time, but there are no general ones. Hence, the yet unsolved problem is how to organize the modeling process across different *layers* and, consequently, the corresponding tool construction and tool adaptation process such that *evolution* due to emerging knowledge is handled *uniformly*, whatever evolution concept is applied. This has to be true, irrespective of whether evolution is before a tool construction process, during the implementation process of a tool, or even at the time when a tool is used.

Models Accounting for Distributed Cooperation

The sketch of the PPM as indicated in Fig. 6.1 would have to be duplicated for any design process carried out in *different companies*. To clarify, let us consider the reactive management case study as an example (see Sects. 3.4 and 6.4). We take the simple scenario of a delegated task to be carried out by a subcontractor company. Then, it is unrealistic to assume a single application management model, corresponding methods, and tools to be applied across the different companies. Indeed, there are different procedures and habits on both sides of the cooperation, possibly also different subdomains (as chemical engineering and plastics processing). Hence, in each of the companies participating in the distributed design process, there must be the possibility to *structure* the *models differently*, to parameterize, and to adapt them correspondingly.

636   M. Nagl

The corresponding modeling problem in the process support case study has been regarded in Sects. 3.1 and 6.2, assuming a *homogeneous* process warehouse for both companies. In Sects. 3.2 and 6.3, we considered the problem in the integrator case study *implicitly*, since integrators can bridge heterogeneity across companies in a design process. There, the integrator is realized in one company and the adaptation to the situation in another company is done via wrappers and corresponding modeling. In Sects. 3.4 and 6.4, we discussed the situation that different models exist on both sides of a cooperation. So, here we have some *explicit* results.

(J) In general, there are different models in the organizations of the partners cooperating in the design process, though there may be some commonalities. Hence, we must be able to explicitly formulate the *commonalities* in the models but also the *differences* for any support concept (cf. Fig. 6.12). This problem is not solved yet. If the differences have to be handled only before a project starts, then this problem is similar to the parameterization problem (H), but it is occurring multiple times. If adaptation to different models of cooperating partners is possible at design process execution (or tool use) time, then we have a multiple evolution problem (see (I)). In any case, distributed cooperation makes the modeling problem harder, as further aspects like independence, contracts, security, knowledge protection etc. are also involved.

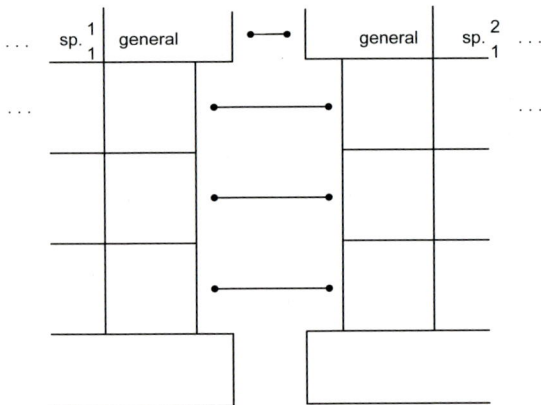

Fig. 6.12. Distributed Models: similar and different

Application to Other Domains

In Sect. 6.1, we have argued that domain knowledge is common to all tools regarded in this volume, but organizational knowledge differs from enterprise to enterprise. In this subsection, the question is how to *change* the *domain*.

Of course, most of the results of this book – and therefore also the modeling results – can be transferred to other domains, say mechanical engineering, electrical engineering, or software development. This means, that the domain knowledge has to be exchanged. In turn, also the organization knowledge has to be exchanged.

The procedure of modeling the other domain is the same as described above. The *parameters* of parameterization are already contained or should be already contained in the application model description. In addition, further determinations have to be added (see vertical integration of above). This serves for the *basic parameterization* of being in another domain or organization.

When moving to another domain, however, also the *other aspects*, namely adaptation to emerging knowledge, distributed models etc. have to be taken into account. So, summing up, there is nothing completely new. The problems are the same as in Chemical Engineering.

(K) Transforming the PPM question and its solutions to another domain means to *aggregate all* the above *modeling problems*, as all appear in any other domain as well. So, the new problem is the collection of all above discussed problems. There is a less and a more ambitious way of transformation.

The less ambitious ways to *start again*, *using* our *knowledge* and our experience about PPM modeling. Then, we have the same problems, as extensively discussed above.

The ambitious solution would use more advanced *reuse techniques* for *transforming* our PPM. Assuming that the above problems were solved, then we could use generic models across different application domains with well-defined instantiations and parameterization mechanisms in order to get a domain-specific PPM. In this case, there would not be a new development by just using modeling knowledge. Instead, there would be a well-defined process by which we get a specific PPM by making use of advanced reuse techniques.

6.5.3 Modeling and New Concepts

We have argued that the part of the PPM below the application layer was exemplarily considered for new support concepts (see (b), (c), and (d) in Fig. 6.1). We assumed a *specific methodology* for realizing any of these new concepts. The methodology chosen was determined by the available knowledge and infrastructure of the research group working on the case study. Therefore, the following open problems can be posted.

(L) If the *new concepts* and tool functionality (experience-based support, consistency of integration, etc.) were realized using *other methodologies* and corresponding infrastructures than those described in this book, would the above *modeling problems* be the same or would they *change*? For example, if we would try to model and realize integrators by some event/trigger/action approach, would that induce other problems?

In some situations, there are also *"light" solutions* to *realize* these support concepts – e.g. batch XML transformators for specific and simple cases of integrators. They have been solved by manual implementation. If we would try to incorporate them into the general modeling approach: would the modeling layers, the problems in these layers, and the transition between layers be the same?

(M) In Sect. 1.1 the above four new support concepts have been introduced to support a design process in chemical engineering. They can also be used in other design or development processes in other engineering disciplines. It is not likely that these concepts are the only possible and reasonable ones for supporting design/development processes. If *further support concepts* are taken into account, would this change any of the modeling problems? Are there new modeling problems to be regarded in addition?

6.5.4 Modeling Infrastructure

All of the above problems deal with modeling aspects. In this subsection, we state further problems, dealing with the *modeling infrastructure* (languages, methods, tools), which is also neither complete nor uniform (see Fig. 6.13).

(N) We have discussed the problem of different perspectives, facets, and degrees of detail in product/process modeling, from an application domain model down to platforms, from explicit models down to implicit code. There is no specification language at hand which covers all these different aspects (*broadband specification language*). Furthermore, there is no language with clear concepts for layering and hierarchies within layers, for the transitions between layers, but also for the other aspects including general or specific determinations, parameterizations etc. Hence, the first open problem is that a suitable broadband specification language is missing. We have used different languages on different layers and for different tool construction processes.

(O) The same holds true for the *methodology* to use such a language. Trivially, this methodology is also missing. The methodology would have to answer general questions, e.g. how to deal with multiple hierarchies, where to use inheritance or other concepts, etc. This general methodology is also missing. We have only specific methods on the application layer and below.

(P) Furthermore, there is no *tool support* at hand supporting the *specifier* when building up a PPM using a broadband modeling language and corresponding methodology.

(Q) Finally, generally usable *submodels*, generic submodels, etc. could be *defined* and *reused*, whenever a PPM is supposed to be built. These general engineering modeling components as well as a method for their definition and reuse are missing. Especially, common mechanisms should be given dealing with subprocess interaction, subproduct integration, and reactivity (see Sect. 1.1).

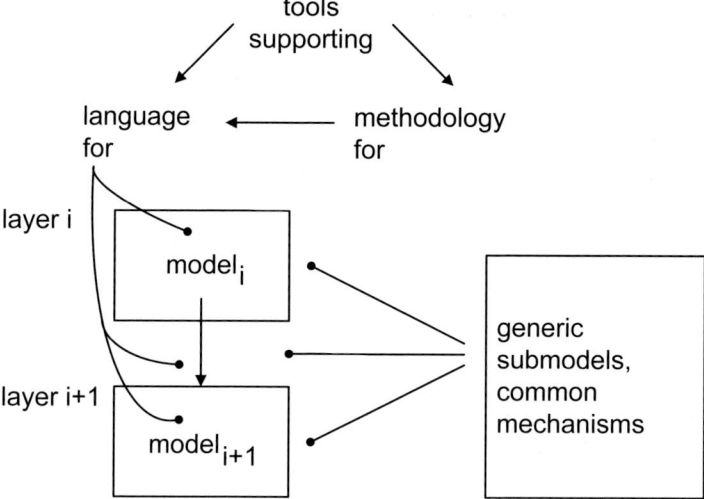

Fig. 6.13. Infrastructure problems: broadband language, methodology, tools, generic submodels/mechanisms

6.5.5 Summary and Conclusion

The *problems* of developing a *PPM*, as introduced in Sect. 1.1, categorized in Sect. 6.1, explained in Sects. 2.6, 6.2 to 6.4, and summarized in this section are of such a *principle character* and deepness that we could not give final answers within the IMPROVE project.

To formulate this statement positively: If we had solved all problems corresponding to the PPM, then we would have deeply understood all phenomena of design processes, we would have been able to describe them formally, we would also have been able to formalize the construction process of integrated environments and, even more, also distribution, parameterization, evolution, and domain transfer. So, we *would* not only *have completely understood* design processes in chemical engineering, but *development processes in general*, as well as the construction of corresponding integrated tool support.

We have got new insight into development processes by regarding the modeling problems of this chapter. We have opened some doors, detected and recognized new rooms. But we have found even more doors to be opened. There is a *long list* of *open problems*.

Nevertheless, dealing with the key question of developing a PPM is also a success story, although we did not get final answers: the question was clearly addressed, the importance of the question was demonstrated, it was identified that it is a big problem, and partial answers were given.

7

Transfer to Practice

The Collaborative Research Center IMPROVE has started and is going to *transfer most of its results to industrial practice* over a period from 2006 to 2009. The majority of IMPROVE subprojects has started a transfer phase. The subprojects are reorganized into *transfer projects*. Most of these transfer projects are financed by German Research Foundation (DFG) in the framework of the transfer center TC 61 "New Concepts and Tools for Process Engineering Practice". This transfer center can be regarded as the fourth phase of IMPROVE.

The individual transfer projects *cooperate* with different *industrial partners* from various industrial sectors, including chemical industry, plastics processing, plant engineering, software industry, and even an insurance company. Due to the dissimilar research interests of these partners, the transfer projects are only loosely connected. Bilateral cooperation will take place where appropriate. But other than the CRC subprojects, the transfer projects do not share a common overall research objective.

It should be pointed out that the transfer projects are the *result* of a consistent *industry-oriented research strategy*, which the CRC followed right from the beginning. Thus, the transfer phase does not constitute an abrupt transition from fundamental research to applied development, but continues the already existing research cooperation between CRC and industry in a different way.

This *chapter* is organized along the transfer projects. Each *section* states the goals of the respective transfer project, the problems to be solved, and the methods to be applied. Moreover, the industrial requirements influencing the direction of research will be discussed.

This chapter deals with another *integration aspect*, extending the technical aspects given in Chaps. 2, 3, 4, and 5, namely the integration between academia and industry. This aspect is more of a political or economical nature.

7.1 Industrial Cooperation Resulting in Transfer

R. Schneider, L. von Wedel, and W. Marquardt

Abstract. This short section is to demonstrate that one of the characteristics of IMPROVE was a permanent exchange of ideas with industrial partners: Ideas were taken up from industry, symposia and workshops were held, spin-offs were founded, etc. So, the transfer center described in this chapter is essentially the result of our long-lasting and continuing cooperation with industry.

7.1.1 Long-Lasting Industrial Collaboration

In the beginning of the CRC project, we held numerous discussion and interview sessions with end-users and experts from the chemical and software industries in order to learn about the particularities of their work processes, to develop realistic *use cases*, and to *specify requirements* for novel software support.

The *implementation* of these requirements into software was again supported by our industrial partners, which helped with practical problems, provided test cases, and evaluated the resulting *prototypes* in industrial settings.

As already indicated in Subsect. 1.1.3, the subproject I1 of IMPROVE had the specific role to act as a clearinghouse, where industrial problems were imported, where results of IMPROVE were exported to industry, and where the management of the IMPROVE cooperations with industry was located. According to this specific task, the main research area of subproject I1 was the investigation of industrial work processes and their improvement.

While some of the industrial *partners collaborated* only temporarily, others were involved over the entire research period of twelve years. For *example*, there was a close and *long-lasting* cooperation on work process modeling between subprojects A1 and I1 of IMPROVE on the one and the companies Air Products and Bayer Technology Services on the other hand. Another example is information and project management, which included subprojects A2, I1, and B4 from IMPROVE, and Degussa as well as Uhde on the industrial side. Finally, there was a cooperation for a number of years between subproject B2 of IMPROVE and the software tool builder innotec.

The cooperation with industry was formalized by the *CRC advisory committee* with the following industrial members: Aventis, Basell, Bayer, BASF, Degussa, innotec, Linde, Shell, Siemens, and Uhde. Regular meetings and workshops were held for the exchange of experience between industrial partners and IMPROVE.

7.1.2 Symposia and Workshops

For a broader audience, we hosted a series of workshops and symposia bringing together participants from academia, software vendors, and end-users. The *objective* of these events was to exchange experiences, to discuss new concepts and ideas, and to obtain feedback from practitioners.

Six *symposia* were held in the years 2000 [396], 2002 [353], 2004 [399], 2005, 2006, and 2007 at Aachen and Berlin. The organization was due to W. Marquardt (RWTH Aachen University) and G. Wozny (TU Berlin). Any of these symposia had about 100 participants, most of them from industry.

The symposia *programs* consisted of oral presentations, panel discussions, and poster sessions. These programs were interesting for both industry and IMPROVE members. Most of the symposia had an associated software exhibition of commercial software tools for engineering design. The results of the symposia were documented in brochures which were distributed to public. Presentation slides of the symposia since 2004 are available online [112].

7.1.3 IMPROVE's Spin-offs

The industry related activities culminated in the formation of two start-ups that are concerned with the development of software tools for the chemical and polymer industries.

aiXtrusion was founded in December 2003. The objective of the company is the development of innovative information systems to support *analysis* in *plastics processing* plants. In particular, aiXtrusion offers a wide variety of simulation applications that permit the study of extrusion processes. The simulation software is easily coupled to measurements from a plant for use as an online simulation.

Beyond process analysis, aiXtrusion offers general solutions for the *integration* of distributed components in *heterogeneous information systems* typical for the polymer processing industries. The software distributed by aiXtrusion has been developed at the Institut für Kunststoffverarbeitung (IKV) at RWTH Aachen University. Further research and development of the software is carried out at IKV, whereas support and maintenance are provided by aiXtrusion.

A second initiative, *AixCAPE*, was founded in 2002 as a consortium of industrial end users of CAPE software. The major objective of this organization is *transfer-oriented research and development* in close cooperation with its industrial members. Transfer is organized in a variety of cooperation opportunities, ranging from short-term consulting services to joint medium-term research projects or a long-term membership in the consortium. Initially, the research results of the Lehrstuhl für Prozesstechnik form the basis of AixCAPE's activities. In the future, other academic collaborators are aimed to be included. The idea is to achieve open technical platforms in which results from various research organizations can be integrated for efficient assessment and use.

Two topics important to the founding members (Atofina, BASF, Bayer, Degussa, Dow, and Shell) have a strong connection to the work carried out in IMPROVE. First, further development in the area of *systematic* management of *mathematical process models* has resulted in a successor project of the ROME model repository (cf. Subsect. 5.3.3). The system, which is now called *MOVE* [1030], has been improved according to industrial requirements. It is better tailored to model management and integration with commercial simulators such as AspenPlus. The technical basis is now the content management system Plone [511].

Second, the *integration* of mathematical models for unit operations into *commercial process simulators* has been carried forward with unit operation and thermodynamic models from research at universities as well as in-house models provided by end users. Background knowledge and experiences gained in the various integration efforts in the IMPROVE project have helped to effectively develop these solutions. From a technical perspective, solutions are based on the independent CAPE-OPEN standard rather than on proprietary interfaces of the various simulators in which the models can be employed.

7.1.4 Transfer Center as a Result of Our Industrial Links

As a consequence of the various initiatives described above, we have been able to identify the *problem areas* where an intensified cooperation by means of transfer projects seems most promising.

There are *seven transfer projects* which build up the transfer center TC 61 "New Concepts and Tools for Process Engineering Practice" (http://se.rwth-aachen.de/research/tb61). This transfer center can be regarded as the fourth phase of IMPROVE.

Table 7.1 gives *organizational details* of these transfer projects, i.e., their title, their code, the subprojects of IMPROVE they originate from, but also the affiliated industrial partners. These transfer projects are *described* in the following sections of this *chapter*.

The cooperation with industry within the transfer center is not a one-way activity. *Industry* is also contributing with a remarkable *financial effort*, as already indicated in Sect. 1.3.

Further workshops are planned for 2008 and 2009 to continue the above series. A bigger event is planned for 2009, however, to disseminate the *results* and to present the *achievements* of the *transfer center* to the public.

7.1.5 Summary

This short section describes the various forms of cooperation of IMPROVE with industrial partners from different fields of industry. Especially, we discussed the *different measures* we have taken in order to install and improve cooperation.

We also shortly described the *transfer center* as a *result* of the cooperations since the beginning of IMPROVE. The remainder of this chapter describes the joint research between the IMPROVE groups and industry.

Table 7.1. Overview of the transfer projects

Title	Code	Originating from IMPROVE subproject	Industrial partners
Ontology-based integration and management of distributed design data (cf. Sect. 7.2)	T1	A2	Evonik Degussa, ontoprise
Computer-assisted work process modeling in chemical engineering (cf. Sect. 7.3)	T2	A1, I1	Air Products, BASF, Bayer Technology Services, Siemens
Simulation-supported workflow optimization in process engineering (cf. Sect. 7.4)	T3	I2, I4	Bayer MaterialScience, ConSense, InfraServ
Management and reuse of experience knowledge in continuous production processes (cf. Sect. 7.5)	T4	B1, C1	aiXtrusion, Fraunhofer FIT, Freudenberg
Tools for consistency management between design products (cf. Sect. 7.6)	T5	B2	innotec
Dynamic process management based upon existing systems (cf. Sect. 7.7)	T6	B4	AMB, innotec
Service-oriented architectures and application integration (cf. Sect. 7.8)	T7	I3	AMB

7.2 Ontology-Based Integration and Management of Distributed Design Data

J. Morbach and W. Marquardt

Abstract. During the design phase of a chemical plant, information is created by various software tools and stored in different documents and databases. These distributed design data are a potential source of valuable knowledge, which could be exploited by novel software applications. However, before further processing, the scattered information has to be merged and consolidated. For this task, semantic technologies are a promising alternative to conventional database technology. This contribution gives an outline of the transfer project T1, which aims at the development of an ontology-based software prototype for the integration and reconciliation of design data. Both ontology and software development will be performed in close cooperation with partners from the chemical and software industries to ensure their compliance with the requirements of industrial practice. The advantages of semantic technologies will be demonstrated by comparing the prototype against a conventional integration solution.

7.2.1 Introduction

During the individual stages of a plant design project, information is created and manipulated by application tools and stored in heterogeneous formats, such as technical documents, CAE systems, simulation files, or asset management tools. The lack of integration between these application tools and data stores creates a significant overhead for the designers, since much time has to be spent on the re-entering of data, the manual reconciliation of overlapping data sets, and the search for information. NIST, the National Institute of Standards and Technology in the U.S, has recently analyzed the efficiency losses resulting from inadequate interoperability among computer-aided design, engineering, and software systems. According to this study, insufficient interoperability causes costs of 15.8 billion dollars in the US capital facilities industries, compared to a hypothetical scenario where the exchange of data and the access to information are not restricted by technical or organizational boundaries [681].

In principle, interoperability between application tools could be achieved if all stakeholders agreed on a shared data model [1035, 1038]. However, none of the standards proposed for the chemical industries has gained wide acceptance so far (cf. discussion in Sect. 2.6). Therefore, several of the major chemical and engineering companies are developing their own solutions to data integration and management. One of these solutions is the *PlantXML* technology [506], which has been established by the Evonik Degussa engineering department to improve the interoperability between their application tools. Data exchange is realized via XML files that comply with an in-house standard specified through XML schemata. Custom-made converters handle the import and export of the

XML files to and from the application tools by mediating between the in-house standard and the internal data models of the application tools.

The section is organized as follows: In Subsect. 7.2.2, three application areas are described which could benefit from a comprehensive information base on top of PlantXML. Subsection 7.2.3 sketches the information model required for such an information base. In Subsect. 7.2.4, the implementation strategy of the transfer project, based on semantic technologies, is discussed. Subsections 7.2.5 and 7.2.6 are about the project's contribution to scientific progress and industrial practice, respectively.

7.2.2 Application Scenario

Currently, PlantXML merely constitutes a convenient solution for data exchange. In a next step, however, additional value can be created by harvesting the information contained in the XML files with appropriate software tools. Three major application areas have been identified for investigation in the transfer project T1:

1. *Automatic detection and elimination of data inconsistencies and design errors.* Typical design errors would be interconnected flanges with different internal diameters or process parameters exceeding the stability limits of some plant equipment.
2. Supporting the project manager by *providing an integrated view on the relevant project data* and by *extracting key figures and performance indicators* from the XML files. Examples of such indicators are the accumulated equipment costs or the percentage of completed sub-tasks.
3. *Automatic generation of a comprehensive and consolidated technical documentation* of the chemical plant for the data handover to the owner.

The transfer project does not intend to develop a fully functional software solution for the above application areas. Instead, the goal is to understand the fundamental issues, to develop a conceptual solution, and to provide a software prototype that demonstrates the potential of the conceptual solution.

As stated before, the software prototype will make use of the information available in the PlantXML format. To this end, appropriate converters will extract selected data from the XML files and assemble them in a *comprehensive information base*, which has the function of integrating and reconciling the distributed information. After such consolidation, the information will be reorganized and transferred into *application-specific information bases*. These are specifically designed to serve the information requirements of *novel application tools* for the three application areas described above. The entire information transformation process is summarized in Fig. 7.1.

Note that the information processing is significantly facilitated by the availability of the PlantXML technology. Without PlantXML, the information required by the novel application tools would need to be extracted from

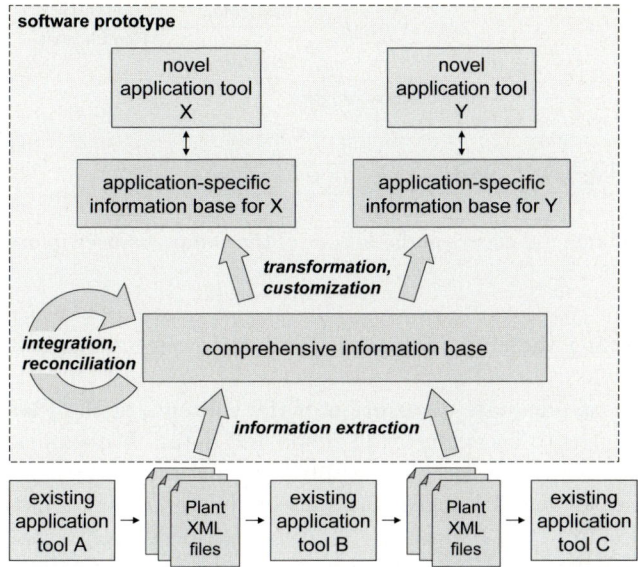

Fig. 7.1. Information transformation process

the proprietary formats of the existing application tools and data stores. In the project setting, however, these information can be directly obtained from the XML files, which can be easily accessed due to their consistent and well-known syntax and semantics. Moreover, the XML files already aggregate and pre-integrate the data, which would otherwise be scattered over a much larger number of disparate sources.

7.2.3 Information Models

As explained in the previous subsection, the scattered information items from the various PlantXML files must be assembled into a comprehensive information base. This requires a *comprehensive information model*, which will be based on the application domain models introduced in Chap. 2, particularly on the domain ontology OntoCAPE presented in Subsect. 2.2.4 and on the document model presented in Sect. 2.3.

As shown in Fig. 7.2, each XML file will be represented by an instance of the class **DocumentVersion**. The version history of the files can be indicated by the relations **hasPredecessor** and **hasSuccessor**, as explained in Subsect. 2.3.3. The **hasContent** relation is employed to denote the contents of the XML files through concepts from the OntoCAPE ontology (cf. Subsect. 2.3.4). A finer-grained description can be obtained by decomposing each file into its structural entities, represented through instances of the **Product** class, and

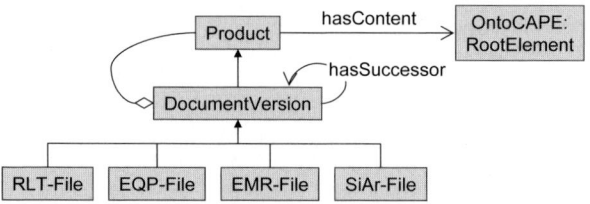

Fig. 7.2. Fundamental classes and relations of the comprehensive information model

by characterizing the content of each **Product** by an adequate concept from OntoCAPE.

To enable an adequate description of the contents of the PlantXML files, OntoCAPE needs to be extended in scope and detail. We expect the changes to take place mainly on the lower, application-oriented layers of OntoCAPE. However, some of these changes might affect the higher levels of the ontology, as well, and could even require the modification of already existing ontology modules. Since OntoCAPE is explicitly designed for reuse (cf. [326]), the effort for implementing these changes is expected to be moderate.

PlantXML defines specific XML schemata for the different phases of a design project: XML-EQP for the design of machines and apparatuses, XML-EMR for the design of instrumentation and control systems, XML-RLT for piping, and XML-SiAr for the design of fittings and safety valves. Four subclasses of **DocumentVersion** are introduced to denote these schemata. For each subclass, the range of the **hasContent** relation is specifically restricted to the corresponding concepts of OntoCAPE.

The four PlantXML schemata are independent of each other but overlap to some extent. These overlaps have to be identified, and multiple occurrences of data have to be merged to obtain a comprehensive, integrated representation of the distributed information. Moreover, interrelations and dependencies between the different schemata must be explicitly represented within the comprehensive information base. A simple but illustrative example is given in Fig. 7.3: Here, both an EQP file and an RLT file contain a data item that denotes a nozzle with the identification number 14. Since both data items refer to the same real-world object, they need to be represented by a single data item in the integrated information base (i.e., **Nozzle #14**). That way, the topological relation between **Vessel #11** represented in the EQP file and **Pipe #23** represented in the RLT file are made explicit. The implementation strategy through which the data consolidation is to be achieved will be described in the next subsection.

Once a comprehensive information base is available, the data can be easily transferred into the application-specific information bases, which meet the requirements of the novel application tools. The project partner Evonik Degussa has defined several use cases to determine the information demands of these

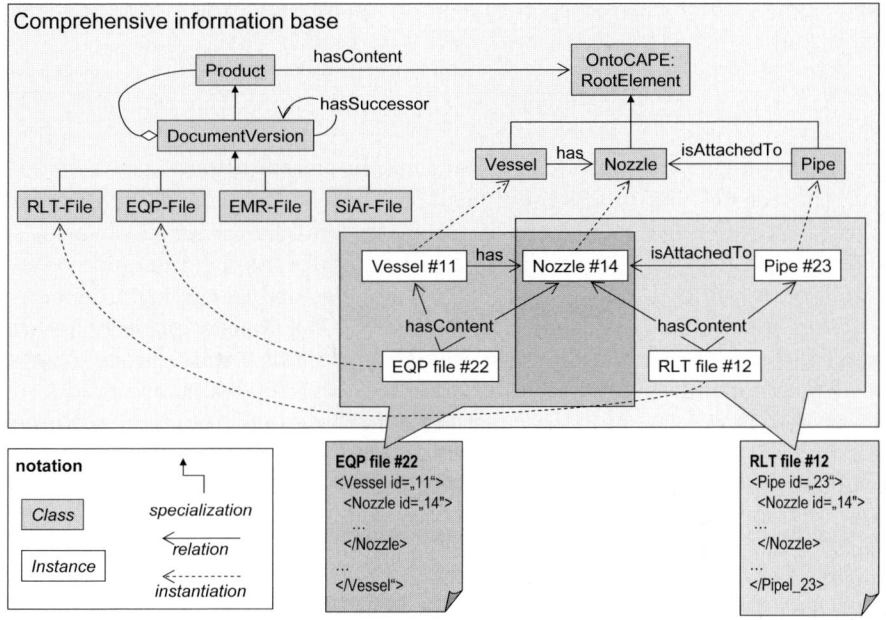

Fig. 7.3. Illustrative example

tools. Based on the requirements specification, *application-specific information models* are currently developed.

7.2.4 Implementation Strategy

We intend to use *semantic technologies* for the realization of both the comprehensive and the application-specific information bases. In this context, the term 'semantic technologies' refers to software systems that use ontologies as internal data models. Since ontologies explicitly specify the semantics of the data, a specialized software component, the so-called *inference engine*, can interpret and reason about the data.

Compared to conventional database technology, semantic systems have several advantages. First to mention is the user-friendly approach to creating internal data models: Semantic systems typically incorporate a graphical modeling environment for the creation of ontologies. Subsequently, these ontologies can be directly used as implementation models for the storage and management of data without further modifications. The underlying program logic and the data structures of the semantic software are completely hidden from the user. That way, the user can intuitively create, extend, and maintain customized information bases even without possessing deep expertise in database design. Ontology engineering is furthermore supported by the inference engine, which checks the ontology for consistency, thus revealing logical errors and ensuring a reliable, validated data model.

Another important advantage of ontology-based systems over conventional database technology is the possibility of partially automating the information integration process. To this aim, two different types of inference mechanisms can be applied: subsumption (i.e., automatic classification) and deduction (i.e., the execution of production rules). The currently available ontology languages and inference engines do not support both inference mechanism simultaneously. Since deduction is particularly useful for merging and consolidating of distributed information [827], it was decided to employ a deductive language (and a compatible inference engine) in the transfer project. Accordingly, the application development system *OntoStudio* was chosen to serve as an implementation basis. OntoStudio, a commercial software product of the project partner OntoPrise, constitutes an design environment for ontologies and semantic applications. Unlike most of the ontology-based systems available today, OntoStudio is scalable and thus suitable for the processing of large ontologies. It relies on the deductive database language F-Logic [779], which allows the representation of ontologies, the definition of mapping rules, and the formulation of queries. F-Logic supports not only the specification of mappings between individual classes and/or instances, but also the definition of general integration rules. Thus, instead of establishing mappings between individual data items, the integration policy can be specified in more general terms. The applicability of a rule to the given data is automatically checked by the inference engine. This approach is more intuitive and less error-prone than conventional database integration techniques (e.g., integration via SQL queries), especially in complex contexts with many relations between the data objects [827].

Both the comprehensive information base and the application-specific information bases will be implemented in the design environment OntoStudio. For the import of PlantXML data, a new interface is to be developed. Ontologies represented in the deductive database language F-Logic are to be used as semantic data models. Information integration and reconciliation will be supported by OntoStudio's built-in inference engine, which will execute mapping and integration rules specified on top of the ontologies. Similarly, the transformation from the comprehensive information base into the application-specific formats will be performed by the inference engine, controlled by conversion rules defined within OntoStudio.

A drawback of semantic technologies is the considerable effort to be spent on ontology engineering. This development effort can be significantly reduced by reusing existing ontologies instead of starting from scratch. Therefore, as already mentioned in Subsect. 7.2.3, the application domain models presented in Chap. 2 will be reapplied in this transfer project. Since OntoStudio supports the import of OWL ontologies, these models can be reused directly.

7.2.5 Intended Contribution to Scientific Progress

The use of ontologies and semantic technologies for the integration of distributed data is a long-standing research issue; an overview on the more recent contributions in this field can be found in [900, 1026, 1031]. Three major approaches can be distinguished, differing in the way in which the semantics of the heterogeneous data sources are described through ontologies:

- According to the *centralized approach*, all information sources are mapped onto a single global ontology. The mappings clarify the semantics of the source objects, thus supporting the identification of semantically corresponding objects.
- In the *decentralized approach*, each information source is described by its own local ontology. Mappings are defined between the local ontologies to identify semantically corresponding terms.
- The *hybrid approach* combines features of the two previous approaches and eliminates most of their drawbacks. Similar to the decentralized approach, each source is described by its own ontology. But this time, the local ontologies are built upon a shared vocabulary, which is provided by a global domain ontology. That way, the local ontologies can be compared more easily, which facilitates their integration. The integration approach to be taken in the transfer project is a variant of the hybrid approach.

To our knowledge, the hybrid approach has never been utilized for the integration of engineering design data. Thus, our research objectives are (1) to evaluate if the hybrid approach is suitable for this application case and (2) to elaborate both the methodology and the ontological models required for its practical usage.

So far, only a few research projects have investigated the use of semantic technologies for information integration in chemical engineering: Weiten and Wozny developed a system to archive the diverse results of research projects (publications, experimental data, mathematical models, etc.) in an integrated fashion [1041–1043]. The system uses ontologies to annotate and link heterogeneous information sources. Similarly, the Process Data Warehouse presented in Subsect. 4.1.5 uses ontologies to describe and interrelate the contents of chemical engineering documents and data stores. Both the PDW and the archival system developed by Weiten and Wozny allow to establish semantic relations between the information sources via content-specific metadata; however, these systems do not merge and reconcile the contents of the sources on a fine-grained level, as it is intended in this project. Moreover, the relations between information sources have to be established manually in those systems, whereas this project proposes to automate the integration process by using integration rules.

Last but not least, the transfer project provides an excellent opportunity for testing the application domain models developed in IMPROVE in an industrial setting. Based on the use case of integrating PlantXML data, two questions will be investigated:

1. Are the application domain models suitable for describing and managing large amounts of real-world data?
2. Can the models be easily adapted to new applications and thus reused without major effort?

Together with our industrial partners, we will evaluate these issues and, if necessary, improve and adapt the models to the requirements of industrial practice. The ultimate goal is to obtain practically validated application domain models, which can then be reused by third parties to build semantic applications in the domain of process engineering.

7.2.6 Intended Contribution to Industrial Practice

In the chemical and process industries, semantic technologies are currently not applied. Commercial software systems for the integration and management of lifecycle data still rely on conventional database technology. Most of these systems are based on a central data warehouse that can import and redistribute data created by application tools. These systems support navigation and retrieval of data and provide some basic data management functions such as version management and change management, but novel functionality as envisioned in Subsect. 7.2.2 is not available today. Typical examples are SmartPlant Foundation [747] from Intergraph and VNET [520] from AVEVA. However, these off-the-shelf solutions can only process the proprietary data formats of the respective vendors and are thus limited to the data created by the vendors' application tools, which normally represent only a minor subset of all data created in the course of a development project. Extending these tools towards the handling of non-proprietary formats is difficult, as one needs to modify the source code and map onto the internal data models of the tools, both of which are poorly documented in most cases.

To overcome these deficiencies, some systems allow the import and export of XML data. But again, as there is no general agreement on a particular XML exchange format, the tools can process only the XML formats propagated by the respective vendors. Examples of such proprietary XML dialects are XMpLant [868], AEX [667], ppdXML [635], and cfiXML [661]. A common data exchange format supported by all established vendors is unlikely to evolve within the near future [506]. As a consequence, many chemical and engineering companies are utilizing in-house technologies such as PlantXML. While these technologies are capable of solving the problem of data exchange, they are less suitable for handling complex integration issues (cf. discussion in Subsect. 7.2.4).

The prototypical software to be developed in this project constitutes a supplement to the existing in-house technologies, providing additional value beyond data exchange (as explained in Subsect. 7.2.2). Even though the project use case is based on PlantXML, we are aiming at a generic solution that can be combined with different in-house technologies and is thus reusable across companies and organizations.

Since semantic technologies are currently not applied in the chemical industry, it is not known if the conjectured advantages of semantic technologies prove true in practical use. As a first benchmark, the reliability of the semantic software and its ability to handle large-scale industrial data sets will be evaluated.

7.2.7 Summary

The transfer project T1 [480] aims at the development of a prototypical software tool for the integration and reconciliation of engineering design data. The implementation will be based on ontologies and semantic technologies; some of the application domain models developed in IMPROVE will be reused for this purpose. As an industrial use case, the project considers the integration of design data available in the PlantXML format. The envisioned tool will extract and merge data from different XML files into a comprehensive information base. Subsequently, the consolidated information can be transformed into application-specific formats for further processing. Prototypical tools for the following application cases are to be realized in this project: detection and elimination of design errors; information provision for the project management; generation of a plant documentation.

7.3 Computer-Assisted Work Process Modeling in Chemical Engineering

M. Theißen, R. Hai, and W. Marquardt

Abstract. The transfer project aims at the integrative modeling, analysis, and improvement of a variety of work processes in the life cycle of a chemical product across disciplinary and institutional boundaries. A methodology is elaborated for the creation of conceptual, coarse grained models of work processes originating from empirical field studies in industry and their subsequent enrichment and formalization for computer-based interpretation and processing.

7.3.1 Introduction

As a continuation of IMPROVE subprojects A1 and I1, the transfer project aims at the integrative modeling, analysis, and improvement of a variety of work processes in the lifecycle of a chemical product across disciplinary and institutional boundaries. Within the project, different types of work processes in four very different phases of the chemical product lifecycle are considered, namely

- the development of chemical products,
- the conceptual design of production processes,
- the specification of operating procedures for the chemical plant, and finally
- the realization of these operating procedures in the real plant.

These work processes do not only differ in the lifecycle phases they address, but are also quite different in nature. The first three examples are *design processes*, whose products are specifications of certain artifacts, whereas the last work process is an *operational process*; its product is the manufactured chemical product itself. The actors of design processes are human beings, typically assisted by certain software tools. In contrast, operational processes are performed by both human beings and computers, i.e., the operational staff and the process control system. As operational processes are largely predetermined and leave limited room for creative decisions on behalf of the actors, they can be automated to a large extent. In contrast, the potential for automating highly creative design processes is low.

Typically, the human actors of an industrial work process originate from *various disciplines*, such as chemistry, chemical engineering, or electrical engineering, and they perform *diverse roles* (e.g., technical, managerial, supervisory). As a result of these different backgrounds, each stakeholder possesses a *personal view and understanding* of the work process. Some actors, in particular those in managerial and supervisory roles, are acquainted with the process as a whole, but are unfamiliar with details of the process. Such detailed knowledge is distributed among the stakeholders. This lack of common understanding hinders collaboration across organizational and disciplinary boundaries and over the lifecycle of the product, and therefore leaves significant room for improvement [791].

A *common understanding* can be achieved by means of simple representations of work processes, which can easily be edited and understood by all stakeholders, independent from their disciplinary and educational backgrounds. Several case studies in cooperation with industrial partners (cf. Sect. 5.1) have proven that the C3 notation for work processes (cf. Subsect. 2.4.4) provides an adequate representation.

C3 is a typical example of a *semi-formal* modeling language: Part of the information comprised in a C3 model is represented in an unambiguous way (e.g., the order of activities as it is defined by control flows), whereas another part is expressed by annotations in natural language (see Subsect. 2.1.1 for a sound definition of the terms *semi-formal* and *formal*).

A drawback of semi-formal models is their limited utility for computer-based applications such as analysis, simulation, or enactment. For this purpose, *formal*, machine-interpretable models are required. However, the creation of formal models is difficult, time-consuming and requires expert skills. Consequently, a methodology is required for the creation of formal work process models by means of an incremental approach with active participation of all stakeholders. This transfer project envisions

- the extension of the modeling procedure and Process Ontology developed in the IMPROVE subprojects A1 and I1 (cf. Sect. 2.4) – which are currently restricted to design processes – to cover *different types* of work processes,
- the creation of *prototypical tools* for work process modeling and for the export of these models in software tools used in certain lifecycle phases, and
- the *validation* of the approach in several case studies performed in cooperation with industrial partners.

The remainder of the section is organized as follows: In Subsect. 7.3.2, the current state of the art in work process modeling is briefly discussed. Subsection 7.3.3 describes the methods applied in the transfer project to reach the goals sketched above as well as various case studies in cooperation with industrial partners for the validation of the approach. Subsection 7.3.4 is about the industrial relevance of the project and the expected benefits for the chemical industries.

7.3.2 State of the Art in Research and Industrial Practice

Currently, different approaches exist for representing and processing work process knowledge in the process industries, depending on the life cycle phase they address. There is no integrated approach supporting the creation of semi-formal models, their incremental formalization, and finally the use of these models for different applications.

Modeling of Design Processes

As argued in Sect. 2.4, no semi-formal representation for design processes is established in the process industries. Instead, notations and models like activity diagrams of the *Unified Modeling Language* (*UML*, [560]), the *Business Process Modeling Notation* (*BPMN*, [873]), or *Event Driven Process Chains* (*EPC*, [842, 949]) are used, although they do not fully meet the requirements induced by the characteristics of design processes. As noted above, the semi-formal C3 notation has proven its potential for filling this gap in a series of industrial case studies.

Moreover, there is no modeling approach for the formal representation of design processes. Despite the above mentioned benefits of formal models of design processes, they are not used yet in industry [401]. This is due to the lack of appropriate modeling languages, but also to the high cost for their creation. Since the knowledge about a design process is distributed among all individuals participating in the process, the creation of a formal model from scratch would require an immense effort. Thus, an incremental procedure is required which starts from a simple semi-formal representation of a design process and which results in a formal model.

Modeling of Operational Processes

The existing formal representations for operational processes such as the languages defined in IEC 61131-3 [748] (e.g., *Sequential Function Chart* (*SFC*), *Block Diagram*) result in very detailed and complex descriptions. As formal process descriptions are indispensable for the automation of chemical processes, several methodologies have been elaborated to simplify their creation and analysis. In [775], class diagrams and state charts of UML-PA, a domain-specific extension of UML, are proposed as intermediate steps before the automatic generation of specifications according to IEC 61131-3. A procedure for the design and analysis of logic controllers for plants is presented in [822], comprising the design of an SFC, its translation into a formal modeling language, and finally the analysis of the formal controller model.

Such methodologies offer valuable support for the specification of operational processes as long as only control and possibly process engineers are involved. An extensive use of these approaches by other stakeholders in preceding phases is impeded by a wide variety of non-trivial documents and formats.

Semi-formal representations can be used as an intermediate step for the generation of a model with the intended level of detail and formalization: First, the operations are specified by chemists or process experts on a conceptual, semi-formal level, and subsequently formalized and refined. Since such semi-formal representations are currently missing, inadequate workarounds like spreadsheets or informal documents are often used in industrial practice.

7.3.3 Objectives and Methods

Extended Modeling Procedure

The *modeling procedure* described in Subsect. 2.4.2 addresses the work processes during the conceptual design of chemical processes. It covers the knowledge acquisition from stakeholders by means of field studies, the creation of conceptual semi-formal work process models, and their subsequent enrichment and formalization. The methodology has been successfully applied in several case studies conducted in industry (cf. Sect. 5.1). This modeling procedure is supposed to be extended

- to cover *different types of work processes* and
- to enable *different applications* based on the models (e.g., simulation or automation of (parts of) the work process).

Fig. 7.4 gives an overview of the extended procedure. It begins with the creation of a *coarse-grained semi-formal* model of the work process considered (1). The purpose of this semi-formal model is a first common understanding of the work process, which is agreed upon by all individuals involved. At this stage, several aspects of the work process may be included in an informal way, for example by means of textual annotations, or they may be neglected completely if they are considered insignificant.

In the next step, *additional aspects* of the work process are added to the model or described in a more precise way (2). Both the choice of aspects and the degree of detail depend on the intended applications of the work process model. For example, if a design process model is supposed to guide an inexperienced designer, behavioral aspects of the design process (such as the sequence of activities to be performed and conditions for their execution) can be expressed in natural language. In contrast, if the work process model is intended, for example, as input specification for a workflow management system, a formal and detailed description of behavioral aspects is required.

In a subsequent analysis phase, the work process model is checked for consistency and completeness (3). *Consistency* checks ensure that the model does not contain contradictory statements about the work process. *Completeness* refers to the aspects which are relevant to the intended applications of the model. Completeness does not mean that all details about an aspect of a work process must be included in the model. Rather, completeness requires that a relevant aspect is modeled on a sufficient level of detail, which complies with the target format of the software application using the work process model.

In case a work process model is intended to be used in a tool processing the work process model for a specific purpose, it must be *transformed* (4) into the target format employed by the candidate software application. The transformed work process model can be further enriched with application-specific content in the target software application.

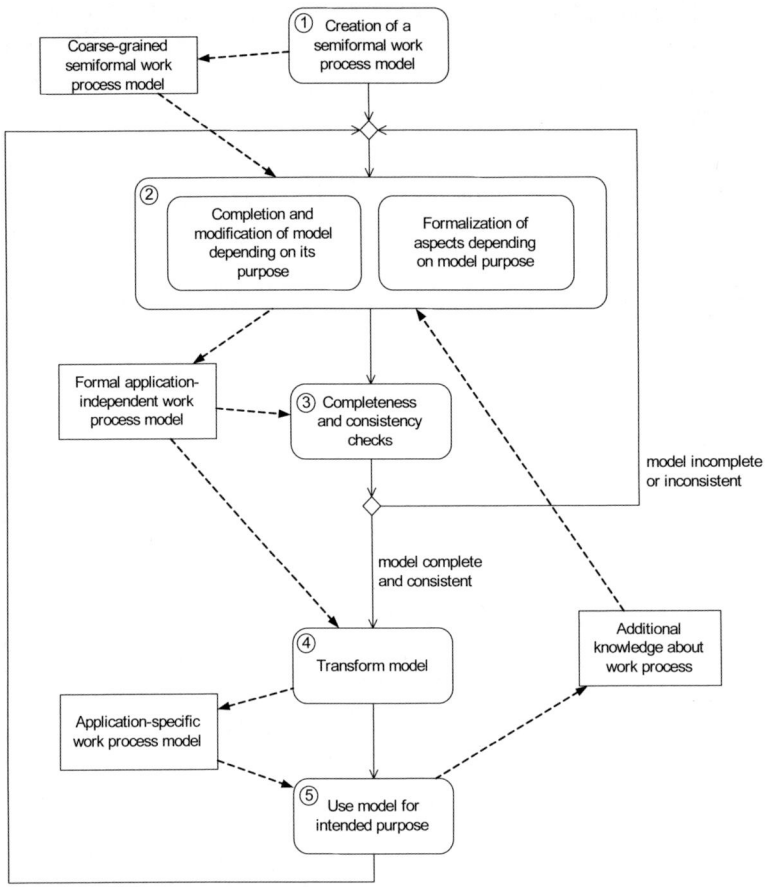

Fig. 7.4. Modeling procedure (in C3 notation, cf. Subsect. 2.4.4)

The *use of the model* in the target application (5) will reveal better insight into the work process considered. For example, simulation studies of an operational process (e.g., emergency shut-down of a plant section) by means of a discrete-event system simulator can reveal deficiencies in the current design of the operating procedure. Based on such findings, the design engineer modifies the work process model and initiates a new iteration of formalizing, checking, transforming, and using the model in the (simulation) application. Note that different tools tailored to different purposes or applications can be integrated into the procedure without difficulties, as long as all application-independent aspects of the work process are included in the model. For example, it is possible to optimize an operational process by means of simulation, and then

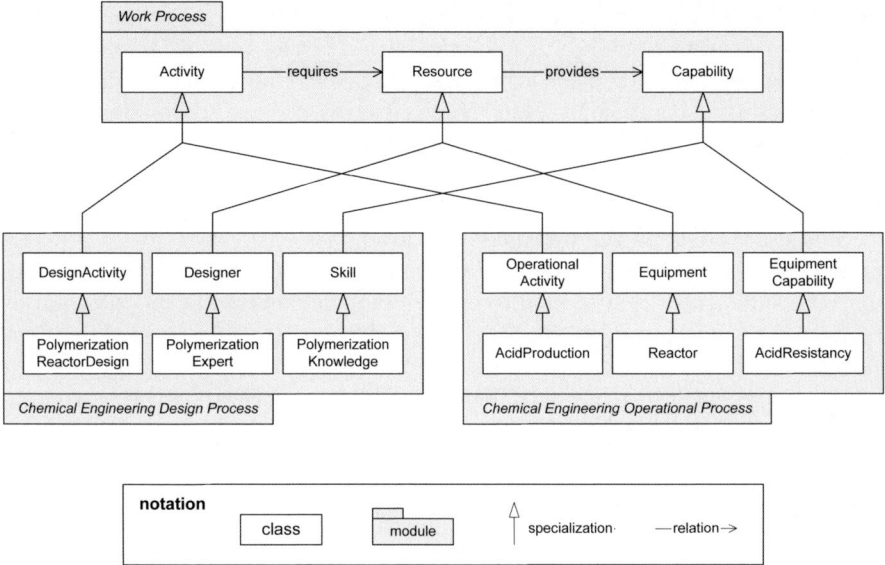

Fig. 7.5. Sketch of the modified Process Ontology

generate sequential function charts as a basis for the specification and implementation of the control and automation system from the same model.

Extended Process Ontology

The modeling procedure described above induces a number of requirements on the Process Ontology, which are not fulfilled by its current version as described in Subsect. 2.4.6. In the following, the most important new features of the Process Ontology are outlined.

The current *Work Process* module of the Process Ontology abstracts from the characteristics of special types of work processes and provides a set of basic elements for modeling a wide range of processes. The *Chemical Engineering Work Processes* module imports both the *Work Process* module and the OntoCAPE product data model, and thus it provides the concepts required for modeling conceptual design processes in chemical engineering and their products (cf. Subsect. 2.6.2).

The extended Process Ontology to be created in this project will have to cover the peculiarities of *different types of work processes*. This implies some minor modifications of the type-independent *Work Process* module (e.g., the replacement of the *Actor* class by a more general *Resource*), but in particular *type-specific modules* are to be created which serve as meta-models for work processes of a certain type. As shown in a simplistic way in Fig. 7.5, the concepts of these modules inherit from the concepts of the general *Work Process* module. For instance, the DesignActivity introduced in the *Chemical Engineering Design Process* module[62] is a specialization of the general Activity within *Work Process*. The execution of a DesignActivity requires the availability of Designers with certain Skills. Thus, these two classes have to be introduced as specializations of Resource and Capability on the common meta-layer. Analogous specializations are introduced in the *Chemical Engineering Operational Process* module.

Process models on the instance layer are intended to be used for *different applications* such as simulation or automation. Thus, the Process Ontology must be sufficiently rich to enable process models to provide the information relevant for the applications. For this purpose, the class definitions of the *Work Process* module and the type-specific modules can be enhanced by a set of attributes. For instance, a ControlFlow can be characterized by the percentage by which the preceding Activity must be completed before the subsequent Activity can start. Exemplary attributes of a Reactor are its volume or its maximum operating pressure.

In principle, the model framework allows to include the entirety of aspects relevant to all possible applications in a single application-independent work process model, provided that its meta-model, i.e. the type-specific module, features all required model elements. This approach is not feasible because it would result in unmanageable complexity of the Process Ontology. Furthermore, a software application tailored to a certain task can offer better support for modeling application-specific aspects of a work process than a general-purpose work process modeling tool which aims at application-independent models. Thus, the application-independent work process model in the work process modeling tool should be restricted to those aspects which are relevant for several applications.

Technical Realization

Like the current Process Ontology, also the extended version will be implemented in the Web Ontology Language (OWL, [546]). Whereas the standard OWL editor Protégé [979] can be used for the modifications and extensions of the Process Ontology itself, a tailored tool is required for the creation of work process models on the instance layer. An extended version of the Workflow Modeling System WOMS (cf. Subsect. 5.1.3), called *WOMS+* in the following, will further on provide an intuitive graphical representation in the proven C3 notation. An outstanding feature of the new tool is its adaptability for different types of work processes, realized by an import functionality for type-specific modules of the Process Ontology.

---

[62] *Chemical Engineering Design Process* corresponds to the *Chemical Engineering Work Processes* module of the current version of the Process Ontology. It has been renamed to emphasize its restriction to design processes.

The output format of WOMS+ is OWL, i.e., work process models are represented by a set of instances of the classes defined in the Process Ontology. Thus, the description logics reasoner RacerPro [918] can conduct the consistency and completeness checks within the modeling procedure.

Model transformations into different applications are realized in *XSLT* (*Extensible Stylesheet Language Transformations*, [602]), which allows to reduce the effort for their implementation to a minimum.

Industrial Case Studies

Cooperation with various partners from the process industries is an essential requirement for the success of the project. The intended long-term cooperation provides access to industrially relevant work processes to the academic research team. Sufficient knowledge about real work processes is a prerequisite for the extension of the Process Ontology as described above. Also, usability of the modeling procedure in general and the modeling tool in particular can only be evaluated in the context of industrial case studies. The cases to be examined in the project address real problems of the industrial partners. They have been carefully selected to cover a wide range of different types of work processes, but also with regard to the anticipated enhancements that can be reached in the project.

A first group of case studies, performed in cooperation with *Air Products and Chemicals, Inc.*, addresses the modeling and improvement of *product and process design processes* for specialty chemicals. In a previous cooperation, models of two concrete design projects have been created. In the transfer project, a generalized model will be elaborated based on the existing models. Given the highly creative character of product and process design, this is by no means a simple combination of the two existing models, but requires a substantial abstraction from the concrete projects. In a subsequent step, the generalized model is transformed into a simulation model for the quantitative analysis of the design process. As time to market is considered a crucial issue for innovative chemical products, simulation studies will in particular address the impact of allocated resources on the cycle time.

A second group of case studies in cooperation with *BASF SE*, *Bayer Technology Services GmbH*, and *Siemens AG* focuses in the first instance on modeling and improving the *design processes during the specification of the operational processes* of a chemical plant. Based on the promising results of a previous study in cooperation with *Bayer Technology Services GmbH*, our fundamental approach for improving these design processes is the *modeling of the operational processes* themselves using WOMS+: The tool will be used by the different experts for describing those aspects of an operational process which are relevant from their point of view (e.g., chemical engineers specify the process steps to be performed and their order of execution). This WOMS+ model is then passed to the subsequent experts, who will further enrich, detail, and formalize the model. The integration of typical application tools (e.g., simulation systems for batch processes or editors for sequential function charts) in the overall design process is supported by adequate model transformations from WOMS+ to the application formats.

7.3.4 Industrial Relevance

Work process modeling is considered to be an important methodology for improving the efficiency and the productivity during the design processes in the chemical product lifecycle. A competitive advantage is often only possible if time to market is reduced and if the development cost for a low-volume product and its associated manufacturing process can be properly controlled. An integrative approach across the variety of work processes in the chemical product lifecycle is indispensable to limit the cost of introducing such a new technology in the industrial environment.

A better understanding of cross-institutional and cross-disciplinary work processes is considered to be of key importance for the improvement of design processes in industry. It is commonly agreed that there is still a lot of room for improvement with respect to the quality of the resulting design as well as to the reduction of elapsed process time. These potential benefits can only be materialized if a suitable work process modeling methodology and a supporting tool in the lifecycle of a chemical product can be provided to the chemical process industries.

The intended incremental refinement approach towards work process models of different levels of detail and different degrees of formalization captures very well the needs of industrial users. This way, semi-formal models can be created for achieving a common understanding in the project teams. On this basis, formal models of work processes can be created at moderate effort, if the benefits of formal work process models are supposed to be exploited. It is highly desirable to assess the potential of formal methods to work processes by computer-aided tools.

As both, modeling methodology and tool are suitable for a wide range of different classes of work processes, only a minimal learning effort is required from potential users to implement a modeling strategy for work processes covering important parts of the chemical product lifecycle. This is a major step towards our long-term objective: a methodology and a tool for interdisciplinary and inter-organizational work process modeling, which can be applied across the chemical product lifecycle in an industrial environment.

7.3.5 Conclusion

The transfer project aims at the integrative modeling, analysis, and improvement of different types of work processes in the lifecycle of a chemical product. To this end, the modeling procedure and the Process Ontology developed in the IMPROVE projects A1 and I1, originally addressing the work processes during conceptual design of chemical processes, are extended to cover several types of design and operational processes.

A novel work process modeling tool is developed to enable any stakeholder in an industrial work process to routinely create and utilize work process models with moderate additional effort. The tool builds on semantic information processing and facilitates the export of the model into other software applications such as simulation systems and process control systems.

Tight cooperation with four industrial partners ensures the practical relevance of the results to be obtained in the project. The success of the transfer project can be measured by means of the following competency questions that check the fulfillment of the project's goals:

- Does the project succeed in the elaboration of an integrative modeling methodology that can be applied to different types of work processes in chemical engineering?
- Is the methodology appropriate for routine application in industry?
- Does the application of the methodology in industry provide a measurable benefit for the user?

An overall indicator for the success of the project will be the rate of acceptance of the newly developed methods and tools in industry.

7.4 Simulation-Supported Workflow Optimization in Process Engineering

B. Kausch, N. Schneider, C. Schlick, and H. Luczak

Abstract. The results of IMPROVE, which are extensively described in this book, are generally interesting for industrial users. In this transfer project, the insights gained in subproject I4 are expanded for the simulation-supported planning of process design and development projects with partners from the chemical and software industries. The planned activities are presented in this section. The goal of the transfer project is the interactive, in parts even automated transformation of work process models – which are created using the participatory C3 modeling method – into workflow models suitable for simulation studies. The required formal representation of workflows is based on Timed Stochastic Colored Petri Nets. That way, the systematic variation of organizational and technological influencing factors and the analysis of organizational effects becomes possible. By providing this simulation method already in the planning phases of design and development projects, precise prognoses for better project management can be achieved.

7.4.1 Introduction

Design and development projects in the chemical industries are characterized by poorly structured creative work processes and a high number of interfaces between the organizational units involved. Therefore, planning labor and resource utilization and the identification of potential bottlenecks is hardly possible using conventional project planning methods. Thus, according to current studies of savings potential in German development departments, a substantial part of research and development investments is not used effectively [700, 724]. Presumably, there is also a significant potential to reduce cycle times [575]. So far, existing approaches for business process reengineering (BPR) and business process modeling (BPM) do not offer adequate support. Given the complexity and limited transparency of these methods and the high costs for the software tools required, they are rarely used in industrial practice. At least, some graphical modeling techniques provide basic support to describe process interdependencies and the collective utilization of resources. However, a profound analysis and evaluation of the complex and multi-layered work processes during design projects in chemical engineering requires more sophisticated approaches. In this transfer project, the usage of discrete-event simulation techniques for workflow management is examined.

So far, the complexity of simulation techniques and the time required for their execution prevent their routine use for the analysis of complex work processes. The high entry costs for simulation-supported project optimization are an additional threshold [925]. To overcome these issues, the C3 method for the participative modeling and analysis of design processes has been developed in IMPROVE (cf. Subsect. 2.4.4). Several case studies have shown that design and development processes in the chemical industries can be collected and displayed quite well and without much effort by means of C3 (cf. Subsect. 5.1.4).

Also, the reference scenario of IMPROVE (the conceptual design of a PA6 process, cf. Subsect. 1.2) has been represented in C3. In a first step, this scenario has been simulated to get a rough approximation. In expert interviews, further attributes influencing the simulation were collected and added to the model. These attributes include, for example,

- additional information required for forecasting the durations of individual activities,
- further conditions for the execution of activities, and
- the expertise (i.e., skills, experience) required for using the tools.

In a final step, a simulation model, implemented as a Petri net, was created according to the transformation rules described Subsect. 5.2.4. This model allows experts to assess workflow management variants in a simulation-based way.

This chapter gives an overview of the simulation of design processes and its current and possible future applications in industry. Therefore, the fields of activity of the industrial partners are described in Sect. 7.4.2. Section 7.4.3 subsumes the goals and strategies which are pursued in this project. Section 7.4.4 describes the concepts for the realization of the goals.

7.4.2 Application in Industrial Process Design

A first scenario taking into account some of the relevant characteristic elements of design processes in chemical engineering was developed from the beginning of IMPROVE in 1997 (cf. Sect. 1.2). It served as a framework for analyzing the interdependencies between individual design activities and the organizational units involved. Thus, tools were developed based on the analysis of the scenario's structure. While these tools supported the cooperation between the designers, they also facilitated an integrative development of the product. This scenario also forms the starting point for the analysis of the connections and influencing factors relevant for supporting the planning of design processes.

These areas of application were not yet fully foreseeable when the scenario was compiled. This contributed to the fact that the scenario had to be expanded in the final phase of IMPROVE in order to even realize a simulation. This extension was done in collaboration with process engineering experts. Besides temporal information about individual activities, it also encompasses the influence of tool usage and the competences of the developers involved. This information can be accounted for in the simulation, though it is not based on a valid foundation. For this reason, it is necessary to collect additional information by modeling new design processes in this transfer project to quantify the dependencies of the design process on diverse influencing factors. The industrial partners represent both process engineering and software engineering fields. In the following, the cooperation between the involved partners will be described.

Cooperation with the Chemical Industries

Two companies with core activities in process and facility development have agreed to join the project as cooperation partners. The continuation of the cooperation with Bayer is, on the one hand, a confirmation of the successful work of IMPROVE. In addition, InfraServ Knapsack participates in the project to ensure the independence of the project results from the specifics of a single company. InfraServ Knapsack is an industrial service provider for the development and operation of chemical processes according to customer specifications.

Both Bayer and InfraServ Knapsack carry out similar projects in their design departments. A project can include conceptual process design as well as basic and detail engineering. For the modeling, simulation, and optimization of chemical processes, several tools like CHEMCAD [592] or Aspen Plus [516] are used. In addition, laboratory experiments may be required to determine reaction parameters or other physicochemical data. A project can also include the planning, construction, operation, and analysis of a pilot plant. Depending on the customer's demands, not all of these project phases are executed.

Specialized knowledge about the work organizational connections and backgrounds is thus available on a fine-grained level throughout the entire duration of a design and development project. Such projects are partially run in networks with other departments or companies; they are based on complex and non-standardized problem solving techniques. A work organizational calculation of the individual activities is necessary in order to fulfill the customer- or order-specific demands and to ensure the effective and efficient cooperation of the different specialists.

The prototypical integrated simulation tools for workflow management of IMPROVE subproject I4 have sparked interest in discussions with managers from the process industries. The demand for constant software-supported methods and tools for simulation-based optimization of design projects in chemical engineering was also emphasized during the workshops organized by IMPROVE (cf. Sect. 7.1).

Cooperation with Software Industry

A research software prototype for work process simulation has been created in subproject I4 (cf. Sect. 5.2). However, the learning effort for potential users is high, and the results must be interpreted in complex procedures independent of the simulation. As a result, the involvement of a software company is indispensable for the success of the transfer project.

ConSense GmbH operates as an innovative and industry-spanning software developer in the field of business process modeling and integrated management systems. The specific experience collected over several years focuses on personalization and on the interactive visualization of complex business processes tailored to the demands of individual employees. A particular emphasis, however, is thereby on the employee-centered and thus often weakly structured process presentation. Due to a tight integration of various information sources in the scope of quality and other management systems and ergonomic design, work organizational connections can more strongly assert themselves through software tools developed by ConSense. ConSense can support the recording of work organizational correlations through software technology. Thus, extending the accepted design process models with mathematical correlations to enable simulation-based analysis of the processes is a noteworthy advancement for the companies as well.

State of the Art

Up to now, concepts of concurrent engineering [521] and BPM have only sporadically been used for planning design projects in the chemical industries. Project planners and divisional directors continuously gain experience relevant for project planning and execution. Such knowledge is the foundation for improving the planning and calculation of future projects, but it is seldom structured and prepared by modern methods of knowledge management. In consequence, decisions during a particular project can often not be reproduced. Also, due to individual preferences of the planners, resources are not used in an optimal manner. Furthermore, the documentation of completed projects is typically restricted to milestone reports.

In consequence, the actual course of a project can thus neither be analyzed in hindsight, nor can it be improved for future projects. Several projects often seem to follow the same course, which is why project plans are regularly reused. Though reusing project plans reduces the effort for project planning, it also prevents the adaptation of the resource requirements to actual specifications and boundary conditions in the company like the available resources or staff. Basically, numerous organizational units, often geographically distributed, are involved in complex design processes. Such issues are rarely considered in the planning phase. Instead, the tasks to be completed by the organizational units are centrally planned, and the available resources are assigned to them.

7.4.3 Goals and Strategies for Efficient Planning of Design Projects

Goals

In order to analyze the existing deficits in the work organizational planning of design processes and to develop measures for their reduction, managers of chemical engineering design projects should be given the opportunity to model the project structure with little effort and to evaluate possible variants of the project progression in simulation experiments. Simulation studies must be performed in the early project phases to have any effect already in the planning phase. It should be possible to discover bottlenecks and weak points in project planning (such as invalid overlaps or parallelizations of tasks or the absence of qualified personnel at specific times) by analyzing project models and simulation results. In this case, the empirical measures of managers and analogies to other projects can be used better than today. The required technical (necessary technical aids for project development) and organizational resources (budget frames, organizational structure, joint sectors) as well as human resource allocation (number of people, qualifications, labor time), all necessary for the project, can be better forecasted. After all, the use of different project resources along with the employment of the available personnel can be arranged in an improved way. The superior purpose of the existing approach in this project is to decrease the cycle time of design and development projects with simultaneous consideration of resources.

Strategies for Improving Design Processes

Based on the results of IMPROVE, further endeavors must be made to accomplish some of the following demands in particular: The participative software-supported modeling of design projects as well as the conduction of experiments based on already constructed project models should be possible with little effort. The trans-departmental integration of process knowledge and experience should be ensured; also an increased planning reliability is desirable. It should be possible to adapt the simulation model to changed constraints (e.g., order situation, resource availability, etc.). The targeted variation of parameters should be possible to change the project organization (e.g., context of tasks, available personnel, available resources, etc.). Using this tool, a project manager should be able to give optimal criteria for the efficient project constellation, and in this sense also determine optimal variants for the management of the project (e.g., the optimal combination of personnel and resources). The simulation results should be available in a task-appropriate and expectation-compliant form such that the managers of design projects can easily use them. A simulation of a running project, and thereby a detailed depiction of the actual influencing variables, should help the managers to choose an improved strategy at certain milestones.

To carry those strategies into effect, some research gaps should be closed which still exist in the area of design process simulation in chemical engineering. First, the relevant influencing factors in project management and development controlling should be identified and weighted empirically. It should be verified if these influencing factors can be modeled as key data and if they can be integrated in the simulation model in the form of output variables. Those parameters of design processes that have a significant influence on the output

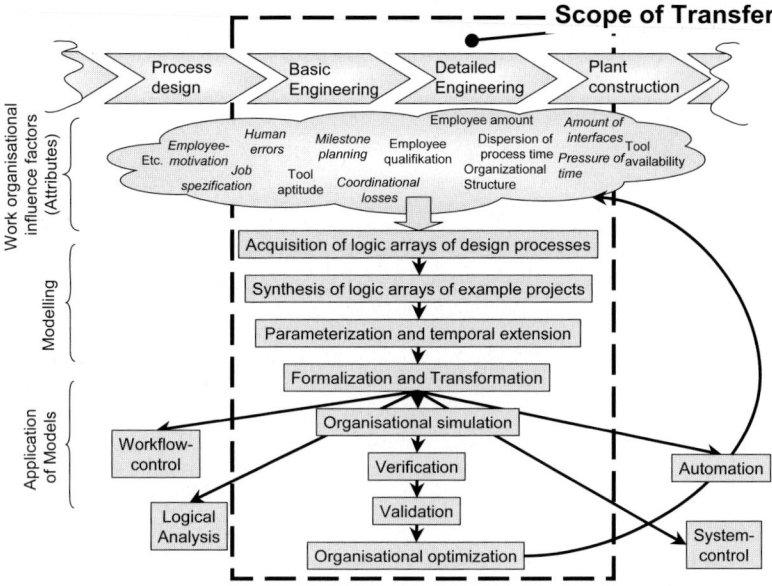

Fig. 7.6. Procedure of application in industry

variables must be gathered empirically. Beyond this, the type and the direction of effect of these influences should be determined. The connections should ultimately be operationalized in a conceptual simulation model. Subsequently, this model must be transformed into a computer model. The computer model must subsequently be verified with respect to the conceptual model. In order to transfer well-founded conclusions about the forecasting quality of the expanded executable simulation model, it must be extensively validated through empirical project data. Figure 7.6 roughly illustrates the described work plan.

7.4.4 Extensions to the Existing Approach and Consortium

The simulation method developed in IMPROVE (cf. Sect. 5.2) is based on correlations between selected influencing factors which have been determined by literature survey and expert discussions in a theoretical rather than empirical way. However, these are just a small excerpt of the success factors relevant for actual design processes. If additional influencing factors are integrated in the simulation for their practical relevance, the models must be validated again.

In cooperation with the industrial partners of the transfer project, the simulation model must be extended to overcome its current restrictions. The most important factor is the extension of staff characteristics, such as the impact of the designers' experience with tools on cycle times. Furthermore, the probability of error must be accounted for depending on the level of practice in different activities; it has a direct influence on the productivity of a particular person due to the probability of iterations during incorrect operations. The individual tasks in the task net – the task net is the part of the simulation model responsible for the integration of different tasks – are still processed by single individual persons. This approach does not, however, match with reality, and must be further developed to reflect the influence of several persons simultaneously working on a task.

Similarly, and in contrast to the currently used normal distribution, the right-skewed β-distribution that shows the trend of the usually underestimated actual processing time is to be used. This is to make allowance for the fact that activities tend to take more time and resources than expected. In the current simulation model, the effectiveness of tools does not depend on the qualifications of the persons who use the tool nor does it depend on the tasks a certain tool is used for. This should be adapted according to actual conditions so that the further development should take these coherences into account.

Furthermore, the characteristics of the information element in C3 must be extended; for instance, it must be possible to decompose an information element in several partial information elements. This way, it will be possible to account for the impact of the status of an information on the upcoming activities. For example, during the conceptual design of a chemical process, simulation studies are performed to evaluate different reactor types. Later on, more detailed simulation studies will provide information about the optimal dimensions of the reactor. However, the first partial information concerning the reactor type is sufficient for a first product inquiry for the reactor; thus, a product inquiry can be started even before the complete information about the reactor is available.

The current simulation model does not account for the information carriers used. In the transfer project, their effect on transfer time and information quality will be accounted for.

In the following, we describe the consortium of industrial partners and the work plan to achieve the required modifications of the existing simulation approach.

Consortium

The involvement of several industrial partners is essential in order to extend, parameterize, and validate the modeling and simulation tool with actual design processes. The following companies will take part in this plan:

- Bayer MaterialScience AG is one of the world's largest producers of polymer and highly valued synthetic materials, and is thereby also an ideal cooperation partner for this project.
- As a service provider in chemical process design, InfraServ GmbH & Co Knapsack KG is a very well-suited cooperation partner. The company has

extensive experience with complex development processes and disposes of a large pool of relevant example processes.
- As a medium-size software company, ConSense GmbH possesses substantial experience in person-based modeling of operating and business processes. The company develops and distributes complete systems for business process modeling, and is thereby an important cooperation partner for the implementation of planned support tools.

Work Plan

At the beginning, representative example processes from the application partners (Bayer and InfraServ) are to be identified and modeled by means of C3. Special consideration is given to the determination of workflow management influencing variables and process parameters in regard to the work topic, work equipment, and also the employee. They are identified and categorized in expert workshops and interviews according to their relevance for workflow management project planning.

Afterward, the structure of the design projects is analyzed to identify recurring routine activities; they are made available in the form of reference process components. Initially, this is done separately for both application partners. In a second step, similarities between the two application partners are identified. The validity and integrity of the collected parameters is increased through this process. In order to represent the design projects created in the first step, including influencing variables and process parameters, an appropriate software environment is identified.

To reduce the degrees of freedom for the succeeding simulation, the relevant influencing variables and process parameters are weighted and finally selected in cooperation with experts of the application partners.

The example projects are modeled in a way similar to workflow models; the identified reference elements are stored in a reference library. Therefore, the existing partial model of the simulation model must be extended with additional process parameters.

In addition, many of the variables that were up to now modeled as input variables should instead be modeled as dependent output variables. To do so, the interdependencies between the parameters must be quantified based on further literature analysis and on surveys of experts. An assistance program for the creation of simulation models will be conceptualized and implemented to support project managers in modeling new concepts of design processes. This contains both the conversion of the C3 model in an executable simulation model as well as the partial automation of test runs for the optimization of workflow management.

To ensure the usability of the entire system for improving the planning periods of design processes, a user interface is developed with regard to software ergonomics. The verification of the extended computer model with respect to the conceptual model is an important step to identify discrepancies between the model and the progression of real projects. Therefore, the model is validated in regard to an actual chemical engineering design process. For this, a design project is generated, modeled, and simulated a posteriori. The results will be statistically tested and analyzed. Possible differences to the conceptual model will be quantified and documented depending on the required modifications. Finally, the results of the individual work packages and the software components will be combined in a modular manner to form the complete system.

7.4.5 Conclusion

Based on the results of IMPROVE, this transfer project aims at closing the gap between work process analysis, the modeling of work organizational interdependencies, and discrete-event simulation by improving workflow management methods. To reach this goal, new methods will be investigated and developed. These methods, created by means of C3, should enable project managers in the chemical industries to create detailed graphical models of design processes interactively. The (semi-)automatic transformation of such models into formal simulation models, represented as timed Petri nets, will be supported. On the basis of these simulation models, variants of the design process can be created, evaluated, and optimized with respect to specific target values given by the project managers.

In order to do so, C3 will be extended with simulation-specific attributes. Furthermore, exemplary design projects will be modeled in cooperation with the industrial partners. By analyzing these processes, frequently occurring routine activities shall be identified and described in form of reference process components. Also, the simulation model for work processes developed in IMPROVE must be conceptually expanded. Therefore, the existing partial models, among other things, should be extended with influence variables and process parameters relevant for workflow management. This conceptual simulation model will be formalized as a Petri net and then transferred into a computer model. The computer model should first be verified in regard to the conceptual model and afterwards be validated on the basis of an actual design process. Furthermore, an assistance system for the creation of simulation models will be conceptualized and developed. This concerns both the implementation of the transformation rules in a computer tool as well as the partial automation of test runs for the evaluation and optimization of design processes. To ensure the usability of the entire system for project managers in the chemical industries, a user interface will be developed with special consideration of software ergonomics.

7.5 Management and Reuse of Experience Knowledge in Extrusion Processes

S.C. Brandt, M. Jarke, M. Miatidis, M. Raddatz, and M. Schlüter

Abstract. Extrusion of rubber profiles, e.g., for the automotive industry, is a highly complex continuous production process which is nevertheless influenced strongly by variability in input materials and other external conditions. As analytical models exist only for small parts of such processes, experience continues to play an important role here, very similar to the situation in the early phases of process engineering studied in CRC IMPROVE. This section therefore describes a transfer research project called MErKoFer conducted jointly with an industrial application partner and a software house founded by former CRC members.

In MErKoFer, results from the CRC projects on direct process support (B1, see Sect. 3.1), process data warehousing (C1, Sect. 4.1), and plastics engineering (A3, see Sect. 5.4) were applied and extended. Specifically, knowledge about extrusion processes is captured by ontology-based traceability mechanisms for both direct process support of extrusion operators, and for process analysis and improvement based on an integration of data mining techniques. The accumulated knowledge assists in ensuring defined quality standards and in handling production faults efficiently and effectively. The approach was experimentally implemented and evaluated in the industrial partner's site, and some generalizable parts of the environment were taken up by the software house partner in their aiXPerience software environment for process automation and process information systems.

7.5.1 Introduction

In the area of plastics processing, the analysis of process data from production plants is continually gaining ground to achieve an improved understanding of the overall process. This is due to the widespread adoption of sensor technology, and the increasing production costs and plant sizes. The data collected during production is, though, usually not systematically evaluated because the context in which it was recorded often gets lost. A great potential for improvement can be found here, by including the contextual information and systematically generating knowledge from such operational data. Thus, the technology transfer project described in this section focuses on supporting the analysis of rubber extrusion by employing knowledge-based methods and on the direct experience-based support of machine operators controlling such production lines.

In the project "MErKoFer – Management of Experience Knowledge in Continuous Production Processes" [373], funded by the German Federal Ministry of Research (BMBF), two major approaches for process support and improvement were combined. The mechanisms of direct process support as developed in the sub project B1 (see Sect. 3.1) were extended for the experience-based support of production line operators. Specifically, the scenario treats the analysis and advancement of production processes for specialized rubber profiles. Additionally, ontological modeling from the sub projects B1 and A3 (see Sects. 4.1 and 5.4, respectively) was used in combination with experience management to generate and improve explicit process understanding. Further research results from projects conducted at the Chair of Informatics 5 were also integrated. The project was realized in cooperation with an industrial partner from the domain of rubber profile extrusion, with applications mostly in the automotive industry. The aiXtrusion GmbH [497], a CRC spin-off from the area of process automation and information systems, contributed solutions to the problems of data integration and pattern identification in process data.

To drive and validate the scenario, the so-called *aiXPerience* environment was established at the industrial partner's site, to record all important influence factors of the production processes. This includes the operators' interactions, process set values, measured and derived process values, product quality and production fault information, and additional information such as environmental conditions. These traces were then analyzed and searched, to allow their direct and situation-based reuse for operator support, and thus to improve the production process itself. Methods of data mining were employed for clustering and detecting characteristic process situations, to find out where and how support functionality can be provided.

The experiences gained from developing and applying the situation-based operator support were also used to gain explicit understanding of and knowledge about the production process itself. As described later in more detail, the processes treated here cannot be effectively described by analytical models or simulations based on physical correlations [942]. Instead, a coherent framework for the analysis of the production processes can be established by combining recorded process traces, available explicit knowledge about the process, and using additional ontological modeling of the process aspects. Data mining, statistics and other mathematical approaches are used to analyze and partially predict the process behavior, and to use this knowledge in process design and improvement by the engineers.

Complementary to the detailed and deep situation-based analysis of the process traces, aggregated analyses based on the concepts of OLAP (Online Analytical Processing) were realized to achieve a broader view of the production processes. Based on the integrated data sources, multi-dimensional cubes were created to aggregate the available information, allowing the well-known operations of slice, dice, and drill down.

To achieve the formalization of the production processes, the Process Data Warehouse (PDW) developed in sub project C1 was integrated into the infrastructure, forming an important part of the aiXPerience environment for the integrated analysis and reuse of process and product artifacts. The Core Ontology of the PDW (see Subsect. 4.1.5) facilitated the definition of products, processes and their characterizations (e.g., category schemata), as well as their linkage to various persistent sources of information. This basic modeling formalism was filled with the explicit knowledge described through a domain

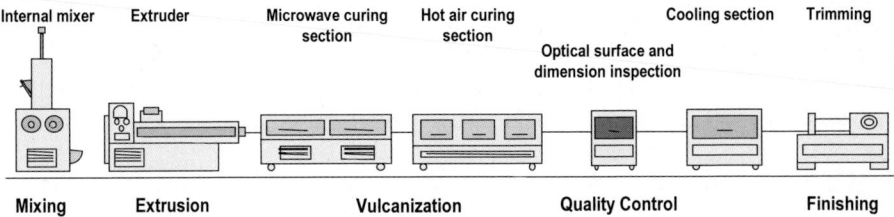

Fig. 7.7. Schematic view of a typical extrusion line for rubber profiles

ontology with relationships and dependencies gained during the analysis of the application domain. Driven by this comprehensive information model, the analysis tools used by the application partner were integrated and extended, and complemented with specialized tools for the classification and visual exploration of large data sets, based on self-organizing feature maps.

The next subsection gives an overview on the issues of using computer science methods for supporting complex production processes in the domain of rubber profile extrusion. Subsection 7.5.3 describes the initial steps followed in the project, i.e., the integrated recording and analysis of the production processes. The following subsection shows how the *production situations* were created, clustered, and analyzed further, to allow extended situation-based operator support. Subsection 7.5.5 describes how the results of the project are used for gaining and applying explicit process understanding. The section closes with a short summary and conclusions.

7.5.2 Supporting and Improving Extrusion Processes

Rubber Profile Extrusion

Figure 7.7 shows a typical rubber profile extrusion line. Several days prior to the extrusion itself, the raw rubber is prepared in an internal mixer, where certain materials such as sulfur or carbon black are added to achieve the required properties. One or more extruders form the head of the production line. There, the rubber is kneaded, heated, intermixed, and then extruded through specially formed dies (final delivery elements) which are responsible for attaining the required profile geometry. A major part of the line is taken by the vulcanization process, usually achieved by near infra red and microwave heating. After optical quality control, the profile is cooled down and cut into pieces of a certain length.

Nowadays, operators for all kinds of production lines have to interact with a large number of different systems for automatic control, process recording, and quality management. On the other hand, documentation tasks within these processes are often partly paper based, e.g., human process manipulations or the administration of raw materials. This especially applies to the domain of rubber profile extrusion, as treated in this section, and partially described in reference to Fig. 7.7. The heterogeneity of these systems and methods does currently not allow integrated processing of the respective data in a reliable and coherent manner [942]. This is also the current state at the site of the project's industrial partner.

In the concrete case of rubber profile extrusion, several additional complications arise that result in very complex production processes. Common approaches for automatic control, e.g., from chemical engineering, do not suffice in this setting due to the complex and non-deterministic properties of these processes, which are mainly caused by the following issues.

- No usable physico-analytical models exist yet for the vulcanization behavior of rubber, especially in the extrusion context. This inhibits the prediction of process behavior.
- It is hardly possible to accurately determine the quality of a profile section before it reaches the end of the production line, as it needs to be examined in its fully vulcanized state. This is usually done by cutting of a small section and examining it manually for surface defects or deviations in the cross-section geometry, or – more recently – by automated visual inspection.
- Due to the length of the production lines (up to 250 m), the time needed for vulcanization and thus, the slow speed of extrusion, the produced profile takes a long time to reach the inspection zone (up to 30 minutes). Thus, the results of operating parameter changes can only be accurately detected after this time.
- The processes react very sensitively to many influence factors, not all of them explicitly known (e.g., the duration and environmental conditions of raw material storage). When changing or correcting the process, this sensitivity has to be taken into account.

Because of these issues, these processes can only be successfully started and controlled by operators with a lot of operational experience. Much variation can be found in the time necessary to achieve a stable production state. At the same time, production costs are mainly determined by the consumption of raw materials. Unfortunately, vulcanized rubber cannot be reused if the final profile does not conform to specification. Therefore, the primary goals of the research described here were to improve the work situation of the machine operators, especially their autonomy, and to reduce the time spent in producing off-spec products, thus saving money and reducing environmental impact.

Operator Support

The approaches used in many other domains for operator support were mostly inapplicable in the project context, as described above. Due to the missing explicit analytical models, no prediction of process behavior and thus expected product quality based on currently measured process values was possible. This also precluded the use of technologies such as soft sensors to alleviate the issue of missing early quality measurements. Thus, many of the characteristics these processes exhibit, resemble those of creative design processes, which form the central topic of the CRC's research:

- personal experience is the primary driving factor for successful process completion;
- it is impossible to "plan" or "predict" the process in detail;
- instead, it is necessary to *react* on unexpected changes;
- there is an obvious need for fine-grained direct process support.

Therefore, the project appeared a good place for transferring some of the results of the CRC IMPROVE into industrial practice. In contrast to development processes, these production processes are executed often, and can be measured quantitatively. This allowed the direct application and validation of the approach in real-world processes, instead of scenario-based evaluation as usually utilized when validating design process support. Thus, the aspects of direct experience-based process support as researched by the sub projects B1 and C1, were to be adjusted and applied in the production context.

Case-Based Reasoning

To enable the direct reuse of experience traces, methods like case-based reasoning (CBR [493]) can be applied. This method is based on using concrete, previously recorded *cases* related to possible solution approaches, instead of using the relations between generalized problem descriptions and conclusions. This way, CBR tries to mimic human problem solving behavior. A new problem is solved by *retrieving* a similar case from historical data or an explicit case base and *reusing* the case and its related solution proposal. Usually, it is necessary to *revise* the solution according to the current problem situation. As a last step, the new solution can be *retained* as a new case. Thus, CBR also enables incremental learning, i.e., starting from a small case base and filling it with manually or automatically recorded solution cases.

Data Mining

For discovering knowledge from large data sets, many different methods and technologies exist for automatically *mining* previously unknown correlations from those data sets. After finding such correlations, the results have to be interpreted to identify concrete knowledge. One common approach for Data Mining is based on discovering *clusters* with common characteristics in the data, and distributing the data items or objects among those clusters. Objects inside one cluster should be as similar as possible, while objects from different clusters need to be dissimilar.

A specialized method for similarity-based visualization of high-dimensional data is formed by self-organizing feature maps (SOM). The data items are arranged on a two-dimensional plane with the aid of neural networks, especially Kohonen nets. Similarity between data items is represented by spacial closeness, while large distances indicate major dissimilarities [968]. At the authors' department, a system called MIDAS had already been developed which combines strategies for the creation of feature maps with the supervised generation of fuzzy-terms from the maps [967].

Process Analysis

It was planned, and accomplished, to use the results of the situation-based process analysis for the creation of some kind of process models for rubber profile extrusion, to replace the not-yet-existing explicit mathematical models. A combined view on the explicit process understanding, the ontology-based domain modeling, and the experiences from situation analysis and operator support was used to support the design and analysis of both existing and newly designed production processes, and thus to improve these processes. Additionally, the experiences from the CRC in design process support were applied here, e.g., using the Process Data Warehouse and the PRIME process-integrated environment. This allowed the same extended application environment to be used both for supporting the production processes, and the design processes that create and change the process specifications.

7.5.3 Recording and Analyzing Production Processes

Considering the problems described above, the approach of experience-based support for creative processes was extended to the support of operation personnel in profile extrusion. In the context of the process analysis of plastics engineering production, a knowledge management system was developed to achieve the aforementioned goals. Ontology-based methods for the explicit management of experience knowledge were combined with innovative methods of data mining and neural networks. The expert knowledge captured in the ontologies was enriched with fine-grained relationships and rules from the analysis activities (data mining, visual data exploration and correlation analysis).

From Operator Support to Process Development

A Stepwise Approach

A prominent goal of the MErKoFer project was the prototypical development of an information system for the support of process data analysis, and experience and knowledge management in rubber profile extrusion. This system was

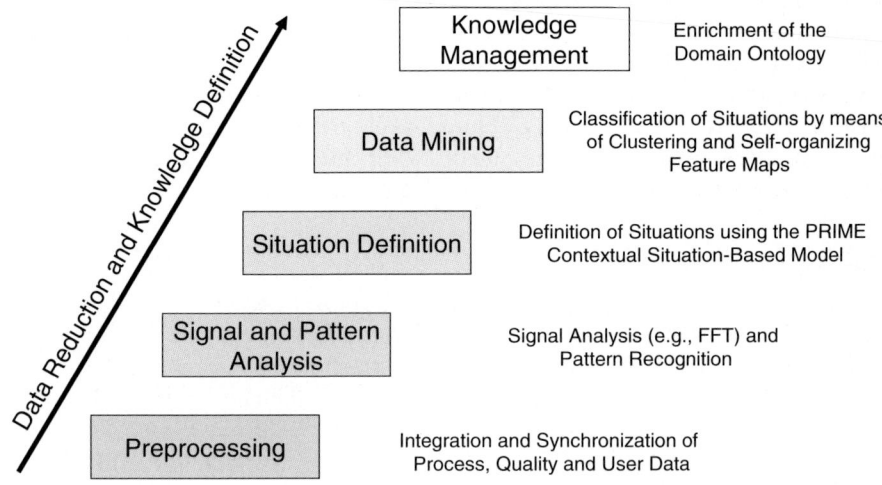

Fig. 7.8. From process and business data to knowledge definition

validated on example scenarios from the project's application partner, and established at the partner's site to provide the process operator with integrated support for his or her tasks. Figure 7.8 gives an overview on some of the necessary steps followed during the design and evolution of the system, starting from the recording and processing of the operational data to the generation and management of explicit process knowledge. As it will be described in the following, the steps cannot be seen as simple and linear as in the figure, but results of the upper steps had to be fed back into the lower ones.

In the following, the various interrelated steps followed during the MErKoFer project are described in more detail, based on the blocks visible in Fig. 7.8. Through increased data reduction, e.g., by the generation of characteristic values and their contextualization, more detailed knowledge is extracted from automatically recorded information.

Preprocessing

The lowest level of the figure is formed by the recording of the process traces, consisting of process values, quality and user data, and other information. To achieve this, it was necessary to integrate the various automatic control data sources, together with additional sources such as environmental conditions. As described in the previous subsection, many of these sources had not been integrated in any way, so that the issue of their heterogeneity had to be addressed first. Also, many important steps had not yet been recorded at all, or only on paper, e.g., material changes, fault situations, countermeasures, and most importantly, quality control data.

In tight relation with the recording, the data had to be analyzed on a technical level, to determine important characteristics, e.g., the importance of various signals, their noise levels, and other factors. Additionally, as some of the sources were – and are – only integrable by offline import, all the data had to be synchronized on common recording times. Thus, about 50 different *signals* were recorded and stored in a joint database.

Signal and Pattern Analysis

The next level in Fig. 7.8 is formed by signal analysis, where various methods were tried and used for recognizing significant positions in the signals. Based on the detection of those positions by, e.g., threshold detection, fourier analysis, or mean value analysis, *deviation patterns* were defined for their classification. By recognizing these patterns on the signals, *situations* were constructed as sets of temporally related deviations. Some important help could already be given to the production line operators here, e.g., by raising alarms in case of obvious deviations.

Situation Definition

Using the PRIME meta model, situations were then defined, based on the detected signal deviations. In an initial step, generic definitions were used to detect a large number of temporarily grouped deviations, called *situation instances*. To form abstract situations out of the instances, methods of data mining had to be employed.

Data Mining

These situation instances, formed by detected deviations on semantically related signals within a short time range, were then mined for correlations, mainly based on two methods. Clustering was used to detect groups of situation instances with common characteristics, while self-organizing feature maps were used to visualize these situation spaces. Based on the visualization, recurring and thus important combinations of signal and deviations were determined. This allowed both the construction of a case base for direct application of recorded situations and their related counter measures in the production process, and for creating explicit knowledge about the process.

Operator Support

Not visible in Fig. 7.8 is the application of the trace and case repository onto the direct support of the production line operators. The aiXPerience system was extended to detect process deviations, to find reference situations matching the current process situation, and thus, to find appropriate countermeasures. These situation-based information and counter-measures are then presented to the operators, who – according to the case based reasoning concept of a *recommender system* [550] – can then decide on how to apply the presented information.

Knowledge Management

During the whole of the project, the MErKoFer domain ontology was developed, extended, and adjusted to experiences gained, and to changing requirements. Based on the Core Ontology of the Process Data Warehouse (PDW) (cf. Subsect. 4.1.5), it was used to integrate the different aspects of production and development processes, production lines, raw materials, products, and process situations. The ontology was both created by, and used for, the determination of important relationships between various domain concepts and objects. For example, production faults and counter measures are both categorized according to appropriate ontological schemata. The content of this ontology, and the knowledge contained therein, forms the highest level in the knowledge definition and discovery structure of Fig. 7.8. Additionally, the explicit knowledge created both during the project, and by running the support system, can be used for analysis and modification of process specifications, or for the creation of new processes, e.g., for new products.

Recording and Preprocessing

Recording Process Values

At the production lines for rubber profile extrusion as installed at the industrial partners site, most process values were already available from automatic control. Yet, these values were only available for direct display, but not recorded in any way. This is partially due to the fact that production lines in this domain usually run without major changes for several decades. Some important process values, e.g., the power consumption of the extruder engines, were not measured at all. Thus, a first important step of the project was the integration and recording of these values. Appropriate hard- and software had to be installed. Additionally, environmental conditions with possible influence on the properties of the raw materials and thus the process behavior had to be recorded, e.g., temperature, humidity and pressure.

At the production line in question, the profiles' surface quality and cross-section geometry deviation was already analyzed and recorded by online optical quality inspection systems. These camera based systems usually mark defective profile sections visibly, to separate them out for the operating personnel [910]. In some cases, this data was analyzed to quantify the amount and kind of failures statistically to control the rate of production faults. For an integrated management of process, user interaction, and fault information, and quantitative quality data, the records of these inspection systems also had to be integrated.

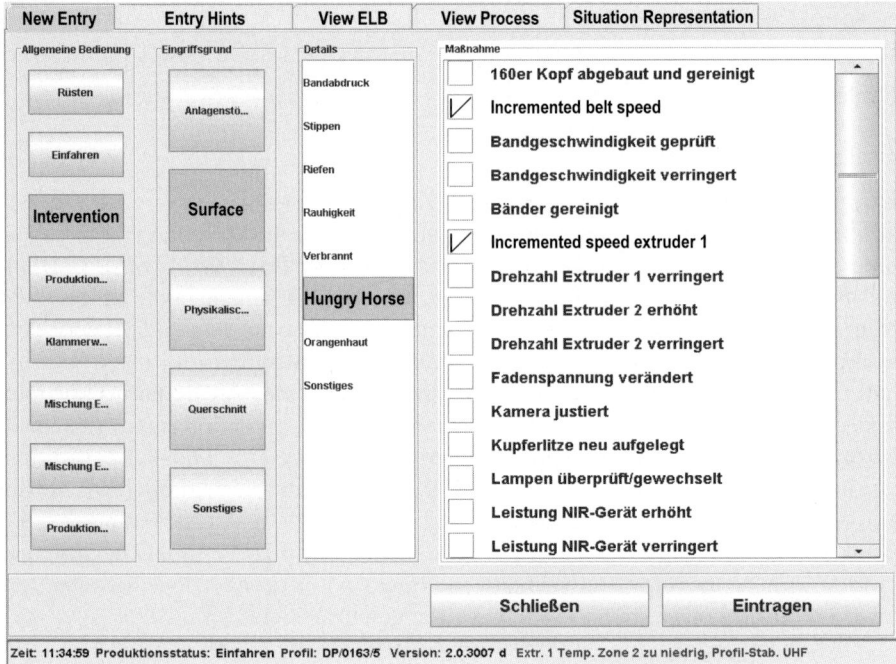

Fig. 7.9. Electronic logbook showing task, cause, error detail, and counter measure

The Electronic Logbook

The machine operators themselves are faced with a bundle of heterogeneous systems which are partially redundant for historical and technical reasons. Many important operator observations and actions, e.g., process faults, interventions, rubber changes, etc., were only recorded manually in a paper-based logbook, and thus unusable for automated trace management. Therefore, the requirements for building a new human machine interface were elaborated in a participatory procedure together with the operators and the engineers. The most important issue was that the "electronic logbook" (ELB) had to be comfortably operated using a touch screen, while wearing gloves. That led to the installation of two pressure sensitive touch panels which are placed at the extruders (the production line's starting point) and the quality control area (the end of the line) as required by the plant's dimension.

The user-interface for the operators was based on the formalized ontology for the production process, as developed so far. The various tasks, interventions, error categories and error details and the available counter measures were all retrieved from this ontology. This allowed to record the user activities during production, and to store this data with its appropriate context to enable context-sensitive data-mining across all available data sources. Together with the logging of current process values this also provided a way to continuously update new situations into the knowledgebase.

In Fig. 7.9, a screenshot of the logbook is shown where a set of countermeasures to remedy a production fault is inserted into the database. On the left of the screen, the list of tasks can be seen. In addition to entering interventions necessary because of detected problems ("Intervention", currently selected), it offers important process phase changes (mounting, starting the run-in, starting production, ending production), and raw material changes. The second column from the left allows to select the cause for the current intervention, "Surface" in the current case because a surface error of the type "Hungry Horse" was detected (i.e., metal inlay bared; third column). The fourth column allows to select one or more counter measures as carried out by the machine operator (Incremented belt speed and Incremented speed extruder 1, here). The time of the intervention may also be selected in an additional dialog, e.g., if an entry needs to be postdated.

Signal Analysis and Pattern Recognition

Analyzing Process Values

It was necessary to establish a set of different steps for signal analysis, pattern analysis and recognition, situation recognition and analysis, and finally, situation-based reuse of the gained experience traces. This was mainly due to the high complexity of the signals and their unknown relationships, and the impossibility of detecting significant changes directly on the complete signal data. In contrast to the fully automated approach initially projected, this allowed to reach high quality for the reference patterns, especially during the final evaluation phase of the project.

The identification of typical patterns and correlations between characteristic process parameters was accomplished in close collaboration with the domain experts using the visual data exploration tool InfoZoom [976] on existing process data. Unfortunately, it was not possible to gain detailed and deterministic information about the quantitative relationships between the various signals and their deviations. Knowledge about the qualitative behavior of the process was available, and independent process specifications for certain nominal values were in use. Yet, knowledge about the quantitative interrelation between process values, fault patterns and production quality was only implicitly known, i.e., in the experience of the operators. For example, when the quality control systems notes a deviation in cross-section geometry, this information is not directly used. Instead, a cut is taken from the finished profile, enlarged, and visually compared with the geometry specification. Thus, it was nearly impossible to find appropriate parameters for deriving product quality and corresponding information about process stability and production quality, from the raw quality inspection data.

Detecting Signal Deviations

The aim of the signal analysis in the MErKoFer project was to determine *significant deviations* of the various signals (i.e., process and other recorded values) from their stable state. Of course, care had to be taken of noise, oscillations, and similar issues. Several methods where therefore applied and evaluated. Partially, Gauss or median filters were used for preprocessing.

- A simple *edge detection* filter was used successfully, especially on set values. Also, some signals such as the infrared reflection in the hot air canals showed a clear "toppling" behavior.
- For some signals, such as the error rate calculated from the surface quality inspection, a simple *threshold detection* was enough to distinguish between "good" and "bad" production states.
- With the help of *polynomial approximation*, especially linear and cubic interpolation, the signal were described in a simplified form.
- The most important and successful analyses were done in the frequency domain, i.e., interesting signal positions were detected after *Fourier-transforming* the signal.

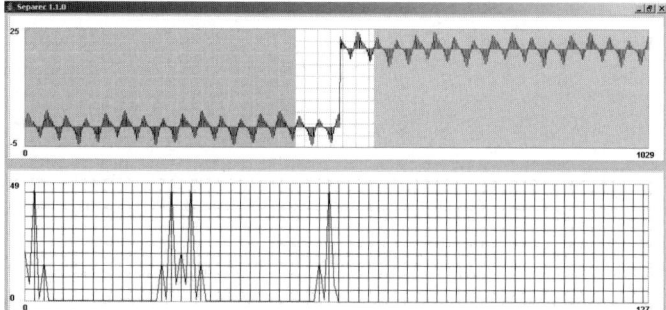

Fig. 7.10. Detection of characteristic signal events (FFT)

Fig. 7.10 shows the application of the Fast Fourier Transformation (FFT) on a sample signal, an extruder temperature curve. On top, the signal can be seen, while the bottom shows its FFT. Three deviations can be seen. Skipping the first deviation, the second one is caused by a change in the oscillation frequency and amplitude, while the third one is caused by a major raise in the temperature value, probably due to some user interaction.

Pattern Recognition

After the detection of significant positions in the signals, i.e., deviations from stationary behavior, it was necessary to determine the type or *class* of deviation. For the more complex signals, *signal patterns* were therefore defined,

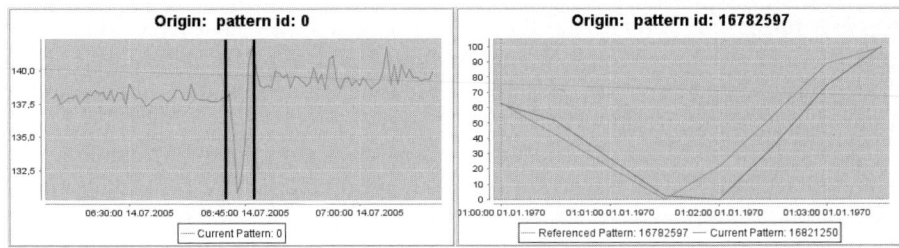

Fig. 7.11. Detection of a signal deviation, and its classification by pattern matching

based on deviation patterns that occurred regularly. These patterns were partially defined explicitly, and partially derived from existing signals, using supervised learning methods. Figure 7.11 shows the detection of a significant deviation on the left, and – on the right – its match against a pattern and thus, its classification. This matching was done using euclidian distance, partially corrected by dynamic time warping, and by using the frequency spectra. Using a large list of reference patterns, grouped into a lower number of signal pattern classes, it was possible to determine and classify the detected signal deviations [468].

Each of these detections and classifications was applied to single signals, only. Therefore, the next step of the project required finding and analyzing *situations*, i.e., sets of temporally and, hopefully, also causally related signal deviations on different signals.

7.5.4 Situation-Based Analysis and Support

The key point about assessing and defining process and product state similar to the machine operators' way, is having objective information about the product quality. In the presented approach, the information from the optical inspection system was used to define characteristic situations based on the profile quality (the kind, distribution and quantity of defects) and the process parameters measured and stored by the automation system. A situation or case is thus characterized, among other things, by the aforementioned profile quality, the kind of profile that is produced, the used rubber-mixture, environmental data like air pressure or humidity, the values and latest progression of physical process parameters like extruder-temperature, power of microwave heating or speed of conveyor-belts and the countermeasures that are taken by the machine operators.

Situations and Situation Instances

A *situation instance* is defined as a set of temporally related events, usually signal deviations or other detected changes, such as interventions or detected product faults, together with the context in which they occurred. They are created to allow the systematic *analysis* and *reuse* of these situations, i.e., to draw conclusions about the expected process behavior in the case of a similar situation occurring in the process.

Three different kinds of situations need to be distinguished in the following. They combine both the results of explicit analysis done by process experts, and the multi-step analysis of the process data done within in the project.

- *Situation definitions* explicitly define various events and their properties as a kind of pattern, e.g.:
  - **rotational speed extruder 1**: constant
  - **engine power consumption extruder 1**: falling
  - **mass pressure extruder 1**: falling
  - Expected behavior:
    power consumption and pressure continue to fall, insufficient mass at extruder die, "hungry horse"
  - Possible counter-measures:
    check rubber feed for extruder 1 (run out of rubber, or wrenched off); raise **rotational speed extruder 1**; turn off infra-read (IR).
- *Situation instances* are automatically determined sets of temporarily related signal deviations, as introduced in the beginning of this subsection;
- *Reference situations* are specially selected from the large number of situation instances to represent certain cases, and to allow to find and characterize these cases, including appropriate counter-measures.

In MErKoFer, the concept of situation was derived from the identically named PRIME concept (see Subsect. 3.1.3 and [371]). The steps described up to now allowed the systematic acquisition of the situation instances, their automated analysis as described in the next subsubsection, and the specific selection of reference situations and appropriate counter-measures. By defining the *cases* as situations with appropriate counter-measures, it was possible to apply *case-based reasoning* (CBR) for the direct support of the production line operators, as described in Subsect. 7.5.2.

Situation Mining

A large number of situation instances was generated by automatically applying the MErKoFer algorithms for signal analysis, deviation detection and pattern recognition onto the available process and quality information. On these situations, two methods of data mining were mainly applied. The first method was used for density-based clustering of the situations, creating groups or clusters of situations to be interpreted by the project participants. The second method used neural networks to create self-organizing feature maps. Thus, it achieved the visualization of the situation instance base, allowing to interactively explore and analyze the situations to determine common vs. distinguishing features.

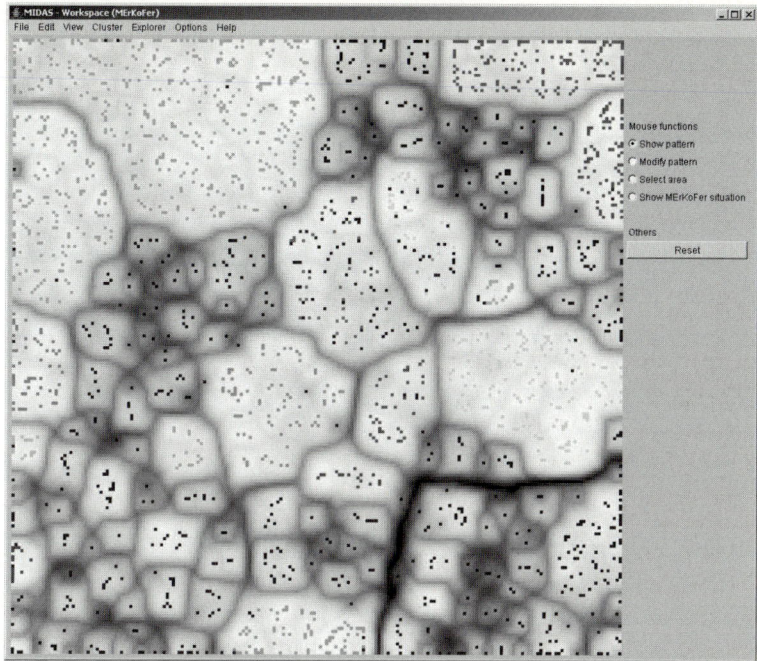

Fig. 7.12. MIDAS feature map with about 1500 situations, colored according to clusters and noise

Clustering

For clustering the situation instances, the density-based method of DBSCAN ("Density-Based Spatial Clustering of Applications with Noise", [662]) was used [466]. Unfortunately, the results where not applicable directly for determining characteristic reference situations. Firstly, due to the many preprocessing steps and the used situation model (vectors with a fixed number of components, and distance penalties for comparing against empty components), the data quality had been massively reduced. Later on in the project, better approaches were therefore researched and applied successfully [468]. Secondly, it was nearly impossible to determine the real correlations of the process parameters from the clustered situations. The main problem was that a situation instance can only be visually represented as a multi-signal time series, e.g., in the process view of the electronic logbook. Thus, comparison of more than two different situations and their composing process values was very difficult. The clustering results were therefore mainly used in combination with the feature maps as described in the following.

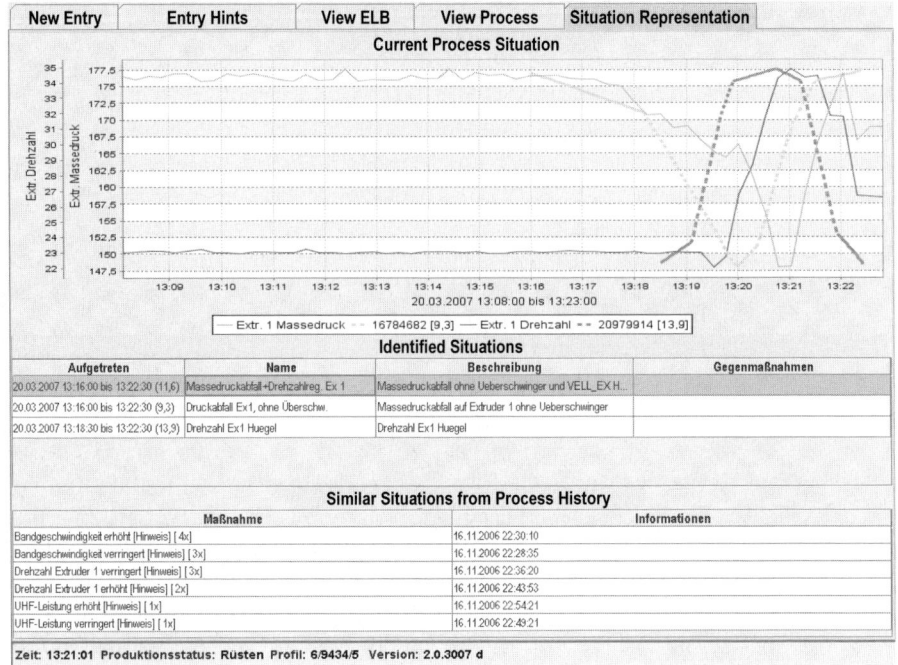

Fig. 7.13. Visualization of situations in the GUI.Dance module

Self-Organizing Feature Maps

Figure 7.12 shows one of the self-organizing feature maps (SOM) created with the multi-strategy tool "MIDAS" (Multi-Purpose Interactive Data Analysis System, see [967]). About 1500 situation instances are placed on a two-dimensional map so that the similarity of the data points is represented by their spatial closeness. Additionally, greyscales are used for achieving an intuitive display, where light "plateaus of similarity" are separated by dark "valleys of differences". This allows to explore the "situation space" in a visual way; by clicking on a situation instance, the respective situation, i.e., its signals, are displayed in the process view of the electronic logbook.

To apply this method onto the situation instances, it was necessary to convert the data items into a vector space. For each situation, the distance to all others could be calculated, based on the signal-wise distance between the recognized patterns. This allowed to apply the FastMap method [664], to create a lower dimensional space that approximately represented the situations' distance matrix. Unfortunately, this preprocessing disallowed to apply the fuzzy-ID3-based rule generation mechanism of MIDAS [969], as no conclusions about the process parameters could be drawn from rules about the generated vector space.

Extended Situation-Based Operator Support

For the direct situation-based support of the production line operators, the electronic logbook (ELB) was extended by a module called GUI.Dance. This module enabled a special visualization to display the currently identified situation to the operators, together with appropriate counter-measures which might help to regain a stable and good production state. In the top half of the situation representation display of the ELB in Fig. 7.13, two process signals are shown, on which significant deviations have been detected (**rotational speed extruder 1** and **mass pressure extruder 1**). As dotted lines, the signal patterns that best matched on these deviations are also plotted. Below, a list of reference situations is shown that fit the current process situation well. Based on these reference situations, a number of counter-measures, and their frequency, is displayed on the bottom.

The operators are immediately able to compare the different situations and assess which of the proposed solutions, or a completely different one, to choose best. While the search for, and the presentation of the situation alternatives can be seen as the *retrieve* step of case-based reasoning (CBR, see Subsect. 7.5.2), the selection and application of an appropriate solution corresponds to the *reuse* aspect. By adapting the recommended course of action and carrying it out, the operator then provides the *revision* step. An important aspect of the system is that the possible actions are neither directly applied onto the process, nor are there any restrictions of the operators about the possible interventions. Instead, the operators' *experience* is supported, but not replaced, by the *recommender* functionality of the support system.

Currently, research is still going on with respect to the best visual representation of the reference situations and the counter-measures. Additionally, aspects such as the Bayes factor (see FIST algorithm in Subsect. 3.1.5 and [80]) are to be integrated, to improve the recommendation quality based on the operators' previous decisions.

7.5.5 Using Recorded Experience for Process Analysis and Development

The MErKoFer Ontology

Originally, it was not planned to develop a full domain ontology for the MErKoFer project. Instead, only coarse-grained extended categorization schemata of various aspects were to be used, e.g., for interventions, error causes, and counter-measures.

As the elaborated domain knowledge gathered from discussions and interviews with the industrial partner grew, it was step by step transformed into comprehensive and fine-grained explicit information model, extending and enriching the Core Ontology from Subsect. 4.1.5. The domain-specific ontology of the industry partner is thus based on the generic Process Data Warehouse

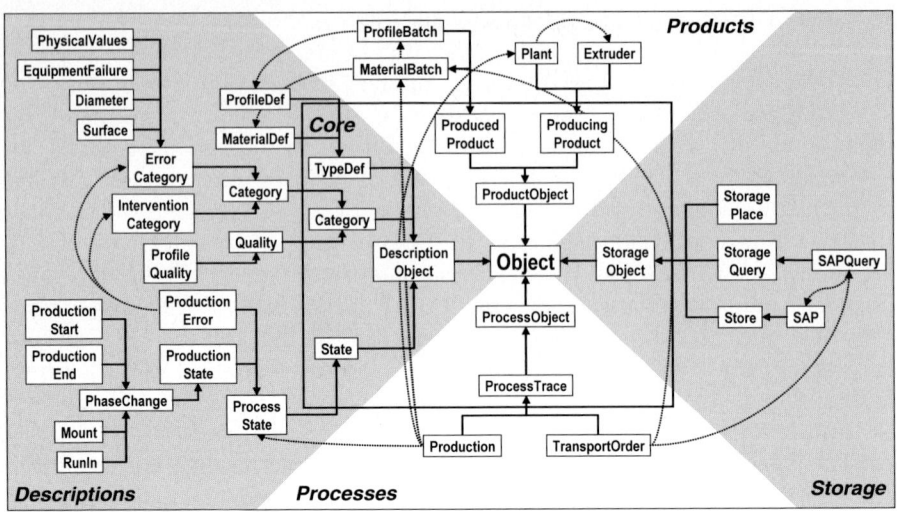

Fig. 7.14. Excerpt of the domain ontology

(PDW) model. It represents a model of the objects and concepts identified in the rubber extrusion domain as well as the relations between them. As a result of that, inside this MErKoFer ontology, the process structure, products and description artifacts are stored as well as instances of the executed production processes themselves, together with all related information. This includes the characteristic situations identified by the described mechanisms.

This MErKoFer ontology is partially shown in Fig. 7.14. In the area of *descriptions*, the most important aspects can be found: profile and material definitions, and the aforementioned categories for phase changes and other tasks, process states, and errors. In the *product* area, the plant and its elements are modeled as *producing products*, in addition to profile and material batches as *produced products*. The *process* area contains the production process itself as primary center for state and material changes, and the transport orders of the material batches which are read from the company's ERP system (SAP R/3) that is part of the *storage* area.

One of the project's main issues, the contextualization of information in form of situations, was achieved much more precisely by considering these explicit correlations and knowledge. Using the domain ontology, the explicitly stored knowledge was thus combined with detailed correlations from the data mining tasks described. For the future, the extension of the model elements will be done by the domain and modeling experts using the generic PDW front-end, while the instances are stored by the automated situation analyzer and the electronic logbook. This allows the easy configuration and parametrization of the operator support system via the modeling front-end of the PDW, and

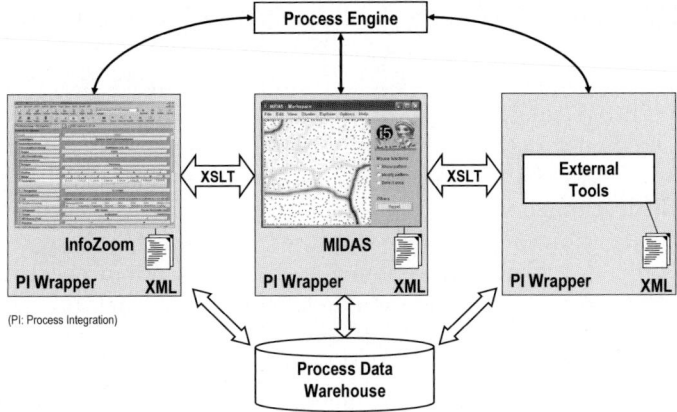

Fig. 7.15. Loose integration of domain-specific tools and the PDW through the PRIME framework

also the transfer of the system onto other, similar application domains, and to different industrial partners.

The PDW has been developed for the support of creative development processes by recording and reusing process and product traces. For the MErKoFer case, it was also used for modeling and supporting production processes. This allows to transfer the knowledge gained by production support into the field of *process design* and *improvement*, e.g., for the design of new processes for new products, or for changing the specifications of existing production processes according to recorded experiences. Additionally, it is possible to integrate the tools of process design with the PDW, thus achieving an integrated repository which supports the complete process lifecycle. This enables a unified view for process designers onto the process from design to execution in a way that is not yet possible. For the support of these design processes, a tight integration of the PRIME environment is required, transferring additional results of sub project B1 (see Subsect. 3.1.5).

Process Integration Using the PRIME Approach

Complementary to the PDW, the Process-Integrated Modelling Environment PRIME can play the role of the middleware infrastructure for the loose coupling of the various participating tools in a coherent process-integrated environment. This environment can serve two important purposes. First, the loose process integration of the interactive tools promotes their effective interoperability and thus, methodically support the users during their tasks. Second, PRIME closely interweaves with the PDW and acts as the mediator for the interpretation of the PDW Core Ontology to tool-specific formats and vice versa. Figure 7.15 gives an overview of the planned environment.

The main goal of this platform is the extraction of knowledge from captured experiences in the domain of profile extrusion, and its storage in the

PDW knowledge repository for long-term use. The processing of the captured experiences requires the use of computer-based tools that provide services to the users to accomplish their tasks. For reasons of efficiency, these tools are process-integrated using the PRIME approach (see Fig. 7.15, "external tools"). From the user's perspective, the tools should be seamlessly integrated with the various process chains and cooperating tools. From a builder's perspective, the integration should be easy, and tight coupling should be avoided in order to remain sufficiently flexible. Because of the large heterogeneity of the tools, the process integration approach has to be tailored according to the needs and limitations of the individual cases. Initially, each tool needs to be evaluated with respect to its provided programming interfaces, in order to find out to which extent it can be process-integrated. Next, appropriate wrappers for each tool have to be built, according to the PRIME generic tool wrapper architecture.

7.5.6 Conclusions

The described transfer project integrates, transfers, and extends results of three sub projects of the CRC 476 IMPROVE (A3, B1, C1) with a focus on experience-based operator support in a plastics engineering application scenario. In this context, the work has been directly validated in industrial practice for improving experience and knowledge management in rubber profile extrusion. Moreover, the coupling of explicit ontology-based knowledge formalizations with innovative and established tools employing data mining and visualization methods has provided further scientific results.

Based on the structured and highly integrated information display, one of the major goals for the industrial partner was achieved, namely, improving the autonomy of the production line operators by offering direct situation-based support derived from the analyzed process traces. Simple and explicit rules are used to notify the production line operators about occurring or imminent process changes in a way not possible previously. Even though the offering of specialized information about the type of deviation and possible countermeasures has not been realized for the day-to-day operations, the available information already indicates the need for interventions to the operators. Integrating and extending the display of the electronic logbook (ELB) with both current and short-term historical process values, provides enhanced support there. By recording and displaying the position of the produced profile visually as well as providing a numeric stability value, additional hints are given. Qualitative information about the process and its quality, derived from the surface and geometry inspection, can directly be used to determine the need for major process corrections or even a full process restart. At the same time, the old paper-based recordings of faults, counter-measures, mixture changes and other interventions, have been replaced by the ELB, allowing simplified execution of, e.g., fault occurrence evaluations, and other analyses necessary from the points of view of both engineering and business.

For the long-term evolution of the research, several aspects have been detected for follow-up. Most importantly, the system is being kept in production, and still extended by ongoing research and implementation work. Especially the aspects of validating and improving the situation detection, of quantifying the long-term waste reduction, and of better integrating the aiXPerience system into the industrial partner's environment, are being worked on. Currently, it is also discussed to extended the system onto further production lines, at least one of them being newly installed. There, the results of explicitly recording and modeling the domain processes and concepts will also be applied for designing both the technical aspects of the new line, and the new processes to be used on it.

7.6 Tools for Consistency Management between Design Products

S. Becker, A. Körtgen, and M. Nagl

Abstract. The results of the IMPROVE subproject B2 (cf. Sect. 3.2) are to be transferred to industry. The corresponding transfer subproject T5 is described in this section. The main goal is to provide a universal integrator platform for the engineering solution Comos PT of our industrial partner innotec. The core results to be transferred are the integration rule definition formalism and the integrator framework including the execution algorithm.

Besides evaluation in practice, the transfer subproject will also deal with major extensions. For instance, the integration rule definition formalism will be extended and repair actions to restore consistency of damaged links will be incorporated into the framework. The transfer is intended to be bidirectional, i.e. our partner's knowledge will influence our research as well.

7.6.1 Introduction

The transfer subproject T5 is a *joint* research *initiative* of RWTH Aachen University, Computer Science 3 and the German software company innotec. It is the transfer activity of the IMPROVE subproject B2.

Innotec is the developer and vendor of *Comos PT*, which is an engineering solution covering the whole engineering life cycle of plants. We are focusing on its use in chemical engineering development processes, though it can be applied to other engineering disciplines as well. The main principle underlying Comos PT is to provide an object-oriented data storage for all engineering data. This is complemented by different tools having domain-specific graphical user interfaces for specific documents.

As Comos PT deals with a lot of different project phases and engineering documents, there is the need to keep the corresponding data consistent. Thus, the methods and tools gained in subproject B2 will be extended and applied to provide a prototype of a *universal integrator platform* for *consistency management* in Comos PT. This prototype will be used to perform the integration of logical documents contained in the Comos PT data model as well as the integration of external data.

While being tailored to support Comos PT and development processes in chemical engineering, a mandatory requirement for the integrator platform is its *adaptability* to *different tools* and other *application domains*.

An important goal of the transfer subproject T5 is to make the results of the CRC subproject B2 (cf. Sect. 3.2) available in an *industrial context*. The *main results* to be transferred are the integration rule definition formalism and the integrator framework, including the rule execution algorithm.

Besides transfer, *additional research* is to be conducted by subproject T5 as well. For instance, the integration rule definition formalism will be extended and support for repair actions dealing with inconsistent relationships will be added to the framework. We expect the application experience of our partner innotec to be a valuable input for our research activities.

The rest of this *section* is structured *as follows*: In Subsect. 7.6.2, some of our past and current cooperations with different industrial partners are described. Subsect. 7.6.3 gives an overview of the system to be realized in the transfer subproject. In Subsect. 7.6.4, selected research activities needed for the realization of the Comos PT integrator platform are described in more detail. Subsect. 7.6.5 concludes this section.

7.6.2 Past Cooperations with Industrial Partners

Cooperation with innotec

The IMPROVE subproject B2 is cooperating with innotec since 2001. Together with the IMPROVE subproject A2, the scenario for the integration of Comos PT and Aspen Plus, as sketched in Subsect. 3.2.1, has been elaborated. This *scenario* was used to discuss the *requirements* for an integrator solution with innotec.

Based on these requirements, an *early version* of the integrator *framework* (cf. Subsect. 3.2.2) has been implemented [27, 251]. It was used to realize a first integrator *tool* between simulation models in Aspen Plus and process flow diagrams in Comos PT (cf. Subsect. 3.2.5). Innotec contributed the wrapper, connecting the framework to Comos PT, and evaluated early prototypes. The final prototypes of the framework and the integrator were limited in functionality and could not easily be adapted to integrate other documents. Nevertheless, they served as proof of concept for our integration approach.

In parallel to the implementation of the framework and the integrator, the *rule definition* language (cf. Subsect. 3.2.3, [39]) was defined. Thus, the integration rules, executed by the integrator, have been defined using a preliminary version of the integration rule editor. During discussions about the scenario with innotec, the visual and UML-based approach for defining rules proved to be very useful.

The prototypes as well as the general approach and the underlying concepts have not only been discussed with our partners at innotec. Additionally, they have successfully been presented to selected *customers* of innotec from the German chemical industry. In multiple *workshops*, presentations and prototype tool demonstrations were given, followed by discussions about further ideas and requirements. We also presented our integration approach at practice-oriented conferences and in corresponding journals [36, 38, 41].

Currently, there are two ongoing projects together with innotec: In the first activity, we are *defining* a set of integration *rules* to import arbitrary columns processing crude oil from Aspen Plus and Pro/II simulation models into process flow diagrams (PFDs) of Comos PT. Here, the current version of the integration rule editor is used for rule definition and consistency check.

The rules are evaluated by generating PROGRES code and executing it using the integration rule evaluation environment IREEN (cf. Subsect. 3.2.5).

Comos PT can be customized using so-called base projects, for instance by defining user-specific types of equipment that can later be used in PFDs. To adapt Comos PT to the needs of a company or a certain project, usually a lot of such definitions are made. Thus, in the second activity, a *tool* is being developed that provides UML views on *base projects* to give an overview to the customizers. This tool will also help defining integration rules in the future, as UML definitions of the equipment types are needed for the document models (cf. Subsect. 3.2.3).

Cooperation with Schwermetall

In another industrial cooperation with the German company Schwermetall [956], we realized an integrator that *collects data* from automation systems of production processes in a *centralized repository*. The data are grouped in a way that for each product leaving the factory its production history is clearly visible. As central repository, Aspen Batch.21 [514] is used. Data sources include different databases, such as those of the process information management system (PIMS) Aspen InfoPlus.21 [515] and some proprietary automation solutions.

The *two* main *problems* to be solved were (a) to handle the heterogeneity of the large number of data sources and (b) to implement a solution, that is flexible enough to be adapted to new process steps, and simple enough to be performed by the system administrator at Schwermetall.

Problem (a) was addressed by writing *light-weight wrappers* (cf. Sect. 5.7) for the data sources. To solve the second problem, the data structure definition file for the Aspen Batch.21 system was extended by additional specifications that *define* the *source* of *data* for each piece of information stored in the Aspen Batch.21 system. The data source definition specifies which wrapper to contact for the data, which function to call, which parameters to pass, etc.

The additional specifications are ignored by the Aspen Batch.21 system but they are *interpreted by* the *integrator* at runtime. Each time a data source signals the availability of new data, all related specifications are executed by the integrator and the collected data is inserted into the Aspen Batch.21 System.

This integrator differs from the other integrators that have been built. Nevertheless, some valuable *further experiences* have been gained, especially concerning the integration of row-oriented data of different databases.

7.6.3 The Concept: An Industrial Integrator Platform

In this subsection, the *functionality* and the *overall architecture* of the integrator platform will be described. As the platform will be evaluated as integration solution for Comos PT, a short overview of the Comos PT system is given first.

Short Introduction to Comos PT

Comos PT is a universal solution for *organizing design* and *maintenance data* for different engineering disciplines. Besides providing a centralized data storage within an underlying database, Comos PT offers specific user interfaces for the manipulation of selected engineering documents, such as process flow diagrams or reactor data sheets for chemical engineering.

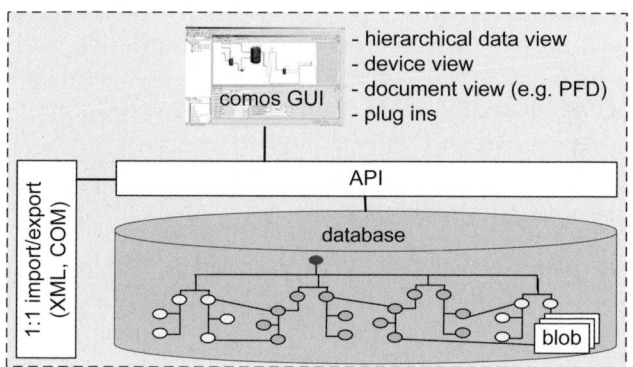

Fig. 7.16. System architecture of Comos PT

We now give a short overview of the internal structure of Comos PT. The coarse-grained overall system architecture is depicted in Fig. 7.16. It is based on a *central database*, which serves as data storage for all Comos PT related data. Most commercial relational databases can be used. The Comos PT *application programming interface* (API) provides an interface to the central database similar to most object-oriented database interfaces. *Objects* in the database can have attributes, which can be structured hierarchically and may include references to other objects. Objects are used in a way that their granularity meets that of real-world objects, e.g. a chemical reactor is represented by a Comos PT object. Objects can be decomposed hierarchically. For instance, a reactor can be further refined by aggregated objects, such as agitators or nozzles. Objects are called devices in Comos PT terminology.

The Comos PT API is used by the Comos PT graphical user interface (GUI) to access the engineering data. The GUI provides a hierarchical overview (tree view) of the devices contained in the database for navigation, as well as detail views on devices. Different types of *graphical* and *textual documents* are provided that can contain references to devices. For instance, process flow diagrams contain references to the devices that are placed on the sheet.

When displaying the sheet, these references are evaluated and the right geometrical form is drawn for each device. Thus, there are no "real" documents stored inside the Comos PT database. Instead, *logical documents* are defined by combining data from different places inside the database.

External data can be imported as large binary objects (BLOB) into the database. To display these data, external applications can be started as *plugins* inside the Comos PT GUI. The plugin mechanism can also be used to extend the GUI with specific functionality. To communicate with Comos PT, the extensions can make use of the Comos PT API.

For some tools, Comos PT contains simple *import* and *export* functionality that translates their data contained in BLOBs inside the database into corresponding Comos PT device structures. The translation works batch-wise without user interaction and performs simple one-to-one mappings only. So far, only BLOBs containing XML files or data of certain applications with COM interfaces (e.g. Aspen Plus) are supported.

Functionality and Realization Overview

The integrator platform is intended to be used for Comos PT related *data integration*, e.g. for keeping simulation models and process flow diagrams consistent. This comprises internal data as well as external data sources. The integrator platform has to fulfill all *requirements* on integration tools as proposed in Subsect. 3.2.1. The key requirements are:

- The behavior of the integrator can be defined by rules. *Rule definition* has to be feasible for the personnel customizing Comos PT.
- Operating the integrator has to be *feasible for the engineers* performing design processes.
- The integration has to be *incremental*. No manual modifications are to be overwritten without asking the user.
- The integration has to be performed *interactively*. The user can select among concurring rules and perform parts of the integration completely manually.
- The integrator platform has to be *tightly integrated* with the *Comos PT* environment from the users' perspective. Nevertheless, it must be easily *adaptable* to *other tools* and *other domains* of applications.

Figure 7.17 illustrates the system *architecture* we have chosen to incorporate our integration framework into Comos PT.

In contrast to the original system architecture of our integrators (cf. Fig. 3.22), Comos PT is connected to the integrator as one single application. Thus, only *one wrapper* instance is needed. The wrapper accesses the Comos PT database via the Comos PT API. It is the wrapper's task to provide logical documents to the integrator and to map them to collections of database objects. The wrapper will be described in more detail in the next subsection.

For the *integration* of *external data*, they have to be imported into the Comos PT database. First, a BLOB containing the external document is created.

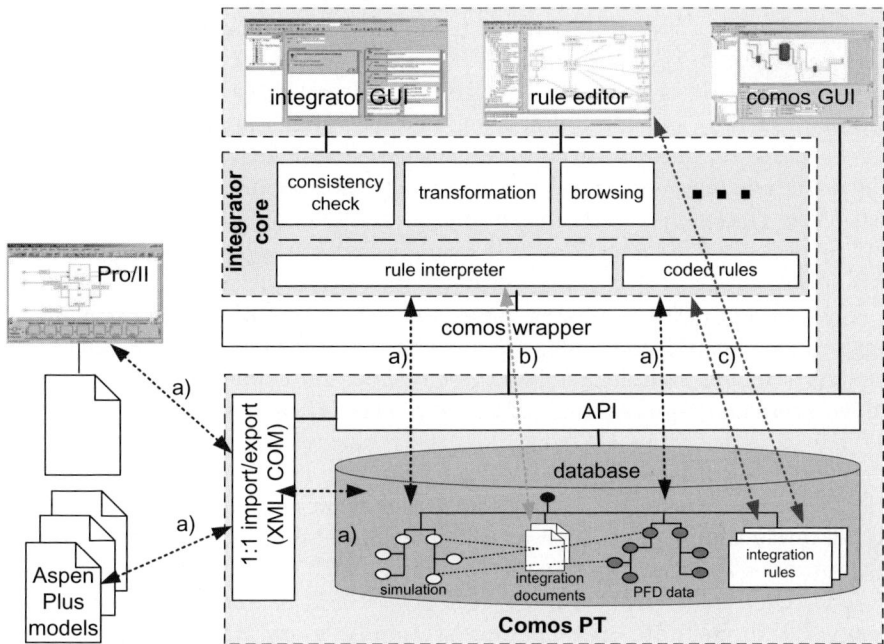

Fig. 7.17. System architecture of the integrator platform for Comos PT

Second, the one-to-one import is used to translate the proprietary document into a corresponding Comos PT data structure. Only then, the integrator performs the complex integration part. The resulting data flow during integration is shown as arrows labeled with **a)** in the figure. To allow the incremental integration of arbitrary data, the current im-/export tool has to be extended to work incrementally and to support additional external applications. Furthermore, the definition of the im-/export behavior has to be facilitated. These extensions will be realized in the transfer subproject as well.

Unlike the original approach, the wrapper does not only provide an interface to the logical documents being integrated. Additionally, it is used to store the *integration documents* in the Comos PT database (cf. arrows labeled with **b)** in Fig. 7.17). As a result, the user does not need to explicitly manage data by himself in the file system. As a further benefit, other parts of the Comos PT system can make *use* of the *links* contained in the integration documents. For instance, they can be used for navigation purposes or to propagate attribute values between related devices using already existing Comos PT functionality (e.g. the so called mapping tables).

The integration *rules* controlling the integrator are stored inside the Comos PT database as well. They are modified by the rule editor and read by the integrator using corresponding wrapper functionality (cf. arrows labeled **c)** in the Figure).

The integrator *core* of the current integrator framework (cf. Subsect. 3.2.2) is used to perform the integration by executing the integration algorithm (cf. Subsect. 3.2.4). It is controlled by the integration rules (cf. Subsect. 3.2.3) contained in the Comos PT database. The basic functionality is already implemented in the integrator core but nevertheless it has to be extended significantly. The most important extensions are the support of repair actions to restore the consistency of damaged links in the integration document, and of new elements in the integration rule formalism. Both will be addressed in the next subsection.

The rule editor as well as the integrator user interface will be executed as plugins in Comos PT. Both have to be adapted to the domain concerning their presentation. The main challenge, thereby, is to provide a *user-friendly view* on the *underlying* complex integrator *concepts*. The tight integration of the rule editor with Comos PT will allow the realization of additional functionality, such as drag and drop between Comos PT flowsheets and integration rules. The rule editor has to support the current formalism for the definition of integration rules as well as the extensions being added in the transfer subproject (see next subsection).

One requirement for the integrator platform is its *tight integration* with Comos PT to facilitate operating integrators by users. Nevertheless, it is intended to keep the overall framework adaptable to other tools and other domains as well. To use the framework with another tool, just the wrapper has to be exchanged. Additionally, the presentation layer of the integrator user interface and the rule editor have to be adapted. To ensure the adaptability, the transfer subproject cooperates with another project at our department performing research concerning data integration in automotive development processes.

7.6.4 Important Extensions to the Original Approach

In this subsection, we describe selected aspects of the transfer subproject, namely those having an *impact* on further *research*.

Comos PT Wrapper

The new Comos PT wrapper we developed deals with *three* new *aspects*:

1. It is responsible not only for accessing the *data* that have *to be integrated* but also for storing all *integration documents* as well as all integration *rules* inside the Comos PT system. To allow other parts of Comos PT to use these data, they cannot be simply stored as BLOB inside the Comos PT database. Instead, they are saved as Comos PT objects, exploiting the possibilities of the Comos PT data model, like attributes and references.
2. Unlike the old wrappers for integrators, the new one does not wrap exactly one specific kind of documents. Instead, it provides a logical document

interface to *arbitrary data* that reside in the Comos PT database. As a result, the wrapper is able to *interpret document definitions* at runtime and map accesses to the logical documents' contents at its interface to specific Comos PT objects. Specifying logical documents and interpreting the specification at runtime leads to interesting problems.

For instance, *intersections* between logical documents have to be *avoided*, at least if they are used for the same run of the integrator. It is not sufficient to simply define logical documents by selecting entire branches of the hierarchical database view of Comos PT, because a logical document can contain information from multiple branches in the tree and a tree branch can contain information that is relevant for multiple logical documents. Even for a specific document type, there are different storing policies. For instance, this is the case for the specialized flowsheet documents PFD and P&ID.

The new wrapper solves this problem with the help of so-called *folder mappings*. When the wrapper is instantiated for a specific flowsheet, it retrieves a folder mapping which maps object types to folders in the tree. Thus, the wrapper determines where to store or to find objects by their type.

Additionally, it has to be figured out *how fine-grained* the specification has to be. While in some cases it could be useful to explicitly include a single attribute into a logical document, this could cause unnecessary overhead in other situations.

3. To facilitate the definition of integration rules, additional *graph constructs*, e.g. paths and derived attributes, are to be supported by the wrapper. Comparable to the definition of logical documents, these constructs have to be defined and interpreted at runtime. This is due to the fact that the Comos PT data model can be customized to the needs of specific users and, thus, it cannot be foreseen e.g. which paths will be needed in the future.

Paths are very useful, e.g. when importing simulation data. The current one-to-one import from Aspen Plus creates Comos PT objects representing simulation streams that reference the connected devices by an attribute containing a string. The string consists of the name of the device and the name of the port the stream is connected to. Thus, in an integration rule containing a stream that is connected to a device, there would be no *visual manifestation* of the connection. Instead, an attribute constraint would have to be included that checks the contents of the string attribute. Using paths, a *global path definition* for such connections could be interpreted by the wrapper and the connection could be visually modeled in the integration rule. Paths are one of the extensions of the rule formalism, which are explained in the following.

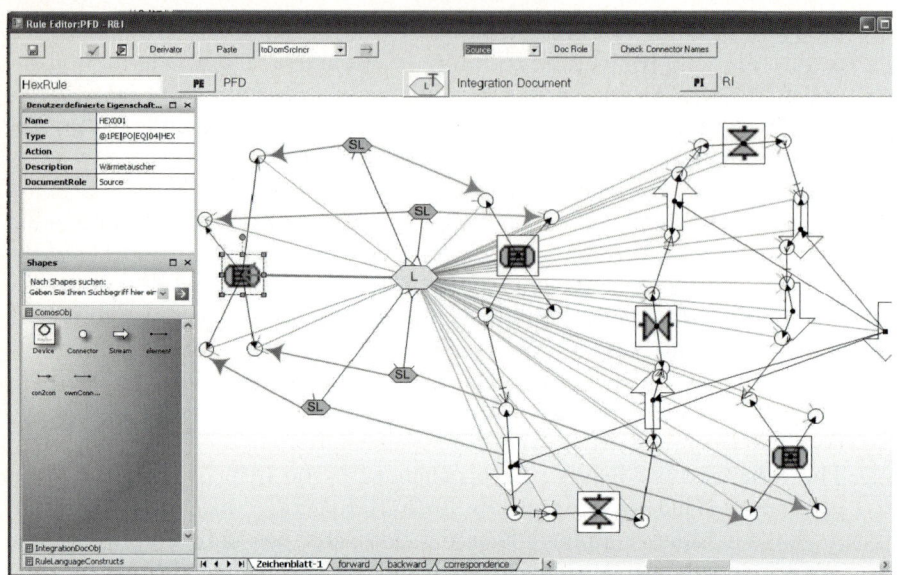

Fig. 7.18. Rule editor to specify rules

Extending the Formalism for Integration Rules

The integration *rule* modeling *formalism* introduced in Subsect. 3.2.3 provides the possibility to specify the behavior of integrators in a clean and consistent way. Nevertheless, it has to be *extended* for two reasons: First, it has to be made more user-friendly by offering specific views. Second, it has to be enriched by additional constructs to support more complex rules.

To facilitate the definition of integration rules for Comos PT customers, *new views* on the rule model have to be provided. For instance, the link type definitions can be hidden from the average user and just be included into an expert user interface. Additionally, a domain-specific visualization using graphical symbols for specific node types is needed. Most important, filters (e.g. supporting edge-node-edge constructs by edges, as in the UPGRADE framework [49]) should be incorporated.

Figure 7.18 shows a screen shot of a *new rule editor* we developed, which is based on the tool Visio from Microsoft. We chose Visio because it brings along basic functionality for designing graphics, it is easy to use, and has a well documented API. *Simplification of rule modeling* is achieved by hiding link type definitions. Thus, a rule modeler starts directly with drawing an integration rule.

For this purpose, a collection of *predefined shapes* is available, which can be dragged and dropped onto the drawing (see shapes window in Figure 7.18). The shapes represent nodes and edges of a rule, each of them has properties, such as document role, type, or action. Shapes representing nodes have in addition a name and description property (see table in Figure 7.18). The type of a node shape can be selected via tree views, which list either link, PFD, or P&ID types. The latter two are retrieved directly from the Comos PT system.

The shapes *look differently* according to their affiliation to a document, which makes their distinction easier and, therefore, more user-friendly. At least *two collections* of predefined shapes have to be loaded for rule modeling: one for shapes which shall be part of the integration document and the other for shapes of the application documents. This separation of shape collections makes it easy to use the rule editor for different documents and domains. For example, the provided domain-specific shapes for flowsheet objects, namely devices, streams, and connectors, are represented as boxes, block arrows, and circles, respectively. Further *customization* to the Comos PT system is realized by loading an image from the Comos PT database when the type of a device shape is set. In the rule depicted in the figure, the devices (boxes) show the representation of heat exchangers in PFD and P&ID of Comos PT.

Additionally, the *creation process* of rules is *simplified* enormously. Drag and drop of devices from the Comos PT tree view as well as copy and paste of devices from a Comos PT flowsheet into the rule editor results in the creation of shapes. If multiple devices inter-connected among each other are selected and copied, the created shapes in the rule editor get inter-connected as well. This *tight integration* allows to reuse already existing templates for transforming certain devices from a PFD into a P&ID from the Comos PT system. These templates are expressed by simple flowsheets containing respective devices. There are functions to support automatic property setting and connection of shapes. A rule checker is implemented to validate rules.

Furthermore, real *extensions* to the *rule model* will be made. The main idea is to incorporate additional constructs already available in current graph rewriting implementations (such as PROGRES [126]) into the rule model. We currently plan to include or have included the following constructs:

- Arbitrary *path definitions* are to be supported, as motivated above. Similar to regular expressions, we plan to support basic operations like alternation and quantification of graph patterns in order to restrict the set of possible paths. Quantification is already realized by the concept of set-valued graph patterns (see below).
- *Negative application conditions* (abbreviated NAC) will be supported. They provide a possibility to decrease the amount of unnecessary user interaction by defining "guards" for the application of certain rules. Incorporating NACs into the integration algorithm (cf. Subsect. 3.2.4) is not straightforward.

Each rule execution potentially produces new nodes in the documents. Thus, it is hard to determine when to *check a NAC*. Especially, it has to be decided whether a rule containing a NAC, that is to be checked at later stages of the algorithm, is to be included into the conflict detection phase

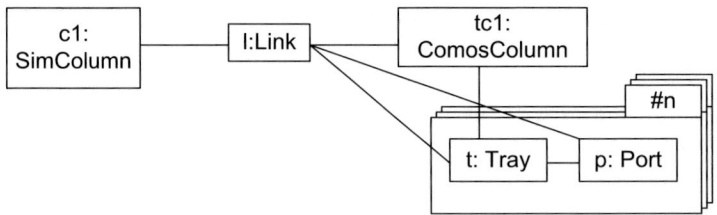

Fig. 7.19. Integration rule with parameter

of the algorithm. As a result, the use of NACs has to be restricted in a way that the rule execution can still be defined in a deterministic way.
- Set-valued nodes are already supported by PROGRES. They are needed for integration rules as well. Especially, the *generalized set-valued patterns* are needed in integration rules. A new construct for modeling *repetitive subgraphs* is realized, which even allows the modeling of successively connected subgraphs [250].
- Additionally, a new construct for modeling *alternative subgraphs* [180] is realized. This is a powerful construct to specify multiple alternative correspondence structures in one rule. For example, there are different options to map a process stream to a pipe which contains different types of valves in different order. All these alternatives can be now specified within *one* rule.
- Concepts for the *parametrization* of integration rules will be defined.: Parameters can be set at *runtime* as well as at *rule definition time*. Figure 7.19 shows an example of a rule with a runtime parameter in combination with a set-valued pattern: A column in the simulation model is mapped to a column in the flowsheet, which has a number of column trays each having a port. The parameter #n controls the number of trays, that are added to the column; it is provided by the user at runtime. Up to now, the parametrization works with attributes occurring in a rule whose values can be retrieved at runtime without user interaction.
  To facilitate rule definition, parameters can be set at definition time as well. This can be compared to the concept of *genericity* as contained in some modern programming languages. We plan to allow different kinds of parameters, ranging from integer values, or node types, to complex graph patterns.
- Additional improvements will be made to the *handling of attributes* during integration.

All extensions affect the integration *algorithm* as well as the *derivation* of forward, backward, and correspondence analysis *rules*. For instance, the existence of paths can be checked, but not all paths can be created in a deterministic way. Thus, the constructs either have to be restricted, or generated forward, backward, and correspondence analysis rules have to be manually post-processed.

To *realize* the above *extensions*, first IREEN will be used to define and evaluate their semantics. Only then, the integration rule editor and the integrator core of the framework will be adapted to support the new functionality.

Concerning all extensions, it is very important to get *feedback* from practice. It has to be made sure that the resulting rule definition formalism is still usable for the average Comos PT customer and that these extensions are really needed.

Repair Actions

After modifications of the integrated documents, already existing *links* in the integration document – that have been created by the execution of rules or manually – can become *invalid*. There are different *reasons* for this:

- *Increments* in source or target documents have been *deleted*, resulting in dangling references from the link.
- *Attribute values* of increments within source or target documents have been *changed* in a way that attribute conditions contained in the integration rule no longer hold.
- *Edges* in source or target documents have been *deleted* or have been assigned a new type. As a result, the pattern matched by the integration rule that created a link does not exist any more.
- A *link* the current link is *depending* on has been *damaged*.

The simplest way to *deal* with a *damaged link* is to delete it and thereby make the remaining increments available again for the execution of other rules. Though possible in general, deleting the link is not a good option. The modification resulting in the damaged link was most probably done on purpose. The link – even if it is damaged – contains valuable information on which parts of the other document may be affected by the modifications. So in most cases just asking the user to resolve the inconsistency manually is a better option than deleting the link.

If the inconsistency has been caused by the deletion of increments, it is possible to propagate the deletion by first *deleting* all remaining *increments* in both documents and then deleting the link. This behavior is restricted to situations, where all increments of a link have been deleted in one of the documents or where one of the dominant increments has been deleted.

Another option is to restore consistency by *removing* the *cause* for the *inconsistency*. For instance, missing increments or edges may be created. This option is desirable only in those cases where the operation causing the damage was carried out accidentally, because it would be undone.

If only some parts of the patterns in source and target documents are missing, it is possible to perform a *pattern matching* for the whole pattern of the rule that created the link using still existing nodes to initialize some of the patterns' nodes. This can be helpful, e.g. if the user first deletes an increment and then recreates it.

Another option is to perform pattern matching of *alternative rules* to the current situation. This means that the altered pattern is to be mapped to different patterns in the corresponding document. If an alternative rule is applied to the new situation, then the corresponding document will be adapted, i.e. the former pattern will be deleted and a new pattern according to the alternative rule will be created.

In general, it cannot be determined automatically which alternative for repairing damaged links is appropriate. So, user *interaction* is *necessary* here as well. The integration tool determines all possibilities for dealing with the inconsistency and let the user decide. As repair actions are performed during an integration run, normal integration rules to transform newly created increments are simultaneously proposed to the user. Thus, repair actions and integration rules are to be presented to the user likewise. This is important, as it is possible that repair actions conflict with integration rules.

We are currently implementing different repair actions listed above. If a repair action is a reasonable option to solve a damaged link, the cause of the damage is determined. As discussed with our industrial partner innotec, in the first place provided repair actions should *re-organize the references* contained in the integration document and let the application documents untouched (conservative strategy). Only if this is not possible, other repair actions should be offered, where the direction of the integration run (forward, backward, or correspondence) is considered.

Document Reuse

In engineering processes, it is common practice to *reuse documents* of preceding projects. This is usually done by searching for mappings between high-level descriptions of the currently designed system and existing documents, e.g. PFD and P&ID documents which describe chemical processes similar to the one to be designed and they contain precise, *refined design decisions*. A mapping between these documents is usually found by detecting name and type equalities of their objects.

To *find* such *mappings* between two documents, the documents have to be created for this purpose, e.g. objects' names have to be chosen according to a specific convention. Because this is not always the case and one cannot assume in general that such conventions are met, it is more appropriate to search the documents for specific patterns. For this purpose, our *integration rule formalism* is well suited. Potential mappings can be specified more precisely, comprising structural aspects as well. Not only one-to-one mappings, even complex patterns in one document, e.g. an arbitrary number of parallel pumps with several fittings, can be specified to be mapped to similar but different patterns in the other document.

We developed *extensions* to our *integration algorithm* to support document reuse. In [230], new document traversal strategies are presented to tackle the problem of combinatorial explosion of possible mappings. Additionally, further extensions were made to the integration rule formalism facilitating its application in practice. For example, we started developing concepts for generic rules which express fuzzy correspondences. For this purpose, alternation and quantification of graph patterns described above are supported.

It is possible to *match* these *patterns* in the documents and *derive* concrete *rules* from these matchings. As we discussed with our industrial partner innotec, it is necessary to keep the reused document and delete unneeded increments from it. There is additional information contained in a flowsheet from a preceding project which cannot be mapped by integration rules, e.g. layout information or additional graphical annotation. If a new flowsheet was generated using the derived rules, the additional information would be lost.

Multiple Document Integration

Currently, the integrator supports pair-wise integration only (cf. Fig. 7.20 a)). If dependencies among *multiple documents* are to be handled, a *master* document has to be identified and pair-wise integrations of the other documents with the master document have to be performed (cf. Fig. 7.20 b)). This is feasible in most cases because there are certain obvious master documents in design processes, e.g. the flowsheet in chemical engineering.

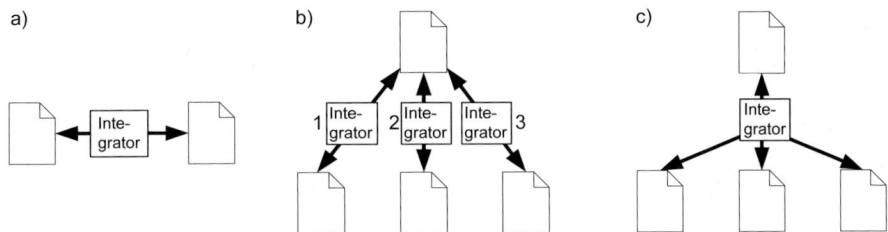

Fig. 7.20. Integration of multiple documents

Up to now, no research has been done on how the single *integrations* have to be *coordinated* with each other in case b). On the one hand, it could be necessary to perform the integrations in an appropriate order. For instance, in the figure integration 2 can have to be performed after integration 1 and 3. On the other hand, an even more fine-grained synchronization mechanism on the basis of single rules can be used. For instance, a certain rule in integration 2 can have to be executed before another rule is executed in integration 1 and so on. For the former case, a simple order of integrations has to be defined. For the latter case, the rule specification has to be enriched by specific constructs for handling inter-integration dependencies.

Another possibility is sketched in Fig. 7.20 c): A single integration tool executes *rules* dealing with all documents *simultaneously*. For this possibility, as well as for the one sketched before, the triple graph grammar approach underlying our integration algorithm has to be extended substantially.

7.6.5 Conclusion

In this section, we gave an overview of the integrator tool transfer activities. We plan to *use* our existing *experience* in industrial cooperations to transfer the results of the IMPROVE subproject B2 to industrial applications. In the last two subsections we discussed still existing challenging research *problems*. Only their solution will make the transfer project a success.

Current practice in industry either relies on simple hand-coded batch transformators for data integration or on manual consistency management. Thus, from the industrial perspective, the main result of this transfer project will be the possibility to *realize integrator tools* at *lower costs* and with improved quality and to increase their use in design processes. This is mostly achieved by separating common implementation parts in the framework from rule specifications that define the behavior of integrators.

However, it is not intended in this transfer subproject to provide complete solutions that are ready for use. Instead, the focus is still on providing *prototypes* that serve as *proof of concept* and as foundation for market-ready implementations.

From the research perspective, the intended *impact* of the transfer subproject is twofold. First, we expect valuable contributions by solving the scientific issues, some of which have been presented in this section. Second, the experience of our industrial partner and the close contact with their customers will provide a benefit for our research activities that could not be achieved without this cooperation.

7.7 Dynamic Process Management Based upon Existing Systems

M. Heller, M. Nagl, R. Wörzberger, and T. Heer

Abstract. In the past funding periods of the CRC 476 the process management system AHEAD has been developed as a research prototype (cf. 3.4). The project T6 aims at transferring the corresponding research results into two different industrial environments. Together with innotec GmbH, we realize a process management system on top of the chemical engineering tool Comos PT. This system will allow for the holistic management of the overall administration configuration (activities, products and resources). Furthermore, we extend our application experience by also considering dynamic business processes. In cooperation with AMB-Informatik, we build a workflow management system that allows dynamic changes of workflows at runtime. This new workflow management system is also built on top of an existing one, which strictly separates build-time from runtime.

7.7.1 Introduction

In IMPROVE the process management system *AHEAD* has been developed, as described in detail in Sect. 3.4. AHEAD's main *characteristics* are (a) the medium-grained representation of the task structure, (b) the coverage and integration of processes, products, and resources at the managerial level, (c) the integration between managerial and technical level, (d) the support for dynamics of design processes, (e) the process management support across companies, and (f) the adaptability of the system to different domains.

Due to these features, AHEAD is *ahead of state of the art systems* for project, workflow, or document management. In particular, these systems cover only parts of the managerial configuration and do not integrate the managerial with the technical level. Common workflow management systems lack the support of dynamic changes within the task structure, which occur frequently in development processes.

Because of these shortcomings, the overall managerial configuration of a process is usually maintained manually. This induces inconsistencies in the administrative configuration, which may lead to, e.g., forgotten tasks or to the use of outdated documents. The risk of inconsistencies emphasizes the *need* for holistic *management support* which can be provided by AHEAD.

The AHEAD system is currently built using the tools PROGRES [414], UPGRADE [49], and GRAS [220], which allow for *rapid prototyping* based on graph transformations. Due to these technical dependencies, AHEAD *cannot easily be transferred* to arbitrary software platforms.

In the *following section*, we describe recent and future efforts concerning the transfer of concepts elaborated within IMPROVE into industrial environments. At first, we discuss two general approaches to transfer a system like AHEAD such that they integrate with existing systems for document,

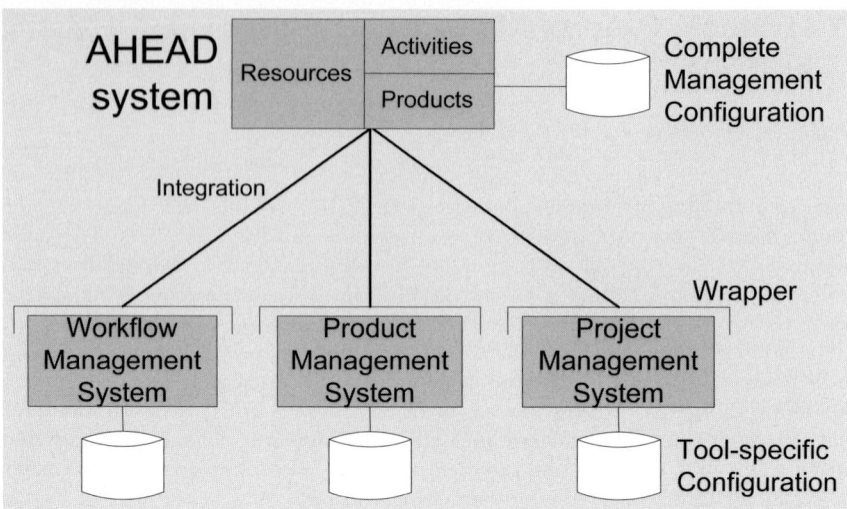

Fig. 7.21. Integration Approach 1 (IA1): AHEAD maintains full management configuration

workflow, or resource management. Thereafter, we depict preliminary work within the AHEAD project to facilitate technology transfer. Next, the implementation of an AHEAD-like system on top of an existing tool for chemical engineering is outlined. This is performed in cooperation with innotec, which develops and sells a tool named Comos PT. In an additional transfer activity together with AMB-Informatik, we show how AHEAD concepts can be applied to the domain of business processes and can be realized on top of an existing system for workflow management.

7.7.2 Strategies for A-posteriori Integration of Management Systems

The transfer into industry aims at the realization of an integrated, *industrial management support system* built up from existing tools. An AHEAD-like system is positioned on top of commercial tools as sketched in Figs. 7.21 and 7.22.

A-posteriori Integration

AHEAD *offers advanced functionality* for the management of dynamic design processes which is currently not available in commercial tools. As discussed in-depth in Sect. 3.4, process management covers several aspects including activity or workflow management, product management, resource management, and additionally time- and cost-based project management. For example, the storage of documents or the support for project planning is done with various commercial tools. They constitute mature solutions for carrying out certain management tasks. Furthermore, users are familiar with these tools. Therefore, it is not feasible to replace these tools by a completely new and monolithic system.

Instead, *AHEAD* and the *commercial tools* are synergistically *combined* into a coherent overall system in a bottom-up fashion resulting in *two advantages*: (1) The commercial tools are embedded in a dynamic process environment. Thus, their functionality can be used in a coordinated manner to achieve new and advanced objectives. (2) AHEAD is extended with new functionality assembled from the commercial tools' functions. This avoids the need to re-implement already existing functionality offered by these tools. By the integration of commercial tools, the AHEAD system can be successfully "embedded" into the industrial technical and managerial overall environment.

The existing and coupled management systems are accessed by *wrappers* which provide an *abstract interface* to these systems. The wrappers comprise a functional interface to call functions and a data interface providing data views. The technical integration, including problems like data format conversion for example, can be encapsulated within the wrappers (cf. [136] and Sects. 5.7 and 7.8).

The functionality offered by commercial tools is often *stripped* in order to concentrate on the main functionality. For example, a product management system may also offer some functionality for managing activities in a design process. We decided not to use this functionality because a more versatile workflow management system is also to be integrated. Consequently, only one dedicated system is used for each management aspect (*separation of concerns*). In particular, activity, product and resource management are handled on the level of existing systems and on the integration level above.

Two Possible Solutions

The *storage* of the *management configuration* of the overall design process can be performed in *two possible ways*: The AHEAD system maintains the full management configuration with all details of the design process as shown in Fig. 7.21 (IA1). Alternatively, AHEAD stores only a part of the overall configuration together with additional integration data, while other parts are maintained in the existing systems and are accessible for AHEAD through views (IA2) as shown in Fig. 7.22. In the latter case, the AHEAD instance has to maintain all data relevant for the coordination of all systems on a more coarse-grained level, while other detailed data are left to be stored within the existing systems on a fine-grained level.

While the *AHEAD* system is used itself as integration *instance* in the first approach, the second approach implements *AHEAD-like functionality* on top of industrial management systems. For both alternatives, it is advantageous to locate AHEAD's *semantical models* of design processes at a *central place*, to guarantee that the semantical submodels fit together. These submodels

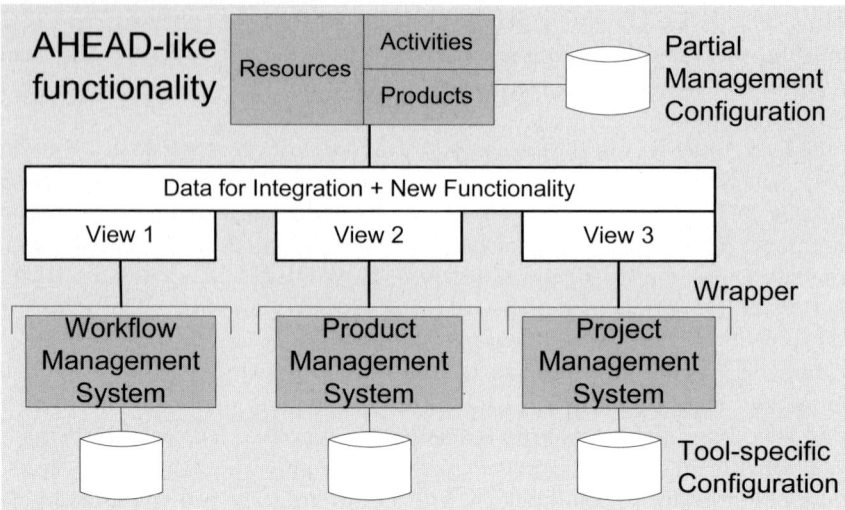

Fig. 7.22. Integration Approach 2 (IA2): AHEAD-like system merely coordinates existing systems

often differ from the models coming with existing systems. So, it cannot be guaranteed in general that all parts of the semantic models in AHEAD can be mapped onto that of existing systems. In some cases, data within the existing systems are not directly available in AHEAD, while in other cases all data are duplicated in AHEAD.

7.7.3 Preliminary Work

In the following subsection, we describe *preliminary work* for the transfer. We show how a *commercial workflow management* system can be used to *execute task nets* that are defined in the AHEAD system. After that, we describe the *integration* of the AHEAD system with *various existing systems* for project planning, email communication, and document management.

Workflow-Based Execution of Dynamic Task Nets

To further investigate the above mentioned integration alternative IA1, where AHEAD maintains the *full management configuration* for the integrated system, we have studied the *integration* of AHEAD with the *commercial workflow management system* COSA from Ley [208]. In this experiment, AHEAD was used for *planning* and *editing* of the overall task net structure, while the *execution* was *delegated* to the workflow management system. AHEAD uses *dynamic task nets* for process modeling, while COSA uses a *Petri net* variant for the same purpose. Thus, the *main problem* for the integration is the mapping of the dynamic task nets into Petri nets and respecting the semantics of dynamic task nets during workflow execution within COSA. For the mapping, the structure and the dynamic behavior of dynamic task nets is mapped into a Petri net by mapping each modeling element of task nets into a small Petri net fragment consisting of places and transitions.

The integration of AHEAD and COSA is achieved in *two steps*. (1) The overall task net instance is stored in a task net description file. A *transformation module* implements this mapping and converts the task net into a COSA net which is stored in a COSA workflow description file. This file is imported into COSA where a corresponding workflow instance is created. (2) AHEAD and COSA are coupled at run-time using a *communication server* in between, similar to the coupling of two AHEAD systems for delegation-based cooperation as described in Subsect. 3.4.5. Both systems *exchange events* to keep each other informed about relevant process changes.

Changes of the process structure cannot be performed in COSA but *only in AHEAD*. The propagation of structural task net changes from AHEAD to COSA at run-time follows a *stop-change-restart-strategy*: the currently executing workflow instance is stopped in COSA after the current state has been stored; the obsolete workflow definition is changed and replaced with a new workflow definition containing the changed process structure; a new workflow instance is created according to the new workflow definition and is restarted after populating it with the stored process state.

The integration of AHEAD and COSA is *limited* with respect to the following aspects. First, the *full dynamic behavior* of task nets is not mapped to COSA; thus, the mapping is only implemented for a restricted part of the AHEAD model. Second, in order for the change propagation strategy to work, the old and new workflow definition of a changed workflow *cannot differ much*; only very limited structural changes can be accommodated to assure that the process state can be restored in the new workflow instance.

An Integrated Management Suite for Development Processes

We now describe preliminary work following *integration approach IA2*, which has been done during a half-year student project. In this project we have combined the AHEAD system with various commercial tools in order to form an integrated management suite. In this case, AHEAD does not maintain the complete management configuration. Parts of the *management configuration* are *distributed* across the commercial tools. For example, some details about documents are not contained in the product submodel of AHEAD but are rather stored in commercial product management systems.

The integrated management system provides the following synergistic functionality for its users reaching beyond the functionalities of the systems involved: (Fig. 7.23):

- The AHEAD system offers a *management* and a *work environment* to project managers and designers, respectively. Additionally, AHEAD pro-

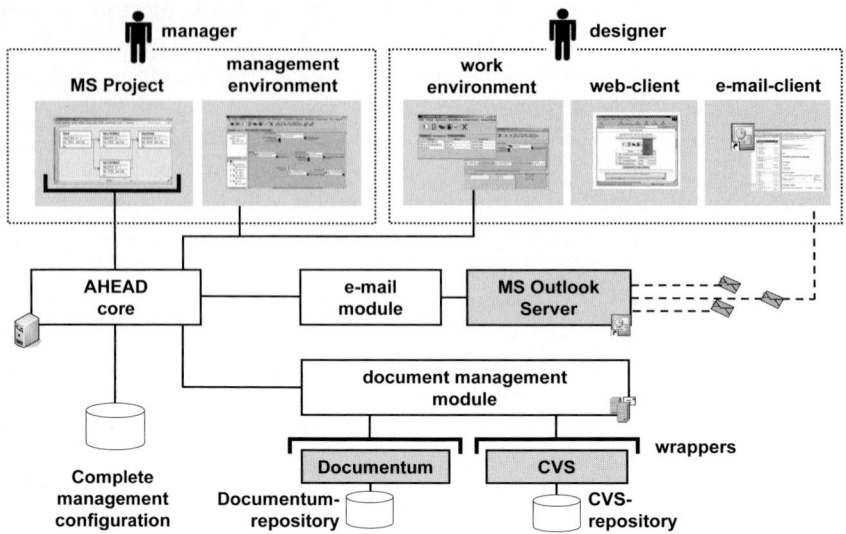

Fig. 7.23. Integrated management solution with AHEAD and commercial management systems

vides a *web-based client* for designers to communicate with AHEAD using a conventional web browser. All user interfaces access the AHEAD core which in turn stores the whole management configuration in its *own database*.

- On the *project management level*, the project manager can additionally use the *commercial tool* MS Project for the management of the design process from a project management perspective. For example, the overall task net can be imported within MS Project as a GANTT- or PERT-chart. The project manager can then use the *project management functionality* to calculate optimal start dates or end dates for the tasks using the critical path method, assigning resources to tasks etc. Finally, the results are exported back to the AHEAD system where the current state is updated.
- An *e-mail based work environment* is offered to project participants who prefer just to receive and to reply to work assignments via a common e-mail-client. An e-mail module of AHEAD realizes the e-mail communication, which is tailored to the commercial e-mail-client MS Outlook. Thus, *work assignments* are displayed as *Outlook activities*; necessary *documents* can also be received from and sent to AHEAD using e-mails. The state of the assigned tasks can be modified, too. An advanced *voting procedure* is also offered for the distribution of work within groups of participants.
- On the *product management level*, a central product management module is available in AHEAD which uses *wrappers* to *different product management systems* to maintain all documents employed throughout the whole

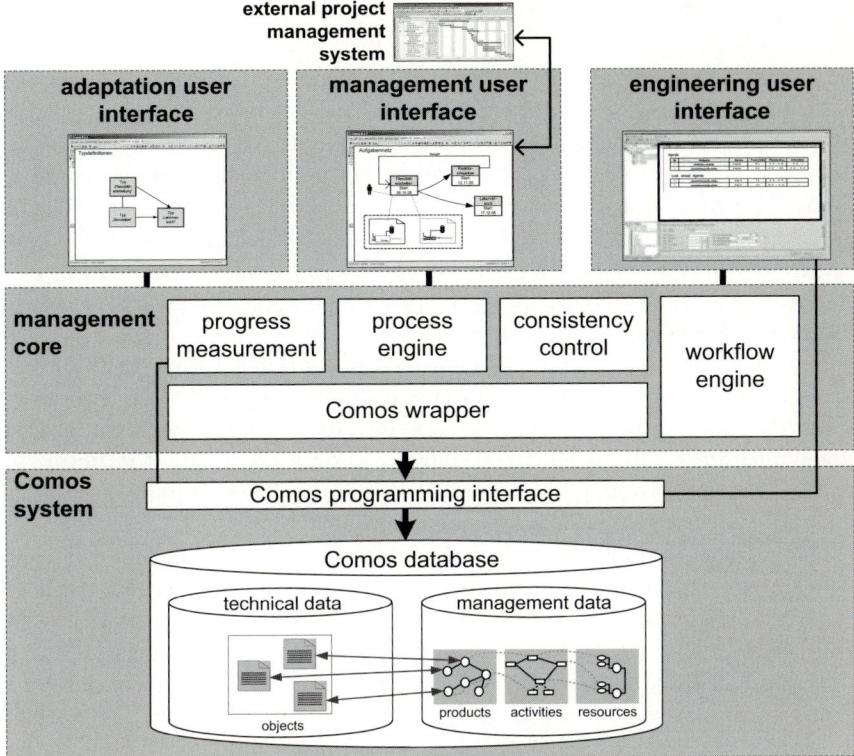

Fig. 7.24. Architecture of the planned management system for Comos PT

design process. This module is responsible for the mapping of logical documents used in the submodel for products in AHEAD to physical documents stored and maintained in the repositories of the coupled product management systems.

7.7.4 Management Tools for Chemical Design Processes

The following subsection gives a description of an integrated *process management system* which is based on an existing product management system according to approach IA2 (cf. Fig. 7.22). Its *realization* will be performed in *cooperation* with *innotec*, a partner in the transfer project T6.

Overview

Starting from the case study of the previous subsection, a process management system will be built to *extend* Comos PT, one of the tools for Computer Aided Engineering (CAE) and Process Lifecycle Management (PLM) supplied by innotec to customers in the process industries (cf. subsection 7.6.3 for a short introduction).

For the time being, Comos PT strongly *focuses* on the *product part* of a design process. The tool is based on an *object-oriented database*, which allows to store all relevant parts of a chemical plant. All stored objects can be enriched with attributes and relationships between each other. Furthermore, Comos PT provides *dedicated views* on these objects (e.g. flow sheets) which show the two-dimensional layout of a (part of a) chemical plant. The support for the *coordination* of engineers is currently *restricted* to administrative attributes of objects, which represent the progress of a certain product according to a certain phase in the design process.

The planned process management system extends Comos PT by functionality for *holistic management* of chemical engineering design processes. It facilitates the coordination of engineers on a medium-grained level by introducing *integrated* submodels which cover not only the *product* structure but also the *task* and *resource* structure of the design process. Figure 7.24 depicts the *architecture* of this process management *system*.

The existing Comos PT system is located on the *lowest layer*. The Comos database stores all technical data within objects. Furthermore, it serves as the *central storage* for *management data* representing products, task, and resources as well as their (overlapping) relationships. Managerial data are strongly related to the technical data but abstract from technical details according to approach IA2 (cf. Fig. 7.22). Access to the Comos PT database can only be realized via the available Comos programming interface.

On the *middle layer*, the management core, an additional Comos wrapper provides an appropriate *abstraction* for accessing integrated management data within the *Comos PT database*. Changes in the integrated *management data*, due to normal progress or occurrence of dynamics, are handled by the *process engine*. Before being committed, changes have to pass a consistency control, which comprises *checks for inconsistencies* regarding the structure of the task net, the execution states of the tasks, as well as for the violation of time constraints. The progress[63] measurement component is used to *measure the progress* of individual tasks and ultimately the progress of the overall design process. Besides the management data, the progress measurement component also has to access the technical data stored in the Comos database, as the internal structure of certain flow diagrams can be used to estimate the effort of related tasks in the design process. Together with innotec GmbH, a workflow engine has been developed based on Microsoft's Windows Workflow Foundation technology. The workflow engine will be used to *automatically execute* parts of the overall design process. Hence, the workflow engine will constitute a part of the management core and will have to be seamlessly integrated with the process engine. Workflows will be started by the process engine, their progress will be measured by the progress measurement com-

---

[63] Note, that this is not related to the PROGRES system mentioned before.

ponent, and their compliance to defined time constraints will be checked by means of the consistency control.

On the *topmost layer* people can access the integrated management data via distinct graphical *user interfaces*. Engineers are notified of assigned tasks by a dedicated *engineering user interface*. Chief Engineers use the *management user interface* to fulfill managerial operations. This interface offers a complete view of the overall management data and provides functions for changing these data in case of dynamics.

Comos PT is mainly used in chemical engineering but can also be used in *other engineering domains*. So the planned process management system is especially designed to be *adaptable* to different domains and to be *integrated* with other tools of innotec. Domain experts adapt the process management system to a certain domain by using the adaptation user interface. This interface allows to declare *new types* of products, resources, activities, and to predefine best-practices as activity patterns. All user interfaces are integrated into the Comos user interface, so that the users do not have to switch between separate applications during their work.

Furthermore, *external project management systems* will be *coupled* with the process management system. For example, *MS Project* offers additional views of the design process like Gantt-charts with marked critical paths.

Problems and Solutions

A-posteriori Integration

The process management system AHEAD has been implemented as a *research prototype* in an a-priori manner. The effort for AHEAD's implementation was reduced by leveraging the high-level graph-based language and compiler PROGRES, the runtime environment UPGRADE, and the graph-oriented database GRAS. PROGRES, UPGRADE and GRAS are themselves research prototypes. Hence, this implementation is *not suitable* for being used in an *industrial environment*.

Instead of porting the current AHEAD implementation directly into the Comos PT environment, AHEAD is *reimplemented* as an extension of Comos PT with restriction to those programming languages and tools that are used within Comos PT. Transferring AHEAD in this way poses two interesting *challenges*:

1. AHEAD's core is specified by a PROGRES specification. This specification comprises integrated submodels for managing products, resources, activities, and definitions of valid operations on the integrated management data as graph transformations. This graph-oriented specification has to be *mapped* to a *common programming language* such as C++, C# or Visual Basic. The resulting code must fulfill two *requirements*: (a) It has to provide an *efficient implementation* of the specified tests and transformations, which are necessary to check the management data for *consistency*

and to perform *complex modifications* to this data, respectively. (b) The code must be *readable and extensible* for developers at innotec. This requirement does not apply for the compiled PROGRES specification that constitutes the AHEAD system.

2. At runtime, AHEAD reads and stores the current managerial configuration via the database system GRAS, which provides a *graph-oriented access* to the stored *data*. The process management system has to read and store management data to the *object-oriented database* of Comos PT by means of the Comos programming interface. The Comos programming interface is restricted, because it just provides a technical view of the stored objects. Therefore, a newly implemented Comos wrapper has to comply with two *requirements*: (a) It has to *bridge* between an object- and a graph-oriented view onto the management data. (b) The Comos wrapper has to provide an *extended view* onto the stored data, which additionally covers management data (resources, activities, and overlapping relationships).

Time Constraints, Expenses, and Staffing

Time constraints, expenses, and staffing do not play a predominant role in the current state of the AHEAD system. Nevertheless, they have to be considered in process management. Widespread *project management systems* like Microsoft Project offer *functions* like the computation of a critical path in an activity network, workload balancing for the staff or calculation of the aggregated expenses for (parts of) a project. Besides these available functions, project management systems offer *additional views* on the current project state, like the activity oriented Gantt-chart or diverse resource-oriented views.

In order to utilize this existing functionality, the process management system for Comos PT can be *coupled* with such existing systems. This can be done similar to the coupling of the AHEAD system with MS Project described in Subsect. 7.7.3.

Although common project management systems provide some useful functions for process management, they *cannot replace* the planned process *management system* of Comos PT since they lack the handling of dynamics and integrated management data. In particular, project management systems are unable to appropriately cope with *unexpected iterations* in the current projects or with *evolving activity structures*. Hence, these systems serve only for executing the described computations or rendering the views mentioned above but are not used as the main environment for process management.

Project management tools offer possibilities to define *time constraints* like deadlines for tasks. But most of these tools do not provide any means to *detect violations* of constraints and they do not *enforce actions* to bring the project back on schedule. Such issues will have to be addressed by our developments. The planned management system will permanently calculate the *current execution state* of the design process and will compare it with the *planned schedule* of the process. In this way, violations of time constraints can be detected. Such violations can be *compensated* by *dynamic changes* of the project schedule at process runtime.

Progress Measurement

As mentioned before, *monitoring* the current execution state of a design process is an important feature of the planned management system. This includes the *measurement of the progress* of tasks and workflows. The integration of the management data with the technical data stored in the Comos database offers the opportunity to *estimate* the *required effort* and the *progress* of tasks based on the internal structure of the created documents.

The flow diagrams created during the design process determine to a certain degree the future tasks. In Comos PT, the *internal structure* of these *documents* is represented by objects. This data can be used for the calculation of the progress of certain tasks. Consequently, the progress is not solely calculated based on estimates of the engineers assigned to the tasks.

The calculated progress measures of elementary tasks have to be *aggregated* and *propagated* upwards in the task hierarchy. In this way, the progress of complex tasks, of project phases and ultimately of the whole project can be estimated. The progress measures help the project manager to decide whether the project is on schedule or whether deadlines will be missed.

Integration with the Workflow Engine

The workflow engine, developed in cooperation with innotec, is already used to *run and monitor workflows*, like the revision of documents or the handover of the final documentation of a chemical plant to the client. The provided *workflow management functionality* should be *used* by the process management system, so that parts of the design process can be *executed automatically* by the workflow engine.

The *integration* of the workflow engine into the management system core has to account for *several issues* such as the mapping of execution states of workflows and other subprocesses, the enforcement of time constraints for a workflow, and the measurement of its progress. Workflows can be started automatically, when the user performs certain actions in Comos. These workflow instances have to be integrated into the overall design process, i.e. their position in the task hierarchy has to be defined, and their relations to other tasks of the process have to be established.

Like all tasks of the design process, workflows have to be executed according to the *project time schedule*. When workflows are executed according to an overall schedule, it is possible to detect work-overload of resources, who participate in multiple workflow instances. The concepts for *progress measurement* of the design process have to be adapted and applied to the management of workflows. The resolution of the aforementioned issues leads to a *seamless integration* of the given workflow engine with the new process management system.

Impact

The implementation of the process management system for Comos PT is interesting from both, the *economic* and the *academic* point of view. The leading position of Comos PT is further improved by a process management system, which allows for the holistic management of the overall design process of a chemical plant.

From the *research perspective* the transfer of the AHEAD system into an industrial environment is challenging w.r.t. several aspects: The process management system has to be realized *on top* of an existing software-system that has not been designed for process management. In contrast to the AHEAD system, graph-based prototyping tools PROGRES, UPGRADE and GRAS cannot be used for its implementation. Instead, the given specification of AHEAD has to be mapped methodically to common programming languages and the graph-based storage of management data must be implemented using an object-oriented database.

Dynamics in design processes have to be addressed with particular emphasis on evolution and iterations of activities, but also on deadlines and expenses.

7.7.5 Dynamic Workflow Management for Business Processes

The following subsubsection describes the realization of a workflow management *system* which provides support for *dynamic business processes*. Following again approach IA2 (cf. Fig. 7.22), this system is based *on top* of an existing workflow management system, which only supports *static* business processes. This objective constitutes the second part of the transfer project T6 and is performed in collaboration with the IT service provider *AMB Generali Informatik Services GmbH*.

Overview

Although AHEAD has been designed within the subproject B4 to capture a multitude of processes, it has always been focusing the management of design processes, while business processes have not been considered in the past. In cooperation with AMB-Informatik, research results of the B4 subproject are *transferred* into the domain of *business processes*. AMB-Informatik is a full-service information technology provider for the insurance group AMB-Generali.

Figure 7.25 shows the coarse grained *architecture* of the planned system. The architecture is divided into three parts, which are described in the following.

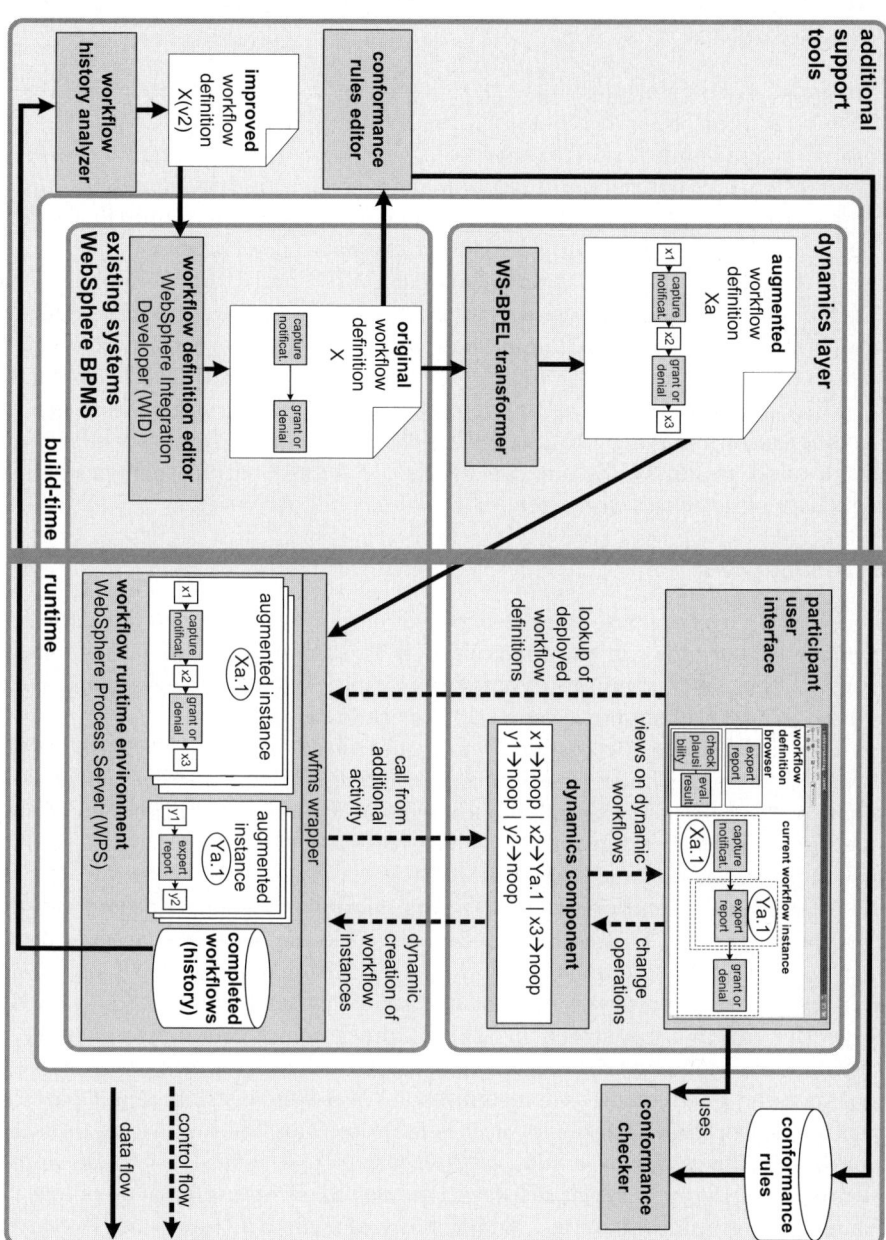

Fig. 7.25. Architecture of the AMB-Informatik workflow management system

Existing Systems

Currently, AMB-Informatik uses IBM's product family WebSphere BPMS in order to support their business processes. WebSphere BPMS [1034] *strictly separates* build-time from runtime of a workflow. Hence, *dynamic business processes* with undeterminable activity sequences *cannot be supported* appropriately. In general, these processes are likely to become dynamic if different people are involved, especially outside of the insurance company.

IBM WebSphere Integration Developer (WID) and IBM WebSphere Process Server (WPS) are the *existing systems* which are relevant for the transfer project. WID can be used at build-time to define workflows consisting of activities that are executed automatically or by human interaction. The workflow definitions are written in the de facto standard "Web Services Business Process Execution Language" (WS-BPEL) [887]. At runtime these workflows can be executed within WPS, but can *neither be altered* within their predefined sequences of activities *nor extended* by additional activities.

Dynamics Layer

The existing systems are extended by a dynamics layer to *facilitate dynamic changes* to workflows in the first place [487]. Roughly speaking, the components of this layer simulate dynamic workflows while the underlying WebSphere BPMS still supports only static workflows.

The distinction between build-time and runtime can also be found in the dynamics layer. At build-time, there is a *WS-BPEL transformer* which adds additional WS-BPEL activities, to an original workflow definition X resulting in an augmented workflow definition Xa. In Fig. 7.25 activities x1, x2, and x3 represent such additional activities.

At runtime, execution of an additional activity triggers a call to the *dynamics component*. This component stores workflow *instances' specific information* about how to handle a call from an additional activity. An additional activity can invoke a newly instantiated workflow instance depending on the respective information stored in the dynamics component. For example, in Fig. 7.25 additional activity x2 invokes workflow instance Ya.1

Since the participant user interface of WPS displays only small parts of the overall workflow in table form, it is inappropriate for supporting dynamic workflow changes. Thus, a new *participant user interface* has to be implemented from scratch which displays the *overall structure* of a workflow instances in a graphical manner. Additionally, workflow instances are displayed in a *condensed manner*, i.e., the additional activities remain hidden and dynamically called workflow instances are displayed inline within the calling workflow instances. Furthermore, *predefined workflow definition* fragments can be selected and inserted by a participant via a *workflow definition browser*. Altogether, the workflow participant experiences a dynamic workflow although the technical basis WPS remains static.

Additional Support Tools

Before a *dynamic change* of a running workflow instance takes effect, the modification is *checked* against certain conformance rules by the *conformance checker tool*. If a check fails, an error message is displayed in the participant user interface, where the change request originated from. Otherwise, the change of the workflow instance takes effect. The conformance rules that are used by the conformance checker originate from a build-time tool named *conformance rules editor*. A workflow modeler can use this editor to define rules concerning dynamic workflow changes like the non-deletability of certain (mandatory) workflow activities.

In order to benefit from the *implicit knowledge* expressed by dynamic workflow changes, completed workflow instances can be analyzed by a *workflow history analyzer* with regard to dynamic changes. A finished workflow instance can be imported into the *workflow history analyzer* and compared with its original workflow definition. Then, it is up to a workflow modeler to *decide* whether a deviation of the instance from its workflow definition is *generic*, and should be part of a *improved workflow definition*, or *special* to the respective instance, and should therefore be kept out of the definition.

Problems and Solutions

Support for Dynamic Changes at Runtime

In contrast to design processes, business processes are more *repetitive* and contain *more static parts* in their overall activity structure. Therefore, major parts of business processes can be *predefined*. Hence, typical dynamic situations are not only evolution of activity structures in a running workflow instance but also *deviations* from prescribed activity structures.

Business processes are likely to be *constrained* by laws or company-specific rules. These rules may demand the execution of certain activities, which therefore must not be removed from a workflow instance.

The planned workflow management system for AMB-Informatik meets both issues. First, it provides support for the *definition at build-time* of workflows and conformance rules via the conformance rules editor. Second, laws or domain specific regulations are *enforced at runtime* by checking dynamic changes of a workflow instance by a conformance checker before these changes actually take effect in the workflow instance.

Process Improvement

As described above, the workflow management system for AMB-Informatik offers functionality for predefining workflows at build-time and for changing workflow instances at runtime. There are *dynamic changes* of *two types*: (1) A dynamic change in a workflow instance might occur as a deviation from the workflow definition due to a *special situation* in a business case that is associated to the workflow instance. (2) A dynamic change in a workflow instance might take place because of an *insufficient workflow definition*.

In the latter case, its is probable that the dynamic change has to be done over and over again for each workflow instance of the corresponding workflow definition. Since completed workflows can be imported into the *workflow history analyzer* and compared to the original workflow definition, a workflow modeler is able to recognize recurring deviations from the workflow definition and can *adopt these deviations* by adding them to the definition. In this way, business processes can be significantly improved.

A-posteriori Integration

Like the process management system for Comos PT, the workflow management system for AMB-Informatik resides on top of an existing system. Hence, it has to be realized in an a-posteriori manner. In contrast to Comos PT, which focuses on product-related aspects of design processes, WebSphere BPMS mainly covers *activity-related parts* of a business process. WebSphere BPMS is therefore used as a *building block* for the new workflow management system for AMB-Informatik. It should be possible to substitute this block.

Impact

The realization of the workflow management system is *beneficial* both in *economic* and *research* respect. The workflow management system empowers AMB-Informatik to *support dynamic business processes*, which WebSphere BPMS does not provide itself. The workflow management system does *not replace* WebSphere BPMS but extends it such that subsequent integration problems can be avoided.

From the *research* perspective, new interesting question emerge by transferring existing concepts for design processes to the domain of business processes. The consideration of dynamics in processes will *shift* from *continuously evolving design processes* to more repetitive but nonetheless *dynamic business processes*.

7.8 Service-Oriented Architectures and Application Integration

Th. Haase and M. Nagl

Abstract. Service-oriented architectures define an architectural style for the construction of a heterogeneous application landscape. By abstracting services, business processes are decoupled from the underlying applications. This section describes how the results of the IMPROVE subproject I3, related to the model-driven development process for wrapper construction, are transferred and extended to the area of business process applications. We present an approach which yields a prototype to formally specify service descriptions and service compositions. This prototype makes it possible to evaluate and explore service-oriented architecture concepts.

7.8.1 Introduction

The role of the subproject I3 of IMPROVE, as described in Sect. 5.7, was to coordinate the software development process of the integrated engineering design environment (cf. Fig. 5.55 on p. 558). Especially, it dealt with the development of a general *framework* for *a-posteriori* integration of existing tools.

Integration was realized on the architectural level. An architecture for the integrated environment was developed on the level of subsystems, where general components were identified. Thereby, a *coordinated development* and *reuse* on the product level were enforced. Additionally, the subproject took care that project-specific components were embedded correctly into the overall environment.

The *architecture* of the *overall environment* describes the "glue" necessary for performing integration. It defines, for example, what kinds of interfaces the tools to be integrated offer, how interfaces are wrapped in order to homogenize them, how tools and wrappers are distributed, how interfaces are accessed via certain middleware techniques, and so on.

Furthermore, to reduce the development effort for building required wrappers, the subproject aimed at *specifying wrappers* using visual models. A corresponding modeling formalism was defined. Based on such models, executable *code* for the wrappers is *generated* and embedded into the general framework.

This section describes how these results are transferred to the area of business application integration in order to apply the approach in an industrial environment. Based on the idea of service-oriented architectures, the modeling formalism to specify wrappers is extended to *model* an integrated *business application* as a loosely coupled set of *interacting services*.

This extension is not only investigated on the conceptual level but also covers a *prototype* implementation for the corresponding modeling environment and its code generator. In this way, a test environment for the *evaluation* and *exploration* of service-oriented concepts in the context of integrated business applications is built.

7.8.2 Basic Notions

In this subsection, the problem of a-posteriori *application integration* in the area of business process support is investigated. Current solutions, representing the state-of-the-art in industry, and their deficits are discussed first. Then, key concepts of the *service-oriented paradigm* are introduced and related to the problem of a-posteriori integration. Expected advantages of the service-oriented approach are finally discussed.

Application Integration

In Sect. 5.7 we argued that current tool *support* in chemical engineering does *not consider* the *overall* design *process*. This situation is due to the fact that tools are provided by different vendors, that they are based on heterogeneous system platforms, proprietary document formats, and different conceptual models of the application domain. The same holds true for the area of business process support [31, 45, 210] as IT landscapes of companies are typically characterized by a portfolio of heterogeneous business application systems.[64]

One approach for the seamless support of the entire supply chain of a company are so-called *ERP systems* (enterprise resource planning) [811]. The basic idea of an ERP system is that of a centralized database, which stores company-wide all relevant data concerning the business processes. This approach fails, because a monolithic ERP system cannot fulfill all conceivable requirements in general [817]. Nowadays, a typical company uses on average 50 different, mission-critical applications for the support of its business processes, as empirical studies have shown [830]. Furthermore, for economic reasons it is not feasible to replace these legacy systems by new applications [973].

As the need for integration emerges in the business domain, so-called *EAI systems* (enterprise application integration)[65] [618] were introduced. An EAI system realizes an abstract communication layer for the mediation of data between heterogeneous applications. Typically, this communication layer aggregates several middleware technologies, e.g. COM/DCOM [846, 847], CORBA [877], Enterprise JavaBeans (EJB) [504], JavaRMI, or Web Services [500], and enables the automatic conversion of the different data formats between them. For EAI systems the same problems as already mentioned in Sect. 5.7 hold: An EAI solution mainly focuses on technical issues, e.g. on a homogenous communication layer above existing middlewares. In contrast to IMPROVE (cf. Sects. 5.7 and 3.2) semantical issues, like data homogenization or data consistency, are not addressed.

---

[64] In contrast to the notion of a-posteriori integration realized *externally* by the means of *wrappers*, as presented in this section, former projects of our department [81–83, 85–87, 89] deal with *internal reengineering* of business applications in order to prepare them for integration.

[65] Another term in this context are B2B systems (business to business application integration) [820]. While EAI focuses on application integration within a company, B2B emphasizes the inter-organizational integration between different companies.

The Service-Oriented Paradigm

The vision of the service-oriented paradigm is to decouple the domain-specific aspects of a business process from its underlying technical realization done by a set of business applications.[66] Decoupling is achieved by a *service layer* consisting of a set of interacting services.

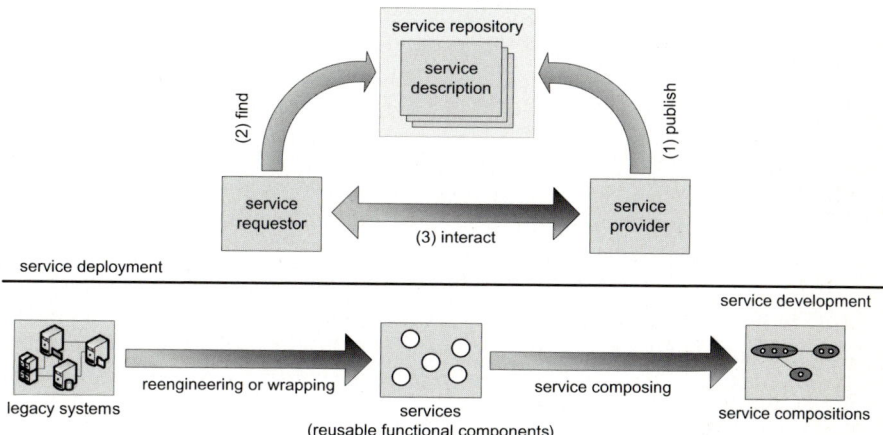

Fig. 7.26. Service deployment and development

Business Perspective

The service layer separates different views on an IT landscape [904]: the *business* perspective, i.e. the *outside* view on a service is that of an atomic unit, well-defined in the terms of a business area, realizing some kind of functionality, and usable within a business process.

In a typical *service-based scenario* [896] (top of Fig. 7.26) the service provider, offering some kind of service, defines a service description of the service and (1) publishes it to a service repository. The service requestor, searching for a service, uses the service repository (2) to retrieve the service description. This enables the service requestor (3) to bind and invoke the service.[67]

---

[66] Strictly speaking, the concept of a service-oriented architecture is nothing else but the application of well-known and established software engineering principles like *separation of concerns* or *information hiding* in the area of IT landscape architectures.

[67] Please note that the terms "service requestor" and "service provider" designate roles, which can represent the same company or different departments within the same company.

Thus, from the business point of view, a service represents a functional module [331] offering an abstract, technically independent interface to access the underlying application systems of the service provider. *Services* are reusable within different business processes and constitute the *building blocks* for defining business processes. According to modified requirements, business processes can be adapted by integrating new services into the business workflow or by substituting services. Therefore, services have to meet certain properties [933]: they have to be (i) self-contained, (ii) coarse-grained, (iii) stateless, (iv) transactional, and (v) potentially asynchronous.

Technical Perspective

The technical perspective or *inside view* on a service refers to its realization and development by the service provider. This level (bottom of Fig. 7.26) deals with the preparation of existing system functionality in order to embed legacy systems into a service-oriented landscape.

Integrating legacy systems can either be done by reengineering or by wrapping. The latter facilitates a transparent integration of legacy systems into a service-oriented environment. Nevertheless, both alternatives aim to realize a functional interface for given systems in the sense of services, which fulfills the required properties stated above.

Furthermore, services can be composed by the service developer to implement a new value-added service. The composed *service aggregates* several basic or composed services, possibly realized by different application systems. From the outside view, i.e. for the service requestor, there is no difference between a basic or a composed service.

Another important issue in this context is the aspect of *data homogenization*. When, for example, several services, realized by different systems, are composed, a common data model is needed. Furthermore, enriching a service description with semantic issues from the business perspective,[68] enables a service requestor to retrieve and to bind services *dynamically* at business process runtime. Such a semantical service description always demands for a common *ontology* of the specific business domain. Following this idea, the data structures of the legacy systems have to be transformed according to this ontology. Therefore, reengineering or wrapping of legacy systems to prepare them for service-orientation also involves the task of data homogenization.

Finally, the service-oriented paradigm is often associated with *Web Services* as the realizing technical infrastructure. This is not necessary [582], since other component technologies [991] like COM/DCOM [846, 847], CORBA [877], or Enterprise JavaBeans [504] are also applicable, as they implement the necessary concepts like (i) implementation transparency, (ii) location transparency, (iii) self-containment, and (iv) an explicit interface independent from its implementation.

---

[68] A semantical service description covers functional as well as non-functional aspects.

7.8.3 A Service-Oriented Integrated Business Application Architecture

This subsection sketches an approach for an integrated *business application architecture*. The architecture is built from a set of existing legacy systems and follows a service-oriented architectural style. We show how the model-driven wrapper development approach (cf. Subsect. 5.7.4) can be applied to the problem of designing such an architecture.

The ideas presented in the following are the results of an ongoing cooperation with an industrial partner, namely the *AMB Generali Informatik Services GmbH* (abbr. AMB-Informatik). AMB-Informatik is the IT service provider for a major affiliated group of insurance companies in Germany.

The *need* for *restructuring* and *integrating* its IT landscape emerges for AMB-Informatik for two reasons: (1) AMB-Informatik aims to substitute client terminal software for accessing legacy mainframe applications by a web-based solution. This requires on the one hand to link existing systems with a new technical infrastructure and on the other hand to reengineer the available user interfaces in order to unify them. (2) The IT landscape of AMB-Informatik grew historically. The incorporation of companies into the affiliated group and the corresponding takeover of their IT systems led to redundancies between the systems, both functional as well as with respect to the databases of the systems.

Architectural Sketch

The *target architecture* aimed at is illustrated in Fig. 7.27. Each layer depends on the one below and realizes the prerequisites for the layer above:

- *Layer 1*: Operational (legacy) systems layer. The basic layer consists of existing custom built business applications. These applications, mainly mainframe systems implemented in COBOL, realize the business functionality. To enable distributed and parallel access to applications, a transaction monitor ensures fault-tolerance, consistency, recovery, and performance balancing.
- *Layer 2*: Technical adapters layer. This layer is responsible to embed legacy systems into the new technical infrastructure. This is done by technical adapters. These components serve as proxies adapting the existing COBOL interfaces to the required new syntax. No kind of semantical homogenization is done on this level. A typical task of an adapter is, for example, the conversion of data types. Furthermore, adapters encapsulate physical distribution and the corresponding communication protocols.
- *Layer 3*: Homogenized components layer. In general, data structures and functionality of different existing applications are not interoperable. Furthermore, not all systems can be restructured internally, or internal reengineering is not feasible for economic reasons. Therefore, layer 3 represents

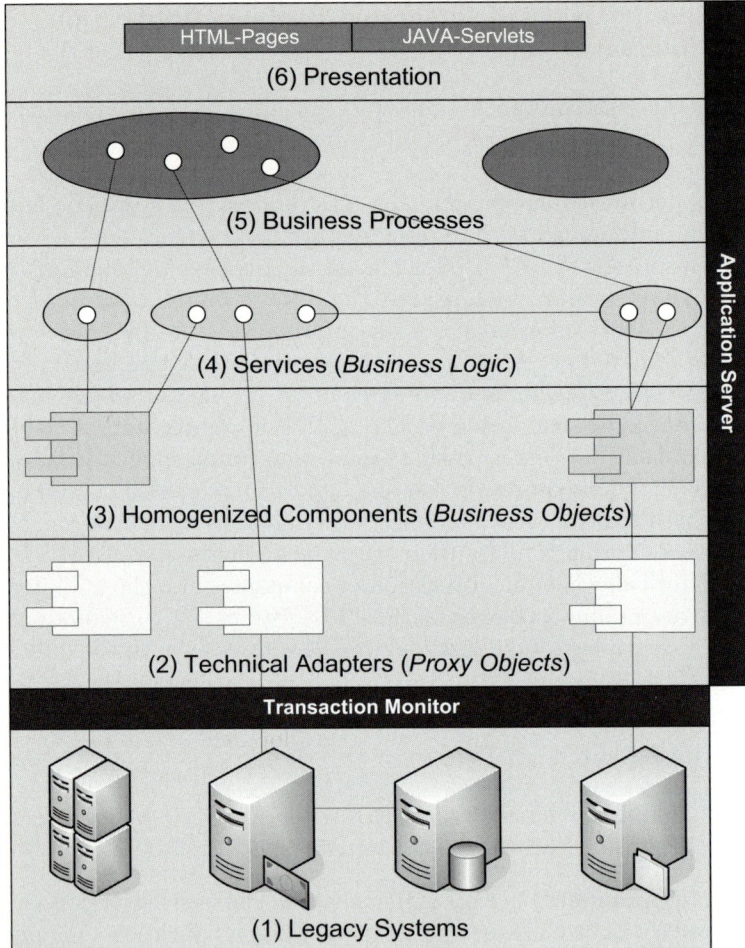

Fig. 7.27. Levels of a service-oriented architecture in the context of a-posteriori business application integration

a homogenization layer. It consists of business objects representing the concepts of the business domain. Their data structures and functions are mapped onto those of the existing applications via technical adapters. With respect to these aspects, a business object is nothing else but a wrapper.
- *Layer 4: Services layer.* The central part of the target architecture is the services layer. Services are built up from (i) an elementary function of a business object, (ii) a composition of elementary functions of one or more business objects, or (iii) a composition of elementary functions and/or other services. Therefore, the service defines the control structure for execution of the composition.

The remaining two layers are out of scope for this section. For completeness, they are shortly mentioned:

- *Layer 5*: *Business processes layer.* This layer is responsible for the choreography of business processes (see also Sect. 7.7). Within a business process, services are usable as atomic units.
- *Layer 6*: *Presentation and access layer.* Finally, user interfaces and the web-based access are realized by this layer.

Application of Available Results

Figure 7.28 relates the layers of the architecture in Fig. 7.27 to the different phases of the *extended* model-driven wrapper development process [138]. While phases (1)–(5) (see Fig. 7.28 left) are already supported by corresponding tools we have developed (cf. Subsect. 5.7.4), phases (6) to (8) are new due to specifying service descriptions and service compositions and generating code for services.

In the following, the relation between the different phases and the layers of the intended architecture will shortly be explained.

We expect technical adapters equipped with a COM interface to exist. This enables to parse the interfaces of legacy systems by the existing parser. Furthermore, a model of the COM component according to our modeling formalism can be generated (cf. *technical wrapper* in Subsect. 5.7.4). In this way, layer 2 is covered by phase 1 of the development process.

Furthermore, the model derived from parsing is *enriched* with *information* received either from exploring (cf. Subsect. 5.7.4) COM components interactively or, alternatively, from extensions specified manually (phase 2). Examples for further specifications in this phase are the renaming of parameter identifiers or the definition of simple function call sequences without any control structures.

Afterwards, the homogenized business objects (layer 3) are specified based upon our modeling formalism. Specification of business objects consists of two modeling phases: (i) Definition of the desired static data structures (phase 3). This is done by an object-oriented class diagram. (ii) Definition of mappings (phase 4) to the data models realized by the adapters (cf. *homogenizing wrapper* in Subsect. 5.7.4). Based on both specifications, the executable program code for business objects can be generated (phase 5).

Finally, the *service* layer is *modeled* (phase 6 and 7 related to layer 4) and, again, the program *code* is *generated* (phase 8). Specifications for service descriptions and service compositions and their transformation into a programming language are not yet covered by our modeling formalism. They are one of the main scientific extensions of the approach, which will be introduced in the next subsection.

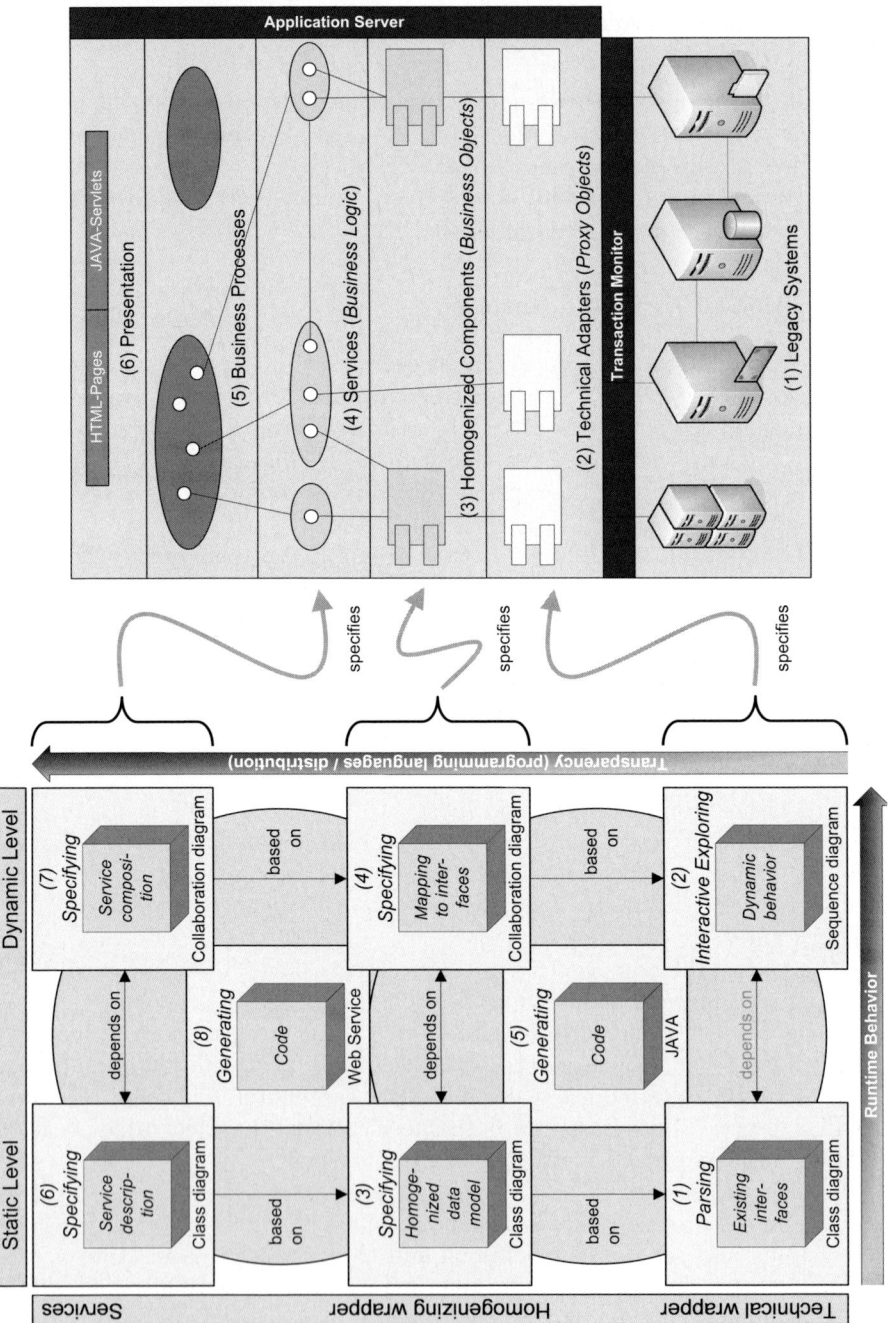

Fig. 7.28. Model-driven wrapper development process related to the layers of a service-oriented architecture

7.8.4 Extensions of Available Results

As stated in the previous subsection, the main goal of our activities is to *extend* our modeling *formalism* and its associated *tools* for the model-based specification of a service-oriented architecture concerning legacy systems. This includes generation of executable program code based upon such specifications. Afterwards, the program code is embedded into a test environment for evaluation and exploration of service-oriented concepts.

The result of these activities is not a "ready-to-use" industrial tool, but a *suitable prototype* allowing an evaluation and exploration of an integrated business administration system. Nevertheless, the resulting prototype has to meet certain requirements, mainly technical ones that enable its usage by our industrial partner.

In the following, several necessary *extensions* of the original approach and, therefore, some challenging problems are presented.

Modifications Regarding Technical Infrastructure

Until now, our approach to model-driven wrapper development deals with tools and applications offering COM interfaces. Conceptually, COM is a *component technology* following the *object-oriented paradigm*. COM components represent subsystems, i.e. a set of classes including attribute and method definitions, relationships between classes, and a set of interfaces to access the subsystem. COM offers an *external description* of application *interfaces* by means of a type library. Such interface descriptions of components included in a type library serve as input for the analyzer to comprehend a component and its various parts syntactically.

As already demanded in the previous subsection, the existence of a COM interface for accessing legacy systems is necessary in order to apply our wrapper construction methodology as well as the corresponding tools. Therefore, the existing application systems, implemented in COBOL, have to be adapted towards this middleware. As the following explanations will show, this approach causes less effort than the alternative adaptation of our tools supporting the wrapper construction methodology.

In a COBOL program, data structures are usually defined by so-called *COBOL copybooks*, which are conceptually comparable to abstract data types and are physically organized as separate files. Existing commercial tools offer the possibility to *automatically transform* COBOL copybooks in so-called *JAVA record classes*, i.e. JAVA classes are generated to represent COBOL data structures one-to-one. Furthermore, the corresponding code to map a COBOL data structure to its associated JAVA record class and vice versa is generated. This code instantiates a JAVA record class from a given instance of a COBOL data structure and propagates changes of the JAVA record class instance to the corresponding COBOL data structure instance. The generated

JAVA classes are then embedded into a corresponding *JAVA-based framework* for accessing COBOL-based mainframe applications via a transaction monitor.

The next step is to equip the JAVA classes with a COM interface. Tools like [629] are able to *automatically* generate a COM interface for a given set of JAVA classes. The required infrastructure code is generated, compiled, and linked to the given code, a type library is generated, and both are embedded into the COM runtime environment.

Using this tool chain, the required COM interfaces to existing COBOL applications are realized without any manual implementation effort. However, only *data aspects* are covered, as a COBOL copybook only describes a data structure. *Functional aspects*, such as a calculation based on some given data can not be handled this way yet.

Formalizing Service Descriptions and Service Compositions

Specifying service descriptions and service compositions causes the main *extensions* of our approach.

The modeling formalism is enriched with concepts to model a service as a functional module, i.e. on the one hand to define the export and the import interface of the service (*service description*), whereas imports are either other services or functions of business objects. On the other hand, the body of the service has to be specified. The body defines the control flow between the imports, i.e. sequences, alternatives, and loops of service or function calls with according execution conditions (*service composition*).

Furthermore, *non-functional properties* of a service can be specified, including (i) the type of communication between services (synchronous vs. asynchronous), (ii) who is allowed to communicate with a service (authentication), or (iii) the predefinition of service alternatives with regard to non-availability of a certain service.

In the case of an *asynchronous service call*, for example, the specifications of both caller and callee have to be completed for event handling. Concerning the callee, the export interface has to contain the signatures of the thrown events and the body definition has to include further specifications regarding under which conditions an event is raised. Analogously, for the import interface of the caller it has to be specified in which events it is interested in. In the body it has to be defined how the caller reacts when a certain imported event is raised.[69]

In addition, the modeling formalism is extended by *analyses* validating on the one hand *syntactical* correct concatenation of services and business object

---

[69] Event handling also affects the business objects layer or the adapters layer, respectively. The business objects layer homogenizes the events thrown by the underlying application systems. However, this is only possible if the interface descriptions of the adapter components include event definitions.

functions such as type conformity of input and output parameters. On the other hand, *structural* properties of service compositions related to runtime behavior are analyzed to detect, for example, unbounded loops or deadlocks[70].

A precise *conceptualization* and a compact *notation* for modeling the aspects mentioned above is challenging. Furthermore, to determine suitable *analyses* for completeness and correctness of the models, so that program code generation is possible, is also not trivial. Both are interesting aspects of the extensions of the modeling formalism to specify service descriptions and service compositions.

As already stated in Sect. 5.7, the modeling formalism for model-driven wrapper development is formulated as a graph rewriting system using the PROGRES language [412, 414]. Consequently, the necessary concepts to model service descriptions and service compositions will be implemented as an *extension* of the PROGRES *specification* shown in Fig. 5.60 on p. 571 and Fig. 5.67 on p. 586.

Transformation to Programming Languages

Further extensions deal with code generation. The generated code allows the execution of specified services. It is embedded into a *test framework* to interactively evaluate and explore the specified services. This test framework is comparable with the tool support to explore existing component interfaces interactively (cf. Subsect. 5.7.4).

We are not aiming to generate code that can be used in the productive environment of our industrial partner. Rather, we are investigating from a scientific point of view how the *concepts* defined by the modeling formalism can be *transformed* to programming languages. The difference to the yet available code generator is the generation of JAVA or C++ code instead of PROGRES code. This change has certain consequences for code generation.

For example, let us regard *associations* between *classes* in an object-oriented programming language like JAVA or C++ in comparison to edges between nodes within the PROGRES language. Whereas PROGRES has a well-defined semantics concerning edges, realized by the PROGRES runtime environment to limit the possible transformation alternatives, for JAVA or C++ several *aspects* have to be considered:

- In general, an association a of object o_1 with object o_2 can be either realized as (i) o_2 is a value of a variable v of o_1 (*variable semantics*) or, alternatively, (ii) v holds a pointer referencing o_2 (*reference semantics*). As o_2 can as well be associated with other objects, variable semantics is in most cases not a feasible solution.[71]

---

[70] Deadlocks can appear in the case of asynchronous, cyclic service calls.
[71] While PROGRES and JAVA make use of reference semantics, C++ offers both alternatives.

- While PROGRES allows bidirectional navigation along edges, this is not possible for pointers in JAVA or C++. Therefore, a bidirectional association a is mapped to an explicit forward pointer a_f and an implicit backward pointer a_b. For that, the generation of additional program code for administrating a_b is necessary.
- An association can be refined by multiplicities. For example, a fixed maximum of associated objects greater than one requires again the generation of additional program code to ensure this constraint during runtime. Analogously, the same has to be done for a fixed minimum greater than zero.

Using *reference semantics* has further *consequences*: When deleting a node in PROGRES, the runtime environment of PROGRES ensures, that all associated edges of that node are deleted as well (outgoing edges as well as incoming edges). In contrast, JAVA uses garbage collection, i.e. all pointers to an object o have to be deleted before deleting o. Thus, the implementation of a deletion operation in JAVA is completely different from PROGRES. Especially, as references to an object o are not known by o, when establishing an association to o, some kind of reference counting has to be generated, so that o can be deleted by deleting the pointers to o.

Transformation to Web Services

Regarding the requirements of our industrial partner (see Subsect. 7.8.3), another topic concerning *code generation* is to realize the specified services as a *Web Service*.[72]

Realizing a specified service as a Web Service includes *several generation steps*:

1. Generation of code for *serializing* a JAVA object, such that it can be encoded as XML and streamed over a network using the SOAP protocol (Simple Object Access Protocol). Especially, it has to be taken into account that the identities of objects are preserved. This is achieved by generating an individual identifier for each object. Using these identifiers, associations between objects can be serialized.
2. Generation of an *interface description*, i.e. a WSDL file (Web Services Description Language), for the service usable by a client to access the service.
3. Generation of a *deployment descriptor*, i.e. a configuration file in order to install the service into the Web Server runtime environment.

While the two latter steps can be realized with the help of existing transformation tools [508], the first step requires further extensions of our code generator.

---

[72] Please note, that the term "service" refers to a concept, which is independent of a specific technique, while the term "Web Service" denotes a concrete realization technique.

Graphical User Interface Extensions

Extensions of the modeling formalism are not limited to the conceptual level. Rather, corresponding tool support has to be implemented to realize a modeling *environment* for *specifying service descriptions* and service *compositions*. This is done by using UPGRADE [48, 49], a JAVA-based framework for building user interfaces for graph-based applications. Based on the PROGRES specification, code is generated and embedded into the UPGRADE framework.

In this way, a first prototype of the tool is developed without any manual implementation. However, the user *interface* of the prototype is *rudimentary* and does not meet usual look&feel standards of current interactive application systems. Therefore, parts of the user interface have to be reworked.

We are not aiming at the development of a tool comparable to a commercial one with respect to properties like robustness, performance etc. Rather, the *prototype* serves for evaluating the *applicability* of the developed concepts by our industrial partner. Nevertheless, the prototype has to be stable enough, such that it can be used for evaluation. This also covers the appropriateness of the user interface.

Extensions of *user interfaces* mainly concern the implementation of different representations to visualize and to edit the static (service description) and the dynamic (service composition) aspects of the specifications. For example, to model service compositions several alternatives are possible: (i) sequence diagrams, (ii) collaboration diagrams, or (iii) a textual notation in some kind of pseudo code. Which of them is the most appropriate alternative can only be determined through experiences using the prototype. The alternative views have to be implemented manually in JAVA and embedded into the UPGRADE framework as well.

Furthermore, the *test environment* for exploring the specified services is *extended* with a *client user interface*, e.g. web pages, to invoke services and to visualize the results returned. Therefore, corresponding HTML code is generated.

7.8.5 Conclusions

In this section, we gave a brief overview over the transfer of the results of the IMPROVE subproject I3 to industrial application. We show, how the model-driven wrapper development process and its associated tool support (cf. Subsect. 5.7.4) can be applied to business application systems. A sketch of an integrated business application architecture following a service-oriented architectural style was drawn. Furthermore, we discussed the adaptation and necessary extensions of the existing approach, namely the technical infrastructure, the modeling formalism, the code generation, and the user interface.

The scientific impact of this transfer project can be characterized by answers to the following questions: (i) Which elements form a service-oriented architecture? (ii) How are they structured on the level of subsystems? (iii) Which aspects have to be considered to define service descriptions and service compositions? (iv) How can the latter be formalized? (v) How can they be represented and edited by visual models? (vi) How can a test driver be generated for service description or service composition, respectively? These basic questions have to be solved in order to allow efficient application of service-oriented concepts in practice.

However, to ensure the applicability of the results in practice, close cooperation with an industrial partner is mandatory. The planned prototype offers an easy-to-use opportunity to evaluate and to explore service-oriented concepts in an industrial environment. In this way, first ideas for restructuring and integrating legacy systems can be identified. Further development is facilitated as the specified models can be regarded as an initial, but nevertheless formal requirements' definition.

8
Evaluation

This chapter gives a *summary* of the *results* contained in this volume by presenting these results under four different perspectives.

The *chapter* is structured as *follows*: In the first section we (a) discuss what IMPROVE has contributed to a better understanding of design or development processes. Then, in the following section, we (b) explain our achievements for new tool functionality/integration. Both sections for (a) and (b) give a review from the academic perspective. The third section describes (c) how we have influenced the industrial state-of-the-art in the process industries on the one, and in the tool vendor industry on the other hand. Finally, in the last section (d) we sketch IMPROVE's success story in academia. These four views are the most characteristic ones for the IMPROVE project.

The summary of our *findings* (what we have achieved, what is still open) on the formal *process/product model* is *not repeated* here. This topic was regarded to be so important that its summary was given in an own chapter (cf. Chap. 6).

The summary *perspectives*, given in the following chapter, do *not coincide* with the *project structure* of Fig. 1.27 and, also, with the chapter structure of this book: For example, our findings on tools come from Chaps. 2, 3, 4, 5, and 7. Our results for process understanding stem from Chaps. 2 and 7, but implicitly also from all other chapters. The same is true for the question, which impact we had on industry. Trivially, the success story in academia relates to the whole book.

8.1 Review from a Design Process Perspective

W. Marquardt

Abstract. This section briefly summarizes and evaluates the results of IMPROVE from an application-oriented perspective. The application domain model, its use for the improvement of design processes, and its implications for the development of design support tools constitute the main focus. We conclude that the model framework has reached a high standard and constitutes a good basis for further refinement and extension. However, the model needs even more validation in industrial case studies. The discussion of this section is given from an academic viewpoint. The industrial relevance of design process results is given in Sect. 8.3.

8.1.1 Introduction

A major working hypothesis of the IMPROVE project has been related to information modeling. More precisely, the *improvement* of *design processes* and their *support* by a suite of integrated software tools *require* a profound *understanding* and *formalization* of the design *process* together with the *results* produced during the various design tasks. This working hypothesis has been largely different from previous work because it explicitly focused on the design process rather than on its results, e.g. the design data and documents. This point of view should naturally lead to software functionality which facilitates work process integration rather than the traditional data and control integration of tools to truly improve the productivity of the design team.

A first objective of the modeling exercise is the *conceptual clarification* of the *application domain*, e.g. of design processes in chemical and plastics engineering. The resulting process-centered information model will not only elucidate the intricate interplay between design tasks, resources used, and data produced. It should, furthermore, serve as the starting point for model extension and refinement to finally result in the set of models required to develop and construct new support tools and tool extensions in a model-based software engineering process. The continuity and consistency of the *suite of models* spanning from the application (and hence presentation) layer to the tool layer is expected to result in both, more efficient software development processes as well as user-oriented support functionality. Since information modeling is a time-consuming task requiring both expert domain knowledge as well as systems engineering and information science skills, the resulting models should be reusable in different yet unknown contexts. This quest for reusability should not alleviate the expressiveness of the model.

This evaluation *section* will briefly reflect on the *achievements* of integrated product and process *modeling* from the perspective of the *application domain*. The first subsection will reflect on the achievements in design process modeling. The following subsection will take the perspective of model validation and will consequently discuss our experiences during the application of the information model for the re-engineering of existing design processes as well as for the construction of design support tools.

8.1.2 Modeling Design Processes

A design *process* is a complex and *multi-faceted* entity. Its modeling has to capture the work processes including the actors and their roles as well as the resources used and the results of the individual work process activities such as the engineering documents produced. The decision processes link design process activities with design data and documents in an intricate manner.

A *divide-and-conquer approach* has been followed during the IMPROVE project to deal with the inherently large complexity of the modeling problem. Consequently, *product data* and *work process* modeling have been pursued largely independently in the first part of the project. Later, *document* modeling has been added not only to complement product data modeling but to also link product data with the activities during the design process and their models. The explicit treatment of the *decision* making process as part of the design process model has not been envisaged in the early phases of the project. However, we had to realize that work process modeling will be at least incomplete or even invalid if decision processes are not considered as an integral part of the modeling. Therefore, decision process modeling has been added in a later project phase.

Some emphasis has been on the *integration* of the *various model parts* to an integrated information model of the application domain. We will next discuss briefly what we have achieved and which open problems we still see. The following paragraphs are referring to Chap. 2 and in particular to Sects. 2.2 to 2.6.

Product Data Modeling

While *product data modeling* in an application domain has focused traditionally on a detailed and complete representation of all domain concepts together with their relations and defining attributes, the IMPROVE approach has been radically different. The major focus has been on the design of an *extensible architectural framework* for product data modeling rather than on a detailed and comprehensive model itself.

This choice of focus has been originating from previous experience in data modeling; it is *explicitly addressing* the objective of *extensibility* and *reusability*. In contrast to a model framework, any comprehensive product data model comprising all details has to be limited to a particular use case and will inevitably lack reusability. Extensibility of the model framework also includes means to support the refinement and detailing of the model to tailor it to a certain application context. Obviously, such a model framework can only be developed if various parts of the model are worked out with considerable level of detail in order to validate the architectural design.

The concepts of *general systems theory* originating from the philosophy of science [578] have proved to provide a powerful set of guiding principles during model framework structuring and development. In addition, the different means to appropriately structure the model, including *modularization* into *partial models* as well as *abstraction* across a number of well-defined *layers*, have proven to be generic and useful for large-scale product data modeling in chemical process engineering and beyond. OntoCAPE, the architectural framework for product data modeling developed during the IMPROVE project is probably the most comprehensive available to date.

Though targeted at the chemical engineering domain, we believe that the modeling framework can also be used in *other engineering domains* because of its systems theory foundation and appropriate structuring. Future work will mainly target at the *extension* of OntoCAPE to cover additional facets of the chemical engineering domain. For example, process control engineering is not adequately covered at this point in time because this sub-domain of chemical engineering has not been in the focus of the IMPROVE project.

Our continuing research efforts on domain information modeling have built on a number of *modeling languages* and *modeling tools* in the past 15 years. Though any model should reflect the understanding of the domain and the intent of the modeler independent of the capabilities of the modeling language, the expressiveness, transparency and even the content covered by the resulting model is determined by the modeling formalism to a very large extent.

Ontology languages have been used in the latest phase of the IMPROVE project. They provide adequate expressiveness and a sound logic basis to be unambiguously processed by some computer program in whatever software application. Besides, their anchoring in semantic web research and development has resulted in a wide distribution and consequently in the continuing development of powerful software tools to simplify model development and use. This technological progress facilitates software development and prototyping.

Document Modeling

Typically, product data models are largely independent of the context of a given design process carried out in a specific institutional environment. In contrast, *design documents* are used in an organization as *carriers* of product data, for example, to *communicate* results of certain design tasks between designers or between designers, project managers and clients, or between humans and software tools. Furthermore, documents are used to *archive* the result of design processes for later use either during plant operation, e.g. to support maintenance or revamping, or during another design project, e.g. to serve as a *blueprint* for a similar plant design. Consequently, documents constitute a certain view on the product data and reflect the nature of the design process as well as organizational aspects.

Our document model covers a *static description* of all the relevant *documents* in process and plant design in chemical engineering including the *type* of the document and its *relation* to other documents. This static description is complemented by a *dynamic* description which covers the evolution of the document over time.

Last but not least, the *content* of the *document* is represented by its internal structure in the sense of a template. In contrast to previous document modeling approaches, the document model is not limited to a description of the document structure but relates its content to the product data. This way, the semantics of the document content can be captured.

The present *document model* can easily be *adapted* and *refined* to reflect the peculiarities of a certain organization. We expect that such an adaptation can be largely restricted to a refinement of the details of the document structure and the content description. We do not expect that the overall structure of the document model will have to be changed. Additional document types can easily be added to the existing document model structure.

Work Process Modeling

The process-oriented approach of IMPROVE calls for a concise *model* of the *design process*. There is an obvious relation to workflow and business process modeling on the one and to the representation of sequential control schemes on the other hand. Many modeling *formalisms* stemming from different scientific disciplines have been *suggested* in the recent past. These languages offer a variety of (largely similar) elements to fulfill the requirements of process modeling.

The *engineering design* process is, however, different from most other work processes: Engineering design is a creative and ill-defined process and can therefore *hardly be captured* by formal languages or even be *prescribed* prior to its execution. Any formalism should therefore be able to cope with the non-determinism and the dynamics inherent to engineering design.

Since design is a social process, which largely lacks any domain theory, *empirical studies* have been carried out in industry first to explore the requirements on the elements of a representation formalism for engineering design processes. These empirical studies led to an extension and adaptation of the *C3 work process modeling language* to engineering design processes. C3 has been designed to support participative modeling of actual (industrial) work processes without stressing a sound formal basis. Though, some of the extensions have been motivated by and specifically address the requirements of engineering design processes, we do not want to strongly argue for a dedicated modeling formalism for engineering design. Rather, these extensions have contributed to evolve C3 into a versatile formalism to be widely used for modeling and analyzing work processes to either *streamline* the design process or to *devise support* software functionalities.

The level of detail and semantic formalization of C3 is not sufficient to allow the interpretation and processing by means of computer software. Therefore, the *Process Ontology* has been developed to *complement* the product data model of OntoCAPE. The Process Ontology follows the same architectural design principles and is based on the same ontology language as OntoCAPE.

The conceptualization of this work process ontology is fully consistent with the one underlying C3. In particular, the elements and connectors of C3 map to classes and relations in the Process Ontology. This *conceptual integration* of both formalisms allows a seamless *transition* from a semi-formal to a *formal representation*. While the former is ideally suited to empirically elucidate an actual work process, the latter targets a tool-based interpretation and processing.

Decision Modeling

Decision making is an inherent part of any work process. In particular, *design decisions* determine and even control the engineering design process. Furthermore, decision making refers to alternative positions and collects arguments for and against each of them. Hence, any integrated product and process data model of an application domain excluding decision making would be incomplete and even qualitatively incorrect. Hence, decision modeling has been *added* to the *modeling project* in IMPROVE.

Decision modeling aims at *two different use cases*: (a) the documentation of the decisions made during a design process constitute the design rationale to be archived for later use, for example, in design revisions or extensions of a built plant; (b) the provision of decision templates to guide the designers during decision making in a certain context of the design process. These two use cases call for modeling capabilities on the instance as well as on the class or template level.

The *Decision Ontology* has been developed in IMPROVE to complement the product data, document and design process models. Its architectural principles are fully compliant with those of OntoCAPE and the Process Ontology. The Decision Ontology is an *extension of DRL*, an existing decision model, to cover the requirements identified in engineering design processes.

In particular, the Decision Ontology not only facilitates *modeling* on the *instance level* but also on the template or *class level*. Modeling on the instance level is involved; its inherent complexity is hardly acceptable to industrial practitioners. Further language constructs for abstraction and aggregation of the complex network of concepts and relations linking decisions with data, documents and work processes are therefore necessary to enhance perceivability and to reduce the effort for the formulation of these models during the design process. Decision templates are an attractive means to offer the designer not only guidance in decision making but also assistance for decision documentation.

Application Domain Model Integration

The *integration* of the *four parts* of the application domain model has been a major objective of our modeling efforts. All model parts have been *imple-*

mented in *OWL* to prepare such an integration on the implementation level. Despite the inherent conceptual complexity and the parallel development of the four model parts by different modelers, a first attempt toward the integration of OntoCAPE, the Document Model, the Process and the Decision Ontology into C^{2}EDM, the so-called *Comprehensive Chemical Engineering Domain Model* has been successfully completed. Though there is still room for improvement and extension to be pursued in on-going and future research, this integrated model seems to be the most comprehensive information model framework for the chemical engineering application domain to date.

OntoCAPE is the core of the integrated model which describes the *concepts* of chemical engineering, the specific application domain of interest in the IMPROVE project. The product *data* in OntoCAPE are well integrated with the Document Model by direct references from the document *content* description to product data elements in OntoCAPE. Likewise, the elements of the document content description link the product model in OntoCAPE to the *decision* and *work process* documentation in the Process and Decision Ontologies.

In contrast to the integration on the instance level, *integration* on the *template* (or class level) needs *further attention*. Work process and decision templates which are connected via document models seem to be an interesting means to effectively support and guide designers during a design project. These template structures are suited to encode best practice knowledge to be reused for improved design process performance and quality assurance. The design of such template hierarchies and their presentation to the designer are largely open issues. This design also has to account for an appropriate representation of the mix of routine and creative tasks in the design process.

Our modeling efforts have been largely targeting at the technical perspective of the design process. More precisely, we have always tried to largely *abstract* the *engineering design process from* the *organizational structure* in the company. It is an open issue whether these dimensions can easily be added by just extending the work process, decision and document models or whether a specific model has to be created to reflect the organizational constraints of the design process. Likewise, our model does not reflect the social dimension of design as identified during a number of field studies [574].

Model Migration to Other Engineering Domains

The architecture of C^{2}EDM and its design principles are not limited to chemical engineering. We expect a straightforward extension to other engineering domains to be easily achievable. The *migration* of the model to *another engineering domain* can be easily accomplished by replacing OntoCAPE by a product data model of the respective domain. The remainder of C^{2}EDM should be reusable as it stands.

8.1.3 Model Validation

The most crucial issue in any kind of modeling project is the validation of the resulting model. In contrast to a specific information model, a *model framework* like C²EDM can not be validated by, for example, assessing the functionality of the software built on the basis of the model. Because of a lacking theory of design the model framework could be validated against, *validation* has to rely on a (very large) *number of use cases* which should span all possible yet unknown contexts the model will be used in.

Obviously, any validation based on use cases is always incomplete and not fully satisfactory because only a relatively *small number* of *real-world problems* can be practically addressed. Furthermore, it is not sufficient to work out complete *software* solutions but their *evaluation* in *field studies* with real industrial design processes would be required in any serious validation effort. Such a validation should be even assisted by an evaluation of the usability of the resulting software as discussed and exemplified on several IMPROVE software prototypes in Sect. 5.6.

It is very *difficult* to *meet* these *validation requirements*, even by a large research project like IMPROVE with close links to industrial practice. This inherent limitation – though rarely mentioned in the literature – is a major obstacle for scientific progress in application domain information modeling.

The following paragraphs *summarize* our efforts to use the information model to improve design processes in industry and hence to contribute to a *validation* of the modeling framework.

Model-Based Re-engineering of Design Processes

The work process *modeling procedure* (cf. Subsect. 2.4.2) and the work process *model* (cf. Sect. 2.4) have been *applied* in a number of *industrial case studies* which were quite different in nature. Participative work process modeling with active participation of all stakeholders in the design process has been confirmed to be decisive for getting a realistic and largely complete representation of the actual work process.

The software *tool WOMS* has proven to be *appropriate* and effective to support the modeling process in an industrial environment. WOMS has different capabilities to present the work process model to the engineer. These views have shown to be helpful to develop, better understand and analyze the work process. However, the complexity of the work process model often hinders transparency. Hence, *additional* structuring and abstraction *mechanisms* have to be developed and integrated in WOMS in on-going research work to solve this problem.

WOMS only provides limited means to analyze the work process. Often a qualitative analysis resulting from a careful inspection of the model by the stakeholders involved is not sufficient. Rather, *quantitative performance measures* are of interest. Such measures can be deduced from Monte-Carlo simulation of the work process models, which can be cast into a discrete-event simulation model using one of the established formalisms (i.e. Petri nets, sequential function charts, etc.).

Simulation requires an *enrichment* of the work process model by *quantitative information* of *different kind*. Examples include the duration of certain activities or the number and type of tools allocated to a certain design task. Such information allows to investigate the time required to accomplish a part of the design process or to study the benefit of employing such tools. Such quantitative data are often hard to get with the desired accuracy in industrial practice.

To reduce the impact of such uncertainty, different case studies have to be carried out for different scenarios. Such *simulation studies* have not only been successfully carried out during IMPROVE but are still in progress during one of the on-going transfer projects. The simulations have been very useful to *analyze* the status quo and to identify concrete measures to *improve* the design process.

All the case studies have revealed that *coarse work process models* like the ones obtained in participative work process modeling by very limited effort already *reveal* the most important *characteristics* and *uncover* the *improvement* potential. Hence, design process modeling is a useful and rewarding investment and largely independent from a later use during the construction of design support software.

The design process knowledge in a tool like *WOMS* is an asset in itself which should be reused in subsequent activities during the design process. Consequently, the modeling tool should provide functionality to *export* the *work process model* to other tools such as discrete-event simulators, workflow engines or even control system design software. This export cannot be confined to providing the model in a standard data format (like XML). Rather, the model has to be refined and extended to fulfill the information requirements of the target application. Furthermore, it has to be mapped to the data format of the target tool which uses the work process model.

First steps toward this objective have been carried out in the context of work process simulation during IMPROVE and an on-going transfer project (cf. Sects. 5.2 and 7.4). A more general approach to the export and reuse of design process knowledge is pursued in another transfer project (cf. Sect. 7.3). There, the work process modeling tool *WOMS* is redesigned and reimplemented using *semantic technologies* to support a gradual extension and refinement of work process models and to map them to the format of a target tool. The modeling tool will be evaluated in three industrial case studies relating to work process simulation, to recipe design for batch processes and to the design of operating procedures and control strategies.

Design Process Models and Tool Implementation

The application domain model forms the upper layer of the integrated product and process model which has been one of the major scientific targets of

IMPROVE. *Model-based software engineering* is supposed to start from such a model and to refine it to ultimately cover all the information required for tool construction. Such a refinement and extension should not be done from scratch for every tool construction project, but should rely on a generic and reusable model which is consistent from application to platform layer.

As argued elsewhere (cf. Chap. 6), this ambitious objective has not been fully achieved by the IMPROVE project. However, a number of case studies have been carried out to show in concrete projects how *model refinement* and *extension* of the application domain model can be successfully carried out to *assist* the *tool construction process*. In particular, the product data and the document models have been successfully employed and thus at least exemplarily practically validated in two software projects during the IMPROVE project targeting at information integration and document management. Further testing of the product data and document models is currently under way in one of the transfer projects (cf. Sect. 7.2). The objective of this project is to demonstrate the use of semantic web technologies and application domain ontologies to integrate and consolidate information across the project lifecycle of process and plant design.

The application domain model has also been employed to *guide* the *development* of *design support* in a less ambitious manner. Rather than aiming at a consistent model-based software development process, parts of the model have been directly used, for example, to design interfaces for data exchange between different simulation and visualization tools in chemical process and extrusion process simulation (cf. Sects. 5.3 and 5.4). The data model has helped the definition of standard data formats to facilitate both the extension and the replacement of tools in an integrated simulation and visualization environment. These applications demonstrate the value of information modeling even in relatively simple data integration projects which are routinely carried out in industrial practice today.

8.1.4 Conclusions

Application domain modeling clarifies understanding and *meaning of concepts* and their relations in an engineering domain. Such a common understanding is an important prerequisite for success in process and any other engineering design activity. Hence, an application domain model is an asset in itself.

Application domain *modeling* is a time-consuming and *expensive effort* which requires an appropriate conceptual foundation to be carried out effectively. The research in IMPROVE has lead to a *general* architectural *framework* for application domain modeling which can serve that purpose. C^2EDM can be directly used, modified, refined and extended in chemical engineering and other engineering domains. There are various levels of sophistication in using the model.

The most ambitious objective pursued in IMPROVE is to use it as *part* of a comprehensive *model-based* software engineering and *tool construction process*. C²EDM provides all the necessary information to serve this purpose. However, as briefly summarized in this section and detailed in other parts of this book, parts of the application domain model can be used in a less ambitious manner to successfully support various aspects of *design process improvement* and design software construction.

8.2 Review from a Tools' Perspective

M. Nagl

Abstract. IMPROVE can also be regarded as a tool integration project with a broad understanding of "integration" (cf. part I of this book). This section is to *summarize* our *findings on tools* by regarding *four* different *views* on these tools: (a) contributions of IMPROVE to a better support of development processes by using tools, (b) lessons learned from tool construction/integration, (c) how tool construction and modeling of development processes interact, and (d) how application-specific or general our results on tools are. The review is restricted to the academic viewpoint. The industrial review for tools – but also for other perspectives – is given in Sect. 8.3.

8.2.1 Introduction

IMPROVE has a *broad understanding* of *integration* (cf. Subsects. 1.1.2 and 1.1.3): Not only system integration based on advanced platforms is regarded. Our understanding of integration is clearly extending the classical integration dimensions [1038]. Especially, we regard enhanced functionality on top of given or even of new tools. The corresponding extensions have been introduced in order to support collaborative development processes or even processes being distributed over different departments or companies. In the preface of Chap. 5 we have summarized all integration aspects regarded in this book.

The main *problems* of *integration*, regarding tool support, have been sketched in Fig. 1.6. There, we find all current gaps of tool support we should bridge by extended tools. The corresponding integration *solutions* are to be found in the main Chaps. 2 to 7 of this book.

This *section* describes the results of IMPROVE on tools from four different perspectives: (a) What contributions has IMPROVE delivered for a better support of design processes in chemical engineering? (b) What have we learned about tool construction/integration in chemical engineering? (c) The whole tool construction/integration approach has to be in close relation with modeling processes on different layers. What benefit did we gain from looking on tool behavior or tool construction from such a modeling perspective? (d) Finally, we regard the question, how specific the tools are, which were presented in this book: Are they only applicable to chemical engineering/plastics processing? Or are the tools or their underlying concepts also applicable to other engincering disciplines? These four *questions* are *addressed* in the following subsections.

This review is given from the academic viewpoint. The industrial review – also for tools – follows in Sect. 8.3. Moreover, in the preface of this chapter we stated that we are not going to repeat the PPM summary of Chap. 6. However, the *interaction* of *modeling* and *tool construction* is discussed in this section. Hence, the *plan* of this section is to summarize all results on tools which are to be found in Chaps. 2, 3, 4, 5, and 7 of this book from an academic perspective.

8.2.2 A Better Support for Design Processes

Design Tools Do Fit Better

A better support for design processes by results of IMPROVE comes from the fact that new tools or new tool functionality on top of given tools are both "derived" from investigating the application *domain* and existing development *processes*. Especially, the deficits and needs in these processes and how to better support the corresponding situations were studied.

As an example, this is to be seen from Sect. 2.4 where a schema for work process modeling is given. All the methodological *hints* "how to do" or the *shortcomings* "what to avoid" are *beforehand identified* from the application side.

Even more, if tools are available, their *usefulness* from the *ergonomy* point of view is also regarded. This was carried out for existing tools, for tools extended within IMPROVE, or new tools built within IMPROVE (see Sect. 5.6).

Both, regarding the needs of a domain and evaluating the ergonomy of tools is related to overcome gaps (a) and (b) of Fig. 1.6. *Tools* according to both perspectives are more useful, as they are *more specific* for the regarded domain or the processes within that domain. They are more semantical in the sense of giving deeper support for one developer in the design process (gap (a)). They are also better suited for the cooperation of different developers in that process (b). All these topics are addressed in Chap. 2. But tools also do *better fit* the ergonomic needs of human developers and their cooperation (see again Sect. 5.6).

New Tool Functionality on Application or Informatics Side

Four *new* informatics concepts were introduced in Sect. 1.1 and discussed in detail in Chap. 3: Direct process support based on developers' experience, consistency control between different developers' results based on underlying structures, direct communication between developers that is related to organized cooperation, and reactive management being able to cover the dynamic situations within design processes. These concepts are new and give valuable support within development processes, as to be seen from the demo description of Sect. 1.2.

Firstly, these concepts are *valuable* for *single developers*: A developer can use his experience, regard the consistency of his results with those having been elaborated earlier, he can find the right discussion partner in case of problems, he can evaluate/manage a change of the process as a manager.

The concepts are also *valuable* for *team activities*: Process chunks of direct process support can also give help for the activities of different developers collaborating on a certain topic. Consistency between the results of different developers is a question for all developers, especially important if they have different roles. Moreover, in some cases the result of a developer may depend on the result of different developers. Direct communication may take place in the form of a conference of several developers, where they discuss a situation to be changed. Changes due to dynamics may come from results of one single person, but usually influences a group.

As explained in detail in Sect. 5.5 these new tool functionalities can again be synergistically integrated. There is no other literature available where the integration of novel collaboration concepts is studied (two-level integration). Hence, *synergy* is definitely a topic of success to be mentioned in this section reviewing our achievements on tools.

Another topic is also remarkable: We studied the aspect of development processes *across* different *departments* of a company or even *different companies* and we developed corresponding tool support. This aspect was addressed on one side in application projects of IMPROVE as, for example, denoting processes by the WOMS tool and discussing them (see Sect. 5.1). This is especially helpful for distributed processes, which are less understood and, therefore, cause many problems in industrial practice. It was also addressed in a more general way, e.g. when studying how the management of dynamic processes looks like, if they include different companies (see Sect. 3.4). Of course, a lot of questions of cross-company support still remain to be studied.

There are further *application-dependent tools*, either invented or developed within or outside of IMPROVE and used therein. In any case, they have been refined due to this use. One such example is the heterogeneous simulation of the chemical overall process (see Sect. 5.3) which unites different single simulations to one aggregated simulation for a whole chemical process. Another topic is a specific simulation approach for plastics processing (see Sect. 5.4) which especially helps to bridge the gap between chemical engineering and plastics processing. Both have been used and extended but not invented within IMPROVE. Other tools are specific for IMPROVE (see e.g. Sects. 5.1 and 5.2).

Again, all the new application-dependent tools, but also all the new informatics concepts and corresponding functionalities of above give better support and *help* to *overcome* the *gaps* of Fig. 1.6: The activities of single developers are better supported: There are new tools available which have not existed before, or given tools have been extended. These tools now better fit the activities of a single human (gap (a), e.g. by using the personal experience of a developer, see Sect. 3.1). The same arguments hold for the cooperation between different developers which is now better supported (see gap (b), e.g. if they are discussing a new chemical process model in a conference, see Chap. 2 and Sect. 3.3). Furthermore, the gap between a group of developers on one hand and the management of the development process on the other hand is better supported (c), for example by using reactive management to solve a problem from the interface of chemical and plastics engineering, see 3.4 and 5. Finally, our investigations specifically addressing the question how cross-company support can be offered, see again gap (c) of Fig. 1.6, have given new results (Sects. 3.4 and 6.4).

8.2.3 Lessons Learned from the Construction/Integration of Tools

There are different kinds of tools described in this book and, also, *different kinds* of *knowledge* how to construct tools. There are (a) end user tools, either developed from the application side or the informatics side, both for bridging the gaps of insufficient current support. There are (b) tools for platforms or wrapping, both facilitating the process of integrating new or given tools. Finally, there are (c) tools supporting the construction process of new tools or tool extensions. Of course, most activities have gone to category (a).

We are going to shortly summarize the *results* of all three *categories*. The reuse aspect of tool construction is summarized later in this section after we have discussed how modeling has influenced tool construction and how general/specific the investigated tools are. It should be remarked again that the tool construction activities are not finished, the ongoing transfer projects are expected to produce further results.

Knowledge How to Construct End User Tools

We can give a short summary, as the functionality of these tools has already been summarized above. Again, *tool* construction and extension *knowledge* on one hand comes from realizing the new informatics concepts (direct process support, ..., reactive management). For either of these functionalities and corresponding tools there is *methodological* knowledge how to realize these tools in a *mechanical* tool construction process. This was explained in Sects. 3.1 to 3.4.

The *synergistical* integration of these new concepts was also addressed. The solution at the moment is a handcrafted one without reuse aspects. Furthermore, *cross-company* support was addressed, mostly in Sect. 3.4, also by giving *first* and not final *solutions*. So, the reuse aspects of synergistical integration need further investigations as well as the support of cross-company processes.

The tool experience of the *application side* is on one hand from tools specifically realized within IMPROVE (as, e.g., WOMS, see Sect. 5.1; or the simulation tool of labor research, see 5.2). On the other hand it is from tools which have been used and refined (ROME, see 5.3; simulation for plastics processing, see 5.4).

Knowledge on Platforms, A-posteriori Integration

The underlying platforms for tools have been introduced in Chap. 4. Sections 4.1 and 4.2 answer the question, which basic services should be provided by a more general platform. There are the *general service* and *homogeneity* aspects of platforms: How do data services and computation services look like, if they are designed to abstract from the diversity of different types of data stores or the heterogeneous computation facilities in a distributed and networked environment?

The second aspect of platforms studied in this Chap. 4 is how to give *specific support* for the tools one layer up of platforms: As an example, we take Sect. 4.1, where specific features are explained in order to extract product or process traces, needed in Sect. 3.1 to bookkeep a developer's activities but also to find out new chunks to be incorporated into direct process support tools in the next step.

Both aspects *facilitate* the *construction* of new *tools*: On one hand general and uniform service functionality is available, such that the tool builder need not use specific services and platforms. On the other hand, specific basic functionality (as trace bookkeeping) for specific tools is given.

Know-how for a-posteriori integration is found in Sect. 5.7. There, we find (a) tools helping to solve the integration problems on top of given tools even in a *distributed* context. The corresponding *architecture modeling tools* especially help to find the *concrete and distributed architecture* for an integrated environment of tools. They help to start from a rather abstract integration situation and to get down to a detailed technical description.

In the same section we also find (b) support for integrating given tools by wrappers. *Wrappers* are *constructed* in some *mechanical* way: Exploration of the interface of given tools, using the found sequences of actions of explorations in order to build new wrapper actions, after specifying their export interface and generating the corresponding realizations.

Knowledge on Tools for Tool Construction

Within Sects. 3.1, 3.2, and 3.4 we also find descriptions of *tools* having been used *for* the *tool construction* process.

In Sect. 3.1 we apply *modeling tools* being based on a common meta model in order to describe the situation to be supported by tools and how this support should happen.

In Sect. 3.2 we find tool support for integrator tool construction on two levels: Firstly, we find tools in order to parameterize integrator tools, i.e. to adapt them to a specific application situation, as e.g. the *correspondence* editor. Secondly, we find *generator* tools, for generating code for a specific integrator specification.

A rather analogous situation is found in Sect. 3.4: Again, there is a *modeling environment* in order to define and adapt design process knowledge for a specific context. So, parameterization is done using this tool. Furthermore, there is another *evolution* tool to extract process knowledge from instance level. As above, generator tools are used to easily get the implementation from high-level specification describing changes of development process situations.

8.2.4 Lessons Learned from Concurrently Modeling and Realizing Tools

This subsection deals with *two aspects*: (1) How application domain modeling influences tool functionality and tool construction and, furthermore, (2) how conceptual modeling drives reuse in the tool construction process. There are limits in our results in direction of a well-understood realization process with reuse, which are also described here.

This subsection is not to repeat the results and open problems of Chap. 6 on modeling. Instead, the *implications* of *modeling* on *tool construction* are to be sketched.

Application Modeling and Tool Functionality

It is new that *application* domain *models* are introduced within the tool construction process. The standard practice is that the tool developer realizes tools, he believes to be useful. Usually, there is only some imagination in form of an implicit and informal application model. In this book we started by *explicitly* defining models on the application side, consisting of a document model, product data model, work process model, and decision model (see Chap. 2).

The *transition* from these models to the necessary input for tool construction, namely an *UI-model*, is not provided. Such an UI-model would consist of a description of tool commands, entities handled by these commands, changes due to commands, as well as UI layout details. Such an extensive UI-model is *not* completely *determined*, as already stated in Chap. 6. So, the transition from application models to tools can only be done manually by adding the information not to be found on the application level.

An *example* for this *transition* can be found in Sect. 6.3 for integrators, where the current steps but also the formal future steps of the transition are described. There, one finds a sketch, how the application domain models could be used and extended in order to end up with an elaborated UI-model describing the behavior of an integrator.

As already stated, the above description is as things should be, not necessarily as things are. The ideal state mainly applies to the three columns of modeling and tool realization sketched in 6.2, 6.3, and 6.4, respectively. For other tools this *coherent modeling*, of the application domain and the conceptual models of tools, does *not apply*. It does also not apply to synergy (Sect. 5.5), or tool integration for a distributed development process, see Sect. 5.7. In the latter cases, tool realization is more done by careful but nevertheless handcrafted programming, or using modeling, however not starting at the application level.

Tool Modeling on Conceptual Level

A formal and abstract description of the internal structure and behavior of tools, in Fig. 6.1 called layer 3 and elsewhere *conceptual modeling for tools*, has two big advantages: (1) It allows the interpretation of this description in order to quickly get to prototyping tools. (2) It is also the basis for more efficient realizations of tools, where the code is generated being based on the formal specification.

If we take the *platform level* into account or the mapping onto that level (see layers 4 and 5 in Fig. 6.1), then we have a higher level for data and computational processes. This makes the *specification* of tool behavior *simpler* and it also facilitates the *generation* process, as more powerful target functionality is at hand.

The state of conceptual *realization modeling* is described in Sect. 6.5, with a long list of open problems to be solved. Nevertheless, it is a *big step* forward what was achieved in IMPROVE, compared to just programming the tools. Especially, it is remarkable that such techniques were applied in a project, where industry was and is involved. Up to now, specification and generation of tools was/is mostly applied in mere academic tool projects.

Trivially, for all the above cases of *minor understanding* – single tools not developed according to the above methodology, synergy of new concepts, integration of tools for distributed development processes – we just *program* and do not specify, interpret, or generate.

Résumé: Modeling and Tool Realization

What we have achieved by introducing different formal modeling levels is currently called *model-driven architectures*, here in this book being applied to tool construction. However, our examples show that the transition between levels cannot be accomplished automatically.

We have integrated model development and tool construction. There are *islands of understanding* of *tool construction* (the three columns of Chap. 6) and other areas where we just use conventional programming in order to get solutions.

The insight we have got corresponds to the *transition* from application domain models to precise tool descriptions, and from there to a complete *conceptual model* describing the details for an executable tool. As mentioned, there are many places, where we just handcraft.

The deep transition *knowledge* we have developed for the intertwined tool modeling and construction process not only applies to *end user* tools in chemical engineering. It can also be used for tools being applied in the *tool construction* process itself, namely for tools adapting tools to a specific context. Deep knowledge on conceptual modeling can also be used for specification, interpretation, and generator tools (see e.g. [334]).

8.2.5 Tools Independent of or Specific to a Domain

The gaps of support we have found and discussed around Fig. 1.6 and the solutions to overcome these gaps are *not specific* to *chemical engineering*. There is a proof for this statement, as different groups involved in this book also worked for supporting design or development processes in other areas, as mechanical engineering, communication engineering, software engineering, and alike.

For the *transfer* of the results of this book to other domains, there are *two possibilities*: (a) We can apply the knowledge to get tools in the other domain. Knowledge is here to be understood in a brought sense, including know how, specification techniques, generators, etc. (b) Transfer might also mean that we can eventually use the same tool, after some kind of adaptation or parameterization.

Transferability of Results to Other Engineering Disciplines

There are results directly transferable to other engineering disciplines. For example, the novel informatics *support concepts* of Sects. 3.1 to 3.4 can also be used for development processes outside of chemical engineering. New or extended tools (on top of existing ones) can be realized using these *support concepts*.

In the same way, the *methodological knowledge* how to implement these tools by using platforms, by using modeling on different levels, by using application knowledge, if available, can be used. This applies to the "islands of understanding" of *tool construction* as well as to direct programming.

Transferability applies to tools of the informatics side, as just described. It also applies to the tool *results* of the *engineering partners* within IMPROVE. Although the tool support is more specific – as mathematical modeling for bridging chemical and plastics engineering, see 5.4, or heterogeneous simulation by aggregating local simulations, see 5.3 – the basic ideas are transferable as well.

Therefore, many of the tool results of this book can be transferred to *other engineering domains*. Others are even directly applicable, as to be seen. So, the book is also addressing readers outside of chemical engineering. This statement is partially verified by looking on the cooperation partners of our technology transfer activities described in Chap. 7.

Generic Tools and Parameterization

There are *general tools* which are not specific for an application domain and, therefore, *applicable* to *any domain*. An example is EXCEL, a broadly used general tool. We find such general tools also as results of IMPROVE. The WOMS tool to denote and discuss work processes is of that type (see Sect. 5.1) as well as the tool for labor capacity simulation (see 5.2).

If some adaptation of a tool to be used in some context is necessary, we distinguish two different forms. On one side, *realization adaptation* is necessary, as to connect to a specific platform, a specific component for data management or storage etc. These topics have to do with porting a tool to another *context*. Thus, these topics are not dependent on an application domain, but on the infrastructure context in which a tool is used.

On the other hand, tools have to be parameterized to fit to a specific *application domain*, but also due to other reasons, as user behavior, specific rules of a company, etc. This is, for example, the case for the reactive management tool of Sect. 3.4. *Parameterization* there means to introduce types of documents, of subprocesses, but also cooperation forms. So, parameterization here means to make the tool suitable in its outside behavior.

Parameterization can be done as a *preparation step* before any use. For the management tool of Sect. 3.4 this was shown by using the modeling environment. The determinations made with this environment are then used to generate a suitable management system. The second possibility is to parameterize a tool during execution time by the end user. We find this *runtime parameterization* as one possibility in Sect. 3.2 on integrators.

Parameterization can have different *granularities*: A tool is "locally" parameterized for a specific user, specific habits in a department, or specific knowledge of a subdomain. Moreover, a tool can also be parameterized for the cooperation of different people from different subdomains. An even more comprehensive form is, if a complete development process is parameterized, implying local parameterizations of used specific tools. The latter might happen also for distributed development processes where subprocesses are carried out in different companies, which needs adaptation to the habits of these different companies.

Any solution for adaptation/parameterization needs *clear interfaces*: When porting a tool, a specific component has to be bound below of a general interface. When parameterizing to a specific context, specific determinations have to be inserted on top of a generic interface.

8.2.6 Summary and Conclusion

There are a lot of *tool results* to be found in this book and the tool construction knowledge, presented here, is rather comprehensive. Tool construction is the second *global goal* of IMPROVE, besides modeling development processes. As we have seen, these two global aspects even do interact.

Summary on Tools

The elaborated state of the art is due to the *comprehensive view* on *tools* and their construction, we have taken in IMPROVE:

- We derive tools which fit to explicit application domain models.

- Tool support addresses all granularities of processes and products: fine-grained work processes of single developers and their corresponding detailed results, fine-grained dependencies between different personal processes and products, medium- and coarse-grained management processes, and the interaction of fine-grained technical and coarse-grained management processes.
- Tool support addresses the different aspects of design processes and products. Let us take the processes as example: WOMS supports denoting work processes and discussions about them on different levels of granularity (see Sect. 5.1). Direct process support allows to use the experience of a developer on a fine-grained technical level (Sect. 3.1), whereas reactive management regards medium-grained processes for the whole project (see Sect. 3.4). The topic of simulation according to resources on the same granularity level is studied in 5.2. Finally, working areas on domain level regard processes of a coarse-grained form (see Chap. 2).
- Even more, support is also given for the parameterization/adaptation of tools (e.g. parameterization before or during execution), not only for using them.
- Tool construction is using reuse techniques. Very elaborated forms of such techniques have been presented.
- The idea of building tools on elaborated platforms was strictly followed.
- Support of distributed development processes was also intensively studied.
- Tool construction was regarded to be integrated with modeling on different levels.
- Transferability of tool results and parameterization of generic tools was in our mind from the very beginning.

Selection of Open Problems

Of course, there are many *open problems* on tools, which IMPROVE delegates to other researchers or research groups. Here is a *small selection*:

- Sects. 3.1 to 3.4 describe new support concepts for cooperation from the informatics perspective. Are there other forms, not addressed here? Also, we find heterogeneous simulation or specific mathematical modeling techniques for evaluating the complete plant, or for regarding specific aspects of plastics processing etc. Are there more application specific tools of that kind? Are there even some more tools which have a general character?
- IMPROVE concentrates on conceptual design and basic engineering. Are all the problems of development processes found in this part of the overall process, or do we find new problems/demands for tool support when regarding later phases of the chemical design process?
- Sect. 6.5 lists a lot of modeling problems still to be solved. These problems are also tool construction problems, as any progress in modeling allows to introduce elaborated reuse techniques for tool construction.

- Of course, the problems of distributed development across different companies are only touched in this book and not completely investigated. So, the essential question of cooperation on one hand and knowledge hiding on the other hand was only partially answered.
- The question of building general tools, which can be parameterized for a specific context, was addressed by different groups within IMPROVE. Can this be made to a uniform approach for tool construction, i.e. not to develop domain- or context-specific tools, but instead general and parameterizable ones?
- IMPROVE can be regarded as a big software tool/integration project. For this project, all support mechanisms studied in this book can be applied as well. So, for example, we could have used the experience of tool developers, integrators for modeling layer transitions, reactive management for the tool projects etc. What would have been our profit on those modified construction processes?

8.3 Review from an Industrial Perspective

W. Marquardt and M. Nagl

Abstract. This short section presents some thoughts on the relevance and the impact of the research work in IMPROVE on industrial practice. Though, the research program has been linked to industrial requirements and has seen a continuous review from industrial colleagues to refocus the research objectives and to assure practical relevance of the IMPROVE research approach, only few concrete results have made it yet into industrial practice. However, the problems addressed in IMPROVE have been receiving significant attention in industry, in particular in the recent past. In this sense, IMPROVE has been addressing a timely and forward-looking fundamental research program, which has come too early to be readily absorbed by industry. However, it will be of significant impact on industrial practice in the future in both, the software as well as the chemical and process industries.

8.3.1 Introduction

This section addresses the question how the research work in *IMPROVE* has had an *impact* on *industrial practice* in process and plant design from the chemical and process engineering as well as from the software vendors' perspectives. We will start this section with a brief discussion on business objectives and the relations between these industries, which obviously act as "constraints" in the process of transferring results from the research in IMPROVE to industrial practice.

The interface between the IMPROVE research project and industry has been complicated because IMPROVE addressed two *different* but interdependent *branches* of *industry* with *different* business *objectives*. The business objective of software tool vendors is to provide functional IT tools and services to their customers, the process engineering companies, and to support and improve the quality of engineering design processes. The business objective of the process engineering companies is to design processes and to construct plants for plant owners and operators with given functional specifications, at target cost and in a given time period. Process engineering software is in the first place considered a cost factor, which has to be justified to management. Though design software environments constitute an indispensable part of the engineering design platform, it is rarely seen as a competitive advantage because it is likewise available to any competitor at similar cost.

Software tool *vendors* will only further develop and *improve* their *software* products if there is either demand for new and improved functionality from their customers or if they can cut down development, maintenance, and deployment costs of their software products. The latter driving force is largely independent of customer requirements but determined by the advances in computer science and software technologies.

Process engineering software is complex in nature and targets a relatively small market. As a consequence, there are only few major product lines available in the market which are widely used in the process industries. The resulting *oligopoly hinders advances* in process engineering software. Given their market power, the software vendors typically try to offer integrated solutions satisfying all the needs of their customers. As a consequence, their software is deliberately not sufficiently open and often only provides low level interfaces for data exchange.

A single customer does typically not have a sufficiently strong market position to enforce software modifications or to even influence the software road map of a vendor. Consequently, *interest groups* have been formed by the *customers* of most major process engineering software packages to use the increased market power to jointly put the screws on the vendors. These interest groups gather and reconcile the sometimes conflicting requests regarding the improvement and further development of the software.

Obviously, this alignment process results in compromises and prevents the development of design support solutions which match the design process of an individual organization and hence contribute to a competitive advantage. Typically, these *requests* are *not addressing* software functionalities from the perspective of the *business process* in the engineering design organizations but rather focus on improvements of single tools from a user's perspective and on their external software interfaces. As a consequence, it is largely accepted that the commercially available design support software determines the way design processes are carried out to a significant extent.

There has been only *little reflection* on fundamental issues of how to manage and organize *design processes across* institutional and geographical *organizational boundaries* in the process industries in the last two decades. In contrast to the automotive, aerospace, and parts manufacturing industries, simultaneous and concurrent engineering have not been raising a significant level of awareness in the process industries.

Still, incremental changes have been continuously applied to the design process to react to the globalization of engineering services, the significant difference in the cost of engineering labor around the globe, and the growing competence level in process design and engineering in the developing countries. This continuing process of *globalizing engineering* skills will *result in* increasing cost pressure and hardening of competition, which is expected to initiate *business process re-engineering* projects in the engineering organizations in the future to improve the design processes in the highly industrialized countries.

In conclusion, the industrial structure has always been posing a substantial *challenge* to *transfer results* from the IMPROVE project to industrial practice and to impact the current status of process engineering design software on the one and the process design process on the other hand. In the following two subsections, we will assess the impact of the IMPROVE project on the software industry and the end-users industries in chemical and process engineering in two separate subsections.

8.3.2 Software Industry

The IMPROVE project has had a lot of *interaction* with software vendors all along during dedicated workshops, project meetings, and scientific conferences. Furthermore, nearly all tool subprojects of IMPROVE had a – more or less long-lasting – connection to industry. Despite these significant efforts, the IMPROVE project has *not* been able to *broadly influence* the developments of the major *vendors* of process engineering software.

The major *scientific objectives* of IMPROVE in the area of software engineering, though timely and relevant from a technological point of view, were, however, *not* always *in line* with the current business objectives and the technological road maps of the software industries. Furthermore, research in IMPROVE resulted in software *prototypes* to illustrate the advanced functionality and to demonstrate the *proof-of-concept*. This software has not been designed to be directly integrated in the software product portfolio of any vendor.

Such a direct transfer of the research results cannot and has not been the research objective of IMPROVE. Rather, such a *transfer requires dedicated* research *activities* involving a software vendor during all phases of the software development cycle, which aims at extending and modifying existing commercial software. Such research activities have been initiated and are described in Chap. 7.

In the following subsection, we will give a few concrete *examples* of how the *results have impacted* the developments in the process engineering software industries.

Integrated Process/Product Model

The objective of IMPROVE to develop an *integrated process and product model*, spanning the application domain, tool functionalities and software platforms with both a product data as well as an engineering workflow perspective, has been by far *too visionary* and forward-looking to get the chance of being picked up in the short-term by the software industries.

According to our knowledge, any kind of formal information model covering at least relevant parts of interest for model-based software construction has been completely lacking in the process engineering software industry when the IMPROVE project has been started and is most likely still not yet employed. Nevertheless, we know of a few *process engineering software vendors* who *have adopted* the IMPROVE strategy *at least to some extent* to *improve* their *software* engineering *processes*. For example, AspenTech has started a major modeling effort to guide the software development and the integration of third party products. This model has been the basis for a complete redesign of the Aspen Engineering Suite. Though the IMPROVE modeling results have not directly been used in AspenTech's project, there have been intensive contacts between IMPROVE researchers and AspenTech's software engineers on information modeling issues.

Information modeling is also a necessary pre-requisite for *data integration* either by means of some central data store or by some data exchange mechanism. Despite an early acknowledgment of the need for a standardized format for process engineering data and various attempts to develop according standardized data models (see Chap. 2), largely independent but often knowing of the IMPROVE activities, these data models have not yet found adequate attention in the process industries. The reasons for this lack of adoption is not completely clear. They seem rather related to company culture than to the maturity of the models or of the software technology employed. It is interesting to note that XML-based data models such as CAEX of ABB or PlantXML of Evonik Degussa have been adopted in industrial applications at least to some extent despite the restriction to product data and the limited applicability due to the lacking semantic foundation.

In conclusion, the *software industry* does *not* yet seem to be *ready* to absorb the more involved modeling concepts and software technologies like those brought forward in the IMPROVE project.

Tool Realization and Integration

The major process engineering software vendors have integrated their product portfolio into engineering design environments. *Tool integration* has been achieved largely manually in time-consuming and expensive software projects. The concepts of IMPROVE have been observed with great interest but have not been applied yet in practice.

The transfer into industrial practice can again only be achieved by concrete joint projects with direct involvement of industry as they are currently carried out. There are three active transfer projects dealing with transferring the *new* and elaborated *support concepts* of Sects. 3.1, 3.2, and 3.4 to make them *available* for *process tools* in *industry*: One is introducing experience-based processes at aiXtrusion, a process automation company for plastics processing (see Sect. 7.5). Two others are carried out with innotec, a tool vending company. On the one hand the idea of integrators is realized in an industrial setting (see Sect. 7.6). On the other hand reactive design process management is introduced as an additional functionality of a suite of tools (see Sect. 7.7).

This is regarded to be a success for IMPROVE. However, these *projects* are *not taking up the long-term philosophy of* IMPROVE's *tool construction process*: (a) Starting from application models to derive UI-models, (b) regarding tool construction as a model-driven process across different layers, (c) using the functionality of a distributed data and process platform within tool construction to improve portability, and (d) that tool construction could and should use elaborated reuse techniques as shown in Chap. 3.

However, these transfer projects took up the idea of IMPROVE that tool integration is more than just having different tools running in the same context. The three transfer projects adopted the idea that *tool integration needs new functionality* invented to improve the quality and the effort of design processes.

Whenever nowadays listening to a talk at a tools' conference the buzzwords *"integration"* and *"model-driven application development"* appear. The different facets of the integration problem and of model-driven development (see Chaps. 3, 4, 5, and 6) have not reached industry. So, the long-term message of IMPROVE was not acknowledged by industry yet. We are optimistic that it will finally be recognized. Both above problems, if a comprehensive solution is aimed at in industry, *will show* the *value of IMPROVE's solutions*.

A-posteriori Integration

Only the smaller companies are interested in *a-posteriori integration of third party products* into an integrated process engineering design environment. In contrast, the major software vendors have always tried to offer their own solutions to their clients to strengthen their market position. Only in a few cases, dedicated interfaces to software of a few selected partners have been implemented. Transparent or even standardized open interfaces between process engineering software components cannot be achieved by a research project like IMPROVE. Rather, a large *consortium* incorporating not only research institutions but the major industrial players *is necessary* to achieve the definition of such standardized interfaces and their enforcement in the longer run.

The *CAPE-OPEN project* may serve for illustration [566]. This project has been funded by the EU in parallel to the IMPROVE project with participation of two IMPROVE research teams. The project successfully defined open interfaces for the major modules of process simulation software. Despite the fairly narrow scope, the financial effort has been substantial. The project could only be successful because of an active participation of the relevant software vendors and their clients who had to sort out not only the technical but also the commercial issues. The academic partners' role has been of an advisory and quality assurance type. Today, a few years after completion of the project, significant efforts are spent by a Co-LaN, a non-for-profit interest group which maintains and extends the standard [997]. Only due to continued efforts of Co-LaN, the CAPE-OPEN *standard* is *implemented* readily by the smaller software companies and only reluctantly by the major vendors.

The *concepts* and *methodologies* for *a-posteriori integration* developed and benchmarked by IMPROVE are of significant importance to those software vendors who offer niche products and hence do not yet have a strong market position. They can benefit significantly from the IMPROVE technologies which had to explicitly address the lack of open interfaces or incomplete knowledge on the data structure and control flows in the available interfaces of the commercially available tools provided by vendors with a strong market position.

The problem of a-posteriori integration does not only appear for tools. It is even more important in the context of *reengineering* of business administration software in the *framework* of a *service-oriented architecture*. Therefore, it is not surprising that the corresponding transfer project is carried out together with the software development branch of an insurance company (see Sect. 7.8).

8.3.3 Chemical and Process Industries

The *understanding* and *reengineering* of existing *process design processes* have been one of the ambitious *objectives* of the *IMPROVE* project. Obviously, actual business processes can hardly be influenced by the results of a research project, which aims at advancing fundamental concepts, methods, and exploratory tools. A change of actual business processes is typically triggered either by economical opportunities or threats. Such changes are often facilitated or even enabled by novel methodologies and tools.

Our *industrial contacts considered* research on design process modeling and improvement as well as process-oriented support tools as an interesting *academic exercise* of little industrial relevance, *when* the *IMPROVE* project *started*. At that time, workflow modeling and support have been introduced in management and business administration in the process industries. Workflow management or process-oriented support functionality has not been considered relevant for engineering design processes beyond the very simple workflow components of document management systems of that time. In the contrary, any monitoring, guidance, or even control functionality had the notion of superintending the well-trained and knowledgeable designers and of constraining their creativity.

Changing Mindset: New Look on Design Processes

The *situation* has significantly *changed* in the *recent past*. The process design and engineering process is considered to take way too much time to be competitive in the global market, the chemical and process industries are facing. A group of industrial and academic experts are currently preparing a conference to take place in summer of 2009 in Tutzing, Germany, to identify all suitable technological and organizational means to *shorten* the *design* and *engineering process* from a product idea to plant start-up by 50 %. This ambitious objective can obviously only be reached if various lines of actions are followed.

One of them is definitely the reengineering of the design process and its support by (work process-oriented) software tools. Established concepts of simultaneous or concurrent engineering are not considered to be sufficient to solve the problem. Rather, a *complete reorganization* of the *design process* emphasizing a sufficiently fine level of granularity is required.

A complete prescription of the design process is not appropriate, rather, some degree of *indeterminism* and *reactivity* has to be accepted to cope with the predictably unforeseeable during the design process. The decision-making power of the individual engineer has to be constrained in those cases where best practice solutions can be reused and where more than one equally good solution is possible. Hence, *de-individualization* and *standardization* have to be carefully enforced where appropriate.

Various measures have to be *regarded*: Design automation as well as automatic design checks have to be put in place to address routine tasks. Supervision and control of the design process as well as proactive risk management requires integrated and consolidated views on the design data. The level of abstraction reflected in these views have to comply with the roles of the various stakeholders in the design process such as the project manager, the task managers as well as the design engineers. The interfaces between the various organizational units participating in the design process need particular attention to avoid time-consuming consolidation of the product data as well as of the design process itself.

The fact that this conference project is industrially driven clearly demonstrates a *new mindset* in the chemical and process engineering. This new mindset is very much *in line with* the *initial objectives of* the research carried out in *IMPROVE*. Many of the concepts, methods, and prototypical tools resulting from the research in IMPROVE form a very good starting point to address and implement a significant reduction of the time needed to build a chemical plant.

In addition to this changing mindset, some concrete developments in better supporting and improving the design process have been or are currently carried out. Two exemplary areas are discussed in the following paragraphs.

Integration and Consolidation of Product Data

The integration and consolidation of the design data during a project has always been an issue in the process industries. *Standardized data models* have been advocated as one building block for product data integration and consolidation. Despite the various substantial efforts reviewed in Chap. 2 of this book, there has been very *little progress for a long time* for several reasons. In particular, some industrialists believed that the work processes and the document styles are significantly different to prevent standardization at reasonable effort. The value of a common data model has been initially underestimated because of the limited exchange of detailed design data across organizational boundaries. In addition to this users' perspective, the early efforts in developing common and standardized product data models have not been successful. The apparent complexity could not be appropriately handled. The scope of standardizing all product data in complete detail was not reasonable, regarding the still immature data modeling methodologies and software technologies available at that time.

Today, there is a great demand for standardized data models not only for integration and consolidation but also for the exchange of data across institutional boundaries during the design process, for electronic procurement of equipment during plant construction and for the handover of the design package from the engineering company to the operator/owner of the plant. *Today*, it is *common sense* that the *standardization of product data* is *technically feasible* and *economically viable* if appropriate industry-driven consortia were formed. IMPROVE has obviously made only a modest contribution of this change in mindset.

The IMPROVE project has neither been in the position to deliver an industrial strength solution to product data standardization but came up with a *modeling framework* which not only comprises a sound starting point for a standardization project but also provides a methodology for compiling reusable data models for data integration and consolidation as well as for tool integration. The first industrial evaluation of the modeling approach is currently undertaken in a transfer project with Evonik Degussa and ontoprise as described in Sect. 7.2.

Design Process Modeling and Its Applications

The design process modeling problem has not been as prominent as the data integration and consolidation problem in the past. *Design process modeling and support* have no track record. Emphasis has always been only on management and administrative work processes which are different from engineering design processes. However, *awareness* for design process modeling and support has been *steadily growing* in the recent past, in particular in those companies who were in contact with the IMPROVE project for a long time.

The IMPROVE project has been able to *convince* a number of industrial *partners* that they should put more emphasis on systematically capturing, improving, and supporting their design processes. Various modeling *case studies* have demonstrated that a better understanding of the actual design processes can be of an enormous value for performance improvements, which often can be implemented at very little cost.

Consequently, design process *modeling and analysis* – either qualitative by simple inspection or quantitative by discrete-event simulation – is considered a simple and convenient means towards design process improvement. Once models describing the actual design process are available, they should be used in various contexts for different tasks during the project. The transfer projects described in Sects. 7.3 and 7.7 address exactly these issues. These projects may be considered a very good starting point to introduce the more sophisticated process support functionalities described in Sects. 3.1 and 3.4 into industrial practice.

8.3.4 Concluding Summary

The following concluding remarks hold true for *industrial recognition* of IMPROVE's results on *tools* by tool *vendors* as well as of its results on a better understanding of design *processes* and their *products* by the *process industries*.

The relevance of the *research agenda* of IMPROVE has *not* fully been acknowledged by industry *when IMPROVE got started*. One reason for this lack of attention has been the business climate of the time. There has not been sufficiently strong economical pressure to reevaluate the design practice and to question the way design support tools have been used. Another reason for the limited attention of industrial practitioners is the focus of IMPROVE on fundamental and long-term research objectives which should result in novel concepts and methodologies rather than in prototypical tools to be demonstrated and evaluated by industry.

The *prototypical tools* developed during IMPROVE have been perfectly serving the needs to demonstrate the power of the novel concepts and methodologies in IMPROVE. However, there has *not* been sufficient functionality built into these prototypes *to facilitate* application and evaluation during some part of an *industrial design process*. Last but not least, the IMPROVE software prototypes did *not* perfectly *fit* the software *infrastructure* in industry. Therefore, major coding effort would have been necessary to overcome this mismatch and to integrate the IMPROVE prototypes with existing industrial tools. An industrial evaluation of the IMPROVE research results would have called for functional prototypes integrated into a specific given industrial software infrastructure. This coding task would have been of a complexity which cannot be dealt with in any university research project even if substantial resources are available.

As a consequence, industrial practitioners could not be fully convinced that the IMPROVE approach delivers the concepts and tool functionality which is necessary to substantially improve actual industrial design processes. The lack of concrete tools industrial partners could "play with" also impeded communication between researchers and practitioners on requirements and possible solutions. The *transfer projects* have been setup in a manner to *address these deficiencies* and to effectively bridge the gap between fundamental research and industrial practice. In all cases, tool functionality is defined and developed in a joint research effort involving academic researchers and industrial practitioners. This tool functionality is put on top of existing process design software to guarantee a seamless integration and use in real design processes.

The changing mindset described in the previous section clearly shows that the *research objectives* of *IMPROVE* have been timely and *forward looking*. As often in the history of science, though to the point, the research results have not been absorbed by industrial practice because the time has not been ripe. The more recent discussions with industrial colleagues, for example during the preparation of the transfer projects or during the preparation of the Tutzing conference, clearly show that the IMPROVE project has been coming too early to result in a revolution in the way industrial process design processes are carried out and supported by information technologies.

Nevertheless, the research *results* documented in this volume will *find their way into industrial application* in due time because they have addressed the right problems and delivered the right concepts and methodologies. The corresponding ideas and concepts have been proved by prototypes, but not in an industrial setting. We are optimistic that this will be done in the next future.

8.4 Review from Academic Success Perspective

M. Nagl

Abstract. This short section aims at evaluating the academic outcome of the CRC IMPROVE from 1997 up to now and also sketches its further outcome as TC 61. We do this by regarding different perspectives: (a) The number and value of publications, (b) the contribution to international conference activities, and (c) how we laid the basis for or accelerated the career of young scientists. Joint activities together with industry and their consequences are discussed in sections 5.1 and 8.3.

8.4.1 Publications Output

Up to now, the members of the CRC IMPROVE or its successor Transfer Center have produced about 485 *publications*, most of them reviewed, which are related to the topic of this book.

Articles

The following table *classifies* the *article publications* of IMPROVE.

Table 8.1. Publications: classifying IMPROVE articles

Scientific Journal Publications	Workshop/Conference Proceedings' Publications	Book Contributions
88	217	84

Most of these publications are *specific* to a subtopic of IMPROVE, e.g. an article on "Conceptual Information Models for Chemical Process Design". So, they are the results of subprojects or of one group. However, in IMPROVE a big number of articles was also written in *cooperation* of two or even more groups. There is hardly another international project with such a big number of joint publications.

Also remarkable is the number of 23 *survey articles* on the IMPROVE approach or its results, produced within this joint project.

Books

Especially, the number of 89 *books* is worth mentioning, many of them Ph.D. theses. Please note, that in the Table 8.2 of this subsection most of the dissertations are not included. Books are only mentioned here, if they appeared as books of an internationally known publishing company. The most comprehensive collective result of IMPROVE in the form of a book is, of course, this volume. The following table gives the list of major book publications.

Table 8.2. Major book publications of IMPROVE

Bayer, B.: Conceptual Information Modeling for Computer-Aided Support of Chemical Process Design, VDI Verlag (2003)

Eggersmann, M.: Analysis and Support of Work Processes within Chemical Engineering Design Processes, VDI Verlag (2004)

Hackenberg, J.: Computer Support of Theory-Based Modeling of Process Systems, VDI Verlag (2006)

Jarke, M./ Lenzerini, M./ Vassiliou, Y./ Vassiliadis, P.: Fundamentals of Data Warehouses, 218 pp., Springer-Verlag (1999), 2nd Ed. (2003)

Jeusfeld, M./ Jarke, M./ Mylopoulos, J.: The Method Engineering Textbook, MIT Press, to appear 2008

Linnhoff-Popien, C.: CORBA – Communication and Management (in German), 370 pp., Springer-Verlag (1998)

Luczak, H./ Eversheim, W.: Telecooperation: Industrial Applications in Product Development (in German), Springer-Verlag (1999)

Luczak, H./ Bullinger, H.-J./ Schlick, C./ Ziegler, J.: Support of Flexible Production by Software-Methods, Systems, Examples (in German), Springer-Verlag (2001)

Nagl, M./ Westfechtel, B. (Eds.): Integration of Development Systems in Engineering Applications – Substantial Improvement of Development Processes (in German), 440 pp., Springer-Verlag (1999)

Nagl, M./ Westfechtel, B.: (Eds.): Models, Tools, and Infrastructures for Supporting Development Processes (in German), 392 pp., Wiley-VHC (2003)

Nagl, M./ Marquardt, W. (Eds.): Collaborative and Distributed Chemical Engineering: From Understanding to Substantial Design Process Support, 871 pp., Springer-Verlag (2008), this volume

Pohl, K.: Process-Centered Requirements Engineering, 342 pp., Research Studies Press, distributed by Wiley & Sons (1996)

Pohl, K.: Continuous Documentation of Information Systems Requirements, Habilitation, RWTH Aachen, Teubner-Verlag (1999)

Schopfer, G.: A Framework for Tool Integration in Chemical Process Modeling, VDI Verlag (2006)

Westfechtel, B.: Models and Tools for Managing Development Processes, Volume 1646 of LNCS, 418 pp., Springer-Verlag (1999)

In addition, there are some further books on projects either being a predecessor project, as [201, 334, 352], or being parallel projects [51, 277] with a more or less close connection to IMPROVE.

8.4.2 Conferences Output

IMPROVE's CRC 476 and TC 61 have contributed to many *international conferences* or *workshops* by giving presentations. Here, we only list those international conferences/workshops which have been heavily influenced by the presentation of IMPROVE results as, for example, there was a complete section of IMPROVE contributions. Moreover, researchers, active in IMPROVE, have organized a lot of further international or national workshops/conferences, having a close connection to IMPROVE.

The following list only contains the *important events*. The industry workshops of IMPROVE in Germany have already been mentioned in Sect. 7.1 and, therefore, are not repeated here.

Table 8.3. Conferences/workshops influenced or organized by IMPROVE

5th Intl. Conf. on Foundations of Computer-Aided Process Design, Breckenridge, CO, USA, 1999 [828]

Symp. on Models, Tools, and Infrastructures for the Support of Development Processes (in German), Aachen, 2002 [353]

8th Intl. Symp. on Process Systems Engineering, Kunming, China [593], 2003

Intl. Workshop on "The Role of Empirical Studies in Understanding and Supporting Engineering Design", Gaithersburg, MD, USA, 2003 [430]

16th European Symp. on Computer-Aided Process Engineering (Escape-16) and 9th Intl. Symp. on Process Systems Engineering, Garmisch-Partenkirchen, 2006 [302]

Intl. Workshop AGTIVE'99 on "Applications on Graph Transformation with Industrial Relevance", Kerkrade, The Netherlands, 1999 [350]

Intl. Workshop AGITVE'03 on "Applications on Graph Transformation with Industrial Relevance", Charlottesville, VA, USA, 2003 [363]

Intl. Workshop AGTIVE'07 on "Applications of Graph Transformation with Industrial Relevance", Kassel, 2007 [416]

GI Conf. INFORMATIK (in German), Aachen, 1997

Seminar on "Scenario Management", Informatics Center Dagstuhl, 1998

Intl. Conf. on Advanced Information Engineering (CaiSE 99), Heidelberg, 1999

Intl. Conf. on Conceptual Modeling (ER 99), Paris, France, 1999

Intl. Conf. on Very Large Data Basses (VDB 2000), Cairo, Egypt, 2000

European Conf. on CSCW (ECSCW 01), Bonn, 2001

Systems Engineering (CaiSE 02), Toronto, Canada, 2002

There are *many other* conferences/workshops influenced or organized by IMPROVE members without having a close relation to IMPROVE. Therefore, they are not given here. The annual reports of IMPROVE groups report on this activities.

8.4.3 Supporting Young Scientists' Career

The most honorable task of university research projects is to *promote* the *career* of young scientists. In the following we concentrate on *two measurements* of career promotions, namely the number of dissertations which have been finished and the number of professorships which have been taken by young researchers of IMPROVE. There are numerous young scientists who have gone to industry and have reached highly-respected positions there. They are not further mentioned in this section, which only deals with academic success.

Doctoral Dissertations

The following table contains the names of *young scientists* who have finished their *Doctoral Dissertation* in connection with IMPROVE. The list also contains the names of some scientists who have successfully finished their "Habilitation", a specific procedure in Germany and some other countries, by which the home university certifies the ability of a young researcher to fulfill the duties of a university professor. The corresponding names are marked by an "H". The list is in chronological order and, within one year, ordered according to the groups of IMPROVE.

Table 8.4. Dissertations/Habilitations of young scientists of/or being related to IMPROVE, by years and groups

Hoff, S., Inf. 4, 1997
Gallersdörfer, R., Inf. 5, 1997
Nissen, H.W., Inf. 5, 1997
Szcurko, P., Inf. 5, 1997
Lohmann, B., LPT 1998
Sattler, U., Inf. 1/LPT, 1998
Reimann, S., IKV, 1998
Stahl, J. IAW, 1998
Krapp, C.-A., Inf. 3, 1998
Westfechtel, B., Inf. 3, 1998, H
Fasbender, A., Inf. 4, 1998
Karabek, R., Inf. 4, 1998
Linnhoff-Popien, C., Inf. 4, 1998, H
Herbst, D., IAW, 1999
Schlick, C., IAW, 1999
Baumann, R., Inf. 3, 1999
Behle, A., Inf. 3, 1999
Cremer, K., Inf. 3, 1999
Gruner, S., Inf. 3, 1999
Winter, A., Inf. 3, 1999

Kesdogan, D., Inf. 4, 1999
Reichl, P., Inf. 4, 1999
Schuba, M., Inf. 4, 1999
v. Buol, B., Inf. 5, 1999
Dömges, R., Inf. 5, 1999
Nicola, M., Inf. 5, 1999
Pohl, K., Inf. 5, 1999, H
Baumeister, M. LPT, 2000
Molitor, R., Inf. 1/LPT, 2000
Depolt, J., IAW 2000
Radermacher, A., Inf. 3, 2000
Meggers, J., Inf. 4, 2000
Trossen, D., Inf. 4, 2000
Haumer, P., Inf. 5, 2000
Kethers, S., Inf. 5, 2000
Klamma, R., Inf. 5, 2000
Bogusch, R., LPT, 2001
Springer, J., IAW, 2001, H
Klein, P., Inf. 3, 2001
Büschkes, R., Inf. 4, 2001

Cseh, C., Inf. 4, 2001
Küpper, A., Inf. 4, 2001
Becks, A., Inf. 5, 2001
Weidenhaupt, K., Inf. 5, 2001
Wolf, M., IAW, 2002
Münch, M., Inf. 3, 2002
Schleicher, A., Inf. 3, 2002
Lipperts, S., Inf. 4, 2002
Stenzel, R., Inf. 4, 2002
Klemke, R., Inf. 5, 2002
Nick, A., Inf. 5, 2002
Bayer, B., LPT, 2003
Foltz, Chr., IAW, 2003
Jäger, D., Inf. 3, 2003
Fidler, M., Inf. 4, 2003
Schoop, M., Inf. 5, 2003, H
Schlüter, M., IKV, 2003
von Wedel, L., LPT, 2004
Bouazizi, I., Inf. 4, 2004
Park, A., Inf. 4, 2004

Thißen, D., Inf. 4, 2004
List, Th., Inf. 5, 2004
Eggersmann, M., LPT, 2005
Gatzemeier, F., Inf. 3, 2005
Marburger, A., Inf. 3, 2005
Imhoff, F., Inf. 4, 2005
Pils, C., Inf. 4, 2005
Hackenberg, J., LPT, 2006
Schopfer, G., LPT, 2006
Böhlen, B., Inf. 3, 2006
Kirchhof, M., Inf. 3, 2006
Meyer, O. Inf. 3, 2006
Becker, S., Inf. 3, 2007
Kraft, B., Inf. 3, 2007
Miatidis, M., Inf. 5, 2007
Haase, Th., Inf. 3, 2008
Heller, M., Inf. 3, 2008
Ranger, U., Inf. 3, 2008

Young Professors

There is also a remarkable number of persons who, after doing research within IMPROVE, got an appointment as a *university professor*. Table 8.5 contains the list in alphabetic and not chronological order.

Table 8.5. IMPROVE has "produced" a number of professors

Stefan Gruner, Senior Lecturer, University of Pretoria, South Africa

Manfred Jeusfeld, Assoc. Professor, University of Tilburg, NL

Claudia Linnhoff-Popien, Full Professor, University of Munich, D

Hans W. Nissen, Professor, University of Applied Sciences Cologne, D

Klaus Pohl, Full Professor, University of Essen, D

Christopher Schlick, Full Professor, RWTH Aachen University, D

Ralf Schneider, Professor, University of Applied Sciences Regensburg, D

Mareike Schoop, Full Professor, University of Hohenheim, D

Andy Schürr, Full Professor, Technical University of Darmstadt, D

Klaus Weidenhaupt, Professor, University of Applied Sciences Niederrhein, Krefeld, D

Bernhard Westfechtel, Full Professor, University of Bayreuth, D

8.4.4 Summary and Conclusion

This short section described the scientific *outcome* of the IMPROVE project. It was mainly done by means of *figures*: counting and classifying publications, conferences, and scientific success by Doctoral Dissertations, Habilitations, and academic positions.

In order to evaluate the *impact* of IMPROVE one would have to regard, how the results have influenced the scientific discussion on development processes in chemical engineering, or engineering in general, and on corresponding tool support and tool construction. This is much harder to evaluate. A rather superficial evaluation would be to regard the number of citations of IMPROVE publications.

Appendices

A.1 Addresses of Involved Research Institutions

Aachener Verfahrenstechnik, Process Systems Engineering
Prof. Dr.-Ing. Wolfgang Marquardt (Speaker substitute of IMPROVE)
Turmstr. 46, D-52064 Aachen

phone: +49 / 241 / 80-94668
fax: +49 / 241 / 80-92326
e-Mail: wolfgang.marquardt@avt.rwth-aachen.de

Chair and Institute of Plastics Processing
Prof. Dr.-Ing. Edmund Haberstroh
Pontstraße 49, D-52062 Aachen

phone: +49 / 241 / 80-93806
fax: +49 / 241 / 80-92262
e-Mail: zentrale@ikv.rwth-aachen.de

Chair and Institute of Industrial Engineering and Ergonomics
Prof. Dr.-Ing. Christopher Schlick
Prof. em. Dr.-Ing. Dipl.-Wirt.Ing. Holger Luczak
Bergdriesch 27, D-52062 Aachen

phone: +49 / 241 / 80-99440
fax: +49 / 241 / 80-92131
e-Mail: info@iaw.rwth-aachen.de

Chair Computer Science 3 (Software Engineering)
Prof. Dr.-Ing. Manfred Nagl (Speaker of IMPROVE)
Ahornstr. 55, D-52074 Aachen

phone: +49 / 241 / 80-21300
fax: +49 / 241 / 80-22218
e-Mail: nagl@informatik.rwth-aachen.de

Chair Computer Science 4 (Communication and Distributed Systems)
Prof. Dr.rer.nat. Otto Spaniol
Ahornstr. 55, D-52074 Aachen

phone: +49 / 241 / 80-21400
fax: +49 / 241 / 80-22220
e-Mail: spaniol@informatik.rwth-aachen.de

Chair Computer Science 5 (Information Systems and Database Technology)
Prof. Dr.rer.pol. Matthias Jarke
Ahornstr. 55, D-52074 Aachen

phone: +49 / 241 / 80-21501
fax: +49 / 241 / 80-22321
e-Mail: jarke@informatik.rwth-aachen.de

Center for Computing and Communication
Dr. rer.nat. Torsten Kuhlen
Seffenter Weg 23, D-52074 Aachen

phone: +49 / 241 / 80-24783
fax: +49 / 241 / 80-22241
e-Mail: kuhlen@rz.rwth-aachen.de

A.2 Members of the CRC 476 and TC 61

In the following, we list the main actors of the CRC IMPROVE. Some of them also appear as authors of the articles of this book. Approximately further 150 students – who either wrote their Master's Thesis on a topic of the IMPROVE project or who where affiliated as student's researchers – contributed to the success of IMPROVE. They do not appear in the following list.

Assenmacher, Ingo, Dipl.-Inform.
Babich, Yuri, Dipl.-Inform.
Baumeister, Markus, Dr.-Ing.
Bayer, Birgit, Dr.-Ing.
Bauer, Lutz, Dipl.-Inform.
Becker, Simon, Dipl.-Inform.
Becks, Andreas, Dr.
Böhlen, Boris, Dr.rer.nat.
Bogusch, Ralf, Dr.-Ing.
Bolke-Hermanns, Helen,
    techn. Angest.
Brandt, Sebastian C., Dipl.-Inform.
Breuer, Marita, Math.-Techn. Ass.
Broll, Wolfgang, Dr.rer.nat.
Bürschgens, Guido,
    Math.-Techn. Ass.
Büschkes, Roland, Dr.
Conrad, Christoph,
    Math.-Techn. Ass.
Cremer, Katja, Dr.rer.nat.
Cseh, C., Dr.
Depolt, Jürgen, Dipl.-Wirt.-Ing.
Docquier, H., Dr.
Dömges, Ralf, Dr.
Eggersmann, Markus, Dr.-Ing.
Fasbender, A., Dr.
Fuß, Christian, Dipl.-Inform.
Foltz, Christian, Dipl.-Ing.
Friedrich, Jutta, Math.-Techn. Ass.
Fritzen, Oliver, Dr.rer.nat.
Geffers, Willi, Dipl.-Inform.
Gerhards, Sascha, Math.-Techn. Ass.
Gruner, Stefan, Dr.rer.nat.
Haase, Thomas, Dipl.-Inform.
Haberstroh, Edmund, Prof. Dr.-Ing.

Hackenberg, Jörg, Dipl.-Ing.
Hai, Ri, Dipl.-Ing.
Haumer, Peter, Dr.
Heer, Thomas, Dipl.-Inform.
Heimann, Peter, Dipl.-Inform.
Heller, Markus, Dipl.-Inform.
Herbst, Detlev,
    Dr.-Ing., Dipl.-Wirt.-Ing.
Hermanns, Oliver, Dr.
Heyn, M., Dr.
Hormes, Jochen, Math.-Techn. Ass.
Hoofe, Markus, Dipl.-Inform.
Jäger, Dirk, Dr.rer.nat.
Jarke, Matthias, Prof. Dr.rer.pol.
Jertila, Aida, Dipl.-Inform.
Jeusfeld, Manfred, Dr., Hauptdozent
Kausch, Bernhard, Dipl.-Ing.
Kesdogan, Dogan, Dr.
Kethers, Stefanie, Dr.
Kirchhof, Michael, Dr.rer.nat.
Klamma, Ralf, Dr.rer.nat.
Klein, Peter, Dr.rer.nat.
Klemke, Roland, Dr.
Kluck, Markus, Math.-Techn. Ass.
Köller, Jörg, Dipl.-Inform.
Körtgen, Anne-Thérèse,
    Dipl.-Inform.
Kraft, Bodo, Dipl.-Inform.
Krapp, Carl-Arndt, Dr.rer.nat.
Krogull, Rainer, techn. Angest.
Krobb, Claudia, Dipl.-Inform.
Krumrück, Elke, Math.-Techn. Ass.
Kuckelberg, Alexander, Dipl.-Inform.
Künzer, Alexander, Dr.-Ing.
Kuhlen, Torsten, Dr.rer.nat.

Kulikov, Viatcheslav, Dipl.-Ing.
Linnhoff-Popien, Claudia,
    Prof. Dr.rer.nat.
Lipperts, Steffen, Dr.rer.nat.
List, Thomas, Dr.rer.nat.
Lohmann, Bernd, Dr.-Ing.
Luczak, Holger,
    Prof.em. Dr.-Ing., Dipl.-Wirt.-Ing.
Lübbers, Dominik, Dipl.-Inform.
Mackau, Dirk, Dipl.-Ing.
Marquardt, Wolfgang, Prof. Dr.-Ing.
Meggers, Jens, Dr.rer.nat.
Meyer, Bernd, Dipl.-Inform.
Meyer, Oliver, Dr.rer.nat.
Meyers, Andrea, Math.-Techn. Ass.
Miatidis, Michalis, Dr.-Ing.
Molitor, R., Dr.
Morbach, Jan, Dipl.-Ing.
Moron, O., Dr.
Münch, Manfred, Dr.rer.nat.
Munz, M., Dr.
Nagl, Manfred, Prof. Dr.-Ing.
Nick, A., Dr.
Niewerth, Christoph,
    Math.-Techn. Ass.
Nissen, Hans W.,
    Prof. Dr.rer.nat.
Oehme, Olaf, Dipl.-Ing.
Oldenburg, Jan, Dr.-Ing.
Peterjohann, Horst, Dipl.-Inform.
Pohl, Klaus, Prof. Dr.rer.nat.
Quix, Christoph, Dr.rer.nat.
Radermacher, Ansgar, Dr.rer.nat.
Ranger, Ulrike, Dipl.-Inf.
Retkowitz, Daniel, Dipl.-Inf.
Sattler, Ulrike, Dr.
Schares, L., Dr.
Scharwächter, Hanno, Dipl.-Ing.
Schleicher, Ansgar, Dr.rer.nat.

Schlick, Christopher, Prof. Dr.-Ing.
Schlüter, Marcus, Dr.-Ing.
Schmidt, Ludger, Dr.-Ing.
Schneider, Nicole, Dipl.-Inform.
Schneider, Ralph, Prof. Dr.-Ing.
Schoop, Mareike, Prof., Ph.D.
Schopfer, Georg, Dr.-Ing.
Schuba, Marko, Dr.
Schüppen, André, Dipl.-Inform.
Schürr, Andy, Prof. Dr.rer.nat.
Sklorz, Stefan, Dipl.-Inform.
Souza, D., Dr.
Spaniol, Otto, Prof. Dr.rer.nat.
Springer, Johannes, Dr.-Ing.
Stenzel, Roland, Dr.
Stepprath, F.-J., Dr.
Stewering, Jörn, Dipl.-Phys.
Theißen, Manfred, Dipl.-Ing.
Thißen, Dirk, Dr.rer.nat.
Töbermann, J.-Christian, Dr.-Ing.
Trossen, Dirk, Dr.rer.nat.
Volkova, Galina, Math.-Techn. Ass.
von den Brinken, Peter, Dr.-Ing.
von Wedel, Lars, Dr.-Ing.
Weidenhaupt, Klaus,
    Prof. Dr.rer.nat.
Weinberg, Tatjana,
    Math.-Techn. Ass.
Weinell, Erhard, Dipl.-Inform.
Westfechtel, Bernhard,
    Prof. Dr.rer.nat.
Wiedau, Michael, Dipl.-Inf.
Wiesner, Andreas, Dipl.-Ing.
Winter, Andreas, Dr.rer.nat.
Wörzberger, René, Dipl.-Inform.
Wolf, Martin, Dipl.-Inform.
Wyes, Jutta, Dr.rer.nat.
Yang, Aidong, Ph.D.

References

R.1 Publications of the IMPROVE Groups[73]

1. Abel, O., Helbig, A., Marquardt, W.: Optimization approaches to control integrated design of industrial batch reactors. In: Berber, R., Kravaris, C. (eds.) Nonlinear model based process control. NATO-ASI Series, pp. 513–551. Kluwer Academic Publishers, Dordrecht (1998)
2. Amin, M.A., Morbach, J.: DAML+OIL to OWL converter (2005), http://www.lpt.rwth-aachen.de/Research/OntoCAPE/daml2owl.php
3. Amin, M.A., Morbach, J.: XML to OWL converter (2006), http://www.lpt.rwth-aachen.de/Research/OntoCAPE/xml2owl.php
4. Armac, I., Retkowitz, D.: Simulation of Smart Environments. In: Proc. of the IEEE Intl. Conf. on Pervasive Services 2007 (ICPS'07), Istanbul, Turkey, pp. 257–266. IEEE Computer Society Press, Los Alamitos (2007)
5. Assenmacher, I., Haberstroh, E., Stewering, J.: Einsatz der Virtuellen Realität in der Kunststofftechnik. WAK Zeitschrift (2007)
6. Babich, Y.: Integration of web services into a QoS aware environment. In: Proceedings of the 3rd International Conference on Computer Science, Software Engineering, Information Technology, e-Business and Applications (CSITeA'04), Cairo, Egypt (2004)
7. Babich, Y., Spaniol, O., Thißen, D.: Service Management for Development Tools. This volume (2008)
8. Baumann, R.: Ein Datenbankmanagementsystem für verteilte, integrierte Softwareentwicklungsumgebungen. PhD thesis, RWTH Aachen University (1999)
9. Baumeister, M.: Ein Objektmodell zur Modellierung und Wiederverwendung verfahrenstechnischer Prozessmodelle. PhD thesis, RWTH Aachen University (2001)

---

[73] For the reason of completeness the publications of this book are also contained in the following bibliography of IMPROVE. So, this bibliography gives a complete view of the publications of the CRC 476 and the TC 61 by the date of appearance of this volume.

10. Baumeister, M., Bogusch, R., Krobb, C., Lohmann, B., Souza, D., von Wedel, L., Marquardt, W.: A chemical engineering data model – website (1998), http://www.lpt.rwth-aachen.de/Research/Completed/veda.php
11. Baumeister, M., Jarke, M.: Compaction of large class hierarchies in databases for chemical engineering. In: Proceedings of BTW'99, Freiburg, Germany, pp. 343–361. Springer, Heidelberg (1999)
12. Baumeister, M., Marquardt, W.: The chemical engineering data model VeDa. Part 1: VDDL – the language definition. Technical Report LPT-1998-01, RWTH Aachen University, Process Systems Engineering (1998)
13. Bausa, J., von Watzdorf, R., Marquardt, W.: Shortcut methods for nonideal multicomponent distillation: 1. simple columns. AIChE Journal 44, 2181–2198 (1998)
14. Bayer, B.: Conceptual Information Modeling for Computer Aided Support of Chemical Process Design. PhD thesis, RWTH Aachen University. Published in: Fortschritt-Berichte VDI: Reihe 3, Nr. 787. VDI-Verlag, Düsseldorf (2003)
15. Bayer, B., Becker, S., Nagl, M.: Model- and rule-based integration tools for supporting incremental change processes in chemical engineering. In: Chen, B., Westerberg, A.W. (eds.) Proceedings of the 8th International Symposium on Process Systems Engineering (PSE 2003), Kunming, China, pp. 1256–1261. Elsevier, Amsterdam (2003)
16. Bayer, B., Bogusch, R., Lohmann, B., Marquardt, W.: Szenariobasierte Analyse von Entwicklungsprozessen. In: Nagl, M., Westfechtel, B. (eds.) Integration von Entwicklungssystemen in Ingenieuranwendungen – Substantielle Verbesserung der Entwicklungsprozesse, pp. 389–401. Springer, Heidelberg (1999)
17. Bayer, B., Eggersmann, M., Gani, R., Schneider, R.: Case studies in design and analysis. In: Braunschweig, B., Gani, R. (eds.) Software Architecture and Tools for Computer-Aided Chemical Engineering, pp. 591–634. Elsevier, Amsterdam (2002)
18. Bayer, B., Marquardt, W.: A comparison of data models in chemical engineering. Concurrent Engineering: Research and Applications 11(2), 129–138 (2003)
19. Bayer, B., Marquardt, W.: Towards integrated information models for data and documents. Computers & Chemical Engineering 28, 1249–1266 (2004)
20. Bayer, B., Marquardt, W.: A conceptual information model for the chemical process design lifecycle. In: Jeusfeld, M., Jarke, M., Mylopoulos, J. (eds.) The Method Engineering Textbook, MIT Press, Cambridge (2007)
21. Bayer, B., Marquardt, W., Weidenhaupt, K., Jarke, M.: A flowsheet centered architecture for conceptual design. In: Gani, R., Jørgensen, S.B. (eds.) Proceedings of the European Symposium on Computer Aided Process Engineering – ESCAPE 11, pp. 345–350. Elsevier, Amsterdam (2001)
22. Bayer, B., Schneider, R.: Third deliverable of T6.1.1 "Support to Lifecycle Models". Technical Report GCO-WP611-InterfaceDescription.DOC, Global CAPE-OPEN (2001)
23. Bayer, B., Schneider, R., Marquardt, W.: Integration of data models for process design – first steps and experiences. Computers & Chemical Engineering 24, 599–605 (2000)

24. Bayer, B., von Wedel, L., Marquardt, W.: An integration of design data and mathematical models in chemical process design. In: Kraslawski, A., Turunen, I. (eds.) Proceedings of the European Symposium on Computer Aided Process Engineering – ESCAPE 13, pp. 29–34. Elsevier, Amsterdam (2003)
25. Becker, S.: Integratoren zur Konsistenzsicherung von Dokumenten in Entwicklungsprozessen. PhD thesis, RWTH Aachen University (2007)
26. Becker, S., Haase, T., Westfechtel, B.: Model-based a-posteriori integration of engineering tools for incremental development processes. Software and Systems Modeling 4(2), 123–140 (2005)
27. Becker, S., Haase, T., Westfechtel, B., Wilhelms, J.: Integration tools supporting cooperative development processes in chemical engineering. In: Proceedings of the 6th World Conference on Integrated Design & Process Technology (IDPT 2002), Pasadena, California, USA, SDPS (2002)
28. Becker, S., Heller, M., Jarke, M., Marquardt, W., Nagl, M., Spaniol, O., Thißen, D.: Synergy by Integrating New Functionality. This volume (2008)
29. Becker, S., Herold, S., Lohmann, S., Westfechtel, B.: A Graph-Based Algorithm for Consistency Maintenance in Incremental and Interactive Integration Tools. Journal of Software and Systems Modeling (2007)
30. Becker, S., Jäger, D., Schleicher, A., Westfechtel, B.: A delegation based model for distributed software process management. In: Ambriola, V. (ed.) EWSPT 2001. LNCS, vol. 2077, pp. 130–144. Springer, Heidelberg (2001)
31. Becker, S., Kirchhof, M., Nagl, M., Schleicher, A.: EAI, Web und eBusiness: Echte Anwendungsintegration macht Aufwand? In: Jähnichen, S. (ed.) Neue Webtechnologien & eBusiness Integration – Proceedings of ONLINE, Congress VI, C630.01–C630.27. ONLINE GmbH (2002)
32. Becker, S., Körtgen, A., Nagl, M.: Tools for Consistency Management between Design Products. This volume (2008)
33. Becker, S.M., Lohmann, S., Westfechtel, B.: Rule execution in graph-based incremental interactive integration tools. In: Ehrig, H., Engels, G., Parisi-Presicce, F., Rozenberg, G. (eds.) ICGT 2004. LNCS, vol. 3256, pp. 22–38. Springer, Heidelberg (2004)
34. Becker, S., Marquardt, W., Morbach, J., Nagl, M.: Model Dependencies, Fine-Grained Relations, and Integrator Tools. This volume (2008)
35. Becker, S., Nagl, M., Westfechtel, B.: Incremental and Interactive Integrator Tools for Design Product Consistency. This volume (2008)
36. Becker, S., Westfechtel, B.: Integrationswerkzeuge für verfahrenstechnische Entwicklungsprozesse. In: Engineering in der Prozessindustrie. VDI Fortschritt-Berichte, vol. 1684, pp. 103–112. VDI-Verlag, Düsseldorf (2002)
37. Becker, S., Westfechtel, B.: Incremental integration tools for chemical engineering: An industrial application of triple graph grammars. In: Bodlaender, H.L. (ed.) WG 2003. LNCS, vol. 2880, pp. 46–57. Springer, Heidelberg (2003)
38. Becker, S., Westfechtel, B.: Integrationswerkzeuge für verfahrenstechnische Entwicklungsprozesse. atp – Automatisierungstechnische Praxis 45(4), 59–65 (2003)
39. Becker, S., Westfechtel, B.: Uml-based definition of integration models for incremental development processes in chemical engineering. In: Proceedings of the 7th International Conference on Integrated Design and Process Technology (IDPT 2003), Austin, Texas, USA, SDPS (2003)

40. Becker, S., Westfechtel, B.: UML-based definition of integration models for incremental development processes in chemical engineering. Journal of Integrated Design and Process Science: Transactions of the SDPS 8(1), 49–63 (2004)
41. Becker, S., Wilhelms, J.: Integrationswerkzeuge in verfahrenstechnischen Entwicklungsprozessen. Verfahrenstechnik 36(6), 44–45 (2002)
42. Becks, A.: Visual Document Management with Adaptable Document Maps. PhD thesis, RWTH Aachen University (2001)
43. Behle, A.: Wiederverwendung von Softwarekomponenten im Internet. PhD thesis, RWTH Aachen University (1999)
44. Behle, A., Kirchhof, M., Nagl, M., Welter, R.: Retrieval of software components using a distributed web system. Journal of Network and Computer Applications 25(3), 197–222 (2002)
45. Behle, A., Nagl, M., Pritsch, E.: Hilfsmittel für verteilte Anwendungssysteme: Erfahrungen aus einigen Projekten. In: Wahlster, W. (ed.) Software-Offensive mit Java, Agenten & XML – Proceedings of ONLINE 2000, Congress VI, C630.01–C630.19. ONLINE GmbH (2000)
46. Böhlen, B.: Specific graph models and their mappings to a common model. In: Pfaltz, J.L., Nagl, M., Böhlen, B. (eds.) AGTIVE 2003. LNCS, vol. 3062, pp. 45–60. Springer, Heidelberg (2004)
47. Böhlen, B.: A Parameterizable Graph Data Base for Development Tools (in German). PhD thesis, RWTH Aachen University, Aachen (2006)
48. Böhlen, B., Jäger, D., Schleicher, A., Westfechtel, B.: UPGRADE: A framework for building graph-based interactive tools. Electronic Notes in Theoretical Computer Science (Proceedings of the International Workshop on Graph-Based Tools (GraBaTs'02), Barcelona, Spain, October 7–8, 2002) 72(2), 113–123 (2002)
49. Böhlen, B., Jäger, D., Schleicher, A., Westfechtel, B.: UPGRADE: Building interactive tools for visual languages. In: Proceedings of the 6th World Multi-Conference On Systemics, Cybernetics and Informatics (SCI 2002), Orlando, Florida, USA, pp. 17–22 (2002)
50. Böhlen, B., Ranger, U.: Concepts for specifying complex graph transformation systems. In: Ehrig, H., Engels, G., Parisi-Presicce, F., Rozenberg, G. (eds.) ICGT 2004. LNCS, vol. 3256, pp. 96–111. Springer, Heidelberg (2004)
51. Bogusch, R.: A Software Environment for Computer-aided Modeling of Chemical Processes. PhD thesis, RWTH Aachen University. Published in: Fortschritt-Berichte VDI: Reihe 3, Nr. 705, VDI-Verlag, Düsseldorf (2001)
52. Bogusch, R., Lohmann, B., Marquardt, W.: Computer-aided process modeling with ModKit. In: Proceedings of the CHEMPUTERS Europe III Conference, pp. 1–15 (1996)
53. Bogusch, R., Lohmann, B., Marquardt, W.: Ein System zur rechnergestützten Modellierung in der Verfahrenstechnik. In: Jahrbuch der VDI-Gesellschaft Verfahrenstechnik und Chemieingenieurwesen, pp. 22–53. VDI-Verlag, Düsseldorf (1997)
54. Bogusch, R., Lohmann, B., Marquardt, W.: Computer-aided process modeling with ModKit. Computers & Chemical Engineering 25, 963–995 (2001)
55. Bogusch, R., Marquardt, W.: A formal representation of process model equations. Computers & Chemical Engineering 21(10), 1105–1115 (1997)

56. Bogusch, R., Marquardt, W.: The chemical engineering data model VeDa. Part 4: Behavioral modeling objects. Technical Report LPT-1998-04, RWTH Aachen University, Process Systems Engineering (1998)
57. Borning, M.: The anonymity service architecture. In: Proceedings of the 10th International Conference on Computer Communications and Networks (IC-CCN'01), Scottsdale, USA, IEEE Computer Society Press, Los Alamitos (2001)
58. Borning, M., Kesdogan, D., Spaniol, O.: Anonymity and untraceability in the internet. IT+TI Informationstechnik und Technische Informatik 43(5), 246–253 (2001)
59. Bouazizi, I.: Proxy Caching for Robust Video Delivery over Lossy Networks. PhD thesis, RWTH Aachen University (2004)
60. Bouazizi, I., Günes, M.: A framework for transmitting video over wireless networks. In: Proceedings of the 6th World Multi-Conference On Systemics, Cybernetics and Informatics (SCI 2002), Orlando, Florida, USA (2002)
61. Brandt, S., Turhan, A.-Y.: Using non-standard inferences in description logics — what does it buy me? In: Proceedings of the KI-2001 Workshop on Applications of Description Logics (KIDLWS'01), Vienna, Austria. CEUR Workshop Proceedings, vol. 44 (2001), http://CEUR-WS.org/Vol-44/
62. Brandt, S.C., Morbach, J., Miatidis, M., Theißen, M., Jarke, M., Marquardt, W.: Ontology-Based Information Management in Design Processes. In: Marquardt, W., Pantelides, C. (eds.) 16th European Symposium on Computer Aided Process Engineering and 9th International Symposium on Process Systems Engineering, Garmisch-Partenkirchen, Germany, July 9–13, 2006. Computer-Aided Chemical Engineering, vol. 21, pp. 2021–2026. Elsevier, Amsterdam (2006)
63. Brandt, S.C., Morbach, J., Miatidis, M., Theißen, M., Jarke, M., Marquardt, W.: Ontology-Based Information Management in Design Processes. Computers & Chemical Engineering 32(1-2), 230–342 (2008)
64. Brandt, S.C., Schlüter, M., Jarke, M.: A Process Data Warehouse for Tracing and Reuse of Engineering Design Processes. In: Proceedings of the 2nd International Conference on Innovations in Information Technology (IIT'05), Dubai, United Arab Emirates (2005)
65. Brandt, S.C., Schlüter, M., Jarke, M.: A Process Data Warehouse for Tracing and Reuse of Engineering Design Processes. International Journal of Intelligent Information Systems 2(4) (2006)
66. Brandt, S.C., Schlüter, M., Jarke, M.: Process Data Warehouse Models for Cooperative Engineering Processes. In: Proceedings of the 9th IFAC Symposium on Automated Systems Based on Human Skill And Knowledge, Nancy, France (2006)
67. Brandt, S., Fritzen, O., Jarke, M., List, T.: Goal-Oriented Information Flow Management in Development Processes. This volume (2008)
68. Brandt, S., Jarke, M., Miatidis, M., Raddatz, M., Schlüter, M.: Management and Reuse of Experience Knowledge in Continuous Production Processes. This volume (2008)
69. Brandt, S.C., Schlüter, M., Jarke, M.: Using Semantic Technologies for the Support of Engineering Design Processes. In: Sugumaran, V. (ed.) Intelligent Information Technologies and Applications, IGI Publishing, Hershey (2008)

70. Braunschweig, B., Fraga, E.S., Guessoum, Z., Paen, D., Piñoll, D., Yang, A.: COGents: Cognitive middleware agents to support e-CAPE. In: Stanford-Smith, B., Chiozza, E., Edin, M. (eds.) Proceedings of Challenges and Achievements in e-business and e-work, pp. 1182–1189 (2002)
71. Braunschweig, B., Irons, K., Köller, J., Kuckelberg, A., Pons, M.: CAPE-OPEN (CO) standards: Implementation and maintenance. In: Proceedings of the 2nd IEEE Conference on Standardization and Innovation in Information Technology, pp. 335–338 (2001)
72. Braunschweig, B., Jarke, M., Köller, J., Marquardt, W., von Wedel, L.: CAPE-OPEN – experiences from a standardization effort in chemical industries. In: Proceedings of the International Conference on Standardization and Integration in Information Technology (SIIT'99), Aachen, Germany (1999)
73. Börstler, J.: Programmieren-Im-Großen: Sprachen, Werkzeuge, Wiederverwendung. PhD thesis, RWTH Aachen University (1993)
74. Börstler, J., Janning, T.: Traceability between requirements engineering and design: A transformational approach. In: Proceedings COMPSAC, pp. 362–368. IEEE Computer Society Press, Los Alamitos (1992)
75. Büschkes, R.: Angriffserkennung in Kommunikationsnetzen. PhD thesis, RWTH Aachen University (2001)
76. Büschkes, R., Noll, T., Borning, M.: Transaction-based anomaly detection in communication networks. In: Proceedings of the 9th International Conference on Telecommunication Systems, Modelling and Analysis, Dallas, Texas, USA, pp. 33–47 (2001)
77. Büschkes, R., Seipold, T., Wienzek, R.: Performance evaluation of transaction-based anomaly detection. In: Proceedings of the 1st International NATO Symposium on Real Time Intrusion Detection, Lisbon, Portugal (2002)
78. Büschkes, R., Thißen, D., Yu, H.: Monitoring and control of critical infrastructures (short paper). In: Proceedings of the 2001 International Conference on Dependable Systems and Networks (DSN 2001), Göteborg, Sweden, pp. B68–B69. IEEE Computer Society Press, Los Alamitos (2001)
79. von Buol, B.: Qualitätsgestützte kooperative Terminologiearbeit. PhD thesis, RWTH Aachen University (2000)
80. Comanns, M.: Werkzeugunterstützung für erfahrensbasierte Wiederverwendung von Prozessspuren. Master's thesis, RWTH Aachen University (2006)
81. Cremer, K.: Using graph technology for reverse and re-engineering. In: Proceedings of the 5th International Conference on Re-Technologies for Information Systems, Klagenfurt, Austria (1997)
82. Cremer, K.: A tool supporting re-design of legacy applications. In: Nesi, P., Lehner, F. (eds.) Proceedings of the 2nd Euromicro Conference on Software Maintenance and Reengineering, pp. 142–148. IEEE Computer Society Press, Los Alamitos (1998)
83. Cremer, K.: Graphbasierte Werkzeuge zum Reverse Engineering und Reengineering. PhD thesis, RWTH Aachen University (1999)
84. Cremer, K., Gruner, S., Nagl, M.: Graph transformation based integration tools: Application to chemical process engineering. In: Ehrig, H., Engels, G., Kreowski, H.-J., Rozenberg, G. (eds.) Handbook on Graph Grammars and Computing by Graph Transformation – Volume 2: Applications, Languages, and Tools, pp. 369–394. World Scientific, Singapore (1999)

85. Cremer, K., Klein, P., Nagl, M., Radermacher, A.: Verteilung von Arbeitsumgebungen und Integration zu einem Verbund: Hilfe durch objektorientierte Strukturen und Dienste. In: Wahlster, W. (ed.) Fortschritte der objektorientierten Softwaretechnologien – Proceedings of ONLINE'96, Congress VI, C.610.01–C.610.23. ONLINE GmbH (1996)
86. Cremer, K., Klein, P., Nagl, M., Radermacher, A.: Restrukturierung zu verteilten Anwendungen: Unterstützung von Methodik durch Werkzeuge. In: Nagl, M. (ed.): Verteilte, integrierte Anwendungsarchitekturen: Die Software-Welt im Umbruch – Proceedings of ONLINE'97, Congress VI, C620.01–C620.25. ONLINE GmbH (1997)
87. Cremer, K., Klein, P., Nagl, M., Radermacher, A.: Prototypische Werkzeuge zur Restrukturierung und Verteilung von Anwendungssystemen. In: Informationstechnik im Zeitalter des Internets – Proceedings of ONLINE'98, Congress VI, C630.01–C630.26. ONLINE GmbH (1998)
88. Cremer, K., Marburger, A., Westfechtel, B.: Graph-based tools for re-engineering. Journal of Software Maintenance and Evolution: Research and Practice 14(4), 257–292 (2002)
89. Cremer, K., Radermacher, A.: Einsatz von Workstations bei der Restrukturierung von zentralistischen Informationssystemen. In: Proceedings Fachtagung Workstations und ihre Anwendung (SIWORK'96), pp. 63–66. Hochschulverlag ETH Zürich, Zürich (1996)
90. Cseh, C.: Flow Control for the Available Bit Rate Service in Asynchronous Transfer Mode Networks. PhD thesis, RWTH Aachen University (2001)
91. Depolt, J.: Kennzahlenbasierte Wirtschaftlichkeitsanalyse von Telekooperation in der Produktentwicklung der Automobilindustrie. PhD thesis, RWTH Aachen University (2000)
92. Dömges, R.: Projektspezifische Methoden zur Nachvollziehbarkeit von Anforderungsspezifikationen. PhD thesis, RWTH Aachen University (1999)
93. Dömges, R., Pohl, K.: Adapting traceability environments to project-specific needs. Communications of the ACM 41(12), 54–62 (1998)
94. Dömges, R., Pohl, K., Jarke, M., Lohmann, B., Marquardt, W.: PROART/CE – an environment for managing chemical process simulation models. In: Proceedings of the 10th European Simulation Multiconference (1996)
95. Eggersmann, M.: Analysis and Support of Work Processes Within Chemical Engineering Design Processes. PhD thesis, RWTH Aachen University. Published in: Fortschritt-Berichte VDI, Reihe 3, Nr. 840, VDI-Verlag, Düsseldorf (2004)
96. Eggersmann, M., Bayer, B., Jarke, M., Marquardt, W., Schneider, R.: Prozess- und Produktmodelle für die Verfahrenstechnik. In: Nagl, M., Westfechtel, B. (eds.) Modelle, Werkzeuge und Infrastrukturen zur Unterstützung von Entwicklungsprozessen, Wiley-VCH, Weinheim (2003)
97. Eggersmann, M., Gonnet, S., Henning, G., Krobb, C., Leone, H., Marquardt, W.: Modeling and understanding different types of process design activities. In: Proceedings of ENPROMER – 3rd Mercosur Congress on Process Systems Engineering – 1st Mercosur Congress on Chemical Engineering, vol. 1, pp. 151–156 (2001)
98. Eggersmann, M., Gonnet, S., Henning, G., Krobb, C., Leone, H., Marquardt, W.: Modeling and understanding different types of process design activities. Latin American Applied Research 33, 167–175 (2003)

99. Eggersmann, M., Hackenberg, J., Marquardt, W., Cameron, I.: Applications of modelling – a case study from process design. In: Braunschweig, B., Gani, R. (eds.) Software Architecture and Tools for Computer-Aided Chemical Engineering, pp. 335–372. Elsevier, Amsterdam (2002)
100. Eggersmann, M., Henning, G., Krobb, C., Leone, H.: Modeling of actors within a chemical engineering work process model. In: Proceedings of CIRP Design Seminar, Stockholm, Sweden, pp. 203–208 (2001)
101. Eggersmann, M., Kausch, B., Luczak, H., Marquardt, W., Schlick, C., Schneider, N., Schneider, R., Theißen, M.: Work Process Models. This volume (2008)
102. Eggersmann, M., Krobb, C., Marquardt, W.: A language for modeling work processes in chemical engineering. Technical Report LPT-2000-02, RWTH Aachen University, Process Systems Engineering (2000)
103. Eggersmann, M., Krobb, C., Marquardt, W.: A modeling language for design processes in chemical engineering. In: Laender, A.H.F., Liddle, S.W., Storey, V.C. (eds.) ER 2000. LNCS, vol. 1920, pp. 369–382. Springer, Heidelberg (2000)
104. Eggersmann, M., Schneider, R., Marquardt, W.: Modeling work processes in chemical engineering – from recording to supporting. In: Grievink, J., van Schijndel, J. (eds.) Proceedings of the European Symposium on Computer Aided Process Engineering – ESCAPE 12, pp. 871–876. Elsevier, Amsterdam (2002)
105. Eggersmann, M., Schneider, R., Marquardt, W.: Understanding the interrelations between synthesis and analysis during model based design. In: Chen, B., Westerberg, A.W. (eds.) Proceedings of the 8th International Symposium on Process Systems Engineering (PSE 2003), Kunming, China, pp. 802–807. Elsevier, Amsterdam (2003)
106. Eggersmann, M., von Wedel, L., Marquardt, W.: Verwaltung und wiederverwendung von modellen im industriellen entwicklungsprozess. Chem.-Ing.-Tech. 74, 1068–1078 (2002)
107. Ehrig, H., Engels, G., Parisi-Presicce, F., Rozenberg, G. (eds.): ICGT 2004. LNCS, vol. 3256. Springer, Heidelberg (2004)
108. Engels, G.: Graphen als zentrale Datenstrukturen in einer Softwareentwicklungs-Umgebung. PhD thesis, University of Osnabrück (1986)
109. Engels, G., Lewerentz, C., Nagl, M., Schäfer, W., Schürr, A.: Experiences in building integrating tools, Part 1: Tool specification. Transactions on Software Engineering and Methodology 1(2), 135–167 (1992)
110. Eversheim, W., Michaeli, W., Nagl, M., Spaniol, O., Weck, M.: The SUKITS project: An approach to a posteriori integration of CIM components. In: Goerke, W., Rininsland, H., Syrbe, M. (eds.) Information als Produktionsmotor. Informatik Aktuell, pp. 494–504. Springer, Heidelberg (1992)
111. Evertz, M.: Gestaltung und Evaluation der Benutzungsschnittstelle eines Administrationssystems für verfahrenstechnische Entwickler. Master's thesis, RWTH Aachen University (2002)
112. Fachgebiet Dynamik und Betrieb technischer Anlagen (TU Berlin), AVT-PT (RWTH Aachen University): Informationstechnologien für Entwicklung und Produktion in der Verfahrenstechnik – Symposiumsreihe – website (2008), http://www.inprotech.de/
113. Fasbender, A.: Messung und Modellierung der Dienstgüte paketvermittelter Netze. PhD thesis, RWTH Aachen University (1998)
114. Fidler, M.: Providing Internet Quality of Service based on Differentiated Services Traffic Engineering. PhD thesis, RWTH Aachen University (2003)

115. Foltz, C.: Softwareergonomische Evaluation des Fließbildwerkzeugs (FBW). Technical report, RWTH Aachen University (2001)
116. Foltz, C., Killich, S., Wolf, M., Schmidt, L., Luczak, H.: Task and information modeling for cooperative work. In: Systems, Social and Internationalization Design Aspects of Human-Computer Interaction, Proceedings of HCI International, Volume 2, New Orleans, USA, pp. 172–176. Lawrence Erlbaum, Mahwah (2001)
117. Foltz, C., Luczak, H.: Analyse und Gestaltung verfahrenstechnischer Entwicklungsprozesse. atp – Automatisierungstechnische Praxis 45(9), 39–44 (2003)
118. Foltz, C., Luczak, H.: Analyzing chemical process design using an abstraction-decomposition space. In: Chen, B., Westerberg, A.W. (eds.) Proceedings of the 8th International Symposium on Process Systems Engineering (PSE 2003), Kunming, China, Elsevier, Amsterdam (2003)
119. Foltz, C., Luczak, H., Schmidt, L.: Representing knowledge for chemical process design using an abstraction-decomposition space. In: Luczak, H., Cakir, A.E., Cakir, G. (eds.) WWDU 2002 – Work With Display Units – World Wide Work. Proceedings of the 6th International Scientific Conference, Berchtesgaden, May 22–25, 2002, pp. 457–462. ERGONOMIC Institut für Arbeits- und Sozialforschung (2002)
120. Foltz, C., Luczak, H., Westfechtel, B.: Use-centered interface design for an adaptable administration system for chemical process design. In: Proceedings of the International Conference on Human-Computer Interaction (HCI International 2003), Crete, Greece, pp. 365–369 (2003)
121. Foltz, C., Reuth, R., Miehling, H.: Telekooperation in der Automobilindustrie – Ergebnisse einer Längsschnittstudie. In: Reichwald, R., Schlichter, J. (eds.) Verteiltes Arbeiten – Arbeiten in der Zukunft, Tagungsband der D-CSCW 2000, pp. 231–242. Teubner, Wiesbaden (2000)
122. Foltz, C., Schmidt, L., Luczak, H.: Not seeing the woods for the trees – empirical studies in engineering design. In: Subrahmanian, E., Sriram, R., Herder, P., Schneider, R. (eds.) The role of empirical studies in understanding and supporting engineering design – Workshop Proceedings, National Institute of Standards and Technology, Gaithersburg, Maryland, USA, April 4 – April 5, 2002, pp. 40–46. DUP Science, Delft (2004)
123. Foltz, C., Schneider, N., Kausch, B., Wolf, M., Schlick, C., Luczak, H.: Usability Engineering. This volume (2008)
124. Foltz, C., Wolf, M., Luczak, H., Eggersmann, M., Schneider, R., Marquardt, W.: Entwurf eines Referenzszenarios zur Analyse und Gestaltung von Entwicklungsprozessen in der verfahrenstechnischen Industrie. In: Gesellschaft für Arbeitswissenschaft e.V (ed.) Komplexe Arbeitssysteme – Herausforderung für Analyse und Gestaltung. Bericht zum 46. Arbeitswissenschaftlichen Kongress der Gesellschaft für Arbeitswissenschaft, Berlin, 15-18 März 2000, pp. 545–548. GfA Press, Dortmund (2000)
125. Foss, B.A., Lohmann, B., Marquardt, W.: A field study of the industrial modeling process. Journal of Process Control 8, 325–337 (1998)
126. Fuss, C., Mosler, C., Ranger, U., Schultchen, E.: The jury is still out: A comparison of AGG, Fujaba, and PROGRES. In: Ehrig, K., Giese, H. (eds.) 6th International Workshop on Graph Transformation and Visual Modeling Techniques, GT-VMT'07. EASST, vol. 6, Electronic Communications of the European Association of Software Science and Technology, Berlin (2007)

127. Gallersdörfer, R.: Replikationsmanagement in verteilten Informationssystemen. PhD thesis, RWTH Aachen University (1997)
128. Gatzemeier, F.H.: CHASID: A semantics-oriented authoring environment. PhD thesis, RWTH Aachen University (2004)
129. Geilmann, K.: Sichtenbasierte Kooperation in einem Prozessmanagementsystem für dynamische Entwicklungsprozesse. Master's thesis, RWTH Aachen University (2005)
130. Gerndt, A., Hentschel, B., Wolter, M., Kuhlen, T., Bischof, C.: Viracocha: An efficient parallelization framework for large-scale CFD post-processing in VE. In: Proceedings of Supercomputing, Pittsburgh, USA (2004)
131. Gruner, S.: Eine schematische und grammatische Korrespondenzmethode zur Spezifikation konsistent verteilter Datenmodelle. PhD thesis, RWTH Aachen University (1999)
132. Gruner, S.: A combined graph schema and graph grammar approach to consistency in distributed modeling. In: Münch, M., Nagl, M. (eds.) AGTIVE 1999. LNCS, vol. 1779, pp. 247–254. Springer, Heidelberg (2000)
133. Gruner, S., Nagl, M., Sauer, F., Schürr, A.: Inkrementelle Integrationswerkzeuge für arbeitsteilige Entwicklungsprozesse. In: Nagl, M., Westfechtel, B. (eds.) Integration von Entwicklungssystemen in Ingenieuranwendungen – Substantielle Verbesserung der Entwicklungsprozesse, pp. 311–330. Springer, Heidelberg (1999)
134. Gruner, S., Nagl, M., Schürr, A.: Integration tools supporting development processes. In: Broy, M., Rumpe, B. (eds.) RTSE 1997. LNCS, vol. 1526, pp. 235–256. Springer, Heidelberg (1998)
135. Haase, T.: A-posteriori Integration verfahrenstechnischer Entwicklungswerkzeuge. Softwaretechnik-Trends (Proceedings of the 5th Workshop on Software Reengineering (WSR 2003), Bad Honnef, Germany, May 7-9, 2003) 23(2), 28–30 (2003)
136. Haase, T.: Semi-automatic wrapper generation for a-posteriori integration. In: Proceedings of the Workshop on Tool Integration in System Development (TIS 2003), Helsinki, Finland, September 1–2, 2003, pp. 84–88 (2003)
137. Haase, T.: Die Rolle der Architektur im Kontext der a-posteriori Integration. Softwaretechnik-Trends (Proceedings of the 6th Workshop on Software Reengineering (WSR 2004), Bad Honnef, Germany, May 3-5, 2004) 24(2), 61–62 (2004)
138. Haase, T.: Model-driven service development for a-posteriori application integration. In: Proceedings of the IEEE International Workshop on Service-Oriented System Engineering (SOSE'07), Hong Kong, China, October 24–26, 2007, pp. 649–656. IEEE Computer Society Press, Los Alamitos (2007)
139. Haase, T.: A-posteriori Integrated Software Systems: Architectures, Methodology, and Tools (in German). PhD thesis, RWTH Aachen University, Aachen (2008)
140. Haase, T., Klein, P., Nagl, M.: Software Integration and Framework Development. This volume (2008)
141. Haase, T., Meyer, O., Böhlen, B., Gatzemeier, F.H.: A domain specific architecture tool: Rapid prototyping with graph grammars. In: Pfaltz, J.L., Nagl, M., Böhlen, B. (eds.) AGTIVE 2003. LNCS, vol. 3062, pp. 236–242. Springer, Heidelberg (2004)

142. Haase, T., Meyer, O., Böhlen, B., Gatzemeier, F.H.: Fire3: Architecture refinement for A-posteriori integration. In: Pfaltz, J.L., Nagl, M., Böhlen, B. (eds.) AGTIVE 2003. LNCS, vol. 3062, pp. 461–467. Springer, Heidelberg (2004)
143. Haase, T., Nagl, M.: Service-oriented Architectures and Application Integration. This volume (2008)
144. Haberstroh, E., Schlüter, M.: Integrierte 1D- und 3D-Simulation von Doppelschneckenextrudern. Plastverarbeiter (2000)
145. Haberstroh, E., Schlüter, M.: Simulating and designing twin screw extruders by using the boundary element method (BEM) in chemical engineering processes. In: Proceedings PPS 17, Montreal, Canada, Polymers Processing Society (2001)
146. Haberstroh, E., Schlüter, M.: The use of modern technologies in the development of simulation software. Journal of Polymer Engineering 21(2-3), 209–224 (2001)
147. Haberstroh, E., Schlüter, M.: Design of twin screw extruders with the MOREX simulation software. In: Proceedings PPS 18, Guimaraes, Portugal, Polymers Processing Society (2002)
148. Haberstroh, E., Schlüter, M.: Integrierte Simulation von Aufbereitungsprozessen. In: Proc. 21st IKV-Kolloquium, Aachen (2002)
149. Haberstroh, E., Stewering, J.: New aspects for the visualisation of compounding processes. In: Proceedings of the 21st Annual Meeting of the Polymer Processing Society, PPS (2005)
150. Haberstroh, E., Stewering, J., Assenmacher, I., Kuhlen, T.: Development Assistance for the Design of Reaction and Compounding Extruders. This volume (2008)
151. Hackenberg, J.: Computer support for theory-based modeling of process systems. PhD thesis, RWTH Aachen University. Published in: Fortschritt-Berichte VDI, Reihe 3, Nr. 860, VDI-Verlag, Düsseldorf (2006)
152. Hackenberg, J., Krobb, C., Marquardt, W.: An object-oriented data model to capture lumped and distributed parameter models of physical systems. In: Troch, I., Breitenecker, F. (eds.) Proceedings of the 3rd MATHMOD, IMACS Symposium on Mathematical Modelling, Vienna, Austria, pp. 339–342 (2000)
153. Hai, R., Heer, T., Heller, M., Nagl, M., Schneider, R., Westfechtel, B., Wörzberger, R.: Administration Models and Management Tools. This volume (2008)
154. Hai, R., Heller, M., Marquardt, W., Nagl, M., Wörzberger, R.: Workflow support for inter-organizational design processes. In: Marquardt, W., Pantelides, C. (eds.) 16th European Symposium on Computer Aided Process Engineering and 9th International Symposium on Process Systems Engineering, Garmisch-Partenkirchen, Germany, July 9–13, 2006. Computer-Aided Chemical Engineering, vol. 21, pp. 2027–2032. Elsevier, Amsterdam (2006)
155. Haumer, P.: Requirements Engineering with Interrelated Conceptual Models and Real-World Scenes. PhD thesis, RWTH Aachen University (2000)
156. Haumer, P., Jarke, M., Pohl, K., Weidenhaupt, K.: Improving reviews of conceptual models by extended traceability to captured system usage. Interacting with Computers 13(2), 77–95 (2000)

157. Heer, T., Retkowitz, D., Kraft, B.: Algorithm and Tool for Ontology Integration based on Graph Rewriting. In: Schürr, A., Nagl, M., Zündorf, A. (eds.) Applications of Graph Transformations with Industrial Relevance. Proceedings of the Third International AGTIVE 2007, Kassel, Germany. LNCS, vol. 5088, pp. 484–490. Springer, Heidelberg (2008)
158. Heer, T., Retkowitz, D., Kraft, B.: Incremental Ontology Integration. In: Proc. of the 10th Intl. Conf. on Enterprise Information Systems (ICEIS 2008), accepted for publication (2008)
159. Heimann, P., Joeris, G., Krapp, C.-A., Westfechtel, B.: A programmed graph rewriting system for software process management. In: Proceedings Joint COMPUGRAPH/SEMAGRAPH Workshop on Graph Rewriting and Computation (SEGRAGRA 1995), Volterra, Italy. Electronic Notes in Theoretical Computer Science, pp. 123–132 (1995)
160. Heimann, P., Joeris, G., Krapp, C.-A., Westfechtel, B.: DYNAMITE: Dynamic task nets for software process management. In: Proceedings of the 18th International Conference on Software Engineering, Berlin, Germany, pp. 331–341. IEEE Computer Society Press, Los Alamitos (1996)
161. Heimann, P., Krapp, C.-A., Nagl, M., Westfechtel, B.: An adaptable and reactive project management environment. In: Nagl, M. (ed.) IPSEN 1996. LNCS, vol. 1170, pp. 504–534. Springer, Heidelberg (1996)
162. Heimann, P., Krapp, C.-A., Westfechtel, B.: An environment for managing software development processes. In: Proceedings of the 8th Conference on Software Engineering Environments, Cottbus, Germany, pp. 101–109. IEEE Computer Society Press, Los Alamitos (1997)
163. Heimann, P., Krapp, C.-A., Westfechtel, B., Joeris, G.: Graph-based software process management. International Journal of Software Engineering & Knowledge Engineering 7(4), 431–455 (1997)
164. Heimann, P., Westfechtel, B.: Integrated product and process management for engineering design in manufacturing systems. In: Leondes, C. (ed.) Computer-Aided Design, Engineering, and Manufacturing – Systems, Techniques and Applications, Volume 4: Optimization Methods for Manufacturing, 2-1–2-47 (2001)
165. Heller, M.: Decentralized and View-based Management of Cross-company Development Processes (in German). PhD thesis, RWTH Aachen University, Aachen (2008)
166. Heller, M., Jäger, D.: Graph-based tools for distributed cooperation in dynamic development processes. In: Pfaltz, J.L., Nagl, M., Böhlen, B. (eds.) AGTIVE 2003. LNCS, vol. 3062, pp. 352–368. Springer, Heidelberg (2004)
167. Heller, M., Jäger, D.: Interorganizational management of development processes. In: Pfaltz, J.L., Nagl, M., Böhlen, B. (eds.) AGTIVE 2003. LNCS, vol. 3062, pp. 427–433. Springer, Heidelberg (2004)
168. Heller, M., Jäger, D., Krapp, C.A., Nagl, M., Schleicher, A., Westfechtel, B., Wörzberger, R.: An Adaptive and Reactive Management System for Project Coordination. This volume (2008)
169. Heller, M., Jäger, D., Schlüter, M., Schneider, R., Westfechtel, B.: A management system for dynamic and interorganizational design processes in chemical engineering. Computers & Chemical Engineering 29(1), 93–111 (2004)
170. Heller, M., Nagl, M., Wörzberger, R., Heer, T.: Dynamic Process Management Based Upon Existing Systems. This volume (2008)

171. Heller, M., Schleicher, A., Westfechtel, B.: A management system for evolving development processes. In: Proceedings of the 7th International Conference on Integrated Design and Process Technology (IDPT 2003), Austin, Texas, USA, SDPS (2003)
172. Heller, M., Schleicher, A., Westfechtel, B.: Graph-based specification of a management system for evolving development processes. In: Pfaltz, J.L., Nagl, M., Böhlen, B. (eds.) AGTIVE 2003. LNCS, vol. 3062, pp. 334–351. Springer, Heidelberg (2004)
173. Heller, M., Schleicher, A., Westfechtel, B.: Process evolution support in the AHEAD system. In: Pfaltz, J.L., Nagl, M., Böhlen, B. (eds.) AGTIVE 2003. LNCS, vol. 3062, pp. 454–460. Springer, Heidelberg (2004)
174. Heller, M., Westfechtel, B.: Dynamic project and workflow management for design processes in chemical engineering. In: Chen, B., Westerberg, A.W. (eds.) Proceedings of the 8th International Symposium on Process Systems Engineering (PSE 2003), Kunming, China, pp. 208–213. Elsevier, Amsterdam (2003)
175. Heller, M., Wörzberger, R.: Management support of interorganizational cooperative software development processes based on dynamic process views. In: 15th International Conference on Software Engineering and Data Engineering (SEDE 2006), Los Angeles, California, July 6–8, 2006, pp. 15–28 (2006)
176. Heller, M., Wörzberger, R.: A management system supporting interorganizational cooperative development processes in chemical engineering. In: 9th World Conference on Integrated Design & Process Technology (IDPT 2006), San Diego, California, USA, 25-30 June 2006, pp. 639–650. SDPS (2006)
177. Heller, M., Wörzberger, R.: A management system supporting interorganizational cooperative development processes in chemical engineering. Journal of Integrated Design and Process Science: Transactions of the SDPS 10(2), 57–78 (2007)
178. Herbst, D.: Entwicklung eines Modells zur Einführung von Telekooperation in der verteilten Produktentwicklung. PhD thesis, RWTH Aachen University (2000)
179. Herzberg, D., Marburger, A.: E-CARES research project: Understanding complex legacy telecommunication systems. In: Proceedings of the 5th European Conference on Software Maintenance and Reengineering (CSMR 2001), pp. 139–147. IEEE Computer Society Press, Los Alamitos (2001)
180. Heukamp, S.: Regelbasierte Werkzeuge zur Unterstützung der Korrespondenzanalyse zwischen Dokumenten. Diploma Thesis, RWTH Aachen University, to appear (2008)
181. Hoff, S.: Mobilitätsmanagement in Offenen Systemen - Leistungsbewertung von Verzeichnisdiensten. PhD thesis, RWTH Aachen University (1997)
182. Imhoff, F.: Objektorientierte Dienste in konvergierenden Kommunikationsnetzen. PhD thesis, RWTH Aachen University (2005)
183. Imhoff, F., Spaniol, O., Linnhoff-Popien, C., Garschhammer, M.: Aachen-Münchener Teleteaching unter Best-Effort-Bedingungen. Praxis der Informationsverarbeitung und Kommunikation (PIK) 3, 156–163 (2000)
184. Janning, T.: Requirements Engineering und Programmieren im Großen. PhD thesis, RWTH Aachen University (1992)

185. Janning, T., Lefering, M.: A transformation from requirements engineering into design – the method and the tool. In: Proceedings of the 3rd International Workshop on Software Engineering and its Applications, pp. 223–237 (1990)
186. Jarke, M.: Experience-based knowledge management: A cooperative information systems perspective. Control Engineering Practice 10(4), 561–569 (2002)
187. Jarke, M., Bubenko, J.A., Rolland, C., Sutcliffe, A., Vassiliou, Y.: Theories underlying requirements engineering: An overview of NATURE at genesis. In: Proceedings of the IEEE Symposium on Requirements Engineering, RE'93, San Diego, California, USA, IEEE Computer Society Press, Los Alamitos (1993)
188. Jarke, M., Fritzen, O., Miatidis, M., Schlüter, M.: Media-Assisted Product and Process Requirements Traceability in Supply Chains. In: Proceedings of the 11th International Requirements Engineering Conference (RE'03), Monterey, USA (2003)
189. Jarke, M., Gallersdörfer, R., Jeusfeld, M.A., Staudt, M., Eherer, S.: ConceptBase – a deductive object base for meta data management. Journal of Intelligent Information Systems 4(2), 167–192 (1995)
190. Jarke, M., Jeusfeld, M.A., Quix, C., Vassiliadis, P.: Architecture and quality of data ware-houses: an extended repository approach. Information Systems (Special Issue on Advanced Information Systems Engineering) 24(3), 229–253 (1999)
191. Jarke, M., Klamma, R.: Metadata and cooperative knowledge management. In: Pidduck, A.B., Mylopoulos, J., Woo, C.C., Ozsu, M.T. (eds.) CAiSE 2002. LNCS, vol. 2348, pp. 4–20. Springer, Heidelberg (2002)
192. Jarke, M., Lenzerini, M., Vassiliou, Y., Vassiliadis, P.: Fundamentals of Data Warehouses, 2nd edn. Springer, Heidelberg (2003)
193. Jarke, M., List, T., Köller, J.: The Challenge of Process Data Warehousing. In: Proceedings of the 26th International Conference on Very Large Databases (VLDB), Cairo, Egypt, pp. 473–483 (2000)
194. Jarke, M., List, T., Weidenhaupt, K.: A process-integrated conceptual design environment for chemical engineering. In: Akoka, J., Bouzeghoub, M., Comyn-Wattiau, I., Métais, E. (eds.) ER 1999. LNCS, vol. 1728, pp. 520–537. Springer, Heidelberg (1999)
195. Jarke, M., Marquardt, W.: Design and evaluation of computer aided modeling tools. In: AIChE Symposium 92, pp. 97–109 (1996)
196. Jarke, M., Mayr, H.C.: Mediengestütztes Anforderungsmanagement. Informatik-Spektrum 25(6), 452–464 (2002)
197. Jarke, M., Miatidis, M., Schlüter, M., Brandt, S.: Media-Assisted Product and Process Traceability in Supply Chain Engineering. In: Proceedings of the 37th Hawaii International Conference on System Sciences (HICSS), Big Island, Hawaii, USA, IEEE Computer Society Press, Los Alamitos (2004)
198. Jarke, M., Miatidis, M., Schlüter, M., Brandt, S.C.: Process-Integrated and Media-Assisted Traceability in Cross-Organizational Engineering. International Journal of Business Process Integration and Management 1(2), 65–75 (2006)
199. Jarke, M., Mylopoulos, J., Schmidt, J.W., Vassiliou, Y.: DAIDA: An environment for evolving information systems. ACM Transactions on Information Systems 10(1), 1–50 (1992)

200. Jarke, M., Pohl, K., Jacobs, S., Bubenko, J., Assenova, P., Holm, P., Wangler, B., Rolland, C., Plihon, V., Schmitt, J.R., Sutcliffe, A.G., Jones, S., Maiden, N.A.M., Till, D., Vassiliou, Y., Constantopoulos, P., Spanoudakis, G.: Requirements engineering: An integrated view of representation, process, and domain. In: Sommerville, I., Paul, M. (eds.) ESEC 1993. LNCS, vol. 717, pp. 100–114. Springer, Heidelberg (1993)
201. Jarke, M., Rolland, C., Sutcliffe, A., Dömges, R. (eds.): The NATURE of Requirements Engineering. Shaker, Aachen (1999)
202. Jarke, M., Rose, T.: Managing knowledge about information systems. In: Proceedings of the ACM SIGMOD International Conference of the Management of Data, Chicago, USA, pp. 303–311. ACM Press, New York (1998)
203. Jeusfeld, M.A.: Update Control in Deductive Object Bases. PhD thesis, University of Passau (1992)
204. Jeusfeld, M.A., Jarke, M., Nissen, H.W., Staudt, M.: ConceptBase. In: Bernus, P., Mertins, K., Schmidt, G. (eds.) Handbook on Architectures of Information Systems, Springer, Heidelberg (1998)
205. Jeusfeld, M.A., Jarke, M., Nissen, H.W., Staudt, M.: ConceptBase. In: Bernus, P., Mertins, K., Schmidt, G. (eds.) Handbook on Architectures of Information Systems, 2nd edn., Springer, Heidelberg (2006)
206. Jäger, D.: Generating tools from graph-based specifications. Information Software and Technology (Special Issue on Construction of Software Engineering Tools) 42(2), 129–139 (2000)
207. Jäger, D.: Modeling management and coordination in development processes. In: Conradi, R. (ed.) EWSPT 2000. LNCS, vol. 1780, pp. 109–114. Springer, Heidelberg (2000)
208. Jäger, D.: Unterstützung übergreifender Kooperation in komplexen Entwicklungsprozessen. PhD thesis, RWTH Aachen University (2002)
209. Jäger, D., Krapp, C.A., Nagl, M., Schleicher, A., Westfechtel, B.: Anpassbares Administrationssystem für die Projektkoordination. In: Nagl, M., Westfechtel, B. (eds.) Integration von Entwicklungssystemen in Ingenieuranwendungen – Substantielle Verbesserung der Entwicklungsprozesse, pp. 311–348. Springer, Heidelberg (1999)
210. Jäger, D., Marburger, A., Nagl, M., Schleicher, A.: EAI heiß nicht Zusammenschalten: Architekturüberlegungen für das verteilte Gesamtsystem. In: Nagl, M. (ed.) B2B mit EAI: Strategien mit XML, Java & Agenten – Proceedings of ONLINE'01, Congress VI, C610.01–C610.33. ONLINE GmbH (2001)
211. Jäger, D., Schleicher, A., Westfechtel, B.: Using UML for software process modeling. In: Nierstrasz, O., Lemoine, M. (eds.) ESEC 1999 and ESEC-FSE 1999. LNCS, vol. 1687, pp. 91–108. Springer, Heidelberg (1999)
212. Jäger, D., Schleicher, A., Westfechtel, B.: AHEAD: A graph-based system for modeling and managing development processes. In: Münch, M., Nagl, M. (eds.) AGTIVE 1999. LNCS, vol. 1779, pp. 325–339. Springer, Heidelberg (2000)
213. Kabel, D.: Entwicklung eines prozeßbasierten Effizienzmodells für Concurrent Engineering Teams. PhD thesis, RWTH Aachen University (2001)
214. Kabel, D., Nölle, T., Luczak, H.: Requirements for software-support in concurrent engineering teams. In: Luczak, H., Cakir, A.E., Cakir, G. (eds.) World Wide Work, Proceedings of the 6th International Scientific Conference on Work With Display Units, Berchtesgaden, Germany, pp. 202–204 (2002)

215. Karabek, R.: Data Communications in ATM Networks. PhD thesis, RWTH Aachen University (1998)
216. Kausch, B., Schneider, N., Schlick, C., Luczak, H.: Integrative Simulation of Work Processes. This volume (2008)
217. Kausch, B., Schneider, N., Schlick, C., Luczak, H.: Simulation-supported Workflow Optimization in Process Engineering. This volume (2008)
218. Kesdogan, D.: Vertrauenswürdige Kommunikation in offenen Umgebungen. PhD thesis, RWTH Aachen University (1999)
219. Kethers, S.: Multi-Perspective Modeling and Analysis of Cooperation Processes. PhD thesis, RWTH Aachen University (2000)
220. Kiesel, N., Schürr, A., Westfechtel, B.: GRAS: a graph-oriented software engineering database system. Information Systems 20(1), 21–51 (1995)
221. Killich, S., Luczak, H., Schlick, C., Weißenbach, M., Wiedenmaier, S., Ziegler, J.: Task modelling for cooperative work. Behaviour & Information Technology 18(5), 325–338 (1999)
222. Kirchhof, M.: Integrated Low-Cost eHome Systems (in German). PhD thesis, RWTH Aachen University, Aachen (2005)
223. Klamma, R.: Vernetztes Verbesserungsmanagement mit einem Unternehmensgedächtnis-Repository. PhD thesis, RWTH Aachen University (2000)
224. Klein, P.: Architecture Modeling of Distributed and Concurrent Software Systems. PhD thesis, RWTH Aachen University (2000)
225. Klein, P., Nagl, M.: Softwareintegration und Rahmenwerksentwicklung. In: Nagl, M., Westfechtel, B. (eds.) Integration von Entwicklungssystemen in Ingenieuranwendungen – Substantielle Verbesserung der Entwicklungsprozesse, pp. 423–440. Springer, Heidelberg (1999)
226. Klein, P., Nagl, M., Schürr, A.: IPSEN tools. In: Ehrig, H., Engels, G., Kreowski, H.J., Rozenberg, G. (eds.) Handbook on Graph Grammars and Computing by Graph Transformation – Volume 2: Applications, Languages, and Tools, pp. 215–265. World Scientific, Singapore (1999)
227. Klemke, R.: Modelling Context in Information Brokering Processes. PhD thesis, RWTH Aachen University (2002)
228. Knoop, S.: Modellierung von Entscheidungsabläufen in der verfahrenstechnischen Prozessentwicklung. Master's thesis, RWTH Aachen University (2005)
229. Kohring, C., Lefering, M., Nagl, M.: A requirements engineering environment within a tightly integrated SDE. Requirements Engineering 1(3), 137–156 (1996)
230. Körtgen, A., Heukamp, S.: Correspondence Analysis for Supporting Document Re-Use in Development Processes. In: Proceedings of the 12th World Conference on Integrated Design & Process Technology (IDPT '08), Taichung, Taiwan, SDPS, to appear (2008)
231. Kossack, S., Krämer, K., Marquardt, W.: Efficient optimization based design of distillation columns for homogenous azeotropic mixtures. Ind. Eng. Chem. Res. 45(24), 8492–8502 (2006)
232. Küpper, A.: Nomadic Communication in Converging Networks. PhD thesis, RWTH Aachen University (2001)
233. Kraft, B.: Semantical Support of the Conceptual Design in Civil Engineering (in German). PhD thesis, RWTH Aachen University, Aachen (2007)

234. Kraft, B., Meyer, O., Nagl, M.: Graph technology support for conceptual design in civil engineering. In: Proceedings of the 9th International EG-ICE Workshop. VDI Fortschritt-Berichte 4, vol. 180, pp. 1–35. VDI-Verlag, Düsseldorf (2002)
235. Kraft, B., Nagl, M.: Parameterized specification of conceptual design tools in civil engineering. In: Pfaltz, J.L., Nagl, M., Böhlen, B. (eds.) AGTIVE 2003. LNCS, vol. 3062, pp. 90–105. Springer, Heidelberg (2004)
236. Kraft, B., Nagl, M.: Graphbasierte Werkzeuge zur Unterstützung des konzeptuellen Gebäudeentwurfs. In: Rüppel, U. (ed.) Vernetzt-kooperative Planungsprozesse im Konstruktiven Ingenieurbau - Grundlagen, Methoden, Anwendungen und Perspektiven zur vernetzten Ingenieurkooperation, Number 3, pp. 155–176. Springer, Heidelberg (2007)
237. Kraft, B., Nagl, M.: Visual knowledge specification for conceptual design: Definition and tool support. Advanced Engineering Informatics 21(1), 67–83 (2007), http://dx.doi.org/10.1016/j.aei.2006.10.001
238. Kraft, B., Retkowitz, D.: Operationale Semantikdefinition für Konzeptuelles Regelwissen. In: Weber, L., Schley, F. (eds.) Proc. Forum Bauinformatik 2005, pp. 173–182. Lehrstuhl für Bauinformatik, Cottbus (2005)
239. Kraft, B., Retkowitz, D.: Graph Transformations for Dynamic Knowledge Processing. In: Robichaud, E. (ed.) Proc. of the 39th Hawaii Intl. Conf. on System Sciences (HICSS'06), Kauai, Hawaii, IEEE Computer Society Press, Los Alamitos (2006)
240. Kraft, B., Retkowitz, D.: Rule-Dependencies for Visual Knowledge Specification in Conceptual Design. In: Rivard, H. (ed.) Proc. of the 11th Intl. Conf. on Computing in Civil and Building Engineering (ICCCBE-XI), Montreal, Canada, ACSE (2006)
241. Kraft, B., Wilhelms, N.: Visual knowledge specification for conceptual design. In: Soibelman, L., Pena-Mora, F. (eds.) Proceedings of the 2005 ASCE International Conference on Computing in Civil Engineering (ICCC 2005), Cancun, Mexiko, pp. 1–14 (2005)
242. Krapp, C.-A.: Parametrisierung dynamischer Aufgabennetze zum Management von Softwareprozessen. Softwaretechnik-Trends 16(3), 33–40 (1996)
243. Krapp, C.A.: An Adaptable Environment for the Management of Development Processes. PhD thesis, RWTH Aachen University (1998)
244. Krapp, C.A., Krüppel, S., Schleicher, A., Westfechtel, B.: Graph-based models for managing development processes, resources, and products. In: Ehrig, H., Engels, G., Kreowski, H.-J., Rozenberg, G. (eds.) TAGT 1998. LNCS, vol. 1764, pp. 455–474. Springer, Heidelberg (2000)
245. Krapp, C.A., Schleicher, A., Westfechtel, B.: Feedback handling in software processes. In: Ehrig, H., Engels, G., Kreowski, H.J., Rozenberg, G. (eds.) Theory and Application of Graph Transformations: 6th International Workshop, TAGT'98, Paderborn, Germany, 1998, November 16–20, 1998, pp. 417–424. Springer, Heidelberg (2000)
246. Krapp, C.-A., Westfechtel, B.: Feedback handling in dynamic task nets. In: Proceedings of the 12th International Conference on Automated Software Engineering, Incline Village, Nevada, USA, pp. 301–302. IEEE Computer Society Press, Los Alamitos (1997)

247. Krobb, C., Hackenberg, J.: Modellierung und Unterstützung verfahrenstechnischer Modellierungsprozesse. In: Informatik 2000, 30. Jahrestagung der Gesellschaft für Informatik (2000)
248. Krobb, C., Lohmann, B., Marquardt, W.: The chemical engineering data model VeDa. Part 6: The process of model development. Technical Report LPT-1998-06, RWTH Aachen University, Process Systems Engineering (1998)
249. Krüppel, S., Westfechtel, B.: RESMOD: A resource management model for development processes. In: Ehrig, H., Engels, G., Kreowski, H.J., Rozenberg, G. (eds.) Theory and Application of Graph Transformations: 6th International Workshop, TAGT'98, Paderborn, Germany, 1998, November 16–20, 1998, Selected Papers, pp. 390–397. Springer, Heidelberg (2000)
250. Körtgen, A.: Modeling Successively Connected Repetitive Subgraphs. In: Schürr, A., Nagl, M., Zündorf, A. (eds.) Applications of Graph Transformations with Industrial Relevance. Proceedings of the Third International AGTIVE 2007, Kassel, Germany. LNCS, vol. 5088, pp. 428–443. Springer, Heidelberg (2008)
251. Körtgen, A., Becker, S., Herold, S.: A Graph-Based Framework for Rapid Construction of Document Integration Tools. In: Proceedings of the 11th World Conference on Integrated Design & Process Technology (IDPT 2007), Antalya, Turkey, SDPS (2007)
252. Kulikov, V., Briesen, H., Grosch, R., Yang, A., von Wedel, L., Marquardt, W.: Modular dynamic simulation of integrated process flowsheets by means of tool integration. Chemical Engineering Science 60(7), 2069–2083 (2005)
253. Kulikov, V., Briesen, H., Marquardt, W.: A framework for the simulation of mass crystallization considering the effect of fluid dynamics. Chemical Engineering and Processing 45, 886–899 (2006)
254. Lefering, M.: Software document integration using graph grammar specifications. Journal of Computing and Information (CD-ROM Journal, Special Issue: Proceedings of the 6th International Conference on Computing and Information, ICCI'94) 1(1), 1222–1243 (1994)
255. Lefering, M.: Integrationswerkzeuge in einer Softwareentwicklungsumgebung. PhD thesis, RWTH Aachen University (1995)
256. Lefering, M.: Realization of incremental integration tools. In: Nagl, M. (ed.) IPSEN 1996. LNCS, vol. 1170, pp. 469–481. Springer, Heidelberg (1996)
257. Lefering, M., Janning, T.: Transition between different working areas: Vertical integration tools. In: Nagl, M. (ed.) IPSEN 1996. LNCS, vol. 1170, pp. 195–207. Springer, Heidelberg (1996)
258. Lefering, M., Kohring, C., Janning, T.: Integration of different perspectives: The requirements engineering environment. In: Nagl, M. (ed.) IPSEN 1996. LNCS, vol. 1170, pp. 178–194. Springer, Heidelberg (1996)
259. Lefering, M., Schürr, A.: Specification of integration tools. In: Nagl, M. (ed.) IPSEN 1996. LNCS, vol. 1170, pp. 324–334. Springer, Heidelberg (1996)
260. Lewerentz, C.: Interaktives Entwerfen großer Programmsysteme: Konzepte und Werkzeuge. PhD thesis, RWTH Aachen University (1988)
261. Licht, T., Dohmen, L., Schmitz, P., Schmidt, L., Luczak, H.: Person-centered simulation of product development processes using timed stochastic coloured petri nets. In: Proceedings of the European Simulation and Modelling Conference, EUROSIS-ETI, Ghent, Belgium, pp. 188–195 (2004)

262. Linnhoff-Popien, C.: CORBA: Kommunikation und Management. Springer, Heidelberg (1998)
263. Linnhoff-Popien, C., Haustein, T.: Das Plug-In-Modell zur Realisierung mobiler CORBA-Objekte. In: Proceedings of the 11th ITG/GI-Fachtagung Kommunikation in Verteilten Systemen, Darmstadt, Germany, pp. 196–209. Springer, Heidelberg (1999)
264. Linnhoff-Popien, C., Lipperts, S., Thißen, D.: Management verfahrenstechnischer Entwicklungswerkzeuge. Praxis der Informationsverarbeitung und Kommunikation (PIK) 1, 22–31 (1999)
265. Linnhoff-Popien, C., Thißen, D.: Assessing service properties with regard to a requested QoS: The service metric. In: Proceedings of the 3rd International Conference on Formal Methods for Open Object-Based Distributed Systems (FMOODS 99), Florence, Italy, pp. 273–280. Kluwer Academic Publishers, Dordrecht (1999)
266. Lipperts, S.: COMANGA – an architecture for corba management using mobile agents. In: Proceedings of the 14th International Conference on Advanced Science and Technology, Chicago, USA, pp. 327–336 (1998)
267. Lipperts, S.: CORBA for inter-agent communication of management information. In: Proceedings of the 5th IEEE International Workshop on Mobile Multimedia Communication, Berlin, Germany, IEEE Computer Society Press, Los Alamitos (1998)
268. Lipperts, S.: Enabling alarm correlation for a mobile agent based system and network management – a wrapper concept. In: Proceedings of the IEEE International Conference On Networks, Brisbane, Australia, pp. 125–132. IEEE Computer Society Press, Los Alamitos (1999)
269. Lipperts, S.: Mobile Agenten zur Unterstützung kooperativer Managementprozesse. In: Beiersdörfer, K., Engels, G., Schäfer, W. (eds.) Proceedings of Informatik'99, Paderborn, Germany, pp. 231–238. Springer, Heidelberg (1999)
270. Lipperts, S.: How to efficiently deploy mobile agents for an integrated management. In: Linnhoff-Popien, C., Hegering, H.-G. (eds.) USM 2000. LNCS, vol. 1890, pp. 290–295. Springer, Heidelberg (2000)
271. Lipperts, S.: On the efficient deployment of mobility in distributed system management. In: Proceedings of the 3rd International Workshop on Mobility in Databases & Distributed Systems (MDDS'2000), Greenwich, UK (2000)
272. Lipperts, S.: Mobile Agent Support Services. PhD thesis, RWTH Aachen University (2002)
273. Lipperts, S., Park, A.S.B.: Managing corba with agents. In: Proceedings of the 4th International Symposium on Interworking, Ottawa, Canada (1998)
274. Lipperts, S., Stenzel, R.: Agents that do the right thing – how to deal with decisions under uncertainty. In: Proceedings of the 6th World Multi-Conference on Systemics, Cybernetics and Informatics (SCI 2000), Orlando, Florida, USA, pp. 77–82 (2000)
275. Lipperts, S., Thißen, D.: CORBA wrappers for a-posteriori management: An approach to integrating management with existing heterogeneous systems. In: Proceedings of the 2nd IFIP International Working Conference on Distributed Applications and Interoperable Systems, Helsinki, Finland, pp. 169–174. Kluwer Academic Publishers, Dordrecht (1999)
276. List, T.: Nachvollziehbarkeit und Überwachung als Vertrauensdienste auf elektronischen Marktplätzen. PhD thesis, RWTH Aachen University (2004)

277. Lohmann, B.: Ansätze zur Unterstützung des Arbeitsablaufes bei der rechnerbasierten Modellierung verfahrenstechnischer Prozesse. PhD thesis, RWTH Aachen University. Published in: Fortschritt-Berichte VDI: Reihe 3, Nr. 531. VDI-Verlag, Düsseldorf (1998)
278. L.P.T.: Process systems engineering at RWTH Aachen University (2006), http://www.lpt.rwth-aachen.de/
279. Luczak, H.: Modelle flexibler Arbeitsformen und Arbeitszeiten. In: Spur, G. (ed.) CIM – Die informationstechnische Herausforderung, Produktionstechnisches Kolloquium, pp. 227–245 (1986)
280. Luczak, H.: Task analysis. In: Handbook of Human Factors and Ergonomics, pp. 340–416. John Wiley & Sons, Chichester (1997)
281. Luczak, H., Cakir, A.E., Cakir, G.: World Wide Work, Proceedings of the 6th International Scientific Conference on Work With Display Units, Berchtesgaden, Germany (2002)
282. Luczak, H., Foltz, C., Mühlfelder, M.: Telekooperation. Zeitschrift für Arbeitswissenschaft 56(4), 295–299 (2002)
283. Luczak, H., Mühlfelder, M., Schmidt, L.: Group task analysis and design of computer supported cooperative work. In: Hollnagel, E. (ed.) Handbook of Cognitive Task Design, pp. 99–127. Lawrence Erlbaum, Mahwah (2003)
284. Luczak, H., Stahl, J.: Task analysis in industry. In: Karwowski, W. (ed.) International Encyclopedia for Industrial Ergonomics, vol. 3, pp. 1911–1914. Taylor & Francis, Abington (2001)
285. Luczak, H., Wolf, M., Schlick, C., Springer, J., Foltz, C.: Personenorientierte Arbeitsprozesse und Kommunikationsformen. In: Nagl, M., Westfechtel, B. (eds.) Integration von Entwicklungssystemen in Ingenieuranwendungen – Substantielle Verbesserung der Entwicklungsprozesse, pp. 403–422. Springer, Heidelberg (1999)
286. Marburger, A.: Reverse Engineering of Complex Legacy Telecommunication Systems. PhD thesis, RWTH Aachen University (2004)
287. Marburger, A., Westfechtel, B.: Tools for understanding the behavior of telecommunication systems. In: Proceedings of the 25th International Conference on Software Engineering (ICSE 2003), pp. 430–441. IEEE Computer Society Press, Los Alamitos (2003)
288. Marquardt, W.: Dynamic process simulation – recent progress and future challenges. In: Arkun, Y., Ray, W.H. (eds.) Proceedings of the Fourth International Conference on Chemical Process Control, Padre Island, Texas, February 17–22, 1991, pp. 131–180 (1991)
289. Marquardt, W.: An object-oriented representation of structured process models. Computers & Chemical Engineering Suppl. 16, S329–S336 (1992)
290. Marquardt, W.: Rechnergestützte Erstellung verfahrenstechnischer Prozeßmodelle. Chem.-Ing.-Tech. 64, 25–40 (1992)
291. Marquardt, W.: Computer-aided generation of chemical engineering process models. International Chemical Engineering 34, 28–46 (1994)
292. Marquardt, W.: Towards a process modeling methodology. In: Berber, R. (ed.) Model-Based Process Control. NATO-ASI Series, vol. 291, pp. 3–40. Kluwer Academic Publishers, Dordrecht (1995)
293. Marquardt, W.: Trends in computer-aided process modeling. Computers & Chemical Engineering 20(6/7), 591–609 (1996)

294. Marquardt, W.: Review from a Design Process Perspective. This volume (2008)
295. Marquardt, W., Nagl, M.: Tool integration via interface standardization? In: Computer Application in Process and Plant Engineering – Papers of the 36th Tutzing Symposium, pp. 95–126. Wiley-VCH, Weinheim (1999)
296. Marquardt, W., Nagl, M.: Arbeitsprozess-orientierte Integration von Software-Werkzeugen zur Unterstützung verfahrenstechnischer Entwicklungsprozesse. In: Engineering in der Prozessindustrie. VDI Fortschritt-Berichte, vol. 1684, pp. 91–101. VDI-Verlag, Düsseldorf (2002)
297. Marquardt, W., Nagl, M.: Arbeitsprozessorientierte Unterstützung verfahrenstechnischer Entwicklungsprozesse. atp – Automatisierungstechnische Praxis 45(4), 52–58 (2003)
298. Marquardt, W., Nagl, M.: Workflow and information centered support of design processes. In: Chen, B., Westerberg, A.W. (eds.) Proceedings of the 8th International Symposium on Process Systems Engineering (PSE 2003), Kunming, China, pp. 101–124. Elsevier, Amsterdam (2003)
299. Marquardt, W., Nagl, M.: Workflow and information centered support of design processes – the IMPROVE perspective. Computers & Chemical Engineering 29(1), 65–82 (2004)
300. Marquardt, W., Nagl, M.: A Model-driven Approach for A-posteriori Tool Integration. This volume (2008)
301. Marquardt, W., Nagl, M.: Review from an Industrial Perspective. This volume (2008)
302. Marquardt, W., Pantelides, C.: 16th European Symposium on Computer Aided Process Engineering and 9th International Symposium on Process Systems Engineering, Garmisch-Partenkirchen, Germany, July 9–13, 2006. Computer-Aided Chemical Engineering, vol. 21. Elsevier, Amsterdam (2006)
303. Marquardt, W., von Wedel, L., Bayer, B.: Perspectives on lifecycle process modeling. In: Malone, M.F., Trainham, J.A., Carnahan, B. (eds.) Foundations of Computer-Aided Process Design. AIChE Symposium Series, vol. 96(323), pp. 192–214 (2000)
304. Meggers, J.: Adaptive admission control and scheduling for wireless packet communication. In: Proceedings of the IEEE International Conference on Networks (ICON'99), Brisbane, Australia, IEEE Computer Society Press, Los Alamitos (1999)
305. Meggers, J.: Adaptiver Videotransport in Mobilfunknetzen. PhD thesis, RWTH Aachen University (2000)
306. Meggers, J., Schuba, M.: Analysis of feedback error control schemes for block based video communication. In: Proceedings of the International Packet Video Workshop (PV'99), New York, USA, IEEE Computer Society Press, Los Alamitos (1999)
307. Meggers, J., Subramaniam, R.: A new feedback error control schemes for block based video communication in packet switched wireless networks. In: Proceedings of the 4th IEEE International Symposium on Computer Communications (ISCC'99), Sharm El Sheik, Red Sea, Egypt, IEEE Computer Society Press, Los Alamitos (1999)
308. Meyer, O.: aTool: Typography as source for structuring texts (in German). PhD thesis, RWTH Aachen University (2006)

309. Mühlfelder, M., Kabel, D., Hensel, T., Schlick, C.: Werkzeuge für kooperatives Wissensmanagement in Forschung und Entwicklung. Wissensmanagement 4, 10–15 (2001)
310. Miatidis, M.: Integrated Experience-Based Support of Cooperative Engineering Design Processes. PhD thesis, RWTH Aachen University (2007)
311. Miatidis, M., Jarke, M., Weidenhaupt, K.: Using Developers' Experience in Cooperative Design Processes. This volume (2008)
312. Miatidis, M., Theißen, M., Jarke, M., Marquardt, W.: Work Processes and Process-Centered Models and Tools. This volume (2008)
313. Michaeli, W., Grefenstein, A.: Engineering analysis and design of twin-screw extruders for reactive extrusion. Advances in Polymer Technology 14(4), 263–276 (1995)
314. Michaeli, W., Grefenstein, A., Frings, W.: Synthesis of polystyrene and styrene copolymers by reactive extrusion. Advances in Polymer Technology 12(1), 25–33 (1993)
315. Michaeli, W., Haberstroh, E., Seidel, H., Schmitz, T., Stewering, J., van Hoorn, R.: Visualisierung von Strömungsdaten aus CFD-Simulationen mit Virtual Reality Techniken. In: Proceedings of the 23rd Internationales Kunststofftechnisches Kolloquium (2006)
316. Münch, M.: Generic Modeling with Graph Rewriting Systems. PhD thesis, RWTH Aachen University (2003)
317. Molenaar, E., Trossen, D.: The multipoint event sharing service (MESS). In: Proceedings of the 13th Information Resources Management Association International Conference (IRMA 2000), Seattle, USA, IRMA, Strasbourg (2002)
318. Molitor, R.: Unterstützung der Modellierung verfahrenstechnischer Prozesse durch Nicht-Standardinferenzen in Beschreibungslogiken. PhD thesis, RWTH Aachen University (2000)
319. Morbach, J., Bayer, B., Yang, A., Marquardt, W.: Product Data Models. This volume (2008)
320. Morbach, J., Hai, R., Bayer, B., Marquardt, W.: Document Models. This volume (2008)
321. Morbach, J., Marquardt, W.: Ontology-Based Integration and Management of Distributed Design Data. This volume (2008)
322. Morbach, J., Theißen, M., Marquardt, W.: An Introduction to Application Domain Modeling. This volume (2008)
323. Morbach, J., Theißen, M., Marquardt, W.: Integrated Application Domain Models for Chemical Engineering. This volume (2008)
324. Morbach, J., Wiesner, A., Marquardt, W.: OntoCAPE 2.0 – a (re-)usable ontology for computer-aided process engineering. In: Proceedings of the European Symposium on Computer Aided Process Engineering – ESCAPE 18, accepted (2008)
325. Morbach, J., Yang, A.: Ontology OntoCAPE (2006), http://www.lpt.rwth-aachen.de/Research/ontocape.php
326. Morbach, J., Yang, A., Marquardt, W.: OntoCAPE – a large-scale ontology for chemical process engineering. Engineering Applications of Artificial Intelligence 20(2), 147–161 (2007)
327. Mylopoulos, J., Borgida, A., Jarke, M., Koubarakis, M.: Telos: Representing knowledge about information systems. ACM Transactions on Information Systems 8(4), 325–362 (1990)

328. Nagl, M.: Graph-Grammatiken: Theorie, Anwendungen, Implementierung. Vieweg, Wiesbaden (1979)
329. Nagl, M.: Characterization of the IPSEN project. In: Madhavji, N., Schäfer, W., Weber, H. (eds.) Proceedings of the 1st International Conference on System Development Environments & Factories, pp. 141–150 (1990)
330. Nagl, M.: Das Forschungsprojekt IPSEN. Informatik Forschung und Entwicklung 5, 103–105 (1990)
331. Nagl, M.: Softwaretechnik: Methodisches Programmieren im Großen. Springer, Heidelberg (1990), 2nd edn. under the title "Modellierung von Software-Architekturen" (to appear, 2009)
332. Nagl, M.: Eng integrierte Softwareentwicklungs-Umgebungen: Ein Erfahrungsbericht über das IPSEN-Projekt. Informatik Forschung und Entwicklung 8(3), 105–119 (1993)
333. Nagl, M.: Software-Entwicklungsumgebungen: Einordnung und zukünftige Entwicklungslinien. Informatik-Spektrum 16(5), 273–280 (1993)
334. Nagl, M. (ed.): IPSEN 1996. LNCS, vol. 1170. Springer, Heidelberg (1996)
335. Nagl, M. (ed.): Verteilte, integrierte Anwendungsarchitekturen: Die Software-Welt im Umbruch – Proceedings of ONLINE'97, Congress VI. ONLINE GmbH (1997)
336. Nagl, M.: Softwaretechnik mit Ada 95: Entwicklung großer Systeme, 6th edn. Vieweg, Wiesbaden (2003)
337. Nagl, M.: From Application Domain Models to Tools: The Sketch of a Layered Process/Product Model. This volume (2008)
338. Nagl, M.: Process/Product Model: Status and Open Problems. This volume (2008)
339. Nagl, M.: Review from Academic Success Perspective. This volume (2008)
340. Nagl, M.: Review from Tools' Side. This volume (2008)
341. Nagl, M.: The Interdisciplinary IMPROVE Project. This volume (2008)
342. Nagl, M., Faneye, O.B.: Gemeinsamkeiten und Unterschiede von Entwicklungsprozessen in verschiedenen Ingenieurdisziplinen. In: Nagl, M., Westfechtel, B. (eds.) Modelle, Werkzeuge und Infrastrukturen zur Unterstützung von Entwicklungsprozessen, pp. 311–324. Wiley-VCH, Weinheim (2003)
343. Nagl, M., Marquardt, W.: SFB 476 IMPROVE: Informatische Unterstützung übergreifender Entwicklungsprozesse in der Verfahrenstechnik. In: Jarke, M., Pohl, K., Pasedach, K. (eds.) Informatik als Innovationsmotor (GI Jahrestagung '97). Informatik Aktuell, pp. 143–154. Springer, Heidelberg (1997)
344. Nagl, M., Marquardt, W.: Übersicht über den SFB IMPROVE: Probleme, Ansatz, Lösungsskizze. In: Nagl, M., Westfechtel, B. (eds.) Integration von Entwicklungssystemen in Ingenieuranwendungen – Substantielle Verbesserung der Entwicklungsprozesse, pp. 217–250. Springer, Heidelberg (1999)
345. Nagl, M., Marquardt, W.: Informatische Konzepte für verfahrenstechnische Entwicklungsprozesse. In: Walter, R., Rauhut, B. (eds.) Horizonte – Die RWTH auf dem Weg ins 21. Jahrhundert, pp. 292–300. Springer, Heidelberg (1999)
346. Nagl, M., Marquardt, W.: Tool integration via cooperation functionality. In: Proceedings of the 3rd European Congress on Chemical Engineering, CD-ROM (2001), Abstract: Chem.-Ing.-Tech. 6, 622 (2001)

347. Nagl, M., Marquardt, W. (eds.): Collaborative and Distributed Chemical Engineering: From Understanding to Substantial Design Process Support. This volume (2008)
348. Nagl, M., Schneider, R., Westfechtel, B.: Synergetische Verschränkung bei der A-posteriori-Integration von Werkzeugen. In: Nagl, M., Westfechtel, B. (eds.) Modelle, Werkzeuge und Infrastrukturen zur Unterstützung von Entwicklungsprozessen, pp. 137–154. Wiley-VCH, Weinheim (2003)
349. Nagl, M., Schürr, A.: Software integration problems and coupling of graph grammar specifications. In: Cuny, J., Engels, G., Ehrig, H., Rozenberg, G. (eds.) Graph Grammars 1994. LNCS, vol. 1073, pp. 155–169. Springer, Heidelberg (1996)
350. Münch, M., Nagl, M. (eds.): AGTIVE 1999. LNCS, vol. 1779. Springer, Heidelberg (2000)
351. Nagl, M., Westfechtel, B.: Das Forschungsprojekt SUKITS. Informatik Forschung und Entwicklung 8(4), 212–214 (1993)
352. Nagl, M., Westfechtel, B. (eds.): Integration von Entwicklungssystemen in Ingenieuranwendungen – Substantielle Verbesserung der Entwicklungsprozesse. Springer, Heidelberg (1999)
353. Nagl, M., Westfechtel, B. (eds.): Modelle, Werkzeuge und Infrastrukturen zur Unterstützung von Entwicklungsprozessen. Wiley-VCH, Weinheim (2003)
354. Nagl, M., Westfechtel, B.: Some notes on the empirical evaluation of innovative tools for engineering design processes. In: Subrahmanian, E., Sriram, R., Herder, P., Schneider, R. (eds.) The role of empirical studies in understanding and supporting engineering design – Workshop Proceedings, National Institute of Standards and Technology, Gaithersburg, Maryland, USA, 4–5 April 2002, pp. 53–64. DUP Science, Delft (2004)
355. Nagl, M., Westfechtel, B., Schneider, R.: Tool support for the management of design processes in chemical engineering. Computers & Chemical Engineering 27(2), 175–197 (2003)
356. Nick, A.: Personalisiertes Information Brokering. PhD thesis, RWTH Aachen University (2002)
357. Nicola, M.: Performance evaluation of distributed, replicated, and wireless information systems. PhD thesis, RWTH Aachen University (1999)
358. Nissen, H.W., Jarke, M.: Repository support for multi-perspective requirements engineering. Information Systems (Special Issue on Meta Modeling and Method Engineering) 24(2), 131–158 (1999)
359. Nissen, H.: Separierung und Resolution multipler Perspektiven in der konzeptuellen Modellierung. PhD thesis, RWTH Aachen University (1997)
360. Norbisrath, U., Armac, I., Retkowitz, D., Salumaa, P.: Modeling eHome Systems. In: Terzis, S. (ed.) Proc. of the 4th Intl. Workshop on Middleware for Pervasive and Ad-Hoc Computing (MPAC 2006), Melbourne, Australia, ACM Press, New York (2006)
361. Park, A.: A Service-Based Agent System Supporting Mobile Computing. PhD thesis, RWTH Aachen University (2004)
362. Park, A.S.B., Lipperts, S.: Prototype approaches to a mobile agent service trader. In: Proceedings of the 4th International Symposium on Interworking, Ottawa, Canada (1998)
363. Pfaltz, J.L., Nagl, M., Böhlen, B. (eds.): AGTIVE 2003. LNCS, vol. 3062. Springer, Heidelberg (2004)

364. Pils, C.: Leistungsorientierte Dienstselektion für mobile Agenten im Internet – Elektronische Staumelder auf der Datenautobahn. PhD thesis, RWTH Aachen University (2005)
365. Pohl, K.: The three dimensions of requirements engineering: A framework and its applications. Information Systems 19(3), 243–258 (1994)
366. Pohl, K.: Process-Centered Requirements Engineering. Research Studies Press, Taunton (1996)
367. Pohl, K.: Continuous Documentation of Information Systems Requirements. Habilitationsschrift, RWTH Aachen (1999)
368. Pohl, K., Dömges, R., Jarke, M.: Towards method-driven trace capture. In: Olivé, À., Pastor, J.A. (eds.) CAiSE 1997. LNCS, vol. 1250, pp. 103–116. Springer, Heidelberg (1997)
369. Pohl, K., Weidenhaupt, K.: A contextual approach for process-integrated tools. In: Jazayeri, M. (ed.) ESEC 1997 and ESEC-FSE 1997. LNCS, vol. 1301, pp. 176–192. Springer, Heidelberg (1997)
370. Pohl, K., Weidenhaupt, K., Dömges, R., Haumer, P., Jarke, M.: Prozeßintegration in PRIME: Modelle, Architektur, Vorgehensweise. In: Proceedings of Softwaretechnik '98, Paderborn, Germany, pp. 42–52 (1998)
371. Pohl, K., Weidenhaupt, K., Dömges, R., Haumer, P., Jarke, M., Klamma, R.: PRIME: Towards process-integrated environments. ACM Transactions on Software Engineering and Methodology 8(4), 343–410 (1999)
372. Pohl, K.R., Dömges, R., Jarke, M.: Decision oriented process modelling. In: Proceedings of the 9th International Software Process Workshop, pp. 203–208 (2001)
373. Raddatz, M., Schlüter, M., Brandt, S.C.: Identification and reuse of experience knowledge in continuous production processes. In: Proceedings of the 9th IFAC Symposium on Automated Systems Based on Human Skill And Knowledge, Nancy, France (2006)
374. Radermacher, A.: Support for design patterns through graph transformation tools. In: Münch, M., Nagl, M. (eds.) AGTIVE 1999. LNCS, vol. 1779, pp. 111–126. Springer, Heidelberg (2000)
375. Radermacher, A.: Tool Support for the Distribution of Object-Based Systems. PhD thesis, RWTH Aachen University (2000)
376. Ramesh, B., Jarke, M.: Toward Reference Models for Requirements Traceability. IEEE Transactions on Software Engineering 27(1), 58–93 (2001)
377. Ranger, U.: Model-driven Development of Distributed Systems using Graph Rewriting Languages (in German). PhD thesis, RWTH Aachen University, Aachen (2008)
378. Ranger, U., Gruber, K., Holze, M.: Defining abstract graph views as module interfaces. In: Schürr, A., Nagl, M., Zündorf, A. (eds.) Applications of Graph Transformations with Industrial Relevance. Proceedings of the Third International AGTIVE 2007, Kassel, Germany. LNCS, vol. 5088, pp. 123–138. Springer, Heidelberg (2008)
379. Ranger, U., Hermes, T.: Ensuring consistency within distributed graph transformation systems. In: Dwyer, M.B., Lopes, A. (eds.) FASE 2007. LNCS, vol. 4422, pp. 368–382. Springer, Heidelberg (2007)
380. Raschka, R.: Entwurf und Gestaltung der Benutzungsschnittstelle einer integrierten Kommunikationsumgebung. Master's thesis, RWTH Aachen University (2003)

381. Reichl, P.: Dynamische Verkehrs- und Preismodellierung für den Einsatz in Kommunikationssystemen. PhD thesis, RWTH Aachen University (1999)
382. Reimann, S.: Entgasung von Polymeren auf gleichlaufenden, dichtkämmenden Doppelschneckenextrudern am Beispiel von Polystyrol. PhD thesis, RWTH Aachen University (1998)
383. Retkowitz, D., Stegelmann, M.: Dynamic Adaptability for Smart Environments. In: Distributed Applications and Interoperable Systems (DAIS 2008), Oslo, Norway. LNCS, vol. 5053, pp. 154–167. Springer, Heidelberg (2008)
384. Sattler, U.: Terminological Knowledge Representation Systems in a Process Engineering Application. PhD thesis, RWTH Aachen University (1998)
385. Schäfer, W.: Eine integrierte Softwareentwicklungs-Umgebung: Konzepte, Entwurf und Implementierung. PhD thesis, University of Osnabrück (1986)
386. Schirski, M., Gerndt, A., van Reimersdahl, T., Kuhlen, T., Adomeit, P., Lang, O., Pischinger, S., Bischof, C.: ViSTA FlowLib – a framework for interactive visualization and exploration of unsteady flows in virtual environments. In: Proceedings of the 7th International Immersive Projection Technologies Workshop and the 9th Eurographics Workshop on Virtual Environments, pp. 77–85. ACM Press, New York (2003)
387. Schleicher, A.: High-level modeling of development processes. In: Scholz-Reiter, B., Stahlmann, H.-D., Nethe, A. (eds.) First International Conference on Process Modelling, pp. 57–73 (1999)
388. Schleicher, A.: Objektorientierte Modellierung von Entwicklungsprozessen mit der UML. In: Desel, J., Pohl, K., Schürr, A. (eds.) Modellierung '99, pp. 171–186. Teubner, Wiesbaden (1999)
389. Schleicher, A.: Formalizing UML-Based Process Models Using Graph Transformations. In: Münch, M., Nagl, M. (eds.) AGTIVE 1999. LNCS, vol. 1779, pp. 341–357. Springer, Heidelberg (2000)
390. Schleicher, A.: Management of Development Processes: An Evolutionary Approach. PhD thesis, RWTH Aachen University (2002)
391. Schleicher, A., Westfechtel, B.: Beyond stereotyping: Metamodeling for the UML. In: Proceedings of the 34th Hawaii International Conference on System Sciences (HICSS), Minitrack: Unified Modeling – A Critical Review and Suggested Future, IEEE Computer Society Press, Los Alamitos (2001)
392. Schleicher, A., Westfechtel, B.: Unterstützung von Entwicklungsprozessen durch Werkzeuge. In: Nagl, M., Westfechtel, B. (eds.) Modelle, Werkzeuge und Infrastrukturen zur Unterstützung von Entwicklungsprozessen, pp. 329–332. Wiley-VCH, Weinheim (2003)
393. Schlick, C.: Modellbasierte Gestaltung der Benutzungsschnittstelle autonomer Produktionszellen. PhD thesis, RWTH Aachen University (2000)
394. Schlüter, M.: Konzepte und Werkzeuge zur rechnerunterstützten Auslegung von Aufbereitungsextrudern in übergreifenden Entwicklungsprozessen. PhD thesis, RWTH Aachen University (2004)
395. Schmidt, L., Luczak, H.: A cognitive engineering approach to computer supported cooperative work. In: Luczak, H., Cakir, A.E., Cakir, G. (eds.) World Wide Work, Proceedings of the 6th International Scientific Conference on Work With Display Units, Berchtesgaden, Germany, pp. 208–210 (2002)
396. Schneider, R.: Workshop des Sonderforschungsbereichs 476 IMPROVE: Rechnerunterstützung bei der Prozessentwicklung. Chemie Ingenieur Technik 73, 275–276 (2001)

397. Schneider, R.: Einsatzmöglichkeiten der Arbeitsprozessmodellierung am Beispiel NA 35. In: Engineering in der Prozessindustrie. VDI Fortschritt-Berichte, vol. 1684, pp. 33–38. VDI-Verlag, Düsseldorf (2002)
398. Schneider, R.: Erfassung und Analyse von Arbeitsabläufen bei Entwicklungsprozessen. Chem.-Ing.-Tech. 74(5), 612 (2002)
399. Schneider, R.: Informationstechnologien für Entwicklung und Produktion in der Verfahrenstechnik. atp – Automatisierungstechnische Praxis 46, 35–37 (2004)
400. Schneider, R., Gerhards, S.: WOMS – a work process modeling tool. In: Nagl, M., Westfechtel, B. (eds.) Modelle, Werkzeuge und Infrastrukturen zur Unterstützung von Entwicklungsprozessen, pp. 375–376. Wiley-VCH, Weinheim (2003)
401. Schneider, R., Marquardt, W.: Information technology support in the chemical process design life cycle. Chemical Engineering Science 57(10), 1763–1792 (2002)
402. Schneider, R., von Wedel, L., Marquardt, W.: Industrial Cooperation Resulting in Transfer. This volume (2008)
403. Schneider, R., Westfechtel, B.: A Scenario Demonstrating Design Support in Chemical Engineering. This volume (2008)
404. Schoop, M.: Business Communication in Electronic Commerce. Habilitation Thesis, RWTH Aachen University (2003)
405. Schoop, M., Jertila, A., List, T.: Negoisst: A negotiation support system for electronic business-to-business negotiation in e-commerce. Data & Knowledge Engineering 47(3), 371–401 (2003)
406. Schoop, M., Köller, J., List, T., Quix, C.: A three-phase model of electronic marketplaces for software components in chemical engineering. In: Schmid, B., Stanoevska-Slabeva, K., Tschammer, V. (eds.) Towards the E-Society, Proceedings of the First IFIP Conference on E-Commerce E-Government, E-Business, Zurich, Switzerland, pp. 507–522. Kluwer Academic Publishers, Dordrecht (2001)
407. Schopfer, G.: A Framework for Tool Integration in Chemical Process Modeling. PhD thesis, RWTH Aachen University. Published in: Fortschritt-Berichte VDI, Nr. 868, VDI-Verlag, Düsseldorf (2006)
408. Schopfer, G., Wyes, J., Marquardt, W., von Wedel, L.: A library for equation system processing based on the CAPE-OPEN, ESO interface. In: Proceedings of the European Symposium on Computer Aided Process Engineering – ESCAPE 15, Elsevier, Amsterdam (2005)
409. Schopfer, G., Yang, A., von Wedel, L., Marquardt, W.: CHEOPS: A tool-integration platform for chemical process modelling and simulation. International Journal on Software Tools for Technology Transfer 6(3), 186–202 (2004)
410. Schüppen, A.: Multimediale Kommunikationsunterstützung. In: Nagl, M., Westfechtel, B. (eds.) Modelle, Werkzeuge und Infrastrukturen zur Unterstützung von Entwicklungsprozessen, p. 373. Wiley, Chichester (2003)
411. Schüppen, A., Spaniol, O., Thißen, D., Assenmacher, I., Haberstroh, E., Kuhlen, T.: Multimedia and VR Support for Direct Communication of Designers. This volume (2008)
412. Schürr, A.: Operationales Spezifizieren mit programmierten Graphersetzungssystemen. PhD thesis, RWTH Aachen University (1991)

413. Schürr, A.: Specification of graph translators with triple graph grammars. In: Mayr, E.W., Schmidt, G., Tinhofer, G. (eds.) WG 1994. LNCS, vol. 903, pp. 151–163. Springer, Heidelberg (1995)
414. Schürr, A., Winter, A., Zündorf, A.: The PROGRES approach: Language and environment. In: Ehrig, H., Engels, G., Kreowski, H.J., Rozenberg, G. (eds.) Handbook on Graph Grammars and Computing by Graph Transformation – Volume 2: Applications, Languages, and Tools, pp. 478–550. World Scientific, Singapore (1999)
415. Schuba, M.: Skalierbare und zuverlässige Multicast-Kommunikation im Internet. PhD thesis, RWTH Aachen University (1999)
416. Schürr, A., Nagl, M., Zündorf, A. (eds.): Applications of Graph Transformations with Industrial Relevance. Proceedings of the Third International AGTIVE 2007, Kassel, Germany. LNCS, vol. 5088. Springer, Heidelberg (2008)
417. Souza, D., Marquardt, W.: The chemical engineering data model VeDa. Part 2: Structural modeling objects. Technical Report LPT-1998-02, RWTH Aachen University, Process Systems Engineering (1998)
418. Souza, D., Marquardt, W.: The chemical engineering data model VeDa. Part 3: Geometrical modeling objects. Technical Report LPT-1998-03, RWTH Aachen University, Process Systems Engineering (1998)
419. Spaniol, M., Klamma, R., Jarke, M.: Data integration for multimedia E-learning environments with XML and MPEG-7. In: Karagiannis, D., Reimer, U. (eds.) PAKM 2002. LNCS (LNAI), vol. 2569, pp. 244–255. Springer, Heidelberg (2002)
420. Spaniol, O., Meggers, J.: Active network nodes for adaptive multimedia communication. In: Yongcharoen, T. (ed.) Intelligence in Networks, Proc. 5th IFIP International Conference on Intelligence in Networks, SmartNet'99 (1999)
421. Spaniol, O., Meyer, B., Thißen, D.: Industrieller Einsatz von CORBA: aktuelle Situation und zukünftige Entwicklungen. Industrie Management 6 (1997)
422. Spaniol, O., Thißen, D., Meyer, B., Linnhoff-Popien, C.: Dienstmanagement und -vermittlung für Entwicklungswerkzeuge. In: Nagl, M., Westfechtel, B. (eds.) Integration von Entwicklungssystemen in Ingenieuranwendungen – Substantielle Verbesserung der Entwicklungsprozesse, pp. 371–386. Springer, Heidelberg (1999)
423. Stahl, J.: Entwicklung einer Methode zur Integrierten Arbeitsgestaltung und Personalplanung im Rahmen von Concurrent Engineering. PhD thesis, RWTH Aachen University (1998)
424. Stahl, J., Luczak, H.: A method for job design in concurrent engineering. In: Seppälä, P. (ed.) From Experience to Innovation, IEA'97 – Proceedings of the 13th Triennial Congress of the International Ergonomics Association, Helsinki, pp. 265–267 (1997)
425. Staudt, M., Jarke, M.: View management support in advanced knowledge base servers. Journal of Intelligent Information Systems 15(4), 253–285 (2000)
426. Steidel, F.: Modellierung arbeitsteilig ausgeführter, rechnerunterstützter Konstruktionsarbeit – Möglichkeiten und Grenzen personenzentrierter Simulation. PhD thesis, TU Berlin (1994)
427. Stenzel, R.: Steuerungsarchitekturen für autonome mobile Roboter. PhD thesis, RWTH Aachen University (2002)
428. Stewering, J.: Auslegung mit Virtual Reality. Plastverarbeiter (2004)

429. Stewering, J.: Erfolgreich Entwickeln und Kunden überzeugen mit Virtual Reality. Plastverarbeiter (2006)
430. Subrahmanian, E., Sriram, R., Herder, P., Schneider, R. (eds.): The role of empirical studies in understanding and supporting engineering design – Workshop Proceedings, April 4–5, 2002. DUP Science, Delft (2004)
431. Szcurko, P.: Steuerung von Informations- und Arbeitsflüssen auf Basis konzeptueller Unternehmensmodelle, dargestellt am Beispiel des Qualitätsmanagements. PhD thesis, RWTH Aachen University (1997)
432. Theißen, M., Hai, R., Marquardt, W.: Computer-Assisted Work Process Modeling in Chemical Engineering. This volume (2008)
433. Theißen, M., Hai, R., Morbach, J., Schneider, R., Marquardt, W.: Scenario-Based Analysis of Industrial Work Processes. This volume (2008)
434. Theißen, M., Marquardt, W.: Decision process modeling in chemical engineering design. In: Proceedings of the European Symposium on Computer Aided Process Engineering – ESCAPE 17, Bucharest, Romania, pp. 383–388. Elsevier, Amsterdam (2007)
435. Theißen, M., Marquardt, W.: Decision Models. This volume (2008)
436. Theißen, M., Schneider, R., Marquardt, W.: Arbeitsprozessmodellierung in der Verfahrenstechnik: Grundlagen, Werkzeuge, Einsatzgebiete. In: Tagungsband 3. Symposium Informationstechnologien für Entwicklung und Produktion in der Verfahrenstechnik, Berlin, Germany (2005)
437. Theißen, M., Hai, R., Marquardt, W.: Design process modeling in chemical engineering. Journal of Computing and Information Science in Engineering 8(1), 011007 (2008)
438. Thißen, D.: Managing distributed environments for cooperative development processes. In: Proceedings of the 6th World Multi-Conference on Systemics, Cybernetics and Informatics (SCI 2000), Orlando, Florida, USA, pp. 340–345 (2000)
439. Thißen, D.: Management of efficient service provision in distributed systems. In: Proceedings of the WSES International Conference on Multimedia, Internet, Video Technologies (MIV 2001), Malta, pp. 140–145. WSES Press (2001)
440. Thißen, D.: Trader-based management of service quality in distributed environments. In: Proceedings of the 5th International Conference on Communication Systems, Africom'01 (2001)
441. Thißen, D.: Flexible service provision considering specific customer resource needs. In: Proceedings of the 10th Euromicro Workshop on Parallel, Distributed and Network-based Processing (Euromicro-PDP 2002), Canary Islands, Spain, pp. 253–260. IEEE Computer Society Press, Los Alamitos (2002)
442. Thißen, D.: Load balancing for the management of service performance in open service markets: a customer-oriented approach. In: Proceedings of the 2002 ACM Symposium on Applied Computing (SAC 2002), Madrid, Spain, pp. 902–906. ACM Press, New York (2002)
443. Thißen, D.: Dienstmanagement für Entwicklungswerkzeuge. In: Nagl, M., Westfechtel, B. (eds.) Modelle, Werkzeuge und Infrastrukturen zur Unterstützung von Entwicklungsprozessen, pp. 363–364. Wiley-VCH, Weinheim (2003)
444. Thißen, D.: A middleware platform supporting electronic service markets. In: Proceedings of the IADIS International Conference on WWW/Internet (ICWI 2004), Madrid, Spain, pp. 1183–1186 (2004)

445. Thißen, D.: Trader-basiertes Dienstmanagement in offenen Dienstmárkten. PhD thesis, RWTH Aachen University (2004)
446. Thißen, D.: Considering qos aspects in web service composition. In: Proceedings of the 11th IEEE Symposium on Computers and Communications (ISCC 2006), Cagliari, Sardinia, Italy, pp. 371–376 (2006)
447. Thißen, D., Linnhoff-Popien, C., Lipperts, S.: Can CORBA fulfill data transfer requirements of industrial enterprises? In: Proceedings of the 1st International Enterprise Distributed Object Computing Workshop, Gold Coast, Australia, pp. 129–137. IEEE Computer Society Press, Los Alamitos (1997)
448. Thißen, D., Neukirchen, H.: Integrating trading and load balancing for efficient management of services in distributed systems. In: Linnhoff-Popien, C., Hegering, H.-G. (eds.) USM 2000. LNCS, vol. 1890, pp. 42–53. Springer, Heidelberg (2000)
449. Thißen, D., Neukirchen, H.: Managing services in distributed systems by integrating trading and load balancing. In: Proceedings of the 5th IEEE International Symposium on Computers and Communications, Antibes, France, pp. 641–646. IEEE Computer Society Press, Los Alamitos (2000)
450. Trossen, D.: Scalable conferencing control service (sccs) - service specification. Technical Report 001-98, RWTH Aachen University, Department of Computer Science 4 (1998)
451. Trossen, D.: Scalable conferencing control service (sccs) - protocol specification. Technical Report 001-99, RWTH Aachen University, Department of Computer Science 4 (1999)
452. Trossen, D.: GCDL: Group communication description language for modeling groupware applications and scenarios. In: Proceedings of the Communication Networks and Distributed Systems Modeling and Simulation Conference, San Diego, San Diego, California, USA (2000)
453. Trossen, D.: Scalable conferencing support for tightly-coupled environments: Services, mechanisms and implementation design. In: Proceedings of the IEEE International Conference on Communication (ICC 2000), New Orleans, USA, IEEE Computer Society Press, Los Alamitos (2000)
454. Trossen, D.: Scalable Group Communication in Tightly Coupled Environments. PhD thesis, RWTH Aachen University (2000)
455. Trossen, D., Kliem, P.: Dynamic reconfiguration in tightly-coupled environments. In: Proceedings of SPIE (International Society for Optical Engineering): Multimedia Systems and Applications II, Boston, USA (1999)
456. Trossen, D., Schüppen, A., Wallbaum, M.: Shared workspace for collaborative engineering. Annals of Cases on Information Technology 4, 119–130 (2002)
457. Trossen, D., Eickhoff, W.C.: Reconfiguration in tightly-coupled conferencing environments. In: Proceedings of the 11th Information Resources Management Association International Conference (IRMA 2000), Anchorage, Alaska, USA, IRMA (2000)
458. Vassiliadis, P., Quix, C., Vassiliou, Y., Jarke, M.: Data warehouse process management. Information Systems (Special Issue on Advanced Information Systems Engineering) 26(3), 205–236 (2001)
459. von Wedel, L.: An Environment for Heterogeneous Model Management in Chemical Process Engineering. PhD thesis, RWTH Aachen University (2004)
460. von Wedel, L., Kulikov, V., Marquardt, W.: An Integrated Environment for Heterogeneous Process Modeling and Simulation. This volume (2008)

461. von Wedel, L., Marquardt, W.: The chemical engineering data model VeDa. Part 5: Material modeling objects. Technical Report LPT-1998-05, RWTH Aachen University, Process Systems Engineering (1999)
462. von Wedel, L., Marquardt, W.: CHEOPS: A case study in component-based process simulation. In: Malone, M.F., Trainham, J.A., Carnahan, B. (eds.) Foundations of Computer-Aided Process Design. AIChE Symposium Series, vol. 96(323), pp. 494–497 (2000)
463. von Wedel, L., Marquardt, W.: ROME: A repository to support the integration of models over the lifecycle of model-based engineering. In: Pierucci, S. (ed.) Proceedings of the European Symposium on Computer Aided Process Engineering – ESCAPE 10, pp. 535–540. Elsevier, Amsterdam (2000)
464. Wallbaum, M., Carrega, D., Krautgärtner, M., Decker, H.: A mobile middleware component providing voice over ip services to mobile users. In: Leopold, H., García, N. (eds.) ECMAST 1999. LNCS, vol. 1629, pp. 552–563. Springer, Heidelberg (1999)
465. Wallbaum, M., Meggers, J.: Voice/data integration in wireless communication networks. In: Proceedings of the 50th IEEE Vehicular Technology Conference (VTC'99), Amsterdam, The Netherlands, IEEE Computer Society Press, Los Alamitos (1999)
466. Wang, B.: Identifikation von charakteristischen Situationen in Betriebsdaten von kontinuierlichen Produktionsprozessen. Master's thesis, RWTH Aachen University (2007)
467. Watzdorf, R., Bausa, J., Marquardt, W.: General shortcut methods for non-ideal multicomponent distillation: 2. complex columns. AIChE Journal 45, 1615–1628 (1998)
468. Weck, M.: Fehlermustererkennung durch Signalanalyse in der Kautschukprofilextrusion. Master's thesis, RWTH Aachen University (2008)
469. Weidenhaupt, K.: Anpassbarkeit von Softwarewerkzeugen in prozessintegrierten Entwicklungsumgebungen. PhD thesis, RWTH Aachen University (2001)
470. Weidenhaupt, K., Bayer, B.: Prozeßintegrierte Designwerkzeuge für die Verfahrenstechnik. In: Beiersdörfer, K., Engels, G., Schäfer, W. (eds.) Proceedings of Informatik'99, Paderborn, Germany, pp. 305–313. Springer, Heidelberg (1999)
471. Weisemöller, I.: Verteilte Ausführung dynamischer Entwicklungsprozesse in heterogenen Prozessmanagementsystemen. Master's thesis, RWTH Aachen University (2006)
472. Westfechtel, B.: A graph-based model for dynamic process nets. In: Proceedings of the 7th International Conference on Software Engineering and Knowledge Engineering SEKE'95, Skokie, Ilinois, USA, pp. 126–130 (1995)
473. Westfechtel, B.: A graph-based system for managing configurations of engineering design documents. International Journal of Software Engineering & Knowledge Engineering 6(4), 549–583 (1996)
474. Westfechtel, B.: Graph-based product and process management in mechanical engineering. In: Ehrig, H., Engels, G., Kreowski, H.-J., Rozenberg, G. (eds.) Handbook on Graph Grammars and Computing by Graph Transformation – Volume 2: Applications, Languages, and Tools, pp. 321–368. World Scientific, Singapore (1999)

475. Westfechtel, B.: Models and Tools for Managing Development Processes. LNCS, vol. 1646. Springer, Heidelberg (1999)
476. Westfechtel, B.: Ein graphbasiertes Managementsystem für dynamische Entwicklungsprozesse. Informatik Forschung und Entwicklung 16(3), 125–144 (2001)
477. Westfechtel, B., Munch, B., Conradi, R.: A layered architecture for uniform version management. IEEE Transactions on Software Engineering 27(12), 1111–1133 (2001)
478. Westfechtel, B., Schleicher, A., Jäger, D., Heller, M.: Ein Managementsystem für Entwicklungsprozesse. In: Nagl, M., Westfechtel, B. (eds.) Modelle, Werkzeuge und Infrastrukturen zur Unterstützung von Entwicklungsprozessen, pp. 369–370. Wiley-VCH, Weinheim (2003)
479. Wiesner, A., Morbach, J., Marquardt, W.: An overview on OntoCAPE and its latest applications. In: AIChE Annual Meeting, Salt Lake City, USA, Nov. 4-9 (2007)
480. Wiesner, A., Morbach, J., Marquardt, W.: Semantic data integration for process engineering design data. In: Proceedings of the 10th International Conference on Enterprise Information Systems (ICEIS), Barcelona, Spain, accepted (2008)
481. Winter, A.: Visuelles Programmieren mit Graphtransformationen. PhD thesis, RWTH Aachen University (1999)
482. Wolf, M.: Groupware zur Unterstützung verfahrenstechnischer Entwicklungsprozesse, Statusbericht des SFB-Teilprojektes I2. Technical report, RWTH Aachen University, Institute of Industrial Engineering and Ergonomics (1998)
483. Wolf, M.: Entwicklung und Evaluation eines Groupware-Systems zur Unterstützung verfahrenstechnischer Entwicklungsprozesse. PhD thesis, RWTH Aachen University (2002)
484. Wolf, M., Foltz, C., Schlick, C., Luczak, H.: Empirical investigation of a workspace model for chemical engineers. In: Proceedings of the 14th Triennial Congress of the International Ergonomics Association and the 44th Annual Meeting of the Human Factors and Ergonomics Society, San Diego, California, USA, 1/612–1/615 (2000)
485. Wolf, M., Foltz, C., Schlick, C., Luczak, H.: Empirische Untersuchung eines Groupware-basierten Unterstützungs-Systems für verfahrenstechnische Entwickler. Zeitschrift für Arbeitswissenschaft 54(3), 258–266 (2000)
486. Wolf, M., Foltz, C., Schlick, C., Luczak, H.: Development and evaluation of a groupware system to support chemical design processes. International Journal of Human-Computer Interaction 14(2), 181–198 (2002)
487. Wörzberger, R., Ehses, N., Heer, T.: Adding support for dynamics patterns to static business process management systems. In: Proceedings of the Workshop on Software Composition 2008. LNCS, vol. 4954, Springer, Heidelberg (2008)
488. Wörzberger, R., Heller, M., Hässler, F.: Evaluating workflow definition language revisions with graph-based tools. In: Electronic Communications of the EASST 6th International Workshop on Graph Transformations and Visual Modeling Techniques (GT-VMT'2007), Braga, Portugal, March 24 – April 1 (2007)

489. Yang, A., Marquardt, W.: An ontology-based approach to conceptual process modeling. In: Barbarosa-Póvoa, A., Matos, H. (eds.) Proceedings of the European Symposium on Computer Aided Process Engineering – ESCAPE 14, pp. 1159–1164. Elsevier, Amsterdam (2004)
490. Yang, A., Morbach, J., Marquardt, W.: From conceptualization to model generation: the roles of ontologies in process modeling. In: Floudas, C.A., Agrarwal, R. (eds.) Proceedings of FOCAPD 2004, pp. 591–594 (2004)
491. Yang, A., Schlüter, M., Bayer, B., Krüger, J., Haberstroh, E., Marquardt, W.: A concise conceptual model for material data and its applications in process engineering. Computers & Chemical Engineering 27, 595–609 (2003)
492. Zündorf, A.: Eine Entwicklungsumgebung für PROgrammierte GRaph ErsetzungsSysteme. PhD thesis, RWTH Aachen University (1995)

R.2 External Literature[74]

493. Aamodt, A., Plaza, E.: Case-based reasoning: Foundational issues, methodological variations, and system approaches. AI Communications 7(1), 39–59 (1994)
494. Abdalla, H.: Concurrent engineering for global manufacturing. International Journal of Production Economics 60–61, 251–260 (1999)
495. Adelberg, B.: NoDoSE – a tool for semi-automatically extracting structured and semistructured data from text documents. In: Proceedings of the 1998 ACM SIGMOD International Conference on Management of Data, Seattle, Washington, USA, pp. 283–294. ACM Press, New York (1998)
496. AixCAPE: Welcome to AixCAPE (2005), http://www.aixcape.org
497. aiXtrusion GmbH: Website (2006), http://www.aixtrusion.de/
498. Akehurst, D., Kent, S., Patrascoiu, O.: A relational approach to defining and implementing transformations between metamodels. Software and Systems Modeling 2(4), 215–239 (2003)
499. Alberts, L.K.: A sharable ontology for the formal representation of engineering design knowledge. In: Gero, J.S., Tyugu, E. (eds.) Formal Design Methods for CAD, pp. 3–32. Elsevier, Amsterdam (1994)
500. Alonso, G., Casati, F., Kuno, H., Machiraju, V.: Web Services: Concepts, Architectures and Applications. Springer, Heidelberg (2004)
501. Ambriola, V., Conradi, R., Fuggetta, A.: Assessing process-centered software engineering environments. ACM Transactions on Software Engineering and Methodology 6(3), 283–328 (1997)
502. Amtsblatt der Europäischen Gemeinschaften Nr. L 156/14-18: EWG 90/270: Richtlinie des Rates vom 29. Mai 1990 über die Mindestvorschriften bezüglich der Sicherheit und des Gesundheitsschutzes bei der Arbeit an Bildschirmgeräten. Amt für Veröffentlichungen der Europäischen Gemeinschaften, Luxemburg (1990)

---

[74] External literature means on one hand literature of other groups and scientist, not being involved in IMPROVE. Furthermore, there are also publications of former IMPROVE members, which appeared after their membership in IMPROVE. Finally, there are also publication of members of IMPROVE which, either from their contents or from their date of appearance, have only a loose or no connection to IMPROVE.

503. Anderl, R., Trippner, D.: STEP – Standard for the Exchange of Product Model Data. Teubner, Wiesbaden (2001)
504. Anderson, G., Anderson, P.: Enterprise JavaBeans Components Architecture. Prentice-Hall, Englewood Cliffs (2002)
505. Andrews, R., Ponton, J.W.: A process engineering information management system using world wide web technology. In: Proceedings of the 3rd International Conference on Foundations of Computer-Aided Operations (1998)
506. Anhäuser, F., Richert, H., Temmen, H.: PlantXML – integrierter Planungsprozess mit flexiblen Bausteinen. atp – Automatisierungstechnische Praxis 46(10), 61–71 (2004)
507. ANSYS, Inc.: CFD flow modeling software & solutions from fluent – website (2007), http://www.fluent.com
508. Apache Software Foundation: Apache Axis Website (2007), http://ws.apache.org/axis
509. Appukuttan, B.K., Clark, T., Reddy, S., Tratt, L., Venkatesh, R.: A model driven approach to model transformations. In: Proceedings of the 2003 Model Driven Architecture: Foundations and Applications (MDAFA 2003), CTIT Technical Report TR-CTIT-03-27, University of Twente, The Netherlands (2003)
510. Arango, G.: Domain Analysis Methods. In: Schaefer, W., Prieto-Diaz, R., Matsumoto, M. (eds.) Software Reusability, pp. 17–49. Ellis Horwood, New York (1994)
511. Aspeli, M.: Professional Plone Development. Packt Publishing, Birmingham (2007)
512. Aspen Technology, Inc.: SPEEDUP, user manual (1995)
513. AspenTech: Polymers Plus User Guide. Release 10.2 (2000)
514. AspenTech: Aspen Batch.21^{TM} – website (2005), http://www.aspentech.com/product.cfm?ProductID=92
515. AspenTech: Aspen InfoPlus.21® – website (2005), http://www.aspentech.com/product.cfm?ProductID=104
516. AspenTech: Aspen Plus® – website (2005), http://www.aspentech.com/product.cfm?ProductID=69
517. AspenTech: Aspen Zyqad^{TM} – website (2005), http://www.aspentech.com/includes/product.cfm?IndustryID=0&ProductID=89
518. AspenTech: AspenTech – website (2005), http://www.aspentech.com
519. AspenTech: Aspen HYSYS – website (2006), http://www.aspentech.com/products/aspen-hysys.cfm
520. AVEVA: VANTAGE Enterprise Net – website (2006), http://www.aveva.com/products/plant/vnet.html
521. Badham, R., Couchman, P., Zanko, M.: Implementing concurrent engineering. Human Factors and Ergonomics in Manufacturing 10, 237–249 (2000)
522. Baggen, R., Hemmerling, S.: Evaluation von Benutzbarkeit in Mensch-Maschine-Systemen. In: Timpe, K.P., Jürgensohn, T., Kolrep, H. (eds.) Mensch-Maschine-Systemtechnik – Konzepte, Modellierung, Gestaltung, Evaluation, pp. 233–284. Symposium Publishing (2000)
523. Balke, W.T., Badii, A.: Assessing web services quality for call-by-call outsourcing. In: Proceedings of the 1st Web Services Quality Workshop, Rome, Italy, pp. 173–181. IEEE Computer Society Press, Los Alamitos (2003)

524. Bañares-Alcántara, R., Lababidi, H.M.S.: Design support systems for process engineering – II. KBDS: An experimental prototype. Computers & Chemical Engineering 19(3), 279–301 (1995)
525. Bañares-Alcántara, R., King, J.M.P.: Design support systems for process engineering – III. design rationale as a requirement for effective support. Computers & Chemical Engineering 21, 263–276 (1996)
526. Bandinelli, S., Fuggetta, A., Ghezzi, C.: Software process model evolution in the SPADE environment. IEEE Transactions on Software Engineering 19(12), 1128–1144 (1993)
527. Bandinelli, S., Fuggetta, A., Grigoli, S.: Process modeling in-the-large with SLANG. In: Proceedings of the 2nd International Conference on the Software Process (ICSP'93), IEEE Computer Society Press, Los Alamitos (1993)
528. Baresi, L., Mauri, M., Pezzè, M.: PLCTools: Graph transformation meets PLC design. Electronic Notes in Theoretical Computer Science 72(2) (2002)
529. Barkmeyer, E.J., Feeney, A.B., Denno, P., Flater, D.W., Libes, D.E., Steves, M.P., Wallace, E.K.: Concepts for automating systems integration. Technical Report NISTIR 6928, National Institute of Standards and Technology, NIST (2003)
530. Barnicki, S.D., Fair, J.R.: Separation system synthesis: A knowledge-based approach. 1. liquid mixture separations. Industrial & Engineering Chemistry Research 29, 421–432 (1990)
531. Barnicki, S.D., Fair, J.R.: Separation system synthesis: A knowledge-based approach. 2. gas-vapor mixtures. Industrial & Engineering Chemistry Research 31, 1679–1694 (1992)
532. Barton, P.I., Pantelides, C.C.: Modeling of combined discrete/continous processes. AIChE Journal 40, 966–979 (1994)
533. Basili, V.R.: The experience factory and its relationship to other quality approaches. Advances in Computers 41, 65–82 (1995)
534. Basili, V.R., Caldiera, G., Rombach, H.D.: The experience factory. In: Marciniak, J.J. (ed.) Encyclopedia of Software Engineering, vol. 1, pp. 469–476. John Wiley & Sons, Chichester (1994)
535. Basili, V.R., Rombach, H.D.: The TAME project: Towards improvement-oriented software environments. IEEE Transactions on Software Engineering 146, 758–773 (1988)
536. Bass, L., Kazman, R.: Architecture-based development. Technical Report CMU/SEI-99-TR-007, Carnegie Mellon University, Software Engineering Institute, SEI (1999)
537. Basu, P.K., Mack, R.A., Vinson, J.M.: Consider a new approach to pharmaceutical process development. Chemical Engineering Progress 95, 82–90 (1999)
538. Batini, C., Lenzerini, M., Navathe, S.B.: A comparative analysis of methodologies for database schema integration. ACM Computing Surveys 18(4), 323–364 (1986)
539. Batres, R., Asprey, S., Fuchino, T., Naka, Y.: A multi-agent environment for concurrent process engineering. Computers & Chemical Engineering 10, 653–656 (1999)
540. Batres, R., Naka, Y.: Process plant ontologies based on a multi-dimensional framework. In: Malone, M.F., Trainham, J.A., Carnahan, B. (eds.) Fifth International Conference on Foundations of Computer-Aided Process Design. AIChE Symposium Series, vol. 96(323), pp. 433–437 (2000)

541. Batres, R., Naka, Y., Lu, M.L.: A multidimensional design framework and its implementation in an engineering design environment. Concurrent Engineering: Research and Applications 7(1), 43–54 (1999)
542. Batres, R., West, M., Leal, D., Price, D., Masaki, K., Shimada, Y., Fuchino, T., Naka, Y.: An upper ontology based on ISO 15926. Computers & Chemical Engineering 31, 519–534 (2007)
543. Batres, R., West, M., Leal, D., Price, D., Naka, Y.: An upper ontology based on ISO 15926. In: Puigjaner, L., Espuña, A. (eds.) European Symposium on Computer-Aided Process Engineering – ESCAPE 15. Computer-Aided Chemical Engineering, vol. 20, pp. 1543–1548. Elsevier, Amsterdam (2005)
544. Bayer, J., Widen, T.: Introducing traceability to product lines. In: Proceedings of the 4th International Workshop on Software Product-Family Engineering (2001)
545. Bechhofer, S., Horrocks, I., Goble, C.A., Stevens, R.: OilEd: A reason-able ontology editor for the semantic web. In: Baader, F., Brewka, G., Eiter, T. (eds.) KI 2001. LNCS (LNAI), vol. 2174, pp. 396–408. Springer, Heidelberg (2001)
546. Bechhofer, S., van Harmelen, F., Hendler, J., Horrocks, I., McGuiness, D.L., Patel-Schneider, P.F., Stein, L.A.: OWL web ontology language reference (2004), http://www.w3.org/TR/owl-ref/
547. Begole, J., Rosson, M.B., Shaffer, C.A.: Flexible collaboration transparency: Supporting worker independence in replicated application-sharing systems. ACM Transactions on Computer-Human-Interaction 6(2), 95–132 (1999)
548. Beßling, B., Lohe, B., Schoenmakers, H., Scholl, S., Staatz, H.: CAPE in process design – potential and limitations. Computers & Chemical Engineering 21, S17–S21 (1997)
549. Benz, M., Hess, R., Hutschenreuther, T., Kümmel, T., Schill, S.: A framework for high quality/low cost conferencing systems. In: Díaz, M., Sénac, P. (eds.) IDMS 1999. LNCS, vol. 1718, pp. 305–320. Springer, Heidelberg (1999)
550. Bergmann, R.: Experience Management – Foundations, Development Methodology, and Internet-Based Applications. Springer, Heidelberg (2002)
551. Bergstra, J.A., Ponse, A., Smolka, S.A.: Handbook of Process Algebra. North-Holland, Amsterdam (2001)
552. Bernstein, P.A., Dayal, U.: An overview of repository technology. In: Proceedings of the 20th VLDB Conference, Santiago, Chile, pp. 705–713 (1994)
553. Bertino, E., Catania, B.: Integrating XML and databases. IEEE Internet Computing 5(4), 84–88 (2001)
554. Bevan, N., Macleod, M.: Usability measurement in context. Behaviour & Information Technology 13(1-2), 132–145 (1994)
555. Beyer, L.T., Holtzblatt, K.: Contextual Design. Morgan Kaufmann, San Francisco (1998)
556. Biegler, L.T., Grossmann, I.E., Westerberg, A.W.: Systematic Methods of Chemical Process Design. Prentice-Hall, Englewood Cliffs (1997)
557. Bieszczad, A., Pagurek, B., White, T.: Mobile agents for network management. In: IEEE Communications Surveys (1998)
558. Bieszczad, J., Koulouris, A., Stephanopoulos, G.: Model.la: A phenomena-based modeling environment for computer-aided process design. In: Malone, M., Trainham, J., Carnahan, B. (eds.) 5th International Conference on Foundations of Computer-aided Process Design, pp. 438–441 (2000)

559. Blass, E.: Entwicklung verfahrenstechnischer Prozesse, 2nd edn. Springer, Heidelberg (1997)
560. Booch, G., Jacobson, I., Rumbaugh, J.: The Unified Modeling Language User Guide. Addison-Wesley, Reading (1999)
561. Borghoff, U., Schlicher, J.: Rechnergestützte Gruppenarbeit. Springer, Heidelberg (1998)
562. Borst, W.N.: Construction of Engineering Ontologies for Knowledge Sharing and Reuse. PhD thesis, University of Twente (1997)
563. Botz, J., Döring, N.: Forschungsmethoden und Evaluation. Springer, Heidelberg (1995)
564. Boyle, J.M.: Interactive engineering systems design: A study for artificial intelligence applications. Artificial Intelligence in Engineering 4(2), 58–69 (1989)
565. Braun, P., Marschall, F.: Transforming object oriented models with BOTL. Electronic Notes in Theoretical Computer Science 72(3) (2003)
566. Braunschweig, B.L., Pantelides, C.C., Britt, H.I., Sama, S.: Process modeling: The promise of open software architectures. Chemical Engineering Progress 96(9), 65–76 (2000)
567. Bray, T., Paoli, J., Sperberg-McQueen, C.M. (eds.): Extensible Markup Language (XML) 1.0. W3C (1998)
568. Brücher, H.: Erweiterung von UML zur geschäftsorientierten Prozessmodellierung. In: Referenzmodellierung 2001, 5. Fachtagung, Dresden, Germany (2001)
569. Brown, D.C., Chandrasekaran, B.: Design problem solving: knowledge structures and control strategies. Morgan Kaufmann, San Francisco (1989)
570. Browning, T., Eppinger, S.D.: Modeling impacts of process architecture on cost and schedule risk in product development. IEEE Transactions on Engineering Management 49(4), 428–442 (2002)
571. Browning, T.R., Eppinger, S.T.: Modelling the impact of process architecture on cost and schedule risk in product development. Technical Report 4050, Massachusetts Institute of Technology, Sloan School of Management (2000)
572. Broy, M.: Software technology – formal models and scientific foundations. Information and Software Technology 41, 947–950 (1999)
573. Broy, M., Slotosch, O.: Enriching the software engineering process by formal methods. In: Hutter, D., Traverso, P. (eds.) FM-Trends 1998. LNCS, vol. 1641, pp. 1–43. Springer, Heidelberg (1999)
574. Bucciarelli, L.L.: Designing Engineers. MIT Press, Cambridge (1994)
575. Bullinger, H.J., Kiss-Preussinger, E., Spath, D. (eds.): Automobilentwicklung in Deutschland - wie sicher in die Zukunft? Chancen, Potenziale und Handlungsempfehlungen für 30 Prozent mehr Effizienz. Fraunhofer-IRB-Verlag, Stuttgart (2003)
576. Bullinger, H.-J., Warschat, J. (eds.): Concurrent Simultaneous Engineering Systems. Springer, Heidelberg (1996)
577. Bundesgesetzblatt I, 1996, 1246: Arbeitsschutzgesetz, Gesetz zur Umsetzung der EG-Rahmenrichtlinie Arbeitsschutz und weiterer Arbeitsschutz-Richtlinien vom 7. August 1996 (1996)
578. Bunge, M.: Ontology II: A World of Systems. Treatise on Basic Philosophy, vol. 4. D. Reidel Publishing Company, Dordrecht (1979)
579. Burmester, S., Giese, H., Niere, J., Tichy, M., Wadsack, J.P., Wagner, R., Wendehals, L., Zündorf, A.: Tool integration at the meta-model level: the Fujaba approach. International Journal on Software Tools for Technology Transfer 6(3), 203–218 (2004)

580. Buschmann, F., Meunier, R., Rohnert, H., Sommerlad, P., Stal, M.: Pattern-Oriented Software Architecture – A System of Patterns. Wiley, Chichester (1996)
581. Carlowitz, B.: Tabellarische Übersicht über die Prüfung von Kunststoffen. Giesel Verlag, Isernhagen (1992)
582. Carney, D., Fisher, D., Morris, E., Place, P.: Some current approaches to interoperability. Technical Report CMU/SEI-2005-TN-033, Carnegie Mellon University, Software Engineering Institute, SEI (2005)
583. Carroll, J.M., Kellog, W.A., Rosson, M.B.: The task-artifact cycle. In: Carroll, J.M. (ed.) Designing Interaction: Psychology at the Human-Computer Interface, pp. 74–102. Cambridge University Press, Cambridge (1999)
584. Casati, F., Ceri, S., Pernici, B., Pozzi, G.: Workflow evolution. In: Thalheim, B. (ed.) ER 1996. LNCS, vol. 1157, pp. 438–455. Springer, Heidelberg (1996)
585. Casati, F., Discenza, A.: Modeling and managing interactions among business processes. Journal of Systems Integration 10(2), 145–168 (2001)
586. Casati, F., Fugini, M., Mirbel, I.: An environment for designing exceptions in workflows. Information Systems 24(3), 255–273 (1999)
587. Castillo, E.: Extreme Value Theory in Engineering. Academic Press, London (1988)
588. Champin, P.-A.: ARDECO: An assistant for experience reuse in computer aided design. In: Proceedings of the Workshop "From Structured Cases to Unstructured Problem Solving Episodes" (2003)
589. Chandrasekaran, B., Josephson, J.R., Benjamins, V.R.: What are ontologies, and why do we need them? IEEE Intelligent Systems 14(1), 20–26 (1999)
590. Chebbi, I., Tata, S., Dustdar, S.: The view-based approach to dynamic inter-organizational workflow cooperation. Technical Report TUV-1841-2004-23, Technical University of Vienna, Information Systems Institute, Distributed Systems Group (2004)
591. Chemie Wirtschaftsförderungs-GmbH: Campus the plastics database (2008), http://www.campusplastics.com/
592. Chemstations, Inc.: Chemcad Version 6 – website (2006), http://www.chemstations.net
593. Chen, B., Westerberg, A.W. (eds.): Proceedings of the 8th International Symposium on Process Systems Engineering (PSE 2003), Kunming, China. Elsevier, Amsterdam (2003)
594. Chidamber, S.R., Kemerer, C.F.: A metrics suite for object oriented design. IEEE Transactions on Software Engineering 20(6), 476–493 (1994)
595. Chiu, D.K.W., Karlapalem, K., Li, Q.: Views for inter-organizational workflow in an e-commerce environment. In: Proceedings of the IFIP TC2/WG2.6 9th Working Conference on Database Semantics, Deventer, The Netherlands, pp. 137–151. Kluwer Academic Publishers, Dordrecht (2003)
596. Cho, S.-H., Eppinger, D.: Product development process modeling using advanced simulation. In: Proceedings of DETC'01, ASME 2001: Design Engineering Technical Conferences and Computers and Information in Engineering Conference, Pittsburgh, USA, pp. 9–12 (2001)
597. Cho, S.-H., Eppinger, D.: A simulation-based process model for managing complex design projects. IEEE Transactions on Engineering Management 52(3), 316–328 (2005)

598. Chou, S.-C., Chen, J.-Y.: Process evolution support in a concurrent software process language environment. Information and Software Technology 41, 507–524 (1999)
599. Christiansen, T.R.: Modeling Efficiency and Effectiveness of Coordination in Engineering Design Teams. PhD thesis, Stanford University (1993)
600. Cimitile, A., de Lucia, A., de Carlini, U.: Incremental migration strategies: Data flow analysis for wrapping. In: Proceedings of the 5th Working Conference on Reverse Engineering (WCRE'98), Hawaii, USA, pp. 59–68. IEEE Computer Society Press, Los Alamitos (1998)
601. CiT – Computing in Technology GmbH: PARSIVAL – website (2007), http://www.cit-wulkow.de/tbcpars.htm
602. Clark, J. (ed.): XSL Transformations (XSLT) 1.00. W3C (1999)
603. Clements, P.C., Northrop, L.: Software architecture: An executive overview. Technical Report CMU/SEI-96-TR-003, Carnegie Mellon University, Software Engineering Institute, SEI (1996)
604. Cohen, G.: The Virtual Design Team: An Object-Oriented Model of Information Sharing in Project Teams. PhD thesis, Stanford University (1992)
605. Cohn, D.: SmartPlant 3D: Changing the future of engineering. Engineering Automation Report 12(10) (2003)
606. Conklin, E.J.: Capturing organisational memory. In: Readings in groupware and computer-supported cooperative work: assisting human-human collaboration, pp. 561–565. Morgan Kaufmann, San Francisco (1993)
607. Conklin, E.J., Yakemovic, K.C.B.: A process-oriented approach to design rationale. Human-Computer Interaction 6(3-4), 357–391 (1991)
608. Conklin, J., Begeman, M.L.: gIBIS: A hypertext tool for exploratory policy discussion. ACM Transactions on Office Information Systems 6, 303–331 (1988)
609. Connolly, D., van Harmelen, F., Horrocks, I., McGuinness, D.L., Patel-Schneider, P.F., Stein, L.A.: DAML+OIL reference description (2001), http://www.w3.org/TR/daml+oil-reference
610. Cook, T.D., Campbell, D.T.: Quasi-experimentation: design & analysis issues for field settings. Houghton Mifflin, Boston (1979)
611. Cook, T.D., Campbell, D.T., Peracchio, L.: Quasi-experimentation. In: Dunnette, M.D., Hough, L.M. (eds.) Handbook of Industrial and Organizational Psychology, vol. 1, 2nd edn., pp. 491–575. Consulting Psychologists Press, Mountain View (1990)
612. Cool, C., Fish, R.S., Kraut, R.E., Lowery, C.M.: Iterative design of video communication systems. In: CSCW'92: Proceedings of the 1992 ACM conference on Computer-supported cooperative work, Toronto, Ontario, Canada, pp. 25–32. ACM Press, New York (1992)
613. Corradini, A., Ehrig, H., Kreowski, H.-J., Rozenberg, G. (eds.): ICGT 2002. LNCS, vol. 2505. Springer, Heidelberg (2002)
614. Corradini, F., Mariani, L., Merelli, E.: An agent-based approach to tool integration. International Journal on Software Tools for Technology Transfer 6(3), 231–244 (2004)
615. COSA: COSA BPM – website (2006), http://www.cosa-bpm.com/COSA_BPM_-_More.html

616. Cugola, G.: Tolerating deviations in process support systems via flexible enactment of process models. IEEE Transactions on Software Engineering 24(11), 982–1001 (1998)
617. Cui, Z., Tamma, V., Bellifemine, F.: Ontology management in enterprises. BT Technology Journal 17, 98–107 (1999)
618. Cummins, F.A.: Enterprise Integration: An Architecture for Enterprise Application and Systems Integration. Wiley, Chichester (2002)
619. Curtis, B., Kellner, M.I., Over, J.: Process modeling. Communications of the ACM 35(9), 75–90 (1992)
620. Cypher, A. (ed.): Watch what I do: Programming by Demonstration. MIT Press, Cambridge (1993)
621. Cziner, K., Hurme, M.: Process evaluation and synthesis by analytic hierarchy process combined with genetic optimization. In: Chen, B., Westerberg, A.W. (eds.) Proceedings of the 8th International Symposium on Process Systems Engineering (PSE 2003), Kunming, China, pp. 778–783. Elsevier, Amsterdam (2003)
622. Daichendt, M.M., Grossmann, I.E.: Integration of hierarchical decomposition and mathematical programming for the sythesis of process flowsheets. Computers & Chemical Engineering 22, 147–175 (1998)
623. Dalton, C.M., Goldfarb, S.: PDXI, a progress report. In: Proceedings of the CHEMPUTERS Europe II Conference (1995)
624. Damm, C., Hansen, K., Thomsen, M., Tyrsted, M.: Tool integration: Experiences and issues in using XMI and component technology. In: Proceedings of TOOLS Europe (2000)
625. Dan, A., Ludwig, H., Pacifici, G.: Web services differentiation with service level agreements (2003), http://www-106.ibm.com/developerworks/webservices/library/ws-slafram
626. Davenport, T.H.: Process Innovation. Harvard Business School Press, Boston (1993)
627. de Lara, J., Vangheluwe, H.: Computer aided multi-paradigm modelling to process petri-nets and statecharts. In: Corradini, A., Ehrig, H., Kreowski, H.-J., Rozenberg, G. (eds.) ICGT 2002. LNCS, vol. 2505, pp. 239–253. Springer, Heidelberg (2002)
628. Derniame, J.-C., Kaba, B.A., Wastell, D. (eds.): Promoter-2 1998. LNCS, vol. 1500. Springer, Heidelberg (1999)
629. Desiderata Software: EZ JCom – website (2007), http://www.ezjcom.com
630. Desrochers, A.A., Al-Jaar, R.Y.: Applications of Petri Nets in Manufacturing Systems. IEEE Computer Society Press, Los Alamitos (1995)
631. Deutsches Institut für Normung eV, Berlin: Benutzer-orientierte Gestaltung interaktiver Systeme (2000)
632. Deutsches Institut für Normung eV, Berlin: Ergonomische Anforderungen für Bürotätigkeiten mit Bildschirmgeräten (2002)
633. Diaper, D. (ed.): Task Analysis for Human-Computer Interaction. Ellis Horwood, New York (1989)
634. DIN Deutsches Institut für Normung e.V. (ed.): Planung einer verfahrenstechnischen Anlage – Vorgehensmodell und Terminologie. Beuth-Verlag, Berlin (2006)
635. DIPPR: ppdXML – Physical Properties Data XML – website (2004), http://www.ppdxml.org/

636. Distributed Management Task Force, Inc. (DMTF): Common information model, CIM (2006), http://www.dmtf.org/standards/cim/
637. Donohoe, P. (ed.): Software Architecture (TC2 1st Working IFIP Conference on Software Architecture, WICSA1), San Antonio, Texas, USA. Kluwer Academic Publishers, Dordrecht (1999)
638. Douglas, J.M., Timmerhaus, K.D.: Conceptual Design of Chemical Processes. McGraw-Hill, New York (1988)
639. Dourish, P., Bly, S.: Portholes: Supporting awareness in a distributed work group. In: Proceedings of the ACM Conference on Human Factors in Computing Systems, New York, USA, pp. 541–547 (1992)
640. Dourish, P., Bly, S.: Culture and control in a media space. In: Proceedings of the 3rd European Conference on Computer-Supported Cooperative Work (ECSCW'93), Dordrecht, The Netherlands, pp. 125–137 (1993)
641. Dowson, M.: Integrated project support with Istar. IEEE Software 4, 6–15 (1987)
642. Dowson, M., Fernström, C.: Towards requirements for enactment mechanisms. In: Warboys, B.C. (ed.) EWSPT 1994. LNCS, vol. 772, Springer, Heidelberg (1994)
643. Dutoit, A.H., McCall, R., Mistrík, I., Paech, B.: Rationale management in software engineering: Concepts and techniques. In: Dutoit, A.H., McCall, R., Mistrík, I., Paech, B. (eds.) Rationale management in software engineering, pp. 1–48. Springer, Heidelberg (2006)
644. Dutoit, A.H., McCall, R., Mistrík, I., Paech, B.: Rationale Management in Software Engineering. Springer, Heidelberg (2006)
645. Eberleh, E., Oberquelle, H., Oppermann, H.: Einführung in die Software-Ergonomie. Gruyter, Berlin (1994)
646. ebXML: Technical architecture specification (2006), http://www.ebxml.org/specs/index.htm
647. Echterhoff, W.: Lernzuwachs und Effizienz. Technical report, Forschungsgruppe für Programmiertes Lernen e. V. (1973)
648. Eckert, C., Pizka, M.: Improving resource managements in distributed systems using language-level structuring concepts. Journal of Supercomputing 13, 33–55 (1999)
649. Edgar, T.F., Himmelblau, D.M., Lasdon, L.S.: Optimization of Chemical Processes. Chemical Engineering Series. McGraw-Hill, New York (2001)
650. Ehrenstein, G.W.: Kunststoff-Schadensanalyse – Methoden und Verfahren. University of Erlangen-Nürnberg, Germany (1999)
651. Ehrenstein, G.W., Drummer, D.: Mechanical behaviour of magnetizable polymers under dynamical load. In: SPE Proceedings ANTEC (2002)
652. Ehrig, H., Engels, G., Kreowski, H.-J., Rozenberg, G.: Handbook on Graph Grammars and Computing by Graph Transformation – Volume 2: Applications, Languages, and Tools. World Scientific, Singapore (1999)
653. Eimer, E.: Varianzanalyse. W. Kohlhammer, Stuttgart (1978)
654. Ellis, C., Gibbs, S., Rein, G.: Groupware – some issues and experiences. Communications of the ACM 34, 39–58 (1991)
655. Elmqvist, H., Mattson, S.E.: Modelica – the next generation modeling language. an international design effort. In: Proceedings of the 1st World Congress on System Simulation, WCSS'97 (1997)

656. Eloranta, E., Hameri, A.-P., Lahti, M.: Improved project management through improved document management. Computers in Industry 43(3), 231–243 (2001)
657. EMC²: EMC Documentum (2006), http://software.emc.com/products/product_family/documentum_family.htm
658. Enders, B.E., Heverhagen, T., Goedicke, M., Tröpfner, P., Tracht, R.: Towards an integration of different specification methods by using the viewpoint framework. Transactions of the SDPS 6(2), 1–23 (2002)
659. Endrei, M., Ang, J., et al.: Patterns: Service-oriented architecture and web services (2004), http://www.redbooks.ibm.com/redbooks/SG246303/wwhelp/wwhimpl/java/html/wwhelp.htm
660. Enhydra.org Community: Enhydra Shark – Java Open Source XPDL workflow, version 1.1-2 (2005), http://www.enhydra.org/workflow/shark/index.html
661. ePlantData: Capital facilities industry XML – website (2004), http://www.cfixml.org
662. Ester, M., Krieger, H.-P., Sander, J., Xu, X.: A Density-Based Algorithm for Discovering Clusters in Large Spatial Databases with Noise. In: Proc. of the 2nd Internat. Conf. on Knowledge Discovery and Data Mining (KDD'96), August 1996, pp. 226–231. AAAI Press, Menlo Park (1996)
663. Fahmy, H., Holt, R.C.: Using graph rewriting to specify software architectural transformations. In: Proceedings of the 15th International Conference on Automated Software Engineering (ASE 2000), Grenoble, France, pp. 187–196. IEEE Computer Society Press, Los Alamitos (2000)
664. Faloutsos, C., Lin, K.-I.: Fastmap: A fast algorithm for indexing, data-mining and visualization of traditional and multimedia datasets. In: Proc. of the 1995 ACM SIGMOD Conf., pp. 163–174 (1995)
665. Feldmann, L.P., Svjatnyj, V.A., Gilles, E.D., Zeitz, M., Reuter, A., Rothermel, K.: Simulationssoftware für eine verteilte parallele Simulationsumgebung für dynamische Systeme. In: Proceedings of the 14th ASIM Symposium, pp. 235–240 (2000)
666. Feng, W.C., Rexford, J.: Performance evaluation of smoothing algorithms for transmitting prerecorded variable-bit-rate video. IEEE Transactions on Multimedia 1(3), 302–313 (1999)
667. FIATECH: Automating equipment information exchange – website (2006), http://www.fiatech.org/projects/idim/aex.htm
668. FIATECH ADI: 15926.org – website (2007), http://15926.org/
669. Finkelstein, A., Kramer, J., Goedicke, M.: ViewPoint oriented software development. In: Proceedings of 3^{td} International Workshop on Software Engineering and its Applications, pp. 337–351. Springer, Heidelberg (1990)
670. Fischer, T., Niere, J., Torunski, L., Zündorf, A.: Story diagrams: A new graph rewrite language based on the unified modeling language and java. In: Ehrig, H., Engels, G., Kreowski, H.-J., Rozenberg, G. (eds.) TAGT 1998. LNCS, vol. 1764, pp. 296–309. Springer, Heidelberg (2000)
671. Fisher, L.: Workflow Handbook 2001. Lighthouse Point (2000)
672. Flores, X., Bonmatí, A., Poch, M., Rodríguez-Roda, I.: Selection of the activated sludge configuration during the conceptual design of activated sludge plants using multicriteria analysis. Ind. Eng. Chem. Res. 44, 3556–3566 (2005)

673. Fowler, M.: UML Distilled – Applying the Standard Object Modeling Language. Addison-Wesley, Reading (1997)
674. Fowler, M.: Patterns of Enterprise Application Architecture. Addison-Wesley, Reading (2003)
675. Frey, W., Lohe, B.: Verfahrenstechnik im Wandel. Chem.-Ing.-Tech. 70, 51–63 (1998)
676. Frühauf, T.: Graphisch-interaktive Strömungsvisualisierung. Springer, Heidelberg (1997)
677. Friesen, J.A., Tarman, T.D.: Remote high-performance visualization and collaboration. IEEE Computer Graphics and Applications, 45–49 (2000)
678. Fritz, H.-G.: Neue Thermoplastische Elastomere. Rezeptierung, Aufbereitung und Werkstoffeigenschaften. Chem.-Ing.-Tech. 67(5), 563–569 (1995)
679. Fuchino, T., Takamura, T., Batres, R.: Development of engineering ontology on the basis of IDEF0 activity model. In: Khosla, R., Howlett, R.J., Jain, L.C. (eds.) KES 2005. LNCS (LNAI), vol. 3681, pp. 162–168. Springer, Heidelberg (2005)
680. Gächter, R., Müller, H.: Taschenbuch der Kunststoff-Additive. Carl Hanser-Verlag, München (1989)
681. Gallaher, M.P., O'Connor, A.C.: Dettbarn Jr., J.L., Gilday, L.T.: Cost analysis of inadequate interoperability in the U.S. capital facilities industry. Technical Report NIST GCR 04-867, National Institute of Standards and Technology, NIST (2004)
682. Gamma, E., Helm, R., Johnson, R., Vlissides, J.: Design Patterns: Elements of Reusable Object-Oriented Software. Addison-Wesley, Reading (1995)
683. Gannod, G.C., Mudiam, S.V., Lindquist, T.E.: An architectural based approach for synthesizing and integrating adapters for legacy software. In: Proceedings of the 7th Working Conference on Reverse Engineering (WCRE'00), Brisbane, Australia (2000)
684. Gao, J.X., Aziz, H., Maropoulos, P.G., Cheung, W.M.: Application of product data management technologies for enterprise integration. International Journal of Computer Integrated Manufacturing 16(7-8), 491–500 (2003)
685. Garlan, D., Monroe, R., Wile, D.: ACME: An architecture description interchange language. In: Proceedings of the 1997 Conference of the Centre for Advanced Studies on Collaborative Research (CASCON'97), Toronto, Ontario, Canada, pp. 169–183 (1997)
686. Garlan, D., Perry, D.E.: Introduction to the special issue on software architecture. IEEE Transactions on Software Engineering 21(4), 269–274 (1995)
687. Gediga, G., Hamborg, K.C., Düntsch, I.: The IsoMetrics usability inventory: an operationalization of ISO 9241-10 supporting summative and formative evaluation of software systems. Behaviour & Information Technology 18(3), 154–164 (1999)
688. Georgakopoulos, D., Prinz, W., Wolf, A.L. (eds.): Proceedings of the International Joint Conference on Work Activities Coordination and Collaboration (WACC'99), San Francisco, CA, USA. ACM SIGSOFT Software Engineering Notes 24(2). ACM Press, New York (1999)
689. Gerber, A., Lawley, M., Raymond, K., Steel, J., Wood, A.: Transformation: The missing link of MDA. In: Corradini, A., Ehrig, H., Kreowski, H.-J., Rozenberg, G. (eds.) ICGT 2002. LNCS, vol. 2505, pp. 90–105. Springer, Heidelberg (2002)

828 References

690. Geyer, W., Richter, H., Fuchs, L., Frauenhofer, T., Daijavad, S., Poltrock, S.: A team collaboration space supporting capture and access of virtual meetings. In: Proceedings of the 2001 International ACM SIGGROUP Conference on Supporting Group Work, Boulder, USA, pp. 188–196 (2001)
691. Gil, N., Tommelein, I.D., Kirkendall, R.: Modeling design development processes in unpredictable environments. In: Proceedings 2001 Winter Simulation Conference (2001)
692. Gill, P.E., Murray, W., Wright, M.H.: Practical Optimization. Academic Press, London (1981)
693. Gilles, E.D.: Network theory for chemical processes. Chem. Eng. Technol. 21(2), 121–132 (1998)
694. Goebel, D.: Modellbasierte Optimierung von Produktionsentwicklungsprozessen. PhD thesis, University of Hannover (1996)
695. Goldsman, D., Nelson, B.L.: Comparing systems via simulation. In: Banks, J. (ed.) Handbook of Simulation, pp. 273–306. John Wiley & Sons, Chichester (1998)
696. Grabowski, H., Lossack, R., Leutsch, M.: A design process model. In: Proceedings of the 3rd International Workshop on Strategic Knowledge and Concept Formation, Sydney, Australia (2001)
697. Greenbaum, J., Kyng, M.: Design at Work. Cooperative Design of Computer Systems. Lawrence Erlbaum, Mahwah (1991)
698. Greenberg, S.: Peepholes: Low cost awareness of ones community. Technical report, University of Calgary, Department of Computer Science (1993)
699. Grefenstein, A.: Rechnergestützte Auslegung von Schneckenreaktoren am Beispiel des dichtkämmenden Gleichdralldoppelschneckenextruders. PhD thesis, RWTH Aachen University (1994)
700. Gröger, M.: Wertschöpfungspotenzial Projektmanagement. REFA-Nachrichten 1, 4–7 (2006)
701. Grisby, D.: OmniORBpy user's guide version 2 (2004), `http://omniorb.sourceforge.net/omnipy2/omniORBpy.pdf`
702. Groeben, N., Scheele, B.: Heidelberger Struktur-Legetechnik. Beltz, Weinheim (1984)
703. Gross, T.: CSCW3: Transparenz- und Kooperationsunterstützng für das WWW. In: Reichwald, R., Schlichter, J. (eds.) Verteiltes Arbeiten – Arbeiten in der Zukunft, Tagungsband der D-CSCW 2000, pp. 37–50. Teubner, Wiesbaden (2000)
704. Grossmann, I., Westerberg, A.: Research challenges in process systems engineering. AIChE Journal 46, 1700–1703 (2000)
705. Gruber, T.R.: A translation approach to portable ontology specifications. Knowledge Acquisition 5(3), 199–200 (1993)
706. Grudin, J.: Evaluating opportunities for design capture. In: Moran, T.P., Carroll, J.M. (eds.) Design Rationale. Concepts, Techniques, and Use, pp. 453–470 (1996)
707. Gulbins, J., Seyfried, M., Strack-Zimmermann, H.: Dokumenten-Management. Springer, Heidelberg (1999)
708. Gunasekaran, A.: Concurrent engineering: A competitive study for process industries. Journal of the Operations Research Society 49, 758–765 (1998)

709. Gutwin, C., Greenberg, S.: The effects of workspace awareness support on the usability of real-time distributed groupware. ACM Transactions on Computer-Human Interaction 6(3), 243–281 (1999)
710. Hacker, W.: Computerization versus computer aided mental work. In: Frese, M., Ulich, E., Dzida, W. (eds.) Psychological Issues of Human-Computer Interaction in the Work Place, pp. 115–130. Elsevier, Amsterdam (1987)
711. Hacker, W.: Allgemeine Arbeitspsychologie. Huber, Bern (1998)
712. Hagen, C., Alonso, G.: Exception handling in workflow management systems. IEEE Transactions on Software Engineering 26(10), 943–958 (2000)
713. Hales, C.: Analysis of the Engineering Design Process in an Industrial Context, 2nd edn. Gants Hill (1991)
714. Hammer, M., Champy, J.A.: Reengineering the Corporation: A Manifesto for Business Revolution. H. Collins (1993)
715. Hammond, J., Koubek, R.J., Harvey, C.M.: Distributed collaboration for engineering design: A review and reprisal. Human Factors and Ergonomics in Manufacturing 11(1), 35–53 (2001)
716. Hao, M.C., Lee, D., Sventek, J.S.: Collaborative design using your favourite 3d applications. In: Proceedings of the 3rd International Conference on Concurrent Engineering Research: Research and Applications, Lancaster, USA, pp. 8–15 (1996)
717. Hao, M.C., Lee, D., Sventek, J.S.: A light-weight application sharing infrastructure for graphic intensive applications. In: Proceedings of the 5th IEEE International Symposium on High Performance Distribution (HPDC-5), Syracuse, New York, USA, pp. 127–131 (1996)
718. Harel, D.: Statecharts: A visual formalism for complex systems. Science of Computer Programming 8(3), 231–274 (1987)
719. Harel, D.: On visual formalisms. Communications of the ACM 31(5), 514–530 (1988)
720. Harmon, P.: G2: Gensym's real-time expert system. Intelligent Software Strategies 9(3), 1–16 (1993)
721. Harold, M., Ogunnaike, B.: Process engineering in the evolving chemical industry. AIChE Journal 46, 2123–2127 (2000)
722. Harris, S.B.: Business strategy and the role of engineering product data management: A literature review and summary of the emerging research questions. Proceedings of the Institution of Mechanical Engineers, Part B (Journal of Engineering Manufacture) 210:B3, 207–220 (1996)
723. Haumer, P., Heymans, P., Jarke, M., Pohl, K.: Bridging the gap between past and future in RE: a scenario-based approach. In: Proceedings of the 4th International Symposium on Requirements Engineering (RE'99), Limerick, Ireland, pp. 66–73 (1999)
724. Hauser, A., Harter, G.: R&D Productivity. Example: Automotive and Health Care/Pharma Industries. Booz, Allen & Hamilton (2003)
725. Hawley, K.: Temporal parts. In: Zalta, E.N. (ed.) The Stanford Encyclopedia of Philosophy (2004), http://plato.stanford.edu/archives/win2004/entries/temporal-parts/
726. Heeg, M.: Ein Beitrag zur Modellierung von Merkmalen im Umfeld der Prozessleittechnik. PhD thesis, RWTH Aachen University (2005)

727. Heinl, P., Horn, S., Jablonski, S., Neeb, J., Stein, K., Teschke, M.: A comprehensive approach to flexibility in workflow management systems. ACM SIGSOFT Software Engineering Notes 24(2), 79–88 (1999)
728. Hensel, T.L.: Die künftige Rolle der Kunststofferzeuger. KU Kunststoffe 90(1), 34–38 (2000)
729. Herder, P., Weijnen, M.: A concurrent engineering approach to chemical process design. International Journal of Production Economics 64, 311–318 (2000)
730. Hermann, H., Burkhardt, U.: Vergleichende Analyse dichtkämmender Gleich- und Gegendrall-Doppelschnecken. Österreichische Kunststoff-Zeitschrift, Sonderheft Kunststoffkolloquium Leoben, 973–984 (1978)
731. Hirsch, D., Inverardi, P., Montanari, U.: Modeling software architectures and styles with graph grammars and constraint solving. In: Donohoe, P. (ed.) Software Architecture (TC2 1st Working IFIP Conference on Software Architecture, WICSA1), San Antonio, Texas, USA, pp. 127–143. Kluwer Academic Publishers, Dordrecht (1999)
732. Holt, R., Winter, A., Schürr, A.: GXL: Towards a standard exchange format. In: Proceedings of the 7th Working Conference on Reverse Engineering (WCRE'00), Brisbane, Queensland, Australia, pp. 162–171 (2000)
733. Hordijk, W., Wieringa, R.: Reusable rationale blocks: Improving quality and efficiency of design choices. In: Dutoit, A.H., McCall, R., Mistrík, I., Paech, B. (eds.) Rationale Management in Software Engineering, pp. 1–48. Springer, Heidelberg (2006)
734. Hornsby, P.R., Tung, J.F., Tarverdi, K.: Characterization of polyamide 6 made by reactive extrusion. i. synthesis and characterization of properties. Journal of Applied Polymer Science 53, 891–897 (1994)
735. Horrocks, I.: Using an expressive description logic: Fact or fiction? In: Proceedings of KR'98, pp. 636–647 (1998)
736. Horrocks, I., Patel-Schneider, P.: Reducing OWL entailment to description logic satisfiability. Journal of Web Semantics 1(5), 345–357 (2004)
737. Hostrup, M., Harper, P., Gani, R.: Design of environmentally benign processes: Integration of solvent design and separation process synthesis. Computers & Chemical Engineering 23, 1395–1414 (1999)
738. Huber, F., Schätz, B., Schmidt, A., Spies, K.: AutoFocus – a tool for distributed system specification. In: Jonsson, B., Parrow, J. (eds.) FTRTFT 1996. LNCS, vol. 1135, pp. 467–470. Springer, Heidelberg (1996)
739. Huber, W., Elting, A.: Abstrakte Muster – Vom Prototypen zum fertigen System. iX –Magazin für professionelle Informationstechnik 10, 106–109 (2002)
740. Huhns, M.N., Singh, M.P.: Service-oriented computing: Key concepts and principles. IEEE Internet Computing 9(1), 75–81 (2005)
741. Humphrey, W.S.: Managing the Software Process. Addison-Wesley, Reading (1990)
742. Huth, C., Erdmann, I., Nastansky, L.: GroupProcess: Using process knowledge from the participative design and practical operation of ad hoc processes for the design of structured workflows. In: Proceedings of the 34th Hawaii International Conference on System Sciences (HICSS), Maui, Hawaii, USA, IEEE Computer Society Press, Los Alamitos (2001)
743. IBM: Rational Rose – website (2006),
http://www.ibm.com/developerworks/rational/products/rose

744. IEEE, Institute for Electrical and Electronics Engineering: IEEE recommended practice for architectural description for software-intensive systems. IEEE Standard 1471–2000 (2000)
745. innotec GmbH: Website (2006), http://www.innotec.de/de/cae/index.php
746. Intergraph: The engineering framework: Integrating the plant information asset throughout the plant life cycle (2004), http://ppm.intergraph.com/library/the-engineering-framework-a4.pdf
747. Intergraph: SmartPlant foundation (2006), http://ppm.intergraph.com/smartplant/foundation/
748. International standard IEC 61131-3: IEC 61131-3, Programmable Controllers – Part 3: Programming Languages (2001)
749. Invensys Process Systems: SimSci-Esscor – simulation software for plant design and optimization (2007), http://www.simsci-esscor.com
750. IONA: Iona Orbix: CORBA for the Enterprise (2006)
751. ISO: ISO 10303 part 231: Process engineering data: Process design and process specifications of major equipment. International Standard (1998)
752. ISO: ISO 10303 Part 231: Process Engineering Data: Process Design and Process Specifications of Major Equipment. Withdrawn Standard (1998)
753. ISO 10303-11: Industrial automation systems and integration – Product data representation and exchange – Part 11: Description methods: The EXPRESS language reference manual. International Standard (2004)
754. ISO 10303-221: Industrial Automation Systems and Integration – Product Data Representation and Exchange – Part 221: Application Protocol: Functional Data and Their Schematic Representation for Process Plants. Standard under Development (2005)
755. ISO 10303-227: Industrial Automation Systems and Integration – Product Data Representation and Exchange – Part 227: Application Protocol: Plant Spatial Configuration. International Standard (2005)
756. ISO 15926-1: Industrial Automation Systems and Integration – Integration of Life-Cycle Data for Process Plants Including Oil and Gas Production Facilities – Part 1: Overview and Fundamental Principles. International Standard (2004)
757. ISO 15926-2: Industrial Automation Systems and Integration – Integration of Life-Cycle Data for Process Plants Including Oil and Gas Production Facilities – Part 2: Data Model. International Standard (2003)
758. ISO 15926-3: Industrial automation systems and integration – Integration of life-cycle data for process plants including oil and gas production facilities – Part 3: Ontology for geometry and topology. Standard under development.(2006)
759. ISO 15926-4: Industrial automation systems and integration – Integration of life-cycle data for process plants including oil and gas production facilities – Part 4: Initial reference data. Standard under development (2005)
760. ISO 15926-7: Industrial automation systems and integration – Integration of life-cycle data for process plants including oil and gas production facilities – Part 7: Implementation methods for data exchange and integration. Standard under development (2005)
761. Isoce, N., Williams, G.B., Arango, G.: Domain modelling for software engineering. In: Proceedings of the 13th International Conference on Software Engineering, pp. 23–29 (1991)

762. Jablonski, S.: Workflow-Management: Entwicklung von Anwendungen und Systemen. dpunkt, Heidelberg (1999)
763. Jablonski, S.: Workflow Management – Modeling Concepts and Architecture. International Thomson Publishing, Albany (1996)
764. Jaccheri, M.L., Conradi, R.: Techniques for Process Model Evolution in EPOS. IEEE Transactions on Software Engineering 19(12), 1145–1156 (1993)
765. Jacobsen, H.A., Krämer, B.J.: A design pattern based approach to generating synchronization adaptors from annotated IDL. In: Proceedings of the 13th International Conference on Automated Software Engineering (ASE'98), Hawaii, USA, pp. 63–72. IEEE Computer Society Press, Los Alamitos (1998)
766. Jahnke, J., Zündorf, A.: Applying graph transformations to database re-engineering. In: Ehrig, H., Engels, G., Kreowski, H.-J., Rozenberg, G. (eds.) Handbook on Graph Grammars and Computing by Graph Transformation – Volume 2: Applications, Languages, and Tools, pp. 267–286. World Scientific, Singapore (1999)
767. Jarke, M., List, T., Nissen, H.W., Lohmann, B., Hubbuch, K.: Bericht zum Workshop verfahrenstechnische Datenbanken. Technical report, Bayer AG (1998)
768. Jarke, M., Nissen, H.W., Pohl, K.: Tool integration in evolving information systems environments. In: Proceedings of the 3rd GI Workshop Information Systems and Artificial Intelligence: Administration and Processing of Complex Structures (1994)
769. Jarke, M., Vassiliou, M.: Foundations of data warehouse quality: An overview of the DWQ project. In: Proceedings of the 16th ACM Symposium on Principles of Database Systems (PODS), Tucson, AZ, USA, pp. 51–61 (1997)
770. Jensen, K.: Coloured Petri Nets: Basic Concepts, Analysis Methods and Practical Use. Volume 1: Basic Concepts. Springer, Heidelberg (1997)
771. Jin, Y., Levitt, R.E.: The virtual design team: A computational model of project organizations. Computational and Mathematical Organization Theory 2(3), 171–195 (1996)
772. Joeris, G., Herzog, O.: Managing evolving workflow specifications. In: Halper, M. (ed.) Proceedings of the International Conference on Cooperative Information Systems (CoopIS'98), pp. 310–321. IEEE Computer Society Press, Los Alamitos (1998)
773. Joeris, G., Herzog, O.: Towards flexible and high-level modeling and enacting of processes. In: Jarke, M., Oberweis, A. (eds.) CAiSE 1999. LNCS, vol. 1626, pp. 88–102. Springer, Heidelberg (1999)
774. Johnson, J.: GUI Bloopers. Morgan Kaufmann, San Francisco (2000)
775. Katzke, U., Vogel-Heuser, B.: Design and application of an engineering model for distributed process automation. In: Proceedings of the 2005 American Control Conference, Portland, USA (2005)
776. Kent, S., Smith, R.: The bidirectional mapping problem. Electronic Notes in Theoretical Computer Science 82(7) (2003)
777. Kerzner, H.: Project Management: A Systems Approach to Planning, Scheduling, and Controlling. John Wiley & Sons, Chichester (1998)
778. Kiepuszewski, B., ter Hofstede, A.H.M., Bussler, C.: On structured workflow modelling. In: Wangler, B., Bergman, L.D. (eds.) CAiSE 2000. LNCS, vol. 1789, pp. 431–445. Springer, Heidelberg (2000)

779. Kifer, M., Lausen, G., Wu, J.: Logical foundations of object-oriented and frame-based languages. Journal of the ACM 42(4), 741–843 (1995)
780. Kim, Y., Kang, S., Lee, S., Yoo, S.: A distributed, open, intelligent product data management system. International Journal of Computer Integrated Manufacturing 14, 224–235 (2001)
781. Kirschner, E.: Running on information. Chemical & Engineering News 75, 15–19 (1997)
782. Kirwan, B., Ainsworth, L.K.: A guide to task analysis. Taylor & Francis, Abington (1992)
783. Klein, R., Anhäuser, F., Burmeister, M., Lamers, J.: Planungswerkzeuge aus Sicht eines Inhouse-Planers. atp – Automatisierungstechnische Praxis 44(1), 46–50 (2002)
784. Klir, G.J.: Architecture of Systems Problem Solving. Plenum Press, New York (1985)
785. Kämpfer, J., Lohmann, B.: Rationalisierung von Planungsprozessen durch die Integration von Informationssystemen. VDI-Berichte 1684 (2002)
786. Königs, A., Schürr, A.: MDI: a rule-based multi-document and tool integration approach. Software and System Modeling 5(4) (2006)
787. Königs, A., Schürr, A.: Tool integration with triple graph grammars – a survey. In: Heckel, R. (ed.) Proceedings of the SegraVis School on Foundations of Visual Modelling Techniques. Electronic Notes in Theoretical Computer Science, vol. 148, pp. 113–150. Elsevier, Amsterdam (2006)
788. Knowledge Based Systems: IDEF0 Function Modeling Method (2006), http://www.idef.com/idef0.html
789. Ko, C.C.W.: Execution Monitoring of Security-Critical Programs in a Distributed System: A Specification-Based Approach. PhD thesis, University of California (1996)
790. Kohlgrüber, K.: Co-rotating Twin-Screw Extruder. Carl Hanser Verlag, München (2007)
791. Konda, S., Monarch, I., Sargent, P., Subrahmanian, E.: Shared memory in design: A unifying theme for research and practice. Research in Engineering Design 4, 23–42 (1992)
792. Kradolfer, M., Geppert, A.: Dynamic workflow schema evolution based on workflow type versioning and workflow migration. In: Proceedings of the International Conference on Cooperative Information Systems (CoopIS'99), Edinburgh, pp. 104–114. IEEE Computer Society Press, Los Alamitos (1999)
793. Krause, F.L., Golm, F., Loske, B., Raupach, C.: Simulation von Produktentwicklungsprozessen. Zeitschrift für wirtschaftlichen Fabrikbetrieb 90(3), 113–115 (1995)
794. Krause, F.-L., Kind, C., Voigtsberger, J.: Adaptive modelling and simulation of product development processes. ANNALS–CIRP 53(1), 135–138 (2004)
795. Krönlof, K. (ed.): Method Integration. Wiley, Chichester (1993)
796. Kummer, O., Wienberg, F., Duvigneau, M., Schumacher, J., Köhler, M., Moldt, D., Rölke, H., Valk, R.: An extensible editor and simulation engine for petri nets: RENEW. In: Cortadella, J., Reisig, W. (eds.) ICATPN 2004. LNCS, vol. 3099, pp. 484–493. Springer, Heidelberg (2004)
797. Kummer, O., Wienberg, M., Duvigneau, M.: Renew-user guide, release 2.0 edition. Technical report, University of Hamburg, Faculty of Informatics, Theoretical Foundations Group (2004)

798. Kunz, W., Rittel, H.: Issues as elements of information systems – working paper no. 131. Technical report, University of California, Berkley, Institute of Urban and Regional Development (1970)
799. Kuraoka, K., Batres, R.: An ontological approach to represent HAZOP information. Technical Report TR-2003-01, Process Systems Engineering Laboratory, Tokyo Institute of Technology (2003)
800. Kushmerick, N.: Wrapper induction: Efficiency and expressiveness. Artificial Intelligence 118(1-2), 15–68 (2000)
801. Lamport, L.: Time, clocks, and the ordering of events in a distributed system. Communications of the ACM 21(7), 558–565 (1978)
802. Laoutaris, N., Stavrakakis, I.: Intrastream synchronization for continuous media streams: a survey of playout schedulers. IEEE Network Magazine 16(3) (2002)
803. Lawrence, P. (ed.): Workflow Handbook. John Wiley & Sons, Chichester (1997)
804. Le Métayer, D.: Describing software architecture styles using graph grammars. IEEE Transactions on Software Engineering 27(7), 521–533 (1998)
805. Leake, D., Hammond, K., Birnbaum, L., Marlow, C., Yang, H.: An integrated interface for proactive, experience-based design support. In: Proceedings of the 6th International Conference on Intelligent User Interfaces (2001)
806. Leal, D.: ISO 15926 "life cycle data for process plant": An overview. Oil Gas Sci. Technol. 60(4), 629–637 (2005)
807. Lee, J.: SIBYL: A qualitative decision management system. In: Winston, P.H., Shellard, S.A. (eds.) Artificial intelligence at MIT expanding frontiers, pp. 104–133. MIT Press, Cambridge (1991)
808. Lee, J., Lai, K.-Y.: What's in design rationale? Human-Computer Interaction 6, 251–280 (1991)
809. Lee, J.: SIBYL: A qualitative decision management system. In: Winston, P.H., Shellard, S. (eds.) Artificial Intelligence at MIT: Expanding Frontiers, pp. 105–133. MIT Press, Cambridge (1990)
810. Lee, J.: Design rationale systems: Understanding the issues. IEEE Expert 12(3), 78–85 (1997)
811. Lee, J., Siau, K., Hong, S.: Enterprise Integration with ERP and EAI. Communications of the ACM 46(2), 54–60 (2003)
812. Levitt, R.E., Cohen, G.P., Kuntz, J.C., Nass, C.I., Christiansen, T., Jin, Y.: The virtual design team: Simulating how organizational structure and information processing tools affect team performance. In: Carley, K.M., Prietula, M.J. (eds.) Computational Organization Theory, Lawrence Erlbaum, Mahwah (1994)
813. Levitt, R.E., Thomsen, J., Christiansen, T.R., Kunz, J., Jin, Y., Nass, C.: Simulating project work processes and organizations: Toward a micro-contingency theory of organizational design. Management Science, Informs 45(11), 1479–1495 (1999)
814. Leymann, F.: Web services: Distributed applications without limits. In: Proceedings of Datenbanksysteme für Business, Technologie und Web (2003)
815. Löffelmann, G., Zgorzelski, P., Ahrens, W.: Produktdatenaustausch auf der Basis standardisierter PROLIST-Merkmalleisten für PLT-Geräte und - Systeme. atp – Automatisierungstechnische Praxis 47, 25–31 (2005)

816. Lieberman, H.: Your wish is my command: Programming by example. Academic Press, London (2001)
817. Light, B., Holland, C.P., Kelly, S., Wills, K.: Best Of Breed IT Strategy: An Alternative To Enterprise Resource Planning Systems. In: Hansen, H.R., Bichler, M., Mahrer, H. (eds.) Proceedings of the 8th European Conference on Information Systems (ECIS 2000), July 03-05, 2000, Vienna, Austria, pp. 652–659 (2000)
818. Lim, W.C.: Effects of reuse on quality, productivity, and economics. IEEE Software 11(3), 23–30 (1994)
819. Linthicum, D.: Enterprise Application Integration. Addison-Wesley, Reading (2000)
820. Linthicum, D.S.: B2B Application Integration: e-Business-Enable your Enterprise. Addison-Wesley, Reading (2001)
821. Liu, D.-R., Shen, M.: Business-to-business workflow interoperation based on process-views. Decision Support Systems 38(3), 399–419 (2004)
822. Lohmann, S., Stursberg, O., Engell, S.: Systematic design of logic controllers for processing plants starting from informal specifications. In: Marquardt, W., Pantelides, C. (eds.) 16th European Symposium on Computer Aided Process Engineering and 9th International Symposium on Process Systems Engineering, Garmisch-Partenkirchen, Germany, July 9–13, 2006. Computer-Aided Chemical Engineering, vol. 21, pp. 1317–1322. Elsevier, Amsterdam (2006)
823. Lunt, T.F.: A survey of intrusion detection techniques. Computers and Security 12(4), 405–418 (1993)
824. Lutz, M.: Programming Python, 2nd edn. O'Reilly, Sebastopol (2001)
825. Maaß, S.: Software-Ergonomie – Benutzer und aufgabenorientierte Systemgestaltung. Informatik-Spektrum 16, 191–205 (1993)
826. MacLean, A., Young, R.M., Bellotti, V.M.E., Moran, T.P.: Questions, options, and criteria: Elements of design space analysis. In: Moran, T.P., Carroll, J.M. (eds.) Design Rationale. Concepts, Techniques, and Use, pp. 53–105. Lawrence Erlbaum, Mahwah (1996)
827. Maier, A., Aguado, J., Bernaras, A., Laresgoiti, I., Pedinaci, C., Pena, N., Smithers, T.: Integration with ontologies. In: Reimer, U., Abecker, A., Staab, S., Stumme, G. (eds.) Professionelles Wissensmanagement – Erfahrungen und Visionen, Beiträge der 2. Konferenz Professionelles Wissensmanagement (2003)
828. Malone, M.F., Trainham, J.A., Carnahan, B. (eds.): Fifth International Conference on Foundations of Computer-Aided Process Design. AIChE Symposium Series, vol. 96(323) (2000)
829. Mark, J.E., Eisenberg, A., Graessley, W.W.: Physical Properties of Polymers. ACS Professional Reference Book. American Chemical Society, Washington, DC (1993)
830. Martin, W.: Strategic Bulletin EAI 2003: Von der Anwendungsintegration zum Prozeßmanagement (2003), http://www.wolfgang-martin-team.net/content/htm/download.htm
831. Martin, D. (ed.) and The OWL Services Coalition: OWL-S: Semantic markup for web services, version 1.1 (2004), http://www.daml.org/services/owl-s/

832. Maurer, F., Dellen, B., Bendeck, F., Goldmann, S., Holz, H., Kötting, B., Schaaf, M.: Merging project planning and web-enabled dynamic workflow technologies. IEEE Internet Computing 4(3), 65–74 (2000)
833. May, A., Carter, C., Joyner, S.: Virtual team working in the european automotive industry: User requirements and a case study approach. Human Factors and Ergonomics in Manufacturing 10(3), 273–289 (2000)
834. Mayer, H.H., Schoenmakers, H.: Integrated use of CAPE tools – an industrial example. In: Malone, M.F., Trainham, J.A., Carnahan, B. (eds.) Fifth International Conference on Foundations of Computer-Aided Process Design. AIChE Symposium Series, vol. 96(323), pp. 466–469 (2000)
835. McCall, R.J.: PHI: a conceptual foundation for design hypermedia. Design Studies 12, 30–41 (1991)
836. McCarthy, J., Bluestein, W.: The Computing Strategy Report: Workflow's Progress. Forrester Research (1991)
837. McCormack, K.: Business process orientation: Do you have it? Quality Progress, 51–58 (2001)
838. McGrath, J.E.: Groups: Interaction and Performance. Prentice-Hall, Englewood Cliffs (1984)
839. McKay, A., Bloor, M.S., de Pennington, A.: A framework for product data. IEEE Transactions on Knowledge and Data Engineering 8(5), 825–838 (1996)
840. McKenaa, T.: Design model of a wiped film evaporator. applications to the devolatilisation of polymer melts. Chemical Engineering Science 50(3), 453–467 (1995)
841. Meeting by Wire: NetMeeting (2006), `http://www.meetingbywire.com/NetMeeting101.htm`
842. Mendling, J., Nüttgens, M.: EPC modelling based on implicit arc types. In: Godlevsky, M., Liddle, S.W., Mayr, H.C. (eds.) Information Systems Technology and its Applications, International Conference ISTA 2003, Ukraine (2003)
843. Mettala, E., Graham, M.H.: The domain-specific software architecture program. Technical Report CMU/SEI-92-SR-009, Carnegie Mellon University, Software Engineering Institute, SEI (1992)
844. Microsoft: Excel – website (2005), `http://office.microsoft.com/excel`
845. Microsoft: Visio – website (2005), `http://office.microsoft.com/visio`
846. Microsoft: Component Object Model (2006), `http://msdn.microsoft.com/library/default.asp?url=/library/en-us/dnanchor/html/componentobjectmodelanchor.asp`
847. Microsoft: DCOM: distributed component object model (2006), `http://msdn.microsoft.com/library/en-us/dnanchor/html/dcom.asp`
848. Mille, A.: From case-based reasoning to traces based reasoning. In: 9th IFAC Symposium on Automated Systems Based on Human Skill And Knowledge, Nancy, France (2006)
849. Miller, R.C., Myers, B.A.: Creating dynamic world wide web pages by demonstration. Technical Report CMU-CS-97-131, Carnegie Mellon University, School of Computer Science (1999)
850. Minenko, W.: The application sharing technology. The X Advisor 1(1) (1995)
851. Misander, P.K.: A document-oriented model of the workflow in an engineering project (2006), `http://CAPE-Alliance.ucl.org.uk/CAPE_Applications_etc/Initiatives_and_Networks/About_CAPENET/CAPENET_Section_Frameset.html`

852. Mittal, S., Araya, A.: A knowledge-based framework for design. In: Proceedings of the 5th National Conference on Artificial Intelligence, pp. 856–865 (1986)
853. Mizrahi, J.: Developing an industrial chemical process: an integrated approach. CRC Press, Boca Raton (2002)
854. Münch, J., Rombach, D.: Eine Prozessplattform zur erfahrungsbasierten Softwareentwicklung. In: Nagl, M., Westfechtel, B. (eds.) Modelle, Werkzeuge und Infrastrukturen zur Unterstützung von Entwicklungsprozessen, pp. 93–106. Wiley-VCH, Weinheim (2003)
855. Moeschlin, O., Grycko, E., Pohl, C., Steinert, F. (eds.): Experimental Stochastics. Springer, Heidelberg (1998)
856. Moran, T.P., Carroll, J.M.: Overview of design rationale. In: Moran, T.P., Carroll, J.M. (eds.) Design Rationale: Concepts, Techniques, and Use, pp. 1–19. Lawrence Erlbaum, Mahwah (1996)
857. Moran, T.P., Carroll, J.M. (eds.): Design Rationale: Concepts, Techniques, and Use. Lawrence Erlbaum, Mahwah (1996)
858. Mostow, J.: Toward better models of the design process. AI Magazine 6(1), 44–57 (1985)
859. Motard, R.L., Blaha, M.R., Book, N.L., Fielding, J.J.: Process engineering databases – from the pdxi perspective. In: Proceedings of the 4th International Conference on Foundations of Computer-Aided Process Design, pp. 142–153 (1995)
860. Mours, M., Flecke, J., Kohlgrüber, K.: Simulation von Zweiwellenextrudern – Einsatz und Grenzen. In: Polymeraufbereitung 2002, pp. 153–174. VDI-Verlag, Düsseldorf (2002)
861. Murata, T., Borgida, A.: Handling of irregularities in human centered systems: A unified framework for data and processes. IEEE Transactions on Software Engineering 26(10), 959–977 (2000)
862. Mylopoulos, J.: Information modeling in the time of revolution. Information Systems 23(3/4), 127–155 (1998)
863. NAMUR – Interessengemeinschaft Prozessleittechnik der chemischen und pharmazeutischen Industrie: NA 35: Abwicklung von PLT-Projekten (2003)
864. Nielsen, J.: Usability Engineering. Morgan Kaufmann, San Francisco (1993)
865. Niles, I., Pease, A.: Towards a standard upper ontology. In: Welty, C., Smith, B. (eds.) Proceedings of the 2nd International Conference on Formal Ontology in Information Systems (FOIS-2001) (2001)
866. Nonaka, I., Takeuchi, H.: The Knowledge-Creating Company. Oxford University Press, Oxford (1995)
867. Noonan, H.: Identity. In: Zalta, E.N. (ed.) The Stanford Encyclopedia of Philosophy (2006), http://plato.stanford.edu/archives/win2006/entries/identity/
868. Noumenon Consulting Limited: Open access to intelligent process plant models (2006), http://www.noumenon.co.uk
869. Noy, N., Rector, A.: Defining n-ary relations on the semantic web (2006), http://www.w3.org/TR/swbp-n-aryRelations/
870. Nuseibeh, B., Finkelstein, A.: Viewpoints: A vehicle for method and tool integration. In: Proceedings of the 5th International Workshop on Computer-Aided Software Engineering (1992)

871. Oberle, D., Volz, R., Motik, B., Staab, S.: An extensible ontology software environment. In: Staab, S., Studer, R. (eds.) Handbook on Ontologies, pp. 311–333. Springer, Heidelberg (2004)
872. Oberquelle, H.: Sprachkonzepte für benutzergerechte Systeme. Informatik-Fachberichte, vol. 144. Springer, Heidelberg (1987)
873. Object Management Group (OMG): Business process modeling notation (BPMN) final adopted specification (2002), http://www.bpmn.org/Documents/OMGFinalAdoptedBPMN1-0Spec06-02-01.pdf
874. Object Management Group (OMG): Meta-object facility (MOF) specification, version 1.4 (2002), http://www.omg.org/technology/documents/formal/mof.htm
875. Object Management Group (OMG): MOF 2.0 query / view / transformations, request for proposal (2002), http://www.omg.org/docs/ad/02-04-10.pdf
876. Object Management Group (OMG): MDA guide, version 1.0.1 (2003), http://www.omg.org/docs/omg/03-06-01.pdf
877. Object Management Group (OMG): Common Object Request Broker Architecture (CORBA): Core Specification, Version 3.0.3 (2004), http://www.omg.org/technology/documents/formal/corba_iiop.htm
878. Object Management Group (OMG): CAD services specification, v1.2 (2005), http://www.omg.org/cgi-bin/doc?formal/2005-01-07
879. Object Management Group (OMG): OCL 2.0 specification (2005), http://www.omg.org/docs/ptc/05-06-06.pdf
880. Object Management Group (OMG): Unified modeling language (UML), version 2.0 (2005), http://www.omg.org/technology/documents/formal/uml.htm
881. Object Management Group (OMG): Domain specifications (2006), http://www.omg.org/technology/documents/domain_spec_catalog.htm
882. Object Management Group (OMG): UML infrastructure specification, version 2.1.1 (2007), http://www.omg.org/cgi-bin/doc?formal/07-02-04
883. Oertel Jr., H. (ed.): Prandtl-Führer durch die Strömungslehre, 10th edn. Vieweg, Wiesbaden (2001)
884. Oestereich, B., Weiss, C., Schröder, C., Weilkiens, T., Lenhard, A.: Objektorientierte Geschäftsprozessmodellierung mit der UML. dpunkt, Heidelberg (2003)
885. Open Applications Group: OAGIS – website (2006), http://www.openapplications.org
886. Oppermann, R., Murcher, B., Reiterer, H., Koch, M.: Software-ergonomische Evaluation. Gruyter, Berlin (1992)
887. Organization for the Advancement of Structured Information Standards (OASIS): Web Services Business Process Execution Language, WSBPEL (2007), http://www.oasis-open.org/committees/tc_home.php?wg_abbrev=wsbpel
888. Orzack, S.H., Sober, E.: A critical assessment of Levins's The strategy of model building in population biology (1966). The Quarterly Review of Biology 68(4), 533–546 (1993)
889. Osswald, A.T.: BEM in der Kunststoffverarbeitung. KU Kunststoffe 89(2), 65–68 (1999)
890. Osswald, T.A., Gramann, P.J.: Polymer processing simulation trends. In: SAMPE, Erlangen, Germany (2001)

891. Ostermann, P.: Security Prosperity; The American Labor Market: How is it changed and what is to do about. Princeton University Press, Princeton (1999)
892. Osterweil, L.: Software processes are software, too. In: Proceedings of the 9th International Conference on Software Engineering, pp. 2–13 (1987)
893. Ouzounis, V., Tschammer, V.: A framework for virtual enterprise support services. In: Proceedings of the 32th Hawaii International Conference on System Sciences (HICSS), IEEE Computer Society Press, Los Alamitos (1999)
894. Pantelides, C., Keeping, B., Bernier, J., Gautreau, C.: Open interface specification: Numerical solvers. Technical Report CO-NUMR-EL-03, The CAPE-OPEN Laboratories Network, CO-LaN (1999), http://www.colan.org/index-33.html
895. Pantelides, C.C., Barton, P.I.: Equation-oriented dynamic simulation: Current status und future perspectives. Computers & Chemical Engineering 17, 263–285 (1993)
896. Papazoglou, M.P.: Service-oriented computing: Concepts, characteristics and directions. In: Proceedings of the 4th International Conference on Web Information Systems Engineering (WISE 2003), Rome, Italy, 10–12 December 2003, pp. 3–12. IEEE Computer Society Press, Los Alamitos (2003)
897. Parnas, D.L.: A technique for software module specification with examples. Communications of the ACM 15(5), 330–336 (1972)
898. Patil, L., Dutta, D., Sriram, R.: Ontology-based exchange of product data semantics. IEEE Transactions on Automation Science and Engineering 3(3), 213–225 (2005)
899. Paton, N.W., Diaz, O.: Introduction. In: Active Rules in Database Systems, pp. 3–27. Springer, Heidelberg (1998)
900. Paton, N.W., Goble, C.A., Bechhofer, S.: Knowledge based information integration systems. Information and Software Technology 42, 299–312 (2000)
901. Patzak, G.: Systemtechnik – Planung komplexer innovativer Systeme. Springer, Heidelberg (1982)
902. Perkins, C.: RTP – Audio and Video for the Internet. Addison-Wesley, Reading (2003)
903. Perkins, C., Hodson, O., Hardman, V.: A survey of packet loss recovery techniques for streaming audio. IEEE Network Magazine 12(5), 40–47 (1998)
904. Perrey, R., Lycett, M.: Service-oriented architecture. In: Proceedings of the 2003 Symposium on Applications and the Internet Workshops (SAINT 2003), Orlando, Florida, USA, 27–31 January 2003, pp. 116–119. IEEE Computer Society Press, Los Alamitos (2003)
905. Perrin, O., Wynen, F., Bitcheva, J., Godart, C.: A model to support collaborative work in virtual enterprises. In: van der Aalst, W.M.P., ter Hofstede, A.H.M., Weske, M. (eds.) BPM 2003. LNCS, vol. 2678, pp. 104–119. Springer, Heidelberg (2003)
906. Peters, M.S., Timmerhaus, K.D.: Plant Design and Economics for Chemical Engineers. McGraw-Hill, New York (1991)
907. Philpotts, M.: An introduction to the concepts, benects and terminology of product data management. Industrial Management & Data Systems 4, 11–17 (1996)

908. PIEBASE – Process Industries Executive for Achieving Business Advantage Using Standards for Data Exchange: PIEBASE Activity Model (1998), http://www.posc.org/piebase/ppam20.pdf
909. PIEBASE – Process Industries Executive for Achieving Business Advantage Using Standards for Data Exchange: PIEBASE – website (2005), http://www.posc.org/piebase
910. PIXARGUS: PIXARGUS GmbH – Automation by Vision (2006), http://www.pixargus.com/
911. Plone Foundation: plone.org – website (2006), http://plone.org/
912. Pohl, K., Böckle, G., van der Linden, F.: Software Product Line Engineering – Foundations, Principles, and Techniques. Springer, Heidelberg (2005)
913. Potts, C., Bruns, G.: Recording the reasons for design decisions. In: Proceedings of the 10th International Conference on Software Engineering (ICSE'88), pp. 418–427. IEEE Computer Society Press, Los Alamitos (1988)
914. Preece, P., Ingersoll, T., Tong, J.: Specification (data) sheets – picture a database. In: Proceedings of the European Symposium on Computer Aided Process Engineering – ESCAPE 4, pp. 467–474 (1994)
915. Proudlove, N., Vadera, S., Kobbacy, K.: Intelligent management systems in operations: A review. Journal of the Operations Research Society 49, 682–699 (1998)
916. PSE: gPROMS – website (2006), http://www.psenterprise.com/gproms/
917. Qiu, J., Knightly, E.: Measurement-based admission control using aggregate traffic envelopes. IEEE/ACM Transactions on Networking 9(2), 199–210 (2001)
918. Racer Systems GmbH & Co. KG: RacerPro – website (2006), http://www.racer-systems.com/products/racerpro/index.phtml
919. Ralyté, J., Rolland, C.: An assembly process model for method engineering. In: Dittrich, K.R., Geppert, A., Norrie, M.C. (eds.) CAiSE 2001. LNCS, vol. 2068, p. 267. Springer, Heidelberg (2001)
920. Ramesh, B.: Factors influencing requirements traceability practice. Communications of the ACM 41(12), 37–44 (1998)
921. Raskin, J.: The Humane Interface – New Directions for Designing Interactive Systems. Addison-Wesley, Reading (2000)
922. Rasmussen, J.: Human Information Processing & Human-Machine Interaction. North-Holland, Amsterdam (1986)
923. Rasmussen, J.: Merging paradigms: Decision making, management, and cognitive control. In: Flin, R., Salas, E., Martin, L. (eds.) Decision making under stress: Emerging paradigms and applications, Ashgate, Aldershot (1997)
924. Rasmussen, J., Pejtersen, A.M., Goodstein, L.P.: Cognitive Systems Engineering. John Wiley & Sons, Chichester (1994)
925. Raupach, H.C.: Simulation von Produktentwicklungsprozessen. PhD thesis, TU Berlin (1999)
926. Rödiger, K.H.: Arbeitsinformatik. In: Luczak, H., Volpert, W. (eds.) Handbuch Arbeitswissenschaft, pp. 176–182. Schäffer-Poeschel, Stuttgart (1997)
927. Regli, W.C., Hu, X., Atwood, M., Sun, W.: A survey of design rationale systems: Approaches, representation, capture and retrieval. Engineering with Computers 16, 209–235 (2000)

928. Reich, Y., Konda, S., Subrahmanian, E., Cunningham, D., Dutoit, A., Patrick, R., Thomas, M., Westerberg, A.W.: Building agility for developing agile design information systems. Research in Engineering Design 11, 67–83 (1999)
929. Reichert, M., Dadam, P.: ADEPT$_{flex}$ – supporting dynamic changes without loosing control. Journal of Intelligent Information Systems 10(2), 93–129 (1998)
930. Reimschüssel, H.K.: Nylon 6 chemistry and mechanisms. Journal of Polymer Science: Macromolecular Reviews (1977)
931. Remberg, C., Wozny, G., Fieg, G., Fett, F.N., Blaich, L.: Entwicklung eines Expertensystems für den Entwurf eines Automatiersungskonzepts für Rektifikationskolonnen. Chem.-Ing.-Tech. 67 (1995)
932. Richter, H., Abowd, G.D., Geyer, W., Fuchs, L., Daijavad, S., Poltrock, S.: Integrating meeting capture within a collaborative team environment. In: Abowd, G.D., Brumitt, B., Shafer, S. (eds.) UbiComp 2001. LNCS, vol. 2201, pp. 123–138. Springer, Heidelberg (2001)
933. Richter, J.-P.: Wann liefert eine Serviceorientierte Architektur echten Nutzen?. In: Liggesmeyer, P., Pohl, K., Goedicke, M. (eds.) Software Engineering 2005, Fachtagung des GI-Fachbereichs Softwaretechnik, Essen, 8.–11.3, 2005. LNI, vol. 64, pp. 231–242. GI (2005)
934. Riegel, J.P., Kaesling, C., Schütze, M.: Modeling software architecture using domain-specific patterns. In: Donohoe, P. (ed.) Software Architecture (TC2 1st Working IFIP Conference on Software Architecture, WICSA1), San Antonio, Texas, USA, pp. 273–301. Kluwer Academic Publishers, Dordrecht (1999)
935. Riggert, K., Terrier, F.: Die Entfernung niedermolekularer Anteile in Polyamid-6-Schmelzen mittels Vakuum. CZ-Chemie-Technik 2(3), 95–99 (1973)
936. Rittel, H.W.J., Webber, M.M.: Dilemmas in a general theory of planning. Policy Sciences 4, 155–169 (1973)
937. Rolland, C.: A comprehensive view of process engineering. In: Pernici, B., Thanos, C. (eds.) CAiSE 1998. LNCS, vol. 1413, p. 1. Springer, Heidelberg (1998)
938. Rose, T.: Visual assessment of engineering processes in virtual enterprises. Communications of the ACM 41(12), 45–52 (1998)
939. Rosemann, M.: Komplexitätsmanagement in Prozeßmodellen – Methodenspezifische Gestaltungsempfehlungen für die Informationsmodellierung. PhD thesis, University of Münster (1996)
940. Rosman, G., van der Meer, K., Sol, H.G.: The design of document information systems. Journal of Information Science 22(4), 287–297 (1996)
941. Rosson, M.B., Carroll, J.M.: Usability Engineering. Academic Press, London (2002)
942. Röthemeyer, F., Sommer, F.: Kautschuk Technologie, 2nd edn. Carl Hanser Verlag, München (2006)
943. Rubin, J. (ed.): Handbook of Usability Testing. Wiley, Chichester (1994)
944. Rumbaugh, J., Blaha, M., Premerlani, W., Eddy, F., Lorensen, W.: Object-Oriented Modeling and Design. Prentice-Hall, Englewood Cliffs (1991)
945. Sahai, H., Agell, M.I.: The Analysis of Variance. Birkhäuser, Basel (2000)
946. Salminen, A., Lyytikäinen, V., Tiitinen, P.: Putting documents into their work context in document analysis. Information Processing and Management 36(4), 623–641 (2000)
947. SAP: SAP R/3® – website (2006), `http://www.sap.com/`

948. Sattler, K.-U.: A framework for component-oriented tool integration. In: Proceedings of the International Conference on Object Oriented Information Systems (1997)
949. Scheer, A.-W.: Aris – Business Process Modeling. Springer, Heidelberg (2006)
950. Schefstroem, D., van den Broek, G. (eds.): Tool Integration – Environments and Frameworks. Wiley, Chichester (1993)
951. Schembecker, G., Simmrock, K.: Heuristic-numeric process synthesis with PROSYN. In: Conference on Intelligent Systems in Process Engineering, pp. 275–278 (1996)
952. Schembecker, G., Simmrock, K.H., Wolff, A.: Synthesis of chemical process flowsheets by means of cooperating knowledge integrating systems. In: Proceedings of the European Symposium on Computer Aided Process Engineering – ESCAPE 4 (1994)
953. Schiemann, B., Borrmann, L.: A new approach for load balancing in high performance decision support systems. In: Liddell, H., Colbrook, A., Hertzberger, B., Sloot, P.M.A. (eds.) HPCN-Europe 1996. LNCS, vol. 1067, pp. 571–579. Springer, Heidelberg (1996)
954. Schäl, T.: Workflow Management Systems for Process Organisations. Springer, Heidelberg (1996)
955. Schmidt-Traub, H., Koester, M., Holtkoetter, T., Nipper, N.: Conceptual plant layout. Computers & Chemical Engineering 22, 499–504 (1998)
956. Schwermetall: Schwermetall Halbzeugwerk – website (2006), http://www.schwermetall.info/
957. Seider, W.D., Seader, J.D., Lewin, D.R.: Process Design Principles – Synthesis, Analysis, and Evaluation. John Wiley & Sons, Chichester (1999)
958. Seidlmeier, H.: Process Modeling with ARIS – A Practical Introduction. Vieweg, Wiesbaden (2004)
959. Sennrath, F., Hermanns, O.: Performance Investigation of the MTP Multicast Transport Protocol. In: Proceedings of the 2nd Workshop on Protocols for Multimedia Systems, Salzburg, Austria, pp. 73–81 (1995)
960. Seppälä, J., Basson, L., Norris, G.A.: Decision analysis frameworks for life-cycle impact assessment. Journal of Industrial Ecology 5, 45–68 (2001)
961. Shanks, G., Seddon, P.: Enterprise resource planning (ERP) systems. Journal of Information Technology 15, 243–244 (2000)
962. Sharp, A., McDermott, P.: Workflow Modeling. Tools for Process Improvement and Application Development. Artech House, Boston (2001)
963. Sheth, A., et al.: NSF workshop on workflow and process automation. ACM Software Engineering Notes 22(1), 28–38 (1997)
964. Shneiderman, B.: Designing the User Interface, 3rd edn. Addison-Wesley, Reading (1998)
965. Silberschatz, A., Galvin, P.B., Gagne, G.: Operating System Concepts. John Wiley & Sons, Chichester (2001)
966. Singhal, S.K., Zyda, M.J.: Networked Virtual Environments: Design and Implementation. ACM Press, New York (1999)
967. Sklorz, S.: A Method for Data Analysis Based on Self Organizing Feature Maps. In: Proceedings of the Internat. Symposium on Soft Computing for Industry (ISSCI'96), Montpellier, France (May 1996)

968. Sklorz, S., Becks, A., Jarke, M.: MIDAS – Ein Multistrategiesystem zum explorativen Data Mining. In: Proceedings of the 2nd Workshop on Data Mining und Data Warehousing als Grundlage moderner entscheidungsunterstützender Systeme (DMDW 1999), pp. 129–143 (1999)
969. Sklorz, S., Jarke, M.: MIDAS: Explorative data mining in business applications. a project description. In: Herzog, O. (ed.) KI 1998. LNCS, vol. 1504, Springer, Heidelberg (1998)
970. Smith, B.: Against idiosyncrasy in ontology development. In: Bennett, B., Fellbaum, C. (eds.) Formal Ontology in Information Systems. Frontiers in Artificial Intelligence and Applications, vol. 150, pp. 15–26. IOS Press, Amsterdam (2006)
971. Smith, M.K., Welty, C., McGuinness, D.L.: OWL web ontology language guide (2004), http://www.w3.org/TR/owl-guide/
972. Sneed, H.M.: Encapsulation of legacy software: A technique for reusing legacy software components. Annals of Software Engineering 9(1) (2000)
973. Sneed, H.M.: Aufwandsschätzung von Software-Reengineering-Projekten. Wirtschaftsinformatik 45(6), 599–610 (2003)
974. Snell, J., Tidwell, D., Kulchenko, P.: Programming Web Services with SOAP. O'Reilly, Sebastopol (2001)
975. Spanoudakis, G., Zisman, A.: Inconsistency management in software engineering: Survey and open research issues. In: Handbook of Software Engineering and Knowledge Engineering, vol. 1, pp. 329–380. World Scientific, Singapore (2001)
976. Spenke, M., Beilken, C.: Visualization of trees as highly compressed tables with infozoom. In: Paton, N.W., Diaz, O. (eds.) Proceedings of the IEEE Symposium on Information Visualization (2003)
977. Sprinkle, J., Agrawal, A., Levendovszky, T., Shi, F., Karsai, G.: Domain model translation using graph transformations. In: Proceedings of the 10th IEEE International Conference on Engineering of Computer-Based Systems (ECBS 2003), Huntsville, AL, USA, April 7– 10, 2003, pp. 159–167. IEEE Computer Society Press, Los Alamitos (2003)
978. Staadt, O.G., Walker, J., Nuber, C., Hamann, B.: A survey and performance analysis of software platforms for interactive cluster-based multi-screen rendering. In: Deisinger, J., Kunz, A. (eds.) Proceedings of IPT 2003, International Immersive Projection Technologies Workshop, ACM Press, New York (2003)
979. Stanford Medical Informatics: The Protégé ontology editor and knowledge acquisition system (2005), http://protege.stanford.edu/
980. Steffen, B., Margaria, T.: METAFrame in practice: Design of intelligent network services. In: Olderog, E.-R., Steffen, B. (eds.) Correct System Design. LNCS, vol. 1710, pp. 390–415. Springer, Heidelberg (1999)
981. Stein, B.: Model construction in analysis and synthesis tasks. PhD thesis, University of Paderborn (2001)
982. Steinmetz, A., Kienzle, M.: e-seminar lecture recording and distribution system. In: Multimedia computing and Networking, San Jose, USA. Proceedings of SPIE (International Society for Optical Engineering), vol. 4312, pp. 25–36 (2001)

983. Stell, J.G., West, M.: A 4-dimensionalist mereotopology. In: Varzi, A.C., Vieu, L. (eds.) Formal Ontology in Information Systems, pp. 261–272. IOS Press, Amsterdam (2004)
984. Strong, D.M., Moller, S.M.: Exceptions and exception handling in computerized information processes. ACM Transactions on Information Systems 13(2), 206–233 (1995)
985. Stroustrup, B.: The C++ Programming Language, 2nd edn. Addison-Wesley, Reading (1991)
986. Störrle, H.: Semantics and verification of data flow in UML 2.0 activities. Electronic Notes in Theoretical Computer Science 127(4), 35–52 (2005)
987. Stuckenschmidt, H., Klein, M.: Integrity and change in modular ontologies. In: Proceedings of the International Joint Conference on Artificial Intelligence (IJCAI'03), Acapulco, Mexico, pp. 900–905. Morgan Kaufmann, San Francisco (2003)
988. Studer, R., Benjamins, V.R., Fensel, D.: Knowledge engineering: Principles and methods. Data & Knowledge Engineering 25(1-2), 161–197 (1998)
989. Subrahmanian, E., Konda, S.L., Dutoit, A., Reich, Y., Cunningham, D., Patrick, R., Thomas, M., Westerberg, A.W.: The n-dim approach to creative design support systems. In: Proceedings of the ASME Design Engineering Technical Conference (1997)
990. Suzuki, M., Batres, R., Fuchino, T., Shimada, Y., Chung, P.W.: A knowledge-based approach for accident information retrieval. In: Marquardt, W., Pantelides, C. (eds.) 16th European Symposium on Computer Aided Process Engineering and 9th International Symposium on Process Systems Engineering, Garmisch-Partenkirchen, Germany, July 9–13, 2006. Computer-Aided Chemical Engineering, vol. 21, pp. 1057–1062. Elsevier, Amsterdam (2006)
991. Szyperski, C.: Component Software: Beyond Object-Oriented Programming. Addison-Wesley, Reading (1997)
992. Taentzer, G., Koch, M., Fischer, I., Volle, V.: Distributed graph transformation with application to visual design of distributed systems. In: Handbook on Graph Grammars and Computing by Graph Transformation – Volume 3: Concurrency, Parallelism, and Distribution, pp. 269–340. World Scientific, Singapore (1999)
993. Tata, S., Chebbi, I.: A bottom-up approach to inter-enterprise business processes. In: Proceedings of the 13th IEEE International Workshops on Enabling Technologies (WETICE 2004), Infrastructure for Collaborative Enterprises, Modena, Italy, June 14–16, 2004, pp. 129–134. IEEE Computer Society Press, Los Alamitos (2004)
994. Teijgeler, H.: InfowebML – OWL-based Information Exchange and Integration based on ISO 15926 (2007), http://www.infowebml.ws/
995. TGL 25000: Chemical engineering unit operations – classification (1974)
996. Thayer, R.H.: Software engineering project management: A top-down view. In: Thayer, R.H. (ed.) Tutorial: Software Engineering Project Management, pp. 15–54. IEEE Computer Society Press, Los Alamitos (1988)
997. The CAPE-OPEN Laboratories Network: CAPE-OPEN standards and supporting documents (2008), http://www.colan.org/
998. Thomas, I., Nejmeh, B.A.: Definitions of tool integration for environments. IEEE Software 9(2), 29–35 (1992)

999. Thomas, M.E.: Tool and Information Management in Engineering Design. PhD thesis, Carnegie Mellon University (1996)
1000. Tichy, W.F. (ed.): Configuration Management. Trends in Software, vol. 2. John Wiley & Sons, Chichester (1994)
1001. To, H.H., Krishnaswamy, S., Srinivasan, B.: Mobile agents for network management: when and when not? In: SAC'05: Proceedings of the 2005 ACM Symposium on Applied Computing, Santa Fe, New Mexico, USA, pp. 47–53. ACM Press, New York (2005)
1002. Traenkle, F., Zeitz, M., Ginkel, M., Gilles, E.D.: PROMOT: A modeling tool for chemical processes. Mathematical and Computer Modelling of Dynamical Systems 6, 283–307 (2000)
1003. Ulich, E.: Arbeitspsychologie. Schäffer-Poeschel, Stuttgart (1991)
1004. Ullrich, H.: Wirtschaftliche Planung und Abwicklung verfahrentechnischer Anlagen. Vulkan-Verlag, Essen (1996)
1005. Underwriters Laboratory: UL 94 flammability testing (2008), http://www.ul.com/plastics/flame.html
1006. Uschold, M., Grüninger, M.: Ontologies: Principles, methods and applications. The Knowledge Engineering Review 11(2), 93–136 (1996)
1007. Uschold, M., King, M., Moralee, S., Zorgios, Y.: The enterprise ontology. The Knowledge Engineering Review 13, 31–89 (1998)
1008. Valette, R., Vergnes, B., Coupez, T.: Multiscale simulation of mixing processes using 3d-parallel, fluid-structure interaction techniques. International Journal of Material Forming (Proc. Symposium MS16: ESAFORM-ECCOMAS Workshop, Paris, France) (2008)
1009. van Daalen, C.E., Thissen, W.A.H., Verbraeck, A.: Methods for the modeling and analysis of alternatives. In: Sage, A.P., Rouse, W.B. (eds.) Handbook of Systems Engineering and Management, pp. 1037–1076. John Wiley & Sons, Chichester (1999)
1010. van Dam, A., Forsberg, A.S., Laidlaw, D.H., LaViola, J.J., Simpson, R.M.: Immersive VR for scientific visualization: A progress report. IEEE Computer Graphics and Applications 20(6), 26–52 (2000)
1011. van der Aalst, W.: Interorganizational workflows: An approach based on message sequence charts and petri nets. Systems Analysis – Modelling – Simulation 34(3), 335–367 (1999)
1012. van der Aalst, W.: Process-oriented architectures for electronic commerce and inter-organizational workflow. Information Systems 24(8), 639–671 (1999)
1013. van der Aalst, W.: Loosely coupled interorganizational workflows: Modeling and analyzing workflows crossing organizational boundaries. Information and Management 37(2), 67–75 (2000)
1014. van der Aalst, W., van Hee, K.: Workflow Management. MIT Press, Cambridge (2002)
1015. van Gigch, J.P.: System Design Modeling and Metamodeling. Plenum Press, New York (1991)
1016. van Schijndel, J., Pistikipoulos, E.N.: Towards the integration of process design, process control, and process operability: Current status and future trends. In: Malone, M.F., Trainham, J.A., Carnahan, B. (eds.) Foundations of Computer-Aided Process Design (FOCAPD'99). AIChE Symposium Series, vol. 323, pp. 99–112. CACHE Publications (2000)

1017. Vandalore, B., Feng, W., Jain, R., Fahmy, S.: A survey of application layer techniques for adaptive streaming of multimedia. Real-Time Imaging 7(5), 221–235 (2001)
1018. VDI: VDI-Richtlinie 5003, Bürokommunikation; Methoden zur Analyse und Gestaltung von Arbeitssystemen im Büro (1987)
1019. VDI: VDI-Richtlinie 2221: Methodik zum Entwickeln und Konstruieren technischer Systeme und Produkte (1993)
1020. VDI: VDI-Richtlinie 3633, Blatt 1, Simulation von Logistik-, Materialfluß- und Produktionssystemen – Grundlagen (1993)
1021. Venkatasubramanian, V., Zhao, C., Joglekar, G., Jain, A., Hailemariam, L., Suresh, P., Akkisetty, P., Morris, K., Reklaitis, G.: Ontological informatics infrastructure for pharmaceutical product development and manufacturing. Computers & Chemical Engineering 30(10-12), 1482–1496 (2006)
1022. Versant Corporation: Versant – website (2007), http://www.versant.com
1023. Vicente, K.J.: Cognitive Work Analysis. Lawrence Erlbaum, Mahwah (1999)
1024. Vinoski, S.: CORBA: Integrating diverse applications within distributed heterogeneous environments. IEEE Communications Magazine, 46–55 (1997)
1025. Virzi, R.A.: Usability inspection methods. In: Helander, M.G., Landauer, T.K., Prabhu, P.V. (eds.) Handbook of Human-Computer Interaction, 2nd edn., pp. 705–715. Elsevier, Amsterdam (1997)
1026. Visser, U., Stuckenschmidt, H., Wache, H., Vögele, T.: Enabling technologies for interoperability. In: Visser, U., Pundt, H. (eds.) Workshop on the 14th International Symposium of Computer Science for Environmental Protection, vol. 20, pp. 35–46 (2000)
1027. Vogel, G.H.: Verfahrensentwicklung: von der ersten Idee zur chemischen Produktionsanlage. Wiley-VCH, Weinheim (2002)
1028. Volkholz, V.: Arbeiten und Lernen. In: Nachhaltige Arbeitsgestaltung, pp. 431–488. Rainer Hampp, Mering (2002)
1029. von Wedel, L.: Model management with MOVE (2005), http://www.aixcape.org/move
1030. von Wedel, L.: Management and reuse of mathematical models in chemical industries with MOVE. In: Braunschweig, B., Joulia, X. (eds.) 18th European Symposium on Computer Aided Process Engineering – ESCAPE 18. Computer-Aided Chemical Engineering, Elsevier, Amsterdam (2008)
1031. Wache, H., Vögele, T., Visser, U., Stuckenschmidt, H., Schuster, G., Neumann, H., Hübner, S.: Ontology-based integration of information – a survey of existing approaches. In: Stuckenschmidt, H. (ed.) Proceedings of the IJCAI-01 Workshop on Ontologies and Information Sharing, pp. 108–117 (2001)
1032. Wagner, K., Aslanidis, S.: Prozessorientierte Nutzung von Erfahrungswissen in den frühen Phasen der Produktentstehung. In: Proceedings of the 4th Conference on Application of Knowledge Management in Industry and Public Administrations – Knowtech, München, Germany (2002)
1033. Wagner, R., Giese, H., Nickel, U.A.: A plug-in for flexible and incremental consistency management. In: Proceedings of the International Conference on the Unified Modeling Language 2003 (Workshop 7: Consistency Problems in UML-Based Software Development), San Francisco, California, USA (2003)
1034. Wahli, U., Avula, V., Macleod, H., Saeed, M., Vinther, A.: Business Process Management: Modeling through Monitoring Using WebSphere V6.0.2 Products. IBM Corp. (2007)

1035. Waligura, C.L., Motard, R.L.: Data management in engineering and construction projects. Chemical Engineering Progress, 62–70 (1977)
1036. Walker, M.B.: Smooth transitions in conversational turn-taking: Implications for theory. Journal of Psychology 110(1), 31–37 (1982)
1037. Wang, W., Haake, J.M., Rubart, J.: A cooperative visual hypermedia approach to planning and conducting virtual meetings. In: Haake, J.M., Pino, J.A. (eds.) CRIWG 2002. LNCS, vol. 2440, pp. 70–89. Springer, Heidelberg (2002)
1038. Wasserman, A.I.: Tool integration in software engineering environments. In: Long, F. (ed.) Software Engineering Environments. LNCS, vol. 467, pp. 137–149. Springer, Heidelberg (1990)
1039. Wasylkiewicz, S., Castillo, F.: Automatic synthesis for complex separation sequences with recycles. In: Gani, R., Jørgensen, S.B. (eds.) Proceedings of the European Symposium on Computer Aided Process Engineering – ESCAPE 11, pp. 591–596. Elsevier, Amsterdam (2001)
1040. Weigand, H., De Moor, A.: A framework for the normative analysis of workflow loops. CAiSE 2001 22(2), 38–40 (2001)
1041. Weiten, M., Wozny, G.: A knowledge based system for the documentation of research concerning physical and chemical processes – system design and case studies for applications. In: Kraslawski, A., Turunen, I. (eds.) European Symposium on Computer Aided Process Engineering – ESCAPE 13, pp. 329–334. Elsevier, Amsterdam (2003)
1042. Weiten, M., Wozny, G.: Advanced information management for process sciences: knowledge-based documentation of mathematical models. International Journal of Internet and Enterprise Management 2(2), 178–190 (2004)
1043. Weiten, M., Wozny, G., Goers, B.: Wege zum Informationsmanagement für interdisziplinäre Forschungsprojekte und die Entwicklung eines prototypischen Systems. Chem.-Ing.-Tech. 74(11), 1545–1553 (2002)
1044. Weske, M.: Formal foundations and conceptual design of dynamic adaptations in a workflow management system. In: Proceedings of the 34th Annual Hawaii International Conference on System Sciences (HICSS), Maui, Hawaii, USA, IEEE Computer Society Press, Los Alamitos (2001)
1045. West, M.: Replaceable parts: A four dimensional analysis. In: Proceedings of the COSIT'03 workshop on fundamental issues in spatial and geographic ontologies (2003)
1046. Westerberg, A.W., Hutchison, W., Motard, R., Winter, P.: Process Flowsheeting. Cambridge University Press, Cambridge (1979)
1047. Westerberg, A.W., Subrahmanian, E., Reich, Y., Konda, S.: Designing the process design process. Computers & Chemical Engineering 21, 1–19 (1997)
1048. Westhaus, U., Droege, T., Sass, R.: DETHERM – a thermophysical property database. Fluid Phase Equilibria (1999)
1049. Whitgift, D.: Methods and Tools for Software Configuration Management. John Wiley & Sons, Chichester (1991)
1050. Whittaker, S., Hyland, P., Wiley, M.: Filochat: Handwritten notes provide access to recorded conversations. In: Proceedings of the Conference on Human Factors in Computing Systems (CHI'94), Boston, USA, pp. 271–277 (1994)
1051. Wikipedia: Need to know (2008), http://en.wikipedia.org/wiki/Need_to_know

1052. Wilcox, A., Weiss, D., Russell, C., Smith, M.J., Smith, A.D., Pooley, R.J., MacKinnon, L.M., Dewar, R.G.: A CORBA-oriented approach to heterogeneous tool integration; OPHELIA. In: Proceedings of the Workshop on Tool Integration in System Development (TIS 2003), Helsinki, Finland, September 1–2 (2003)
1053. Wilding, W., Rowley, R., Oscarson, J.: DIPPR project 801 evaluated process design data. Fluid Phase Equilibria (1998)
1054. Willumeit, H., Gediga, G., Hamborg, K.-C.: IsoMetricsL: Ein Verfahren zur formativen Evaluation von Software nach ISO 9241/10. Ergonomie & Informatik 27, 5–12 (1996)
1055. Wodtke, D., Weissenfels, J., Weikum, G., Kotz-Dittrich, A., Muth, P.: The MENTOR workbench for enterprise-wide workflow management. In: Proceedings of the ACM SIGMOD International Conference on the Management of Data, Tucson, Arizona, USA, pp. 576–579. ACM Press, New York (1997)
1056. Wolisz, A., Tschammer, V.: Performance aspects of trading in open distributed systems. Computer Communications 16(5), 277–287 (1993)
1057. Workflow Management Coalition: Workflow management application programming interface (interface 2&3) specification (1998)
1058. Workflow Management Coalition: Terminology & glossary. Technical Report WFMC-TC-1011, Workflow Management Coalition (1999)
1059. Workflow Management Coalition: Workflow process definition interface – XML process definition language (XPDL), version 1.0 (2002), http://www.wfmc.org/standards/XPDL.htm
1060. World Wide Web Consortium (W3C): Extensible markup language, XML (2006), http://www.w3.org/XML/
1061. World Wide Web Consortium (W3C): Web services activity (2006), http://www.w3.org/2002/ws
1062. Wozny, G., Gutermuth, W., Kothe, W.: CAPE in der Verfahrenstechnik aus industrieller Sicht – Status, Bedarf, Prognose oder Vision?. Chem.-Ing.-Tech. 64(8), 693–699 (1992)
1063. YAWL Foundation: YAWL yet another workflow language (2006), http://yawlfoundation.org
1064. Zantout, H., Marir, F.: Document management systems from current capabilities towards intelligent information retrieval: An overview. International Journal of Information Management 19(6), 471–484 (1999)
1065. Zhao, C., Hailemariam, L., Jain, A., Joglekar, G., Venkatasubramanian, V., Morris, K., Reklaitis, G.: Information modeling for pharmaceutical product development. In: Marquardt, W., Pantelides, C. (eds.) 16th European Symposium on Computer Aided Process Engineering and 9th International Symposium on Process Systems Engineering, Garmisch-Partenkirchen, Germany, July 9–13, 2006. Computer-Aided Chemical Engineering, vol. 21, pp. 2147–2152. Elsevier, Amsterdam (2006)
1066. Zhao, C., Jain, A., Hailemariam, L., Joglekar, G., Venkatasubramanian, V., Morris, K., Reklaitis, G.: A unified approach for knowledge modeling in pharmaceutical product development. In: Marquardt, W., Pantelides, C. (eds.) 16th European Symposium on Computer Aided Process Engineering and 9th International Symposium on Process Systems Engineering, Garmisch-Partenkirchen, Germany, July 9–13, 2006. Computer-Aided Chemical Engineering, vol. 21, pp. 1929–1935. Elsevier, Amsterdam (2006)

1067. Zhao, C., Joglekar, G., Jain, A., Venkatasubramanian, V., Reklaitis, G.V.: Pharmaceutical informatics: A novel paradigm for pharmaceutical product development and manufacture. In: Puigjaner, L., Espuña, A. (eds.) European Symposium on Computer-Aided Process Engineering – ESCAPE 15. Computer-Aided Chemical Engineering, vol. 20, pp. 1561–1566. Elsevier, Amsterdam (2005)
1068. Zhao, W., Olshefski, D., Schulzrinne, H.: Internet quality of service: an overview. Technical Report CUCS-003-00, Columbia University (2000)
1069. Ziegler, J.: Eine Vorgehensweise zum objektorientierten Entwurf graphisch-interaktiver Informationssysteme. Springer, Heidelberg (1996)
1070. Ziegler, J.: Modeling cooperative work processes – a multiple perspectives framework. International Journal of Human-Computer Interaction 15(2), 139–157 (2002)
1071. Zülch, G.: Zeitwirtschaftliche Voraussetzungen für die simulationsunterstützte Planung von Produktionssystemen. REFA-Nachrichten 2, 4–11 (2004)
1072. Zülch, G., Jagdev, H.S., Stock, P. (eds.): Integrating Human Aspects in Production Management. Springer, Heidelberg (2004)
1073. Zope Corporation: Zope.org – website (2006), http://www.zope.org/

Author Index

Assenmacher, Ingo, 268, 493

Babich, Yuri, 401
Bayer, Birgit, 93, 111
Becker, Simon, 224, 519, 612, 696
Brandt, Sebastian C., 369, 675

Eggersmann, Markus, 126

Foltz, Christian, 527
Fritzen, Oliver, 369

Haase, Thomas, 555, 727
Haberstroh, Edmund, 268, 493, 519
Hai, Ri, 111, 433, 621, 656
Heer, Thomas, 621, 711
Heller, Markus, 300, 519, 621, 711

Jarke, Matthias, 185, 369, 519, 605, 675
Jäger, Dirk, 300

Körtgen, Anne-Thérèse, 696
Kausch, Bernhard, 126, 451, 527, 666
Klein, Peter, 555
Krapp, Carl-Arndt, 300
Kuhlen, Torsten, 268, 493
Kulikov, Viatcheslav, 477

List, Thomas, 369
Luczak, Holger, 126, 451, 527, 666

Marquardt, Wolfgang, 3, 83, 93, 111, 126, 153, 169, 433, 477, 519, 605, 612, 643, 647, 656, 743, 764

Miatidis, Michalis, 185, 605, 675
Morbach, Jan, 83, 93, 111, 169, 433, 612, 647

Nagl, Manfred, 3, 61, 224, 300, 519, 555, 593, 612, 621, 629, 696, 711, 727, 753, 764, 774

Raddatz, Marcus, 675

Schleicher, Ansgar, 300
Schlick, Christopher, 126, 451, 527, 666
Schlüter, Marcus, 493, 675
Schneider, Nicole, 126, 451, 527, 666
Schneider, Ralph, 39, 126, 433, 621, 643
Schüppen, Andre, 268
Spaniol, Otto, 268, 401, 519
Stewering, Jörn, 493

Tackenberg, Sven, 451
Theißen, Manfred, 83, 126, 153, 169, 433, 605, 656
Thißen, Dirk, 268, 401, 519

von Wedel, Lars, 477, 643

Weidenhaupt, Klaus, 185
Westfechtel, Bernhard, 39, 224, 300, 621
Wolf, Martin, 527
Wörzberger, René, 300, 621, 711

Yang, Aidong, 93

Printing: Mercedes-Druck, Berlin
Binding: Stein+Lehmann, Berlin

Lecture Notes in Computer Science

Sublibrary 2: Programming and Software Engineering

For information about Vols. 1– 4536
please contact your bookseller or Springer

Vol. 5142: J. Vitek (Ed.), ECOOP 2008 – Object-Oriented Programming. XIII, 694 pages. 2008.

Vol. 5140: J. Meseguer, G. Roşu (Eds.), Algebraic Methodology and Software Technology. XIII, 432 pages. 2008.

Vol. 5136: T.C.N. Graham, P. Palanque (Eds.), Interactive Systems
Design, Specification, and Verification. IX, 311 pages. 2008.

Vol. 5119: S. Kounev, I. Gorton, K. Sachs (Eds.), Performance Evaluation: Metrics, Models and Benchmarks. X, 323 pages. 2008.

Vol. 5095: I. Schieferdecker, A. Hartman (Eds.), Model Driven Architecture – Foundations and Applications. XIII, 446 pages. 2008.

Vol. 5091: B.P. Woolf, E. Aïmeur, R. Nkambou, S. Lajoie (Eds.), Intelligent Tutoring Systems. XXI, 832 pages. 2008.

Vol. 5089: A. Jedlitschka, O. Salo (Eds.), Product-Focused Software Process Improvement. XIV, 448 pages. 2008.

Vol. 5079: M. Alpuente, G. Vidal (Eds.), Static Analysis. X, 379 pages. 2008.

Vol. 5063: A. Vallecillo, J. Gray, A. Pierantonio (Eds.), Theory and Practice of Model Transformations. XII, 261 pages. 2008.

Vol. 5060: C. Rong, M.G. Jaatun, F.E. Sandnes, L.T. Yang, J. Ma (Eds.), Autonomic and Trusted Computing. XV, 666 pages. 2008.

Vol. 5055: K. Al-Begain, A. Heindl, M. Telek (Eds.), Analytical and Stochastic Modeling Techniques and Applications. XI, 323 pages. 2008.

Vol. 5052: D. Lea, G. Zavattaro (Eds.), Coordination Models and Languages. X, 347 pages. 2008.

Vol. 5051: G. Barthe, F.S. de Boer (Eds.), Formal Methods for Open Object-Based Distributed Systems. X, 259 pages. 2008.

Vol. 5048: K. Suzuki, T. Higashino, K. Yasumoto, K. El-Fakih (Eds.), Formal Techniques for Networked and Distributed Systems – FORTE 2008. XII, 341 pages. 2008.

Vol. 5047: K. Suzuki, T. Higashino, A. Ulrich, T. Hasegawa (Eds.), Testing of Software and Communicating Systems. XII, 303 pages. 2008.

Vol. 5030: H. Mei (Ed.), High Confidence Software Reuse in Large Systems. XII, 388 pages. 2008.

Vol. 5026: F. Kordon, T. Vardanega (Eds.), Reliable Software Technologies – Ada-Europe 2008. XIV, 283 pages. 2008.

Vol. 5025: B. Paech, C. Rolland (Eds.), Requirements Engineering: Foundation for Software Quality. X, 205 pages. 2008.

Vol. 5020: J. Barnes, Ada 2005 Rationale. IX, 267 pages. 2008.

Vol. 5016: M. Bernardo, P. Degano, G. Zavattaro (Eds.), Formal Methods for Computational Systems Biology. X, 538 pages. 2008.

Vol. 5014: J. Cuellar, T. Maibaum, K. Sere (Eds.), FM 2008: Formal Methods. XIII, 436 pages. 2008.

Vol. 5007: Q. Wang, D. Pfahl, D.M. Raffo (Eds.), Making Globally Distributed Software Development a Success Story. XIV, 422 pages. 2008.

Vol. 5002: H. Giese (Ed.), Models in Software Engineering. X, 322 pages. 2008.

Vol. 4989: J. Garrigue, M.V. Hermenegildo (Eds.), Functional and Logic Programming. XI, 337 pages. 2008.

Vol. 4970: M. Nagl, W. Marquardt (Eds.), Collaborative and Distributed Chemical Engineering. XII, 851 pages. 2008.

Vol. 4966: B. Beckert, R. Hähnle (Eds.), Tests and Proofs. X, 193 pages. 2008.

Vol. 4954: C. Pautasso, É. Tanter (Eds.), Software Composition. X, 263 pages. 2008.

Vol. 4951: M. Luck, L. Padgham (Eds.), Agent-Oriented Software Engineering VIII. XIV, 225 pages. 2008.

Vol. 4949: R.M. Hierons, J.P. Bowen, M. Harman (Eds.), Formal Methods and Testing. XIII, 367 pages. 2008.

Vol. 4937: M. Dumas, R. Heckel (Eds.), Web Services and Formal Methods. IX, 169 pages. 2008.

Vol. 4922: M. Broy, I.H. Krüger, M. Meisinger (Eds.), Model-Driven Development of Reliable Automotive Services. XVIII, 183 pages. 2008.

Vol. 4916: S. Leue, P. Merino (Eds.), Formal Methods for Industrial Critical Systems. X, 251 pages. 2008.

Vol. 4909: I. Eusgeld, F.C. Freiling, R. Reussner (Eds.), Dependability Metrics. XI, 305 pages. 2008.

Vol. 4906: M. Cebulla (Ed.), Object-Oriented Technology. VIII, 204 pages. 2008.

Vol. 4902: P. Hudak, D.S. Warren (Eds.), Practical Aspects of Declarative Languages. X, 333 pages. 2007.

Vol. 4899: K. Yorav (Ed.), Hardware and Software: Verification and Testing. XII, 267 pages. 2008.

Vol. 4888: F. Kordon, O. Sokolsky (Eds.), Composition of Embedded Systems. XII, 221 pages. 2007.

Vol. 4880: S. Overhage, C.A. Szyperski, R. Reussner, J.A. Stafford (Eds.), Software Architectures, Components, and Applications. X, 249 pages. 2008.

Vol. 4849: M. Winckler, H. Johnson, P. Palanque (Eds.), Task Models and Diagrams for User Interface Design. XIII, 299 pages. 2007.

Vol. 4839: O. Sokolsky, S. Taşiran (Eds.), Runtime Verification. VI, 215 pages. 2007.

Vol. 4834: R. Cerqueira, R.H. Campbell (Eds.), Middleware 2007. XIII, 451 pages. 2007.

Vol. 4829: M. Lumpe, W. Vanderperren (Eds.), Software Composition. VIII, 281 pages. 2007.

Vol. 4824: A. Paschke, Y. Biletskiy (Eds.), Advances in Rule Interchange and Applications. XIII, 243 pages. 2007.

Vol. 4821: J. Bennedsen, M.E. Caspersen, M. Kölling (Eds.), Reflections on the Teaching of Programming. X, 261 pages. 2008.

Vol. 4807: Z. Shao (Ed.), Programming Languages and Systems. XI, 431 pages. 2007.

Vol. 4799: A. Holzinger (Ed.), HCI and Usability for Medicine and Health Care. XVI, 458 pages. 2007.

Vol. 4789: M. Butler, M.G. Hinchey, M.M. Larrondo-Petrie (Eds.), Formal Methods and Software Engineering. VIII, 387 pages. 2007.

Vol. 4767: F. Arbab, M. Sirjani (Eds.), International Symposium on Fundamentals of Software Engineering. XIII, 450 pages. 2007.

Vol. 4765: A. Moreira, J. Grundy (Eds.), Early Aspects: Current Challenges and Future Directions. X, 199 pages. 2007.

Vol. 4764: P. Abrahamsson, N. Baddoo, T. Margaria, R. Messnarz (Eds.), Software Process Improvement. XI, 225 pages. 2007.

Vol. 4762: K.S. Namjoshi, T. Yoneda, T. Higashino, Y. Okamura (Eds.), Automated Technology for Verification and Analysis. XIV, 566 pages. 2007.

Vol. 4758: F. Oquendo (Ed.), Software Architecture. XVI, 340 pages. 2007.

Vol. 4757: F. Cappello, T. Herault, J. Dongarra (Eds.), Recent Advances in Parallel Virtual Machine and Message Passing Interface. XVI, 396 pages. 2007.

Vol. 4753: E. Duval, R. Klamma, M. Wolpers (Eds.), Creating New Learning Experiences on a Global Scale. XII, 518 pages. 2007.

Vol. 4749: B.J. Krämer, K.-J. Lin, P. Narasimhan (Eds.), Service-Oriented Computing – ICSOC 2007. XIX, 629 pages. 2007.

Vol. 4748: K. Wolter (Ed.), Formal Methods and Stochastic Models for Performance Evaluation. X, 301 pages. 2007.

Vol. 4741: C. Bessière (Ed.), Principles and Practice of Constraint Programming – CP 2007. XV, 890 pages. 2007.

Vol. 4735: G. Engels, B. Opdyke, D.C. Schmidt, F. Weil (Eds.), Model Driven Engineering Languages and Systems. XV, 698 pages. 2007.

Vol. 4716: B. Meyer, M. Joseph (Eds.), Software Engineering Approaches for Offshore and Outsourced Development. X, 201 pages. 2007.

Vol. 4709: F.S. de Boer, M.M. Bonsangue, S. Graf, W.-P. de Roever (Eds.), Formal Methods for Components and Objects. VIII, 297 pages. 2007.

Vol. 4680: F. Saglietti, N. Oster (Eds.), Computer Safety, Reliability, and Security. XV, 548 pages. 2007.

Vol. 4670: V. Dahl, I. Niemelä (Eds.), Logic Programming. XII, 470 pages. 2007.

Vol. 4652: D. Georgakopoulos, N. Ritter, B. Benatallah, C. Zirpins, G. Feuerlicht, M. Schoenherr, H.R. Motahari-Nezhad (Eds.), Service-Oriented Computing ICSOC 2006. XVI, 201 pages. 2007.

Vol. 4640: A. Rashid, M. Aksit (Eds.), Transactions on Aspect-Oriented Software Development IV. IX, 191 pages. 2007.

Vol. 4634: H. Riis Nielson, G. Filé (Eds.), Static Analysis. XI, 469 pages. 2007.

Vol. 4620: A. Rashid, M. Aksit (Eds.), Transactions on Aspect-Oriented Software Development III. IX, 201 pages. 2007.

Vol. 4615: R. de Lemos, C. Gacek, A. Romanovsky (Eds.), Architecting Dependable Systems IV. XIV, 435 pages. 2007.

Vol. 4610: B. Xiao, L.T. Yang, J. Ma, C. Muller-Schloer, Y. Hua (Eds.), Autonomic and Trusted Computing. XVIII, 571 pages. 2007.

Vol. 4609: E. Ernst (Ed.), ECOOP 2007 – Object-Oriented Programming. XIII, 625 pages. 2007.

Vol. 4608: H.W. Schmidt, I. Crnković, G.T. Heineman, J.A. Stafford (Eds.), Component-Based Software Engineering. XII, 283 pages. 2007.

Vol. 4591: J. Davies, J. Gibbons (Eds.), Integrated Formal Methods. IX, 660 pages. 2007.

Vol. 4589: J. Münch, P. Abrahamsson (Eds.), Product-Focused Software Process Improvement. XII, 414 pages. 2007.

Vol. 4574: J. Derrick, J. Vain (Eds.), Formal Techniques for Networked and Distributed Systems – FORTE 2007. XI, 375 pages. 2007.

Vol. 4556: C. Stephanidis (Ed.), Universal Access in Human-Computer Interaction, Part III. XXII, 1020 pages. 2007.

Vol. 4555: C. Stephanidis (Ed.), Universal Access in Human-Computer Interaction, Part II. XXII, 1066 pages. 2007.

Vol. 4554: C. Stephanidis (Ed.), Universal Acess in Human Computer Interaction, Part I. XXII, 1054 pages. 2007.

Vol. 4553: J.A. Jacko (Ed.), Human-Computer Interaction, Part IV. XXIV, 1225 pages. 2007.

Vol. 4552: J.A. Jacko (Ed.), Human-Computer Interaction, Part III. XXI, 1038 pages. 2007.

Vol. 4551: J.A. Jacko (Ed.), Human-Computer Interaction, Part II. XXIII, 1253 pages. 2007.

Vol. 4550: J.A. Jacko (Ed.), Human-Computer Interaction, Part I. XXIII, 1240 pages. 2007.

Vol. 4542: P. Sawyer, B. Paech, P. Heymans (Eds.), Requirements Engineering: Foundation for Software Quality. IX, 384 pages. 2007.